Understanding

Zoonotic
Diseases

Join us on the web at

agriculture.delmar.com

Understanding
Zoonotic
Diseases

JANET AMUNDSON ROMICH, DVM, MS

THOMSON

DELMAR LEARNING ™

Australia   Brazil   Canada   Mexico   Singapore   Spain   United Kingdom   United States

Understanding Zoonotic Diseases

Janet Amundson Romich, DVM, MS

Vice President, Career Education Strategic Business Unit:
Dawn Gerrain

Director of Learning Solutions:
John Fedor

Acquisitions Editor:
David Rosenbaun

Managing Editor:
Robert L. Serenka, Jr.

Product Manager:
Christina Gifford

Editorial Assistant:
Scott Royael

Director of Production:
Wendy A. Troeger

Production Manager:
Mark Bernard

Project Manager:
Joan Conlon

Technology Project Manager:
Sandy Charette

Director of Marketing:
Wendy E. Mapstone

Marketing Manager:
Gerard McAvey

Marketing Coordinator:
Jonathan Sheehan

Art Director:
Joy Kocsis

Cover Design:
Studio Montage

Cover Images:
Tick courtesy CDC/James Gathany; chicks courtesy USDA; other images courtesy CDC

Text Design:
Rose Design

Library of Congress Cataloging-in-Publication Data

Romich, Janet Amundson.
  Understanding zoonotic diseases/ Janet Amundson Romich.
    p. cm.
  Includes bibliographical references and index.
  ISBN 1-4180-2103-2
  1. Zoonoses. I. Title.
  SF740.R66 2008
  616.9'59—dc22
                              2007027596

NOTICE TO THE READER

Publisher does not warrant or guarantee any of the products described herein or perform any independent analysis in connection with any of the product information contained herein. Publisher does not assume, and expressly disclaims, any obligation to obtain and include information other than that provided to it by the manufacturer.

The reader is expressly warned to consider and adopt all safety precautions that might be indicated by the activities herein and to avoid all potential hazards. By following the instructions contained herein, the reader willingly assumes all risks in connection with such instructions.

The Publisher makes no representation or warranties of any kind, including but not limited to, the warranties of fitness for particular purpose or merchantability, nor are any such representations implied with respect to the material set forth herein, and the publisher takes no responsibility with respect to such material. The Publisher shall not be liable for any special, consequential, or exemplary damages resulting, in whole or part, from the readers' use of, or reliance upon, this material.

Contents

APPENDICES

GLOSSARY   667

INDEX   675

INTRODUCTION

Bird flu, handling chickens, and migratory birds. SARS, airplanes, and civet cats. *Toxoplasma*, pregnant women, and litter boxes. Rabies, bats, and children. Natural disasters, flooding, and the spread of disease. These are just a few sound bites that may be heard on a daily basis regarding the risks involved in daily life. It is impossible to read through a newspaper, watch television, or listen to the radio without encountering something about an emerging infectious disease. Of the world's new and emerging diseases, approximately 75% are zoonotic. Zoonotic diseases include those directly transmitted from animals to humans (such as rabies); those indirectly acquired by humans through ingestion, inhalation, or contact with infected animal products, soil, water, or other environmental surface that has been contaminated with animal waste or a dead animal (such as *Escherichia coli* food contamination); and those present in an animal reservoir, but requiring an arthropod (mosquito, flea, tick, or fly) to transmit the disease to humans (such as Lyme disease). Emerging infectious diseases that are zoonotic make the role of veterinary health care professionals extremely important in preventing both animal and human disease. Veterinary health care professionals have a great opportunity to disseminate information to the general public about zoonotic diseases as well as clarify misinformation that is circulating among the general population.

It is difficult to acquire current and relevant knowledge of emerging infectious diseases; however, it is equally difficult to gather and interpret new information regarding ancient infectious diseases. Consider tuberculosis, a bacterial respiratory disease believed to have been present in our primate ancestors. It was not until 1882 when a diagnostic stain was discovered to demonstrate the organism in clinical samples. Antibiotic treatment for tuberculosis was not available until 1943 to help physicians treat cases of tuberculosis. In industrialized countries that were treating people with tuberculosis, the incidence of tuberculosis declined until the mid-1980s when cases began to increase as a result of high rates of immigration from tuberculosis ravaged countries and HIV infection. The development of multiple-drug–resistant strains of the bacterium causing tuberculosis has caused new concerns regarding the ancient disease and its treatment. The development of a tuberculosis vaccine currently in clinical trials may help prevent some cases of tuberculosis, but will it prevent infection with some of the multiple-drug-resistant strains? The organism is still found in many wild animals that may spread the disease among the human population and events, such as war, political unrest, and poverty, can alter immunity in people, affect health care accessibility, and limit drug delivery in certain areas of the world.

In addition to directly producing human disease, zoonotic diseases can also affect the food chain and produce infection among other animals. Brucellosis is a bacterial disease that causes undulating fever in people characterized by weakness, loss of appetite, chills, headache, back pain, and intermittent (undulating) fever. Brucellosis can be transmitted to people from infected cows and goats through contaminated milk and cheese making food handling procedures like pasteurization and refrigeration important in controlling this disease. Brucellosis may be endemic in buffalo in Yellowstone National Park, which produced a fear among cattle ranchers in Wyoming and Montana that their cattle would contract brucellosis from infected buffalo. Spread of brucellosis from buffalo to cattle can produce severe economic loss for those involved in the beef industry.

The environment can also affect zoonotic diseases. The World Health Organization (WHO) reports that global warming has affected 40% of the world's ecosystems. Changing ecosystems have resulted in an increase in tropical disease and the spread of insect vectors typically found in southern regions to northern regions that are now warmer than before. Deforestation in Africa has been associated with the spread of Ebola virus, present in wild primates (chimpanzees) that live in the affected jungles. The increase in Lyme disease in the eastern United States is associated with population growth, deforestation of the hardwood forests, and urban sprawl. Natural disasters such as droughts, hurricanes, and floods have also caused an increase in certain zoonotic diseases such as vibriosis and leptospirosis.

Culture has also played a role in development of zoonotic diseases. Domestication of animals increased the intermingling of animals and humans. Once animals were domesticated, humans had more contact with animal feces, animal secretions, and animal products such as meat and hides. This increased animal contact has allowed for the transmission of zoonotic diseases such as cowpox, swine flu, and salmonellosis. Practices within certain cultures have also been responsible for the spread of zoonotic diseases. Handling of brain tissue in some ritualistic ceremonies has resulted in spread of diseases such as prion disease (kuru). Ingestion of undercooked meat containing tapeworm cysts has resulted in spread of diseases such as cysticercosis.

Human disease and advancements in treatment of human diseases has also increased the incidence of zoonotic diseases. Pet ownership has both emotional and health benefits for people with a variety of diseases; however, the increase in immunosuppressive conditions in people such as HIV infection and cancer chemotherapy has caused concern among human physicians for their patients who have pets. Diseases caused by *Toxoplasma*, *Cryptosporidium*, *Giardia*, and *dermatophytes* are just a few examples of zoonotic diseases seen more commonly in immunosuppressed people, making routine physical examination of animals (especially parasite assessment) and client education of owners extremely important veterinary functions.

Veterinary health care professionals play a key role in maintaining human health in a variety of ways. One way is the acquisition of knowledge. Knowledge allows the veterinary team to be aware of the potential dangers associated with animals, their care, and their use. Disseminating this knowledge to others allows them to become enlightened about animal diseases and their role in human disease. A second way to maintain human health is by minimizing exposure to zoonotic diseases through the use of proper personal protective equipment, proper animal handling techniques, and effective personal hygiene measures which all help combat contraction of zoonotic diseases. Thirdly, being able to recognize the clinical signs of zoonotic diseases in animals and contacting the proper authorities if warranted are keys to preventing the spread of zoonotic disease. It is the hope of this textbook to provide veterinary health care professionals with the tools needed to successfully identify, understand, and control zoonotic diseases. This textbook and the accompanying learning materials make the process of understanding, applying, and staying current of the changes of zoonotic diseases as straightforward as possible. This textbook offers brief but thorough explanations of zoonotic diseases, their routes of transmission, epizootiologic importance, clinical signs in animals and people, diagnosis in animals and people, treatment in animals and people, and control measures. Summaries, review questions, and case studies throughout the text and online material provide the learner with many ways to examine and to learn the material presented in this textbook. The online material provides presentation materials, critical thinking questions, and a variety of tools to help in both instruction and understanding of zoonotic diseases.

ORGANIZATION OF THE TEXTBOOK

This textbook begins with an introduction to the history of zoonotic disease including a review of terminology, disease transmission, agencies monitoring diseases, and job safety regarding zoonotic diseases. Immune topics and diagnostic procedures used in understanding zoonotic diseases is covered following the introduction. The remainder of the text is divided into chapters on bacterial zoonoses, tick-borne bacterial zoonoses, fungal zoonoses, parasitic zoonoses, viral zoonoses, and prion diseases that begin with an overview of each type of organism. Within these chapters are sections on zoonotic diseases caused by that category of organism. Each zoonotic disease section includes a historical overview of the disease, explanation of the causative agent, public health significance of the disease, followed by transmission, pathogenesis, clinical signs, diagnostic procedures, treatment, and control of each disease. Each chapter has a summary and review questions to help learners test their knowledge and quickly review the material. Tables offer condensed information that is easily accessible for the learner to comprehend and refer to when reviewing the material. Images and line art help clarify more difficult concepts and reinforce clinical signs and disease processes. The variety of presentation formats enhances content comprehension and retention.

FEATURES AND BENEFITS

- Chapter objectives and key terms are listed at the beginning of each chapter to guide students in understanding the organization of that chapter.
- Descriptions of the different microbes that cause zoonotic diseases are within the text so that students do not have to reference other textbooks or resources.
- A thorough explanation of the immune response to the different types of microbes causing zoonotic disease.
- Clear explanations and illustrations of diagnostic tests used in identifying zoonotic diseases.
- A summary and chapter review questions are provided for each chapter to allow learners to review and track their understanding of a topic.
- Case studies are provided at the end of many chapters to give students a chance to apply the information they have just learned.
- Easy-to-use tables and charts help students organize diseases and disease processes.
- Straightforward illustrations and photos help students comprehend difficult concepts.

- Appendices provide students with quick references throughout the course and in clinical settings.
- A glossary at the end of the text gives students a resource in which to quickly look up key terms used in describing zoonotic diseases.
- An Online Companion is available to both students and instructors. To view the online supplements, go to www.agriculture.delmar.com

Acknowledgments

Special thanks to the following people who helped review this text and answered many questions regarding zoonotic diseases throughout its development.

Darwin R. Yoder, DVM, MS
Sul Ross State University
Alpine, TX

Karl M. Peter, DVM
Foothill College—Veterinary
    Technology
Los Altos Hills, CA

David L. Berryhill, Ph.D.
North Dakota State University
Fargo, ND

Mary O'Horo Loomis, DVM
SUNY—Canton
Canton, NY

Regina J. Brotherton, DVM
Johnson College
Scranton, PA

Terry D. Canerdy, DVM
Murray State University
Murray, KY

Bonnie Ballard, DVM
Gwinnett Technical College
Lawrenceville, GA

James Meronek, DVM
UW College of Agriculture and Life
    Sciences
Madison, WI

**Claire B. Andreasen, DVM, PhD,
Diplomate ACVP**
Director Pathology Laboratory Services
Iowa State University
Ames, IA

Laura Lien, BS, CVT
Moraine Park Technical College
Fond du Lac, WI

Lois Morrison
The University of Georgia
Athens, GA

**Robert Garrision, DVM, MS,
Diplomate ABMM**
Schoolcraft, MI

Simon T. M. Allard, MS, PhD
EPICENTRE Biotechnologies
Madison, WI

About the Author

Dr. Janet Romich received her Bachelor of Science degree in Animal Science from the University of Wisconsin–River Falls, and her Doctor of Veterinary Medicine and Master of Science Degree from the University of Wisconsin–Madison. Currently, Dr. Romich teaches at Madison Area Technical College in Madison, Wisconsin where she has taught and continues to teach a variety of science-based courses. Dr. Romich was honored with the Distinguished Teacher Award in 2004 for use of technology in the classroom, advisory and professional activities, publication list, and fundraising efforts. She is a member of the Biosafety Committee for a bio-pharmaceutical company, an IACUC member for a hospital research facility, and an advisory board member for a distance learning veterinary technician program. Dr. Romich authored the textbooks, *An Illustrated Guide to Veterinary Medical Terminology with Interactive CD-ROM*, 2E and *Fundamentals of Pharmacology for Veterinary Technicians*, as well as serving as a co-author on *Delmar's Veterinary Technician Dictionary*. Dr. Romich remains active in veterinary practice through her relief practice, where she works in both small- and mixed-animal practices.

1

Zoonotic Disease History

Objectives

After completing this chapter, the learner should be able to

- List classic examples of zoonoses
- Describe how zoonoses became prevalent in human society
- Differentiate between biological and mechanical vectors
- Describe ways that zoonotic disease can be transmitted from animals to humans
- Differentiate between endemic, enzootic, sporadic, epidemic, epizootic, pandemic, and panzootic diseases

Key Terms

acute disease	enzootic disease	indirect transmission	panzootic disease	vector
biological vector	epidemic disease	infectious disease	prevalence	vertical transmission
chronic disease	epidemiology	latent disease	reservoir	zoonoses
communicable disease	rate	mechanical vector	reverse zoonoses	
contagious disease	epizootic disease	morbidity rate	sentinel	
direct transmission	fomites	mortality rate	source of infection	
emerging zoonoses	horizontal	noncommunicable	sporadic disease	
endemic disease	transmission	disease	subacute disease	
	incidence	pandemic disease	transmission	

OVERVIEW

Animals, both domestic and wild, are subject to a great variety of diseases. Many of these diseases are confined to a particular animal species; however, some animal disease agents may cause human disease. Diseases that are naturally transmitted from animals to humans are termed **zoonoses** (singular is zoonosis). Zoonotic infections are indigenous, or native, to a particular animal and they may or may not produce clinical illness in the animal.

More than 150 zoonoses are known, but fewer than half of them are clinically significant. Classic examples of zoonoses are the plague, also known as "The Black Death" during the Middle Ages, and tuberculosis, also known as consumption because this disease consumes one's entire body. In addition to well-established zoonoses, new zoonoses continue to be discovered including Lyme disease (discovered in the late 1970s) and hantavirus pulmonary syndrome (discovered in

> Zoonoses are diseases that are naturally transmitted from animals to humans. The term **reverse zoonoses** or anthroponoses refer to diseases naturally transmitted from humans to animals.

the United States in 1993). The West Nile virus outbreak in New York City in the summer of 1999 infected a large number of birds (mostly crows) and humans who developed flu-like symptoms from the bite of virus-infected mosquitoes.

Emerging zoonoses are zoonotic diseases caused either by apparently new agents or by previously known agents appearing in places or in species in which the disease was previously unknown. For example, closely-related strains of some viruses may mutate into "new" viruses that cause human disease. Such is the case with influenza, a virus that causes human respiratory disease. Closely-related strains of the same virus infect birds, horses, and pigs. When these viral strains replicate in different animal hosts, different selective pressures may favor different mutant forms of the virus. When diverse strains of a virus infect the same animal, genetic recombination can produce a "new" virus that might have enhanced virulence (or ability to cause disease) for humans. Scientists have found that a single gene may have been responsible for the devastating virulence of the virus that caused the 1918 Spanish influenza outbreak that killed an estimated 20 to 40 million people. Public health professionals continue to be concerned about the role that existing and emerging zoonoses play on the human and animal populations.

THE HISTORY OF ZOONOSES

To understand animal origins of human disease the evolution of infectious organisms needs to be considered. Diseases that have infected humans can be traced to our ancient ancestors. These diseases followed humans through the evolutionary process and continue to infect humans today. Prior to the agricultural revolution (about 10,000 years ago) humans lived in small, hunter-gatherer groups. These groups moved frequently and were isolated from neighboring groups of people. Their diets consisted mainly of plants and small amounts of animal protein. The nomadic nature of these people limited their accumulation of sewage and garbage that in turn limited the number of rodents and the number of disease-causing organisms to which hunter-gatherers were exposed.

When people altered their lifestyles with the domestication of plants and animals, the spread of disease became easier. Food production and the ability to store food for later use allowed people to cease roaming and to live in communities. The domestication of animals by 6000 B.C. resulted in the close contact of animals and people. Crowded barnyards contained animal feed and animal waste, which were optimal environments for insects and rodents that in turn lead to fruitful places for the breeding and **transmission** (transfer) of diseases. These barnyards were the first breeding grounds for diseases such as rinderpest and cowpox in cattle as well as sources of disease transmission such as tuberculosis from consumption of unpasteurized milk.

As surpluses in agricultural products such as grain and meat began to accumulate, worldwide trade routes developed. Worldwide trade routes became avenues for the spread of disease. Diseases prevalent in Europe, such as the plague, were spread to Asia by the transit of flea-infested materials. When Christopher Columbus reached the New World his men brought diseases such as smallpox and measles to the native people of the West Indies and the Americas. The spread of pathogenic organisms no longer depends upon explorers searching for new territories to conquer. International flights, transcontinental railroads, and other modes of international travel allow tourists, business people, military personnel, and refugees to potentially carry diseases from one locale to another. The shipment of goods via ships and trucks can transport rodents and insects carrying disease-causing organisms from one country to the next. International trade of laboratory animals, pets, and livestock also creates the potential for pathogenic organisms to travel from one place to another.

With this background, the story of zoonotic disease and its impact on animal and human disease unfolds. The importance of understanding how animals play a role in the transfer of pathogens, how they serve as a source of infection for people, how they harbor microbes in the environment, and how they can serve as a disease **sentinel** (a domestic animal host for a particular disease that is placed at various locations to determine the potential for human exposure to a particular disease) are all key concepts to understanding zoonoses and their role in human health.

ORIGINS OF ZOONOSES

Infectious agents need a place to reside if they are going to survive and be spread within an environment. A **reservoir** is an animate or inanimate object that serves as a long-term habitat and focus of dissemination for an infectious agent. Animals can serve as reservoirs for certain human pathogens, although soil, water, and plants can also serve as reservoirs. A reservoir is not the same as a source of infection. A **source of infection** is an individual or object from which an infection is actually acquired. In some cases the reservoir and the source of infection are the same (as is the case with syphilis in which the reservoir and the source of infection are the human body) and in other cases the reservoir and the source of infection are different (as is the case with *Cryptosporidium parvum* in which the reservoir may be cattle and the source of infection is water).

Types of Reservoirs

Reservoirs may be living or nonliving. Living reservoirs include humans, animals, and arthropods (insects, arachnids, and crustaceans). Humans are the most important reservoirs of human infectious diseases. Infected humans may show symptoms of the disease and pass it to another human or they may harbor a microbe, but not show symptoms of disease; however, they have the ability to transfer the microbe to another human/animal (carrier).

Wild and domestic animals also serve as reservoirs of human disease. Animals provide an environment for microbes quite similar to the human body. For some zoonotic pathogens, several animal species can serve as a reservoir. For example, most mammals, including cats, dogs, bats, skunks, and raccoons, can be a reservoir for rabies. Humans become infected with the rabies virus when they come into contact with the saliva of a rabid animal, usually through an animal bite. Pathogens can also pass from animal reservoir to humans when people consume or handle contaminated animals or animal products as is the case with the consumption of pork contaminated with the worm *Trichinella spiralis*.

Arthropods can also transmit zoonotic diseases. Arthropods consist of insects (flies, fleas, and mosquitoes), arachnids (ticks, mites, and spiders), and crustaceans (crayfish and crabs). A classic example of an arthropod spreading disease is biting insects that have previously bitten an infected animal passing along a zoonotic disease. *Yersinia pestis*, the bacterium that caused the plague, was harbored in rodents, but was transmitted to humans by fleabites. When arthropods transmit disease they are referred to as vectors. The term **vector** is any live animal that transmits an infectious agent from one host to the next; however, this term is typically used to describe arthropods. Vectors are placed into two categories: biological and mechanical.

> A vector is an organism that does not cause disease itself but which spreads infection by conveying pathogens from one host to another.

- **Biological vectors** actively participate in a pathogen's life cycle, serving as a place where the pathogen multiplies or completes its life cycle. Biological vectors spread infectious agents to its host by biting (injecting infected saliva into blood, defecating around the bite wound, or regurgitating blood into the bite wound), aerosol formation, or touch. Mosquitoes, fleas, ticks, and tsetse flies are good examples of biological vectors.

■ **Mechanical vectors** are not necessary to the life cycle of the pathogen and are passive participants in the transmission of disease. Mechanical vectors may spread disease when their external body parts become contaminated through contact with the pathogen. The pathogen is subsequently transferred to the human or animal indirectly by an intermediate such as contaminated food or directly by contact between the contaminated body part and a mucous membrane or skin surface. Houseflies and cockroaches are good examples of mechanical vectors.

Nonliving reservoirs include air, soil, dust, food, milk, water, and **fomites** (objects that are able to transfer disease organisms). Fomites (fomes and fomite are singular forms of the term) are also referred to as vehicles. Direct contact with a zoonotic microbe by way of a nonliving reservoir can transmit animal diseases to humans. Consider the transmission of *Toxoplasma gondii* by the handling of cat feces in a litterbox. Contaminated water, such as that containing the protozoan *Cryptosporidium parvum*, is another route of zoonotic disease transmission via a nonliving reservoir. Air contaminated with *Bacillus anthracis* endospores from animal hides serves as another example of transmission of zoonoses to humans by nonliving reservoirs.

The Effect of Animal Reservoirs

An animal reservoir greatly affects the pattern of disease. New forms of a disease can develop in an animal reservoir. Animal reservoirs can make diseases harder to control. An illustration of how animal reservoirs complicate disease eradication is seen with yellow fever. A virus that is spread through mosquito bites causes yellow fever. Yellow fever ravaged the continents of North America, South America, and Africa until it was discovered that mosquitoes transmitted the virus. Knowing how the disease was spread in these continents allowed for its eradication through the establishment of mosquito control programs. The same approach to eradication of yellow fever in Panama failed in Central America during the construction of the Panama Canal because the virus was found in another animal reservoir, the monkey. Monkeys in the jungle surrounding the Panama Canal site constituted an immense animal reservoir population that made eradication of yellow fever impossible for this region.

TRANSMISSION OF ZOONOSES

The routes of disease transmission are many and varied. Zoonotic pathogens can be transmitted between animals and humans by either direct or indirect transmission (Figure 1-1). **Direct transmission** is the immediate transfer of an agent from a reservoir to a susceptible host. **Indirect transmission** is the transfer of an infectious agent carried from a reservoir to a susceptible host. Disease transmission can be further classified as horizontal or vertical. **Horizontal transmission** is the spread of disease through a population from one infected individual to another. Horizontal transmission can be either direct or indirect. **Vertical transmission** means the disease is spread from parent to offspring via the placenta, sperm, milk, or ovum. Vertical transmission from parent to offspring is always direct; however, a pregnant woman can contract a zoonotic disease directly from an animal then pass that disease to the fetus vertically (indirect transmission from the original animal; however, the disease is spread directly from mother to fetus). Figure 1-1 summarizes the types of zoonotic disease transmission and provides examples of each type.

Direct transmission

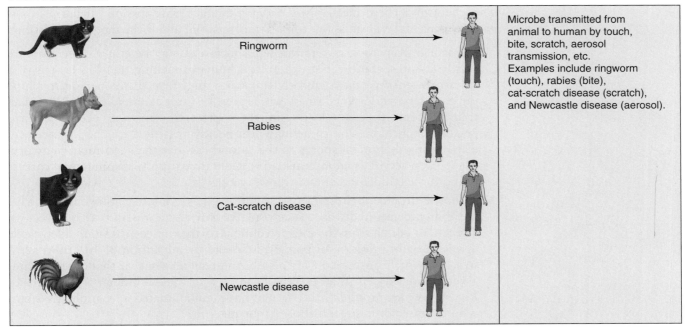

Ringworm

Rabies

Cat-scratch disease

Newcastle disease

Microbe transmitted from animal to human by touch, bite, scratch, aerosol transmission, etc. Examples include ringworm (touch), rabies (bite), cat-scratch disease (scratch), and Newcastle disease (aerosol).

Indirect transmission

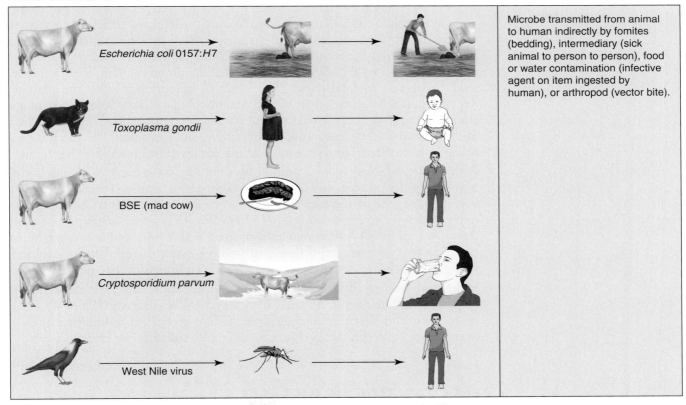

Escherichia coli 0157:H7

Toxoplasma gondii

BSE (mad cow)

Cryptosporidium parvum

West Nile virus

Microbe transmitted from animal to human indirectly by fomites (bedding), intermediary (sick animal to person to person), food or water contamination (infective agent on item ingested by human), or arthropod (vector bite).

Figure 1-1   Types of zoonotic disease transmission.

Direct and indirect transmission of zoonotic disease can occur by a variety of mechanisms. These mechanisms include contact transmission, airborne transmission, placental transmission, fomite transmission, and arthropod transmission:

- Contact transmission occurs through touch and may be either direct or indirect in its presentation. For example, a human touching the skin lesions of a ringworm-positive cat frequently transmits ringworm. If the person touching the cat's lesions develops ringworm lesions, it is an example of direct contact transmission. An example of indirect contact transmission occurs when a person touches the skin of a ringworm-positive cat then touches another cat transferring the fungal spores to the second cat that then contracts ringworm (the person serves as an intermediary). Contamination from an animal carcass is another example of contact transmission.
- Airborne droplets of respiratory secretions may contain pathogens and have the ability to spread disease via aerosol. Airborne transmission can be direct via a sneeze or cough from the infected animal to an uninfected animal. Newcastle disease, a viral infection of poultry, is spread by inhalation of infectious aerosols. Airborne transmission can also be indirect via a sneeze that contaminates a ventilation system that eventually infects an animal/human. Inhalation of *Legionella pneumophila* bacteria that have contaminated air-handling systems is an example of indirect airborne transmission.
- Placental transmission occurs from mother to offspring and is a form of direct transmission. The placenta is an organ formed by maternal and fetal tissue that separates the blood of the fetus from the mother, yet permits diffusion of dissolved nutrients and gases to the fetus. Because the fetal and maternal circulation is separated to some degree, the placenta is a relatively effective barrier against microbes entering the fetal circulation from the mother's circulation. However, some microbes can cross the placenta, enter the umbilical vein, and spread by fetal circulation to fetal tissue. The result of microbes crossing the placenta depends upon the gestation stage and the microbe involved. Serious neurologic complications can occur with placental transmission of zoonotic organisms such as *Listeria monocytogenes* and *Toxoplasma gondii* in which the mother contracted the disease from an animal and placentally transmitted it to her fetus.
- Fomites (or vehicles) such as animal bedding that have become contaminated by blood, saliva, urine, feces, vomit, exudates, respiratory secretions, or milk may be a source of indirect transmission of infectious agents. Bedding contaminated with *Escherichia coli 0157:H7* can be a route of indirect fomite transmission. Fecal-contaminated food and water and contaminated blood products or parenteral injections (substances given directly into the bloodstream) are other examples of indirect fomite transmission of organisms.
- Arthropods such as fleas, mosquitoes, lice, ticks, mites, and flies can indirectly transmit infectious agents. A variety of viral encephalitis diseases such as West Nile encephalitis and Eastern equine encephalitis are spread via mosquitoes and are examples of indirect arthropod transmission (also called vector transmission).

CLASSIFICATION OF DISEASES

Every disease, whether zoonotic or not, can also be classified based on its severity and longevity. **Acute diseases** are diseases that develop rapidly but last only a short time. The common cold in humans is an example of an acute disease. **Chronic diseases** are diseases that develop slowly, usually with less severe clinical signs, and are continual or recurrent. Tuberculosis and leprosy are examples of chronic diseases. **Subacute diseases** have severities and durations somewhere between acute and chronic diseases (they do not come on as rapidly as acute diseases and do not persist as long as chronic diseases). Examples of subacute diseases are cytomegalovirus in people and some forms of anthrax in animals. **Latent diseases** are diseases in which a pathogen remains inactive for long periods of time before becoming active as is the case with herpes viral infections of animals and humans.

Diseases can also be classified based on their degree of infectivity. **Infectious disease** is a disease that is acquired from an infected host. Infectious diseases are also known as **communicable diseases**. Examples of communicable diseases are tuberculosis and influenza. Easily transmitted communicable diseases are known as **contagious diseases** as is the case with chickenpox in people. **Noncommunicable diseases** are not spread from one host to another and diseased individuals do not serve as a source of contamination for others. Tooth decay and tetanus are examples of noncommunicable diseases.

EPIDEMIOLOGY AND ZOONOSES

The effects of diseases on a community involve the field of **epidemiology**. An epidemiologist studies the factors that determine the frequency and distribution of disease within populations. Some factors epidemiologists study include microbe virulence, portals of microbe entry and exit, and the course of disease. Epidemiologists are also concerned about disease surveillance. Disease surveillance involves the statistical analysis (collecting, analyzing, and reporting) of data on the rates of disease prevalence, mortality, and morbidity. Table 1-1 defines statistical terms used in the field of epidemiology.

> Epidemiologists are concerned with the who, what, where, when, why, and how of infectious disease.

The Frequency of Disease

Epidemiologists monitor statistics to determine the frequency of a particular disease in a given population. There are several stages of disease in a population: endemic, sporadic, epidemic, and pandemic. Endemic means native to the population; therefore, **endemic diseases** are diseases that are always present within a population of a particular geographic area. **Enzootic diseases** are conditions affecting animals of a specific geographic area. An enzootic disease is constantly present in a specific animal community, but only occurs in a small number of cases. In some parts of the United States, plague is enzootic in prairie dogs and rodents but is not endemic in people. Plague is only seen occasionally in humans in the United States. When there are a few isolated cases of a disease, such as plague in humans, seen in widespread areas in an unpredictable manner the disease is termed **sporadic**. Epidemic means visiting the population; therefore, an **epidemic disease** is a disease with a sudden onset and widespread outbreak within a group. When an epidemic disease enters a new population that has few defenses against it, the disease causes acute signs in all ages. In the early 1980s the annual incidence of measles in the United States was between 1,000 and 3,000 cases. By the late 1980s the number of cases of measles had reached nearly 25,000 and was considered to be an epidemic. Widespread disease in populations of animals other than humans is referred to as being **epizootic**. Epizootic diseases spread rapidly, simultaneously affecting a large number of animals in a region. Foot-and-mouth disease (FMD) is an example of a disease that can be epizootic. Pandemic means all the population; therefore, a **pandemic disease** is a disease that is a widespread epidemic and generally involves the spread across

Table 1–1   Statistical Terms Used in Epidemiology

Term	Definition	Example
Rate	The ratio of the number of individuals in a particular category to the total number of individuals in the population being studied.	The pulmonary form of anthrax has a 100% fatality rate.
Incidence[*] (also called the **morbidity rate**)	The number of *new* cases of a disease in a defined population over a specific time period as compared to the general healthy population in a certain time period (typically 1 year). Number of new cases ÷ total number of individuals × 100. Think of incidence as the rate of acquiring a disease or condition during a certain period or the general health of a population by giving information about the sick rate.	In a population of 100, 10 people contract ringworm. The incidence is 1 in 10 or 10 ÷ 100 × 100 = 10%. The incidence of ringworm in this population is 10 cases per 100 population. The morbidity rate of stray dogs with leptospirosis is 40%.
Prevalence	The total number of *existing* cases in a population compared to the entire population. Total number of existing cases in a population ÷ total number of individuals in a population × 100. Think of the prevalence as the number of people/animals with a certain disease at a particular time or a snapshot view of the disease. Prevalence is not a rate and is the measure of a disease in a population because it includes all who have been diagnosed in prior years as well as the current year.	In a population of 500 people, 100 individuals have hantavirus infection. The prevalence is 100 ÷ 500 × 100 = 20%.
Mortality Rate	The ratio of individuals who die from a particular disease compared with those individuals in a population.	The mortality rate in people who contract *Listeria monocytogenes* from animals is 20%. The mortality rate of pneumonia is 7.5 per 1000 animals (if the population is 100,000 than there would be 750 deaths).

[*]Chronic diseases like diabetes mellitus can have low incidence but high prevalence because prevalence is a sum of past incidence.

continents. An example of a pandemic disease is the influenza outbreak of 1918 in which 20 to 40 million people died worldwide, including 500,000 people in the United States. **Panzootic disease** is widespread epizootic disease. An example of a panzootic disease is the H5N1 avian influenza outbreak in birds that began in 2003 in parts of Southeast Asia and spread to birds in other parts of the world.

AGENCIES MONITORING ZOONOSES

Public health agencies strive to prevent the spread of zoonotic disease and to identify and eliminate any that do occur. These agencies educate the public about how the diseases are transmitted and explain proper sanitation procedures, identify and attempt to eliminate reservoirs and sources of infection, carry out measures to isolate sick animals/humans, and help in the treatment of ill animals/people. These public health agencies have helped to eradicate smallpox and greatly reduce the incidence of poliomyelitis in many parts of the world. Examples of these agencies include:

- The World Health Organization (WHO) is a specialized agency within the United Nations. WHO was founded in 1948 and its missions are to promote cooperation for health care among nations, carry out disease control and eradication programs, and improve the quality of human/animal life. When epidemics occur, WHO sends out teams of epidemiologists to investigate the outbreak and to assist in bringing the outbreak under control. WHO works at the control (reducing the incidence and/or prevalence of a disease), elimination (reduction of case transmission to a predetermined very low level), and eradication (achieving a status where no further cases of that disease occur anywhere and where continued control measures are unnecessary) of diseases.

■ The Centers for Disease Control and Prevention (CDC) is a United States federal agency under the control of the U.S. Department of Health and Human Services and its function is to assist state and local health departments in all aspects of epidemiology. The CDC was first established as the Communicable Disease Center in Atlanta, Georgia in 1946 and its focus was on communicable diseases (at that time malaria and typhus). The current role of the CDC is "to promote disease prevention and health promotion goals that will foster a safe and healthful environment where health is protected, nurtured, and promoted." One branch within the CDC is the National Center for Infectious Disease (NCID) and its mission is "to prevent illness, disability, and death caused by infectious disease in the United States and around the world." Some zoonotic diseases, known as nationally notifiable diseases, must be reported to the CDC and include examples such as salmonellosis, shigellosis, tuberculosis, and Lyme disease. The CDC also prepares a weekly publication entitled *Morbidity and Mortality Weekly Report* (MMWR), which contains information and statistics about infectious outbreaks in the United States and other parts of the world.

■ The National Institutes of Health (NIH), a division of the U.S. Department of Health and Human Services, is the primary federal agency for conducting and supporting medical research. The NIH was founded in 1887 with the creation of the Laboratory of Hygiene at the Marine Hospital in Staten Island, New York. NIH publishes guidelines for the care of animals in research facilities as well as the prevention of disease transmission among animals and between animals and humans. NIH funds research on a variety of disease topics including zoonoses and ways to prevent their spread. NIH also educates the public and physicians about the role of zoonoses and human health through providing information such as the role of pets and the immunocompromised person.

ZOONOSES AND JOB SAFETY

It is imperative that veterinary professionals be aware of the different types of zoonotic disease and ways to prevent their spread. The Occupational Safety and Health Administration (OSHA), a division of the U.S. Department of Labor, was developed to "make the place of employment free from recognized hazards that are causing or are likely to cause death or serious physical harm." The most common injury among workers at veterinary clinics is animal bites that occur when trying to restrain animals. Animal bites can transmit zoonoses; therefore, proper technique when handling animals is critical.

Contact with blood, stool, laboratory cultures, and infected animals may all serve as zoonotic disease sources. OSHA requires each workplace to conduct a hazard assessment for each job to determine the level of protective equipment and training required for each task. Examination gloves, masks, safety goggles, aprons, work shoes or boots, eyewash stations, and other protective equipment may be needed when handling certain specimens. Specific safety equipment needed for a particular zoonotic disease will be covered under that section.

Veterinary professionals are also required to report diseases that may affect human public health. Being able to recognize the transmission of zoonoses and to identify means of prevention as well as the education about zoonoses to the public are important functions of the veterinary staff. Both federal and state laws regulate which diseases are reportable and the government agency to which they must be reported. At the federal level, the U.S. Department of Agriculture's (USDA) Animal and Plant Health Inspection Service (APHIS) requires licensed veterinary personnel to report diseases that pose a significant threat to human health. Similar guidelines may exist at the state level depending upon the state involved. Individual state guidelines are available through the state's Department of Food and Agriculture or similar agency.

Review Questions

Multiple Choice

1. Zoonoses
   a. have always been a problem for humans because of our close association with animals.
   b. became more of a problem when the nomadic nature of humans increased.
   c. increased with the alteration of human lifestyles with the domestication of plants and animals.
   d. are easily contained within one geographic area as a result of the isolation of one country to the next.

2. A domestic animal that serves as a host for a particular disease and is placed at various locations to determine the potential for human exposure to a particular disease is a
   a. reservoir.
   b. source.
   c. carrier.
   d. sentinel.

3. The bubonic plague is caused by the bacterium *Yersinia pestis*. This bacterium is found in rodents. When fleas take a blood meal from infected rodents, the bacteria multiply in the flea. What type of vector is the flea in this case?
   a. biological
   b. mechanical
   c. horizontal
   d. vertical

4. The effects of diseases on a community involve what field of study?
   a. invertebrate disease transmission
   b. biology
   c. entomology
   d. epidemiology

5. What is the name of the weekly CDC publication that contains information and statistics about infectious outbreaks in the United States and other parts of the world?
   a. *Morbidity and Mortality Weekly Report* (MMWR)
   b. *CDC Weekly Updates* (CDC-WU)
   c. *U.S. Department of Health and Human Services Weekly* (US-DHHSW)
   d. *Incidence of Mortality Weekly* (IMW)

6. The statistic that tells about the general health of a population is the
   a. ratio.
   b. incidence.
   c. prevalence.
   d. mortality rate.

7. The statistic that provides a snapshot view of a particular disease is the
   a. ratio.
   b. incidence.
   c. prevalence.
   d. mortality rate.

8. The immediate transfer of an agent from reservoir to a susceptible host is known as
   a. direct transmission.
   b. indirect transmission.
   c. horizontal transmission.
   d. vertical transmission.

9. When a person gets infectious agents from a drinking glass, the drinking glass is known as a
   a. biological vector.
   b. reservoir.
   c. vector.
   d. fomite.

10. An animate or inanimate object that serves as a long-term habitat and focus of dissemination for an infectious agent is known as a
    a. reservoir.
    b. source of infection.
    c. vector.
    d. fomes.

Matching

11. _____ zoonoses

12. _____ vector

13. _____ sporadic diseases

14. _____ vertical transmission

15. _____ horizontal transmission

16. _____ endemic diseases

17. _____ enzootic diseases

18. _____ epidemic diseases

19. _____ epizootic diseases

20. _____ panzootic diseases

A. diseases that are always present within a population of a particular geographic area

B. a live animal that transmits an infectious agent from one host to the next

C. conditions affecting or are peculiar to animals of a specific geographic area

D. diseases naturally transmitted from animals to humans

E. diseases with a sudden onset and widespread outbreak within a group

F. widespread epizootic disease

G. isolated cases of a disease seen in widespread areas in an unpredictable manner

H. spread of disease from one parent to offspring

I. spread of disease through a population from one infected individual to another

J. widespread diseases in populations of animals other than humans

Case Studies

Identify the following case studies as endemic, enzootic, sporadic, epidemic, epizootic, pandemic, or panzootic.

21. In spring 1993, cryptosporidiosis (a diarrheal disease caused by the protozoan *Cryptosporidium parvum*) affected more than 400,000 people in Milwaukee, Wisconsin. The oocysts (thick-walled structures in which protozoa develop and are transferred to new hosts) of *Cryptosporidium parvum* were present in cattle feces that were washed off Wisconsin dairy farms into Lake Michigan. The water of Lake Michigan provides Milwaukee with its drinking water. Although the lake water had been treated with chlorine the tiny oocysts were not killed and passed through the filters that were being used in Milwaukee. Thus the oocysts were present in the city's drinking water and people who drank city water became infected. This outbreak caused the death of more than 100 immunocompromised people.

22. The first documented evidence of human immunodeficiency virus (HIV) infection in humans can be traced to an African serum sample collected in 1959 (it is possible that HIV infection occurred before this date). HIV is thought to have been transferred to humans from nonhuman primates. In the 20-year period from 1981 to 2001 WHO stated that acquired immunodeficiency syndrome (AIDS) has become the most devastating disease humankind has ever faced. More than 60 million people have become infected with HIV and an estimated 22 million people have died of AIDS.

23. Foot-and-mouth disease (FMD) is a viral, infectious disease of domestic and wild cloven-hoofed animals. The disease is characterized by vesicular lesions and erosions of the epithelium of the mouth, nares, muzzle, feet, teats, and udder. Morbidity and mortality are the highest in the young; dairy breeds are particularly susceptible to FMD infection. The virus may be spread via aerosols, usually when animals are in close proximity. In 2001, many countries were experiencing outbreaks of FMD. Only the continents of Australia, Antarctica, and North America were free of FMD.

References

Burton, G., and P. Engelkirk. 2004. *Microbiology for the Health Sciences*, 7th edition. Philadelphia, PA: Lippincott, Williams & Wilkins, pp. 182–193.

Glossary of epidemiology terms. 2002. http://www.atsdr.cdc.gov/glossary.html (accessed February 28, 2005).

Ingraham, J., and C. Ingraham. 2004. *Introduction to Microbiology: A Case History Approach*, 3rd edition. Pacific Grove, CA: Thomson Brooks Cole, pp. 364–5.

Kobasa, D., A. Takada, K. Shinya, et al. 2004. Enhanced virulence of influenza A viruses with the haemagglutinin of the 1918 pandemic virus. *Nature* 431:703–7.

Shapiro, L. 2005. *Pathology and Parasitology for Veterinary Technicians*. Clifton Park, NY: Thomson Delmar Learning, pp. 99–107.

Talaro, K., and Talaro, A. 1993. *Foundations in Microbiology*. Dubuque, Iowa: William C. Brown Publishing, pp. 351–354.

Zoonoses: The Natural History of Disease. San Diego Natural History Museum Field Guide, http://www.sdnhm.org/fieldguide/zoonoses/index.html (accessed February 25, 2005).

Chapter 2

Principles of Immunity and Diagnostic Techniques

Objectives

After completion of this chapter, the learner should be able to

- Differentiate between innate and acquired immunity
- Describe different types of innate immunity
- Describe how fever is induced
- Describe the steps involved in inflammation
- Describe the steps involved in phagocytosis
- Describe the different types of acquired immunity
- Differentiate the types of T lymphocytes and their functions
- Differentiate the types of immunoglobulins and their functions
- List examples of cytokines and their functions
- Describe humoral immunity
- Describe cell-mediated immunity
- Differentiate the primary and secondary immune responses
- Describe immunodiagnostic tests
- Describe diagnostic techniques involving genetic material

Key Terms

acquired immunity
agglutination test
cell-mediated immunity
complement fixation
complement system
cytokines
enzyme-linked immunosorbent assay (ELISA)

fever
fluorescent antibody techniques
humoral immunity
immunoassays
inflammation
inflammatory peptides
innate immunity
interferons

lactoferrin
lactoperoxidase
lymphokines
monokines
natural killer cells
opsonization
phagocytosis
polymerase chain reaction (PCR)
precipitation reaction

primary immune response
pyrogen
radioimmunoassay (RIA)
reverse transcriptase-polymerase chain reaction (RT-PCR)
secondary immune response

sensitivity
specificity
vasoactive amines
Western blot test

OVERVIEW

Animals and humans have survived on earth for hundreds of thousands of years because they have developed naturally occurring nonspecific defense mechanisms against pathogens as well as complex interactions among different types of immune cells and cellular secretions that target specific pathogens. The ability of any animal species to resist foreign invaders and recover from disease is a result of both innate and acquired immunity. **Innate immunity** is often thought of as "inborn immunity" and is a mechanism of defense that does not depend upon

Innate immunity is nonspecific immunity that consists of a set of disease-resistance mechanisms that are not specific for a particular antigen. Acquired immunity is specific immunity that displays a high degree of specificity as well as the property of memory.

prior exposure to an infectious agent to be effective. Innate immunity consists of nonspecific defense mechanisms that come into play immediately or within hours of the appearance of the antigen (molecule that triggers an immune response) in the body. Innate immunity is different for each species. The fact that dogs do not get human immunodeficiency virus (HIV) is a result of innate immunity. **Acquired immunity**, sometimes referred to as adaptive immunity, is often thought of as the immunity acquired as one goes through life and is specific to a particular foreign infectious agent, requires time to develop, and occurs more quickly and vigorously upon second exposure to that particular agent. It is the combination of these two types of immunity that work in conjunction with each other to protect living organisms from becoming diseased.

INNATE DEFENSE MECHANISMS

There are many different innate defense mechanisms that are nonspecific and help protect the body from pathogens. Innate defense mechanisms can be as simple as a physical barrier to prevent initiation of infectious agents into the body and can become more complex as is the case with substances such as complement and cytokines. An animal's defenses are sometimes categorized as occurring in stages; the first two lines of defense are nonspecific and the third line of defense is specific.

Anatomic and Physiologic Properties

Animals and humans are constantly in the process of defending themselves against microbial invaders. The first line of defense against infectious agents includes any barrier that prevents an organism from entering the body. This first line of defense limits access to the internal tissues and organs of the body. These first lines of defense include:

- *Anatomic properties.* Intact (unbroken) surfaces serve as an anatomic barrier to infectious agents because few pathogens can penetrate unbroken skin or mucous membranes. Hard outer surfaces like skin have an outer layer of dead epithelial cells that have been cornified and keratinized (the epithelial cells are compacted, cemented together, and have an insoluble protein in them called keratin). Desquamation (flaking) of skin also helps skin rid the animal or human of potential pathogens. Soft outer surfaces such as the lining of the digestive, urinary, and respiratory tracts are usually protected by a layer of mucus, which lubricates the surface and helps dislodge particles from it. Mucus producing cells rapidly divide, are constantly produced, and are constantly released from mucous membranes taking bacteria adhering to them out of the body. These soft surfaces may have other adaptations to further protect the animal or human. Other adaptations include hairs and cilia, which can help trap particles and keep them from entering the body or they may propel entrapped particles either cranially out of the body or caudally into another body part. For example, the epithelial lining of the gastrointestinal tract has cilia and mucus. If a foreign particle gets trapped in this cilia and mucus, it can either be expelled cranially out of the esophagus or caudally into the stomach where it can be destroyed by the stomach acid.
- *Physiologic properties.* The wide variety of surfaces found in living organisms has unique properties that protect an animal or human from disease.
  - Skin provides an anatomical barrier against microbe invasion, but it also has physiologic properties that help the skin resist pathogens. These physiologic

properties include dryness, acidity, and temperature of the skin; these inhibit the growth of many microbes. Oil produced by sebaceous glands contains fatty acids that are toxic to some pathogens. Sweat produced by sweat glands flushes microbes from pores and skin surfaces and contains the enzyme lysozyme, which degrades part of bacterial cell walls. Sweat also contains salts, urea, and lactic acid that discourage microbe growth.

■ Mucus produced at mucous membranes contains many substances that can kill or inhibit the growth of bacteria. Lysozyme, **lactoferrin** (a protein that binds iron that is needed by all pathogens), and **lactoperoxidase** (an enzyme that produces highly reactive superoxide radicals that are toxic to bacteria) are all examples of substances found in mucus.

■ Many body systems have pH levels that can alter the rate of microbe growth. Hydrochloric acid production in the stomach produces an acid environment that helps retard the growth of some ingested microbes. The small intestine contains digestive enzymes and bile (an alkaline substance) that destroy microbes. In the reproductive systems, vaginal fluid has a low pH and semen contains an antimicrobial chemical that inhibits bacterial growth. Tears and saliva contain lysozyme, a chemical that can degrade bacterial cell walls and has a basic pH to hinder bacterial growth.

■ Gravity's role in disease prevention is a result of its ability to hinder organisms from gaining a foothold in many body systems. Urination flushes microbes from the urethra and helps reduce the number of microbes that can colonize the urinary tract. Peristalsis and expulsion of feces help remove microbes from the intestine. Lacrimation flushes the eye's surface with tears and carries irritants from the eye.

■ Normal flora or microflora is a complex mixture of microbes (bacteria, fungi, protozoa, and viruses) that reside on or within an animal. The animal body provides a favorable habitat for microbes because it is a constant source of nourishment and moisture, relatively stable pH and temperature, and extensive surfaces on which to live. Animals acquire normal flora at birth and it may fluctuate to some extent during the animal's life. The uterus is normally sterile during embryonic development until prior to an animal's birth at which time the fetus is exposed to microbes with the breaking of the fetal membranes. At this time exposure to many microbes occurs and continues with the process of birth through the vagina. The nature of the intestinal normal flora depends initially upon the type of milk (either mother's milk or milk replacer) being consumed, but in time will be influenced by the environment, feed, and contact with other animals. Normal flora protects an animal from infection by transient microbes because these nonresident microbes must compete with the normal flora for space and nutrients of a particular body area. Many times there are just not enough space and nutrients to go around. Normal flora may also produce substances that help kill transient bacteria thus protecting the environment of normal flora. Body systems that do not contain normal flora such as the nervous system and blood can be especially vulnerable to infections once a microbe gets into these areas.

Cellular and Chemical Protection

Microbes that are able to penetrate the first line of defense are usually destroyed by nonspecific responses. These responses are cellular and chemical in nature and act rapidly at both the local and systemic levels once the first line of defense has been broken. The second line of defense includes:

■ *Fever*. Many animals have specific normal body temperatures. An elevation of body temperature above the normal range is referred to as a **fever**. Normal body temperature is maintained by the hypothalamus, a control center located in the brain. Initiation of fever occurs when a **pyrogen** (a fever-stimulating chemical) resets the hypothalamic thermostat to a higher level. This resetting of the internal thermostat signals muscles to increase heat production and peripheral arterioles to decrease heat loss through vasoconstriction. Pyrogens may be produced in the body (endogenous pyrogen) or they may be produced outside the body (exogenous pyrogen). Interleukin-1 is an example of an endogenous pyrogen because it is produced by activated macrophages found in the body. Bacteria, viruses, parasites, and fungi are all examples of exogenous pyrogens. The benefits of fever include the inhibition of replication of certain temperature-sensitive microbes, the stimulation of white blood cells to destroy microbes, the increase in metabolism of certain cells, the increase in phagocytosis, the reduction in iron available for replication of bacteria, and the enhancement of the effects of interferon. The disadvantages of fever include increased heart rate, increased demand for calories, seizures, and dehydration. Figure 2-1 provides an example of fever induction.

■ *Complement*. The **complement system**, named because it complements the immune reaction, is a group of approximately 30 different proteins found in blood plasma in an inactive form. Complement is an important defense against bacteria and some fungi. The proteins in the complement system interact with one another in a step-wise manner known as the complement cascade. The complement system is activated by the classical pathway (named because it was discovered first) when complement proteins come in contact

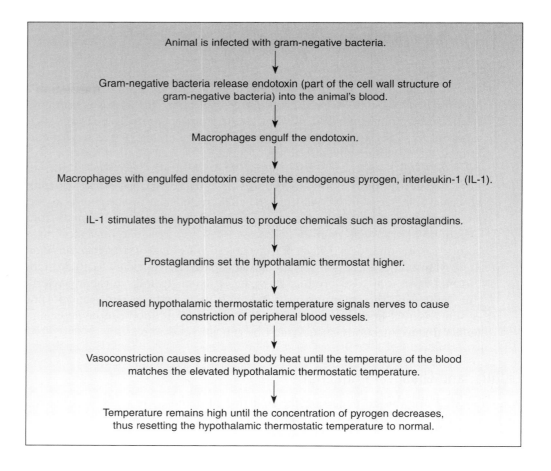

Figure 2-1 Example of fever induction.

with a foreign substance. A series of reactions follow in a sequential manner and as each complement protein is activated, it activates the next complement protein, until the final protein is activated. The classical pathway depends upon an antibody attaching to an antigen; therefore, it responds to the specific immune system covered later in this chapter. In the alternate pathway, the first protein is activated spontaneously in blood and binds to foreign cell surfaces. This binding initiates a cascade of activations, ultimately destroying the foreign cell. The main function of the alternate pathway is rapid lysing of foreign cells, especially viruses and gram-negative bacteria, in the absence of specific immunity. In addition to destroying foreign cells, the complement system also enhances the inflammatory process and phagocytosis.

■ *Interferons.* **Interferons** are a group of glycoproteins released by a variety of cells in response to invasion by intracellular parasites (including viruses) and other stimuli. There are three known types of interferon: alpha (α) a product of lymphocytes that is induced by infection with viruses, bacteria, and other agents; beta (β) a product of fibroblasts, epithelial cells, and macrophages in response to viruses; and gamma (γ) a product of T lymphocytes and natural killer cells that function in immune regulation. Alpha and beta interferon are important in the nonspecific immune response, whereas gamma interferon is important as part of the specific immune response to antigens. Virus-infected cells that produce interferons are unable to save themselves from destruction; however, the interferon produced by these cells attaches to membranes of surrounding cells and prevents viral replication from occurring in those cells. Thus, interferon inhibits the spread of infection. Interferons are not virus-specific (they are effective against a variety of viruses, not just a particular type) but are species-specific (they are effective only in a particular species and ineffective in another species). In addition to cell protection, alpha interferon produced by T lymphocytes also activates a subset of cells called **natural killer cells** (NK cells). NK cells are nonspecific cytotoxic cells that work against any foreign antigen or abnormal cell, but particularly well against tumor and virus-infected cells. NK cells kill animal cells by releasing various cytotoxic molecules, some of which create holes in the target cell's membrane causing cell lysis, whereas others enter the target cell and fragments its nuclear DNA.

> Interferons interfere with viral replication.

■ *Inflammation.* **Inflammation** is a reaction to any traumatic event in the body. Local injury, irritation, microbe invasion, or toxin release are all examples of traumatic events in the body that can trigger the inflammatory response. Once the initial traumatic event has occurred, a chain reaction takes place at the site of damaged tissue, calling beneficial cells and fluids into the injured area. Some of the earliest changes occur in the vasculature near the damaged tissue. These changes are controlled by the nervous system and **cytokines** (chemical mediators) released by blood and tissue cells. The initial reflex response to damaged tissue is vasoconstriction (narrowing of the blood vessels). Vasoconstriction only lasts for a few seconds or minutes and is rapidly followed by vasodilation (widening of the blood vessels). Vasodilation is caused by vasoactive agents such as histamine and prostaglandin released from damaged cells. Vasodilation causes increased blood flow to the damaged area, which in turn causes the redness and heat associated with inflammation. Vasodilation also causes the endothelial cells lining the capillaries to stretch and form gaps through which blood components can leak into the extracellular spaces. The exudate (fluid) that escapes is typically plasma and it accumulates in the tissues causing edema (local swelling and hardness because of exudate accumulation). The accumulation of fluid (dilutes toxic substances), cells, and

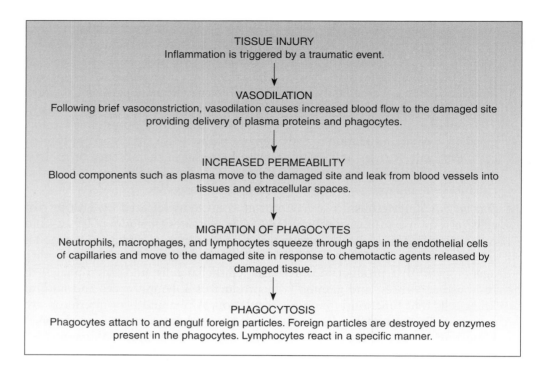

TISSUE INJURY
Inflammation is triggered by a traumatic event.

↓

VASODILATION
Following brief vasoconstriction, vasodilation causes increased blood flow to the damaged site providing delivery of plasma proteins and phagocytes.

↓

INCREASED PERMEABILITY
Blood components such as plasma move to the damaged site and leak from blood vessels into tissues and extracellular spaces.

↓

MIGRATION OF PHAGOCYTES
Neutrophils, macrophages, and lymphocytes squeeze through gaps in the endothelial cells of capillaries and move to the damaged site in response to chemotactic agents released by damaged tissue.

↓

PHAGOCYTOSIS
Phagocytes attach to and engulf foreign particles. Foreign particles are destroyed by enzymes present in the phagocytes. Lymphocytes react in a specific manner.

Figure 2-2   Steps involved in inflammation.

cell debris signals the infiltration of neutrophils (a phagocytic white blood cell) to the area. After a period of time, more slowly reacting phagocytic cells such as monocytes and macrophages come to the damaged site. Clearing of fluid, cellular debris, dead neutrophils, and damaged tissue is done by the macrophages. Lymphocytes react by producing antibodies or kill intruders directly (see specific immune response described later in this chapter). Figure 2-2 summarizes the steps involved in inflammation.

■ *Phagocytosis.* Animal cells must be able to recognize when a substance does not belong in the body. The recognition of nonself is important in the process of **phagocytosis,** the engulfment of an invading particle and abnormal cells (such as dead cells) via invagination of the cell membrane. Phagocytosis begins when phagocytes migrate to the needed site as a result of chemical attraction. This chemical attraction is called chemotaxis. A variety of cells produce chemotactic agents during the complement cascade and during inflammation. Phagocytes move across a concentration gradient (they move from an area of low concentration of chemotactic agents to an area of high concentration of chemotactic agents) and are attracted to the site where they are needed. Different chemotactic agents attract different leukocytes (white blood cells). There are several categories of cells capable of phagocytosis. Types of phagocytes include:

■ *Granulocytes.* Granulocytes are leukocytes that originate from stem cells in the bone marrow and have prominent cytoplasmic granules that can be seen after staining with the appropriate dyes. Granulocytes are divided into neutrophils, eosinophils, and basophils. Neutrophils and eosinophils are phagocytic. Neutrophils are the most abundant and efficient circulating phagocyte and provide the first line of phagocytic defense in an infection. Neutrophils react early in the inflammatory response to bacteria and a high neutrophil count in blood (neutrophilia) is a common finding with bacterial infections. Eosinophils are attracted to sites of parasitic infections and antigen-antibody reactions and an increase in eosinophils (eosinophilia) indicates

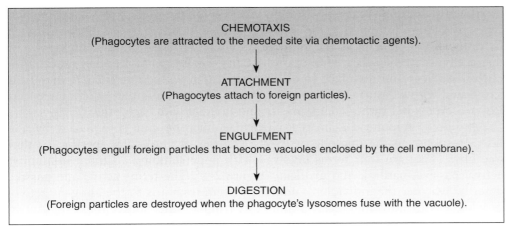

Figure 2-3   Steps involved in phagocytosis.

certain types of parasitic infections or allergic reactions. Basophils are not phagocytic and are involved in allergic and inflammatory reactions.

- *Monocytes.* Monocytes are leukocytes that originate from stem cells in the bone marrow and enter the peripheral blood stream. As monocytes leave the blood and spread through tissues they differentiate into active phagocytes called macrophages. Monocytes become macrophages in lymph nodes, spleen, lungs, and nervous system; they become Kupffer cells in the liver; they become alveolar macrophages in the air sacs of the lungs, and they become microglial cells in the central nervous system (CNS). Macrophages are extremely efficient phagocytes and engulf foreign particles such as cellular secretions, dead leukocytes, erythrocytes (red blood cells), and tissue cells. Macrophages live longer than neutrophils and have the ability to replicate. Macrophages also play an important role in the specific immune response.

Phagocytosis continues when the phagocyte attaches to the foreign particle. Phagocytes can only ingest foreign particles to which they can attach. For this attachment to occur sometimes a process called **opsonization** is needed. Opsonization is a process that facilitates phagocytosis by the deposition of opsonins such as antibodies or complement fragments that coat the surface of foreign particles. This coating of the surface of foreign particles facilitates the recognition and engulfment of the foreign particle because the phagocyte possesses surface receptors for antibodies and complement fragments, thus allowing the phagocyte to attach to the foreign particle. Following attachment of the phagocyte to the foreign particle, the phagocyte then surrounds the particle with pseudopodia causing fusion of the phagocyte and foreign particle. The foreign particle is then engulfed (this process may be referred to as endocytosis). When a phagocyte engulfs a particle, the particle becomes a vacuole enclosed by the cell membrane and is destroyed by the phagocyte's lysosomes. Lysosomes are membrane-bound cellular structures that are filled with digestive enzymes that destroy the foreign particle when they fuse with the vacuole and release these enzymes. Figure 2-3 summarizes the steps involved in phagocytosis.

ACQUIRED DEFENSE MECHANISMS

When innate defense mechanisms fail to protect an animal from a foreign particle, the acquired defense mechanisms need to be activated. An acquired immune response is activated by a specific foreign particle called an antigen. Antigens are usually proteins and are usually (but not always) foreign to the host. There are two types of acquired immunity, the humoral and cellular, which are both mediated by different components of the immune system and function in the elimination of distinct types of microbes. Humoral immunity is based on antibodies found both on cell surfaces and dissolved in blood and lymph, whereas cellular immunity is associated with cell surfaces. The two types of acquired immunity do not occur as isolated events but rather communicate and interact with each other.

Acquired immunity against a particular microbe can be induced by the host's response to the microbe or by the transfer of antibodies specific for that microbe. There are four terms involved with understanding acquired immunity: active, passive, natural, and artificial (Figure 2-4). The terms active and passive describe whether the individual's immune system responds to the antigen (active) or whether the individual receives immunity from another source (passive). Active immunity occurs when an individual's own body makes activated lymphocytes to a particular antigen; therefore, the host's immune system plays an active role in responding to the antigen. Active immunity takes days to weeks to develop and lasts for a long time because memory cells have been produced. Examples of active immunity include the body responding to antigens during an infection or via vaccination. Passive immunity occurs when the immune components develop in another animal and are transferred to an individual who was not previously immune; therefore, the recipient becomes immune without having been exposed to or having responded to that particular antigen. Passive immunity provides immediate protection to a specific antigen but lasts for only a short time because memory cells were not produced. Examples of passive immunity include transfer of antibodies across the placenta or the introduction of antibodies in antisera. The terms natural and artificial refer to how the immunity is obtained. Natural immunity occurs when the immunity is acquired unintentionally through everyday living. Artificial immunity occurs when deliberate action is taken to acquire the

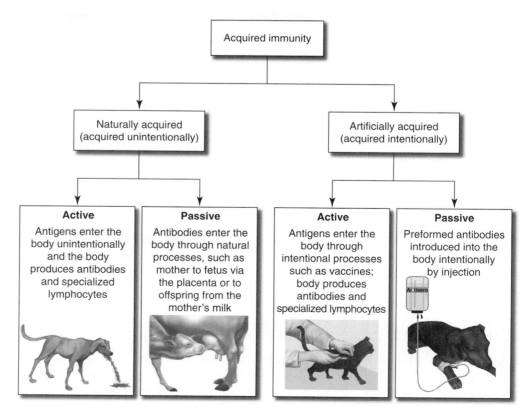

Figure 2-4 Types of acquired immunity.

immunity such as getting a vaccine. Combining these terms provides four types of acquired immunity: naturally acquired active immunity (the body responding to antigens that enter unintentionally such as during infection), artificially acquired active immunity (the body responding to antigens that are intentionally introduced into the body such as vaccines), naturally acquired passive immunity (the unintentional transfer of antibodies from mother to offspring across the placenta or through colostrum/breast milk), and artificially acquired passive immunity (the intentional injection of antitoxins or antisera (obtained from immune individuals) into an animal). Figure 2-4 provides examples of acquired immunity.

> An antigen is any substance that will stimulate an immune response.

Components of Acquired Immunity

The basis for acquired immunity is the recognition of self versus nonself. Nonself recognition involves proteins imbedded in the cell surface called major histocompatibility complexes (MHC). There are two classes of MHC proteins: class I is found on the surface of almost all cells, whereas class II is found on certain cells that have a role in the immune system. MHC proteins are recognized by two basic types of molecules: antibodies and cells that have T-cell receptors.

Lymphocytes are types of leukocytes involved in acquired immunity. There are two main categories of lymphocytes known as B and T lymphocytes and they function differently. Both types of lymphocytes originate in the bone marrow from the same basic stem cell, but develop into two distinct types. Maturation of B lymphocytes is believed to occur in the bone marrow (or some believe they mature in lymphoid tissue of the gut), whereas T lymphocytes mature in the thymus gland. The process of maturation commits each B or T lymphocyte to one specific type. Both types of lymphocyte then migrate to precise areas in lymphoid tissue until they are needed. B lymphocytes have antibody molecules in their surface and give rise to plasma cells. Plasma cells actively secrete antibodies into blood. T lymphocytes have surface receptors that bind antigens. Table 2-1 describes a variety of different types of T lymphocytes. When lymphocytes become activated they are stimulated to move from a stage of recognition where they bind with particular antigens to a stage where they proliferate and differentiate into cells that function to eliminate antigens.

Antibodies are proteins called immunoglobulins that are produced when B lymphocytes become sensitized to a specific antigen, multiply, and mature into plasma cells. The rate of antibody production is extraordinary; one functional plasma cell can make approximately 2,000 antibody molecules per second for about the first 5 days of its activation. The basic structure of an antibody molecule resembles the letter Y

Table 2–1   Types of T Lymphocytes and Their Functions

Type of Lymphocyte	Function
Helper T cell (also known as CD4+ or T_H)	Assists in immune response and secrete substances that stimulate B lymphocytes.
Cytotoxic T cell	Binds tightly to target cell and secretes a protein that causes pores to form in the foreign cell membrane.
Suppressor T cell (also known as CD8+)	Inhibits B lymphocytes and the immune response.
Memory T cell	Remembers the specific antigen and stimulates a faster and more intense response if the same antigen is presented.

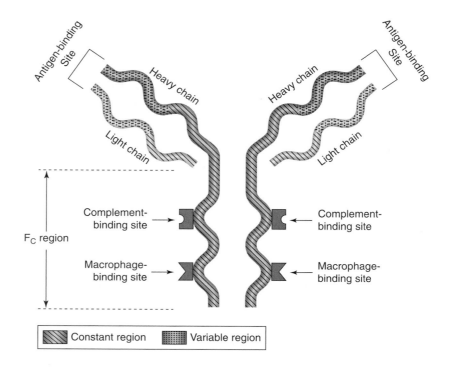

Figure 2-5 Antibody structure.

Each immunoglobulin molecule is shaped like the letter Y and consists of two identical heavy chains and two identical light chains held together by disulfide bonds. Five different kinds of heavy chains form the immunoglobulin's stem (known as the F_C region). The arms of the Y terminate in variable regions to form two antigen-binding sites. Because the amino acid sequences of the two light chains of an antibody molecule are identical (as are the two heavy chains) the binding sites are identical.

and consists of two identical light polypeptide chains, two identical heavy polypeptide chains, two antigen-binding sites, and an F_C region (Figure 2-5). The stem region of the Y, known as the F_C region, is always constant and consists of macrophage-and complement-binding sites. The F_C region is the portion of the antibody that allows it to bind to cells such as neutrophils, macrophages, basophils, and mast cells that possess surface receptors able to recognize this region. Attached to the stem region are two projections that produce the arms of the Y. Each arm of the Y consists of a light and heavy polypeptide chain. The light polypeptide chains contain fewer amino

Table 2–2   Immunoglobulins and Their Functions

Type	Function
Immunoglobulin M (IgM)	M is for macro; therefore, IgM is a huge molecule with a great capacity to bind antigen. It is the first antibody produced in response to antigen. It circulates in blood and is too large to cross the placenta.
Immunoglobulin G (IgG)	IgG is produced by memory cells responding for the second time to a particular antigen. It is the most prevalent immunoglobulin. It circulates in blood and is the only immunoglobulin to cross the placenta.
Immunoglobulin A (IgA)	IgA is a secretory immunoglobulin found in the mucous and serous secretions of the salivary glands, nasal membrane, mammary tissue, lung, urinary tract, and reproductive tract. IgA provides local immunity to the gastrointestinal, respiratory, and urogenital systems and is passed to offspring via nursing.
Immunoglobulin D (IgD)	IgD is found in small amounts in plasma and may play a role in immune suppression.
Immunoglobulin E (IgE)	IgE stimulates an inflammatory response and is involved in allergic and parasitic conditions.

acids than the heavy polypeptide chains; therefore, they are shorter and lighter in weight than the heavy polypeptide chains. The end of each arm contains pockets called antigen-binding sites that are varied in shaped to accommodate a wide variety of antigens. This variability to the end of the antibody is because of the presence of a variable region in which the amino acid composition is varied from one clone of B lymphocyte to another. The rest of the light and heavy polypeptide chain is known as the constant region because its amino acid composition does not vary greatly from one antibody to another. Figure 2-5 shows the basic structure of an antibody.

There are five different classes of antibody, referred to as IgM, IgG, IgA, IgD, and IgE. The class of antibody determines the role of the antibody in the immune response, but not the antigen that it recognizes. Table 2-2 summarizes the classes of immunoglobulins and their functions.

Antibodies can destroy foreign particles in a variety of ways (Figure 2-6). Some ways include:

- Opsonization. Foreign particles become coated with IgG so that they can be more easily recognized by phagocytes. When the foreign particles become coated macrophages are stimulated to engulf the particle.

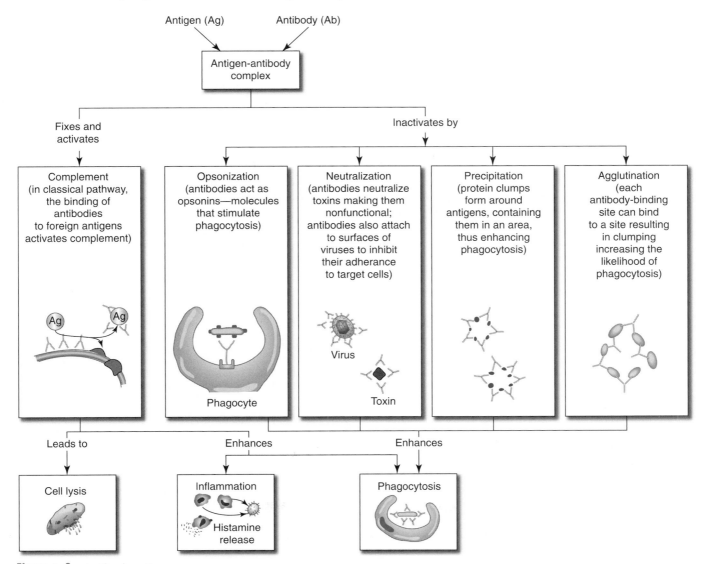

Figure 2-6 Antibody action.

- *Neutralization*. IgG and IgM can neutralize some viruses and toxins secreted by bacteria, whereas IgA neutralizes toxins in digestive and respiratory secretions. Neutralization masks the dangerous parts of bacterial toxins and viruses. Immunoglobulins can bind to a virus' envelope and prevent the virus from attaching to host cells thus preventing it from functioning normally.
- *Complement activation*. The classical pathway of the complement cascade is activated by bound antibody. Bound antibody fixes and activates complement, which leads to antigenic cell lysis.
- *Precipitation*. Antibodies produce clumping of proteins around antigen so that the antigens are contained within an area. This containment enhances the process of phagocytosis.
- *Agglutination*. Antibodies produce clumping of red blood cells or microbial cells because each basic antibody has two antigen-binding sites. Each antigen-binding site can attach to two antigenic determinants at once resulting in several antibody molecules binding with two cells. Agglutination enhances the chance of phagocytosis and hinders the activity of phagocytic organisms.

Cytokines are protein hormones that serve as chemical communicators helping the body mount cooperative mechanisms against foreign particles. There are several different types of cytokines such as **monokines** (cytokines produced by monocytes and macrophages), **lymphokines** (cytokines produced by lymphocytes), **inflammatory peptides** (cytokines produced by neutrophils), and **vasoactive amines** (cytokines produced by platelets and mast cells). Some cytokines act during inflammation and allergic reactions, whereas others function as part of the acquired immune system. Table 2-3 lists a variety of cytokines.

Humoral Immunity

Humoral immunity or antibody-mediated immunity involves the production of antibodies (Figure 2-7). Humoral immunity is most effective against bacteria, viruses located outside of body cells, and toxins. When antigen is introduced into the body it is taken up by antigen-presenting cells (APC) such as macrophages. Macrophages digest the antigen and the antigen binds to the MHC receptor. The combined antigen-MHC molecule is then displayed on the macrophage's

Table 2–3 Some Types of Cytokines and Their Functions

Type	Source	Function
Interleukin-1	Monokine produced by activated macrophages	Activates B and T lymphocytes; mediates inflammation
Interleukin-2	Lymphokine produced by helper T cells	Growth factor for B and T lymphocytes; enhances cytotoxic effects of nonkiller (NK) cells
Interferon	Lymphokine produced by helper and suppressor T cells	Activates macrophages; promotes B- and T-cell differentiation; activates neutrophils and NK cells
Tumor necrosis factor	Monokine produced by activated macrophages	Mediator of inflammation

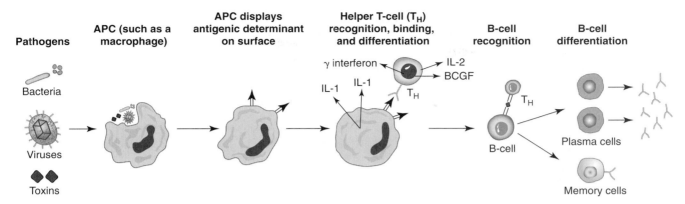

Figure 2-7 Humoral immunity.

The body mounts the humoral immune response against exogenous antigens (toxins or microorganisms that replicate outside of body cells). For exogenous antigens, an antigen-presenting cell (APC), such as a macrophage, internalizes the invading pathogen and enzymatically breaks down the pathogen's antigenic molecules. The antigenic determinants are inserted into MHC molecules and are displayed outside the APCs plasma membrane. Macrophages secrete interleukin-1 to stimulate helper T cells. Helper T cells recognize and bind to the MHC and then differentiate and secrete interleukin-2, B cell growth factor (BCGF), and gamma interferon. B lymphocytes are stimulated by BCGF and the helper T cell binds to the B cell and becomes activated. The B cells then differentiate into antibody-producing plasma cells and long-lived memory cells.

surface. The macrophage secretes interleukin-1, which stimulates helper T cells (APC present antigen to all helper T cells; however, only those helper T cells that have binding sites complementary to the presented antigen will be activated). The appropriate helper T cells attach to the antigen-MHC molecules and then they divide and secrete interleukin-2. The helper T cells also secrete chemicals such as B-cell growth factor that reach a B lymphocyte and activate it. The activated B cell will divide to produce a clone of identical B cells. The majority of these B-cell clones matures into antibody-producing plasma cells and expels antibodies for several days until the plasma cell dies. Each plasma cell makes only one antibody type that binds to that specific antigen that initiated its production. Each plasma cell can make thousands of antibody molecules per second. Plasma cells have high activity levels and are thus short-lived (most die within a few days of activation); however, antibodies can remain in body fluids for months.

Small amounts of activated B cells initially formed by B-cell proliferation do not become antibody-producing plasma cells but rather become memory B cells. Memory B cells are long-lived cells with receptors to the specific antigen that triggered their production. Memory B cells persist in lymphoid tissue for months or years. These memory B cells are a reserve of antigen-sensitive cells that become active and are able to respond rapidly should the antigen enter the body at a later time. Memory B cells can proliferate and differentiate rapidly into plasma cells without requiring interaction with an APC. These newly differentiated plasma cells produce large amounts of antibody within a few days. Once the antigen is under control, suppressor T cells inhibit antibody production.

Cell-Mediated Immunity

Cell-mediated immunity is used when intracellular pathogens or abnormal body cells are present because antibodies are unable to enter cells. Cell-mediated immunity involves the interactions of many cell types and cytokines (Figure 2-8). Although cell-mediated immunity does not involve antibody production, antibodies produced during humoral immunity may play a role in some cell-mediated

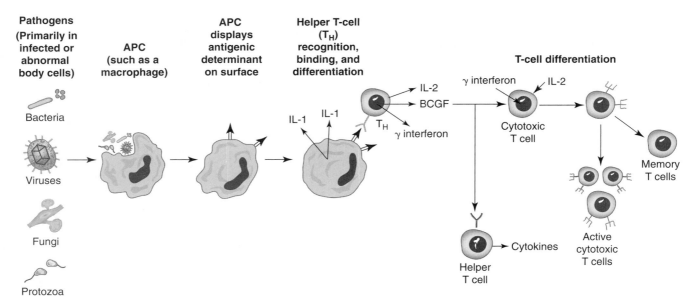

Figure 2-8 Cell-mediated immunity.

The body mounts the cell-mediated immune response against endogenous antigens (cells or microorganisms produced by pathogens that replicate inside the body cells). For endogenous antigens, an antigen-presenting cell (APC), such as a macrophage, internalizes the invading pathogen and enzymatically breaks down the pathogen's antigenic molecules. The antigenic determinants are inserted into MHC molecules and are displayed outside the APC's plasma membrane. Macrophages secrete interleukin-1 to stimulate helper T cells. Helper T cells recognize and bind to the MHC and then differentiate and secrete interleukin-2, B cell growth factor (BCGF), and gamma interferon. Interleukin-2 and gamma interferon produced by the helper T cell stimulates inactive cytotoxic T cells which in turn produce more activated cytotoxic T cells and memory cells. Activated cytotoxic T cells produce cytotoxins such as perforins that destroy the antigen. Other cells divide and become memory T cells that can respond more quickly (without APC interaction) to the antigen on its subsequent presentation and produce cytotoxic T cells. Helper T cells continue to remain active producing cytokines.

responses. Cell-mediated immunity can be described as four distinct types: delayed type hypersensitivity, cytotoxic T-cell response, natural killer cell responses (which are nonspecific), and immediate hypersensitivity. The cytotoxic T-cell response will be described here. Cell-mediated immunity is initiated when a macrophage engulfs and digests an antigen. Antigenic fragments are displayed on the surface of the macrophage. A helper T cell binds to one of the antigenic parts being displayed on the macrophage surface and the helper T cell produces interleukin-2 and gamma interferon. Interleukin-2 and gamma interferon activate those cytotoxic T cells that have T-cell receptors for the antigen. Most activated cytotoxic T cells differentiate into more cytotoxic T cells, whereas a few differentiate into memory T cells that persist for months or years in lymphoid tissue. The "daughter" cytotoxic T cells produce interleukin-2 and become self-stimulating (they no longer need an APC or helper T cell for activation). Cytotoxic T cells have vesicles containing cytotoxins that form channels in the target cell through which enzymes signaling cell death can be transferred. Once the target cell is destroyed the cytotoxic T cell moves to another infected cell.

Activated T cells that have become memory T cells remain inactive until subsequent contact with the same antigen. When memory T cells encounter the same antigen they respond immediately (without interaction with APC) by differentiating to cytotoxic T cells. Because the number of memory T cells is greater than the original number of T cells that recognized the antigen during the initial exposure, the secondary cell-mediated response is much more rapid and effective.

Figure 2-9 summarizes the interaction of the humoral and cell-mediated immune response.

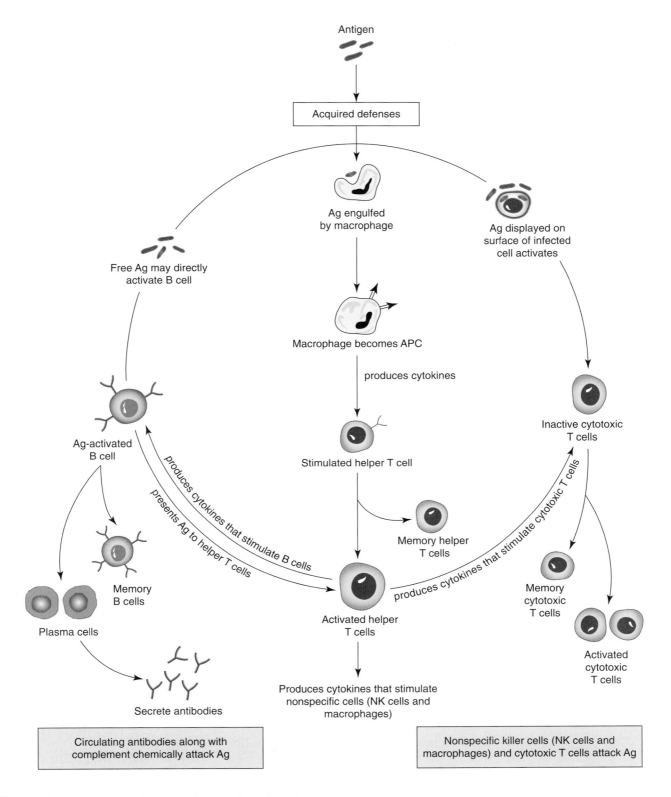

Figure 2-9   Interaction of the humoral and cell-mediated immune response.

Primary and Secondary Immune Responses

The initial action of the immune system to a particular antigen is the **primary immune response**. The primary immune response occurs the first time a specific antigen is identified by either lymphocyte. Following initial exposure to an antigen there is a delayed primary response in antibody production. This delay, known as the lag phase, is a result of the time needed to process antigen. In time, B cells differentiate into plasma cells capable of producing antibody. The primary response typically takes 10 to 14 days for relatively small amounts of antibodies to be produced. As antigen is destroyed, the number of antibodies in blood declines because the plasma cells die marking the end of the primary response. The antigen-stimulated B cells that did not differentiate into plasma cells now become memory B cells. Memory B cells remain dormant in lymphatic tissue until subsequent exposure by the same antigen. Activated memory B cells will differentiate rapidly (approximately 1 to 3 days) and will produce large amount of antibodies. This more rapid response and increased production of antibody following subsequent exposure to the same antigen occurs because so many more cells are able to recognize and respond to the antigen. This rapid recognition and response on subsequent exposure to antigen is called the **secondary immune** or anamnestic **response** (ana- means against and mimneskein means to call to mind).

Vaccine Theory and Immune Response

Primary and secondary immune responses play a role in vaccination theory. When an animal receives the first injection of vaccine, the primary immune response takes place. In time, antibody levels are present in the animal as well as memory cells (the amount of time varies for different types of vaccine). When the animal receives the second injection of vaccine, the secondary immune response takes place. Because memory cells to that specific antigen are present in the animal's body, a more rapid and intense reaction against the antigen occurs resulting in higher antibody levels. Therefore, booster vaccines are given weeks to months apart to raise antibody levels, thus providing adequate levels of antibody protection. Figure 2-10 depicts the primary and secondary humoral immune response.

Another value of booster vaccines is to ensure that maternal antibodies that may be present in young animals are not blocking antibody production in the vaccinated animal. Young animals receive passive immunity through placental transfer of antibodies, consumption of colostrum (antibody-rich milk produced by the mother in the first days following parturition), or if immunoglobulin-containing products are given to them in the first hours of life. Maternal antibodies protect them from infectious agents in the short term; however, they interfere with or delay the young animal's ability to protect itself from infectious agents in the long term. Maternal antibodies treat vaccine antigens like real antigens and inactivate them before they get a chance to stimulate the young animal's immune system. Maternal antibodies have varying durations as a result of both the individual and the infectious agent they protect against. This variable rate of maternal antibody decline makes the timing of vaccine administration difficult. For most infectious agents, maternal antibody concentrations fall to nonprotective levels by 2 to 3 months of age; however, any residual maternal antibody could make antibody production in young animals unresponsive for additional weeks or months. Some maternal antibodies can persist in young animals until 6 months of age, thus preventing immune responses to some antigens for a prolonged period of time.

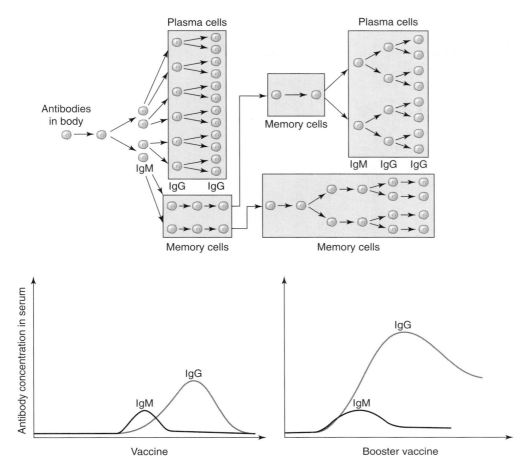

Figure 2-10   Primary and secondary immune response to vaccine.

When initially vaccinated for a disease, the primary immune response produces relatively small amounts of antibody and it may take days or weeks to have sufficient levels of antibody to clear the infection. When a booster vaccine is given, the secondary immune response activates a reserve of antigen-sensitive memory cells that quickly proliferate and differentiate into plasma cells without interaction with antigen-presenting cells (APCs). The newly differentiated plasma cells produce large amounts of antibody within a few days resulting in inactivation of the antigen before it can cause disease.

IMMUNODIAGNOSTICS

Diagnosing infectious disease can be done in a variety of ways. Culturing bacteria on a variety of agars (solid growth media) and performing staining procedures can aid in the identification of bacterial pathogens. Observing parasite eggs or larvae in stool examinations can be a fast and inexpensive way to identify parasitic pathogens. However, sometimes these tests may take longer than desired or the isolation of some pathogens may be impossible with current technology. Thus, numerous tests have been developed that take advantage of the patient's immune response in the direct or indirect identification of infectious agents. Although some of the tests described below have provided the opportunity to identify infectious agents or an animal's exposure to them, there may be problems that exist with some of these tests. **Specificity** is the property of a test to detect only a certain antibody or antigen and not react with an unrelated antibody or antigen (Figure 2-11A). **Sensitivity** is the capability of a test to detect even very small levels of antibody or antigen for which the test was developed (Figure 2-11B). If tests are not very specific nor very sensitive, false-positive or false-negative results can occur. For example, false-positive results may occur when two related agents have antigenic properties similar enough to cross react with antibodies produced against the other. False-positive results typically occur with tests that do not have a high level of specificity. False-negative results may occur when the level of antibody is not high enough to provoke a positive result yet the animal was exposed to the agent

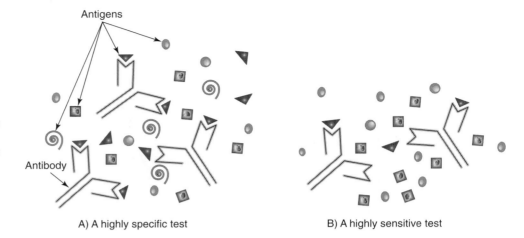

Antigens

Antibody

A) A highly specific test

B) A highly sensitive test

Figure 2-11   Specificity and sensitivity in immune testing. Specificity and sensitivity are terms used to describe immune testing. A) Specificity is the property of a test to detect with extreme exactness only one type of antigen or antibody. B) Sensitivity is the property of a test to detect antigen or antibody at very small levels.

in the past. False-negative results typically occur with tests that do not have a high level of sensitivity.

Immunodiagnostic procedures are laboratory procedures that help diagnose infectious disease by the detection of either antigens or antibodies in clinical specimens. Detection of antigen in a clinical specimen serves as an indication that a particular agent is present in the patient, providing direct evidence that the patient is infected with that organism. Detection of antibodies directed against a particular agent serves as indirect evidence of infection with that organism. The presence of antibodies to a particular organism may indicate exposure to an antigen by past infection or vaccination or may indicate current infection. Because the presence of antibodies may indicate several possible explanations and takes approximately 10 to 14 days to develop, the presence of antigens provides the best evidence of infection.

Some ways to increase the value of antibody tests are to specifically test for IgM antibodies and to utilize paired sera tests. Measuring IgM antibodies is valuable because IgM is the first antibody to appear and is short-lived; therefore, its presence indicates current infection or recent exposure. Paired sera tests involve the collection of one serum sample (the acute sample) during the acute stage of the disease and another serum sample (the convalescent sample) approximately 2 weeks after the acute stage. A significant rise in antibody levels between the acute and convalescent samples demonstrates that the patient was actively producing antibodies against the organism during the 2-week period.

Agglutination Tests

Agglutination occurs when antibodies (also called agglutinins) cross-link with insoluble antigens (also called agglutinogens) to form visible clumps (Figure 2-12A). In **agglutination tests**, the antigens are whole cells such as bacteria or red blood cells (Figure 2-12B). In some agglutination tests, special agglutinogens have been made by attaching inert particles onto the antigen. Sometimes the inert particles are latex beads (known as latex agglutination tests) or red blood cells that react with viral antigens (known as viral hemagglutination tests).

Precipitation in Agar Gel

Precipitation reactions occur when soluble antigen (also called precipitinogen) is made insoluble by an antibody (also called a precipitin). When optimal proportions

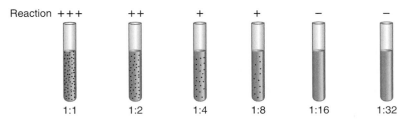

A) Agglutination occurs when antibodies cross-link with insoluble antigens to form visible clumps.

Reaction +++   ++   +   +   −   −

1:1   1:2   1:4   1:8   1:16   1:32

B) Tube agglutination tests use patient's serum that is serially diluted with saline. The number of antibodies are halved in each subsequent tube. An equal amount of the antigen is added to each tube. After incubation and centrifugation, each tube is examined for clumping. A titer is determined and is defined as the dilution of the last tube in the series that shows agglutination (in this example, the titer is 1:8).

Figure 2-12 Agglutination testing.

A) Agglutination involves clumping of whole cells
B) Tube agglutination involves serial dilutions of patient serum that are incubated and examined for clumps to determine the level of antigen or antibody present

of antibody and antigen are reached, a precipitate forms. Precipitates are easily disrupted in liquid media therefore most precipitation reactions are carried out in agar gels (Figure 2-13). Agar gels are soft enough to allow antibody and antigen to freely diffuse, yet are firm enough to hold the antibody-antigen precipitate in place.

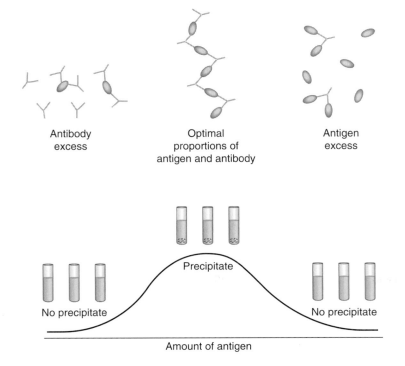

Antibody excess

Optimal proportions of antigen and antibody

Antigen excess

Precipitate

No precipitate

No precipitate

Amount of antigen

Figure 2-13 Precipitation in agar gel.

When antigen and antibody are mixed in proper proportions, they form large lattice-like macromolecules called precipitates. Precipitation is when a solution of soluble antigen is mixed with specific antibody that forms an insoluble precipitate consisting of antigen-antibody complexes.

Complement Fixation

Complement fixation tests are methods to determine whether complement has been bound to an antigen-antibody complex (Figure 2-14). When complement is bound it is termed fixed and is inactive. Some antibodies bound to the surface antigens of cells have the ability to combine with complement and to make complement inactive. Sample serum is heated to destroy naturally present complement and the sample is incubated with the antigen of interest in the presence of added complement, red blood cells, and anti–red blood cell antibodies. If antibody is present in the sample serum, components of complement bind to the antigen-antibody complex. Complement is then unavailable to bind with the anti–red blood cell antibodies that have bound to the red blood cells and therefore cannot lyse the red blood cells (no

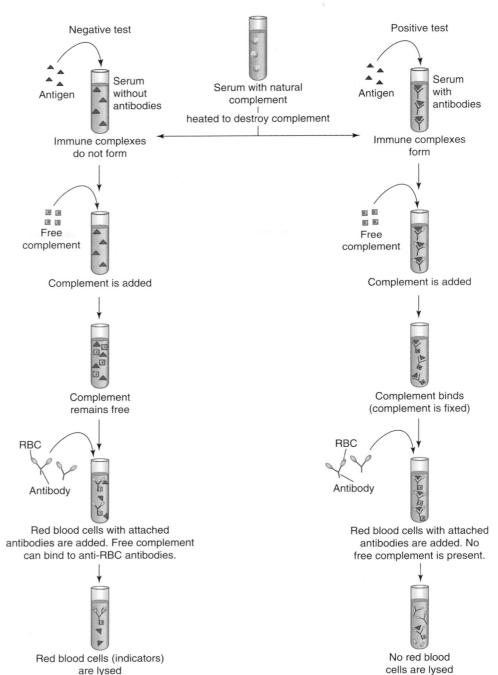

Figure 2-14   Complement fixation.

Complement fixation testing uses a patient's serum that is heated to destroy naturally present complement. Known amounts of the desired antigen and complement are added and animal red blood cells and antibody against them (anti–red blood cell [RBC] antibodies) are added. A positive test occurs when antibody against the antigen are present and antigen-antibody complexes bind to complement which then cannot bind with anti-RBC antibodies to lyse the red blood cells. A negative test occurs in the absence of antibody leaving free complement to bind to anti-RBC antibody attached to the red blood cells which are then lysed.

lysis = positive result). If antibody is not present in the sample serum, the components of complement remain free and inactivated. Complement is then available to bind with the anti–red blood cell antibodies that have bound to the red blood cells and can lyse the red blood cells (lysis = negative result). Lysis of the sensitized red blood cells indicates that complement is activated because it was not fixed (bound) in an antigen-antibody complex because antibody was not present in the sample serum.

Fluorescent Antibody Techniques

Fluorescent antibody techniques use the properties of dyes such as fluorescein and rhodamine to emit visible light in response to ultraviolet radiation (Figure 2-15). Fluorescent antibodies can be used for direct testing or indirect testing of antigen. In direct testing, an unknown antigen is fixed to a slide and exposed to a fluorescent antibody solution for a known antigen. If the antibodies are specific for the unknown antigen, they will bind to it. The slide is then rinsed to remove unbound antibodies and observed with a fluorescent microscope. Fluorescing cells indicate the presence of antigen-antibody complexes and a positive test result. In indirect testing, the fluorescent antibodies are antibodies to a portion of another antibody. A known antigen is then combined with the sample serum. If the sample serum has antibodies to the known antigen, the antibody will bind to the antigen. The fluorescent antibody solution (the antibody to the antibody in the sample serum) is then added and rinsed. Fluorescing cells indicate that the fluorescent antibodies have combined with the antibodies in the sample serum providing a positive test result. In a negative test, no fluorescing complex will be seen.

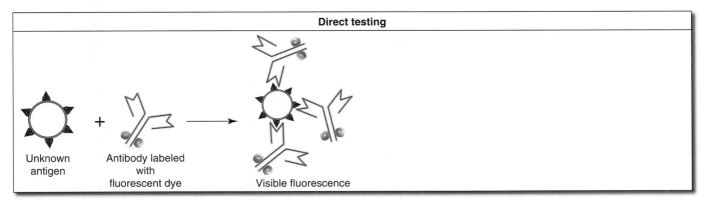

Figure 2-15 Fluorsecent antibody testing.

A) Direct fluorescent antibody testing uses unidentified antigen that is directly tagged with fluorescent antibody. Visible fluorescence is seen. B) Indirect fluorescent antibody testing uses known antigen that binds unknown antibody. When a second antibody with fluorescent dye is added it binds to the first antibody and visible fluorescence is seen.

Immunoassays

The most sensitive immunologic test methods that detect antibodies or antigens are those that have a molecular tag that is easy to detect at extremely low concentrations. Examples include immunoassays and Western blotting.

Immunoassays are extremely sensitive methods that accurately and rapidly identify either antigen or antibody. Ways of detecting antigen or antibody include radioactive isotope labels, enzyme labels, or electronic sensors. **Radioimmunoassay (RIA)** is a technique in which antibodies or antigens are labeled with a radioactive isotope that can be used to detect small amounts of a corresponding antigen or antibody. In RIA the labeled substance competes with its natural, unlabeled one for a reaction site. Large amounts of bound radioactive component indicate that the sample did not have the substance for which the sample was being tested (in other words, it did not have to compete for the reaction site with its natural form). The level of radioactivity is measured with an isotope counter or autoradiograph. **Enzyme-linked immunosorbent assay (ELISA)** is an easy to perform technique in which antibody is linked to an antigen and antibody is linked to an enzyme-antibody complex that produces a color change (Figure 2-16). In direct ELISA, a known antibody is exposed to a sample. If the sample contains the specific antigen, the antibody will bind to it. An enzyme-antibody complex that can react with the antigen is then added. Bound antigen will attract enzyme-antibody and keep it in place. A substrate to the enzyme is then placed in the test. If an enzyme

ELISA method (direct)

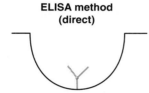

Direct ELISA testing detects antigen levels in a sample. Antibody specific for that antigen is coated onto the wall of the test well.

If antigen is present in the sample, it will bind to the antibody on the wall of the test well.

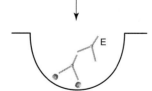

A second antibody with an attached enzyme specific for the same antigen is added. If antigen is not bound to original antibody, the second antibody is washed away.

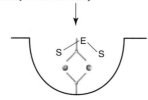

Substrate is added. If enzyme is present the substrate is converted to a product producing a color change. No color change means the enzyme was washed away and is a negative test.

ELISA method (indirect)

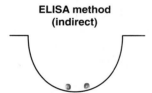

Indirect ELISA testing detects antibody in a sample. The antigen specific for that antibody is coated onto the wall of the test well.

Sample is added and if antibody is present, it binds to the antigen coating the test well.

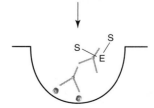

An anti-antibody with an attached enzyme is added. If the first antibody did not bind, the second antibody can not bind and is washed away.

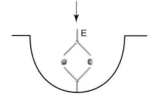

Substrate is added. If enzyme is present the substrate is converted to a product producing a color change. No color change means the enzyme was washed away and is a negative test.

Figure 2-16 Direct and indirect ELISA testing.

Enzyme-linked immunoabsorbant assay (ELISA) testing may be used to detect the amount of either antigen or antibody in a sample. Direct ELISA testing detects antigen in a sample; indirect ELISA testing detects antibody.

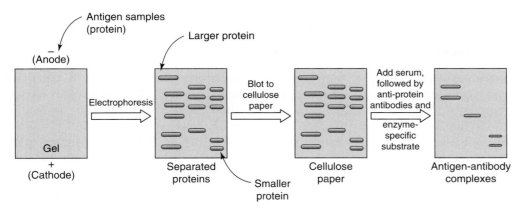

Figure 2-17　Western blot technique.

The Western blot test is used for detecting antibodies against multiple antigens in a mixture. Antigens in solution are placed in wells and separated by gel electrophoresis. Each protein in solution is resolved into a single band, producing a pattern of protein bands. Each of the protein band patterns is transferred to a cellulose paper by absorbing the solution into absorbent paper and cutting it into strips. Each strip is incubated with a test solution. The strips are washed, an enzyme-labeled anti-antibody solution is added, the strips are washed again, and exposed to the enzyme's substrate.

has been affixed to antigen, it combines with the substrate and releases a colored dye. Any color development is a positive test; lack of color is a negative test. In indirect ELISA, an antigen specific for the antibody being measured is adsorbed to the surface of the test plate. The sample is then added. If the sample contains antibodies to the antigen the two will bind. An enzyme-antibody reagent to the antibody being measured is added to the test plate. The substrate to the enzyme is then added. Any color development indicates that all the components reacted because antibody was present in the sample.

> Direct ELISA tests for antigen; indirect ELISA tests for antibody.

Blotting Tests

Blot tests are used to detect deoxyribonucleic acid (DNA), ribonucleic acid (RNA), or protein using electrophoresis (separation of ionic molecules based on their rate of migration on a medium such as paper or gel that is then stained and quantified). Blot tests used to detect DNA or RNA are covered in the next section. **Western blot tests** separate proteins by electrophoresis, transfer them to membranes, and identify them through the use of labeled antibodies specific for the protein of interest (Figure 2-17). Test strips containing antigen are incubated with patient serum and if antibodies are present they will bind to their corresponding antigen. Unbound antibody is removed by washing. Colored bands appear in the positions where antigen-specific antibodies are present. Western blot tests are commonly used to detect antibodies to specific parts of electrophoretically separated antigenic components. Western blot tests are often used to confirm the specificity of antibodies detected by ELISA screening tests.

DIAGNOSTIC TECHNIQUES INVOLVING GENETIC MATERIAL

Recombinant DNA technology is a collection of procedures for manipulating genetic material in vitro. The tools of recombinant DNA technology are used in a variety of basic techniques to multiply, identify, isolate, and sequence the nucleotides of genes.

Southern Blot Tests

Blot tests are used to detect DNA, RNA, or protein (covered previously) using electrophoresis. Southern blot tests (developed by Ed Southern in 1975) denature specimen DNA, treat the DNA with restriction enzymes resulting in DNA fragments, and then separate single-stranded DNA fragments via electrophoresis. These fragments are then blotted to a membrane that has radiolabeled single-stranded DNA fragments with sequences complementary to those being sought. The presence of double-stranded DNA bearing radiolabel detected by radiography indicates a positive test.

Polymerase Chain Reaction

Polymerase chain reaction (PCR) is a technique available since 1985 that is used to amplify and analyze DNA (Figure 2-18). PCR is becoming a valuable tool in disease detection because it rapidly increases the amount of DNA in a sample allowing for possible detection of infection from a single gene copy. PCR allows a single DNA molecule to be detected in a group of other molecules and to make unlimited copies of the DNA. In PCR, the DNA to be amplified is denatured, primed, and replicated by a polymerase enzyme that can function at high temperatures. When the DNA sample is denatured it is heated in a machine called a thermocycler. Heating separates the strands of DNA exposing each base. The sample is then cooled. During priming, synthetic short DNA strands attach at the ends of the test DNA strands to promote replication. Priming prepares the two DNA strands for synthesis. In the last phase heat-stable DNA polymerase enzyme and nucleotides are added to make complementary strands of DNA. These steps are repeated until multiple copies of the original DNA are produced. PCR can go through many cycles within 2 to 3 hours, making it a valuable tool for diagnosing disease.

    Reverse transcriptase-polymerase chain reaction (RT-PCR) is used when the nucleic acid of interest is RNA rather than DNA (when the virus of interest is an

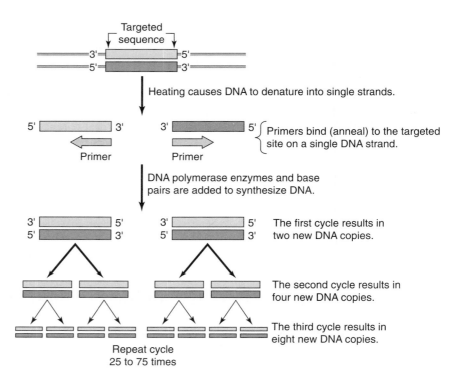

Figure 2-18 Polymerase chain reaction.

Polymerase chain reaction (PCR) is a technique used to produce a large amount of identical deoxyribonucleic acid (DNA) molecules in a laboratory setting. PCR is a repetitive process that alternately separates and replicates DNA by denaturing, priming, and extension of this material.

Table 2–4   Phases of Acquired Immunity

Phase	Action
Recognition	Exposure to a specific antigen resulting in selective activation and expansion of those lymphocytes with antigenic receptors specific for that antigen.
Activation	Activation of lymphocytes by binding of antigen
Effector	Lymphocytes that have been specifically activated by antigen perform their functions that lead to elimination of antigen (such as secretion of antibodies in humoral immunity or destruction of cells in cell-mediated immunity).
Decline	Elimination of cells from the body.
Memory	Generation of memory cells that provide lasting protection from a specific antigen.

RNA virus). RT-PCR modifies the PCR procedures to include conversion of RNA to DNA using reverse transcriptase in the initial steps.

SUMMARY

An animal's ability to resist foreign invaders and recover from disease is a result of both innate and acquired immunity. Innate immunity is a defense mechanism that does not depend upon prior exposure to an infectious agent to be effective, whereas acquired immunity is the protection acquired through life and is specific to a particular foreign invader. Types of innate immunity include physical and chemical barriers, fever, complement, interferons, inflammation, and nonspecific cellular responses such as phagocytosis. Acquired immunity involves the activation of T and B lymphocytes against a particular antigen. Humoral immunity involves the production of antibodies, whereas cell-mediated immunity involves the interaction of many cell types and cytokines. Table 2-4 shows a summary of the phases of acquired immunity.

Immunodiagnostic testing involves detecting disease via antigen-antibody recognition. Some examples of immunodiagnostics include agglutination tests, precipitation reactions, complement fixation, fluorescent antibody techniques, Western blot tests, and immunoassays such as radioimmunoassay and enzyme-linked immunosorbent assays. Recombinant DNA diagnostic techniques include Southern blot tests, PCR, and RT-PCR.

Review Questions

Multiple Choice

1. Regulatory T cells ensure that the immune response is effective but not destructive. What two T cells are considered regulatory T cells?
   a. memory and suppressor
   b. suppressor and cytotoxic
   c. cytotoxic and helper
   d. suppressor and helper

2. Cyclosporine is a drug that inhibits the production of interleukin-2 and is used to prevent rejection of transplanted organs. Why would cyclosporine be effective in preventing transplant rejection?
   a. it inhibits interleukin-2, a substance that stimulates B-cell production
   b. it inhibits interleukin-2, a substance that stimulates T-cell production
   c. both of the above
   d. none of the above

3. Humans are not able to contract fowl cholera. This is an example of
   a. artificially acquired active immunity.
   b. naturally acquired active immunity.
   c. innate immunity.
   d. acquired immunity.

4. What type of immunity can be referred to as antigen-specific immunity?
   a. active
   b. acquired
   c. passive
   d. innate

5. Activated macrophages produce what endogenous pyrogen?
   a. interferon
   b. interleukin-1
   c. interleukin-2
   d. immunogen

6. Alpha interferon activates what subset of cells?
   a. cytotoxic T lymphocytes
   b. cytokines
   c. pyrogens
   d. natural killer cells

7. One advantage to acquired immunity is that
   a. it takes time for it to occur.
   b. it produces memory cells.
   c. it produces antibodies instead of activated cells.
   d. it activates cells instead of producing antibodies.

8. Indirect immunodiagnostic testing involves the identification of
   a. antigen.
   b. antibody.
   c. T lymphocyte.
   d. B lymphocyte.

9. Because it is the first antibody to be produced in response to an antigen, its presence can be used to indicate current infection. What antibody is this?
   a. IgG
   b. IgM
   c. IgA
   d. IgE

10. What diagnostic test method amplifies and analyzes DNA in its identification of antigen?
    a. agglutination test
    b. precipitation test
    c. ELISA test
    d. polymerase chain reaction test

Short Answer

11. Describe five physiologic properties of surfaces that protect animals from infection.

12. What is the complement system and what are some advantages to having it?

13. Differentiate between cell-mediated and humoral immunity with regard to the types of antigens they eradicate.

14. What type of test would give you the greatest chance of false-positive results, a test with low specificity or one with low sensitivity? Why?

15. Give two examples of situations where the active response occurs, yet the animal does not develop signs of disease.

References

Bauman, R. W. 2004. *Microbiology*. San Francisco, CA: Pearson Benjamin Cummings.

Bennett, M. Indiana State University. Inflammation and the immune response. 2006. http://web.indstate.edu/mary/PATHOPHY.htm (accessed March 17, 2005).

Black, J. 2005. *Microbiology: Principles and Explorations*, 6th ed. Hoboken, NJ: John Wiley & Sons.

Burton R., and P. Engelkirk. 2004. *Microbiology for the Health Sciences*, 7th ed. Baltimore, MD and Philadelphia, PA: Lippincott, Williams & Wilkins.

Cowan, M., and K. Talaro. 2006. *Microbiology: A Systems Approach*. New York, NY: McGraw-Hill.

Decker, J. University of Arizona Veterinary Science and Microbiology. Immunity rules. 2006. http://microvet.arizona.edu/Courses/MIC419/Tutorials/immunityrules.html (accessed March 17, 2005).

Dorresteyn Stevens, C. 2003. *Clinical Immunology and Serology: A Laboratory Perspective*, 2nd ed. Philadelphia, PA: F.A. Davis.

Leboffe, M., and B. Pierce. 2005. *A Photographic Atlas for the Microbiology Laboratory*, 3rd ed. Englewood, CO: Morton Publishing Company.

Lunn, D. P. 2006. Baby Boosters. https://www.netpets.org/horses/healthspa/boost.html (accessed March 15, 2005).

Roberts L., and J. Janovy. 2005. *Foundations of Parasitology*, 7th ed. New York, NY: McGraw-Hill.

Shimeld, L. A. 1999. *Essentials of Diagnostic Microbiology*. Albany, NY: Delmar Publishers.

Turgeon, M. L. 2003. *Immunology and Serology in Laboratory Medicine*, 3rd ed. St. Louis, MO: Mosby.

University of Arizona—The Biology Project. Immunology. 2004. http://www.biology.arizona.edu/immunology/immunology.html (accessed March 17, 2005).

3

Bacterial Zoonoses

Objectives

After completing this chapter, the learner should be able to

- Describe how living organisms are classified
- Differentiate between prokaryotic and eukaryotic cells
- Describe properties unique to bacteria
- Identify the appearance of bacteria microscopically and by colony growth
- Briefly describe the history of specific bacteria causing zoonotic disease
- Describe the causative agents of specific bacterial zoonoses
- Identify the geographic distribution of specific bacterial zoonoses
- Describe the transmission, clinical signs, and diagnostic procedures of specific bacterial zoonoses
- Describe methods of controlling bacterial zoonoses
- Describe protective measures professionals can take to prevent transmission of bacterial zoonoses

Key Terms

bacteria
binary fission
capnophil
coliform
endotoxin
enterotoxin

eukaryote
exotoxin
facultative
microaerophil
nomenclature
obligate

prokaryote
taxonomy
thermophil
xenodiagnosis

OVERVIEW

Life on earth is incredibly diverse with new species of all types of organisms still being discovered. New technologies and scientific discoveries are allowing organisms to be classified and reclassified based upon new criteria that attempt to identify relationships among living things. Classifying organisms in an orderly fashion allows the scientific community to understand and communicate about a variety of life forms in a clear and concise manner. Describing a rabbit as "the animal with fur and short tail" can cause one person to think of a rabbit, another person thinks of a Manx cat, and yet another person thinks of a grizzly bear. By knowing classification schemes of living organisms, one can clearly understand the roles that various animals and types of microbes play in zoonotic diseases.

THE TAXONOMIC SCHEME

To effectively study living things and their relationships to each other it is necessary to understand the orderly system of classifying organisms known as **taxonomy**. In the 4th century B.C., Aristotle was the first person to group all organisms as either plants or animals. Although a variety of different classification schemes have been developed since Aristotle's day, the basic rules and **nomenclature** (naming process) that are used today have been around since the 1700s. In 1735 the Swedish scientist Carolus Linnaeus recognized that confusion would result from having several common names for the same organism. He spent many years giving scientific names to a variety of different organisms. Linnaeus developed the system of classifying organisms, called the taxonomic system, that divides every organism into seven basic categories. These seven categories are kingdom, phylum, class, order, family, genus, and species (some people use eight categories with domain being more diverse than kingdom). The seven levels in this system are arranged into descending ranks with kingdom being the most diverse and species being the most specific.

The Taxonomic Hierarchy

Kingdoms are the largest and most general taxon (category) in the Linnaean system of classification. In 1969, Robert Whittaker of Cornell University divided all life (except viruses) into five kingdoms based on cellular organization and nutritional patterns. Table 3-1 shows these five kingdoms and their characteristics.

Kingdoms contain smaller groups called phyla. Each phylum (singular form of phyla) is divided into several classes; each class is divided into several orders; each order is divided into several families; each family is divided into several genus groups; and each genus is divided into several species. Each sequential division contains more specific descriptions about living things, thus by the time an organism is given its species name little confusion exists about its characteristics. Table 3-2 shows the classification of the house cat, human, and *Escherichia coli* bacteria.

Table 3-1   The Five Kingdom System

Kingdom	Cellular Organization	Nutrition Type	Examples
Monera	■ Prokaryotic ■ Unicellular	■ Varies	Bacteria
Fungi	■ Eukaryotic ■ Unicellular or multicellular	■ Heterotroph (cannot make its own food) ■ Absorptive	Molds, yeasts, mushrooms, smuts, rusts
Protista	■ Eukaryotic ■ Unicellular	■ Heterotroph (a few autotrophic species exist)	Protozoa
Animalia	■ Eukaryotic ■ Multicellular	■ Heterotroph ■ Ingestive	Invertebrates, vertebrates
Plantae	■ Eukaryotic ■ Multicellular	■ Autotroph (can make its own food)	Plants, mosses, ferns

Table 3-2   Classification of the House Cat, Human, and *E. coli* Bacteria

Category	House cat	Human	E. coli bacteria
Kingdom	Animalia	Animalia	Monera
Phylum	Chordata	Chordata	Proteobacteria
Class	Mammalia	Mammalia	Gamma proteobacteria
Order	Carnivore	Primate	Enterobacteriales
Family	Felidae	Hominidae	Enterobacteriaceae
Genus	*Felis*	*Homo*	*Escherichia*
Species	*domestica*	*sapien*	*coli*

Nomenclature refers to the naming of organisms. Binomial nomenclature uses two parts for each organism's scientific name. The binomial nomenclature system assigns a genus and species name to each organism and allows scientists to communicate with a common language. Writing scientific names involves capitalizing the genus name and beginning the species name with a lower case letter. Both names are either italicized or underlined.

PROKARYOTIC VERSUS EUKARYOTIC MICROBES

There are two basic types of cells that differ greatly in their size and cellular organization. These two cell types are called prokaryotic and eukaryotic. **Prokaryotic** cells usually have a single, circular chromosome located in a part of the cytoplasm called the nucleoid (nuclear region). The prokaryotic chromosome is not surrounded by a nuclear membrane. **Eukaryotic** cells have a membrane-bound nucleus. Table 3-3 shows a comparison of prokaryotic cells and eukaryotic cells.

INTRODUCTION TO BACTERIA

Bacteria were among the earliest life forms on Earth billions of years ago. Many believe that more complex cells developed as free-living bacteria took up residence in other cells, eventually becoming the organelles in eukaryotic cells. Bacteria are often viewed as harmful because they cause human and animal disease; however, certain bacteria are beneficial, such as those that produce antibiotics and those that live symbiotically as normal flora of some body system of animals. Bacteria are extremely important organisms because of their adaptability and capacity for rapid growth and reproduction.

The Kingdom Monera contains bacteria. **Bacteria** are prokaryotic, unicellular, and reproduce asexually by a process called binary fission. Bacteria usually

Table 3-3   **Comparison of Prokaryotic and Eukaryotic Cells**

Characteristic	Prokaryotic	Eukaryotic
Average size of cells	0.20–2.0 μm in diameter	10–100 μm in diameter
Nucleus	No nuclear envelope or nucleoli	Membrane-bound, nucleoli present
Location/type of genetic material	Single, circular chromosome in cytoplasm; some have plasmids	Multiple, linear chromosomes in nucleus; other DNA in organelles
Membrane-bound organelles	Not present	Present (examples include mitochondria and endoplasmic reticulum)
Flagella	Hollow, made of protein, attached by basal body	Complex, 9+2 arrangement of microtubules
Glycocalyx	Exists as capsule or slime layer	Exists in animal cells
Cell wall	Usually present; many contain peptidoglycan	Present in plant cells, no peptidoglycan
Plasma membrane	No carbohydrates, most lack sterols	Sterols and carbohydrates present
Cytoskeleton	Not present	Present
Ribosomes	70S	80S (70S in organelles)
Cell division	Binary fission	Mitosis
Sexual reproduction	Transfer of DNA fragments by conjugation, transformation, or transduction	Involves meiosis

have one circular chromosome (some bacteria such as *Borrelia burgdorferi*, the causative agent of Lyme disease, has a linear chromosome) and in some cases small, circular molecules of DNA called plasmids. Table 3-4 and Figure 3-1 summarize other structures found in bacteria.

What Do Bacteria Look Like to the Naked Eye?

Bacteria can be grown outside of the host by providing the nutrients and growth factors needed by that particular organism. Bacteria are routinely grown on blood agar plates (BAP) because this agar supports the growth of most bacteria, but there are other specialized agars that allow for differentiation and selection of certain bacteria as well. When growing bacteria, their appearance on culture media varies depending upon which bacteria they are; however, most appear as colonies on the surface of the agar. Depending upon the bacterium, the colonies may appear mucoid, dry, flat, depressed, pinpoint, small, or a variety of other descriptions. Figure 3-2 shows bacteria growing on a blood agar plate.

Biochemical tests are also available for identifying specific bacteria. These tests, in addition to colony growth, allow for identification of most bacteria.

Table 3–4   Structures Found in a Typical Prokaryotic Cell

Structure	Characteristics	Function
Glycocalyx (capsule or slime layer)	Gelatinous polysaccharide polypeptide layer	Surrounds the cell wall May protect against phagocytosis and dessication (drying) Aids in adherence to surface
Fimbriae and pili	Short, thin, hollow appendages attached to the cell wall	Fimbriae—attachment to surfaces Pili—conjugation
Flagella	Long, thin, hollow structures consisting of a filament, hook, and basal body	Flagella rotate to push the cell Attach to the cell wall
Axial filaments	Similar to flagella but wrapped around the cell, associated with spirochetes	Provides motility to spirochetes
Cell wall	Two types, gram-positive and gram-negative	Surrounds the cell membrane and protects cell from environmental stress Contains peptidoglycan
Cell membrane	Selectively permeable, phospholipid bilayer and protein	Surrounds cytoplasm and contains enzymes involved in metabolic reactions
Cytoplasm	Gelatinous matrix located inside the cell membrane	Made of water and organic and inorganic molecules
Ribosomes	70S, contain rRNA and protein	Site of protein synthesis
Nucleoid	Contains the bacterial chromosome (DNA molecule)	Area in the cytoplasm where the main chromosome is located
Plasmids	Small, circular, extrachromosomal DNA molecules	Found in some cells in addition to the main chromosome
Inclusions	Reserve deposits of various materials found in the cytoplasm	Examples include sulfur granules and metachromatic granules
Endospores	The dormant, resistant stage of some bacteria (6 genera of gram-positive bacteria; 2 genera of medical significance)	Assist survival in adverse conditions
Capsule	Protective structure outside the cell wall in some bacteria	Protects against or delays phagocytosis

What Do Bacteria Look Like Under the Microscope?

Bacteria are extremely small (measured in micrometers) and are transparent. In order to view them under the microscope, they need to be stained. The most common stain used is a differential stain called the Gram stain. The Gram stain is based on the fact that some bacteria have a different cell wall structure than

A)

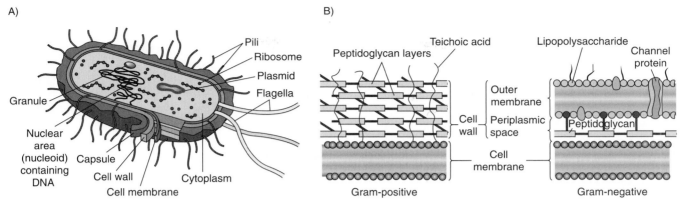

B)

Figure 3-1 A) Structures found in a typical prokaryotic cell (described in Table 3-4). B) Gram-positive and gram-negative cell wall structures. Gram-positive bacteria have cell walls with a thick peptidoglycan layer that usually contains teichoic acid. The thick peptidoglycan layer is closely attached to the outer surface of the gram-positive cell membrane. In contrast, gram-negative bacteria have cell walls with a thin peptidoglycan layer without teichoic acid. Gram-negative cell walls have lipopolysaccharide, an outer membrane, and a periplasmic space (contains toxins and enzymes to protect the bacterium).

others and this difference allows some cells to retain the stain crystal violet. Bacteria described as gram-positive retain the crystal violet stain and appear dark purple, whereas gram-negative bacteria do not retain the crystal violet stain and appear pink. Gram-negative bacteria have lipids in their cell walls that prevent crystal violet stain from entering the cell. When gram-negative bacteria are rinsed with a decolorizer such as acetone, the lipids in their cell walls are removed resulting in open pores. These open pores allow the counterstain safranin to pass through the cell wall. The pink color of safranin is retained by gram-negative bacteria making them appear pink. The Gram stain and a variety of other stains allow visualization of bacteria under the microscope. Figure 3-3 describes the Gram stain process.

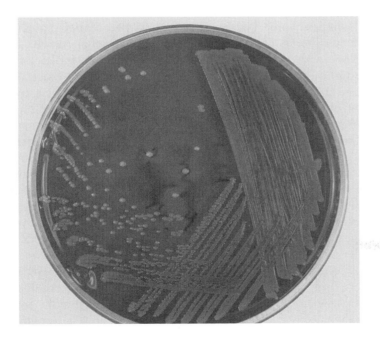

Figure 3-2 Colony growth on blood agar plate (BAP). Each colony originated from a single bacterium.

(Courtesy Shimeld)

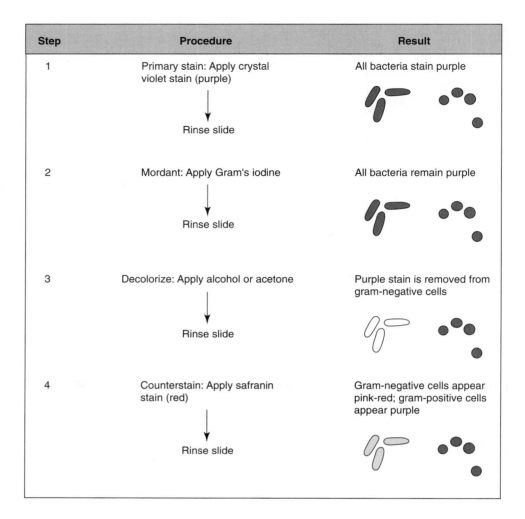

Step	Procedure	Result
1	Primary stain: Apply crystal violet stain (purple) ↓ Rinse slide	All bacteria stain purple
2	Mordant: Apply Gram's iodine ↓ Rinse slide	All bacteria remain purple
3	Decolorize: Apply alcohol or acetone ↓ Rinse slide	Purple stain is removed from gram-negative cells
4	Counterstain: Apply safranin stain (red) ↓ Rinse slide	Gram-negative cells appear pink-red; gram-positive cells appear purple

Figure 3-3 The Gram stain. The Gram stain procedure is a differential staining method using two stains to differentiate bacteria based on cell wall structure. 1. Smear is covered with crystal violet (primary stain). 2. Stain is removed and smear is covered with iodine (mordent, which is a chemical that helps a stain adhere to the cell and intensify color). 3. Smear is washed with decolorizer to remove excess, unbound stain. 4. The decolorizer is washed off and safranin (counterstain) is used on the smear. The counterstain stains structures that were unstained with the primary stain. The stain is then removed by rinsing with water.

There are thousands of different types of bacteria, but there are basically a few different shapes. Some bacteria are rod-shaped (called bacilli); others are shaped like little spheres (called cocci); others are comma-shaped (called vibrio); others are spiral in shape (called sphirochetes [tight] or spirilla [loose] depending upon the tightness of their spirals). Some bacterial cells exist as individuals, whereas others group together to form pairs (diplo-), chains (strepto-), clusters (staphylo-), or other arrangements. Figures 3-4, 3-5A, and 3-5B illustrate bacterial morphology.

Some bacteria, namely the gram-positive rods *Bacillus* and *Clostridium*, have the ability to produce structures called endospores. Endospores, commonly called spores, are protective structures that form when unfavorable conditions are present. Vegetative cells (bacteria that are metabolically active) of the genera *Bacillus* or *Clostridium* can form endospores when nutrients are exhausted or other conditions become unfavorable for growth. When an endospore-forming species stops growing, it starts forming endospores and when favorable conditions return, the endospores germinate to produce new vegetative cells (one endospore germinates into only one vegetative cell). Endospores are extremely resistant to drying, heat, radiation, and chemicals (such as alcohols and bleach). Endospores can be seen microscopically and can form either centrally, subterminally (near one end), or terminally (at one end). Figure 3-6 illustrates endospore formation.

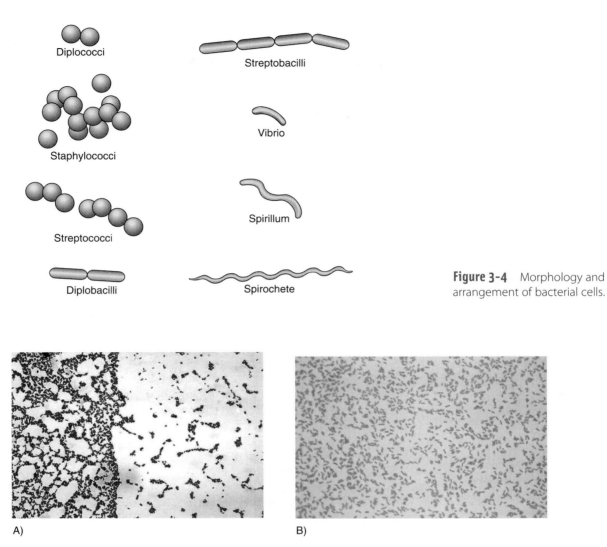

Figure 3-4 Morphology and arrangement of bacterial cells.

Figure 3-5 Gram-positive versus gram-negative bacteria. A) A photomicrograph of gram-positive cocci. (Courtesy of CDC/Dr. Richard Facklam). B) A photomicrograph of gram-negative bacilli. (Courtesy of CDC).

How Do Bacteria Reproduce?

Prokaryotic cell reproduction is quite simple compared to eukaryotic cell reproduction. Prokaryotic cells reproduce by **binary fission**, where one cell (the parent cell) splits in half to become two daughter cells identical to the parent. Prior to replication the chromosome is duplicated so that each daughter cell possesses the same genetic information as the parent.

The time it takes binary fission to occur is called the generation time. The generation time varies from one bacterial species to another and is also dependent upon growing conditions such as pH, temperature, and availability of nutrients. Under ideal conditions some bacteria can replicate every 20 minutes.

How Do Bacteria Obtain Nutrients?

Chemical nutrients such as carbon, hydrogen, oxygen, and nitrogen are needed for bacterial growth. Bacteria obtain these nutrients from a variety of environmental

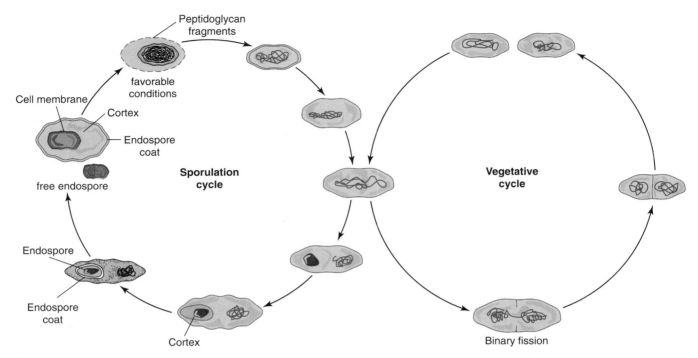

Figure 3-6   Endospore formation.

Sporulation is the formation of endospores. Sporulation occurs in a few bacilli-shaped bacteria such as those in the *Bacillus* and *Clostridium* genera.

sources. Bacteria that use inorganic carbon source (CO_2) as their sole carbon source are called autotrophs. Bacteria that break down organic molecules (such as proteins, carbohydrates, amino acids, and fatty acids) are heterotrophs and acquire these nutrients from other organisms.

Some bacteria make their own food from sunlight. Other bacteria absorb food from the material they live on or in. Some of these bacteria produce by-products such as iron or sulfur. The microbes that live in the gastrointestinal tract absorb nutrients from digested food.

How Do Bacteria Cause Infection?

Not all bacteria cause disease. Some bacteria such as normal flora bacteria are needed by the host to maintain health of a body system. Colonization is the persistence of microbes in a body site without causing disease. Normal flora is an example of colonization. Normal flora and the host are an example of a symbiotic relationship that is mutualistic (both organisms benefit) or commensal (no harm is done to the host) rather than parasitic.

Microbes capable of causing disease are pathogens. Microbial pathogenicity is the biochemical mechanism by which microbes are able to cause disease. Infection is like a miniature battle between bacteria and the host; the bacteria are trying to remain present and multiply, whereas the host is trying to prevent bacteria from gaining access. The term infection is used to describe the persistence or multiplication of a pathogen (disease-causing organism) on or within the host. Disease occurs when an infection causes clinical signs in the host. Infection and disease are dependent on host factors as well as virulence of the microbe.

Pathogenic bacteria have certain virulence factors that help them cause disease. The first step in pathogenesis of bacterial infection is for the bacteria to reach the site of interest and to remain there. Some ways bacteria can get to and remain in an area are with the aid of fimbria and flagella. Fimbria are hair-like structures on bacteria that enable them to attach to certain body sites and help them not get washed away by body secretions. For example, *E. coli* produces fimbria that attach to the epithelial lining of the urinary tract allowing the bacteria to remain in the urinary bladder without getting flushed away. Flagella are structures on bacteria that help propel them from one area to another. The flagella help bacteria reach a body site where they can survive and multiply; therefore, flagella have a function in pathogenicity.

Invading microorganisms must also survive phagocytosis in order to cause disease. Some pathogens avoid direct contact with a phagocyte by producing a slippery mucoid capsule. This capsule inhibits the chemical identification of bacteria by the phagocyte. If the pathogen is not recognized as foreign by phagocytes, it has time to replicate in the body and cause disease. Pathogens that produce the thickest capsules are some of the most virulent.

The next challenge for pathogens is to compete with normal flora. One way bacteria compete with normal flora is by producing toxic compounds that cause harm to their host producing clinical signs such as vomiting, diarrhea, paralysis, pain, or fever. Some bacteria produce toxins wherever they grow and can cause illness even without being in the affected area as is the case with some forms of food poisoning. Another way bacteria compete with normal flora is by invading the host cells especially in areas such as the gastrointestinal tract and other body systems with increased numbers and varieties of normal flora. For example, *Salmonella enterica* serovar Typhimurium (commonly called *S. typhimurium*) will destroy the cells of the intestine causing them to release their cell contents that are used by bacteria as nutrients. The result in the host is diarrhea due to intestinal cell damage that alters absorption of fluid and nutrients, against the result for the pathogen is being lost from the intestine with the stool.

ANTHRAX

Overview

Anthrax, also known as black bane, fifth plague, and woolsorter's disease, has been a documented disease since biblical times as described in the Old Testament. The very severe plague cast upon the Pharaoh's cattle in the book of Exodus was believed to be anthrax, especially because the endospores that are produced by this bacterium can persist for many years in soil similar to that found in the Nile valley. In the 19th century, anthrax was known in Europe as woolsorter's and ragpicker's disease because these groups of people caught the endospores from the fibers and hides they were handling. The first cases of anthrax in the United States were reported in Louisiana in the early 1700s.

Anthrax, named directly from the Greek word *anthrax* meaning coal, is named after the cutaneous black "coal-like" lesions it can cause. Anthrax is caused by inhalation, skin exposure, and gastrointestinal absorption and is primarily a disease of herbivores with humans contracting the disease from infected animals. Anthrax is caused by *Bacillus anthracis*, which was the first bacterium shown to cause disease. In 1877, Robert Koch, a Polish scientist, helped launch the science of bacteriology when he became interested in this disease after several local cows died suddenly without a true explanation as to the cause of death. Koch believed that *B. anthracis* was the cause of the outbreak, but he did not know how the bacterium was transmitted and the mechanism by which it caused disease. Koch began his investigation by studying blood from dead cattle and found numerous rods and threads in it that did not appear in healthy cattle blood. He then used blood from cattle that had died of anthrax and applied the blood to open cuts on mice. Each mouse that came in contact with the infected blood died. To determine the impact of *B. anthracis* inside an animal, Koch cultured parts of swollen spleens from the infected mice and liquid from inside infected cattle eyes and watched as bacteria grew and formed the unusual rods and threads in his laboratory. He then put the samples in water and sunlight, and found that if placed in ordinary water, the rods and threads separated and disappeared. In sunlight only, the rods and threads were killed. These observations led Koch to believe that the bacterium must remain in the body to maintain its virulence.

As anthrax spread through sheep flocks in France in 1877, Louis Pasteur worked on a vaccination for this disease and by 1881 a successful anthrax vaccine for livestock was developed. Further preventative measures against anthrax were strengthened by the 1920s law that required testing of shaving brushes that were made of horse or pig bristles.

As a result of the ability of *B. anthracis* to produce endospores, anthrax has long been studied as a potential biological weapon. In 1925, the Geneva Protocol banned bacteriologic warfare as part of the World War I peace treaties with ratification occurring prior to World War II (except by the United States and Japan). A resolution calling on all United Nations members to ratify the ban was rejected in 1952 and all major nations maintained extensive facilities for producing and testing germ warfare (including anthrax). In 1969, a secret military test range in Utah accidentally released nerve gas killing hundreds of sheep, which lead President Nixon to renounce chemical and biological weapons in the United States. In 1975, the United States finally ratified the international ban on chemical and biological weapons. In 2001, bioterrorist-related cases of cutaneous and inhalation anthrax were seen in the United States prompting the desire for vaccination against the disease and for prophylactic antibiotics.

> The risk of getting anthrax remains low in developed countries with modern animal husbandry and industrial hygiene.

Causative Agent

Anthrax is an acute infectious disease caused by *B. anthracis* bacteria. *B. anthracis* are very large, nonmotile, gram-positive, encapsulated, endospore-forming bacilli. *B. anthracis* has two main virulence factors: the presence of a capsule and production of toxins.

- The capsule helps the bacillus avoid engulfment by phagocytes, thus allowing for establishment of infection in the animal/person. All virulent strains of *B. anthracis* form a capsule and are known as S or smooth variants. S variants produce mucoid or "smooth" colonies when grown on agar. R or rough variants do not produce the capsule and are relatively avirulent. The ability to produce capsules can be transferred to nonencapsulated *B. anthracis* via plasmid transfer.
- Anthrax toxins consist of three types including:
  - Protective antigen (PA), also known as factor II, is a protein that binds to select cell receptors in the target tissue, which in turn forms a channel that permits the other factors to enter those cells.

- Edema factor (EF), also known as factor I, is a toxin that converts adenosine triphosphate (ATP) to cyclic adenosine monophosphate (cAMP). As cAMP increases cellular edema occurs in the target tissue. Build-up of fluid surrounding the lungs can inhibit immune function and can be fatal.

- Lethal factor (LF), also known as factor III, is a toxin that is believed to inhibit phagocytosis by neutrophils and release cytokines. LF can kill infected cells or prevent them from working properly.

B. anthracis has the ability to form endospores and these endospores form in the middle of the cells and may persist for long periods in dry products such as feed, contaminated objects, or in soil. The endospores revert to the vegetative (reproductive) form when environmental conditions are optimal, including warmer seasons when temperature is above 60° F and there is heavy rainfall. Flooding allows bacteria to accumulate at the ground surface of low-lying areas and drought conditions favor the development of endospores. *B. anthracis* is typically found in neutral to mildly alkaline soil (pH 6 to 8.5). Figure 3-7 shows *B. anthracis* endospores.

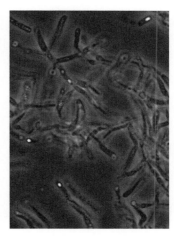

Figure 3-7 *Bacillus anthracis* endospores seen using phase contrast microscopy.

(Courtesy of CDC/Larry Stauffer, Oregon State Public Health Laboratory)

Epizootiology and Public Health Significance

Anthrax occurs worldwide and is seen most commonly in agricultural regions without adequate veterinary public health programs such as South and Central America, Southern and Eastern Europe, Asia, Africa, and the Middle East. International figures are not exact because of difficulty in reporting cases in developing nations. Anthrax is endemic in Africa and Asia.

In animals, anthrax occurs sporadically throughout the United States mainly in cattle, bison, and deer herds. In humans, the incidence of anthrax in the indigenous United States averages less than one case per year. From 1955 to 1994, there were 235 human cases of anthrax in the United States (average 5 per year) with 20 of them being fatal. In the United States there are recognized areas of infection in South Dakota, Nebraska, Arkansas, Mississippi, Louisiana, Texas, and California; small areas exist in a number of other states. Anthrax is rarely seen in the United States and when seen occurs most frequently in farmers, herdsmen, butchers, veterinarians, and in wool, tannery, and slaughterhouse workers. U.S. cases of human anthrax were between 40 and 50 in 1952, but since 1962 have numbered less than 10 per year (except for the bioterrorism-related cases in 2001 and 2002). During September 2001 through November 2001, an outbreak of intentionally spread anthrax in the United States caused five deaths and a total of 22 infections (18 confirmed and 4 suspicious). Anthrax is identified by the Centers for Disease Control and Prevention (CDC) as being capable of causing death and disease in large enough numbers to devastate cities or developed areas. According to a World Health Organization (WHO) estimate, the release of 50 kilograms of anthrax over a city of 5 million people would result in 250,000 deaths, with 100,000 patients dying before receiving treatment.

Preventing and treating anthrax includes vaccination and the use of antibiotics. The U.S. Department of Defense has given more than 2 million anthrax vaccinations to more than 500,000 military personnel. The cost of the vaccine is approximately $18 for a complete immunization series (at $3 per dose). The vaccine cost is higher if it is given by individual clinicians rather than as part of mass public vaccinations. Anthrax treatment using ciprofloxacin for 60 days costs about $700 (2005 values).

Anthrax is typically a disease of herbivores.

Transmission

B. anthracis is a **facultative** (capable of adapting to different conditions) organism whose cycle of vegetative growth and endospore formation occur in soil. Infection with *B. anthracis* occurs when animals grazing on contaminated pasture ingest endospores. The pathogenic bacillus is returned to the soil in animal excrement or when the animals die. Once in the soil the bacterium sporulates and the soil remains a long-term reservoir of infection for animals.

B. anthracis is typically transmitted between animals via ingestion of endospore-contaminated water or pasture in areas where previously infected animals lived. Ingestion of infected feedstuffs, such as bloodmeal or bonemeal, has also been implicated as a cause of infection. During an epidemic, insect transmission of disease may occur, but is not usually significant. Some animals, such as pigs, dogs, cats, mink, and captive wild animals, have acquired the disease from consumption of contaminated meat.

B. anthracis is typically transmitted to people by infected animals and their products (fur, skin, bonemeal). *B. anthracis* most frequently enters the body via skin abrasions, injuries, or blemishes resulting in cutaneous anthrax. Airborne infection (via shearing of infected sheep or handling of infected hides) and gastrointestinal infection (via consumption of contaminated meat or milk) can also lead to infection.

Pathogenesis

B. anthracis owes its pathogenicity to the formation of a capsule and production of toxins. The capsule protects the bacillus from bactericidal components in blood and prevents phagocytosis of the organism. The capsule plays an important role in establishment of infection because the organism is allowed to increase in number as a result of the slowing of phagocytosis by neutrophils. Toxin production plays an important role both in establishment of infection and in the later stages of disease where the toxins are responsible for the clinical signs observed in infected animals. Toxins produced by *B. anthracis* can have both short antiphagocytic activity and leukocidal effects on white blood cells. The PA toxin binds to cell receptors in target tissue causing exposure of a binding site. This exposed binding site combines with either EF to form edema toxin or binds with LF to form lethal toxin. Edema toxin causes cellular edema in target tissue, whereas lethal toxin may inhibit phagocytosis and may cause release of tumor necrosis factor and interleukin-1.

Once anthrax bacilli are contracted, they multiply at the site of the lesion. Phagocytic cells migrate to the site, but are unable to engulf bacilli as a result of the presence of their capsules. If some bacilli are engulfed they can resist killing and digestion by producing toxins that impair phagocytic activity and can be lethal to leukocytes. Depending upon the port of entry, bacteria and their toxins cause cutaneous, pulmonary, or gastrointestinal disease. Systemic anthrax results from the spread of these forms.

There is considerable variation in susceptibility to anthrax among animal species. The infectious dose of anthrax also varies widely among animal species with mice only requiring five bacteria to initiate disease, whereas rats require 10^6 bacteria to cause disease. This variation among animal species may be the result of a particular species resistance to the toxin produced or to the ability of the bacterium to establish disease in that species. Figure 3-8 depicts the pathogenesis of anthrax in humans.

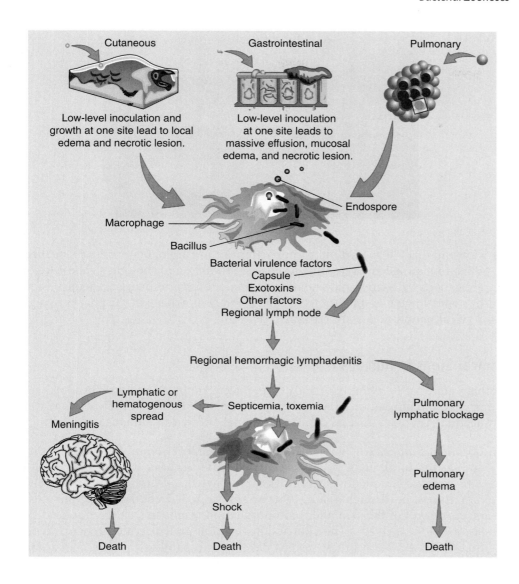

Figure 3-8 Anthrax in humans may present in three forms: cutaneous, gastrointestinal, and pulmonary (systemic anthrax results from spread of these forms). In cutaneous anthrax, the endospore germinates in macrophages of the skin and toxin production results in raised skin lesions. In gastrointestinal anthrax, contaminated meat is ingested and endospores are typically deposited in the upper gastrointestinal tract where they germinate producing localized lymphadenopathy, edema, and sepsis following development of an oral or esophageal ulcer. In pulmonary anthrax, the bacillus endospores are deposited into the alveolar space where macrophages will lyse and destroy most spores. Surviving endospores may reach lymph nodes where they return to the vegetative state and replicate in the lymph nodes, releasing toxins and producing hemorrhage, edema, and necrosis. If any form of anthrax is left untreated, the infection may become systemic.

Clinical Signs in Animals

In animals there are three forms of anthrax including:

- *Peracute form.* The peracute form of anthrax occurs mainly in ruminants especially cattle, sheep, and goats. Signs in affected animals may include ataxia, dyspnea (difficulty breathing), trembling, and seizures. Sometimes sudden death is the only sign observed in these animals (Figure 3-9).

- *Acute form.* In cattle and sheep the first sign of acute anthrax may be high fever and excitement followed by stupor, respiratory depression, cardiac depression, seizures, and death. Milk production is reduced and abortion may be seen in pregnant animals. Animals may also hemorrhage from the mouth, nose, and anus. Splenomegaly may be evident and may appear as blackberry jam on necropsy. The acute form of anthrax in swine may present as sudden death or may present as progressive swelling of the throat, which may cause suffocation. In horses, the acute form of anthrax presents as fever, chills, severe colic, anorexia, muscular weakness, bloody diarrhea, and swelling in the neck, lower abdomen, and genital area.

> In animals, anthrax typically presents as an acute fatal septicemia with splenomegaly and hemorrhagic infiltration of the subcutaneous tissue.

Figure 3-9 A beef carcass from an animal that has died of anthrax. The carcass bloats soon after death as a result of rapid decomposition.
(Courtesy of USDA)

- *Chronic form*. The chronic form of anthrax in ruminants produces localized, subcutaneous edema typically in the neck, thorax, and shoulders. These lesions are caused by bacteremia and are not analogous to human cutaneous lesions. In swine, lingual and pharyngeal edema and hemorrhage of the pharyngeal and cervical lymph nodes are typically seen.

Clinical Signs in Humans

There are no reported cases of human-to-human transmission of anthrax.

Humans contract *B. anthracis* mainly from infected animals and their products such as fur, skin, bloodmeal, or bonemeal. The symptoms of anthrax vary with the route of infection. There are several different forms of anthrax in humans including:

- *Cutaneous anthrax*. Cutaneous anthrax is the most common form of anthrax and occurs in about 95% of cases. Cutaneous anthrax has an incubation of 2 to 5 days. Cutaneous anthrax occurs when *B. anthracis* endospores enter nonintact skin through a laceration or abrasion. The endospore germinates in macrophages of the skin and toxin production results in small, red, raised skin lesions (Figure 3-10). The skin lesions enlarge and ulcerate or may become fluid filled. By 7 days these lesions develop into painless, black eschars (dark, sloughing scab) surrounded by edema and blood vessels. The eschars dry and fall off in about 1 to 2 weeks. If left untreated, the infection can become systemic producing symptoms of fever, lethargy, vomiting, and hypotension. If the portal of entry of the bacterium is on the neck or face and if septicemia or meningitis develops, the prognosis is poor. Mortality from untreated cutaneous anthrax is 10% to 20%; in treated cases mortality is about 1%.

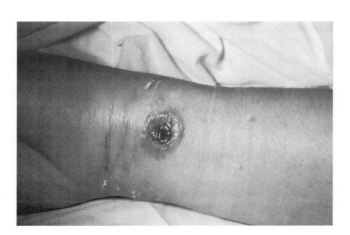

Figure 3-10 A cutaneous anthrax lesion.
(Courtesy of CDC/Dr. Philip S. Brachman)

- *Pulmonary (inhalation) anthrax.* Pulmonary anthrax is rare in the United States and was historically associated with woolsorters at industrial mills. Pulmonary anthrax has an incubation of 1 to 14 days and typically develops in the textile and tanning industries among workers handling contaminated animal wool, hair, and hides. More recent cases of pulmonary anthrax in the United States are associated with biological warfare. Pulmonary anthrax occurs when aerosolized *B. anthracis* endospores (8,000 to 50,000 endospores needed to initiate disease) enter the respiratory tract via inhalation and deposit in the alveolar space. Macrophages will lyse and destroy some endospores and infection in the lung is rare. Surviving endospores reach the mediastinal and peribronchial lymph nodes where germination may occur up to 60 days postinfection. After germination the anthrax bacilli replicate in the lymph nodes, release toxins, and hemorrhage, edema, and necrosis occurs. Inhalation anthrax produces biphasic clinical symptoms. In the first stage of the disease symptoms resemble those of a cold (fever, dyspnea, cough, headache, chills). As endospores travel to the lungs (secondary stage), symptoms such as dyspnea, sweating, shock, cyanosis, and hypotension may appear (Figure 3-11). Rapid death may occur in 1 to 3 days following the onset of symptoms. Mortality rates without treatment are greater than 95%, and even humans treated with antibiotics rarely survive unless they are treated immediately after the onset of symptoms.

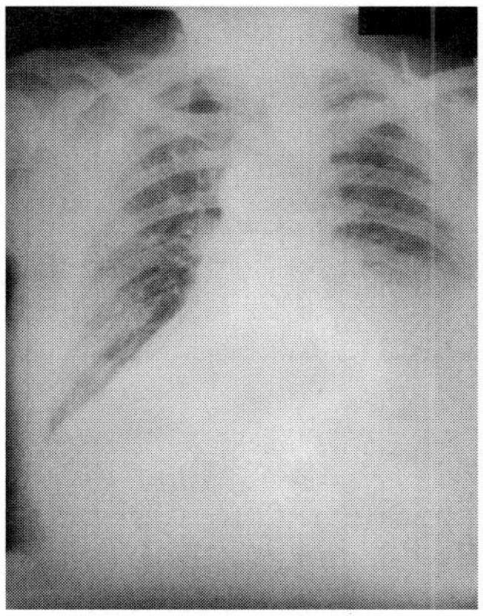

Figure 3-11 Chest radiograph of a person with pulmonary anthrax.
(Courtesy of CDC)

- *Gastrointestinal anthrax.* Gastrointestinal anthrax is associated with ingestion of contaminated meat and produces two distinct syndromes (oral-pharyngeal and abdominal). When *B. anthracis* endospores are deposited in the upper gastrointestinal tract, they germinate producing localized lymphadenopathy, edema, and sepsis following development of an oral or esophageal ulcer. Dysphagia (difficulty eating/swallowing) and dyspnea may also occur. When *B. anthracis* endospores are deposited in the lower gastrointestinal tract, they germinate producing primary intestinal lesions. Symptoms include nausea, abdominal pain, vomiting, and lethargy that may progress to bloody diarrhea or sepsis. The abdominal form of gastrointestinal anthrax is more common than the oral-pharyngeal form. Gastrointestinal anthrax is rare; however, cases have been reported in Asia and Africa.

- *Systemic anthrax.* Systemic infections caused by *B. anthracis* arise from hematogenous spread of bacteria from the cutaneous, pulmonary, and gastrointestinal forms. Spread from the cutaneous route is most common. Systemic infection usually produces meningitis; however, spread of disease from primary infection sites is rare. Symptoms of meningitis include fever, fatigue, neck pain, nausea, behavioral changes (such as agitation) and neurologic signs (Figure 3-12). Systemic anthrax is usually fatal.

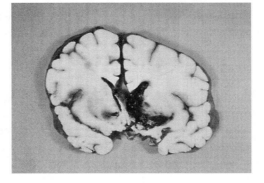

Figure 3-12 A brain section revealing an interventricular hemorrhage as a result of toxins released from *Bacillus anthracis* bacteria.
(Courtesy of CDC)

Diagnosis in Animals

The carcass of any animal suspected of having anthrax should not be necropsied because the anthrax bacilli form endospores when exposed to oxygen. Rigor mortis in these animals is typically absent or incomplete.

Diagnosis of anthrax in animals via clinical signs is difficult; therefore, laboratory confirmation is needed. Blood collected from suspect animals should be shipped to reference laboratories in leak-proof containers or as a dried specimen on either a sterile swab or dried blood smear. Tissue samples should be shipped refrigerated or frozen in leak-proof containers. All specimens shipped should be labeled as "suspected anthrax" so that proper biosafety handling measures can be followed. Common laboratory methods for identifying *B. anthracis* include stained blood smears, bacterial culture, animal inoculation, and serologic tests. Blood smears are typically stained with Gram stain, polychrome methylene blue stain (also called McFadyean's stain; a rapid stain that stains rods blue surrounded by pink capsular material), or Giemsa stain (demonstrates the encapsulated bacillus). Organisms stain gram-positive when stained from young cultures, but can stain gram-variable or gram-negative with age. On Gram stain endospores may be seen and appear clear because they do not retain the crystal violet or safranin stain. Endospores will stain green if the sample is stained with a malachite green spore stain using heat; the vegetative cell will appear red when counterstained with safranin.

B. anthracis bacteria can be cultured using routine media (blood agar, chocolate agar) under aerobic or anaerobic conditions and special media such as PLET (polymixin-lysozyme EDTA-thallous acetate) and bicarbonate agar. On blood agar, *B. anthracis* grows as nonhemolytic, medium-large, gray, flat, irregular colonies with swirling, comma-shaped projections. *B. anthracis* is differentiated from other gram-positive rods on culture by its lack of hemolysis and motility and by growth on phenylethyl alcohol blood agar.

Guinea pigs and mice are occasionally used to diagnose cases of anthrax by inoculation of blood or tissue from an anthrax suspect (**xenodiagnosis**). *B. anthracis* is much more pathogenic for guinea pigs and mice than other *Bacillus* species typically causing death within 24 hours. If the animal has died within 48 hours following inoculation, blood smears and touch preps of splenic tissue can demonstrate the characteristic large, encapsulated bacilli.

Serologic tests available for identification of *B. anthracis* include enzyme-linked immunoabsorbant assay (ELISA) or indirect hemagglutination antibody (IHA) tests that identify titers of antibodies directed against the capsular antigen, protective antigen, lethal factor, or edema factor (quadrupling of the titer is diagnostic). Fluorescent antibody (FA) is also available for identification of the bacterium in tissue and culture. Polymerase chain reaction (PCR) methods that amplify markers of *B. anthracis* have also been developed.

Diagnosis in Humans

Anthrax in people is diagnosed differently depending on the form of anthrax. Cutaneous anthrax is diagnosed when exudate from skin lesions is swabbed or aspirated and the fluid stained or cultured. Skin biopsy and immunohistochemical staining is also used to diagnose cutaneous anthrax. Pulmonary anthrax is diagnosed by radiographic lesions of the thorax that show symmetrical mediastinal widening and by computed tomography (CT) scans that show enlarged hilar lymph nodes, pleural effusion, and airway edema. Confirmation of pulmonary anthrax is by staining and culturing of sputum specimens. Gastrointestinal anthrax is diagnosed by staining and culture of fecal samples, vomitus, or hemorrhagic fluid from body cavities. Serologic testing of samples for anthrax confirmation includes PCR, ELISA, IHA, and FA. In some countries an anthraxin skin test is used to confirm cases of anthrax. The anthraxin skin test involves subdermal injection of commercially produced chemical extract of an attenuated strain of *B. anthracis*. A positive test indicates cell-mediated immunity. The accuracy of positive test results increases with the duration of the disease.

Treatment in Animals

Anthrax is highly fatal requiring early and vigorous treatment. All sick animals should be isolated and treated with antibiotics (the bacterium is susceptible to many antibiotics including oxytetracycline, erythromycin, and sulfonamides) and all healthy animals in the herd and on surrounding farms should be immunized.

In addition to antibiotic treatment and vaccination, controlling the spread of anthrax includes the following:

- notification of the appropriate regulatory offices
- enforcement of quarantine
- prompt disposal of dead animals, manure, bedding, and other contaminated material by cremation or deep burial
- isolation of sick animals and removal of healthy animals from the contaminated area
- disinfection of stables and equipment
- improved sanitation

Treatment in Humans

Prior to October 2001, the treatment and prophylaxis for human cases of anthrax was penicillin; however, concern for genetically engineered penicillin-resistant anthrax strains prompted the CDC to recommend the use of other antibiotics. For patients with severe cases of anthrax, corticosteroid and intravenous (IV) antibiotic treatment is recommended. Pulmonary anthrax patients typically received a multidrug regimen of either ciprofloxacin or doxycycline (doxycycline is not used in patients with meningitis as a result of poor drug penetration to the central nervous system [CNS]) for 60 days and another antibiotic such as rifampin, vancomycin, or an aminoglycoside. Cases of gastrointestinal and cutaneous anthrax are treated with ciprofloxacin or doxycycline for 60 days followed by amoxicillin or amoxicillin/clavulanic acid. Despite early treatment, people infected with pulmonary, gastrointestinal, or meningeal anthrax have a very poor prognosis.

Management and Control in Animals

Prevention of anthrax is attained by annual vaccination of all grazing animals in an endemic area and by implementation of control measures during outbreaks. Vaccines for anthrax composed of killed bacilli and/or capsular antigens do not produce significant immunity; therefore, the current vaccine is a live vaccine. The Sterne's vaccine, which uses the Sterne strain of *B. anthracis*, produces sublethal amounts of toxin allowing for antibody production in animals and is approved for horses, cattle, sheep, and pigs. Vaccination should be done 2 to 4 weeks before the season when outbreaks may be expected in the area. Animals should not be vaccinated within 2 months of anticipated slaughter. Because it is a live vaccine, antibiotics should not be administered within 1 week of vaccination. Animals surviving naturally-acquired anthrax are immune to reinfection and second attacks are extremely rare. Permanent immunity to anthrax appears to require antibodies to both the toxin and capsule.

B. anthracis may survive for 20 to 30 years in dried cultures and remains viable in soil for many years. Freezing temperatures have little effect on the bacillus; however, endospores can be destroyed by autoclaving at standard conditions, by boiling for 30 minutes, or by exposure to dry heat at 140°F (60°C) for 3 hours. Most chemical disinfectants must be used in high concentrations over a long period of time to be effective. Cremation or deep burial (at least 6 feet or 1.8 meters) in lime (calcium oxide) is recommended for disposal of the carcasses of animals that died of anthrax.

Management and Control in Humans

Anthrax is a reportable disease.

A vaccine consisting of protective antigen of an avirulent, noncapsulated strain of *B. anthracis* has been used to protect U.S. military personnel and others at risk of infection such as people who work with imported animals hides, furs, bone, meat, wool, animal hair, and equipment used in grooming these animals. Multiple doses are given (three subcutaneous injections given 2 weeks apart followed by three additional subcutaneous infections given at 6, 12, and 18 months) and an annual booster is required to maintain protective immunity. The vaccine should only be used in healthy people 18 to 65 years of age and its safety in pregnant women has not been established. Passive vaccines that deliver antibody directed against protective antigen are being investigated.

Summary

B. anthracis, the causative agent of anthrax, is a rod-shaped, gram-positive, endospore-forming bacterium that typically infects herbivores such as cattle, sheep, and horses. Human disease may be contracted by handling contaminated hair, wool, hides, flesh, blood, and excretions of infected animals and from manufactured products such as bonemeal, as well as by purposeful dissemination of endospores. Infection is introduced through scratches or abrasions of the skin, wounds, inhalation of endospores, eating insufficiently cooked contaminated meat, or by flies. *B. anthracis* endospores are very stable and may remain viable for many years in soil and water. Anthrax endospores were weaponized by the United States in the 1950s and 1960s prior to the termination of the old U.S. offensive program. *B. anthracis* is easy to cultivate and endospore production is readily induced with endospores being highly resistant to sunlight, heat, and disinfectants.

Anthrax presents as three clinical syndromes in animals: peracute, acute, and chronic. The peracute form of anthrax occurs mainly in ruminants, especially cattle, sheep, and goats, with signs including ataxia, dyspnea, trembling, seizures, and sudden death. The acute form of anthrax occurs in cattle and sheep and presents with high fever and excitement followed by stupor, respiratory depression, cardiac depression, seizures, and death. Animals may also hemorrhage from the mouth, nose, and anus. Splenomegaly may be evident and may appear as blackberry jam on necropsy. The acute form of anthrax in swine may present as sudden death or may present as progressive swelling of the throat, which may cause suffocation. In horses, the acute form of anthrax presents as fever, chills, severe colic, anorexia, muscular weakness, bloody diarrhea, and swelling in the neck, lower abdomen, and genital area. The chronic form in ruminants produces localized, subcutaneous edema typically in the neck, thorax, and shoulders. The chronic form of anthrax in swine presents with lingual and pharyngeal edema and hemorrhage of the pharyngeal and cervical lymph nodes.

Anthrax presents as three distinct clinical syndromes in humans: cutaneous, pulmonary, and gastrointestinal disease. The cutaneous form occurs most frequently on the hands and forearms of persons working with infected livestock beginning as a papule followed by formation of a blister-like fluid-filled vesicle. The vesicle dries and forms a coal-black scab. Pulmonary anthrax, known as woolsorter's disease, is a rare infection contracted by inhalation of the endospores. It occurs mainly among workers handling infected hides, wool, and furs. Gastrointestinal anthrax is contracted by the ingestion of insufficiently cooked meat from infected animals. In humans, the mortality of untreated cutaneous anthrax ranges up to 25% and in pulmonary and gastrointestinal anthrax, the fatality rate is almost 100%.

Diagnosis of anthrax in animals via clinical signs is difficult; therefore, laboratory confirmation is needed. Common laboratory methods for identifying *B. anthracis* include stained blood smears, bacterial culture, animal inoculation, and serologic tests. Diagnosis of anthrax in humans includes blood smears, bacterial culture, serologic tests, and clinical tests such as thoracic radiographs (for identification of pulmonary anthrax). Treatment of anthrax in both animals and humans includes antibiotics that need to be administered early in the disease course to be successful. Prevention of anthrax includes vaccination of animals using the Sterne's vaccine and selective vaccination of high-risk humans with vaccine developed from attenuated strains of *B. anthracis*.

BITE WOUNDS

Overview

Approximately half of the U.S. population will incur an animal bite sometime in their life. Every year, about 330,000 people are seen in emergency rooms in the United States for dog bites; about 4% of these cases are hospitalized and approximately 20 deaths are estimated to occur in the United States from animal bites. Dog bites compose approximately 80% of animal bite wounds and are typically seen in children (peak incidence is between the ages of 5 and 9 years). Dog bites typically occur on the extremities and an estimated 4% to 25% of dog bite wounds become infected. Cat bites occur more commonly in women with half of all victims of cat bites older than 20 years of age. Cat bites also typically occur on the extremities. Because cats have thin, sharp teeth they tend to cause puncture wounds (about 85% are puncture wounds). Approximately 30% to 50% of cat bite wounds become infected. The first symptoms associated with cat bite wounds typically occur within 12 hours following the bite. The first symptoms associated with dog bite wounds typically occur approximately 24 hours following the bite. The focus in this section will be dog and cat bite wounds.

> A large dog can exert more than 450 psi of pressure with its jaws, causing significant injury and tissue devitalization to a bite area.

Causative Agent

The microbes causing infection following dog and cat bites may consist of normal flora from the animals' mouths or from the skin of the person/animal being bitten. At least 30 different infectious agents have been reported to be transmitted from dog or cat bites and most infections as a result of dog and cat bites consist of multiple bacteria (a median of five isolates per infected wound). About 50% of dog bites and 60% of cat bites involve both aerobic and anaerobic bacteria. Bacteria found in bite wounds include:

- *Pasteurella* spp. are the most common pathogens found in dog and cat bite wounds (typically *P. multocida* and *P. canis*). Wounds containing *Pasteurella* bacteria tend to show signs of infection more rapidly than wounds containing other bacteria. *Pasteurella* bacteria are commonly found in abscesses and non-purulent wounds. *P. multocida* is normal oral flora in 50% to 70% of healthy cats and is found in about 45% of all cat bite wounds. Other animals that transmit *Pasteurella* bacteria through their bites are dogs, horses, sheep, and pigs.

- *Staphylococcus aureus* bacteria are commonly isolated from nonpurulent wounds. Animals that transmit *St. aureus* bacteria through bite wounds are dogs, cats, horses, camels, pigs, lizards, and rodents.

- *Streptococcus mitis* bacteria are commonly isolated from nonpurulent wounds. Animals that transmit *Str. mitis* are dogs and cats; other species of *Streptococcus* are spread through the bites of horses, camels, pigs, simians, squirrels, and birds.

- *Moraxella* spp., *Corynebacterium* spp., and *Neisseria* spp. are commonly isolated aerobic bacteria from the bite wounds of dogs, cats, horses, simians, rodents, and squirrels.

- *Bergeyella zoohelcum* (formerly known as *Weeksella zoohelcum*) is an uncommon zoonotic pathogen that causes acute cellulitis from dog and cat bites.

- *Capnocytophaga canimorsus* and *C. cynodegmi* are normal flora of the canine and feline mouth. These bacteria may cause local wound infections after a dog or cat bite and may lead to sepsis, meningitis, and disseminated coagulopathy. *Capnocytophaga* bacteria are also associated with rabbit bites.

- *Fusobacterium* spp., *Bacteroides* spp., *Porphyromonas* spp., and *Prevotella* spp. are anaerobic bacteria cultured from dog, cat, horse, camel, simian, rodent, bird, and reptile wounds, but they are rarely cultured alone.

Epizootiology and Public Health Significance

Animal bite wounds are seen throughout the world and are common in the United States (about 1% of all emergency room visits are related to animal bites). In the United States an estimated 1 to 3 million animal bites occur annually with approximately 80% to 90% from dogs, 5% to 15% from cats, and 2% to 5% from rodents (the remainder coming from other small animals, such as rabbits and ferrets, farm animals, monkeys, and reptiles). Internationally, it is difficult to get accurate numbers as a result of the variety of animals inflicting animal bites including large cats (tigers, lions, and leopards), wild dogs, hyenas, wolves, crocodiles, and other reptiles. In England and Wales, 200,000 people seek medical care for dog bites; in France, 500,000 people seek medical care for dog bites; and in Germany, 35,000 people are bitten by dogs. Most bites seen worldwide are from domestic dogs. Young children tend to get bitten by dogs, whereas adult women are more frequently bitten by cats. Other groups of people at risk include veterinary professionals, animal keepers, breeders, and trainers. In developing countries, dog bites carry a high risk of rabies infection.

Animal bites result in approximately 0.4% to 1.5% of all emergency room visits with an annual cost of approximately $100 to $165 million dollars in health care expenses and lost income. Most bites are the result of a family's own pet or a neighbor's animal. Because 58% of all households in the United States having at least one pet, the potential public health significance of animal bites is enormous.

Capnocytophaga spp. and *Bergeyella zoohelcum* bacteria can be transmitted to people through close contact with animals as well as through a bite wound.

Transmission

Most of these organisms are transmitted by bite wounds, but can also be transmitted through close animal contact such as the animal licking an open wound of a person/animal (especially immunocompromised individuals).

Pathogenesis

The pathogenesis of bite wounds is variable depending on the type of animal that produces the bite. Dog bites typically cause crushing-type wounds, because their rounded teeth and strong jaws cause injury to deeper tissues such as bones, vessels, tendons, muscle, and nerves. The pointed teeth of cats usually cause puncture wounds and lacerations that may inoculate bacteria into deep tissues. Other animals such as monkeys and herbivores can cause injury from bite wounds as a result of the trauma they cause followed by infection.

Wound infections from dog and cat bites can cause abscess formation, septic arthritis, osteomyelitis, endocarditis, and CNS infections. The incubation period from animal bites varies with the animal producing the bite, the organism or organisms causing the infection, and health factors of the person/animal being bitten. Patients who present within 8 to 12 hours of a bite typically show local lesions without significant signs of local inflammation. Those patients that present after 12 hours may present with localized cellulitis, pain at the bite site, discharge, and enlarged lymph nodes. If septicemia develops, signs include fever, chills, vomiting, diarrhea, abdominal pain, lethargy, dyspnea, and headache. Immunocompromised people (splenectomized people, alcoholics, people on corticosteroids or chemotherapy) may develop more severe signs associated with animal bites and signs may include endocarditis, pneumonia, meningitis, peripheral gangrene, and shock.

> In general, the better the vascular supply and the easier the wound is to clean (i.e., laceration versus puncture), the lower the risk of infection. Bites on the hand have a high risk for developing infection because of the relatively poor blood supply in the hand and difficulty of adequately cleansing the wound.

Clinical Signs in Animals

Bite wounds inflicted from one animal to another are usually caused by fighting.

- *Cats.* Intact male cats fight more than neutered male cats, which fight more than female cats. Feline fight wounds typically occur on the face, legs, back, tail, and rump. The most common complication of fight wounds is infection because cat bites create small puncture wounds in the skin that quickly close and are difficult to find in affected animals. Microbes from the biting cat's mouth enter the other animal's skin and multiply rapidly. An abscess will form if loose skin surrounds the bite site. Cellulitis may develop on less fleshy areas, such as the foot or tail (Figure 3-13). Both abscesses and cellulitis trap pus, causing swelling and pain. If a cat is bitten by a cat with feline leukemia virus (FeLV) or feline immunodeficiency virus (FIV), it could contract these viruses. Transmission of rabies is also a concern when any animal is bitten by a cat.

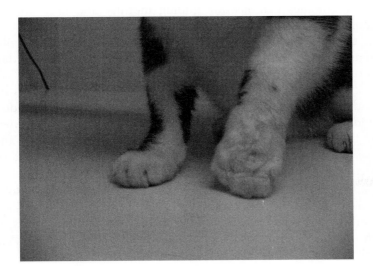

Figure 3-13 Cat bite wound on the distal front limb of a cat. Note the degree of cellulitis associated with a wound inflicted from the bite of another cat.

- *Dogs.* Intact male dogs fight more than neutered male dogs. Female dogs tend to fight with other female dogs. Dog bite wounds tend to cause more trauma from ripping or damaging of tissue followed by infection once the microbes from the oral cavity of the dog enter the wound and multiply. Abscesses and cellulitis are common findings following dog bite wounds. If a dog is bitten by an animal with rabies virus it could contract this illness.

Clinical Signs in Humans

Clinical signs of animal bite wounds in people depend on the area in which the bite occurs. Hands are common locations of bite wounds in people. The skin is thin over most of the hand, offering little protection (especially over the joints). Hand wounds are prone to soft tissue, joint, and tendon sheath infection (Figure 3-14). Complications of animal bite wounds include cellulitis, septic arthritis, osteomyelitis, and sepsis.

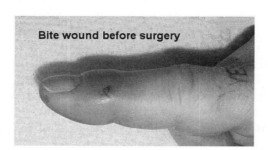

Figure 3-14   Lateral view of a cat bite wound on human finger.
(Courtesy of www.ThePetCenter.com)

Diagnosis in Animals

A tissue sample obtained surgically or at necropsy could show the causative organism with final confirmation based on bacterial culture. Samples are typically obtained via swabbing of the wound area. Diagnosis of bacterial pathogens causing infection following animal bites varies with the disease-causing microbe. Identification of specific bacteria such as *Pasteurella*, *Staphylococcus*, and *Streptococcus* is covered under separate sections; those covered here are only found in bite wounds.

- *Capnocytophaga* spp. are gram-negative, fusiform-shaped bacteria with one rounded end and one tapered end. The species found in dogs may be curved. *Capnocytophaga* spp. are slow growers and are identified by culture on enriched media (blood agar), where after 48 to 72 hours the colonies are small- to medium-size, opaque, shiny, nonhemolytic pale beige or yellow in color. The colonies demonstrate gliding motility and may produce swarming similar to *Proteus* spp. This bacterium grows best at 35°C to 37°C in an aerobic environment containing 5% to 10% CO_2 or anaerobically (they are known as **capnophiles** or bacteria that require additional carbon dioxide). The species of *Capnocytophaga* found in the oral cavity of dogs are *C. canimorsus* and *C. cynodegmi* and can be differentiated from each other via biochemical tests with both being oxidase-positive, catalase-positive, and indole-positive; *C. cynodegmi* is positive for esculin hydrolysis, whereas *C. canimorsus* is variable on this agar. Other species of *Capnocytophaga* are oxidase- and catalase-negative. Various selective media and PCR tests have been developed for identification of this bacterial genus.
- *Bergeyella zoohelcum* are gram-negative, short, straight, nonendospore forming, aerobic rods. This bacterium does not grow on MacConkey agar and on blood agar produces yellow colonies that are circular, shiny, sticky, and smooth. *B. zoohelcum* is oxidase, catalase, and indole-positive. PCR tests are available for identifying this bacterium.

Diagnosis in Humans

Diagnosis in humans is the same as in animals.

Treatment in Animals

Bite wounds in animals are typically cleaned, irrigated with saline under high pressure, and debrided (any necrotic tissue is removed). Antibiotics given within 24 hours of obtaining the wound will often prevent infection. Fight wounds and infections usually heal within a few days with proper treatment. More aggressive treatment may be needed if the therapy is started after infection has set in. Evaluation of rabies status needs to be performed when animals are presented with bite wounds; evaluation of other infectious diseases is dependent on the species. Treatment of carrier animals is not warranted.

Treatment in Humans

People are treated with a variety of antibiotics for bite wounds. A complete history should include questioning whether the person is immunocompromised. The person's tetanus immunization status needs to be checked as well as rabies immunization status of the animal. Any simian (monkey, apes, and humans) bite should have viral cultures done for *Herpesvirus simiae*. Wounds are typically cleaned, irrigated with saline under high pressure, and the necrotic tissue debrided. Infected wounds seen within 24 hours of being bitten could be sutured following irrigation and debridement. If the wound is over 24 hours old it should be left open. Wounds in areas that typically develop infection are usually left open no matter when they present for medical care. Antibiotics are typically prescribed for 7 to 14 days unless the wounds are complicated by joint or bone involvement, which warrants 3 to 6 weeks of treatment.

Management and Control in Animals

Almost half of the U.S. population will be bitten by an animal in their lifetime. In some areas animal leash laws and other animal ordinances try to limit the general public's exposure to animals. Required vaccination for rabies can help limit the spread of this virus secondary to a bite wound. Species specific vaccines may limit the spread of other diseases between animals of the same species.

Management and Control in Humans

There are an estimated 52 to 68 million dogs and 57 million cats kept as pets in the United States. Educating people about the risk of infection following an animal bite is important in patients receiving prompt treatment of animal bite injuries. Teaching children to be cautious around unfamiliar animals and to properly handle animals is essential in preventing animal bites in people. Immunocompromised people need to be especially careful around animals.

Summary

Almost half of the U.S. population will be bitten by an animal in their lifetime. The microbes that can cause infection following dog and cat bites may consist of normal flora from the animals' mouths or from the skin of the person/animal being bitten. At least 30 different infectious agents have been reported to be transmitted from a dog or cat bite and most infections as a result of dog and cat bites consist of multiple bacteria. Bacteria found in bite wounds include *Pasteurella* spp., *Staphylococcus aureus*, *Streptococcus mitis*, *Moraxella* spp., *Corynebacterium* spp., *Neisseria* spp., *Bergeyella zoohelcum* (formerly known as *Weeksella zoohelcum*), *Capnocytophaga canimorsus* and *Capnocytophaga cyno-degm*, *Fusobacterium* spp., *Bacteroides* spp., *Porphyromonas* spp., and *Prevotella* spp. Wound infections from dog and cat bites can cause abscess formation, septic arthritis, osteomyelitis, endocarditis, and CNS infections. The incubation period from animal bites varies with the animal producing the bite, the organism or organisms causing the infection, and health factors of the person/animal being bitten. Animal lesions from bites tend to be isolated to abscesses, whereas human lesions can range from abscesses to septicemia to CNS infection. Educating people about medical care regarding animal bites and ways to avoid bites is crucial in lowering the incidence of animal bites. Rabies immunization status and tetanus immunization status should also be determined in treating human cases of animal bites. Methods of identifying particular bacteria vary with each bacterium.

BRUCELLOSIS

Overview

Brucellosis, also known as Bang's disease and contagious abortion in animals, and Malta fever, Mediterranean fever, and undulant fever in humans, is a disease named after Sir David Bruce, an English army surgeon who identified the cause of this disease in 1887. Bruce found the causative agent of brucellosis, *Bacillus melitensis*, in the spleens of British soldiers who died of undulant fever on the Mediterranean island of Malta. Several years later the infection in these British soldiers was traced to the soldier's drinking contaminated goat's milk. In 1897, *Brucella abortus* was isolated and identified from an aborted bovine fetus by Danish veterinarian, Dr. Fredrick Bang. The infection in cattle became known as Bang's disease and was eventually proven to be ubiquitous in many animals. Brucellosis is one of the most serious diseases of livestock because of the damage it causes, including decreased milk production, weight loss, loss of young, infertility, and lameness.

Brucella spp. was the first microbe that the United States chose to develop as a biological weapon. It was considered as a biological weapon by the United States in World War II until the time of the destruction of its stockpile in the 1970s. The reasons it was chosen as a biological weapon include its low lethality, ease of manufacture, susceptibility to sunlight, and its ability to be spread by aerosol dispersion or by contamination of food or milk. It has the advantage of being debilitating to people without being fatal. Under optimal storage conditions it has a half-life of a few weeks. Field tests with live bacteria were performed in the early 1950s and it was effectively disseminated in 4-pound bombs. In 1954, it became the first biological agent developed by the old U.S. offensive biological weapons program with field testing on animals beginning soon afterwards. By 1955, the United States was producing *Br. suis*–filled cluster bombs for the U.S. Air Force at the Pine Bluff Arsenal in Arkansas. Development of *Brucella* as a biological weapon was halted in 1967, and President Nixon later banned development of all biological weapons on November 25, 1969.

Causative Agent

Brucellosis is a contagious bacterial disease that typically affects cattle and bison (*Br. abortus*), swine (*Br. suis*), dogs (*Br. canis*), and sheep and goats

(*Br. melitensis*). *Br. neotomae* (from desert wood rats), *Br. ovis* (mainly from sheep), and *Br. maris* (from dolphins) have not been isolated from people. Brucellosis only occasionally affects horses, and cats are relatively resistant to *Brucella* infections. Depending upon how an animal contracts brucellosis, a different species of *Brucella* may be causing the infection. For example, pigs, sheep, and goats that are in contact with infected cattle can be infected with *Br. abortus*. Dogs that ingest placentas from farm animals may be infected with *Br. abortus*, *Br. suis*, and *Br. melitensis*.

Brucella spp. are gram-negative coccobacilli that infect the placenta, uterus, and fetus, causing abortion in females, and infect the testes and accessory sex glands, causing orchitis and accessory sex gland infection in males. Brucellosis can cause infertility in both sexes. *Br. abortus* and *Br. canis* cause mild disease in humans, whereas *Br. suis* and *Br. melitensis* can be fatal.

Epizootiology and Public Health Significance

Brucellosis has a worldwide distribution and can affect a variety of animals, including reindeer in Alaska and Siberia, camels in the Middle East, and livestock throughout the world.

Concentrations of brucellosis can be seen in Europe, Africa, India, Mexico, and Central and South America. In the United States and Europe brucellosis is uncommon as a result of its elimination from cattle herds. In unvaccinated herds, infection spreads quickly causing many abortions; however, after the first exposure cattle typically develop antibodies and subsequent gestations and lactations appear normal.

Br. melitensis in sheep and goats represents the most important source of brucellosis in humans. *Br. melitensis* is not enzootic in the United States, Canada, northern Europe, Australasia, or Southeast Asia, but is prevalent in Latin America, the Mediterranean area, Central Asia and, especially, in the countries around the Arabian Gulf. Humans are infected by the handling of animals during the birthing process and the consumption of raw milk and milk products, especially fresh soft cheeses.

Br. suis affects both sexes of swine, causing infertility, abortion, orchitis, and bone and joint lesions. The prevalence is generally low except in parts of South America and Southeast Asia. *Br. suis* occurs in areas in which pigs are kept, including the southeastern United States and Australia where populations of feral swine are heavily infected. Human infections with *Br. suis* occur in people handling pigs on farms and during slaughtering and processing feral and domestic swine.

Bovine brucellosis, caused by *Br. abortus*, has been eradicated from Canada, Japan, northern Europe, and Australia. Cases in humans tend to be sporadic and are acquired by drinking unpasteurized milk, by working with infected cattle at a slaughter facility, by attending infected parturient cattle, and by accidental inoculation with live vaccine.

Br. canis infection in humans tends to occur in dog handlers because close, frequent contact seems to be necessary for transmission.

In the United States, the frequency of brucellosis is related to the number of infected animals. Infected animals are rare in the United States and pasteurization of milk has eliminated that mode of transmission. Occupational exposure (cattle-workers, veterinarians, slaughterhouse workers) is the main transmission route in the United States. The incidence is approximately 200 cases per year or 0.04 per 100,000 persons. People with brucellosis in the United States are primarily found in Texas, California, Virginia, and Florida. Worldwide the frequency of brucellosis varies across nations but is higher in places where handling of animal products and dairy products is less stringently monitored.

Transmission

Brucellosis is commonly transmitted to susceptible animals by direct contact with infected animals or with an environment that has been contaminated with discharges from infected animals.

Brucella spp. are facultative, intracellular bacteria that are able to establish infection because the virulent strains are able to survive inside phagocytes. Animal brucellosis is transmitted by contact, by mechanical vectors such as contaminated food, water, and excrement, or by ingestion of bacteria present in large numbers in aborted fetuses and uterine discharges. Cattle can contract brucellosis from contaminated feed or water, licking of contaminated genitals or aborted fetuses, venereal transmission from infected bulls during natural copulation or artificial insemination, or through mucous membranes, lacerations, and rarely intact skin. Replacement cattle or bison that are infected or have been exposed to infection prior to purchase or when wild animals or animals from an affected herd mingle with a brucellosis-free herd are common ways that cattle become infected with *Brucella*. Goats and sheep can contract brucellosis from ingestion of bacteria and through conjunctiva, vaginal, and subcutaneous wounds. In sheep, transmission occuring between rams is especially common during mating season when healthy rams acquire the infection by servicing ewes previously serviced by infected rams. Swine contract brucellosis by animal-to-animal contact, usually through ingestion of infected material and sexually transmitted fluids like semen. In dogs, transmission is congenital or venereal or by ingestion of infective materials. Humans get brucellosis by direct contact with infected animals or their secretions. People who handle diseased animals can be infected through a break in the skin, across mucous membranes such as conjunctiva, or by inhalation. Drinking infected milk is an important mode of transmission because *Brucella* bacteria concentrate in mammary glands of infected animals. Pasteurization kills *Brucella* spp. and has helped decrease cases of brucellosis where milk is pasteurized. Veterinarians may become infected as a result of accidental vaccination of themselves while vaccinating animals. Figure 3-15 summarizes the transmission of *Brucella* bacteria.

Pathogenesis

Person-to-person spread of brucellosis is rare.

Once in the body, *Brucella* spp. are engulfed by neutrophils and are carried in the lymphatic fluid to the lymph nodes draining the infected area. The infected neutrophils release bacteria into the blood and bacteria localize in certain organs such as the liver, spleen, bone marrow, and kidney. The affected organs vary with the animal/human and the species of *Brucella*. The gross lesions seen in an animal are subtle and rarely diagnostic. In cows, placental lesions include edema, necrosis, and a brownish odorless discharge. In aborted bovine fetuses, edema and bronchopneumonia may be seen. In cows, mammary glands and supramammary lymph nodes may show diffuse inflammation, whereas in bulls the scrotum becomes enlarged and thick connective tissue may compress the testes. In swine, *Br. suis* causes the formation of white nodules on the uterus of females and testes of males, and similar lesions in the spleen, liver, kidney, lymph nodes, and bone of both sexes. In sheep, brucellosis causes edema and inflammation of the epididymis in rams, necrosis of the placenta in ewes, and inflammatory changes in the lung, liver, lymph nodes, spleen, and kidneys of lambs. *Br. canis* causes uterine and placental lesions in bitches, orchitis in males, and bronchopneumonia in pups. In humans, the lung, spleen, liver, CNS, bone marrow, and synovium are more frequently affected with disease manifestation reflecting this distribution.

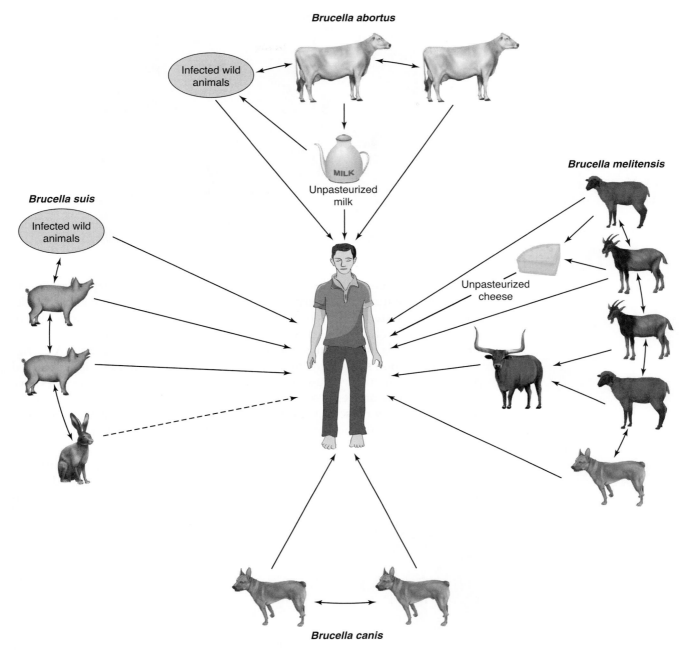

Figure 3-15 Cycle of *Brucella* infection.

Clinical Signs in Animals

There is no distinct appearance to animals with brucellosis. The signs of brucellosis are spontaneous abortion and inability to conceive in females and inflammation of sex organs in male animals. In susceptible animals, primarily cattle, swine, and goats, brucellosis causes infertility and death. The incubation period of brucellosis is variable ranging from 2 weeks to 1 year or longer (the typical length is 30 to 60 days). Some animals will abort prior to developing a positive reaction to the diagnostic test making determination of an incubation period difficult. Some infected animals never abort. Species specific clinical manifestations are described as follows.

> Brucellosis in animals typically causes late-term abortion in females and inflammatory lesions in the male reproductive tract.

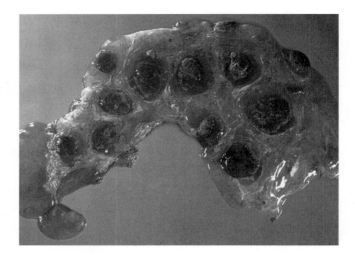

Figure 3-16 Bovine placenta with lesions due to *Brucella* infection.

(Courtesy of National Security Education Program)

- *Cattle.* The severity of the clinical signs of brucellosis is dependent upon the immune status of the herd or animals. In nonvaccinated pregnant cattle, abortion rates are high (up to 90%) and abortions occurring after the fifth to sixth month of pregnancy causing autolysis of the fetus is a common finding. *Brucella* bacteria also colonize the mammary glands causing mastitis. A carrier state is a common sequela in cows leading to a reduction in milk yield, retained placenta, and metritis. Figure 3-16 shows a placenta with *Br. abortus* lesions. In bulls, orchitis and epididymitis are common in *Brucella* infected animals. Bilateral or unilateral scrotal swelling may persist for a considerable amount of time resulting in necrosis of the testes and infertility. *Brucella* organisms may also be isolated from arthritic joints and lameness.

- *Sheep and goats.* In rams the first sign of brucellosis is the deterioration of semen quality. Scrotal edema and inflammation (Figure 3-17), epididymitis, fever, and an increased respiratory rate may also be seen in rams. Abortion or the birth of weak or stillborn lambs is only rarely seen in ewes. In goats, abortion during late pregnancy, birth of weak kids, and mastitis are the most common findings with brucellosis. Fever, weight loss, and diarrhea may also be seen in affected goats.

Figure 3-17 *Brucella* lesion on a ram scrotum.

(Courtesy of Simon Peek, BVSC, MRCVS, PhD, DACVIM)

- *Swine.* In sows irregular estrous cycles, infertility, and abortion typically in the third month may be seen with brucellosis. Orchitis, testicular necrosis, sterility, lameness, and incoordination may be observed in boars. Heavy mortality may be seen with piglets.

- *Dogs.* In bitches the main clinical sign is abortion in the last trimester of pregnancy, typically between 40 and 60 days of gestation. Prolonged vaginal discharge typically follows the abortion and repeated abortions in subsequent pregnancies are common. Stillbirths and conception failures are also seen in females. In affected males, epididymitis, orchitis, scrotal dermatitis, and prostatitis are frequently observed. Brucellosis is an important differential diagnosis for diskospondylitis in dogs of either sex.

- *Horses.* In horses, *Br. abortus* or *Br. suis* can be found in the bursa of the neck and withers, muscles, tendons, and joints. Brucellosis is not common in horses.

> Brucellosis is incapacitating rather than fatal (fatality is about 2%).

Clinical Signs in Humans

Brucella spp. are able to establish an infection by surviving phagocytosis and are passed from the lymph to blood and then to organs throughout the body, mainly the liver, spleen, and bone marrow. In humans, the incubation period is typically 5 to 60 days (or longer) and the most prominent symptoms are weakness, loss of appetite, chills, headache, back pain, and intermittent (undulating) fever. Brucellosis persists for weeks to months if left untreated but is seldom fatal in humans. In the chronic form, symptoms may persist for years, either continuously or intermittently, and may assume an undulant nature with periods of normal temperature between acute attacks. Chronic infection can damage joints and the spinal cord.

Diagnosis in Animals

Tissues infected with *Brucella* spp. do not provide pathognomonic (distinctive of a disease) findings. In some fetuses, pneumonia may be found. In females the placenta is edematous, wheras in males inflammation of the testes may be found. Histologic samples include the uterus which shows nodular inflammatory thickening and abscessation. The placenta of affected ruminants may show firm, yellow-white plaques in the cotyledons of the placenta. Arthritis and vertebral body necrosis may be found in lame swine and dogs. Testicular necrosis and scrotal inflammation may be seen in males of all species. In fetuses, findings consistent with pneumonia may be found.

> Abortions seen with brucellosis may be described as a "storm" of abortions. The storm of abortions is the high number of abortions seen when the disease is introduced into a herd or flock, followed by a period of resistance during which abortions do not occur.

Diagnosis in animals consists of bacteriologic or serologic identification. Bacteriologic identification involves the Gram stain and cellular morphology. *Brucella* is faintly staining, tiny, gram-negative coccobacilli. They are nonendospore forming and lack capsules or flagella; therefore, they are nonmotile. To ensure safety, all tiny gram-negative coccobacilli should be processed in a biosafety cabinet until *Brucella* can be ruled out. *Brucella* bacteria are aerobes but some species require an atmosphere with added CO_2 (5% to 10%). Multiplication is slow at the optimum temperature of 37°C and enriched medium is needed to support adequate growth. *Brucella* colonies become visible on suitable solid media in 2 to 3 days, but should be incubated at least 21 days before discarding the samples. Many laboratories now rely on commercial identification systems to identify *Brucella* or send suspect samples to a reference lab.

Serum agglutination and ELISA tests are standard for diagnosing bovine cases of brucellosis. Antibodies to the bacteria are present in the blood serum, vaginal mucus, seminal fluid, and milk and serve as important diagnostic factors. ELISA tests have been developed for detecting antigens in vaginal discharge and antibodies in milk and serum. Standard plate or tube agglutination tests are used to identify positive animals within a herd. In cattle serum dilutions of 1:100 or above for unvaccinated animals and 1:200 for vaccinated animals between 3 and 9 months of age are considered positive (the animals are classified as reactors). In goats a titer of 1:100 in any goat in a herd is positive and all goats at 1:50 and 1:25 are considered reactors and culled. In sheep a variety of tests may be used. A complement fixation test is run on ram serum and bacterial culture is done on semen or aborted tissue. FA staining is another highly specific diagnostic test in sheep. In swine, identification of *Brucella* organisms is done with the brucellosis card test. In dogs, qualitative agglutination tests such as latex agglutination are used as a screen to diagnose brucellosis and quantitative agglutination tests such as slide agglutination or agar gel immunodiffusion (AGID) are used to definitively diagnose the disease.

Diagnosis in Humans

Brucellosis diagnosis is primarily dependent on clinical suspicion, adequate history of possible exposure including travel, and isolation of the organism. Clinical presentation can be highly variable and focal lesions may present decades after exposure to the bacterium making brucellosis difficult to diagnose. An unequivocal diagnosis requires isolation of the organism using blood culture as the method of choice. Specimens need to be obtained early in the disease and cultures may need to be incubated for up to 28 days to ensure accurate identification of the organism. Isolation rates of only 20% to 50% are reported even from experienced laboratories and failure to grow the organism is common, especially in cases of *Br. abortus* infection. Commercial identification systems are hampered by the small amount of CO_2 produced during the incubation of organisms. Culture from bone marrow, blood, and affected organs may be successful. Presumptive identification of cultures based on colony morphology and slide agglutination with specific antiserum should be followed by further work in a reference facility. Molecular techniques for identification of *Brucella* organisms are being developed.

Serology is the preferred method of laboratory diagnosis in humans, but the interpretation of results is difficult. The standard serum agglutination test (SAT) and a modified Coombs' (antiglobulin) test have been used to identify *Brucella* organisms, whereas ELISA tests have been used to differentiate between specific IgM and IgG antibodies. The use of antibody tests has been limited because the species of *Brucella* have many common antigens and cross-reacting antigens are seen with many gram-negative bacteria. In addition, cases of brucellosis are often investigated late in their course and rising antibody titers may be missed. The variability of individual responses and the frequency of subclinical infections make the interpretation of single high titers difficult. PCR allows for a rapid diagnosis, but the technique has not been standardized.

Treatment in Animals

Treatment of infected animals is not attempted because animals may recover from the disease signs but do not clear the infection. Efforts to control the disease are aimed at eradication. Dogs may be isolated and management procedures such as individual cages may be attempted; however, these animals remain a source of infection for others.

Treatment in Humans

Humans are treated with antibiotic combinations for 4 to 6 weeks with doxycycline and rifampin (adults) or trimethoprim-sulfamethoxazole and rifampin (children).

Management and Control in Animals

Measures for prevention and control of brucellosis include vaccination of calves, periodic testing of bulk milk from farms, blood testing of adults, and slaughtering of infected animals. The level of enzootic disease can be reduced by intensive use of live, attenuated vaccines (*Br. abortus* RB51 in cattle, *Br. melitensis* strain Rev. 1 for sheep and goats). Detection of infected herds (by skin tests in sheep; serologic tests on milk or blood samples taken at sale or slaughter in cattle) and individual animals (by serologic tests) also reduces cases of brucellosis. Finally, the elimination of infected animals by slaughter effectively reduces the source of *Brucella* infections.

In the United States a federal program for brucellosis eradication called the Cooperative State Federal Brucellosis Eradication Program has existed since 1934. This program has minimum standards (called the Uniform Methods and Rules) for states to achieve eradication. States are deemed brucellosis free when none of their cattle or bison is found to be infected for 12 consecutive months under an active surveillance program. There are different class statuses for rates of infection (Class A <0.25% of herds are infected; Class B 0.26% to 1.5% of herds are infected; Class C >1.5% of herds are infected). In June 2000, 44 U.S. states, plus Puerto Rico and the U.S. Virgin Islands, were brucellosis free.

There are two surveillance procedures used to locate infection without having to test every animal: milk from dairy herds is checked two to four times annually by testing a sample from creameries or the bulk milk tank and blood tests on animals upon change of ownership.

- Milk is tested by the brucellosis ring test (BRT) in which milk from each cow in a herd is pooled and a sample taken for testing. A suspension of stained, killed *Brucella* organisms is added to the milk and if any cow is positive a bluish ring forms at the cream line.
- Animals (cattle and bison) are tested using market cattle identification (MCI). MCI involves using U.S. Department of Agriculture (USDA)-approved numbered tags (backtags) placed on the shoulders of adult breeding animals being marketed and blood samples are collected from these animals at livestock markets and slaughter facilities. If a sample reacts, it is traced to the backtag number of the herd and the herd owner is contacted by a state or federal animal health official to arrange for herd testing. All eligible animals in the herd are tested at no cost to the owner. At slaughter, eligible animals are all cattle and bison 2 years of age or older except steers and spayed heifers. At market, eligible animals are all beef cattle and bison older than 24 months of age and all dairy cattle older than 20 months of age except steers and spayed heifers. Pregnant and postparturient heifers are tested regardless of age. Eligible animals for herd tests include all cattle and bison older than 6 months of age except steers and spayed heifers. MCI provides a means of determining the brucellosis status of animals marketed from a large area and eliminates the need to round up all animals from a herd for testing. Blood collected can either be tested by blood agglutination tests or brucellosis card tests.
- Blood agglutination detects nonspecific antibodies to *Brucella* using serum taken from each animal and mixing it with a test fluid containing killed *Brucella* organisms (antigen). If the organisms agglutinate, the test is positive.
- The brucellosis card test, also known as the Rose Bengal test, is a compact test kit in which serum on a white card has *Brucella* antigen added to it. The test is read 4 minutes after the serum and antigen are mixed. Agglutination is a positive result. False-positive results can be seen with the card test especially

as a result of residual antibody in calves from vaccination, colostral antibody in calves, and cross-reaction with other bacteria.

Control of brucellosis in cattle and bison is through vaccination. Vaccination is about 65% effective in preventing cattle from becoming infected by an average exposure to *Brucella*. RB51 is the newer, live, attenuated vaccine strain used today (it was licensed in February 1996). Strain 19 *Brucella* vaccine was the original vaccine developed in 1941 that caused postvaccination reactions in cattle such as abortions and localized inflammation at the vaccine injection site. Unlike the strain 19 vaccine, RB51 does not stimulate the same type of antibodies that could be produced by actual infection causing confusion on interpretation of standard diagnostic tests. The organism in the RB51 vaccine is cleared from blood within 3 days and is not present in nasal secretions, saliva, or urine and the organism is not spread from vaccinated to nonvaccinated cattle. The vaccine is safe in all cattle older than 3 months of age. Live vaccines can only be administered by an accredited veterinarian or state or federal animal health official. In case of human exposure, strain RB51 is sensitive to a range of antibiotics used in the treatment of human brucellosis, but is resistant to rifampin and penicillin.

Guidelines for vaccinating cattle include:

- Female dairy calves are vaccinated between 4 and 8 months of age. Female beef calves are vaccinated between 4 and 10 months of age. At the time of vaccination, a tattoo is applied in the right ear that identifies the animal as an official vaccinate and identifies the year in which the vaccination took place. RV/5 would indicate that the calf was vaccinated with RB51 (RV) in 2005 (5). A brucellosis tag is also placed in the right ear of the vaccinate.
- Vaccination is not done in pregnant animals because of the risk of vaccine-induced abortion.
- Males are not vaccinated because the live vaccine may cause bacteria to colonize the male reproductive tract resulting in venereal spread during coitus. Any vaccinated male is neutered.

Swine brucellosis is controlled through serologic testing, inspection at slaughter, and tracing infections back to the farm of origin. Swine are not vaccinated for brucellosis.

The cost of maintaining the brucellosis eradication program is offset by the financial savings to the livestock and dairy industries. Losses from lowered milk production, aborted calves, and reduced breeding efficiency have decreased from $400 million in 1952 to less than $1 million today. The number of infected herds has dropped from 124,000 (1956) to 700 (1992) to 6 (2000) to 7 (2004).

Management and Control in Humans

Human brucellosis is an occupational disease among farmers, veterinarians, slaughterhouse workers, meat packers, laboratory workers, and others who come in direct contact with infected animals or their products (raw meat or unpasteurized dairy products). Individuals who are occupationally exposed can be protected by wearing impermeable clothing, rubber boots, gloves and face masks, and by practicing good personal hygiene. Adequate containment of the organisms to reduce aerosol spread in a laboratory setting is essential. Failure of laboratory kits to identify *Brucella* spp. quickly and accurately has also caused infection in unsuspecting laboratory workers.

Cases of human brucellosis are usually caused by *Br. melitensis* in travelers to areas such as Mexico and the Mediterranean region where this organism is

Brucellosis in free-ranging bison in Yellowstone National Park and Grand Teton National Park threatens the brucellosis status of livestock herds in that area. State and federal agencies are working toward containing the spread of brucellosis from bison to domestic livestock.

prevalent, and by the importation of infected (unpasteurized) dairy products. Pasteurization of milk and other dairy products is effective in protecting people from brucellosis. Unpasteurized milk and cheese is still available in foreign countries such as France and can serve as a source of infection to foreign travelers.

Brucellosis is a less significant problem in the United States where approximately 200 human deaths from this disease occur annually compared with more than 500,000 human deaths per year worldwide. Eradication of brucellosis from domestic animals has greatly reduced the threat of disease to humans in the United States and several other countries. No widely accepted vaccines for humans have been developed.

Summary

Brucellosis is a costly, contagious, zoonotic disease that can cause considerable damage in livestock. The organisms of greatest concern are *Br. abortus*, *Br. melitensis*, and *Br. suis*. Brucellosis in animals typically affects the reproductive organs and udder, causing abortion, inflammation of the reproductive organs, and mastitis. Bacteria are shed in milk and are found in aborted fetuses, placentas, and reproductive discharges. Signs of brucellosis in animals are unremarkable and are typified by late-term abortions in females and inflammatory lesions in the male reproductive tract. Human brucellosis also presents with vague signs such as headache, backache, and an undulating fever. Brucellosis is identified with bacterial culture and serologic tests. Eradication programs in livestock, aimed at testing milk and animals, have greatly reduced the incidence of brucellosis in the United States. Vaccination of cattle has also greatly reduced the incidence of brucellosis.

> Brucellosis is a reportable disease in the United States and state and federal health authorities must be notified within 7 days of diagnosis.

CAMPYLOBACTERIOSIS

Overview

Campylobacteriosis, also known as vibriosis and vibrionic abortion in animals, is a zoonotic disease caused by the genus of bacteria *Campylobacter*, a gram-negative curved rod-shaped bacterium (*kampter* is Greek for bend or angle and *bakterion* is Greek for little rod). In 1886, Theodor von Escherich (of *E. coli* fame) observed organisms resembling *Campylobacter* in the stool of children with diarrhea. In 1913, two English veterinarians named McFaydean and Stockman identified campylobacters in fetal tissues of aborted sheep and named them *Vibrio fetus*. Since that time, campylobacters have been recovered in blood samples of children with diarrhea and stool samples of patients with diarrhea. For many years, bacteria of the genus *Campylobacter* were thought to be *Vibrio* organisms and it was not until the 1970s that the name of *Vibrio* bacteria was changed to *Campylobacter*. In 1972 the development of selective growth media allowed laboratories to test stool specimens for *Campylobacter*.

Campylobacteriosis is one of the most common bacterial causes of human diarrheal disease in the United States and the enteric disease in people is caused by *Ca. jejuni*. Campylobacteriosis produces symptoms in people that range from loose stools to dysentery that commonly present as diarrhea, fever, and abdominal cramping. *Ca. jejuni* can also produce bacteremia, septic arthritis, and can be a trigger for Guillain-Barre syndrome (a neuromuscular paralyzing disease). *Ca. coli*, *Ca. lari*, and *Ca. upsaliensis* are also associated with human disease. A variety of *Campylobacter* species inhabit the gastrointestinal tract of animals such as poultry, dogs, cats, sheep, and cattle, as well as the reproductive organs of several animal species. *Ca. jejuni* is found in many foods of animal origin and ingestion of raw agricultural products is implicated in many cases of human infection.

Causative Agent

Campylobacter are pleomorphic, helical (curved), gram-negative rods that are commonly referred to as having a "gull-winged" appearance. They have a long flagellum at one or both ends of the cell that may be several times the length of the cell and is responsible for its rapid motility (Figure 3-18). This motility contributes to *Campylobacter*'s ability to colonize and infect the intestinal mucosa. *Campylobacter* spp. are **microaerophilic** (optimal growth occurs with reduced oxygen levels at about 5% to 7%) and capnophilic (optimal growth occurs with increased carbon dioxide at about 10%). *Ca. jejuni* and *Ca. coli* are **thermophilic** species of *Campylobacter* and grow better at about 42°C. Cultures are typically incubated for 42 to 78 hours and grow as fine, pinpoint colonies on blood agar. Most pathogenic species of *Campylobacter* are oxidase- and catalase-positive.

Campylobacter spp. have been isolated from dogs, cats, hamsters, ferrets, nonhuman primates, rabbits, swine, sheep, cattle, birds, and wildlife. There are more than 13 species of *Campylobacter*, but only some are pathogenic to animals or are zoonotic in nature. Strains of *Campylobacter* that cause disease include:

> *Ca. jejuni* grows best at about 42°C, which is approximately the same temperature as the body temperature of chickens.

- *Ca. jejuni* was first identified as a human diarrheal pathogen in 1973 and is the most frequently diagnosed bacterial cause of human gastroenteritis in the United States. *Ca. jejuni* is a commensal in the intestinal tract of many species of domestic animals, including poultry, dogs, cats, cattle, goats, pigs, sheep, mink, and ferrets. *Ca. jejuni* subspecies *jejuni* is found in animals, whereas *Ca. jejuni* subspecies *doylei* is found in children with diarrhea and in gastric biopsies from adults. Some strains of *Ca. jejuni* produce a heat-labile (unstable in the presence of heat) **enterotoxin** thought to be responsible for causing diarrhea.

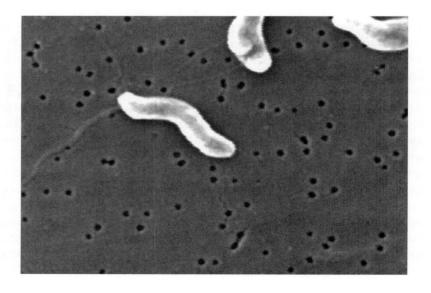

Figure 3-18   Scanning electron micrograph of the gram-negative *Campylobacter jejuni* bacterium. Note the long flagellum that aids in its motility.
(Courtesy of CDC/Dr. Patricia Fields, Dr. Collette Fitzgerald)

- *Ca. coli* is a commensal in the intestinal tract of poultry, swine, and humans. It can cause enteritis in humans and piglets. Dogs and pigs can be carriers of *Ca. coli*. *Ca. coli* produces a heat-labile toxin and is difficult to differentiate from *Ca. jejuni* (hippurate hydrolysis is one way to differentiate these two species).

- *Ca. upsaliensis* is found in the feces of both healthy and diarrheic dogs and cats. Dogs are the major reservoir of *Ca. upsaliensis* worldwide. This species of *Campylobacter* can also be found in the feces of healthy children. *Ca. upsaliensis* causes gastroenteritis (acute, watery diarrhea), septicemia, and abscesses in humans.

- *Ca. hyointestinalis* is a species of *Campylobacter* mainly found in swine, but can also be found in cattle, hamsters, and deer. In humans *Ca. hyointestinalis* causes gastrointestinal disease (watery or bloody diarrhea and vomiting).

- *Ca. fetus* is a major veterinary pathogen causing reproductive problems in ruminants. There are two subspecies of *Ca. fetus* found worldwide: *Ca. fetus* subspecies *fetus* and *Ca. fetus* subspecies *venerealis*. *Ca. fetus* subspecies *venerealis* is found in the prepuce of bulls and the genital tract of cows and heifers producing both infections and carriers in these animals. *Ca. fetus* subspecies *venerealis* mainly causes infertility (irregular estrous cycles, resorption of embryos, and infected semen) and may cause low levels of abortions in cattle. This bacterium is spread through coitus (natural or through artificial insemination). *Ca. fetus* subspecies *venerealis* can cause septicemia in humans. *Ca. fetus* subspecies *fetus* causes abortion mainly in sheep, but also cattle and pigs producing both infected and carrier animals. Contaminated tissue such as feces, uterine discharge, and aborted fetuses and membranes are sources of infection with secondary sources being scavenger birds. In humans *Ca. fetus* subspecies *fetus* causes gastroenteritis, septicemia, abortion, and meningitis. Both subspecies are believed to cause zoonotic disease; however, direct and indirect transmission of these bacteria from animals to humans has not been proven.

- *Ca. lari* is found in birds (mainly seagulls), poultry, dogs, and river and seawater animals. In humans, *Ca. lari* rarely causes gastroenteritis and septicemia and its identification is sometimes confused with *Ca. jejuni*.

- *Ca. mucosalis* is found in pigs with swine proliferative enteritis and can be cultured from the oral cavity and intestinal content of healthy pigs. *Ca. mucosalis* is not believed to cause human disease.

Epizootiology and Public Health Significance

Campylobacter is the most common cause of acute infectious diarrhea in developed countries and is increasing worldwide particularly during the warm summer months in temperate climates (with a secondary peak occurring in the late fall in the United States). In the United States about 2 million cases of gastroenteritis caused by *Campylobacter* occur annually. The peak incidence of this disease is found in children younger than 1 year of age and in young adults. Other people at high risk for developing campylobacteriosis are those with frequent contact with animals. In developing countries *Campylobacter* infections can be endemic with symptomatic disease occurring in young children and persistent carriers found in adults. *Ca. jejuni* accounts for greater than 80% of all gastroenteritis cases caused by *Campylobacter*. *Ca. coli* accounts for only 2% to 5% of total cases of campylobacteriosis in the United States; however, it accounts for a higher percentage of cases in developing countries. *Ca. fetus* infections in animals occur worldwide and can be an unrecognized cause of abortion on farms for long periods of time.

In response to the increase in cases of campylobacteriosis, the CDC began a national surveillance program in 1982 to understand the number of cases and the spread of *Campylobacter*. In 1996 a more detailed program was initiated to learn the frequency of this disease and the risk factors for acquiring it. In addition to the CDC, the USDA conducts research on how to prevent the *Campylobacter* infection in chickens (in 1999 up to 88% of poultry in supermarkets tested positive for *Campylobacter*). No vaccines are currently available for *Campylobacter* spp.

Transmission

Ca. jejuni is spread to humans by oral ingestion of contaminated food or water or by contact with the excretions of infected animals (poultry and cattle are the main sources of human infection). *Campylobacter* bacteria can survive at 40°C in feces and milk for up to 3 weeks, in water for 4 weeks, and in urine for 5 weeks. In animals *Campylobacter* can be shed in feces for at least 6 weeks postinfection. Person-to-person transmission via the fecal-oral route has also been documented, as well as person to animal (kittens and puppies).

> In contrast to other bacteria that can cause foodborne gastroenteritis, *Campylobacter* does not multiply in food.

Ca. jejuni can be transmitted between animals in a variety of ways. Many chicken flocks have *Campylobacter* carriers and the bacterium can be easily spread from bird to bird through a common water source or through contact with infected feces. During slaughter an infected bird can transfer *Campylobacter* from the gastrointestinal tract to the meat. *Campylobacter* is also present in the giblets (especially the liver) and the skin.

Transmission in cattle may occur from unpasteurized milk from a cow that has *Campylobacter* in her udder, from *Campylobacter* manure-contaminated milk, and from contaminated surface water and mountain streams.

Ca. fetus is spread between animals via coitus, artificial insemination, peneal contact, and contaminated bedding.

Pathogenesis

> A very small number of *Campylobacter* bacteria (fewer than 500) can cause clinical disease in humans.

Campylobacter bacteria have adhesins (that help with attachment to the mucosa), cytotoxins, and **endotoxins** that appear to help them colonize and invade the jejunum, ileum, and colon, producing hemorrhagic lesions that trigger inflammation. Motility by means of a polar flagellum also contributes to *Campylobacter*'s ability to colonize the intestinal mucosa. *Campylobacter* cells adhere to the intestinal lining by receptors on its cell wall. Bacteria then penetrate the intestinal mucosa by burrowing, causing ulcerative lesions in the gastrointestinal tract. Pathology appears to involve a heat-labile enterotoxin (*Ca. jejuni* enterotoxin (CJT) that stimulates a secretory diarrhea) but the understanding of *Campylobacter*'s pathogenicity is not completely understood.

Clinical Signs in Animals

Many animals maintain *Campylobacter* spp. without producing clinical signs. *Campylobacter* can cause diarrhea typically 1 to 7 days after exposure in a variety of animals including dogs and cats (particularly puppies and kittens), calves, sheep, ferrets, mink, hamsters, and nonhuman primates. These animals can also be carriers of the disease. Younger animals are more likely to acquire this infection and shed the organism. Most species infected with *Campylobacter* develop mild to moderate watery diarrhea that may contain mucus, bile, or blood. Some animals such as ferrets, hamsters, and Guinea pigs can develop proliferative lesions of the gastrointestinal tract. *Campylobacter* can also cause hepatitis in poultry and abortion in ruminants. Reproductive lesions found in ruminants include diffuse, mucopurulent endometritis, infiltration of the uterine mucosa with white blood cells, and necrotic and autolytic changes of the placenta. Most aborted fetuses are resorbed, but if passed vaginally they will appear macerated and may contain necrotic liver lesions.

Clinical Signs in Humans

Campylobacter produces acute gastrointestinal disease in humans presenting with diarrhea (with or without blood), abdominal pain, lethargy, and fever. The incubation period is 1 to 7 days. The degree of diarrhea can range from loose stools to profuse watery diarrhea producing bowel movements ten or more times daily. Although campylobacteriosis is self-limiting, symptoms in people may last a week or longer. A potential sequela to campylobacteriosis is Guillain-Barre syndrome (GBS), an autoimmune neuropathy that can occur in approximately 1 out of 1,000 cases of *Campylobacter* infection. GBS is most frequently found with *Ca. jejuni* infections that produce an acute inflammatory demyelinating neuropathy from cross-reaction of antibodies to *Ca. jejuni* antigens with Schwann cells or myelin. Twenty to 40% of all cases of GBS are preceded by *Ca. jejuni* infection. *Ca. fetus* subspecies *fetus* is most commonly associated with systemic infections particularly in immunocompromised people and range from bacteremia, arthritis, septic abortion, and meningitis.

Diagnosis in Animals

Tissue samples obtained from biopsy or aborted fetuses containing *Campylobacter* bacteria may show ulcerated intestinal mucosa with crypt abscesses and infiltration of the tissue with neutrophils, monocytes, and eosinophils. In swine, hamsters, and ferrets proliferative lesions are seen in the intestinal mucosa.

Diagnosis of *Campylobacter* is made via fecal culture from liquid stool or rectal swabs. Samples should be examined within a few hours and transported in semi-solid transport media such as Cary-Blair transport medium or Wang's medium. It is important for fecal samples not to be exposed to oxygen. *Campylobacter* grows well on CAMPY-BAP (a selective media with antibiotics in a *Brucella* agar base with sheep blood) or Skirrow's agar (a selective media with peptone and soy protein base agar with lysed horse blood and antibiotics). *Campylobacter* is slow-growing and produces pinpoint colonies when incubated in a microaerophilic atmosphere (5% to 7% oxygen, 5% to 10% carbon dioxide). *Ca. jejuni* and *Ca. coli* are thermophilic (grow better at higher temperatures) and require 42°C to 43.5°C for 24 to 72 hours for growth. A filtration method can also be used with a nonselective medium, but is not as sensitive as direct culture. Direct examination under the microscope reveals gram-negative "gull-wing" rods. Darting motility using a wet mount preparation is seen using phase-contrast or darkfield microscopy. Direct examination is best done on samples collected during the acute stage of clinical diarrhea when large numbers of the organism are being shed in the feces. Heat-stable or heat-labile antigen tests are available to check rising antibody titers.

Diagnosis in Humans

Stool cultures are used in humans as they are in animals. Blood cultures are routinely set up for humans in addition to fecal cultures. In humans latex agglutination kits and complement fixation tests are also available, but have provided variable results as a result of the large number of *Campylobacter* serovars. PCR tests for identification of *Campylobacter* have been developed.

Treatment in Animals

Treatment of *Campylobacter* infection in animals depends on the severity of the disease and the zoonotic potential. Antibiotics such as erythromycin, gentamicin, and doxycycline are effective against *Campylobacter* bacilli in both animals and humans (penicillins are ineffective against *Campylobacter*) and should be prescribed for adequate amounts of time (anywhere from 7 to 28 days of treatment). Animals will continue to shed the bacteria in the stool despite antibiotic treatment; therefore, follow-up fecal cultures are important after antibiotic treatment. Fluid and electrolyte replacement is important in animals that are dehydrated, especially young animals.

Treatment in Humans

Campylobacter infections in humans are typically self-limiting within 2 to 5 days (up to 10 days in some cases). Fluid and electrolyte replacement is important in humans with more severe cases of *Campylobacter* gastroenteritis. Antibiotics should only be used when diarrhea is persistent or recurs. Antibiotics may not shorten the duration of disease but can shorten the length of bacterial shedding. Antibiotics effective against *Campylobacter* include erythromycin, gentamicin, and doxycycline. Resistance to ciprofloxacin has been demonstrated in some strains of *Campylobacter*.

Management and Control in Animals

Control of diarrhea-causing bacteria is discussed in the chapter on *Escherichia coli* and those precautions apply to preventing *Campylobacter* outbreaks. The use of antibiotics, particularly fluoroquinolones, in feed has been associated with drug-resistant strains of *Campylobacter* in poultry. Within 2 years of the 1995 approval of fluoroquinolone use in poultry the rate of domestically-acquired human cases of fluoroquinolone-resistant campylobacteriosis increased. The use of the fluoroquinolone antibiotic enrofloxacin (Bayril® 3.23%) has been banned for use in poultry by the FDA in July 2005.

Management and Control in Humans

Prophylaxis of *Campylobacter* infections is aimed at domestic animal reservoirs and interrupting transmission of the bacterium to people. Most human cases of *Campylobacter* are caused by fecal contamination of food making proper food hygiene a key factor in preventing campylobacteriosis. Proper food hygiene involves proper cooking of pork and poultry (170°F for breast meat and 180°F for thigh meat), preventing cross-contamination between utensils and cutting surfaces and foods, and drinking pasteurized milk and treated water. Proper hand washing after handling animals (livestock, wildlife, and pets) is also important in controlling spread of this bacterium from animals to people. Physicians and laboratories who diagnose campylobacteriosis in people should report their findings to the local health department.

Summary

Campylobacter are gram-negative bacteria that normally reside in the intestines of animals; however, some species appear to be pathogenic in humans and animals. Some species inhabit the reproductive system of animals. *Ca. jejuni* and *Ca. coli* are the two species some commonly associated with infections in humans and are usually transmitted by contaminated food, milk, or water. Poultry and cattle are the main sources of human campylobacteriosis. *Ca. jejuni* causes about 2 million cases of gastroenteritis annually in the United States and is the number one cause of gastroenteritis in the United States. *Ca. jejuni* is typically spread to humans by oral ingestion of contaminated food or water or by contact with the excretions of infected animals. *Campylobacter* can cause diarrhea typically 1 to 7 days after exposure in a variety of animals including dogs and cats (particularly puppies and kittens), calves, sheep, ferrets, mink, hamsters, and nonhuman primates; however, many animals do not show clinical signs while harboring this bacterium. These animals can also be carriers of the disease. *Campylobacter* produces acute gastrointestinal disease in humans, producing diarrhea (with or without blood), abdominal pain, lethargy, and fever. The incubation period in humans is also 1 to 7 days. Diagnosis of *Campylobacter* is made via fecal culture from liquid stool or rectal swabs. In humans latex agglutination kits and complement fixation tests are also available. Treatment of *Campylobacter* infection in animals depends on the severity of the disease and the zoonotic potential, and may include antibiotics such as erythromycin, gentamicin, and doxycycline. These antibiotics are also effective against *Campylobacter* bacilli in humans. The use of antibiotics, particularly fluoroquinolones, in feed has been associated with drug-resistant strains of *Campylobacter* in poultry and these drugs have been banned for use in poultry by the FDA. Prophylaxis of *Campylobacter* infections is aimed at domestic animal reservoirs and interrupting transmission of the bacterium to people. Using proper food hygiene techniques is a key factor in preventing campylobacteriosis.

CAT-SCRATCH DISEASE

Overview

Cat-scratch disease (CSD) is a bacterial disease caused by *Bartonella henselae* and is spread through cat bites or scratches. Forty percent of cats are estimated to carry *Ba. henselae* at some time during their lives with kittens more likely to be infected than adult cats. The clinical syndrome of CSD was first documented by the Parisian physician, Robert Debré, in the early 1930s, but the etiologic agent has only recently been confirmed. For almost 50 years, a variety of microbial agents (including *Pasteurella* and *Chlamydia*) have been suspected as the causes of CSD. In 1983, Douglas Wear at the Armed Forces Institute of Pathology (AFIP) described the features of the cat-scratch agent using Warthin-Starry silver stain. By 1992 the name *Afipia felis* was proclaimed to be the agent causing CSD (*Afipia* is an acronym for the Armed Forces Institute of Pathology and *felis* refers to the vector of this infection). Some problems existed with the identification of *Af. felis* including the inability of other laboratories to isolate *Af. felis* from CSD patients and that CSD patients did not mount an immune response to *Af. felis* antigen. When human immunodeficiency virus (HIV) became more prevalent in the 1990s, a newly recognized disease called bacillary angiomatosis was recognized. Bacillary angiomatosis lesions contained bacillus bacteria visualized using Warthin-Starry stain. Since bacillary angiomatosis and CSD tissue sections contained bacillus organisms indistinguishable from each other many scientists believed that bacillary angiomatosis may be disseminated CSD in immunocompromised people. In time, polymerase chain reaction (PCR) amplification of ribosomal DNA was used to examine this bacterium and it became apparent that this organism was not *Af. felis* but was similar to *Rochalimaea quintana* (the agent of trench fever). The CSD causing bacterium was *Rochalimaea*-like and was named *Ro. henselae* after Diane Hensel, a microbiologist who isolated several of these bacteria. Shortly after this, sera from CSD-suspect patients were evaluated for the new *Ro. henselae* antibodies and 88% were positive. Several cat fleas combed from these bacteremic cats were also positive for *Ro. henselae*. Serologic data now existed that suggested that *Ro. henselae* was associated with CSD. In time, genotypic evaluation of members of the genus *Rochalimaea* demonstrated that they are related to *Bartonella* spp. and the genus designation of *Bartonella* is now applied to all species of the old genus *Rochalimaea*. *Ba. henselae* may progress to other diseases such as bacillary angiomatosis, bacillary peliosis, and Perinaud's oculogranular syndrome (a small sore on the conjunctiva, redness of the eye, and swollen lymph nodes in front of the ear). CSD should not be referred to as bartonellosis, a disease caused by *Ba. bacilliformis*.

Causative Agent

| It is generally accepted that most cases of CSD are caused by *Ba. henselae* with a small percentage being caused by *Af. felis* and *Ba. clarridgeiae*. |

The bacterium that causes most cases of CSD is *Ba. henselae*, a short, slightly curved, gram-negative, oxidase-negative, aerobic, fastidious rod. *Bartonella* bacteria grow on chocolate agar, but not on blood agar or MacConkey agar. Many *Bartonella* species require long incubation periods in a humid, 37°C environment with increased levels of carbon dioxide. Bacteria belonging to the genus *Bartonella* have gone through extensive name changes and were once grouped with *Rochalimaea* genus bacteria with members of the family Rickettsiae. *Bartonella* and *Rochalimaea* are now one genus which currently includes 16 species; however, only 5 species are currently considered to cause disease in humans. All species of *Bartonella* are found in animals, but do not cause disease in animals (there are some reports that *Ba. vinsonii* causes endocarditis, lameness, and granulomatous lymphadenitis in dogs). *Bartonella* organisms are of a concern especially in immunocompromised people and recognized as causing clinical syndromes in immunocompromised and immunocompetent people.

Currently, there are six *Bartonella* species that infect humans: *Ba. quintana, Ba. henselae, Ba. elizabethiae, Ba. clarridgeiae, Ba. vinsonii,* and *Ba. bacilliformis*. Other than *Ba. henselae* causing CSD, the other major disease-causing species of *Bartonella* are *Ba. bacilliformis* and *Ba. quintana*. *Ba. bacilliformis is* the causative agent of bartonellosis, an often fatal disease causing fever, severe anemia, joint pain, and chronic skin infections that is transmitted by *Phlebotomus* flies (sand flies) in Peru, Ecuador, and Columbia. *Ba. quintana* is the causative agent of both trench fever (a louse-transmitted disease that varies from being asymptomatic to producing severe headaches, fever, and bone pain) and bacillary angiomatosis (a vascular disease of the skin and lymph nodes). Other species cause chronic asymptomatic bacteremia in a wide variety of mammalian hosts ranging from deer, wildcats, cattle, wild rodents, and the Norwegian rat. The full zoonotic potential of these species is unknown.

Epizootiology and Public Health Significance

CSD occurs only in humans and is not a disease of animals. CSD can be seen worldwide with *Ba. henselae* being endemic in Europe, Africa, Australia, and

Japan. In the United States about 25,000 cases of CSD occur annually with 80% of those occurring in children and adolescents. In temperate climates, higher rates of CSD are reported in the autumn and winter (peak between September and March), which can be attributed to the seasonal breeding of the domestic cat. The highest levels of seropositive cats are seen in the southeastern states, coastal California, Hawaii, and the Pacific Northwest. There is only one genotype of *Ba. henselae* reported in North America. There are at least two genotypes of *Ba. henselae* reported in Europe.

The incidence of patients discharged from hospitals in the United States with a diagnosis of CSD is between 0.77 and 0.86 per 100,000 people per year. The estimated incidence of disease in outpatients is about 9 per 100,000 people per year. The estimated annual health care cost of CSD is thought to be more than $12 million.

Transmission

The reservoir of *Ba. henselae* is domestic cats, which do not show clinical signs of disease but are bacteremic for extended periods of time. *Ba. henselae* has been isolated from bacteremic cats, with transmission among cats believed to be via the cat flea (*Ctenocephalides felis*). In regions with particularly high humidity and warm temperature, 40% to 70% of cats, especially feral cats, are carriers of *Ba. henselae*. Although other *Bartonella* species are transmitted by arthropod vectors, it is unlikely that the cat flea is involved directly in human infection, but plays a role in amplifying the bacteria in cats. Transmission of *Ba. henselae* to humans is believed to be via dried infected flea feces via the claws of cats by scratching (Figure 3-19).

> Person-to-person transmission of CSD has not been documented; dogs have been implicated in 5% of CSD cases.

Pathogenesis

Bartonella infections have both intracellular and extracellular phases that may coexist causing a variety of clinical presentations in people. Typically, CSD causes regional lymphadenopathy in those lymph nodes that drain the inoculation site (area of the scratch in which bacteria are introduced) with the most common lymph nodes involved found in the upper extremities. Low-grade fever and

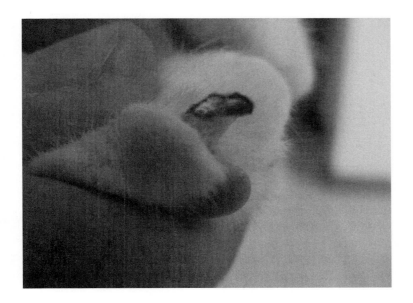

Figure 3-19 Infected feces on the nail of cat claws is the main route of transmission of CSD.

lethargy occurs in up to 50% of patients with lymphadenopathy. Approximately 40% to 60% of patients report a cutaneous inoculation site of about 0.5 to 1.0 centimeters at the site of a cat scratch or bite. Skin lesions typically develop in 3 to 10 days after injury and progress to lymphadenopathy in one to two weeks. As the patient becomes bacteremic many organ systems become infected including the central nervous system, eyes, lungs, and bones. The affects on individual body systems include:

- Lymph nodes become enlarged in 1 to 2 weeks following exposure and are tender and occasionally fluid filled;
- Central nervous system involvement occurs in approximately 2% to 3% of patients abruptly and typically 1 to 6 weeks after lymphadenopathy occurs. CNS signs include disorientation, confusion, and seizures with deterioration to coma in some cases;
- Ocular involvement includes painless, unilateral vision loss with ophthalmo-scopic examination revealing edematous optic discs and exudates surrounding the macula;
- Pulmonary involvement is rare developing in approximately 1 to 5 weeks after lymphadenopathy and presenting as pneumonia and pleural thickening and/or effusion;
- Angiomatosis involves angiogenesis in the skin, regional lymph nodes, and internal organs such as the liver, spleen, bones, and lungs. Angiomatosis typically occurs in immunocompromised people and the disease is termed bacillary angiomatosis. Lesions of the skin and lymph nodes will appear red and may ulcerate;
- Liver involvement is characterized by microscopic blood-filled cysts and is termed bacillary peliosis. Bacillary peliosis may also involve the spleen and lymph nodes. Patients with bacillary peliosis typically have symptoms such as fever, chills, and hepato- or splenomegaly.

Clinical Signs in Animals

Infected cats do not show signs of disease, but are bacteremic. *Ba. vinsonii* may cause endocarditis, lameness, and granulomatous lymphadenitis in dogs.

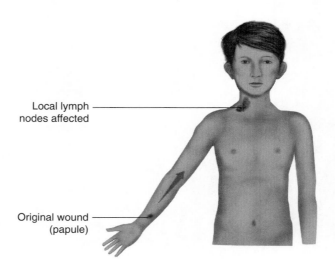

Figure 3-20 CSD causes a cutaneous lesion (papule) followed by regional lymphadenopathy.

Local lymph nodes affected

Original wound (papule)

Clinical Signs in Humans

A cat scratch is typically the primary lesion with CSD and this lesion serves as the portal of entry for the bacteria (Figure 3-20). The skin lesion usually appears as an erythematous (red) papule (Figure 3-21). Following an incubation of 1 to 2 weeks, CSD usually presents as a self-limiting lymphadenopathy associated with a cat scratch or bite. The most common initial finding is subacute regional lymph node enlargement that may develop purulent discharge in 10% to 15% of cases. Generalized lymphadenopathy with CSD is rare. In up to 25% of cases systemic involvement may occur which includes ocular involvement, encephalopathy, hepatitis, and hepatosplenic infection.

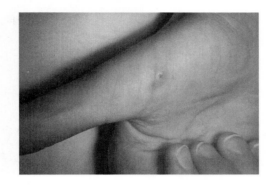

Figure 3-21   Cat-scratch disease lesion on the thumb. (Courtesy of CDC/Emory University, Dr. Sellers)

Diagnosis in Animals

Diagnosis in cats is not done because this bacterium does not cause disease in cats.

Diagnosis in Humans

The tissues infected with *Bartonella* bacteria vary in people and biopsies are typically not used in people to diagnose CSD. If biopsies are taken, the early stages of disease provide the better chance for microscopic detection of bacteria. Microscopy with Warthin-Starry silver stain and immunocytochemistry of blood smears is rarely positive as a result of the low density of bacteria in the blood. Diagnosis of CSD includes identifying the following clinical conditions: regional lymphadenopathy, contact with animals, and primary skin lesions. Positive culture results are optimal in diagnosing this disease, but may not occur because *Ba. henselae* is difficult to culture and may take up to 6 weeks to grow. Either serology or PCR tests are considered to be the best methods of detection. Anti-*Ba. henselae* IgG and IgM indirect fluorescence assay (IFA) or enzyme-linked immunosorbent assays (ELISA) tests exist for the diagnosis of CSD. IFA tests are up to 100% sensitive and ELISA tests for IgM are approximately 95% sensitive (Figure 3-22). Genetic variation occurs amongst *Ba. henselae* strains, which may explain the inconsistency of some diagnostic techniques and cross-reaction within the genus making species identification difficult. A separate serogroup (Marseilles) has been reported in a seronegative patient with CSD, and *Ba. clarridgeiae* and *Af. felis* have the potential to cause the disease which may also complicate the diagnosis of CSD. PCR tests are also available and typically coupled with serologic and histologic tests to increase the accuracy of diagnosis.

>   CSD is the most common cause of regional lymphadenopathy in children and young adults.

Histopathologic findings from involved lymph nodes depends on the stage of infection: early in the disease lymphoid hyperplasia is seen; as the disease progresses granulomas with areas of central necrosis are seen in the lymph nodes; late in the disease areas of necrosis in the lymph nodes start to coalesce. *Bartonella* bacteria can be seen on properly stained sections.

Skin tests can also be performed by preparing antigen by heating pus from affected lymph nodes and injecting 0.1 mL of the antigen intradermally in

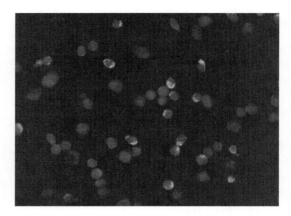

Figure 3-22 IFA test used to diagnose CSD. Magnification 400X.
(Courtesy of BION Enterprises, Park Ridge, IL)

suspected patients. A positive result is at least 5 mm of induration after 48 and 72 hours. Skin testing is controversial as a result of the risk of disease transmission.

Treatment in Animals

Cats are not ill with this disease; therefore, no treatment is used in animals. Treatment of carrier animals is not warranted.

Treatment in Humans

The majority of CSD cases resolve spontaneously and do not require antibiotic treatment. Antipyretics and analgesics may be needed to reduce fever and pain in patients. In complicated CSD, treatment with trimethoprim-sulfamethoxazole, ciprofloxacin, doxycycline, or azithromycin is recommended as a result of their ability to achieve high intracellular concentrations. Aspiration of enlarged, fluid-filled lymph nodes may be recommended to relieve pain; however, incision and drainage leaves a scar and may spread the bacterium from one location to another.

Management and Control in Animals

There is no current management of bacteremic cats to control CSD. Flea control may help limit transmission between cats and reduce bacterial load in the reservoir host.

Management and Control in Humans

Prevention of CSD includes teaching people, especially children, how to handle pets gently to avoid scratches. Any scratch or bite from animals, especially cats, should be thoroughly washed immediately and monitored to make sure it resolves and that lymphadenopathy does not occur. Cats should not be allowed to lick open wounds. People who are immunocompromised, especially people with AIDS, have an increased risk of developing CSD and any skin lesion should be examined by medical professionals. Immunocompromised peoples should especially avoid contact with feral, flea-infested cats and kittens.

Summary

Cat-scratch disease (CSD) is a bacterial disease caused by *Ba. henselae*. Most people with CSD acquire the infection via a cat bite or scratch; the bacterium is spread amongst cats by fleas. Cats that carry *Ba. henselae* do not show signs of disease; people with CSD have swollen lymph nodes, fever, headache, fatigue, and anorexia. CSD may progress to a variety of different diseases including ocular involvement, encephalopathy, hepatitis, hepatosplenic infection, and bacillary angiomatosis. *Ba. henselae* is difficult and time consuming to culture; therefore, either serology or PCR tests are considered to be the best methods of detection. The majority of CSD cases resolve spontaneously and do not require antibiotic treatment. Flea control may help limit transmission between cats and reduce bacterial load in the reservoir host. Teaching people, especially children, how to handle pets gently to avoid scratches is also important in preventing CSD. Immunocompromised people should especially avoid contact with feral, flea-infested cats and kittens.

CLOSTRIDIAL INFECTIONS

Overview

Clostridium bacteria are a large genus of anaerobic bacteria responsible for diseases such as botulism, tetanus, gas gangrene, and pseudomembranous colitis. Although there are over 100 species of *Clostridium*, only a few species are implicated in serious disease. *Clostridium* comes from the Greek word *kloster* meaning spindle and was thus named as a result of its large, almost rectangular morphology. This large genus of bacteria has varied habitats including soil, sewage, vegetation, organic debris, and as commensals of the intestinal tract of humans and animals. *Clostridium* bacteria were first identified by Louis Pasteur in 1861 (the first identified species was *Cl. butyricum*).

Cl. botulinum consists of seven members (four harmful) and was first discovered in 1793 when there were scattered cases of muscle weakness and respiratory failure linked to eating sausage in Germany, Scandinavia, and Russia. The word botulism is derived from the Latin word for sausage, *botulus*. In 1897, van Ermengem found the *Clostridium* organism in contaminated ham and discovered that the organism would only multiply in oxygen-free containers. He also discovered that *Cl. botulinum* produced a powerful toxin that could be inactivated by heat and the bacterium would not produce the toxin at all if the food was too salty. Botulism is a disease that is seen most frequently with home canning products, with cases peaking in the United States in the 1930s (about 35 cases per year). *Cl. botulinum* releases a gas that causes the tin cans used in food canning to bulge. Refrigeration and food preservation techniques (the use of nitrites in hot dogs and cold cuts is to prevent botulism) have made foodborne botulism rare today. More recent botulism outbreaks in the United States include a 1977 outbreak when 59 people got sick after eating improperly preserved jalapeño peppers, a 1993 outbreak when a restaurant in Georgia used contaminated canned cheese sauce, and another 1993 outbreak that occurred in a restaurant in Texas that served contaminated dip. *Cl. botulinum* can also cause infant botulism (when infants ingest the endospores and the toxin forms in the intestines in the absence of mature gastrointestinal normal flora) and wound botulism (where the bacterium multiplies in a wound and produces toxin which is absorbed into the bloodstream).

Horses are especially sensitive to tetanus and are used for commercial production of tetanus antitoxin.

Cl. tetani is the causative agent of tetanus, which comes from the Greek word *tetanos* meaning to stretch. Tetanus has been recognized since the time of Hippocrates in the 5th century B.C. In ancient Egypt, dung was a favored medicine to treat wounds. During wartime, lacerations and burns contaminated with *Cl. tetani* are a breeding ground for tetanus. Tetanus was not fully understood until the late 19th century when the Japanese bacteriologist Shibasaburo Kitasato demonstrated the damage done by this bacterium's toxin by injecting bacteria into the tips of mouse tails and chopping their tails off within an hour of injection. Even with this prompt "treatment," the mice died from the toxin. The discovery of tetanus toxoid (produced by inactivating the toxin with formaldehyde or other means) occurred during World War I when it was injected into French cavalry horses. Additionally, small amounts of toxin could be injected into animals enabling them to produce an antitoxin that could be used on humans. The discoveries of long-term immunization with tetanus toxoid and injection of tetanus antitoxin within a day or two following injury virtually wiped out tetanus. Most U.S. tetanus cases today are in people over fifty who are not up-to-date on their vaccinations. *Cl. tetani* also causes neonatal tetanus which results in hundreds of thousands of infant deaths in Asia and Africa where umbilical cords are cut with contaminated instruments or are packed with dirt.

Cl. perfringens is the cause of food poisoning (especially in cooked beef and poultry) and gas gangrene (named from the Greek term *gangraina* meaning eating sore). The Latin term *perfringere* means to break up and this bacterium is believed to be named because it produces necrotizing enzymes. Around 1892, Welch, Nuttall, and other scientists isolated a gram-positive anaerobic bacillus from gangrenous wounds. This organism was originally named *Bacillus aerogenes capsulatus*, then *Bacillus perfringens*, then *Cl. welchii*, and finally *Cl. perfringens*. Gas gangrene is typically associated with war injuries. In World War I, 6% of open fractures and 1% of open wounds were complicated with gas gangrene; in World War II these cases decreased to 0.7%; in the Korean War they decreased to 0.2%; in the Vietnam War they decreased to 0.002%; and were nonexistent during the war in the Falkland Islands in 1982.

Cl. difficile, an opportunistic bacterium of the gastrointestinal tract and potential zoonotic agent, is named because it is so difficult to culture. *Cl. difficile* causes antibiotic-associated diarrhea and pseudomembranous colitis in humans. Healthy people do not get *Cl. difficile* since it can be part of the normal gastrointestinal flora; however, people who have undergone gastrointestinal surgery, have serious underlying illness, or are immunocompromised are at risk of developing disease from this bacterium. Horses can develop *Cl. difficile*-associated diarrhea and may pose a zoonotic risk for this disease.

Causative Agent

The genus *Clostridium* consists of large, **obligate** (strictly) anaerobic, gram-positive, endospore-forming bacilli that are found in almost all anaerobic habitats in nature where organic compounds are present (soil, aquatic sediments, and the gastrointestinal tracts of animals). Their resistance to environmental changes is a result of their ability to produce endospores and their virulence is a result of the vegetative cell's secretion of toxins. *Clostridium* bacteria produce oval to spherical endospores that swell the vegetative cell and the cells occur in single, paired, or chain arrangements. The main pathogenic species of *Clostridium* in humans are:

■ *Cl. botulinum.* This species of *Clostridium* forms subterminal endospores and is commonly found in soil, decaying vegetation, and lake and pond sediments. The gastrointestinal tracts of birds, mammals, and fish may occasionally contain *Cl. botulinum.* Its endospores are able to survive improper food canning techniques and as the vegetative cell germinates they release neurotoxins into the container. Different strains of *Cl. botulinum* produce one of seven distinct neurotoxins (identified as A, B, C1, D, E, F, and G). In the United States type A is the most common cause of botulism (62% of cases). Types A, B, and E are most important in botulism in people; type B toxin causes botulism in horses and foals in the eastern United States; C1 in most animal species (wild ducks, pheasants, chickens, mink, cattle, and horses); D in cattle; type F has been responsible for two outbreaks in humans; type G (isolated from soil in Argentina) is not known to have been involved in any outbreak of botulism. Botulism neurotoxins affect the peripheral nervous system.

■ *Cl. tetani.* This species of *Clostridium* is typically found in heavily-manured soils and in the gastrointestinal tracts and feces of many animals. There are 11 strains of *Cl. tetani* distinguished from each other on the basis of flagellar antigens. *Cl. tetani* produces terminal endospores within a swollen end giving it a tennis-racket appearance (Figure 3-23). *Cl. tetani* produces tetanospasmin, a neurotoxin that disrupts nerve impulses to muscles. This toxin is produced by vegetative cells and is released on cell lysis, which occurs naturally during germination.

■ *Cl. perfringens.* *Cl. perfringens* is found in soil and dust, manure, in lakes and streams, on skin, and in the gastrointestinal tract. *Cl. perfringens* is nonmotile (most other species are motile) and produces subterminal endospores. This bacterium produces at least 12 different toxins and antigens and is divided into types A to E based on differences in the toxins. Important toxins include alpha (lethal toxin that is also hemolytic and necrotizing), beta (lethal toxin that is responsible for inflammation of the intestine and loss of mucosa), epsilon (lethal toxin that is necrotizing), itoa (lethal and necrotizing), delta (lethal and hemolyzing), kappa (lethal toxin that is a proteolytic enzyme that breaks down collagen), lambda (proteolytic enzyme that attacks hemoglobin), and

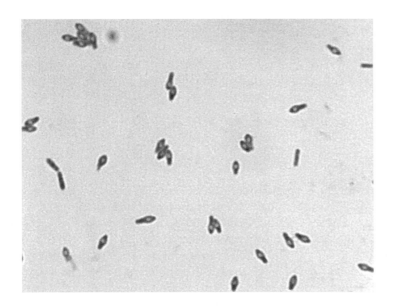

Figure 3-23 Photomicrograph of *Clostridium tetani* demonstrating its terminal endospores.
(Courtesy of CDC)

Table 3-5   Selected Species of *Clostridium* and Their Properties

Clostridium Species	Disease	Spore Location
Cl. perfringens	Gas gangrene in humans; food poisoning in humans; enteric disease in animals	Subterminal
Cl. novyi	Type A: gas gangrene in humans, cattle, and sheep; big head in rams Type B: black disease in sheep (occasionally cattle) Type C: osteomyelitis in water buffalo	Subterminal
Cl. septicum	Gas gangrene in humans Malignant edema in cattle, sheep, horses, pigs, and other animals	Subterminal
Cl. tetani	Tetanus in humans and animals	Terminal
Cl. botulinum	Botulism in humans and animals; termed limberneck in birds	Subterminal
Cl. difficile	Colitis (antibiotic associated in people; *Cl. difficile* associated diarrhea in horses) Pseudomembranous colitis in humans; ileocecitis in laboratory animals	Subterminal
Cl. bifermentans	Gas gangrene in humans and animals	Subterminal
Cl. sporogenes	Gas gangrene in humans and animals; enterotoxemia in rabbits	Subterminal
Cl. histolyticum (formerly known as *Cl. novyi* type D)	Gas gangrene in humans and animals; bacillary hemoglobinuria or red water disease in sheep and cattle	Subterminal

mu (hydrolyzes hyaluronic acid). The toxin is deemed lethal as tested by injection in mice.

- *Cl. difficile*. This organism has a subterminal endospore and produces two toxins. Toxin A is an enterotoxin because it causes fluid accumulation in the intestines and toxin B is a cytopathic toxin that is extremely lethal. These two toxins destroy the intestinal lining causing diarrhea. In minor infections these lesions are self-limiting; in serious cases a life-threatening pseudomembranous colitis occurs in which large sections of the colon wall slough, potentially perforating the colon, leading to massive internal infection by fecal bacteria and possible death.

Selected *Clostridium* bacteria are summarized in Table 3-5.

Epizootiology and Public Health Significance

Thirty grams of pure botulism toxin could kill every U.S. citizen.

Clostridium spp. are ancient organisms and have worldwide distribution. *Clostridium* bacteria are found in soil and any other environment where organic compounds are found. Certain species may be more prevalent in some parts of the world and will be described with each disease.

Annual U.S. cases of botulism are about 25 with a 20% mortality rate (mortality rates used to be over 60% prior to the establishment of critical care). Typically the first person who contracts botulism in an outbreak has 25% mortality with subsequent cases being more rapidly diagnosed and treated carrying a 4% mortality rate. About 100 cases of infant botulism are reported annually in the United States. Since 1994, the use of black tar heroin by chronic IV drug users has led to a dramatic increase in wound botulism in the United States.

Over 1 million cases of tetanus occur annually worldwide, mostly in underdeveloped countries with inadequate health care and vaccination protocols. The mortality rate of tetanus is 50%; neonatal tetanus (usually from umbilical cord infection) has a 90% mortality rate. In the United States tetanus is uncommon as a result of the vaccine.

Gas gangrene is uncommon in the United States with only 1,000 to 3,000 cases occurring in the United States annually. The mortality rate with gas gangrene is about 40% even with therapy.

Cl. difficile infections among hospitalized patients are increasing. Studies from Canada indicate that between March 2000 and April 2004 the number of cases doubled (from about 3,300 to 7,000) with a 60% increase in deaths associated with *Cl. difficile* infection. Exact case numbers in the United States are not documented, but this increase is likely occurring in this country as well. Other than *Cl. difficile* infections which appear to be rising, clostridial diseases are uncommon in the United States.

> The occurrence of *Cl. tetani* in the soil (and the resulting incidence of tetanus in man and horses) is higher in the warmer parts of the various continents.

Transmission

Clostridial infections can be transmitted in a variety of ways including:

- Ingestion of preformed toxin in vegetable and meat-based foods for *Cl. botulinum*.
- Contamination of wounds or puncture of object contaminated with *Cl. tetani*, *Cl. perfringens*, and less commonly *Cl. septicum* and *Cl. novyi*.
- Colonization of gastrointestinal tract with toxin-producing bacterium for *Cl. botulinum* (infant botulism).
- Person-to-person spread for *Cl. difficile*.
- Potential zoonotic spread via fecal contaminated equipment for *Cl. difficile*.

Pathogenesis

Botulism

In foodborne botulism the toxin is ingested with contaminated food in which endospores have germinated and vegetative cells have multiplied. Foodborne botulism is a form of intoxication because the ingested food contains preformed toxin. In the upper gastrointestinal tract the toxin is absorbed, passes into the blood, and in time reaches the peripheral neuromuscular junction. The toxin binds to the presynaptic membrane and blocks the release of acetylcholine, a neurotransmitter required for muscle stimulation by a nerve. Botulism occurs from ingested uncooked foods and commonly is associated with home canning. Clinical signs of botulism occur 18 to 36 hours after toxin ingestion and include weakness, dizziness, and gastrointestinal signs like diarrhea and vomiting. Neurologic signs develop soon afterwards and include blurred vision, weakness of skeletal muscles, and respiratory paralysis.

> The three most poisonous substances known are botulism toxin, tetanus toxin, and diphtheria toxin.

Infant botulism is caused by infection with *Cl. botulinum* and typically occurs in infants 5 to 20 weeks of age that have been exposed to solid foods containing endospores. *Cl. botulinum* establishes itself in the intestines of infants prior to development of competent normal flora. Bacteria produce toxin in the gastrointestinal tract which causes signs such as constipation and generalized weakness.

Tetanus

Most tetanus cases result from puncture wounds or lacerations that become contaminated with endospores that germinate in tissue and produce toxin. Infection is typically localized causing only minimal inflammation; however, because *Cl. tetani* is a strict anaerobe the endospores cannot germinate to vegetative cells unless the tissues are necrotic and have poor blood supply (lowers the oxygen tension). The tetanospasmin toxin is produced during cell growth and released upon cell lysis (occurs during vegetative growth and the production of endospores). The toxin moves along nerve pathways from the wound to the peripheral nerve endings. It travels within the axon to the spinal cord. In the spinal cord the toxin binds to specific sites on spinal neurons that are responsible for inhibiting skeletal muscle contraction. The tetanospasmin blocks the release of neuroinhibitors needed to regulate muscle contraction causing the muscles to contract uncontrollably. Larger muscle groups are initially affected such as the muscles of the jaw and then back (causing arching), arms, and legs. Respiratory paralysis is often the cause of death in severe cases.

Gas gangrene

Gas gangrene, also known as myonecrosis, is caused by a few species of *Clostridium* with *Cl. perfringens* being the most common etiologic agent. *Cl. perfringens* produces alpha toxin which causes red blood cell rupture, edema, and tissue damage. There are two forms of gas gangrene: anaerobic cellulitis in which previously injured necrotic tissue is infected with *Cl. perfringens* producing toxin and gas (infection remains localized and does not spread into healthy tissue) and myonecrosis in which toxins are produced in damaged tissue and are diffused into nearby healthy tissue (resulting in continued spread of disease as more tissue is destroyed producing gaseous bacterial waste products).

Pseudomembranous colitis

Cl. difficile produces two toxins: toxin A is an enterotoxin (causes fluid accumulation in the intestines) and toxin B is a cytopathic toxin (causes cell death). Typically following antibiotic treatment or in immunocompromised people, *Cl. difficile* endospores germinate in the intestine allowing it to predominant the normal flora. The toxins and enzymes produced by this bacterium produce hemorrhagic necrosis of the intestinal mucosa. In serious cases large sections of the colon will slough potentially causing bowel perforation and internal infection by fecal bacteria.

Clinical Signs in Animals

Clostridial infections in animals vary with the species of *Clostridium* and animal species. Some examples of resulting infections from select *Clostridium* species include (others can be found in Table 3-5):

- *Cl. botulinum*. This species of *Clostridium* causes botulism, a rapidly fatal motor paralysis in animals. The usual source of the toxin is decaying carcasses or contaminated vegetable materials such as decaying grass, hay, grain, or spoiled silage. In animals toxicoinfectious botulism occurs when *Cl. botulinum* grows in tissues of a living animal and produces toxins. Predisposing factors for the development of toxicoinfectious botulism include gastric ulcers, liver necrosis, navel and lung abscesses, skin and muscle wounds, and necrotic lesions of the gastrointestinal tract. The incidence of botulism in animals is unknown, but it is relatively low in cattle and horses, more frequent in chickens, and high in wild waterfowl (between 10,000 and 50,000 birds are lost annually in the

United States, with losses reaching 1 million or more during outbreaks in the western United States). The zoonotic potential of botulism is minimal.

- Most affected birds are ducks, although loons, geese, and gulls also are susceptible. In birds, *Cl. botulinum* is commonly found in the gastrointestinal tract of poultry and wild birds and in litter, feed, and water in broiler chicken houses. Intoxications are sporadic in poultry, but massive mortality has occurred in waterfowl in western North America. Clinical signs in poultry and wild birds are similar and include flaccid paralysis of the legs, wings, eyelids, and neck (hence the term limberneck) (Figure 3-24), with paralytic signs progressing cranially from the legs to the wings, neck, and eyelids. In affected waterfowl, neck paralysis can lead to drowning.

Figure 3-24 Botulism in birds is often referred to as limberneck as a result of the flaccid paralysis of the neck muscles.
(Courtesy of Canadian Cooperative Wildlife Health Centre)

- Dogs, cats, and pigs are comparatively resistant to most types of botulism toxin when administered orally. Botulism occurs sporadically in dogs as a result of type C toxin; botulism has not been reported in cats. Clinical signs of botulism include progressive motor paralysis, altered vision, difficulty in chewing and swallowing, and generalized weakness.

- Most cases of bovine botulism occurs in South Africa, where phosphorus-deficient cattle chew any bones they find on the range and intoxication occurs if these bones and flesh came from an animal that had been carrying type D strains of *Cl. botulinum*. As little as one gram of dried flesh from an infected carcass may contain enough toxin to kill mature cattle. Type C strains also cause botulism in cattle (this form is rare in the United States with only a few cases reported from Texas and referred to as loin disease). Cattle may drool, be unable to urinate, have dysphagia, and may be in sternal recumbency that progresses to lateral recumbency just before death. Clinical signs in cattle resemble milk fever, but the cows do not respond to calcium therapy. Death results from respiratory or cardiac paralysis. There are no characteristic lesions with botulism in cattle.

- Botulism in sheep has been reported in Australia, in sheep eating carcasses of rabbits and other small animals found on the range.

- Botulism in horses often results from forage contaminated with type C or D toxin. Foals with botulism, also known as shaker foal syndrome, are usually less than 4 weeks old and may be found dead without signs. The most common sign is progressive symmetrical motor paralysis; other signs include stilted gait, muscular tremors, dysphagia, constipation, mydriasis, frequent urination, dyspnea with extension of the head and neck, tachycardia, and respiratory arrest. Death occurs in foals within 24 and 72 hours after the onset of clinical signs.

- Botulism in unvaccinated mink occurs when mink eat raw meat diets. Mink botulism is typically caused by type C strains that have produced toxin in chopped raw meat or fish. Many mink are typically found dead within 24 hours of exposure to the toxin, whereas others show varying degrees of paralysis and dyspnea. Necropsy findings are nonspecific.

- *Cl. tetani.* Tetanus toxemia in animals is caused by a neurotoxin produced by *Cl. tetani* in necrotic tissue. Most mammals are susceptible to tetanus, although dogs, cats, and birds are relatively resistant. Horses are the most sensitive of all animal species. Tetanus in animals has a worldwide distribution, although in some areas, such as the northern Rocky Mountains of the United States, the bacterium is rarely found in the soil and tetanus is rare. Tetanus is seen following a puncture wound, a surgical procedure such as docking or castration, and occasionally the point of entry cannot be found because the wound is small or healed. The classic clinical sign is muscle rigidity and even minor stimulation of the affected animal may trigger muscle spasms. Muscle spasms affecting the larynx, diaphragm, and intercostal muscles may lead to respiratory failure. Involvement of the autonomic nervous system results in cardiac arrhythmias, tachycardia, and hypertension. Fever is rare, but may occur. The incubation period averages 10 to 14 days but can vary from 1 to several weeks (dogs and cats often have a longer incubation period). In most animals the first sign is localized stiffness, typically of the masseter muscles, muscles of the neck, hindlimbs, and the region of the infected wound, with generalized stiffness becoming pronounced a few days later with tonic spasms and hyperesthesia becoming evident. Hyperreflexia, general spasms, and difficulty in prehension and mastication of food (hence the common name lockjaw) are commonly seen.

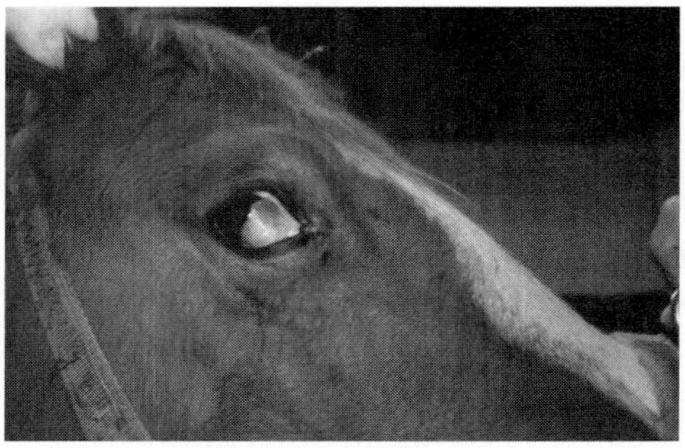

Figure 3-25 Horses with tetanus commonly present with a prolapsed third eyelid. (Courtesy of Simon Peek, BVSC, MRCVS, PhD, DACVIM)

  - In horses, clinical signs include erect ears, stiff and extended tail, dilated anterior nares, prolapsed third eyelid, difficulty walking, turning, and backing, extension of the head and neck as a result of muscle spasms, a "sawhorse" stance, sweating, increased heart rate, rapid breathing, and congestion of mucous membranes (Figure 3-25).

  - Sheep, goats, and pigs present with similar signs to horses and often fall to the ground and exhibit opisthotonos when startled. Mentation is not affected.

  - In dogs and cats, localized tetanus presents as stiffness and rigidity in a limb with a wound. The stiffness progresses to the opposite limb and may advance cranially. Generalized tetanus in dogs and cats is similar to other animals except they may have a partially open mouth with the lips drawn back (as seen in humans).

- *Cl. perfringens.* This species of *Clostridium* is widely distributed in the soil and the gastrointestinal tract of animals. Five types of isolates (A, B, C, D, and E) have been identified.

  - Type A strains of *Cl. perfringens* are commonly found as part of the normal intestinal flora of animals and do not have as destructive toxins as some other strains. *Cl. perfringens* type A produces necrotic enteritis in poultry and dogs (causing massive destruction of the villi and coagulation necrosis of the small intestine), colitis in horses, and diarrhea in pigs.

  - Type B and C strains of *Cl. perfringens* produce highly necrotizing and lethal β toxin (responsible for the severe intestinal damage) and cause severe enteritis, dysentery, toxemia, and high mortality in young lambs, calves, pigs, and foals. *Cl. perfringens* type C also causes enterotoxemia in adult cattle, sheep, and goats. Lamb dysentery is caused by *Cl. perfringens* type B in lambs up to 3 weeks of age (lambs stop nursing, become listless, and remain recumbent and commonly develop blood-tinged diarrhea with death occurring in a few days; some lambs may die before signs appear); calf enterotoxemia is caused by *Cl. perfringens* types B and C in well-fed calves up to 1 month of age (acute diarrhea, dysentery, abdominal pain, convulsions, and opisthotonos resulting in death in a few hours or in less severe cases recovery over a period of several days); pig enterotoxemia is caused by *Cl. perfringens* type C in piglets during the first few days of life (acute illness within a few days of birth with diarrhea, dysentery, reddening of the anus, and a high mortality rate typically within 12 hours); foal enterotoxemia is caused by *Cl. perfringens* type B in foals in the first week of life (acute dysentery, toxemia, and rapid death); and goat enterotoxemia caused by *Cl. perfringens* type C in adult goats (death is typically the only sign).

- Type D strain of *Cl. perfringens* causes an enterotoxemia of sheep (and to a rare extent goats and cattle). The main predisposing factor of this disease is ingestion of excessive amounts of feed or milk in the very young and grain in feedlot lambs (from where it gets it common name overeating disease). As starch intake increases it provides a suitable environment for bacterial growth and toxin production. This disease is commonly seen in lambs younger than 2 weeks of age or in weaned lambs on feedlot. The most common clinical sign is sudden death in the best conditioned lambs; in some cases excitement, incoordination, and seizures may be seen prior to death. Adult sheep may be affected occasionally showing similar signs.

- Type E strain of *Cl. perfringens* has not been linked to animal disease.

- *Cl. difficile*. *Cl. difficile* causes acute, sporadic gastrointestinal disease of horses characterized by diarrhea and colic. Clinical signs in horses range from mild and self-limiting to rapidly fatal. The factors that cause the disease are unknown; however, alteration in the normal flora allowing excessive multiplication of these bacteria, which are capable of producing toxins that cause intestinal damage and systemic effects, is believed to play a key role. Dietary change and antibiotic therapy likely play a role in disease development. Other host factors that may determine whether disease develops include age, immunity, and presence or absence of intestinal receptors for the clostridial toxins. *Cl. difficile* has been identified in foals with diarrhea, but not in healthy foals. Foals are more frequently affected, but adult horses may also get the disease. Clinical signs include dead horses without any signs, abdominal pain, diarrhea with or without blood, dehydration, toxemia, and shock. *Cl. difficile* has emerged in recent years as a cause of diarrhea and edema of the colon in neonatal swine (1- to 7-day-old pigs).

Clinical Signs in Humans

Clinical signs of botulism include weakness, dizziness, blurred vision, dry mouth, dilated pupils, constipation, and abdominal pain followed by progressive paralysis that ultimately affects the diaphragm. Death results from the inability to inhale. People remain alert when they have botulism. Babies with infant botulism excessively cry, have constipation, and fail to thrive. Paralysis and death with infant botulism is rare. Wound botulism is a rare disease and occurs when the endospores get into an open wound and multiply in an anaerobic environment (Figure 3-26). The symptoms produced are similar to those described above, but may take up to 2 weeks to appear.

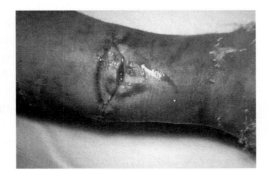

Figure 3-26 Wound botulism on the arm of a person with a bone fracture.
(Courtesy of CDC)

Figure 3-27 Baby displaying muscle rigidity produced by *Clostridium tetani* neurotoxin. (Courtesy of CDC)

Clinical signs of tetanus include the initial and diagnostic sign of tightening of the jaw and neck muscles, sweating, drooling, and back spasms (Figure 3-27). As the disease progresses irregular heart rate, changes in blood pressure, excessive sweating, and spasms spreading to other muscles such as those of the arms, fists, and feet. Death can occur as a result of the person's inability to exhale.

Clinical signs of gas gangrene include intense pain at the site of infection (as a result of tissue swelling and necrosis), the presence of gas in the tissues (as a result of bacterial waste products) (Figure 3-28), shock, kidney failure, and death.

Clinical signs of pseudomembranous colitis include a self-limiting explosive diarrhea in mild cases and in severe cases sloughing of the colon wall, intestinal perforation, sepsis, and possible death.

Figure 3-28 Gas gangrene causing necrosis of the fingers. (Courtesy of CDC/Dr. Jack Poland)

Diagnosis in Animals

Diagnosis of clostridial infections varies with the species of *Clostridium*.

- Botulism is diagnosed by clinical signs, eliminating other causes of motor paralysis, analyzing feed, detecting toxin in the blood (by mouse inoculation tests or ELISA testing), Gram stain, and anaerobic tissue culture.
- Tetanus is diagnosed by clinical signs and history of recent trauma, demonstrating the presence of tetanus toxin in serum, and demonstration of the bacterium in Gram-stained smears, and by anaerobic culture when a wound is apparent.
- *Cl. perfringen*s infections can be diagnosed based on necropsy lesions (hemorrhagic enteritis with ulceration of the mucosa is the major lesion in all species), Gram-stained smears of intestinal contents, and toxin detection.

■ *Cl. difficile* infections can be diagnosed based on lesions (necrotizing enterocolitis, loss of colonic and cecal mucosal epithelial cells, and thrombosis in capillaries of the intestinal mucosa), demonstration of toxins in feces or intestinal fluid, and demonstration of large numbers of the bacterium on Gram stain and through anaerobic culture.

Diagnosis in Humans

Diagnosis in people is similar to animals. Proper collection and transport of specimens for anaerobic culture are essential. Material for anaerobic culture is best obtained by tissue biopsy or fine needle aspirate. Swabs expose the specimen to drying, contamination, and retention of the organism to the swab fibers (although oxygen-free transport swabs are available). Some clostridial diseases (such as foodborne *Cl. perfringens* and *Cl. botulinum* and enteritis caused by *Cl. difficile*) must be sent to a public health laboratory for confirmation. Suitable anaerobic media include anaerobic blood agar, chopped meat broth, thioglycollate broth, and cycloserine cefoxitin fructose agar (CCFA), which is selective for *Cl. difficile*. Anaerobic incubation can be obtained by using anaerobe jars, holding jars, or an anaerobe chamber. Commercial identification methods are also available (Figure 3-29).

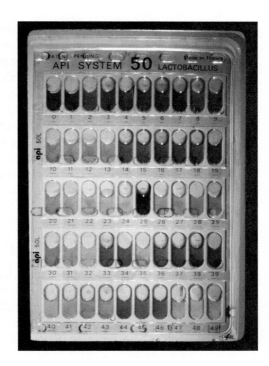

Figure 3-29 Commercial identification method used to identify gram-negative bacteria. (Courtesy of CDC/Dr. Gilda Jones)

Treatment in Animals

Treatment of *Clostridium* infections includes the early use of antibiotics and antitoxin if available, removal of contaminated feed or material, and disinfection. Altering the anaerobic environment of contaminated wounds through debridement and antisepsis is also recommended.

The incidence of botulism in animals is relatively low in cattle and horses, somewhat frequent in chickens, and high in wild waterfowl. Probably 10,000 to 50,000 birds are lost annually in the United States with ducks being the bird most affected. Dogs, cats, and pigs are comparatively resistant to all types of botulinum toxin when administered orally. Botulism is treated with botulinum antitoxin in ducks and mink with type C antitoxin; however, such treatment is rarely used in cattle. Treatment with guanidine hydrochloride may ease paralysis caused by the toxin; however, it is not used extensively.

Tetanus is treated with curariform agents, tranquilizers, or barbiturate sedatives, in conjunction with tetanus antitoxin in horses. Draining and cleaning of wounds is also important in treating cases of tetanus. Keeping animals in a quiet, darkened area and assisting with ambulation are also important in the nursing care of animals with tetanus.

Treatment of *Cl. perfringens* infections is usually ineffective in animals because of the severity of the disease; however, specific hyperimmune serum and oral administration of antibiotics may be helpful.

Treatment of *Cl. difficile* in horses is oral metronidazole.

Treatment in Humans

Treatment of *Clostridium* infections in humans revolves around the proper use of antibiotics and antitoxin. Botulism is treated with antibiotics, antibodies against the botulism toxin (so that new toxin doesn't bind to neurons; any prior toxin binding is irreversible), and repeated washing of the intestinal tract to remove the bacterium. Tetanus is treated with antibiotics, passive immunization with antitoxin (binds to and neutralizes toxin before it can bind), cleansing of the wound to remove endospores, and active immunization with tetanus toxoid (stimulates the formation of antibodies that neutralize the toxin). As with botulism, bound tetanus toxin cannot be reversed. Sometimes sedatives or tranquilizers may be given to ease the spasms as well as keeping the person in a dark, quiet place. Gas gangrene is treated with antibiotics, administration of antitoxin, surgical debridement of the wound to eliminate the anaerobic environment, and sometimes hyperbaric oxygen treatment. Clostridial food poisoning is usually self-limiting. Treatment of *Cl. difficile* involves discontinuing the implicated antibiotic (mild cases) or treatment with vancomycin or metronidazole (severe cases where about one third of patients will relapse).

Management and Control in Animals

Control of *Clostridium* infections in animals includes proper disposal of carcasses, correcting any dietary deficiencies, avoiding overeating, removing contaminated grass or feed, proper surgical technique following strict guidelines for asepsis, using clean needles, and vaccination if available. Vaccines are available for cattle and sheep (a variety of combinations for *Cl. chauvoei*, *Cl. septicum*, *Cl. novyi*, *Cl. sordelii*, *Cl. perfringens* types C and D, *Cl. tetani* (toxoid and antitoxin), *Cl. haemolyticum*), mink (toxoid for botulism), horses (*Cl. botulinum* type B and *Cl. tetani* [toxoid and antitoxin]), and pigs (*Cl. perfringens* and *Cl. tetani* toxoid). In horses (who are especially sensitive to tetanus) the toxoid should be annually; mares should be vaccinated during the last 6 weeks of pregnancy and the foals vaccinated at 5 to 8 weeks of age, then repeated annually. If horses incur a deep wound after immunization, another injection of toxoid may be given to increase circulating antibody levels and repeated in 30 days. If the horse has not been immunized previously, it should be treated with tetanus antitoxin, which usually provides passive protection for up to 2 weeks. In high-risk areas, foals may be given tetanus antitoxin immediately after birth and every 2 to 3 weeks until they are 3 months of age, at which time they can be given toxoid. Vaccination of lambs, calves, and pigs depends on the prevalence of tetanus in the area.

Cl. difficile infections in horses can be reduced by judicial use of antibiotics (antibiotics such as metronidazole and chloramphenicol for oral administration is recommended for high-risk horses).

Management and Control in Humans

Clostridial infections in people can be controlled in a variety of ways. Botulism can be controlled by proper home canning techniques, preventing endospore germination by refrigeration or establishing an acidic environment, or destroying the toxin by heating to at least 80°C for at least 20 minutes. Infant botulism can be prevented by not feeding infants honey until they are over 1 year of age. Wound botulism can be avoided with proper wound cleaning. Tetanus can be controlled by vaccination with tetanus toxoid (part of the DTP or DT vaccine) as per CDC recommendations of three doses during the first year of life, a booster in about one year, a booster when entering elementary school, followed by a booster every 10 years of life. Gas gangrene infections are hard to control because the organism is so prevalent in the environment. Myonecrosis can be controlled with proper cleaning of wounds. Foodborne *Cl. perfringens* infections can be prevented through refrigeration of food and reheating of food to destroy any toxin present. Hot foods should be served immediately or held above 140°F. Refrigerated foods should be stored in small containers and reheated to 165°F prior to serving. *Cl. difficile* infections can be prevented with proper and limited use of antibiotics and proper hygiene to avoid nosocomial infections. Zoonotic potential can be reduced through the use of barrier precautions (gloves, gowns, and boots), personal hygiene, and disinfection of equipment with an appropriate disinfectant such as 5% to 10% bleach.

Summary

Clostridial infections are caused by bacteria in the genus *Clostridium*—large, strictly anaerobic, gram-positive, endospore-forming bacilli that are found in almost all anaerobic habitats in nature where organic compounds are present. Their resistance to environmental changes is a result of their ability to produce endospores and their virulence is a result of the vegetative cell's secretion of toxins. The main pathogenic species of *Clostridium* in humans are *Cl. botulinum*, *Cl. tetani*, *Cl. perfringens*, and *Cl. difficile*. *Clostridium* spp. are ancient organisms and have worldwide distribution. *Clostridium* bacteria are found in soil and any other environment where organic compounds are found. Clostridial infections can be transmitted in a variety of ways including ingestion of preformed toxin in vegetable and meat-based foods, contamination of wounds or puncture wounds, colonization of the gastrointestinal tract with toxin-producing bacteria, person-to-person spread, and potential zoonotic spread via fecal contaminated equipment. The pathogenesis of clostridial disease revolves around toxins and toxin production.

Clostridial infections in animals vary with the species of *Clostridium* and the animal species. Some examples of resulting infections from select *Clostridium* spp. include *Cl. botulinum* (causes botulism, a rapidly fatal motor paralysis in animals), *Cl. tetani* (causes tetanus toxemia in animals as a result of a neurotoxin produced in necrotic tissue that results in muscle rigidity and spasms), *Cl. perfringens* (causes wound contamination and gastroenteritis in animals), and *Cl. difficile* (causes acute, sporadic gastrointestinal disease of horses characterized by diarrhea and colic). In people, clostridial infections cause botulism (signs include weakness, dizziness, blurred vision, dry mouth, dilated pupils, constipation, and abdominal pain followed by progressive paralysis that ultimately affects the diaphragm), tetanus (signs include the initial and diagnostic sign of tightening of the jaw and neck muscles, sweating, drooling, and back spasms, progressing to irregular heart rate, changes in blood pressure, excessive sweating, and spasms spreading to other muscles such as those of the arms, fists, and feet), gas gangrene (signs include intense pain at the site of infection (as a result of tissue swelling and necrosis), the presence of gas in the tissues (as a result of bacterial waste products), shock, kidney failure, and death), and pseudomem-branous colitis (signs include a self-limiting explosive diarrhea in mild cases and in severe cases sloughing of the colon wall, intestinal perforation, sepsis, and possible death). Clostridial infections are typically diagnosed via clinical signs, toxin detection, Gram stain, and anaerobic tissue culture. Treatment of clostridial infections includes the early use of antibiotics and antitoxin if available, removal of contaminated feed or material, and disinfection. Altering the anaerobic environment of contaminated wounds through debridement and antisepsis is also recommended. Control of clostridial infections in animals includes proper disposal of carcasses, correcting any dietary deficiencies, avoiding overeating, removing contaminated grass or feed, proper surgical technique following strict guideline for asepsis, and vaccination if available. Vaccines are available for cattle and sheep (a variety of combinations for *Cl. chauvoei*, *Cl. septicum*, *Cl. novyi*, *Cl. sordelii*, *Cl. perfringens* types C and D, *Cl. tetani* (toxoid and antitoxin), *Cl. haemolyticum*, mink [(toxoid for botulism)], horses (*Cl. botulinum* type B) and *Cl. tetani* (toxoid and antitoxin)), and pigs (*Cl. perfringens* and *Cl. tetani* toxoid). In people, botulism can be controlled by proper home canning techniques and by not feeding infants honey until they are older than 1 year of age. Wound botulism can be avoided with proper wound cleaning. Tetanus can be controlled by vaccination with tetanus toxoid. Gas gangrene infections can be controlled with proper cleaning of wounds. Foodborne *Cl. perfringens* infections can be prevented through refrigeration of food and reheating of food to destroy any toxin present. *Cl. difficile* infections can be prevented with proper and limited use of antibiotics and proper hygiene to avoid nosocomial infections. Zoonotic potential can be reduced through the use of barrier precautions, personal hygiene, and disinfection of equipment with an appropriate disinfectant.

ERYSIPELOID

Overview

Erysipeloid is a disease of traumatized skin consisting of fever, vesicles of the hands and feet, and inflammation of the mucous membranes caused by *Erysipelothrix rhusiopathiae* (from the Greek words *erythros* for red, *pella* for skin, *thrix* for hair). The taxonomic name *Er. rhusiopathiae* literally means erysipelas thread of red disease. *Er. rhusiopathiae* (formerly known as *Er. insidiosa*) was first isolated by Koch in 1878. In 1886 it was described by Loeffler as the etiologic agent of swine erysipelas. Erysipeloid was first described in humans by William Morrant Baker in 1873. In 1909, Rosenbach isolated the bacterium from a human patient with localized cutaneous lesions thus establishing it as a human pathogen. Rosenbach used the term erysipeloid to avoid confusion with the cutaneous lesions of human erysipelas (caused by *Streptococcus pyogenes*). Joseph Victor Klauder in 1917 published the first account in English.

Erysipeloid is known by a variety of other names such as Baker-Rosenbach syndrome, Klauder's syndrome, Rosenbach's erysipeloid, crab dermatitis, ectodermosis erosiva pluriorifacialis, fish-handlers' disease, and swine erysipelas in man. The eponym Rosenbach's disease is used in describing the mild form of the disease, whereas Klauder's form is used to describe a syndrome of severe systemic involvement. Erysipeloid appears most commonly in kitchen workers, butchers, fishermen, and other persons coming in contact with contaminated meat, animal products, or animal carcasses.

Causative Agent

Er. rhusiopathiae is a straight or slightly curved, thin, nonmotile, nonendospore forming, gram-positive bacillus that is facultatively anaerobic. There is no capsule or flagellum. There are two forms of this bacterium based on cellular morphology. Bacteria isolated from tissues during acute infection or from smooth (S) colonies (small, circular, transparent colonies on BAP with a smooth glistening surface and edge) are straight or slightly curved small rods that may occur in short chains. Bacteria from older cultures or rough (R) colonies (larger, flatter, colonies on BAP with a matte surface and jagged edge) tend to become filamentous and may be confused with fungal mycelia. The S form prefers alkaline environments, whereas the R form prefers acidic environments.

There are 24 serotypes (designated 1 through 24) of *Er. rhusiopathiae* based on heat-stable antigens. No correlation has been shown to exist between the serotype and the manifestation of the septicemic, urticarial, or endocardial forms of the disease. *Er. rhusiopathiae* is the only pathogenic species in the genus *Erysipelothrix* (recently a nonpathogenic species *Er. tonsillarum* has been identified as normal flora in the tonsils of healthy pigs and in surface waters but is not considered zoonotic).

> Infection by *Er. rhusiopathiae* in humans is known as erysipeloid. Erysipelas is the name given to an infection in animals caused by the bacterium *Er. rhusiopathiae*. Erysipelas in humans is an acute *Streptococcus* bacterial skin infection (historically known as St. Anthony's fire as a result of its red lesions).

> Over 50 animal species may be infected with *Er. rhusiopathiae*, but it is especially common in domesticated pigs. Adult pigs, and especially nursing sows, are more susceptible than others.

Epizootiology and Public Health Significance

Infection with *Er. rhusiopathiae* occurs worldwide in a variety of animals, especially hogs (the major reservoir of this bacterium), but is also found in sheep, horses, cattle, chickens, crabs, fish, dogs, and cats.

Erysipeloid is an occupational disease seen more commonly among farmers, butchers, cooks, housewives, and fishermen. The infection is more likely to occur during the summer or early fall. Since it is not a reportable disease in the United States there are limited statistics on its prevalence.

Transmission

Both animals and humans are infected with *Er. rhusiopathiae* through contact with infected animals, fish, or their products allowing bacteria to gain access through skin wounds and abrasions (including mucous membrane damage during insemination). In endemic areas, pigs are exposed naturally to the organism when they are young and have maternal antibodies (thus they have some immunity and do not show overt clinical signs). Infection may occur in animals by ingestion of contaminated foodstuffs (particularly cannibalism of infected carcasses). Rarely, human infections have been reported to occur through dog bites or the consumption of contaminated meat. Rodents may serve as reservoirs. Insect vectors and ticks may transmit the bacteria mechanically.

Er. rhusiopathiae is shed in the feces of infected (and perhaps carrier) animals contaminating the soil. This bacterium may survive for long periods depending on the environmental temperature and soil pH. Seasonal changes in climate (especially cold, rainy weather) have been associated with disease outbreaks.

> *Er. rhusiopathiae* is excreted by infected animals and survives for short periods in moist soil.

Pathogenesis

Er. rhusiopathiae typically enters the skin through scratches or open wounds. In the skin, the bacterium produces enzymes that dissect tissues allowing the organism to move through these dissected areas. *Er. rhusiopathiae* also produces a hyaluronidase (the significance of which is not known) and a neuraminidase (its level of activity correlates with virulence). The neuraminidase breaks links in neuraminic acid located on the surface of host cells helping the bacterium invade tissues. Two adhesive surface proteins (RspA and RspB) are also produced by this bacterium that help the microorganism bind to biotic (living) and abiotic (nonliving) surfaces.

There are several forms of the disease and these forms may occur separately, in sequence, or together. The erysipeloid lesions are the result of thrombotic vasculitis.

> *Er. rhusiopathiae* can survive in meat even after smoking, pickling, or salting.

> *Er. rhusiopathiae* is carried in the pharynx of subclinically infected pigs and shed in feces, urine, and oronasal secretions of up to 30% of pigs.

Clinical Signs in Animals

Erysipelas occurs in many animals; however, the main reservoirs are pigs. Clinical signs vary in different animal species and include:

- *Swine. Er. rhusiopathiae* can be found in 30% to 50% of healthy swine (typically in the pharynx, feces, urine, or oronasal secretions). The bacterium can also be isolated from the environment. There are three basic forms of erysipelas in swine that represent different progressive stages of the disease.
  - *Acute septicemia.* Pigs with acute septicemia may die suddenly without clinical signs. Acute septicemia occurs most frequently in finishing pigs (100 to 200 pounds). Clinical signs include fever (104°F to 108°F), stiff gait (walking up on their toes), anorexia, vomiting, and lying in sternal recumbency separately rather than piling in groups. Hemorrhages may be found in a variety of organs. Mortality is 0% to 100% and death may occur up to 6 days after the first signs of illness. Acutely affected pregnant sows may abort and suckling sows may stop milk production. Untreated pigs may progress to the other stages.

  - *Cutaneous form* (also known as diamond skin disease). Skin discoloration varies from widespread erythema (redness) and purplish discoloration of the ears, snout, and abdomen, to diamond-shaped skin lesions anywhere on the body, but particularly on the abdomen (Figure 3-30). Skin lesions become raised and firm to the touch within 2 to 3 days of illness. In untreated cases, skin lesions become necrotic and large areas of skin can separate. In addition to skin lesions, lymph nodes are usually enlarged, the spleen is swollen, and the lungs are congested. Petechiael hemorrhages may be found in the kidneys and heart.

Figure 3-30 Skin lesions caused by *Erysipelothrix rhusiopathiae* in this pig. Skin discoloration varies from widespread erythema of the ears, snout, and abdomen, to diamond-shaped skin lesions anywhere on the body. (Courtesy of USDA)

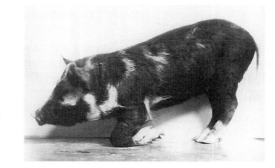

  - *Chronic arthritis.* This is the most common form of chronic infection producing mild to severe lameness (Figure 3-31). Affected joints tend to become visibly enlarged and firm. Mortality in chronic cases is low, but growth in these animals is slowed.

Figure 3-31 Arthritis causing lameness in a pig with erysipelas. (Courtesy of USDA)

- *Endocarditis.* Heart lesions usually result in large, irregular masses on the mitral valve. In long standing cases, hypertrophy of affected ventricles occurs in an attempt to keep blood flow at adequate levels.

- *Birds*. Erysipelas or fowl erysipelas occurs worldwide in poultry of all ages typically as an acute septicemia. Signs consist of acute death without any other clinical signs. Outbreaks occur suddenly with a few birds dying one day followed by increasing mortality on subsequent days. Mortality rates may range from <1% to 50% depending on vaccination status of the flock. Prior to death, some birds may appear droopy, with an unsteady gait. Chronic clinical disease in a flock is rare and if occurs produces cutaneous lesions and swollen hocks. Turkeys may develop vegetative endocarditis without showing clinical signs and may die suddenly. Clinical signs in chickens include general weakness, depression, diarrhea, decreased egg production in hens, and sudden death. Erysipelas may affect the fertility of the male and may contribute to production losses. Generalized or diffuse darkening of the skin is common. At necropsy the liver and spleen are usually enlarged, friable, and mottled. Turkeys are the most frequent poultry species affected, but outbreaks have also occurred in chickens, ducks, and geese.

> Although the acute septicemic form of swine erysipelas may cause death, the greatest economic loss probably occurs from the chronic, nonfatal forms of the disease.

- *Sheep*. *Er. rhusiopathiae* in sheep occurs as an extension of a focal cutaneous infection typically around the hoof resulting in laminitis (adults) or as arthritis (lambs). In adults, infection begins after dipping in a solution with bacteriostatic activity that has become contaminated with *Er. rhusiopathiae*. When animals are dipped, small skin abrasions occur near the hoof and fetlock joint when their legs scrape against the sides of the vat resulting in lameness of one or more legs. The affected leg appears normal except that the hoof and pastern are hot and painful. In sheep clinical signs appear 2 to 4 days after dipping with most sheep recovering spontaneously in 2 to 4 weeks. In lambs *Er. rhusiopathiae* enters the body through wounds that occur during docking and castration. Lambs develop septicemia which leads to joint infection (the entry site does not appear infected). The characteristic lesion in lambs is an acute, nonsuppurative (without pus) arthritis manifested by heat, pain, and slight joint swelling (typically involving the hock, stifle, elbow, and carpus). In lambs the clinical signs appear 9 to 19 days after the operation (docking or castration) with all cases developing in a 5-day period. Morbidity is 10% to 50%.

> Erysipelas can occur in sheep flocks that have been artificially inseminated 4 to 5 days prior to an episode of death without clinical signs.

- *Calves*. Calves can develop arthritis similar to lambs.
- *Reptiles, amphibians, marine mammals, and fish*. Erysipelas has also been reported in reptiles, amphibians, marine mammals, and fish and typically presents without clinical signs (reptiles, amphibians, fish) or as peracute, acute, or chronic forms typically having septicemia and skin lesions (marine mammals). The organism has been isolated from the surface slime of fish, which may serve as a source of infection for other species.

Clinical Signs in Humans

The disease caused by *Er. rhusiopathiae* in man is erysipeloid and presents itself in three ways:

- *Localized cutaneous form (also known as erysipeloid of Rosenbach)*. The most common and least severe form of erysipeloid is the localized form that manifests itself cutaneously. Typical lesions are clearly defined bright red to purple lesions with smooth shiny surfaces that slowly expand over a few days to develop sharp or curvaceous borders. Tiny blisters may be present. Skin lesions produce local burning, intense pruritus or pain at lesion sites, and may be warm and tender to the touch. Most skin lesions occur on the hands, forearms, or any other exposed area of the body and leave a brownish discoloration on the skin when resolving. Occasionally patients may experience mild fever, regional lymphadenopathy, chills, and malaise. This form is self-limiting.

- *Generalized (diffuse) cutaneous form.* Multiple, well-demarcated lesions appear on various parts of the body. Lesions appear as plaques with an advancing border and central clearing. Cellulitis may develop extending proximal to the initial infection site.
- *Septicemic form.* Rarely, a severe systemic form of erysipeloid may develop where skin lesions may not be apparent. If skin lesions are present they appear as localized areas of swelling surrounding a necrotic center or may present as several follicular, erythematous papules. In the septicemic form other organs are infected, such as the heart (endocarditis), brain, joints, and lungs. People with systemic disease may experience symptoms such as chills, fever, headache, cough, joint pain, and weight loss.

Diagnosis in Animals

In swine, the diamond-shaped skin lesions are diagnostic for erysipelas. At necropsy, demonstration of the bacterium in stained smears or cultures confirms the diagnosis. In chronic arthritis cases, organisms may not be cultured making diagnosis difficult. Typical necropsy lesions include large, irregularly shaped masses on the mitral valve projecting into the lumen of the left ventricle, thickened joint capsules with folding synovial membranes, and cutaneous lesions that appear rectangular in shape.

In all animal species, *Er. rhusiopathiae* can be definitively identified via Gram stain, culture, and serology. *Er. rhusiopathiae* is a gram-positive, short rod with long filaments. It can be isolated on BAP (blood agar plates) from spleen, kidney, and long bone samples of acutely sick animals and from the tonsils and lymph nodes of many apparently normal animals. Its colonies may present as large and rough or small, smooth, and translucent on BAP with alpha hemolysis with both forms after prolonged incubation. Bacterial culture may require specialized media (such as Fletcher, Stuart, Ellinghausen combined with neomycin to control growth of other bacteria). *Er. rhusiopathiae* is the only catalase-negative, gram-positive, nonendospore forming rod that produces hydrogen sulfide when inoculated on triple sugar iron (TSI) agar.

Serology is available for identification of *Er. rhusiopathiae* but can prove unreliable. A rising titer in an agglutination test (with controls) is available as well as a complement fixation test. Identification can be made by fluorescent antibody staining or a mouse protection test.

Diagnosis in Humans

Laboratory diagnosis of *Er. rhusiopathiae* requires culture of the organism (using blood, CSF, or urine). If blood cultures are negative, an aspirate of the center or edge of the cellulitis lesion may be used to isolate the bacterium. Bacterial culture on BAP and specialized media may be needed and was previously described. Serologic conversion by the microagglutination (MA) test requires a fourfold or greater rise in titer between the acute and convalescent sample for diagnosis. The MA test is difficult to perform and is usually done by reference laboratories. Several rapid serologic tests such as the indirect hemagglutination assay (IHA) have been developed that are reliable and commercially available. Many feel serologic tests for *Er. rhusiopathiae* are not reliable. A PCR test has also been developed.

Treatment in Animals

- *Swine.* In treating acute cases in an unvaccinated swine herd, antiserum may be administered to protect uninfected pigs. Penicillin, cephalosporin, or tetracycline antibiotics may also be given to exposed pigs in an attempt to provide prophylaxis. Penicillin and erythromycin are effective in acutely affected pigs (with or without concurrent antiserum administration). Treatment of swine with chronic infection is ineffective and these animals should be culled.
- *Birds.* Penicillin is the drug of choice for treating birds along with a full dose of erysipelas bacterin. Broad-spectrum antibiotics like erythromycin are also effective. Vaccination with a bacterin helps protect those birds in the flock not yet infected. Neither antibiotic therapy nor vaccination eliminates the carrier state.
- *Sheep.* Lambs are treated with penicillin as early as possible in the disease course. Most adult sheep recover spontaneously in 2 to 4 weeks.
- *Cattle.* Penicillin and cephalosporin antibiotics have been used in cattle to treat *Erysipelothrix* infections. Antibiotic withdrawal times need to be adhered to in food-producing animals.
- *Reptiles, fish, and marine mammals.* Penicillin and cephalosporin antibiotics are used to threat *Erysipelothrix* infections in these animals.

Treatment in Humans

In humans cutaneous infection resolves spontaneously in 3 to 4 weeks after disease onset. Antibiotics such as penicillin and cephalosporin will speed resolution of the disease. *Er. rhusiopathiae* is resistant to vancomycin and its failure may be used to diagnose this type of infection. Invasive disease with septicemia and endocarditis requires IV penicillin for 4 to 6 weeks, whereas milder forms can be treated with oral antibiotics.

Management and Control in Animals

Ways to reduce the incidence of erysipelas in animals includes elimination of carriers, good sanitation practice including prompt waste removal, and establishment of a vaccination program. Individual species variation includes:

- *Swine.* Killed bacterins are used to vaccinate pigs in the United States, whereas live-culture strains of low virulence are used in other countries. The formalin-killed, aluminum-hydroxide-adsorbed bacterin provides immunity that protects growing pigs from acute disease until market age. An oral vaccine of low virulence is also available. Young breeding stock should be vaccinated twice at the recommended interval, and then revaccinated every 6 months or after each litter. Vaccination of pregnant sows is not recommended. Although vaccination raises the level of immunity it does not provide complete protection and acute cases may develop after periods of stress. Antigenic variation exists between bacterial strains, so a vaccine may not be effective against all wild strains of the bacterium.
- *Birds.* Both inactivated and live vaccines are available for use in turkeys. The use of bacterins in flocks used for meat is useful but labor intensive and should be given every 2 to 4 months in breeding stock. The use of live vaccines administered in the drinking water does not require handling each bird; however, it cannot be guaranteed that all birds are inoculated.

- *Sheep*. There is not a vaccine commonly used in the United States for preventing erysipelas in sheep. In Australia there is a vaccine that is used to vaccinate ewes prior to lambing (the vaccine is given once and then repeated 4 to 6 weeks later, and at least 4 weeks prior to lambing in previously unvaccinated sheep). Annual boosters are recommended. Lambs are vaccinated prior to mulsing (extensive tail docking where a dinner plate size area of skin is removed by the tail). Prevention includes using copper sulfate in dipping wash to prevent bacterial growth and using strict antiseptic techniques when docking and castrating lambs.
- *Cattle*. There is not a vaccine for preventing erysipelas in cattle.
- *Marine mammals*. There is a commercially available bacterin for marine mammals such as dolphins. The bacterin is followed by a modified live vaccine in 6 months. Annual revaccination is recommended. The vaccine is typically given in the dorsal musculature using a long needle to assure that the vaccine is placed in the muscle.

Er. rhusiopathiae is not readily destroyed by the usual laboratory disinfectants (it may survive in litter or soil for various lengths of time) making disinfection of premises difficult. It is inactivated by a 1:1000 concentration of bichloride of mercury, 0.5% sodium hydroxide solution, 3.5% liquid cresol, or a 5% solution of phenol.

Management and Control in Humans

Preventing *Er. rhusiopathiae* infection in people includes avoiding direct contact with animal tissues, animal secretions/excretions, and/or contaminated soil. People handling animals should wear gloves as barrier protection. Frequent hand washing with an antiseptic soap and prompt medical treatment of cuts and abrasions can also prevent erysipeloid.

Summary

> Erysipeloid is a not a nationally reportable disease in the United States; however, many states require notification of erysipeloid.

Er. rhusiopathiae is a gram-positive, facultatively anaerobic, nonmotile rod found worldwide that can survive in water, soil, decaying organic matter, slime on the bodies of fish, and carcasses, even after processing. *Er. rhusiopathiae* usually enters the body through traumatized skin and in animals causes swine erysipelas, nonsuppurative arthritis in lambs, postdipping lameness in sheep, and acute septicemia in poultry. In humans the infection may be localized, generalized, or systemic and is termed erysipeloid. Erysipeloid should not be confused with erysipelas in man, a form of cellulitis caused by *Streptococcus pyogenes*. This bacterium is typically diagnosed with Gram stain and culture. Serologic tests are available providing variable results. Treatment consists of antibiotics such as penicillin, cephalosporins, tetracyclines, and erythromycin. Vaccines are available for swine, poultry, and marine mammals. People can avoid contracting the organism by avoiding direct contact with animal tissues, animal secretions/excretions, and/or contaminated soil.

E. COLI INFECTION

Overview

Escherichia coli, more commonly known as *E. coli*, is a **coliform** bacterium (coliforms are gram-negative normal intestinal flora which ferment lactose within

48 hours) originally known as *Bacterium coli commune*. *E. coli* belongs to the family Enterobacteriaceae and was first isolated and described in 1885 by the German bacteriologist and pediatrician, Theodor von Escherich. Most *E. coli* strains are harmless to animals and humans. Their ability to multiply quickly has helped the scientific community through their use in the biotechnology field; however, their ability to multiply quickly has also resulted in the production of disease especially in the young and old.

Historically, *E. coli* has been associated with many disease forms including enteritis, cystitis, and meningitis. Infantile diarrhea, known by a number of synonyms including griping in the guts, cholera infantum, and summer diarrhea, was a term used to describe the clincial signs of an infection that had been noted for a number of centuries. Over the past four centuries, infantile diarrhea was a major problem worldwide, with high morbidity and mortality. The isolation of *E. coli* from cases of infantile summer diarrhea had already been noted as early as 1889 when it was suggested that there were both pathogenic and nonpathogenic strains of *E. coli*. The term enteropathogenic *E. coli* (EPEC) was introduced in 1955 to describe strains of *E. coli* implicated epidemiologically with infantile diarrhea.

Even though *E. coli* is a single species of bacteria, there are many different strains of the species including the dangerous pathogen *E. coli* O157:H7. *E. coli* O157:H7 is a mutation discovered in 1982 that has at least 62 subtypes and causes an acute bloody diarrhea. Enterohemorrhagic *E. coli* infection, also called hemorrhagic colitis that may lead to hemolytic uremic syndrome (HUS), is an infectious disease caused by the microbe *E. coli* O157:H7. It is found in feces and fecal-contaminated meat (contaminated meat is the primary source of infection). When milk, cider, water, sawdust, and air come in contact with cattle feces they too may become contaminated with bacteria. Drinking or swimming in sewage-contaminated water may also be a source of pathogenic *E. coli* as well as some vegetables rinsed or watered with contaminated water. Since the first outbreak in 1982, there have been several *E. coli* O157:H7 outbreaks in the United States: in June and July 1997, *E. coli* O157:H7 infection outbreaks occurred in Michigan and Virginia associated with the consumption of alfalfa sprouts grown from the same seed lot; in October 1996, there were simultaneous outbreaks in California, Colorado, and Washington linked to the consumption of unpasteurized apple juice; in July 2002, 28 reported infections in Colorado were all linked to consuming contaminated beef products produced and later recalled by ConAgra Beef Company; in the Spring and Fall of 2000, an outbreak of *E. coli* O157:H7 was reported among school children in Pennsylvania and Washington when they became infected after having direct contact with animals during farm visits with their families and/or school. Many of the infections resulted in the development of HUS. HUS was first described in 1955 by a Swiss pediatrician and is now the leading cause of kidney failure in children (approximately 10% of all hemorrhagic colitis cases end in HUS and nearly 5% of all HUS cases lead to death). *E. coli* O157:H7 causes 90% of all HUS cases and it is the only known cause of HUS in children.

Causative Agent

E. coli are motile, gram-negative, rod-shaped bacteria of the family Enterobacteriaceae (Figure 3-32). All Enterobacteriaceae, including the genus *Escherichia*, have three kinds of major antigens: somatic or cell wall (O), surface or envelope (K), and flagellar (H) (Figure 3-60). Some strains of *E. coli* are virulent pathogens and are distinguished from each other by two of the types of antigens they produce (these antigens are then assigned numbers). One of these antigens is designated O for outer membrane and the other is designated H for flagellum. Some *E. coli* antigens such as O157, O111, H8, and H7 are associated with virulence. Virulent strains have genes

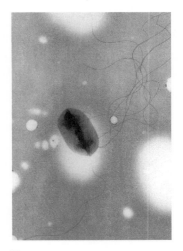

Figure 3-32 Electron microscopic image of *E. coli* with flagella.

(Courtesy of CDC/Elizabeth H. White, M.S.)

Table 3-6 The Different Types of _E. coli_ and Their Properties

Type of _E. coli_	Human Disease	Animal Disease	Description of Diarrhea	Virulence Factors
*Enterohemorrhagic (EHEC)	Hemorrhagic colitis, hemolytic uremic syndrome, thrombotic thrombocytopenic purpura	Hemorrhagic colitis	Bloody or nonbloody	Adherence causing A/E lesions, enterotoxin production that may enter the bloodstream and bind to endothelial cells
*Enteropathogenic (EPEC)	Enteritis in infants (infantile diarrhea)	Enteritis	Watery	Adherence causing A/E lesions
Enterotoxigenic (ETEC)	Choleriform enteritis, traveler's diarrhea, infant diarrhea	Enteritis in newborn and young animals	Watery, may be bloody	Adherence, enterotoxin production that stimulates hypersecretion
Enteroinvasive (EIEC)	Dysentery-like colitis	?	Bloody or nonbloody	Adherence, mucosal invasion
Enteroaggregative (EAggEC)	Chronic enteritis	?	Watery	Adherence, enterotoxin production

*zoonotic

E. coli can survive and grow in the presence or absence of oxygen making it a useful experimental bacterial model in the laboratory.

for fimbriae, adhesions, and a variety of **exotoxins** that allow these strains to colonize animal tissue and cause disease. Pathogenic strains of _E. coli_ can adhere to animal cells, invade tissue, and produce toxin. The set of virulence factors that a pathogenic strain possesses determines whether it is classified as enterohemorrhagic, enteropathogenic, enterotoxigenic, enteroinvasive, or enteroaggregative (Table 3-6).

- Enterotoxigenic strains of _E. coli_ are the primary cause of traveler's diarrhea and infant diarrhea in developing countries. These strains produce two enterotoxins: one heat-labile toxin (destroyed by heat) and the other heat-stable toxin (not destroyed by heat). The heat-labile toxin is similar to the toxin that causes cholera because it stimulates intestinal secretion and fluid loss. This strain is typically spread by poor sanitation in developing countries.

- Enteroinvasive strains of _E. coli_ cause dysentery similar to shigellosis. These bacteria produce proteins that allow invasion into cells; however, they are not as virulent as _Shigella_ bacteria and larger numbers of bacteria are needed to initiate infection. This strain causes an inflammatory disease that involves invasion and ulceration of the large intestinal mucosa.

- Enteropathogenic strains of _E. coli_ cause diarrhea in infants. This strain attaches to mucosal cells of the small intestine and destroys the microvilli of the intestinal epithelium.

- Enteroaggregative strains of _E. coli_ cause chronic watery diarrhea in infants as a result of their adherence to animal cells most likely by pili. This strain is also believed to produce hemolysis-like toxins, although the exact mechanism of pathology is unknown.

- Enterohemorrhagic strains of _E. coli_ produce bloody diarrhea that may lead to a life-threatening condition called hemolytic uremic syndrome (HUS). The virulence of _E. coli_ O157:H7 is a result of the production of shiga-like toxin (also called Vero toxin) that this strain of *E. coli* may have acquired through plasmid transfer from a strain of *Shigella*. This strain of *E. coli* is resistant to the acidity of gastric secretions and is able to attach to intestinal cells and disrupt their cell structure.

Epizootiology and Public Health Significance

E. coli is widespread in nature (including soil and water) and is found worldwide with some endemic areas present in developing countries. Most strains are commensals of the gastrointestinal tract of animals and help protect against infection by enteric pathogens. *E. coli* is important to animal health because it manufactures vitamins B$_{12}$ and K from undigested food in the large intestine. In animals, particularly ruminants, *E. coli* does not cause disease (except in the very young) and animals tend to be carriers and excretors of this bacterium. In terms of global public health, enteropathogenic *E. coli* is the most widespread diarrhea-causing *E. coli*. However, since 1982, enterohemorrhagic *E. coli* has emerged as a cause of sporadic or epidemic disease in North and South America, Europe, Asia, and Africa. The annual incidence of enterohemorrhagic *E. coli* infection in the United States is approximately 8 cases per 100,000 people, with cases peaking between June and September. In some states *E. coli* O157:H7 is the second or third most frequent cause of diarrhea (more frequent than *Shigella* and *Yersinia*). In United States patients with bloody diarrhea, 40% of cases are caused by *E. coli* O157:H7.

An estimated 2,100 people are hospitalized annually in the United States from *E. coli* O157:H7 infections. The illness is often initially misdiagnosed resulting in more expensive diagnostic procedures and therapies once the disease is properly identified. Patients who develop HUS often require prolonged hospitalization, dialysis, and long-term follow-up care.

> Because *E. coli* is a prominent normal intestinal bacterium, it is used as an indicator bacterium to monitor fecal contamination in water, food, and dairy products. *E. coli*, like other coliforms, are present in large number, are more resistant, and are easier and faster to detect than true pathogens. If a certain number of coliforms are found in a water sample, it is deemed unsafe to drink.

Transmission

E. coli is spread by fecal-oral transmission with some serotypes being species specific, whereas others are not. The organism is excreted in nasal and oral secretions, urine, and feces before clinical signs appear; therefore, animals most likely acquire the bacterium from environmental sources such as manure, manure-contaminated objects, or other objects contaminated by secretions. Bacteria enter the animal's body through the nasal and oropharyngeal mucosa, the intestinal mucosa, or via the umbilicus and umbilical veins. In groups of calves, transmission is by direct nose-to-nose contact, urinary and respiratory aerosols, or as the result of navel-sucking or fecal-oral contact. People tend to acquire *E. coli* from ingestion of meat that is contaminated during the slaughtering process and is inadequately cooked or from cross-contamination of food during its preparation. The reservoir of this pathogen appears to be mainly cattle and other ruminants and is transmitted to humans primarily through consumption of contaminated foods, such as raw or undercooked ground meat products and raw milk. Fecal contamination of water and cross-contamination during food preparation (contaminated surfaces and kitchen utensils) will also lead to *E. coli* infection. Foods implicated in outbreaks of *E. coli* O157:H7 include undercooked hamburgers, dried cured salami, unpasteurized fresh-pressed apple cider, yogurt, cheese and milk, and fruits and vegetables contaminated with feces from domestic or wild animals at some stage during cultivation or handling. Waterborne transmission has been reported, both from contaminated drinking water and from recreational waters. *E. coli* can survive for months in manure and water-trough sediments. Person-to-person contact (infected people that do not wash their hands after using the toilet or diapering children) is an important mode of transmission through the oral-fecal route and bacteria can be shed for about one week in adults and even longer in children. Farms and other venues where people come into direct contact with farm animals and inappropriate hygiene after contact with these animals is another source of *E. coli* O157:H7 infection. In a few cases, *E. coli* O157:H7 is transmitted by direct contact

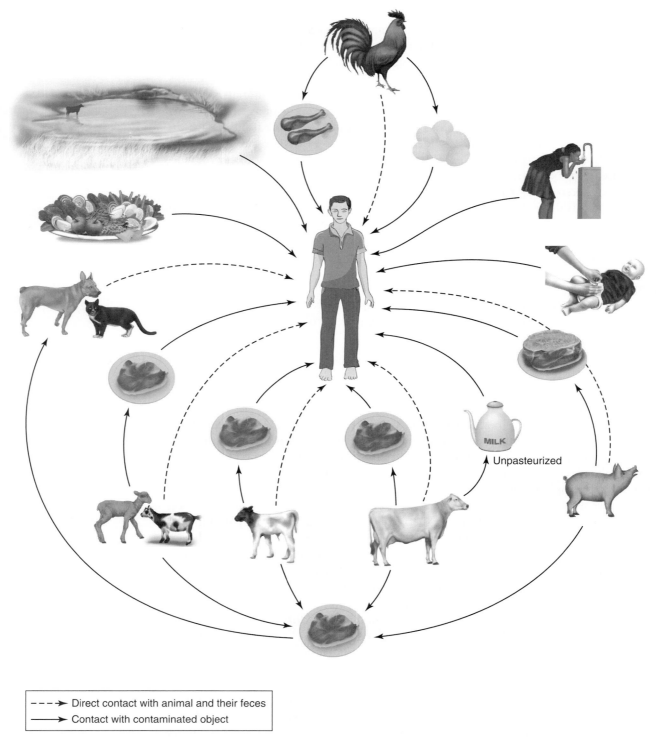

- - - → Direct contact with animal and their feces
———→ Contact with contaminated object

Figure 3-33 Transmission routes for *E. coli* from animals to humans.

with infected cattle or horses to humans. The infective dose of *E. coli* O157:H7 in people is 10^1 to 10^2 CFU (colony forming units), which indicates high infectivity. Figure 3-33 shows the various transmission routes for *E. coli*.

Pathogenesis

E. coli infection in animals is sometimes known as colibacillosis and is a major cause of bacterial diarrhea in young ruminants especially calves. Two distinct types of diarrheal disease are produced in calves by two different strains of this bacterium.

- Enterotoxigenic *E. coli* has two virulence factors associated with the production of diarrhea: antigens on fimbria (shorter filaments on the bacterial cell) and production of enterotoxin. Fimbrial antigens enable *E. coli* to attach to and colonize the villi (microscopic raised portions) of the small intestine resulting in villous atrophy. The loss of mature cells at the tips of the villi results in a decrease in villous height (and subsequent decrease in the surface area for absorption) and in the loss of digestive enzymes produced by these cells (Figure 3-34). Villous atrophy decreases the capacity of the intestinal mucosa to absorb fluids and nutrients and results in malabsorptive diarrhea. Most *E. coli* strains in calves commonly possess either K99 (F5) or F41 fimbrial antigens or both. Enterotoxigenic *E. coli* also produce an enterotoxin (Sta) that affects intestinal ion and fluid secretion to produce a noninflammatory secretory diarrhea. This enterotoxin stimulates hypersecretion by activating enzymes and by producing a net secretion of sodium and chlorine.

- Enteropathogenic *E. coli* cause diarrhea by adhering to the intestine producing a lesion that results in cellular damage of microvilli at the site of bacterial attachment (these lesions are known as attachment-and-effacement or A/E lesions). These lesions decrease enzyme activity and alter ion transport in the intestine. Some enteropathogenic *E. coli* produce verotoxin, which may be associated with a more severe hemorrhagic diarrhea. Infections with verotoxin-producing enteropathogenic *E. coli* result in accumulation of fluid in the large intestine and extensive damage to the large intestinal mucosa (edema, hemorrhage, and erosion and ulceration of the mucosa) which results in blood and mucus in the intestinal lumen. The infection most frequently occurs in the cecum and colon, but the distal small intestine can also be affected. Inflammation is also a major component of the pathology produced by enteropathogenic *E. coli*. Inflammation leads to vascular and lymphatic damage and to structural damage of the intestinal crypts and villi.

Most human food-poisoning *E. coli* bacteria do not cause clinical illness in animals (except in young animals).

Meat package labels are required to have cooking instructions indicating proper temperatures to kill foodborne bacteria.

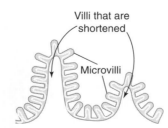

A) Cross-section of normal intestine

B) Cross-section of intestine showing villous atrophy. Shortened villi decrease the amount of surface area for absorption of nutrients

Figure 3-34 Normal intestinal villi compared with villi atrophy.

Enterohemorrhagic *E. coli* causes hemorrhagic colitis in humans. Enterohemorrhagic *E. coli* have the ability to produce Shiga toxins or verotoxins and the ability to cause attachment-and-effacement lesions. Enterohemorrhagic *E. coli* is resistant to the acidity of gastric secretions allowing its survival in the stomach and passage into the small intestine where it attaches to the intestinal cells causing structural changes in these cells. Shiga toxins and verotoxins have the ability to cause severe intestinal cell damage and if they enter the bloodstream they can bind to endothelial cells of blood vessels causing platelet aggregation (which can result in a consumptive thrombocytopenia), microinfarcts, red blood cell damage (leading to anemia), and hypertension. The development of HUS is a result of endothelial cell injury that triggers a cascade of events resulting in microvascular lesions and microthrombi that occlude arterioles and capillaries. In HUS microthrombi are confined to the kidney and in thrombocytopenia microthrombi occur throughout the microcirculation including areas of the brain, skin, intestines, skeletal muscle, pancreas, spleen, adrenal glands, and heart.

Clinical Signs in Animals

Animals that develop clinical manifestations of *E. coli* include:

- *Young ruminants. E. coli* is commonly found in the feces of healthy young ruminants and whether or not *E. coli* leads to clinical disease depends on the virulence of strain, the immune status of the animal (especially in relation to the success or failure of passive transfer of colostral antibodies), and environmental factors such as stress and poor diet. Newborn calves, lambs, and kids develop diarrhea from *E. coli* that can lead to dehydration and death (Figure 3-35). Diarrhea as a result of enterotoxigenic (K99-bearing) *E. coli* occurs in calves younger than 3 to 5 days of age, is sudden in its onset, and rapid in its clinical course (3 to 8 hours). Profuse amounts of diarrhea are passed, and calves quickly become depressed, dehydrated, and recumbent. Body temperature is usually normal or lower than normal, but may be increased. Hypovolemic shock and death may occur in 12 to 24 hours (mortality can be close to 100%). Disease produced by enteropathogenic *E. coli* typically occurs in calves from 4 days to 2 months of age producing bloody and/or mucoid diarrhea. The clinical course of enteropathogenic *E. coli* is short. Septicemia from *E. coli* may occur during

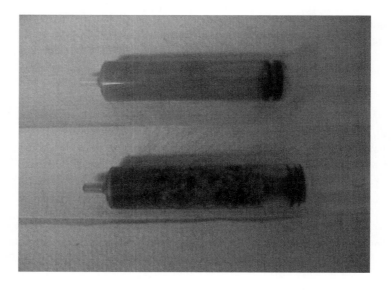

Figure 3-35 Syringes containing fecal samples from calves with scours. Upper syringe contains watery sample in a 1-day-old calf; lower syringe contains bloody diarrhea from a 5-day-old calf.

Figure 3-36   Normal, healthy cattle can transmit pathogenic strains of *E. coli.*

the first week of life (typically between 2 and 5 days of age) in calves and is seen in lambs less than one week of age. Septicemia in young ruminants causes peracute death. Chronic disease can occur for up to 2 weeks of age with localization of infection causing polyarthritis, meningitis, and convulsions. The disease is usually sporadic and is more commonly seen in dairy calves than beef calves.

- *Adult ruminants.* Healthy adult ruminants may be carriers of *E. coli,* periodically excreting the organism in feces (Figure 3-36). Stress (such as parturition) may increase the excretion of *E. coli* making the disease more likely to be seen in ruminants around freshening. Contaminated calving areas and infection of the udder and perineum of the dam can lead to infection in offspring.

- *Poultry.* In poultry, *E. coli* causes an acute fatal septicemia or subacute pericarditis and air sacculitis (Figure 3-37). Pathogenic strains most commonly seen in poultry are 01, 02, and 078 serotypes. Poultry facilities contain large numbers

Acute *E. coli* septicemia in neonatal calves, sometimes called colisepticemia and septicemic colibacillosis, may be rapidly fatal.

Figure 3-37   Infected air sacs from a bird with *E. coli* infection.
(Courtesy of Noah's Arkive/Ted Leighton)

of *E. coli* bacteria through fecal contamination and initial exposure to pathogenic *E. coli* may occur in the hatchery from infected or contaminated eggs. Systemic infection from *E. coli* usually requires predisposing environmental factors such as poor air quality or co-infection with other diseases and occurs when large numbers of pathogenic *E. coli* enter the bloodstream from the respiratory tract or intestine. Bacteremia can lead to septicemia and death, or infection to other organs. Clinical signs in poultry are nonspecific and vary with age, organs involved, and concurrent diseases. Young birds dying of acute septicemia have few lesions other than hepatomegaly and splenomegaly. Birds that survive septicemia may develop air sacculitis, pericarditis, and hepatitis. Air sacculitis is a classic lesion of colibacillosis. Sporadic lesions seen with *E. coli* infection include pneumonia, arthritis, osteomyelitis, and salpingitis.

- *Pigs.* Specific serotypes of *E. coli* can affect healthy, rapidly growing nursery piglets producing a condition called edema disease (also known as bowel edema or gut edema). Edema disease is a peracute toxemia that causes edema of the gastric and intestinal submucosa. Edema disease is caused by four serotypes of enteropathogenic *E. coli*: O138:K81:NM, O139:K12:H1, O141:K85a,b:H4, and O141:K85a,c:H4. Piglets acquire this infection during nursing from a contaminated sow or after weaning as a result of fecal-oral contact, feed, or short-distance aerosol transmission. Clinical signs range from rapid death to CNS signs such as ataxia and recumbency. Periocular edema, swelling of the facial region, open-mouth breathing, and anorexia are common. Edema disease usually occurs 1 to 2 weeks postweaning and typically involves the healthiest animals in a group. Morbidity levels are approximately 30% to 40% and mortality levels may be 50% to 90%.

- *Rabbits.* Colibacillosis in rabbits is caused by enteropathogenic strains of *E. coli*. Normal healthy rabbits do not have *E. coli* of any strain associated with their gastrointestinal tract; however, rises of intestinal pH can cause *E. coli* to colonize the gut causing attachment-and-effacement lesions. Intestinal lesions lead to diarrhea and possibly death. Two types of colibacillosis are seen in different ages of rabbits: in 1- to 2-week old rabbits a severe yellowish diarrhea develops that results in high mortality and in 4- to 6-week old weaned rabbits diarrhea develops in which the intestines are fluid-filled causing death in 5 to 14 days or leaving rabbits stunted and unthrifty.

- *Dogs and cats.* The role of *E. coli* as a primary intestinal pathogen in small animals is unclear. Several studies have suggested that the incidence of enterotoxigenic *E. coli* diarrhea is extremely low in dogs (less than 1% to 2% of diarrhea cases). There is some evidence that enteropathogenic *E. coli* plays a role in diarrhea in cats.

Clinical Signs in Humans

E. coli produces three types of infections in humans depending upon the virulence of the bacterium: intestinal diseases (gastroenteritis), urinary tract infections (UTI), and neonatal meningitis. UTIs and neonatal meningitis are nonzoonotic.

- Five types of *E. coli* cause diarrheal diseases in humans: enterotoxigenic *E. coli* (ETEC), enteroinvasive *E. coli* (EIEC), enterohemorrhagic *E. coli* (EHEC), enteropathogenic *E. coli* (EPEC), and enteroaggregative *E. coli* (EAggEC).
  - ETEC are an important cause of diarrhea in infants and travelers in underdeveloped countries or regions of poor sanitation. Clinical signs vary from minor discomfort to severe diarrhea without fever.

- EIEC cause a dysentery-like diarrhea with fever.
- EPEC produce watery diarrhea and are an important cause of traveler's diarrhea in Mexico and North Africa.
- EAggEC are associated with persistent nonbloody diarrhea in young children.
- EHEC are typically represented by a single strain O157:H7 (although it is believed there are more EHEC strains that cause disease), which causes a diarrheal syndrome producing large amounts of bloody discharge and no fever. Diarrhea caused by this strain can be fatal especially in children developing acute kidney failure as a result of HUS. HUS develops when *E. coli* O157:H7 enters the bloodstream through the intestinal wall and begins to release shiga-like toxin (SLT) causing kidney failure and the hemolysis of red blood cells. Red blood cell hemolysis leads to brain hemorrhaging, uncontrolled bleeding, and the formation of clots in the bloodstream.
- *E. coli* cause 90% of the urinary tract infections (UTI) in anatomically normal, unobstructed urinary tracts of humans. *E. coli* colonize the feces or perineal region and ascend the urinary tract to the urinary bladder.
- *E. coli* strains invade the blood stream of human infants from the nasopharynx or gastrointestinal tract and are carried to the meninges causing neonatal meningitis. Neonatal meningitis affects 1 out of every 2,000 to 4,000 infants in the United States.

Diagnosis in Animals

Tissue samples obtained from biopsy or surgical excision containing *E. coli* bacteria can be Gram stained and cultured for diagnostic identification. Several samples should be taken in the early stages of diarrhea from untreated animals to obtain the best culture results. *E. coli* grows well on blood agar, chocolate agar, and MacConkey agar. On MacConkey agar *E. coli* produces flat, dry, pink colonies (pink colonies are produced as a result of its ability to ferment lactose). *E. coli* also grows on selective agars such as Hektoen enteric (HE) agar, xylose-lysine-deoxycholate (XLD) agar, and *Salmonella-Shigella* (SS) agar. Slow fermentation of sorbitol on sorbitol MacConkey (SMAC) agar is the basis of microbiologic identification of *E. coli* O157: H7 (colonies grow colorless on SMAC plates, whereas other *E. coli* bacteria grow pink on SMAC plates). Latex agglutination tests using O157-specific serum are also available.

Blood results may show a moderate leukocytosis and neutrophilia early in the disease with a marked leukopenia seen in the later stages of disease. Joint fluid may show an increase in protein and inflammatory cells with bacteria evident on microscopic examination. A deficiency of circulating IgG may be seen in calves using zinc sulfate or total protein estimation. Demonstration of the bacteria in blood or tissue can be seen using Gram-stain techniques.

The best clinical diagnostic test for *E. coli* in animals is necropsy of untreated, acutely affected animals. Examination of intestinal mucosa for diagnostic lesions and for the presence of bacteria is the optimal way to diagnose disease associated with the attachment-and-effacement strains of *E. coli*. The diagnostic value of a necropsy diminishes quickly with time after death as a result of autolysis of lesions. Edema in the gastric submucosa, fibrin strands in the peritoneal cavity, and serous fluid in the pleural and peritoneal cavity help diagnose *E. coli* infection in necropsied swine. Birds will show respiratory lesions consistent with pneumonia and air sacculitis. In rabbits the intestine will be fluid-filled with petechial hemorrhages on the serosal surface.

Diagnosis in Humans

Routine stool cultures will allow growth of *E. coli*, but because *E. coli* are normal fecal flora, laboratories must be advised to check for pathogenic *E coli* when a sample is submitted. Most 0157:H7 isolates do not ferment sorbitol; therefore, cultivation of specimens on sorbitol MacConkey (SMAC) medium is warranted (and is required by the CDC in all cases of bloody diarrhea submitted to human laboratories). SMAC uses sorbitol instead of lactose as the primary carbohydrate, which *E. coli* O157:H7 does not ferment (it grows as colorless colonies on SMAC plates). Confirmation requires identification of presumptive isolates with specific antiserum. Rapid enzyme immunoassays (EIA) for *E. coli* 0157:H7 have been developed but are not yet widely used clinically. All infants with suspected sepsis should have specimens of blood, urine, and cerebrospinal fluid sent for culture and Gram stain prior to initiating antimicrobial therapy to detect possible *E. coli* infections.

Treatment in Animals

Treatment of *E. coli* infection in animals varies with the affected species and the severity of clinical signs. In calves, fluid and electrolyte replacement are essential to counteract severe dehydration seen in young animals. Antibiotic therapy is initiated using an antibiotic with efficacy against gram-negative bacteria because culture and sensitivity testing takes too long to wait for these results. Despite aggressive treatment mortality in calves is high. In poultry treatment involves antibiotic treatment along with controlling predisposing infections and environmental factors. In swine antibiotics administered via the drinking water prior to disease outbreak in a herd may be helpful in controlling the spread of *E. coli*. Management techniques such as altering feeding practices and adding high levels of fiber to the diet have also been utilized in *E. coli* infections in swine. In rabbits antibiotics may be helpful in mild cases of *E. coli* infections. In severe cases, affected rabbits are culled and the facility is disinfected.

Treatment in Humans

Symptoms associated with *E. coli* O157:H7 infections are typically self-limiting within 5 to 10 days. In more severe cases, fluid and electrolyte therapy is the main treatment for diarrhea caused by *E. coli* because it helps correct any dehydration that may have developed as well as contributes to kidney perfusion which may help prevent HUS development. Antibiotic therapy rarely is indicated and should not be instituted in cases of enterohemorrhagic *E. coli* as a result of the potential risks of developing HUS. Antidiarrheal agents should not be given to patients with *E. coli* O157:H7 infections because they may prolong the clinical and bacteriologic course of disease. Cases of HUS are treated with kidney dialysis.

Management and Control in Animals

There are many causes of diarrhea in young animals and total prevention in large facilities may not be obtainable; however, limiting exposure to the infectious agent and ensuring adequate immune status in young animals is critical. Principles that apply to newborns and young animals in all herds/flocks include isolating diseased animals from healthy ones, practicing good general hygiene, providing good nutrition to dams and young animals, assuring that newborns receive colostrum within a few hours after birth, and by vaccinating dams or newborns if available for that species. Vaccination of pregnant dams with *E. coli* vaccine can control enterotoxigenic colibacillosis in calves. Monoclonal K99 *E. coli* antibody is commercially available for oral administration to calves immediately after birth serving as an effective substitute for the K99-specific antibody in the colostrum of vaccinated cows and as a supplement to calves who receive adequate levels of colostrum. Clinical trials began in 2002 for a cattle vaccine against *E. coli* 0157:H7. In poultry, commercial bacterins are available for breeder hens and chicks providing some protection against specific *E. coli* serotypes. Vaccines are not currently available for swine and rabbits.

Management and Control in Humans

Most human cases of *E. coli* are caused by fecal contamination of food, making proper hygiene critical in preventing these infections. To kill *E. coli* O157:H7, the contaminated material must be cooked at 165°F or higher. Annually in the United States, approximately 73,000 people are infected with *E. coli* O157:H7 toxicity and 61 people die. The prevention of infection requires control measures at all stages of the food chain, from agricultural production on the farm to processing, manufacturing, and preparation of foods in both commercial establishments and the environment. At the national level the number of cases of *E. coli* contamination may be reduced by implementing strategies for ground beef (i.e., preslaughter screening of animals to reduce large numbers of pathogens in the slaughtering environment, proper sanitation of the facility, adequate hygiene measures, irradiation of the product) (Figure 3-38). Prevention of contaminated raw milk includes the education of farm workers in principles of good hygienic practices and pasteurization or irradiation of milk. Water areas

Figure 3-38 Contaminated meat is the number one cause of *E. coli* infection in people. (Courtesy of USDA)

and drinking sources also need to be protected from animal wastes to prevent their contamination.

Guidelines are established to educate food handlers in the handling of food for public consumption. Food handlers should follow the Recommended International Code of Practice, General Principles of Food Hygiene contained in the Joint FAO/ WHO Food Standards Programme (updated in 2001). Important recommendations in this document include cooking meat thoroughly so that at least the center of the food reaches 165°F. Outbreaks of *E. coli* involving fresh fruits and vegetables in recent years have resulted in guidelines for growing and harvesting fruits and vegetables in a document called *Codex Code of Hygienic Practice for Fresh Fruits and Vegetables* (adopted in 1969). It is believed that some vegetables, especially sprouts, are contaminated as seeds in the field or during harvesting, storage, or transportation. During the germination process, low levels of pathogens present on seeds may quickly reach levels high enough to cause disease. By following procedures outlined in this document the risk of developing disease from fruit and vegetable consumption is lowered. The recommendations in this code include the sanitary use of irrigation water, the sanitary disposal of human and animal wastes, sanitary harvesting of crops, and proper processing techniques (using equipment and product containers that do not pose a health threat and adopting processing techniques that remove unfit product and do not contain microbe, insect, or chemical contamination). Preventing cross-contamination of food is covered in the salmonellosis section.

E. coli O157:H7 infection is nationally reportable and is reportable in most U.S. states. HUS is also reportable in most states. A nationally notifiable disease is one for which regular, frequent, timely information on individual cases is necessary to prevent and control the disease. Each year the list is agreed upon and maintained by the Counsel of State and Territorial Epidemiologists (CSTE) and the CDC. Diseases that are nationally notifiable may or may not be designated by a given state as notifiable (reportable) by legislation or regulation at the local or state level (as a result of cost/benefit factors such as incidence versus cost to maintain state records in addition to using the national notifiable list). The CDC currently has six surveillance systems for obtaining information about *E. coli* O157:H7 that serve different purposes and provide information on various features of the organism's epidemiology including:

- Public Health Laboratory Information System (PHLIS) is a laboratory-based surveillance system that collects data about many infections, including *E. coli* O157:H7. Cases confirmed by culture and verified at the state public health laboratory are reported to the PHLIS and information about the infection is reported to the CDC by the state.

- National Electronic Telecommunications System for Surveillance (NETSS) is a physician-based surveillance system that records both laboratory-confirmed and clinically suspected cases of all nationally notifiable diseases, including *E. coli* O157:H7. *E. coli* O157:H7 infections and other surveillance data collected by NETSS are published weekly in the CDC *Morbidity and Mortality Report* (MMWR).

- The Foodborne Diseases Active Surveillance Network (FoodNet) is a surveillance system for identifying culture-confirmed foodborne infections including *E. coli* O157:H7. In addition to monitoring the number of *E. coli* O157:H7 infections, investigators monitor laboratory techniques for isolation of bacteria, determine foods associated with illness, and administer questionnaires to people to better understand trends in the eating habits of Americans.

- National Molecular Subtyping Network for Foodborne Diseases Surveillance (PulseNet) is a national network of public health laboratories that perform pulsed-field gel electrophoresis (PFGE) on certain foodborne bacteria, including *E. coli* O157:H7. PFGE, a type of DNA fingerprinting, provide patterns that are submitted to CDC and can be compared with others in a large database to help determine if individual infections are related or if an outbreak is occurring.

- National Antimicrobial Resistance Monitoring System (NARMS) is a surveillance system that monitors antimicrobial resistance of *E. coli* O157:H7 and selected other bacteria that cause human illness.
- Foodborne Outbreak Detection Unit of the CDC monitors outbreaks of foodborne disease. Each year epidemiologists report the results of outbreak investigations to the CDC.

Summary

E. coli are gram-negative bacteria that normally reside in the intestines of animals and humans. Most strains are harmless and part of the intestinal normal flora; however, several strains produce toxins that can cause diarrhea. The strain of *E.coli* of most concern is *E. coli* O157:H7 that is acquired by eating contaminated food. *E. coli* O157:H7 bacteria live in the intestines of some healthy cattle and meat can become contaminated with feces containing this microbe during the slaughtering process. Careless food handling, drinking contaminated water, swimming in contaminated lakes, exposure to infected farm animals, and inappropriate personal hygiene after using the toilet or diapering children are other ways of spreading this bacterium. *E. coli* infections in animals, sometimes called colibacillosis, tend to produce severe diarrheal disease in young animals. In infected people, signs range from mild diarrhea to bloody, severe diarrhea with abdominal cramps. Culturing of bacteria and specialized laboratory techniques are used to identify *E. coli* bacteria and to identify them by type. Colibacillosis in animals is treated with antibiotics and fluid therapy; in people enterohemorrhagic *E. coli* generally resolves in 5 to 10 days without treatment and the use of antibiotics and antidiarrheal drugs is contraindicated. The spread of *E. coli* infection can be limited by isolating diseased animals from healthy ones, practicing good general hygiene, providing good nutrition to dams and young animals, assuring that newborns receive colostrum within a few hours after birth, and by vaccinating dams or newborns if a vaccine is available for that species. *E. coli* infections in people can be prevented by through cooking of meat, proper food handling procedures, and adequate hygiene practices. Infection with *E. coli* O157:H7 is a reportable disease in most states and a variety of agencies monitor outbreaks caused by this bacteria.

GLANDERS

Overview

Glanders, also known as malleus (named for its devastating effect on horses), farcy (Latin for sausage which describes the cutaneous lesions associated with this form of the disease), and droes, is an acute to chronic infectious disease caused by the bacterium *Burkholderia mallei* (formerly known as *Pseudomonas mallei* and *Actinobacillus mallei*) and one of the oldest known diseases (it was first described by Hippocrates in 425 B.C.). *Glandulus* is Latin for little nut, which describes the subcutaneous nodules that form, ulcerate, and discharge pus with this disease. Glanders is primarily a fatal skin and respiratory disease affecting horses, donkeys, and mules, but it can also be naturally contracted by goats, dogs, and cats (glanders is technically the respiratory form of the disease and farcy is the cutaneous form of the disease). Human infection is characterized by pustular skin lesions, multiple abscesses, respiratory tract necrosis, pneumonia, and sepsis. Other than one case in a laboratory worker in 2000, glanders has not been seen in the United States since 1945. Glanders occurs rarely and sporadically among laboratory workers and those in direct and prolonged contact with infected, domestic animals; however, there have never been reports of any epidemics of human disease. Sporadic cases continue to occur in Asia, Africa, South and Central America, and the Middle East.

Glanders is believed to have been deliberately spread by the Central Powers in World War I to infect Russian horses and mules in Eastern Europe resulting in the slowing of troop, supply, and artillery convoys. During and after World War I human cases of glanders increased in Russia. The Japanese infected horses, civilians, and war prisoners with *Bu. mallei* at the Pinfang China Institute during World War II. In 1943–1944, the United States studied *Bu. mallei* as a potential biological weapon, but as a result of the low transmission rates to humans from infected horses its use as an infectious weapon was abandoned.

Causative Agent

In nature, *Bu. mallei* is only found in infected hosts; it is not found in water, soil, or plants.

Bu. mallei bacteria are small, gram-negative, nonmotile, aerobic bacilli (some sources say coccobacilli) that grow in 48 hours on blood agar (growth is accelerated with the addition of 1% to 5% glucose and/or 5% glycerol). On blood agar the colonies appear white, semitranslucent, and viscid (older colonies appear yellow). *Bu. mallei* also grows on MacConkey agar. Its oxidase status is variable.

Epizootiology and Public Health Significance

Glanders is rarely seen in the United States, but is endemic in Africa, Asia, the Middle East, and Central and South America. Globally, naturally-acquired cases of glanders are rare and if they occur are sporadic. People who are at greater risk of contracting this disease are people who work with clinical samples or have close contact with horses. The fatality rate in humans with the septicemic form of glanders is about 95% in untreated cases and greater than 50% in treated cases. The fatality rate in humans with localized disease is 40% in untreated cases and 20% in treated cases. There have not been reported human epidemics of glanders.

Transmission

Bu. mallei is found in nasal and skin exudates of infected animals. The bacterium is commonly spread among animals by ingestion of contaminated food or water, but can also be spread by inhalation, discharges of actively infected animals and subclinical carriers, or through skin abrasions and the conjunctiva. Carnivores can be infected after consumption of contaminated meat. After ingestion, bacteria invade the intestinal wall and localize in the lungs, skin, nasal mucosa, and other viscera. *Bu. mallei* is also spread on fomites such as harnesses, grooming equipment, and water troughs. *Bu. mallei* can survive for up to 30 days at room temperature.

Bu. mallei is transmitted to humans by direct contact with infected animals which enables bacteria to enter the body through the skin and mucosal surfaces of the eyes and nose. Ingestion and inhalation are also possible routes of transmission. Cases of human-to-human transmission have been reported (two suggested cases of sexual transmission and several cases in family members who cared for infected patients). *Bu. mallei* is inactivated by heat, light, drying, and a variety of chemicals.

Pathogenesis

Once in the host, *Bu. mallei* produce toxins such as pyocyanin (blue-green pigment that interferes with energy production via the electron transfer system), lecithinase (causes cell lysis by degrading lecithin of cell membranes), collagenase, lipase, and hemolysin. These toxins interfere with cellular functions causing cell death in affected areas.

Clinical Signs in Animals

In animals there are several forms of glanders including:

- *Acute form.* Signs of the acute form of glanders in horses, mules, and donkeys include high fever, cough, dyspnea, thick nasal discharge, and deep ulcers on the nasal mucosa (Figure 3-39). Ulcers that heal appear star-shaped. Submaxillary lymph node enlargement and pain as well as thickening of facial lymphatic vessels are seen with acute glanders. Secondary skin infections may occur. Affected animals typically die in 1 to 2 weeks.
- *Chronic form.* The chronic form of glanders produces more vague signs of coughing, unthriftiness, weight loss, an intermittent fever, and purulent nasal discharge (usually from one nostril). Ulcers and nodules on the nasal mucosa, enlarged submaxillary lymph nodes, joint swelling, and edema of the legs are also seen. The chronic form slowly progresses and may be fatal.
- *Latent form.* The latent form of glanders shows few signs other than nasal discharge and labored breathing. Lesions may be found in the lung.

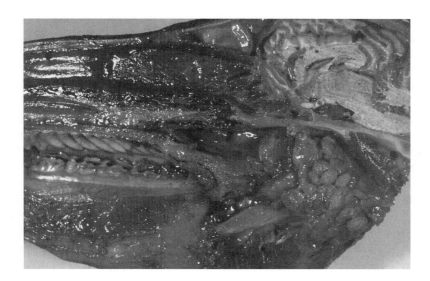

Figure 3-39 Glanders in horses, mules, and donkeys presents with nodules, abscesses, and ulcers in mucous membranes, skin, and lymph nodes.

(Courtesy of Dr. Daniel Harrington, Prof. Emeritus, Dept. Comparative Pathology, School of Vet. Medicine, Purdue University)

Clinical Signs in Humans

Bu. mallei is usually associated with infections in laboratory workers or those with close and frequent contact with infected animals, such as veterinarians, animal caretakers, slaughterhouse workers, and laboratory personnel. The symptoms of glanders vary with the route of infection. There are several different forms of glanders in humans including:

- *Localized form. Bu. mallei* enter nonintact skin through a laceration or abrasion producing local infection that leads to nodules, abscesses, and ulcers in mucous membranes, skin, and lymph nodes. A mucopurulent, blood-tinged discharge may be seen from the mucous membranes. The incubation period is 1 to 5 days. If the infection spreads, a pustular rash, abscesses in internal organs, and pulmonary lesions may be seen.
- *Pulmonary form.* Aerosolized *Bu. mallei* that enter the respiratory tract via inhalation may cause pulmonary infections such as pneumonia, pulmonary abscesses, and pleural effusions. A cough, fever, dyspnea, and mucopurulent discharge may be seen. The incubation period is 10 to 14 days.
- *Septicemia.* When *Bu. mallei* are disseminated in the bloodstream the disease is usually fatal within 7 to 10 days. Septicemia affects many organs of the body such as skin, liver, and spleen. Signs include fever, chills, muscle pain, and chest pain that develop rapidly.
- *Chronic infection.* In chronic forms of glanders, multiple abscesses, nodules, or ulcers can be observed in the skin, liver, and spleen. The chronic form of glanders in humans is known as farcy.

Diagnosis in Animals

Pathologic findings for glanders vary with the form of disease. Bloody, purulent discharge is usually present in the nasal cavity and paranasal sinuses (this discharge is highly infectious). In equine, there may be ulcers, nodules, and stellate (star-shaped) scars in the nasal cavity, trachea, pharynx, larynx, and skin. In all forms of the disease, pathologic findings are characterized by poorly encapsulated pyogranulomas that may spread locally or disseminate along lymphatics. In the pulmonary form there may be signs of bronchopneumonia with enlarged bronchial lymph nodes and variable numbers of pyogranulomatous lesions scattered throughout the lung tissue. In the septicemic form the lungs, liver, spleen, and kidneys may have firm, round, gray nodules. In the nasal form, nodular lesions that ulcerate the nasal septum mucosa are seen. In the cutaneous form, multiple, pyogranulomatous nodules occur along lymphatics. These pus-filled lesions become enlarged and are known as farcy pipes. These lesions may rupture releasing tenacious exudate and bacteria.

Glanders can be identified via culture, animal inoculation, the mullein test, or serology in animals. *Bu. mallei* may be cultured from a lesion or blood sample. *Bu. mallei* is a gram-negative rod that grows as small, white, semitranslucent, viscous colonies on blood agar. PCR testing is available to identify this bacterium to the species level.

Animal inoculation into guinea pigs is known as the Straus reaction. Guinea pigs are injected intraperitoneally with infectious material from affected animals. In positive cases, Guinea pigs develop peritontitis involving the scrotal sac. The testes will become enlarged, painful, and necrotic in Guinea pigs injected with material from affected animals.

The intracutaneous mallein test is the standard test for identifying *Bu. mallei*. Mallein is a product formed from the lysis of *Bu. mallei* that contains both endotoxins and exotoxins. Animals that are infected with *Bu. mallei* are allergic to mullein and following inoculation with mullein will exhibit local and systemic hypersensitivity reactions similar to tuberculin tests. Mallein (0.1 mL) is usually injected intrapalpebrally into the dermis of the lower eyelid. Positive cases will show edema of the eyelids, conjunctivitis, photophobia, and pain within 12 to 72 hours. The test is typically read 48 hours after injection. Other mullein tests include an ophthalmic test with injection of mullein into the conjunctival sac and a subcutaneous test. It should be noted that animals inoculated with mullein may produce a humoral serologic reaction that may be permanent if the animals undergoes repeated mullein testing.

A variety of serologic tests are available for identification of *Bu. mallei*, including complement fixation, ELISA, indirect hemagglutination, immunoelectrophoresis, and immunofluorescence. In horses, the most accurate and reliable tests are complement fixation (not valid when used in donkeys or mules) and ELISA.

Diagnosis in Humans

Glanders in people is diagnosed in the laboratory by isolating *Bu. mallei* from blood, sputum, urine, or skin lesions. Gram stain may reveal small, gram-negative bacilli, which stain irregularly with methylene blue or Wright stain, and may demonstrate a safety pin, bipolar appearance. Organisms can be grown on blood agar as described above. Animal inoculation in Guinea pigs and serology tests such as agglutination and complement fixation tests are also available. Agglutination tests may be positive after 7 to 10 days, but a high background titer and cross-reactivity makes interpretation difficult. Complement fixation tests are more specific and are considered positive for glanders if the titer is equal to or greater than 1:20.

Treatment in Animals

A variety of antibiotics are effective against *Bu. mallei*; however, treatment is typically not recommended as a result of the potential spread to humans and the development of asymptomatic carriers.

Treatment in Humans

Bu. mallei is usually sensitive to a variety of antibiotics including tetracyclines, ciprofloxacin, streptomycin, novobiocin, gentamicin, imipenem, ceftrazidime, and the sulfonamides. Resistance to chloramphenicol has been reported. Treatment may be ineffective especially in cases of septicemia.

Management and Control in Animals

In glanders-endemic areas, routine testing and euthanasia of positive animals have been successful in eradication of this disease from some areas. In endemic areas, congregation of animals, communal feeding, and communal watering sites should be avoided. *Bu. mallei* is sensitive to heat, drying, and common disinfectants; however, it can survive for long periods in warm, moist environments. When outbreaks occur, bedding, feed, stalls, and harness equipment need to be disinfected and susceptible species removed from the environment and isolated. There is no animal vaccine for glanders.

Management and Control in Humans

In endemic countries, prevention of glanders in humans involves identification and elimination of the infection in the animal population. Transmission of *Bu. mallei* in the health-care setting can be prevented by using routine blood and body fluid precautions. There is no human vaccine for glanders.

Summary

Glanders is an acute to chronic infectious disease caused by the bacterium *Bu. mallei* (formerly known as *Pseudomonas mallei* and *Actinobacillus mallei*). *Bu. mallei* bacteria are small, gram-negative, nonmotile, aerobic bacilli that are only found in infected hosts and not in water, soil, or plants. *Bu. mallei* is found in nasal and skin exudates of infected animals. The bacterium is commonly spread among animals by ingestion of contaminated food or water, but can also be spread by inhalation, discharges of actively infected animals and subclinical carriers, or through skin abrasions and the conjunctiva. After ingestion, bacteria invade the intestinal wall and localize in the lungs, skin, nasal mucosa, and other viscera. *Bu. mallei* is transmitted to humans by direct contact with infected animals, which enables bacteria to enter the body through the skin and mucosal surfaces of the eyes and nose. In animals there are several forms of glanders including the acute form, chronic form, and latent form. *Bu. mallei* in people is usually associated with infections in laboratory workers or those with close and frequent contact with infected animals, such as veterinarians, animal caretakers, slaughter house workers, and laboratory personnel. There are several different presentations of glanders in humans including the localized form, pulmonary form, septicemia, and chronic infection. Glanders can be identified via culture, animal inoculation, the mullein test, or serology in animals and culture, animal inoculation, or serology in humans. Treatment of glanders in animals is usually not recommended as a result of the zoonotic nature of the disease. Antibiotic treatment in humans may or may not be successful.

LEPROSY

Overview

Leprosy is a disease that derives its name from the Greek word *lepros*, which means scaly, rough, or mangy. The ancient Greeks described a scaly disease among its people, but this was probably not leprosy as we know it today, but rather psoriasis. Aretaeus the Cappadocian accurately described leprosy in the second century A.D., but he referred to it as leontiasis because of the facial deformity involved with the disease (*leonis* is Latin for lion, which described the bone abnormalities of people with the disease and is now used to describe the appearance of people with Paget's disease). Leprosy was rampant in Europe from around A.D. 1200 until the Black Death era (bubonic plague) in the fourteenth century, when the loss of human hosts probably halted the spread of the bacterium that caused the disease. Leprosy has been known as Hansen's disease since the 1970s when it was named for Armauer (Gerhard) Hansen, a Norwegian bacteriologist who first diagnosed and described the causative agent (*Mycobacterium leprae*) of the disease in 1873. Today, about 2 to 3 million people worldwide have leprosy which is usually nonfatal but permanently disfiguring.

Causative Agent

Mycobacterium leprae, like other *Mycobacterium* organisms, are gram-positive (though weakly staining), acid-fast bacilli that are nonendospore forming, obligate aerobes that have an unusual waxy cell wall that affects many of its properties.

> Glanders is a reportable disease.

My. leprae differs from other *Mycobacterium* spp. in that it is a strict parasite that has not been grown in artificial media or human tissue culture and it is the slowest growing species of *Mycobacterium*. *My. leprae* replicates within host cells in globi (large packets) at an optimal temperature of 30°C.

Epizootiology and Public Health Significance

Humans are the principle reservoir of leprosy; however, *My. leprae* is found in nine-banded armadillos, chimpanzees, and mice. These animals can serve as a source of infection for people; however, this is difficult to confirm as a result of the long incubation period of the bacterium in humans and the inability to eliminate a human source of infection in an endemic area. Approximately 12 million people worldwide have leprosy, with the highest prevalence in Asia, Africa, Latin America, and Oceania followed by India, the Philippines, Korea, and some South American countries. The highest burden of leprosy cases is concentrated in six countries: India, Brazil, Myanmar, Madagascar, Nepal, and Mozambique. In the United States, about 2,500 known cases of leprosy exist (found mainly in immigrants); however, naturally-arising cases have been found in Texas, Louisiana, Hawaii, and Puerto Rico. In 1999, 108 new cases of leprosy occurred in the United States. Worldwide, approximately 107,000 new cases are identified annually. An estimated 2 to 3 million people globally are permanently disabled as a result of leprosy. Leprosy tends to be found in areas of low socioeconomic standards. The form of the disease, whether lepromatous leprosy or tuberculoid leprosy, also varies regionally. Asia and North and South America tend to have lepromatous leprosy, whereas Africa tends to have tuberculoid leprosy.

My. lepraemurium can cause disease in rats, mice, and cats similar to leprosy; however, this organism is not zoonotic.

Transmission

Leprosy was once believed to be an exclusively human disease; however, research has demonstrated that this infection and disease occur naturally in wild animals. The origin of the animal infection is unknown. Animals may have contracted *My. leprae* from a human source since these bacilli may remain viable in dried nasal secretions for about 7 days. Since armadillos are in close contact with soil this may be a source of infection if it is contaminated with organisms from nasal secretions. Leprosy in animals can be seen at high levels in certain locations indicating that armadillos may pass the infection to another armadillo either via inhalation of or direct contact with the bacillus. *My. leprae* has been found in milk, making maternal milk another possible transmission route.

Armadillos harbor a species of *Mycobacterium* indistinguishable from *My. leprae* making the transmission of this organism to people possible.

In people, leprosy is believed to be transmitted directly into the skin through contact with a person with leprosy, by mechanical vectors, or by inhalation of respiratory droplets. There have been cases of human leprosy attributed to a person's handling of armadillos and eating their meat and hunting and cleaning armadillos (without known contact with humans with leprosy). About one-third of all leprosy patients in the United States have had contact with armadillos. *My. leprae* has a very low level of virulence, so most people that come in contact with the organism do not develop clinical disease.

Pathogenesis

Once *My. leprae* enter the body, they are phagocytosed by macrophages, which destroy the bacilli. In some cases the macrophage response is slow or weak, which allows the bacilli to survive and grow slowly in the lesions. Localized skin and nerve lesions are granulomatous (tumor-like masses, with active fibroblasts

My. leprae can survive for about 7 days outside the body. This survival time has supported the belief that armadillos acquired the disease from people with lepromatous leprosy whose secretions came in contact with soil.

and capillaries, containing macrophages surrounded by mononuclear cells), inflammatory processes. The incubation period is typically 2 to 5 years (ranges from 3 months to 40 years). In untreated cases, the bacilli replicate in skin macrophages and Schwann cells of peripheral nerves, causing several different outcomes.

Clinical Signs in Animals

Clinical signs of leprosy in animals vary. Skin lesions vary from mild, self-healing lesions to severe, destructive lesions. Dermal nerves can be invaded by these bacilli causing nerve damage. Many organisms can be found in the macrophages of lymph tissue, spleen, and liver. *My. leprae* prefers cool parts of the body such as the foot pads of mice and bodies of armadillos, which have a body temperature of 30°C to 35°C. Armadillos experimentally inoculated with *My. leprae* had dissemination of the organisms in lymph glands, liver, spleen, lungs, bone marrow, and meninges (Figure 3-40).

Figure 3-40   Armadillos are a source of infection and research model for *Mycobacterium leprae.*
(Courtesy of CDC/Dr. Charles Shepard)

Clinical Signs in Humans

Clinical signs of leprosy in humans depend on the form of the disease involved.

- Tuberculoid leprosy is characterized by localized lesions of the skin and nerves that are often asymptomatic. The skin lesions are few in number (one to three), asymmetric, and shallow containing few bacilli. If the lesions involve the dermal nerves these nerves may become damaged resulting in local loss of pain perception. This form of leprosy has fewer complications and is more easily treated. Tuberculoid leprosy results in destruction of the organism by cellular immunity (and therefore produce low serum antibody titers).

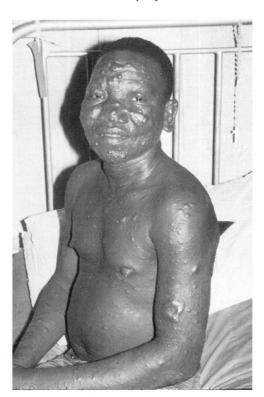

Figure 3-41   A man with lepromatous leprosy.

- Lepromatous leprosy is characterized by many symmetrical skin lesions of varying sizes. Mucosa of the upper respiratory tract, lymph nodes, liver, spleen, and testicles may also be involved. This form of leprosy is responsible for the disfigurations commonly associated with the disease. Because these bacilli prefer the cooler regions of the body such as the nose, ears, eyebrows, and chin the organisms grow there (Figure 3-41). As replication of bacteria continues, the face of the person develops folds and granulomatous thickenings called lepromas. Advanced cases of lepromatous leprosy may cause a loss of sensitivity that results in self-mutilation, secondary infections, and blindness. Lepromatous leprosy activates humoral immunity; therefore, antibody levels are high. Although antibody levels are high, they do not protect the person from developing extensive skin damage as a result of *My. leprae*.

- Borderline leprosy is intermediate between the above two forms and can progress to either form depending on the patient's treatment and immune status. The borderline leprosy cases can result in early damage to nerves that control the hands and feet as well as sensory nerve damage leading to trauma and loss of fingers and toes.

Diagnosis in Animals

Large numbers of acid-fast bacilli are seen in affected organs such as skin, lymph nodes, spleen, and liver. Smears of skin lesions and nasal secretions may also demonstrate acid-fast bacilli. Bacterial culture for *My. leprae* is not possible; therefore, diagnosis is based on clinical signs and histopathology. In vivo culture on mouse foot pads can be performed with animal and human samples. ELISA testing of IgM to *My. leprae* has been used in monitoring armadillos in Texas and Louisiana.

Diagnosis in Humans

Leprosy in humans is diagnosed by clinical symptoms, histopathology, and patient history. A tissue biopsy sample obtained surgically or a tissue smear will show acid-fast bacilli. Tissue biopsies will demonstrate granulomatous, inflammatory lesions. Bacterial culture is unsuccessful. An additional test used in humans is the lepromin test (leprosy skin test). Patients with tuberculoid leprosy mount a cell-mediated immune response that will result in a positive lepromin test. Patients not infected with *My. leprae* or those with compromised immune systems fail to mount a cell-mediated immune response that results in a negative lepromin test.

Treatment in Animals

Treatment of leprosy in animals is not documented.

Treatment in Humans

People with leprosy are treated with the antileprosy drug dapsone (DDS), which is bacteriostatic (often treatment lasts for life). Resistance to dapsone has been reported; therefore, multidrug therapies (MDT) are used to treat human leprosy. Rifampicin is typically used in these combination therapies because it eliminates contagious organisms in 1 to 2 weeks; thereby eliminating the need for isolation of patients in leprosariums. In the United States leprosy patients are treated at the National Hansen's Disease Center in Louisiana (operated by the U.S. Public Health Service).

Management and Control in Animals

Armadillos are being tested in Texas and Louisiana to monitor the level of leprosy in this wildlife source of infection. Other animals do not pose a threat to human health.

Management and Control in Humans

Control of leprosy in humans is based on early detection and antibiotic treatment. The WHO is sponsoring vaccine trials for killed leprosy bacilli. Armadillo-derived killed *My. leprae* vaccines have been studied since 1984. The use of BCG vaccine (the tuberculosis vaccine) has also been attempted in the control of leprosy and is about 40% to 50% effective. The use of BCG plus killed *My. leprae* vaccines is being evaluated. Multiple vaccine studies have also been performed in India; however, a widely used vaccine is not available to date. Proper hygiene after handling armadillos or having contact with people with leprosy can also help control the spread of leprosy. The main control strategy is multidrug therapy, which has caused a remarkable decline in leprosy cases. Drugs to treat leprosy are supplied free through some charitable foundations and campaigns to educate people about leprosy through case detection and public awareness are helping control the disease. Continued research is needed to discover more rapid and simplified diagnosis of leprosy, simplified treatment regimens of shorter duration, and prevention of disability in affected patients.

Summary

Leprosy is caused by a slow-growing, acid-fast bacterium called *Mycobacterium leprae*, which is related to *My. tuberculosis* and *My. bovis*. Leprosy is transmitted human to human by respiratory secretions (nasal discharge and respiratory droplets) and direct contact with skin lesions. A natural reservoir of infection is the nine-banded armadillo, which may serve as an important source of infection and research model. Clinical signs of leprosy in animals vary. Skin lesions vary from mild, self-healing lesions to severe, destructive lesions. In humans, lesions from leprosy are not fatal, but disfiguring. Common lesions involve skin and nerve endings leading to skin thickening and folding and loss of sensation. *My. leprae* does not grow in culture, making diagnosis difficult. In animals, large numbers of acid-fast bacilli are seen in affected organs such as skin, lymph nodes, spleen, and liver. Smears of skin lesions and nasal secretions may also demonstrate acid-fast bacilli. Leprosy in humans is diagnosed by clinical symptoms, histopathology, and patient history. Animals are not treated for leprosy. Multidrug treatments and development of vaccines are being used in humans with leprosy.

LEPTOSPIROSIS

Overview

Leptospirosis, also known in people as Weil's disease (named after German physician Adolph Weil, who first described the disease as infecious jaundice in 1886), is a disease caused by finely coiled spirochetes (*leptos* is Greek for fine or slender). Leptospirosis is known by a variety of names that typically reflect the sources of infection or species of bacteria that causes the disease. Swineherder's disease (named because people contracted the disease from handling infected animals); rice-field fever, cane-cutter fever, swamp fever, and mud fever (contracted from contaminated water) are names that reflect the source of infection. Icterohemorrhagic fever (caused by serovar *L. icterohaemorrhagiae*) and Canicola fever (caused by *L. canicola)* are names that reflect the species of bacteria causing the disease.

Leptospirosis has historically been found in miners, farmers, fish handlers, and sewer workers exposed to bacteria from infected animals or contaminated water. During World War I, soldiers living in trenches came down with leptospirosis, most likely from contamination of the environment with animal urine, especially rat urine. In 1942 soldiers training at Fort Bragg, North Carolina came down with a disease producing high fever, headache, and rash, which was later traced to swimming in ponds and streams contaminated by livestock urine. The largest recorded outbreak of leptospirosis in the United States was in Illinois in 1998 during a triathlon event that sickened 110 participants. *Leptospira* bacteria are ubiquitous, making leptospirosis one of the world's most widespread zoonotic diseases. It is most common where the climate is warm and humid, soils are alkaline, and there is abundant surface water.

Causative Agent

Leptospira bacteria are gram-negative, aerobic spirochetes (tight coils) with hooked ends (Figure 3-42). *Leptospira* spp. are motile by means of axial filaments and do not have external flagella. The nomenclature system used to classify leptospires has been revised; the traditional system divided the genus into two species: the pathogenic *L. interrogans* and the nonpathogenic *L. biflexa*. These species were divided further into serogroups, serovars, and strains (based on shared antigens), which gave the species *L. interrogans* more than 250 serovars. Prior to DNA-DNA hybridization, the classification system divided *Leptospira biflexa* into five species and recognized the variability within the classic *L. interrogans* species, dividing it into seven named and five unnamed species as follows:

- *L. interrogans*
- *L. weilii*
- *L. santarosai*
- *L. noguchi*
- *L. borgpetersenii*
- *L. inadai*
- *L. kirschner*
- *Leptospira* species 1, 2, 3, 4, and 5.

Leptospira bacteria are now divided into 17 genomospecies based on DNA-DNA hybridization which added *L. alexander* to the list of pathogenic *Leptospira* organisms.

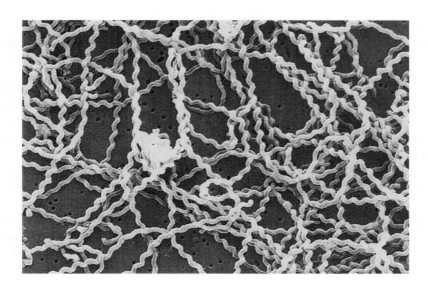

Figure 3-42 *Leptospira* organisms are spirochetes as seen on this scanning electron micrograph (SEM).
(Courtesy of CDC/Rob Weyant)

Animals exposed to *Leptospira* bacteria will develop immunity (even if they do not show signs of the disease) but only to the particular serotype to which they have been exposed.

Leptospira spp. infect many types of mammals including rats, mice, dogs, cats, cattle, pigs, squirrels, raccoons, bats, deer, foxes, rabbits, goats, birds, frogs, snakes, fish, and mongooses. Animals are critical to the maintenance of *Leptospira* in a particular location. The reservoir host animals of the different *Leptospira* species and serogroups vary from region to region, do not typically cause disease in these hosts, and individual animals may carry multiple serovars. *L. interrogans* has multiple serovars found in animals including *L. interrogans* serovar *canicola* in dogs; *L. interrogans* serovar *icterohaemorrhagiae* in rats; *L. interrogans* serovar *hardjo* in cattle, *L. interrogans* serovar *grippotyphosa* in voles, raccoons and other small mammals; and *L. interrogans* serovar *bratislava* in pigs, rats, and other small mammals. The pathogenic serovars of *Leptospira* do not replicate outside of an animal host. *Leptospira* bacteria can persist and evade the immune system in the renal tubules of these animals without causing clinical signs of disease, and they can be excreted in the urine for prolonged periods of time. Rats are the most common source of infection for humans worldwide; in the United States the most significant sources of infection for humans are dogs, followed by livestock, followed by rodents, followed by wild mammals.

Epizootiology and Public Health Significance

Leptospirosis is endemic worldwide and is an occupational hazard for people who work outdoors or with animals (farmers, sewer workers, veterinarians, fish workers, dairy farmers, or military personnel) and is a recreational hazard (campers or those who participate in outdoor sports in contaminated lakes and rivers). In humans, the disease is seen more frequently during summer and fall.

Leptospirosis most commonly occurs in temperate or tropical climates. One hundred to 200 cases of leptospirosis occur in the United States annually, mainly in the Southeast and Hawaii, where water temperatures are optimal for the bacterium's reproduction.

The incidence of leptospirosis is highest in tropical areas, especially following heavy rainfall, which increases risks for waterborne infection and drives rodents into urban dwellings. In 2000, there was a massive outbreak associated with flooding in Thailand in which over 5,000 people were infected and over 180 people died. In 1998, 9 workers at the University of Missouri swine facility developed leptospirosis from handling infected swine and drinking or smoking while working with the animals. In the United States, leptospirosis is most likely underdiagnosed and underreported. The reported annual incidence ranged from 0.02 to 0.04 cases per 100,000 persons from 1985 to 1994, prompting recommendations to remove leptospirosis from the list of notifiable diseases. Internationally high-risk areas include the Caribbean islands, Central and South America, Southeast Asia, and the Pacific islands. Reporting of leptospirosis in these areas is flawed and frequently the disease gains public attention when outbreaks occur in association with natural disasters, such as flooding. Leptospirosis can reach a mortality rate of 10%. Leptospirosis is diagnosed most commonly in adult males as a result of occupational and recreational exposures.

In animals, outbreaks of mastitis and a significant decrease in milk production can be seen in dairy herds infected with *Leptospira*. In both dairy and beef herds, decreased calving percentage as a result of abortions and high death rate in calves may constitute considerable economic loss. The cost of leptospirosis vaccination is an affordable way to prevent disease.

Leptospirosis continues to re-emerge as a notable source of morbidity and mortality in the Western Hemisphere. The largest recorded outbreak in the continental United States (110 cases in a group of 775 people who participated in triathlons, which included swimming in a lake) occurred in June and July 1998 in Illinois. Significant increases in incidence were also reported from Peru and Ecuador (following heavy rainfall and flooding) in 1998, and Thailand has also reported a rapid increase in incidence of leptospirosis between 1995 and 2000.

Transmission

The bacterium infects a variety of wild and domestic animals that excrete the organism in their urine or in the fluids excreted during parturition. Humans are exposed to the organism when they come in contact with water or soil contaminated by infected animals. *Leptospira* bacteria multiply in fresh water, damp soil, vegetation, and mud. Flooding following heavy rainfall helps spread the organism because, as the flood water saturates the environment, *Leptospira* present in the soil pass directly into surface waters. *Leptospira* bacteria can enter the body either through nonintact skin and mucous membranes (including the conjunctiva) or ingestion of contaminated water. Humans are susceptible to infection with a variety of serovars.

Leptospira spp. localize in the kidneys and are shed in the urine for long periods of time. In both animals and humans, *Leptospira* bacteria are transmitted via contaminated urine (direct contact with urine or through contact with contaminated water and soil) followed by invasion of the organisms across mucosal surfaces or nonintact skin. Under favorable conditions, *Leptospira* spp. can survive in fresh water for as long as 16 days and in soil for as long as 24 days. Human-to-human transmission is rare.

Veterinary personnel (veterinarians, veterinary technicians, veterinary assistants, and kennel workers) should handle urine specimens and urine contamination as if they contained *Leptospira* bacteria by wearing gloves and a face shield.

Pathogenesis

Leptospirosis occurs in two phases: the leptospiremic phase and the immune phase.

- *Leptospiremic, acute, or early phase.* During this phase the bacterium enters the animal or human via mucous membranes or breaks in the skin. Once the organism, which is highly motile, penetrates the mucous membranes or damaged skin, it enters the bloodstream and distributes throughout the body. Bacteria that reach the kidneys multiply there and are excreted in the urine. During this phase bacteria can also enter the cerebrospinal fluid. Clinical signs at this stage of the disease include high fever, weakness, anorexia, and vomiting.

- *Immune, subacute, or second phase.* During this phase the blood infection resolves and the fever lessens as a result of circulating IgM antibodies. Clinical signs during this phase include milder fever, labored breathing, increased thirst, icterus, reluctance to rise, and signs of lumbar or abdominal pain (in animals) and milder fever, headache secondary to meningitis, and kidney and liver involvement resulting in jaundice and anemia (in humans).

Research suggests that *Leptospira* spp. produce an endotoxin (hemolysin) and lipase as possible causes of its pathogenicity. The true mechanism of host tissue injury, however, remains unclear and likely involves a complex set of interactions. The most common pathologic finding in cases of leptospirosis is vasculitis of capillaries in every affected organ system resulting in loss of red blood cells and fluid through enlarged capillary spaces. This leads to secondary tissue injury and most likely accounts for many of the clinical signs associated with leptospirosis.

Transmission of leptospirosis can occur following bites by mice, rats, or hamsters, when these animals void urine at the time of the bite.

Clinical Signs in Animals

Leptospirosis is a contagious disease that may occur without clinical signs or may result in a variety of disease conditions.

- Leptospirosis in dogs is also known as canine typhus, Stuttgart disease, and infectious jaundice and is caused by *L. interrogans* serovar *canicola* or

The source of *Leptospira* infection in farm animals is through pastures, drinking water, or feed, when contaminated by infected urine. Infection may also occur as a result of contact with infected uterine discharges and aborted fetuses.

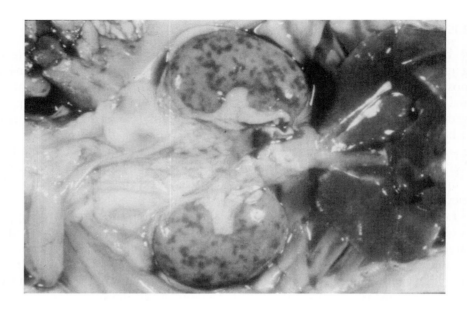

Figure 3-43 Petechial hemorrhages (pinpoint hemorrhages as a result of vascular damage) as seen on these kidneys are commonly seen on the skin of dogs with leptospirosis. Vascular damage to organs, especially of the kidneys, is common with leptospirosis.
(Courtesy of Shapiro)

L. interrogans serovar *icterohaemorrhagiae* (although atypical canine infection with *L. interrogans* serovar *grippotyphosa*, *L. interrogans* serovar *bratislava*, and/or *L. interrogans* serovar *pomona* is seen in dogs). Dogs of any age can contract leptospirosis; however, it is most common in young males. The incubation period is 5 to 15 days and many canine cases of *Leptospira* infections are subclinical. When clinical signs are observed the classical presentation includes fever, anorexia, vomiting, and hepatic disease and oral hemorrhages (petechiae and ecchymoses) (Figure 3-43). This acute form of hepatic and hemorrhagic disease is most often associated with infection with *L. interrogans* serovar *icterohaemorrhagiae*. Within days, the temperature will drop, breathing becomes labored, thirst increases, and reluctance to stand from a sitting position may be noted. Salivary secretions may thicken and become blood tinged and swallowing becomes difficult. Bloody vomit and feces may be seen if hemorrhagic gastroenteritis develops. Once signs of acute renal failure are observed (oliguria or anuria), the prognosis for recovery worsens. Fatality rates are about 10% and in fatal cases death usually occurs 5 to 10 days after disease onset.

- Leptospirosis in cattle is also known as redwater of calves and is caused primarily by *L. interrogans* serovar *hardjo*, *L. interrogans* serovar *pomona*, and *L. interrogans* serovar *grippotyphosa*. In 50% of young calves infected with leptospirosis, hemolytic icterus and hemoglobinuria (producing clear-red or port-wine colored urine) occurs and mortality rates range from 5% to 15%. In calves, additional clinical signs include fever, anorexia, dyspnea, and anemia. A common manifestation of *Leptospira* infection in adult cattle is abortion with retention of the placenta occurring most often in the third trimester, and the fetuses are generally autolyzed and icteric. Approximately 30% of cows will reabort during the next pregnancy. Stillbirths, birth of weak calves, rapid drops in milk production, and atypical mastitis can also occur with *Leptospira* infections. Cows will occasionally show evidence of systemic disease with fever, icterus, and hemoglobinuria.

- Leptospirosis in sheep and goats occurs less frequently than in cattle and is caused primarily by *L. interrogans* serovar *pomona* and *L. interrogans* serovar *hardjo*. Clinical signs in sheep and goats are similar to those described in mature cattle and calves.

- Leptospirosis in swine is primarily caused by *L. interrogans* serovar *pomona* (they may also be infected with *L. interrogans* serovar *grippotyphosa*, *L. interrogans* serovar *canicola*, and *L. interrogans* serovar *icterohaemorrhagiae* serovars) and typically presents as either inapparent disease, febrile reactions lasting 3 to 4 days, or reproductive disease (later term abortions or birth of weak neonates). Occasionally, typical signs of systemic leptospirosis such as fever, icterus, hemorrhages, and death will develop in young pigs. *L. interrogans* serovar *pomona* can also be introduced to cattle herds following the animals' exposure to infected pigs onto the property.

- *Leptospira* infection in horses is characterized by fever, dullness, anorexia, and icterus. Abortion may occur several weeks after the fever and chronic uveitis months later. The etiologic agent of recurrent uveitis (moon blindness) is *L. interrogans* serovar *pomona*, which has been isolated from the eyes of afflicted horses and may progress to blindness. Uveitis typically develops 12 or more months after infection and is an immune-mediated disorder. It has been suggested that the basis for this immune reaction is antigenic cross-reactivity between a *Leptospira* protein and a protein in the equine cornea. *L. interrogans* serovar *pomona* is a major cause of abortion in horses and can cause liver disease in the equine fetus.

- *Leptospira* infection and disease is rare in cats and if present resembles the disease seen in dogs.

> Leptospirosis in cattle, pigs, sheep, and goats is characterized by fever, depression, anemia, and abortion; in horses the disease produces an ocular infection; in dogs the disease causes a severe kidney infection.

Clinical Signs in Humans

People with leptospirosis may present without any symptoms of disease or may present with a wide range of symptoms such as high fever, headache, chills, muscle aches, vomiting, jaundice (10% of cases), abdominal pain, diarrhea, or a rash. The incubation period is 2 days to 4 weeks (average is about 10 days). Following the incubation period, leptospirosis begins abruptly typically producing fever as the first symptom. Leptospirosis may be biphasic: the first phase produces fever, chills, headache, muscle aches, vomiting, or diarrhea and the second phase produces more severe systemic disease. Direct tissue injury from invasion of *Leptospira* bacteria and their toxins characterize the acute phase. Symptoms resolve when systemic proliferation of bacteria ends. The patient may recover from the first phase but become ill again. If the second phase occurs, it is more severe and the person may develop kidney or liver failure or meningitis. This phase is also called Weil's disease and is most often a result of infection with the serovar *L. interrogans* serovar *icterohaemorrhagiae*. The second phase is characterized by increasing antibody titers and inflammation of affected organ systems. Aseptic meningitis and renal dysfunction are commonly seen in the second phase. Symptoms may persist for 6 days to more than 4 weeks. Without treatment, recovery may take several months and kidney damage, meningitis, liver failure, and respiratory distress as a result of pulmonary hemorrhage may occur. Mortality rates are 10%.

Diagnosis in Animals

Biopsies of tissues infected with *Leptospira* spp. present in a variety of different ways. Membranes such as oral mucosa, pleura, and peritoneum will have petechial hemorrhages. Grossly the liver appears swollen and yellowish with petechiae. Hepatocytes will shrink and the microscopic architecture of the liver will lose its organization. In all affected species, the kidney may have areas of petechiae and renal tubules are severely altered (they become swollen, granular, and vacuolated). In time the kidneys may contain grayish to white lesions that may be scattered in the cortex (bovine) or concentrated at the corticomedullary junction (canines). The spleen and lymph nodes are grossly enlarged and may contain areas of edema and hemorrhage. Other organs, such as the heart, urinary bladder, pancreas, and lung, may also show areas of hemorrhage and edema. In pigs, abortion and birth of stillborn pigs may be seen. In horses, ocular tissue is affected and bacteria are not present in the urine. Microscopic lesions of the equine eye include alterations to the anterior uvea with congestion of the iris and ciliary body including lymphocyte infiltration. Corneal vascularization is also seen in horses.

Leptospirosis in animals is diagnosed using multiple testing methods. Darkfield microscopic examination of urine may show the spirochete; however, absence of the organism does not rule-out *Leptospira* infection. Bacterial culture of the *Leptospira* organisms in blood (early in the course of disease) or urine (later in the course of disease) should be performed. Heparinized blood is inoculated into semisolid media tubes enriched with rabbit serum or bovine serum albumin. Urine should be inoculated as soon as possible because the acidity of the urine may harm the spirochetes. All cultures are incubated at room temperature and kept in the dark for up to 6 weeks. Microscopic agglutination (MA) testing is the standard serologic test used to identify *Leptospira* bacteria. MA uses live cells and pools of bacterial antigens that contain many different serotypes for identification of agglutination. If antibody is present in the sample it will agglutinate with its specific antigen providing a positive result. The presence of agglutination is examined under darkfield microscopy. Other serologic tests include FA or antigen-capture ELISA identification. PCR tests are also available.

Diagnosis in Humans

Laboratory diagnosis of leptospirosis requires culture of the organism (using blood, CSF, or urine) or demonstration of serologic conversion by the microagglutination (MA) test. Bacterial culture is relatively insensitive and requires specialized media (such as Fletcher, Stuart, Ellinghausen combined with neomycin to control growth of other bacteria). The MA test is difficult to perform making the use of reference laboratories important. Serodiagnosis of leptospirosis requires a fourfold or greater rise in titer between the acute and convalescent sample using the MA test. More recently, several rapid, simple serologic tests such as the indirect hemagglutination assay (IHA) have been developed that are reliable and commercially available. Histopathology using silver staining techniques for identification of the organism in tissues can be done on tissue biopsies.

Treatment in Animals

Dogs require aggressive antibiotic therapy as well as supportive therapy such as fluid administration for renal failure. Penicillin is given IV initially to eliminate the leptospiremia, followed by 3 weeks of oral doxycycline to eliminate the leptospiruria. Other treatments include tetracycline, streptomycin, and dihydrostreptomycin. Cattle, small ruminants, horses, and swine may be treated with similar antibiotics and treatment is successful if initiated early in the disease course.

Treatment in Humans

In humans leptospirosis is treated with antibiotics, such as doxycycline or penicillin, which are effective if given early in the course of the disease. Intravenous antibiotics may be required for people with more severe symptoms. Supportive treatment including fluid therapy and renal dialysis is also helpful in treating leptospirosis.

Vaccine-induced titers in dogs rarely exceed 1:400 and generally decrease to approximately 1:100 by 2 to 3 months after vaccination.

Management and Control in Animals

The prevalence of leptospirosis in dogs in the United States and Canada has increased since 1983. Dogs at greatest risk are male herding and other working dogs and hounds; however, leptospirosis has also been documented in house pets (with German shepherd dogs having the highest rate of reporting). Vaccination using multivalent (including multiple serovars) vaccines is used. In dogs, two vaccine forms are available: the original vaccine, which contained only the *L. interrogans* serovar *canicola* and *L. interrogans* serovar *icterohaemorrhagiae* serovars; more recent vaccines are also affective against *L. interrogans* serovar *grippotyphosa* and *L. interrogans* serovar *pomona*.

Cattle should receive annual vaccination. Management methods used to control exposure to *Leptospira* bacteria include rodent control, fencing of cattle from contaminated water, separation of cattle from swine and wildlife, and selection of replacement stock for leptospirosis-negative herds.

Early in a swine outbreak, swine abortions can be prevented by early treatment and vaccination of all sows in the herd and by separation of age groups.

Management and Control in Humans

Currently, no vaccine is available to prevent leptospirosis in humans. Travelers are advised to consider preventive measures such as wearing protective clothing and footwear and minimizing contact with potentially contaminated water. Antibiotic prophylaxis with doxycycline may be warranted when traveling to some areas.

> Vaccinated dogs and livestock are protected from clinical disease, but they can still persistently shed *Leptospira* organisms in their urine serving as a source of infection for humans.

Summary

Leptospirosis is a zoonotic disease caused by a spirochete that is common worldwide, especially in tropical countries with heavy rainfall. Infected rodents and other wild and domestic animals pass *Leptospira* bacteria in their urine, which can directly infect a person or can contaminate an environmental source that can transmit the organism. *Leptospira* bacteria can live for a long time in fresh water, damp soil, vegetation, and mud. Flooding after heavy rainfall helps spread the bacteria in the environment. *Leptospira* bacteria can enter the body through broken skin and mucous membranes or by ingestion of contaminated food or water, including water swallowed during water sports. Once in the bloodstream, bacteria can reach all parts of the body and cause signs and symptoms of illness. Signs of disease vary with animal species and include fever, muscle aches, acute renal failure, abortion, and liver disease. Leptospirosis in people may be biphasic: the first phase produces fever, chills, headache, muscle aches, vomiting, or diarrhea and the second phase produces more severe systemic disease. Leptospirosis is common in tropical countries where people have regular contact with fresh water and animals. The disease is probably underdiagnosed in the United States (about 50 to 150 cases are reported annually). Leptospirosis in animals and humans is diagnosed using multiple testing methods including darkfield microscopic examination of urine, bacterial culture of the *Leptospira* organisms in blood (early in the course of disease) or urine (later in the course of disease), microscopic agglutination (MA) testing, and serologic tests including FA, or antigen-capture ELISA identification. PCR tests are also available. Leptospirosis is treatable with antibiotics and supportive care. Vaccines are available for dogs, cattle, and swine, but not for humans.

> Preventative measures for controlling leptospirosis include vaccination of livestock and dogs, rodent control measures, and preventing recreational exposures (such as avoiding swimming in freshwater ponds).

> Leptospirosis is a not a nationally reportable disease in the United States; however, many states require notification of leptospirosis.

LISTERIOSIS

Overview

Listeriosis, also known as circling disease, listeriasis, and listerellosis, is caused by the gram-positive motile bacteria in the *Listeria* genus. The genus *Listeria* contains the two pathogenic species: *Listeria monocytogenes* and *Li. ivanovii* (*Li. ivanovii* subspecies *ivanovii* and subspecies *londoniensis* are rare agents of human disease) and the four apparently apathogenic or rarely pathogenic species *Li. innocua*, *Li. seeligeri*, *Li. welshimeri*, and *Li. grayi*. Listeriosis is most commonly caused in animals and humans by *Listeria monocytogenes* (there are 13 serovars); however, *Li. ivanovii* is also associated with animal disease. *Listeria* bacteria were first cultured by Murray, Webb, and Swann in 1926 from Guinea pigs and rabbits with hepatic necrosis. *Listeria* is named for Joseph Lister, the English surgeon who pioneered antiseptic surgery and the species *monocytogenes* is used for this bacterium's effect on monocytes. *Li. monocytogenes* has been recognized as a human pathogen for approximately 60 years, but its role as a foodborne pathogen was not known until relatively recently. Listeriosis produces flu-like symptoms (headache, fever, and diarrhea) and meningitis/encephalitis in humans and encephalitis, abortions, and septicemia in animals. Listeriosis occurs primarily in newborn infants, elderly patients, and immunocompromised people. *Li. monocytogenes* may be present in the intestinal tract of 1% to 10% of humans and has been found in at least 37 mammalian species and at least 17 species of birds and some species of fish, shellfish, and insects.

Causative Agent

Li. monocytogenes organisms are gram-positive, nonendospore-forming, motile, pleomorphic bacilli that are sometimes arranged in short chains (Figure 3-44). *Li. monocytogenes* are considered foodborne pathogens because the majority of these infections are associated with the consumption of contaminated food (mainly unpasteurized dairy products and meat); however, a woman can pass *Li. monocytogenes* to her baby during pregnancy and farmers/veterinarians/butchers can develop *Listeria* skin infections by touching infected calves or poultry. *Li. monocytogenes* is relatively resistant to drying, freezing, and heat. It will grow at a wide range of temperatures (1°C to 45°C) and can multiply at refrigeration temperatures on contaminated foods. It can survive at pH range 3.6 to 9.5 and a pH greater than five favors its growth. In susceptible humans, an infective dose of *Li. monocytogenes* is less than 1,000 organisms.

Virulence and resistance factors associated with *Li. monocytogenes* include:

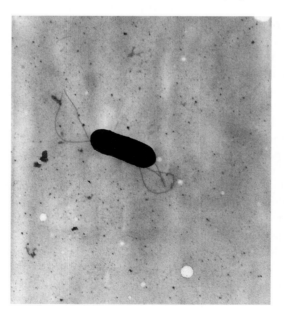

Figure 3-44 An electron micrograph of *Listeria monocytogenes.*
(Courtesy of CDC/Dr. Balasubr Swaminathan; Peggy Hayes)

- The ability to grow at low temperatures. *Listeria* can grow and remain viable in certain foods at freezing and refrigeration temperatures. *Li. monocytogenes* is usually killed by cooking or pasteurization.

- Resistance to drying. *Li. monocytogenes* can survive for months in food, bedding (straw and shavings), and soil.

- Motility. *Listeria* bacteria have flagella at room temperature, but do not produce flagella at human body temperature (37°C). This allows for bacterial existence and spread outside of the host environment.

- Adherence and invasion. *Listeria* organisms produce chemical substances to aid in their adhesion, invasion, and survival in host cells.
  - A membrane protein called internalin mediates invasion of the host cell by *Listeria* bacteria.
  - Act A, a gene product, promotes the lengthening of actin (part of the host cell cytoskeleton) on the bacterial cell surface. Actin sheets function to move bacteria across the cytoplasm to the cell surface where pseudopods form.
  - Listeriolysin O (LLO) is a hemolytic and cytotoxic substance that helps *Listeria* bacteria survive within host cells.
  - Other hemolysins such as phosphatidylinositol-specific phospholipase C (PI-PLC) and phosphatidylcholine-specific phospholipase C (PC-PLC) disrupt membrane lipids.

Epizootiology and Public Health Significance

Li. monocytogenes is found worldwide and is widely distributed in the environment. *Li. monocytogenes* can be found in the feces of farm animals, domestic animals, wild animals, birds, and insects and it has been found in fecal-contaminated soil, water, sewage, and animal feed. Five out of every 100 people carry *Li. monocytogenes* asymptomatically in their intestinal tract. Human infections peak in July and August, whereas animal infections peak between February and April.

In the United States, the estimated annual incidence of listeriosis is approximately 7.4 cases per million people (approximately 2,500 cases are reported each year since 1997, with 500 of them being fatal). Internationally the estimated annual incidence of listeriosis is approximately 4 cases per million people in Canada with lower incidences reported in Australia, England, and Denmark. In susceptible groups, the overall mortality rate is 20% to 30% reaching as high as 70% in untreated cases. The overall death rate for listeriosis is 26%.

Transmission

The main reservoirs of *Li. monocytogenes* are soil and the intestinal tract of animals. Animals can carry the bacterium without appearing ill and can contaminate foods of animal origin such as meats and dairy products via feces, milk, and uterine discharges. The bacterium has been found in a variety of raw foods, such as uncooked meats and vegetables, as well as in processed foods that become contaminated after processing, such as soft cheeses and cold cuts. Vegetables become contaminated from the soil or from manure used as fertilizer. *Listeria* is killed by pasteurization and cooking; however, in certain ready-to-eat foods such as hot dogs and deli meats, contamination may occur after cooking but before packaging (Figure 3-45). *Listeria* is mainly spread through ingestion, but can also be spread

> The seasonal use of silage as livestock feed is frequently followed by an increased incidence of listeriosis in animals.

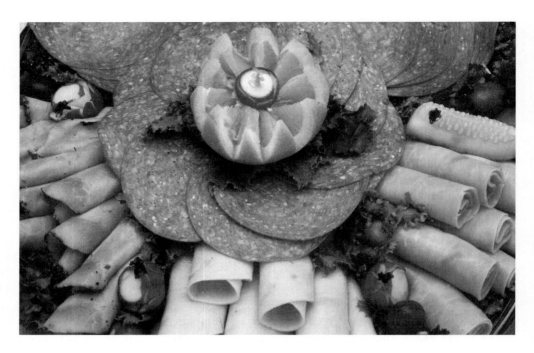

Figure 3-45 *Listeria monocytogenes* is a foodborne pathogen that can be found on processed meat such as cold cuts and hot dogs.
(Courtesy of USDA)

by inhalation or direct contact. Venereal transmission might also be possible. In ruminants, consumption of contaminated silage or other feed is the typical route of transmission, whereas in humans, ingestion of contaminated raw meat and fish, unpasteurized dairy products, and undercooked vegetables are the main routes of transmission.

Vertical transmission is the main route of infection for newborn human infants and ruminants. Vertical transmission is either transplacentally or from an infected birth canal. Humans can also be infected after handling infected animals during calving, lambing, or necropsies.

Pathogenesis

Li. monocytogenes is typically ingested with raw, contaminated food (humans) or contaminated silage (ruminants). *Li. monocytogenes* bacteria are facultative intracellular parasites that bind to epithelial cells of the gastrointestinal tract or macrophages. This binding of *Listeria* bacteria triggers phagocytosis. After engulfment the bacteria produce a pore-forming protein called listeriolysin O that forms a hole in the host cell's cytoplasm before the bacterium is destroyed by a lysosome. *Listeria* bacteria can then survive and replicate in the host cell's cytoplasm avoiding identification by the host's immune system. *Listeria* bacteria then transfer themselves to nearby cells without having to leave the originally-infected host cell through a mechanism of movement through pseudopods. *Li. monocytogenes* activates the host cell's actin molecules to stiffen and lengthen, which pushes the bacterium to the host cell's surface, forming a pseudopod (false foot or extension). A nearby macrophage or epithelial cell phagocytizes the pseudopod and *Listeria* bacteria tunnel out of the cell that engulfed it and continue to survive as an intracellular parasite in the new host cell. Once inside a host cell, this bacterium can enter the blood stream and can then be transported to a variety of body systems (most commonly the central nervous system (brain and spinal cord) and placenta). Figure 3-46 illustrates how *Listeria* bacteria avoid the host's immune system.

Direct zoonotic transmission of *Li. monocytogenes* between infected animals and humans is relatively uncommon.

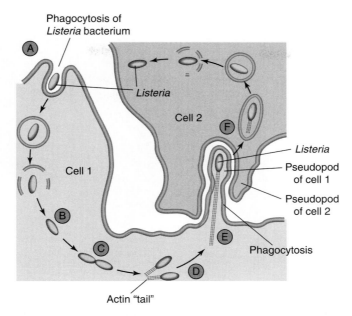

Figure 3-46   *Listeria* bacteria cause disease by avoiding the host's immune system. After phagocytosis (A), the intracellular bacterium escapes into the cytoplasm before it can be destroyed by the host cell's lysosome (B). The bacterium multiplies in the cytoplasm of the phagocyte (C). The *Listeria* bacterium then fuses with the host cell's actin molecules to form a tail of actin filaments (D) that push the bacterium to the cell's surface where if forms a pseudopod (E). A nearby macrophage or epithelial cell then phagocytizes the pseudopod (F) and the *Listeria* bacterium infects a new cell.

Clinical Signs in Animals

A wide variety of animal species, such as domestic and wild mammals, birds, fish, and crustaceans, carry *Li. monocytogenes* as commensals in the gastrointestinal tract. Clinical disease is mainly seen in ruminants, but may also be seen in rabbits, pigs, dogs, cats, poultry, and pet birds.

> Once inside host cells, *Listeria* bacteria can hide from immune responses and become inaccessible to certain antibiotics.

- *Ruminants*. In ruminants the incubation period for *Li. monocytogenes* is 10 to 21 days and can cause encephalitis, abortions, and septicemia in cattle, sheep, and goats. *Li. monocytogenes* can affect all ages of ruminants and may appear as an epizootic disease in feedlot cattle or sheep.
  - In cases of encephalitis, infected animals become solitary and anorexic followed by neurologic signs including leaning against stationary objects, circling in one direction, facial paralysis with profuse salivation, torticollis, strabismus, incoordination, and head pressing (Figure 3-47). Neurologic signs are usually unilateral. The disease course is short in small ruminants (1 to 4 days in sheep and goats with death in 1 to 2 days) and more chronic in cattle (4 to 14 days). Encephalitis is usually not seen in ruminants before the rumen becomes functional and is most commonly seen in 1- to 3-year-old animals. The mortality rate in the neurologic form of the disease is 70% in sheep and 50% in cattle.
  - The reproductive form of the disease produces placental infection, abortions and stillbirths that occur late in gestation in animals displaying no other clinical signs except fever and anorexia. Retained placentas and metritis may develop in some animals. Herd morbidity rates are typically about 30% with the abortion rate as high as 20% in sheep and cattle.
  - The septicemic form of the disease is usually seen in newborns and young ruminants with typical signs being fever, anorexia, and death.

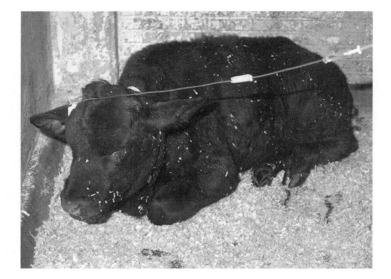

Figure 3-47 Signs of listeriosis in cattle include leaning against stationary objects, circling in one direction, facial paralysis with profuse salivation, incoordination, and head pressing.
(Courtesy of Noah's Arkive, Dr. Wayne Crowell)

- *Rabbits. Li. monocytogenes* in rabbits typically causes abortion during late pregnancy, encephalitis (rare), and sudden death. Animals will display non-specific clinical signs such as anorexia, being obtunded (severe depression or unresponsiveness), and weight loss.
- *Pigs. Li. monocytogenes* infection is rare in pigs, but may occur as the septicemic form in young piglets, with death occurring within 3 to 4 days.
- *Dogs and cats. Li. monocytogenes* is rare in dogs and cats. Clinical signs include being obtunded, anorexia, vomiting, and diarrhea. In dogs septicemia and neurologic signs may mimic rabies. In cats encephalitis and septicemia may occur.
- *Birds. Li. monocytogenes* infection is rare in birds, but may occur in young animals. Affected birds usually have the septicemic form with clinical signs of being obtunded, lethargy, diarrhea, and emaciation. Death can occur without clinical signs. In some birds (geese) encephalitis may also be seen. The incubation period for *Li. monocytogenes* in turkeys is 16 hours to 52 days.

Clinical Signs in Humans

Humans become infected with *Li. monocytogenes* following the consumption of contaminated food (contaminated milk, raw meat, cold cuts, vegetables, seafood, etc.) with symptoms of infection appearing anywhere from 3 to 70 days (the median incubation period is about 21 days). Newborns infected during the birthing process can develop symptoms a few days to a few weeks after birth. Most healthy people do not develop noticeable disease symptoms when infected with *Li. monocytogenes* but may develop gastrointestinal symptoms such as fever, headache, nausea and vomiting, lethargy, and diarrhea. The incubation period for gastroenteritis caused by *Li. monocytogenes* is 1 to 2 days.

Listeriosis is a serious problem in pregnant, newborn, elderly, and immunocompromised people. Pregnant women experience mild, flu-like symptoms such as fever, muscle aches, upset stomach and intestinal problems or may be asymptomatic. These women recover from the disease, but the infection can cause miscarriage, premature labor, septicemia in the newborn, and stillbirth. Abortions typically occur during the third trimester.

Newborns can contract listeriosis in utero (transplacental) or during vaginal delivery from bacteria in the vagina (vertical transmission). There are two types of disease processes in newborns: early-onset disease (a serious septicemia present at birth that usually causes the baby to be born prematurely) and late-onset disease (a disease of full-term babies that become infected during childbirth resulting in meningitis about 2 weeks after birth). Babies infected during the pregnancy that become septic may develop systemic infection called granulomatosis infantisepticum. Babies with late-onset disease have a better chance of surviving the infection. Approximately half of the newborns infected with *Li. monocytogenes* die from the illness.

Immunocompromised adults are at risk of developing meningitis or less frequently septicemia from *Listeria* infections. Symptoms of listerial meningitis occur about four days after the flu-like symptoms and include fever, personality change, incoordination, tremors, muscle contractions, seizures, and loss of consciousness. Other conditions that have been reported in people are endocarditis, brain abscesses, ocular infections, hepatitis, peritonitis, arthritis, and rarely pneumonia.

Diagnosis in Animals

Listeria bacteria cause minimal if any lesions in animals with the encephalitic form of listeriosis. The CSF may be turbid and congestion of meningeal vessels and softening of the medulla oblongata may be seen at necropsy. The septicemic form of listeriosis may show necrotic foci in internal organs such as the liver. In the placental form aborted fetuses may be autolyzed with focal necrosis of the liver, lung, or spleen. The placental cotyledons may also be necrotic. In birds, myocardial necrosis and pericarditis may be seen with petechial hemorrhage in the proventriculus (gizzard) and heart.

Listeriosis can be diagnosed by culture of the bacterium from affected organs (placenta, fetus, uterine discharge, CSF, or blood depending upon the form of disease present). Necropsy samples may include the liver, kidney, spleen, or brain tissue (pons or medulla). Milk and feces can also be cultured; however, *Li. monocytogenes* can be found in clinically normal animals. Samples are typically plated on blood agar and incubated at normal conditions and with a cold enrichment technique (incubating in cooler temperatures prior to normal incubation conditions). Cold enrichment techniques may require up to 3 months for growth. *Li. monocytogenes* grow as small, white, smooth, transparent, beta-hemolytic colonies on blood agar. Biochemical tests such as the CAMP test and hippurate test are also used to identify *Li. monocytogenes*. Commercial rapid identification techniques such as ELISA, immunofluorescence, immunochromatography, and PCR are available. Serology is not routinely used for diagnosis as a result of cross-reactions with enterococci and *Staphylococcus aureus*. Antibodies against listeriolysin O can be used, but the methods have not been standardized.

Diagnosis in Humans

In human newborns, lesions associated with early-onset listeriosis are typical of sepsis (microabscesses or granulomas) or meningitis. Late-onset listeriosis frequently presents with purulent meningitis. A rash with small, pale nodules is histologically characteristic of granulomatosis infantisepticum. Other people with *Listeria* infections usually have an underlying immunodeficiency or are immunocompromised and lesions vary depending on the degree of immune suppression and the organs affected.

The only way to diagnose listeriosis in people is to isolate *Li. monocytogenes* from blood, CSF, or reproductive tissue (stool samples are not valid because up to 10% of the human population can be carriers of *Li. monocytogenes*). Identification methods are the same for humans as described for animals; identification of *Li. monocytogenes* in the food source strengthens the diagnosis.

Treatment in Animals

Listeriosis can be treated using a variety of antibiotics (tetracyclines, sulfonamides, and penicillin), but must be given at high doses and early in the disease. Supportive therapy may also be needed. Animals with neurologic signs usually die despite treatment.

Treatment in Humans

Listeriosis in people is treated with antibiotics (ampicillin or trimethoprim-sulfamethoxazole). Because the bacteria live intracellularly, treatment may be difficult and the treatment periods prolonged. The cure rate can be low as a result of the infection occurring in the young, elderly, and immunocompromised. Human patients are often hospitalized for treatment and monitoring. Other drugs may be given to the patient to relieve pain and fever.

Management and Control in Animals

The risk of ruminants developing listeriosis can be reduced by feeding good quality, low pH silage. Corn silage ensiled before it is too mature is likely to have a low pH, which discourages the replication of *Li. monocytogenes*. Spoiled or moldy silage should be discarded. Rodent control can also decrease the spread of *Li. monocytogenes*. Quarantine of new animals and clinically affected animals is important in containing and preventing infection as are removal of the placenta and fetus following abortion. There is currently no vaccine commercially available for listeriosis.

Management and Control in Humans

Prevention of human listeriosis revolves around food safety. High-risk people (young, elderly, and immunocompromised) should thoroughly cook all food from animal sources, wash raw vegetables, and avoid drinking or eating unpasteurized dairy products. Avoiding cross-contamination during food preparation by frequent hand washing and washing of food-preparation equipment is also important. Immunocompromised people should avoid eating hot dogs, luncheon meats, or deli meats, unless they are reheated until steaming hot. *Li. monocytogenes* can be shed up to 10% to 30% of subclinically infected people, so proper hygiene is important in prevent disease transmission.

In 2006, the FDA approved bacteriophage treatment of processed meat in an effort to control *Li. monocytogenes* contamination of these products.

In the 1980s, the U.S. government began testing processed meats and dairy products for *Li. monocytogenes*. The FDA and the Food Safety and Inspection Service (FSIS) can legally prevent food shipment or order food recalls if any *Listeria* bacteria are detected in food. In 1996, the CDC began a nationwide foodborne disease surveillance program called FoodNet, which determined that the hospitalization rate for listeriosis (94%) was higher than for any other foodborne illness and that *Listeria* bacteria reached the blood and CSF in 89% of cases, a higher percentage than in any other foodborne illness. The National Center for Infectious Diseases (NCID) studies listeriosis in several states to help measure the impact of prevention activities, recognize trends in disease occurrence, and assist local health departments in investigating outbreaks.

Summary

Listeriosis is a disease caused by infection with *Li. monocytogenes* and *Li. ivanovii* with the latter being a rare cause of zoonotic disease. *Li. monocytogenes* organisms are gram-positive, nonendospore-forming, motile, pleomorphic bacilli that are sometimes arranged in short chains. *Li. monocytogenes* are considered foodborne pathogens but can be passed from mother to fetus during pregnancy and farmers/veterinarians/butchers can develop *Listeria* skin infections by touching infected calves or poultry. Virulence and resistance factors associated with *Li. monocytogenes* include the ability to grow at low temperatures and resistance to drying. *Li. monocytogenes* is found worldwide and is widely distributed in the environment. The main reservoirs of *Li. monocytogenes* are soil and the intestinal tract of animals. Clinical disease is mainly seen in ruminants, but may also be seen in rabbits, pigs, dogs, cats, poultry, and pet birds. Clinical signs in animals range from encephalitis, abortions, and septicemia in cattle, sheep, and goats; abortion during late pregnancy, encephalitis (rare), and sudden death in rabbits; septicemia in young piglets; anorexia, vomiting, and diarrhea in dogs and cats; and lethargy, diarrhea, and emaciation in birds. Most healthy people do not develop noticeable disease symptoms when infected with *Li. monocytogenes* but may develop gastrointestinal symptoms such as fever, headache, nausea and vomiting, lethargy, and diarrhea. Listeriosis is a serious problem in pregnant, newborn, elderly, and immunocompromised people. Pregnant women experience mild, flu-like symptoms such as fever, muscle aches, upset stomach, and intestinal problems or may be asymptomatic. These women recover from the disease, but the infection can cause miscarriage, premature labor, septicemia in the newborn, and stillbirth. Babies infected during the pregnancy develop sepsis may develop systemic infection called granulomatosis infantisepticum; babies with late-onset disease have a better chance of surviving the infection. In immunocompromised adults meningitis or less frequently septicemia are the main forms of disease.

Listeriosis can be diagnosed by culture of the bacteria from affected organs. Biochemical tests and commercial rapid identification techniques (ELISA, immunofluorescence, immunochromatography, and PCR) are available. Treatment of listerosis involves the use of antibiotics that need to be administered early in the disease and at high doses. Prevention of listeriosis revolves around feeding/consumption of uncontaminated food products.

MELIOIDOSIS

Overview

Melioidosis, also called Whitmore's disease and pseudoglanders, is a collective term for infection caused by the soil bacterium *Burkholderia pseudomallei* (formerly known as *Pseudomonas pseudomallei* and *Malleomyces pseudomallei*). Whitmore first described and isolated the causative organism of melioidosis in 1912 from a drug addict in Rangoon, Burma. The causative agent of melioidosis was classified for many years within the *Pseudomonas* genus; however, in 1992, it was reclassified along with *P. mallei* and four other species, to a new genus named after the U.S. microbiologist Walter Burkholder. Melioidosis is clinically similar to glanders disease, but the ecology and epidemiology of melioidosis are different from glanders. Melioidosis is predominately found in contaminated water and soil in tropical regions; especially in Southeast Asia (it is endemic in this region). Melioidosis also affects birds and mammals such as sheep, goats, horses, pigs, and cattle. Both humans and animals acquire the disease from the soil and surface water; zoonotic transmission to humans from contact with infected animal fluid is extremely rare.

Causative Agent

In many countries, *Bu. pseudomallei* bacteria are so prevalent that it is a common contaminant found when culturing bacteria in laboratory settings.

Bu. pseudomallei bacteria are aerobic, straight, slender, gram-negative, bipolar-staining, motile rods and are isolated from wet soils, agricultural soils, streams, pools, stagnant water, and rice paddy fields. *Burkholderia* bacteria grow well on routine culture media such as blood, chocolate, and MacConkey agar. *Bu. pseudomallei* have been shown to form an extracellular polysaccharide capsule in response to low pH.

Epizootiology and Public Health Significance

Melioidosis is endemic in Southeast Asia and Oceania (the region between the Tropic of Cancer and the Tropic of Capricorn) occurring mainly in Vietnam, Laos, Cambodia, Burma, Thailand, Malaysia, and northern Australia. It has a seasonal peak during the rainy season. Within the tropics, there are two areas where melioidosis is in extremely high numbers: the Northern Territory in Australia and some northeastern provinces of Thailand. Almost all cases of melioidosis have been diagnosed in temperate climates, with the exception of an outbreak in France in the mid-1970s. The average annual incidence of melioidosis in the Northern Territory of Australia between 1989 and 1998 was 16.5 per 100,000 people. Melioidosis has become an important cause of morbidity and mortality in foreign troops fighting in Southeast Asia. Melioidosis is rare in the United States and tends to be found more frequently in IV drug users.

Transmission

Transmission occurs by direct contact with contaminated soil and water. Humans and animals acquire the infection by inhalation of dust, ingestion of contaminated water, consumption of food (meat, milk, cheese) from infected animals, and contact with contaminated soil especially through skin abrasions (cuts and burns), and contamination of war wounds. Person-to-person transmission has occurred, but it is rare. Animals are rarely the direct source of infection.

Pathogenesis

Bu. pseudomallei is a facultative, intracellular pathogen that can survive within phagocytic cells and macrophages. The presence of a capsule in some forms of *Bu. pseudomallei* may permit prolonged survival within phagocytes. Once in the host, *Bu. pseudomallei* produce toxins such as pyocyanin (blue-green pigment that interferes with energy production), lecithinase (causes cell lysis), collagenase, lipase, and hemolysin. The exact roles these toxins play are unknown.

Acute infection typically exhibits localized necrosis and often presents as a painful nodule at the site of inoculation. Regional lymphadenitis is also seen with some lymph nodes producing yellow odorless pus. As bacteria enter the blood, the localized forms may progress to hematogenous melioidosis involving many organs (most frequently the lungs, liver, and spleen). Acute pulmonary suppurative disease may follow inhalation of the bacterium, but is more frequently from hematogenous spread.

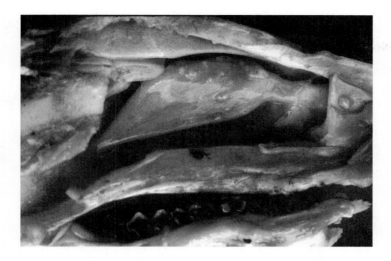

Figure 3-48 Melioidosis produces multiple raised pale nodules (abscesses) on the nasal mucosa on this goat's nasal turbinates.
(Courtesy of National Institute of Animal Health, Japan)

Clinical Signs in Animals

Many animal species are susceptible to melioidosis including sheep, goats, horses, swine, cattle, dogs, cats, birds, tropical fish, and a variety of wild animals. Laboratory animals, such as Guinea pigs, hamsters, and rabbits, are highly susceptible. Clinical signs vary with the site of the lesion and may mimic many other diseases. In sheep and goats, lung abscess and pneumonia-like signs are common (Figure 3-48). In horses, nervous disorders, respiratory signs, and gastrointestinal signs (colic and diarrhea) are common. In pigs, splenic abscesses are common.

Clinical Signs in Humans

The incubation period for melioidosis is not clearly defined and can range from 2 to 3 days to many years because the organism can remain quiescent for long periods of time. Melioidosis in people is classified as acute, subacute, and chronic. Because most people exposed to *Bu. pseudomallei* do not become ill, there is also a recognized subclinical infection classification.

- Acute infection occurs in the monsoonal wet seasons of the various endemic regions and presents with symptoms that have been present for less than 2 months. Acute melioidosis occurs in three forms: localized skin infection that may spread to regional lymph nodes; lung infection with high fever, headache, chest pain, and coughing; and a septicemic form that is characterized by disorientation, dyspnea, severe headache, and skin eruptions on the head or trunk. The septicemic form may be rapidly fatal.
- Subacute infection is characterized by suppurative lesions most frequently found in the lungs as abscesses. The liver may also be involved and will demonstrate solitary or multiple abscesses.
- Chronic infection involves a variety of organs (typically joints, viscera, lymph nodes, skin, brain, liver, lung, bones, and spleen). Chronic melioidosis is often characterized by osteomyelitis and pus-filled abscesses in the skin, lungs, and other organs.

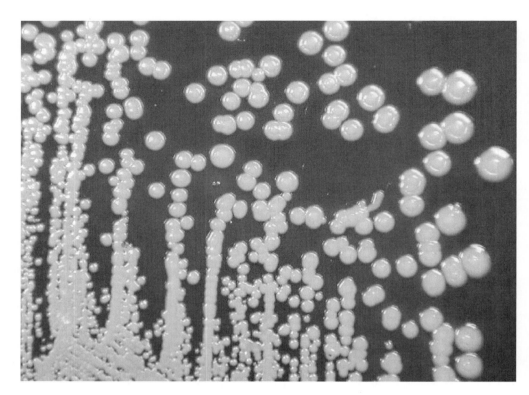

Figure 3-49   Bacterial culture is one way to diagnose melioidosis. *Burkholderia pseudomallei* is a gram-negative aerobic bacterium and is shown here growing on sheep blood agar.

(Courtesy of CDC/Larry Stauffer, Oregon State Public Health)

Diagnosis in Animals

Melioidosis can be identified via routine staining, culture, and agglutination tests using specific antisera. *Bu. pseudomallei* is a gram-negative rod with bipolar staining. This bacterium grows well on blood (smooth and mucoid to dry and wrinkled colonies), chocolate, and MacConkey agar (colorless colonies because it does not ferment lactose) (Figure 3-49). Biochemical tests can also help in the identification of *Bu. pseudomallei*. Complement fixation and indirect hemagglutination assays on serum are also available.

Diagnosis in Humans

Melioidosis should be suspected based on the patient's history, especially travel, occupational exposure to infected animals, or a history of intravenous drug use. Diagnosis is confirmed by staining, culture, biochemical tests, and serology. *Bu. pseudomallei* can be cultured from the person's sputum, blood, or tissue fluid from abscesses. When stained with methylene blue, *Bu. pseudomallei* displays a characteristic bipolar or "safety pin" configuration. Isolation is achieved by using routine culture media such as blood, chocolate, and MacConkey agars. Selective media, such as modified Ashdown's broth, are usually required for culturing respiratory tract specimens to ensure reliable isolation. *Bu. pseudomallei* may require 48 to 72 hours of incubation and may be easily overgrown in mixed cultures on nonselective media. The colonies are typically wrinkled, purplish, and emit a musty odor. Biochemical tests include oxidase (positive) and production of gas from nitrate. Blood tests, including complement fixation (CF) tests and hemagglutination tests, help confirm the diagnosis. Serologic testing in endemic areas is limited by the high numbers of latent seropositivity rates. Several PCR tests have also been developed, but are not used clinically at this point.

Treatment in Animals

Some antibiotics (such as tetracycline, sulfonamides, and kanamycin) are effective against *Bu. pseudomallei*; however, once treatment is discontinued many animals relapse.

Treatment in Humans

Bu. pseudomallei is resistant to many antibiotics; however, it is sensitive to ceftazidime, penicillin, and amoxicillin-clavulanate. High doses of antibiotics for 6 to 12 months are recommended. Surgery to remove abscesses may be needed in some cases. Approximately 5% of all cases recur following antibiotic therapy.

Management and Control in Animals

There is no animal vaccine for melioidosis. Since animals are rarely the source of direct infection, controlling their role in disease transmission is limited.

Management and Control in Humans

There is no human vaccine for melioidosis. Preventing infection in endemic areas is difficult because contact with contaminated soil is so common. People with immune compromising diseases (such as diabetes mellitus) and skin lesions should avoid contact with soil and standing water. Wearing boots while doing agricultural work can prevent infection through the feet and distal legs. In health care settings, using universal precautions can prevent transmission. Prompt cleansing of scrapes, burns, or other open wounds in people who live in endemic areas is important. Avoiding needle sharing among drug addicts can also prevent the spread of melioidosis.

Summary

Melioidosis, caused by the gram-negative rod-shaped bacterium *Bu. pseudomallei*, is an important disease in Southeast Asia and northern Australia. *Bu. pseudomallei* is found in the environment predominantly in wet soils. The disease can be found in a variety of animals including sheep, goats, horses, swine, cattle, dogs, cats, birds, tropical fish, and a variety of wild animals. Laboratory animals, such as Guinea pigs, hamsters, and rabbits, are highly susceptible. Clinical signs in animals vary with the site of the lesion and may mimic many other diseases. In people it typically infects adults with an underlying predisposing condition, mainly diabetes mellitus. Melioidosis in people presents as an acute, subacute, or chronic form. Acute melioidosis occurs in three forms: localized skin infection that may spread to regional lymph nodes; lung infection with high fever, headache, chest pain, and coughing; and a septicemic form that is characterized by disorientation, dyspnea, severe headache, and skin eruptions on the head or trunk. Subacute infection is characterized by suppurative lesions most frequently found in the lungs as abscesses. Chronic infection involves a variety of organs (typically joints, viscera, lymph nodes, skin, brain, liver, lung, bones, and spleen). Melioidosis can be identified via routine staining, culture, biochemical tests, and serology. Melioidosis should be suspected based on the patient's history, especially travel, occupational exposure to infected animals, or a history of intravenous drug use. Treatment in animals and people includes the use of antibiotics; however, once antibiotics are discontinued there is a risk of relapse. There are currently no vaccines for melioidosis. Preventing exposure is difficult because the bacterium is commonly found in endemic environments.

ORNITHOSIS

Overview

Ornithosis, also known as parrot fever, chlamydiosis, and psittacosis, is a bacterial disease of people caused by *Chlamydophila psittaci* (formerly known as *Chlamydia psittaci*). The term psittacosis comes from the Greek *psittakos*, which means parrot and the term ornithosis comes from the Greek word *ornis*, which means bird. Outbreaks of "ornithosis" have occurred in birds other than psittacines (parrots and parrot-like birds such as budgerigars and cockatiels); therefore, the term ornithosis is more accurate than psittacosis. In birds, *Ch. psittaci* infections are referred to as avian chlamydiosis (AC). Ornithosis is a worldwide zoonosis that is carried latently in wild and domesticated birds becoming active under stressful conditions such as overcrowding. Documented cases of ornithosis have been known for more than one hundred years (it was identified for the first time in 1879), but the clinical aspects of the disease were not known during the pandemic of 1929 to 1930 when it was believed that only psittacines could transmit the disease. This pandemic occurred as an atypical and often severe pneumonia in humans and because of its believed origin was termed psittacosis. This pandemic resulted from the shipment of large numbers of amazon parrots from Argentina to a variety of locations around the world. As a result of this pandemic, the United States and other countries banned the importation of birds (this ban was partially lifted in the United States in 1967 and completely removed in 1973). These outbreaks stimulated research of this organism, which included identification of minute basophilic particles in blood and tissue from infected birds. The relationship of these particles with ornithosis was concluded by Bedson, who also described these organisms as obligate intracellular parasites with bacterial affinity (a concept that was not accepted for another 30 years). This bacterium was known at one time as *Bedsonia*; the term *Chlamydia* (*chlamus* is Greek for cloak) did not appear in the literature until 1945. In 1965 it became evident that chlamydiae are not viruses and for many years was the only bacterial order that had just one family and one genus. In the 1990s new diagnostic techniques have resulted in identification of over 40 chlamydial strains and the splitting of the Chlamydiaceae family into two genera, *Chlamydia* and *Chlamydophila*. The name *Chlamydophila* was given to the group of bacteria that were "like chlamydia" by Hans Truper and Johannes Storz after the order Chlamydiales was established in 1999.

Since the 1929-1930 pandemic *Ch. psittaci* has been found in over 100 species of birds including pet psittacine birds, pigeons, doves, poultry, birds of prey, and shore birds. Ornithosis in people typically causes influenza-like symptoms that can lead to severe pneumonia. Prior to the development of tetracycline antibiotics, ornithosis was a serious human disease with an estimated mortality of 20%. Clinical signs of AC are typically inapparent and shedding of the organism is common. The disease is widespread in birds.

Causative Agent

Ornithosis is caused by a group of closely related, gram-negative, aerobic, nonmotile, coccoid bacteria that are obligate intracellular bacteria. Chlamydial cells are unable to carry out energy metabolism and are dependent on eukaryotic host cells

Chlamydial infections are endemic in avian and mammalian populations.

Table 3-7   Divisions in the Family Chlamydiaceae

Genus *Chlamydia*

Species	Host	Properties
Chlamydia trachomatis	Humans	■ Most common sexually transmitted bacterial pathogen in United States ■ Causes ocular and venereal infections in people
Chlamydia suis	Swine	■ Pneumonia, conjunctivitis, and polyarthritis in pigs ■ Most likely is endemic in pigs
Chlamydia muridarum	Mouse, hamsters	■ Genital infection in murine species ■ Closely related to *Chlamydia trachomatis*

Genus *Chlamydophila*

Chlamydophila psittaci	Humans, birds (six avian serovars and two mammalian serovars)	■ Flu-like disease in people ■ Respiratory disease, diarrhea, and polyuria in birds
Chlamydophila pneumoniae	Hamsters, horses, humans, other mammals	■ Pneumonia, bronchitis, and sinusitis in animals ■ Pneumonia and possibly atherosclerosis in people
Chlamydophila pecorum	Mammals	■ Encephalomyelitis and endometritis in cattle ■ Pneumonia and conjunctivitis in sheep ■ Polyarthritis in ruminants ■ Urogenital disease in koalas
Chlamydophila felis	Cats	■ Conjunctivitis in cats ■ Atypical pneumonia in humans
Chlamydophila caviae	Guinea pigs	■ Conjunctivitis in Guinea pigs ■ Highly specific for Guinea pigs
Chlamydophila abortus	Ruminants	■ Abortion and stillbirths as a result of colonization of the placenta ■ Probably endemic in ruminants ■ Reports of abortion and respiratory disease in humans

to supply them with ATP (cellular energy). Originally, they were categorized into their own order (Chlamydiales) with one family (Chlamydiaceae) with one genus (*Chlamydia*) with four species (*C. trachomatis*, *C. psittaci*, *C. pneumoniae*, and *C. pecorum*). In 1999, Everett recommended that the genus *Chlamydia* be divided into two genera (*Chlamydia* and *Chlamydophila*) containing nine species. The genera of *Chlamydia* and *Chlamydophilla* are summarized in Table 3-7.

The life cycle of *Chlamydophila* has a biphasic cycle with two distinct forms (intracellular and extracellular). The large and fragile intracellular form, also known as the initial body or reticulate body, is the reproductive stage. The metabolically inert extracellular form, also known as the elementary body, is the infectious, nonreplicating form. Infectious elementary bodies start the cycle by

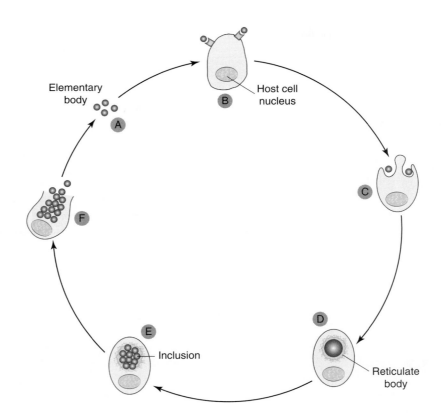

Figure 3-50 The life cycle of *Chlamydiae*. Elementary bodies (infectious stage) (A) attach to specific host cell receptors (B). Elementary bodies are endocytosed and reside in a membrane-bound vesicle of the phagocyte (C). The elementary bodies reorganize into a large reticulate body (reproductive stage) within a vacuole (D). Reticulate bodies multiply by binary fission within a structure called an inclusion (E). Reticulate bodies transform back into elementary bodies that are released from the cell via exocytosis or host cell lysis (F).

attaching to host cell membranes and gain access into the host cell via endocytosis. Once inside the cell, the bacterium remains within an enlarging intracellular vacuole in order to avoid damage by lysosomes. Elementary bodies will differentiate into metabolically active reticulate bodies during the first few hours. Reticulate bodies will multiply using the host cell's energy and nutrients. After multiple sets of divisions, reticulate bodies will transform back to elementary bodies. The infectious elementary bodies are released into the cytoplasm by exocytosis or host cell lysis and can initiate new cycles in new host cells. This cycle typically takes 48 to 72 hours. When an infected animal defecates, sneezes, or coughs it releases these infective elementary bodies into the environment. The intracellular and extracellular phases of *Chlamydophila* are illustrated in Figure 3-50.

All strains of Chlamydiaceae share an identical genus-specific antigen in the lipopolysaccharide of their cell wall but differ in the composition of other cell-wall antigens, which accounts for the different serotypes. There are three bacteria mainly responsible for human disease (*C. trachomatis*, *C. pneumoniae*, and *Ch. psittaci*), which differ in respect to their antigens, host cell preference, antibiotic susceptibility, and morphology. *Ch. psittaci* is the most important zoonotic bacterium in this group and will be the focus of this section.

Epizootiology and Public Health Significance

Avian chlamydiosis is a subclinical, acute, subacute, or chronic disease of wild and domestic birds worldwide and is characterized by respiratory, digestive, or systemic infection. Inapparent infections are common and a high percentage of avian populations are positive. Birds are the natural hosts of *Ch. psittaci* and this bacterium has been detected in over 130 bird species (57 psittacine species). AC is mainly a disease of nestlings and young birds, particularly in colonial nesting

The elementary body is the infectious form of the bacterium, which grows into the reticulate body (the vegetative phase).

species, domestic poultry (turkeys, pigeons, and ducks), caged birds (primarily psittacines), and raptors. The incidence of disease is high following capture as a result of the stresses of poor nutrition, overcrowding, and physiologic stress. This disease is most often diagnosed in birds owned for less than six months. Different strains of *Ch. psittaci* are also found in other animal species such as Guinea pigs, mice, sheep, goats, cattle, and horses. AC occurs worldwide as a result of its presence in the worldwide bird population. From 1988 through 2003, the CDC received 935 reported cases of avian chlamydiosis, which is probably an under-representation of the disease.

Since 1996, fewer than 50 confirmed human cases of ornithosis have been reported annually in the United States. According to the CDC, about 70% of infected people had contact with infected pet birds. People at greatest risk of contracting ornithosis include bird owners, pet shop employees, veterinarians, and people with compromised immune systems (no person-to-person cases have ever been reported).

Transmission

Ch. psittaci is excreted in the feces and respiratory secretions of infected birds where it can remain infectious in the environment for months (elementary bodies are resistant to drying and are the form found in the environment). Apparently healthy birds can shed the organism intermittently and shedding can be activated by stress. Spread of *Ch. psittaci* occurs via inhalation or ingestion of air- or dust-borne organisms. Humans become infected when exposed to infected birds (when processed at dressing plants, working with birds, or handling a single pet bird). Infection in people usually occurs after inhalation of bacteria that have been aerosolized from dried feces or respiratory secretions of infected birds. Other infection routes for humans include mouth-to-beak contact and handling infected birds' plumage and tissues. Strains of *Ch. psittaci* found in other species causing infection of the reproductive tract can be spread to humans and animals via exposure to reproductive fluids and placentas of infected animals.

Ch. psittaci can survive in the environment for several months.

Pathogenesis

After *Ch. psittaci* are inhaled, they are deposited in the alveoli. Some bacteria are engulfed by alveolar macrophages and carried to regional lymph nodes. From the lymph nodes they are disseminated through the body and grow within cells of the reticuloendothelial system (macrophages of the liver, spleen, and bone marrow; tissue macrophages, and circulating monocytes). *Ch. psittaci* initiate infection when elementary bodies attach to the microvilli on the host's mucosal epithelial cells and are engulfed by endocytosis. Elementary bodies within endosomes of the host cell's cytoplasm differentiate into metabolically active, noninfectious reticulate bodies that replicate and form many infectious, metabolically inactive elementary bodies. Newly formed elementary bodies are released from the host cell via cell lysis or exocytosis and the cycle begins again.

Clinical Signs in Animals

Ch. psittaci has been isolated from a variety of animals and the clinical presentation varies with the animal species. Affected species include:

- *Birds.* The duration between exposure to *Ch. psittaci* and onset of disease ranges from 3 days to several weeks. Active disease can appear without identifiable

Birds with avian chlamydiosis either present with acute disease (upper respiratory signs, anorexia, lethargy, and green feces), chronic disease (sick, unthrifty bird with poor feather coat), or as an asymptomatic chronic carrier (appear normal with no signs of disease).

exposure making incubation periods difficult to assess. Birds may present with acute disease with such signs as weight loss, air sacculitis, transient anorexia, yellow to green urates, nasal and ocular discharges, conjunctivitis, sinusitis, inactivity, ruffled feathers, weakness, and loose feces. Many birds are emaciated when presented for examination. Some birds may simply appear unthrifty (chronic disease) or may not show any signs (chronic carrier).

- *Livestock.* Ch. psittaci has been isolated from fecal samples of clinically normal cattle, goats, sheep, and pigs making the gastrointestinal tract an important reservoir and source for disease transmission. Some strains may cause abortions, whereas others cause pneumonia, polyarthritis, encephalomyelitis, or conjunctivitis. Enteritis caused by *Ch. psittaci* has been documented in newborn calves.

- *Cats.* Feline pneumonitis caused by *Ch. psittaci* is a less frequent cause of respiratory disease in cats. *Ch. psittaci* infections characteristically produce low-grade conjunctivitis and infected cats sneeze and have a fever occasionally. Convalescent cats may go through several disease relapses.

Clinical Signs in Humans

Ornithosis presents itself in humans in a variety of ways depending on host and microbial factors, route of transmission, and intensity of exposure. Onset of clinical signs typically follows a 5- to 14-day incubation period. Clinical signs vary from inapparent illness to systemic illness with severe pneumonia. Symptomatic infection usually has an abrupt onset of fever, chills, headache, lethargy, and muscle pain. A nonproductive cough may be accompanied by dyspnea. Splenomegaly and a nonspecific rash may also be seen. If the bacterium affects organ systems other than the respiratory system, it tends to produce endocarditis, myocarditis, hepatitis, arthritis, conjunctivitis, and encephalitis.

Diagnosis in Animals

Birds with AC may have hepatomegaly, splenomegaly, air-sac changes (opacification, thickening, caseation), and fibrous pericarditis on necropsy. The liver may have necrotic areas with inflammatory cell infiltrates (Figure 3-51). The air sacs may be thickened by fibrinous exudates and the lungs may be congested and edematous. Enlargement and discoloration of the spleen or liver are seen with chronic

> Inapparent chlamydial infection is the most frequently seen form in mammals and birds.

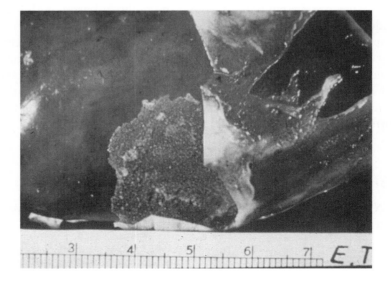

Figure 3-51 Sheets of fibrinous exudate partially cover the capsular surface of the liver in this bird with AC.

(Courtesy of Armed Forces Institute of Technology (AFIP), Education Branch, Division of Research and Education, Dept. of Pathology, Washington, DC)

infections. Lesions are usually absent in latently infected birds, even though the bacterium is often being shed.

Due to the variety of clinical presentations and presence of latently infected carriers, there is not a single diagnostic test that can reliably determine infection with *Ch. psittaci*. History and clinical signs are very important factors in diagnosing AC. In acute disease, there are greater numbers of infective organisms and the diagnosis is easier. The organism can often be identified in impression smears of affected tissues stained by Giemsa or Gimenez stains. Confirmation requires cell culture isolation and identification of *Ch. psittaci* via submission of cloacal, choanal, and conjunctival swabs from live birds, and tissues (liver, spleen, and serosal membranes) from dead birds. Only laboratories with Biosafety level 3 biohazard containment facilities can culture this bacterium.

Serology is almost always used to confirm *Ch. psittaci*. Antibodies may or may not be detectable depending on the stage of infection, whether the bird has been treated, and the test used. Interpretation of single serum titers is difficult making paired or sequential samples or samples from several birds in a population beneficial. Multiple samples collected for 3 to 5 days are recommended for detection of intermittent shedding. Available antigen-antibody test methods include complement fixation, latex agglutination, ELISA, and immunofluorescence (Figure 3-52). PCK tests are also available. Microimmunofluorescence (MIF) using chick yolk sacs is a more recent test developed for identification of *Ch. psittaci*.

Live birds being screened for *Ch. psittaci* might not shed bacteria daily; therefore, serial species should be collected for 3 to 5 consecutive days and pooled before being cultured (to reduce costs).

Diagnosis in Humans

Laboratory diagnosis of *Ch. psittaci* in humans involves the use of MIF and PCR. Cultivation methods are restricted to state public health laboratories.

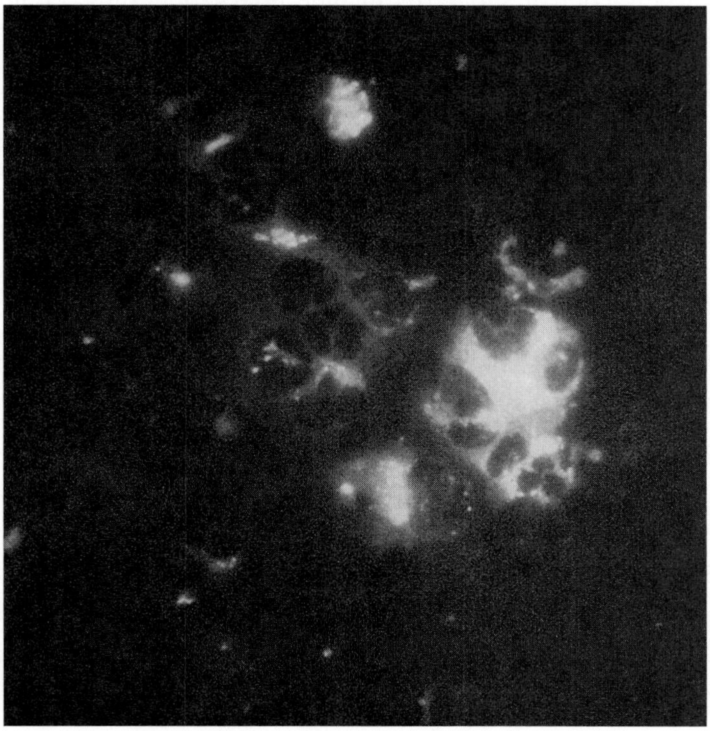

> Freezing, improper handling, and certain transport media can affect the viability of *Ch. psittaci*; refrigeration, placing specimens in sealed plastic bags or other containers, and prompt delivery are preferred methods of shipment.

Figure 3-52 Fluorescent antibodies are used to diagnose *Chlamydia psittaci* infections. (Courtesy of CDC/Dr. Vester Lewis)

Treatment in Animals

Tetracyclines are the antibiotics of choice for treating *Ch. psittaci*. This group of antibiotics is effective against actively multiplying *Ch. psittaci* organisms and in acutely affected birds it rapidly controls shedding. Birds used to be treated by using chlortetracycline (CTC) in the feed of all imported psittacines during the 30-day quarantine (formerly required by law in the United States). Currently approved therapy for bird flocks includes CTC premix in food for 45 days (large psittacines), CTC-containing pelleted feeds, or hulled millet impregnated with CTC. Doxycycline is effective but also requires 45 days of treatment. Prolonged treatments using high levels of antibiotic may not completely eliminate latent infection and shedding may recur. Tetracycline antibiotics are the treatment of choice for livestock infected with *Ch. psittaci* and cats with respiratory disease caused by *Ch. psittaci*.

Treatment in Humans

Tetracycline antibiotics are the treatment of choice for *Ch. psittaci* in people. Relapses are common and treatment should continue for at least 10 to 14 days after the fever is controlled. Macrolide antibiotics such as erythromycin may be used in children and pregnant women.

Management and Control in Animals

There is not a vaccine approved in the United States for *Ch. psittaci* in birds or livestock. In cats some of the upper respiratory vaccine combinations contain either chick-embryo- or cell-line-origin *Ch. psittaci*. A single parenteral dose is recommended for cats older than 12 weeks of age, whereas younger kittens that were vaccinated younger than 12 weeks of age should be revaccinated when they reach 16 weeks. All cats should be revaccinated annually.

Controlling the introduction and spread of *Ch. psittaci* in an avian population is critical. Specific control standards include quarantine and examination of all new birds (these records should be kept for at least one year to help identify sources of infected birds and potentially exposed people); isolation of newly acquired, ill, or exposed birds to minimize cross-contamination; isolation and treatment of affected and contact birds (including stress reduction and maintaining proper nutrition); testing birds before they are boarded or sold; thorough cleaning and disinfection of premises; and practicing preventative husbandry (such as positioning cages to prevent the transfer of fecal material, feathers, and food from one cage to another). *Ch. psittaci* is susceptible to most disinfectants such as 1:10 bleach solution and 70% alcohol. Applying a detergent antiseptic to wet the feathers of dead birds can also reduce exposure to this bacterium.

Ch. psittaci is relatively rare in poultry. Quarantine practices for pet birds do not mandate treatment of *Ch. psittaci*; however, treatment is common. AC is not considered an exotic disease because it is found in wild and captive bird populations in North America. AC is a reportable disease in the United States and specific quarantine and treatment regimens must be carried out for active cases.

Management and Control in Humans

Ornithosis is an important occupational hazard in the poultry and pet bird industries. Avoiding exposure to affected birds (wearing face masks and protective clothing, proper sanitation and hygiene, etc.), treatment of infected birds, and/or quarantine of imported birds can prevent this disease.

Summary

Ornithosis, also known as parrot fever, chlamydiosis, and psittacosis, is a bacterial disease mainly of birds and people caused by *Ch. psittaci* (the disease in birds is called avian chlamydiosis) *Ch. psittaci* are gram-negative, aerobic, nonmotile, coccoid bacteria that are obligate intracellular bacteria. The life cycle of *Chlamydophila* has a biphasic cycle with two distinct forms (the reproductive, intracellular stage and the infectious, nonreproductive, extracellular stage). *Ch. psittaci* is excreted in the feces and respiratory secretions of infected birds where it can remain infectious in the environment for months. Apparently healthy birds can shed the organism intermittently and shedding can be activated by stress. Spread of *Ch. psittaci* occurs via inhalation or ingestion of air- or dustborne organisms. After *Ch. psittaci* are inhaled, they are deposited in the alveoli and are disseminated from the lymph nodes through the body and grow within cells of the reticuloendothelial system.

Ch. psittaci has been isolated from a variety of animals and the clinical presentation varies with the animal species. In birds the acute disease may present with signs such as weight loss, air sacculitis, yellow to green urates, conjunctivitis, sinusitis, and ruffled feathers. Some birds may simply appear unthrifty (chronic disease) or may not show any signs (chronic carrier). *Ch. psittaci* has been isolated from fecal samples of clinically normal cattle, goats, sheep, and pigs, making the gastrointestinal tract an important reservoir and source for disease transmission. Ornithosis presents itself in humans in a variety of ways and can vary from inapparent illness to systemic illness with severe pneumonia. This organism can often be identified in stained impression smears of affected tissues, cell culture, and serology. Antigen-antibody test methods and PCR methods are available. Treatment of *Ch. psittaci* typically consists of tetracycline antibiotics in both animals and humans. Controlling the introduction and spread of *Ch. psittaci* in an avian population is critical in managing this disease. Ornithosis continues to be an important occupational hazard in the poultry and pet bird industries.

Large-scale commercial importation of psittacine birds from foreign countries ended in 1993 with the implementation of the Wild Bird Conservation Act.

PASTEURELLOSIS

Overview

Pasteurellosis, a bacterial disease of various animals caused by *Pasteurella multocida*, can cause local or systemic human disease, mainly after people are bitten or scratched by infected animals. *P. multocida* also causes a variety of diseases in animals and is named after Louis Pasteur, the French chemist who made numerous contributions to microbiology including the discovery of anaerobes, vaccines, and methods for controlling the spread of infectious organisms. Pasteurellosis in cattle was first described in 1878 by Bollinger in Germany. The causative agent of pasteurellosis was isolated by Kitt in 1885 during a period in history that also saw the discovery of the microorganisms causing fowl cholera (Pasteur 1880) and swine plague (Loeffler 1886). Ruappe, a German pathologist, noted similarities in the diseases pasteurellosis, fowl cholera, and swine plague thereby proposing that they share the collective name *Bacillus septipaemiae haemorrhagicae*. In 1896, Kmae shortened the name to *Bacillus bovispotlous* and in 1900 Ligniers described the whole group by the name *Pasteurella*. The species name *multocidum* was given to this bacterium by Lehown and Neumann and did not appear in the literature until 1899. In 1937, Rosenbusch and Merchant's named the organism *P. multocida*. Multocida comes from the Latin *multi* meaning many and *cidere* meaning to kill. There are many species of *Pasteurella* and as bacterial identification techniques have improved, some bacteria formerly classified as *Pasteurella* organisms have been renamed. In 1999, the organism *P. hemolytica* was renamed *Mannheimia hemolytica*. This name change was based on taxonomic differences of this organism from other closely related organisms, in particular *P. multocida*. These differences were identified and described by Dr. Mannheim in 1974.

Causative Agent

P. multocida is a small, facultatively anaerobic, oxidase-positive, gram-negative, nonendospore forming rod that exhibits bipolar staining. *Pasteurella* bacteria are part of the normal oral, respiratory, genital, and gastrointestinal flora in a variety of wild and domestic animals. *P. multocida* has a strongly hydrophilic capsule that protects it from dehydration and makes it more resistant to phagocytosis. *P. multocida* grow well on blood agar as convex, smooth, gray, nonhemolytic colonies that may be rough or mucoid in appearance. There are three subspecies of *P. multocida*:

- *P. multocida* subspecies *multocida* is the most common strain that causes disease in domestic animals.
- *P. multocida* subspecies *septica* has been recovered most frequently from cats and dogs, and infrequently humans.
- *P. multocida* subspecies *gallicida* is most frequently isolated from birds and occasionally pigs.

Epizootiology and Public Health Significance

P. dagmatis is a commensal of the pharynx of dogs and cats and has been isolated from animal bite wounds in humans.

Pasteurellosis occurs worldwide in wild and domestic animals as a normal inhabitant of the pharynx of birds and many mammals. Sources of human infection include pets (cats, dogs, birds, rabbits, and Guinea pigs), livestock (cattle, pigs, and sheep), and wild and zoo animals (buffalo, deer, and lions). Other species of *Pasteurella* can cause human disease such as *P. dagmatis*, *P. canis*, *P. stomatis*, *P. cabali*, and *P. hemolytica* (now known as *Mannheimia hemolytica*); however, *P. multocida* is the most common.

Because pasteurellosis is typically acquired by bite and scratch wounds, the risk of contracting pasteurellosis is good especially since there are more than 100 million dogs and cats in the United States (half of all Americans will be bitten in their lifetime). Approximately 5% of dog bites and 30% of cat bites become infected. Refer to the chapter on animal bite wounds for more statistics on animal bites.

Transmission

P. multocida causes opportunistic disease in animals. Stress is an important factor in the breakdown of respiratory defense mechanisms allowing *P. multocida* to invade lung tissue causing pneumonia. Transmission among animals can occur via bite and scratch wounds from infected animals, ingestion, and inhalation.

Transmission of *P. multocida* to humans occurs primarily by bite and scratch wounds of infected animals. Respiratory droplets and consumption of infected meat are rare routes of transmission.

Pathogenesis

P. multocida bacteria are introduced in humans and animals via a bite or scratch wound producing local inflammation and swelling of regional lymph nodes. From the lymph nodes infection spreads to the joints, bones, and other lymph nodes.

Figure 3-53 Snuffles in a rabbit caused by *Pasteurella multocida.*
(Courtesy of USDA)

Complications include cellulitis, abscesses, and osteomyelitis. Cat bites are particularly bad as a result of the small, sharp, penetrative characteristics of cat teeth.

In addition to bite wound injuries, *P. multocida* in animals causes a variety of clinical disease such as hemorrhagic septicemia and shipping fever of cattle, fowl cholera, swine plague, snuffles in rabbits, and pneumonia in sheep. Many times *P. multocida* is a secondary invader following stress or immune system weakening by another infectious agent. Viruses are believed to alter alveolar macrophage function resulting in decreased clearance of inhaled bacteria. The host response of sending increased numbers of neutrophils to the area of infection causes enzyme release when these neutrophils are lyzed contributing to lung tissue damage. This damage also allows *P. multocida* bacteria to gain a foothold and multiply.

Clinical Signs in Animals

Many animal species are susceptible to *Pasteurella* infection including sheep, goats, swine, cattle, and rabbits. The diseases in these animals include:

- Bronchopneumonia in all ages of sheep and goats especially lambs and kids. Bronchopneumonia of the cranioventral lung lobes is caused by both *P. multocida* and *Mannheimia hemolytica* (formerly known as *P. hemolytica* and the more common cause). Edematous bright red lungs and pericardial effusion are seen on necropsy.
- Shipping fever in cattle, a severe respiratory disease associated with *Pasteurella* bacteria generally occurring in younger animals following shipping.
- Snuffles (rhinitis) and pneumonia in rabbits (Figure 3-53).
- Torticollis, commonly known as wry neck, in rabbits as a result of middle or inner ear infections (Figure 3-54).

Figure 3-54 Torticollis (wry neck) in a rabbit as a result of *Pasteurella multocida* infection.
(Courtesy of USDA)

- Abscesses in a variety of animals especially rabbits.
- Fowl cholera in birds that produces clinical signs such as rapid death, fever, anorexia, mucus discharge, ruffled feathers, diarrhea, tachypnea, and localized swelling of wattles, joints, and footpads. In the peracute form, fowl cholera is one of the most virulent and infectious diseases of poultry.
- Abortion in cattle, sheep, and goats.
- Respiratory disease in dogs and cats.
- Atrophic rhinitis (along with other bacteria) and bronchopneumonia in swine.
- Meningitis in dogs.

Clinical Signs in Humans

The incubation period for pasteurellosis varies from 2 to 14 days depending on the route of entry. Wound infections show redness, swelling, and pain. Cellulitis and abscess formation in the skin and subcutaneous tissue may follow. Other disease complications include respiratory infections (mainly upper respiratory infections like sinusitis rather than lower respiratory tract infections), endocarditis, and rarely meningitis. Septicemia is uncommon and usually indicates an underlying medical condition in affected people.

Diagnosis in Animals

Pasteurellosis can be identified via routine staining such as Gram stain (bipolar staining gram-negative rods) and culture methods of samples obtained via swabs or washes. Bipolar staining bacteria can be demonstrated in Wright-stained blood smears of animals with septicemia. *P. multocida* grow well on blood agar as convex, smooth, gray, nonhemolytic colonies that may be rough or mucoid. Biochemical testing can also be used to identify *P. multocida*. It is indole positive, urea negative, catalase positive, and ferments mannitol, sucrose, and maltose. Indirect fluorescent antibody tests and ELISA tests are also available. Serologic tests are rarely used for diagnosis of fowl cholera.

Diagnosis in Humans

Pasteurellosis in people is diagnosed via Gram stain, special stain (such as Wright stain), culture, and biochemical tests as described in animals.

Treatment in Animals

Antibiotic treatment should be initiated as soon as possible to be effective against *P. multocida*. The best basis for antibiotic choice is culture and sensitivity results; however, ceftiofur, oxytetracycline, and tylosin have been effective. Removal of environmental stressors is also important in the treatment of pasteurellosis.

Treatment in Humans

P. multocida infections in humans are typically treated with beta-lactam antibiotics (such as amoxicillin or amoxicillin-clavulanic acid). Resistance to macrolides has been documented. Any bite wound should be thoroughly cleansed and irrigated. Debridement of nonviable tissue can reduce the risk of infection. Rabies and tetanus status should be determined in patients with animal bite wounds.

The risk of multiple bacteria in a bite wound must be considered in cases of animal bite injuries.

Management and Control in Animals

There is an animal vaccine for pasteurellosis in cattle (both *P. multocida* and *Mannheimia hemolytica*; however, there are many strains of these bacteria that may or may not be covered in all vaccines) and in birds (either bacterins containing aluminum hydroxide or oil as adjuvant prepared from multiple serotypes or live vaccine). Correcting poor ventilation, overcrowding, poor nutrition, failure of passive transfer, commingling of animals from various farms, avoiding animal-to-animal bites and scratches, and minimizing transportation are all ways to control pasteurellosis. Vaccinating cattle for other respiratory diseases that may predispose them to contracting *Pasteurella* bacteria is also helpful in reducing cases of pasteurellosis.

Management and Control in Humans

There is no human vaccine for pasteurellosis. Proper hygiene after handling animals and avoiding animal bites and scratches is indicated.

Summary

Pasteurellosis, caused by *P. multocida*, is a disease found worldwide and people are typically infected through animal bites. *Pasteurella* bacteria are part of the normal oral, respiratory, genital, and gastrointestinal flora in a variety of wild and domestic animals. *P. multocida* is a small, facultatively anaerobic, oxidase-positive, gram-negative, nonendospore forming rod that exhibits bipolar staining. *P. multocida* causes opportunistic disease in animals with stress being a predisposing factor. Transmission among animals can occur via bite and scratch wounds from infected animals, ingestion, and inhalation. Transmission of *P. multocida* to humans occurs primarily by bite and scratch wounds of infected animals. Pasteurellosis in animals presents as bronchopneumonia in all ages of sheep and goats; shipping fever in cattle; snuffles, pneumonia, and wry neck in rabbits; abscesses in a variety of animals especially rabbits; fowl cholera in birds; abortion in ruminants; respiratory disease in dogs and cats; atrophic rhinitis and bronchopneumonia in swine; and meningitis in dogs. In humans wound infections are red, swollen, and painful. Disease complications include cellulitis, abscesses, respiratory infections, endocarditis, and rarely meningitis. Pasteurellosis can be identified in animals and people via routine staining and culture methods of samples obtained via swabs or washes. Biochemical testing can also be used to identify *P. multocida*. Indirect fluorescent antibody tests and ELISA tests are also available. Antibiotic treatment should be initiated as soon as possible to be effective against *P. multocida*. Removal of environmental stressors and thoroughly cleansing and irrigation of wounds are also important in treating pasteurellosis. There is an animal vaccine for pasteurellosis in cattle and birds. Correcting poor environmental and management practices, preventing commingling of animals from various farms, and minimizing transportation are all ways to control pasteurellosis. There is no human vaccine for pasteurellosis. Proper hygiene after handling animals and avoiding animal bites is indicated.

PLAGUE AND YERSINIOSIS

Overview

Plague

Plague, also known as the Black Death, the great pestilence, and the bubonic plague, is a disease first described in the Old Testament of the Bible and later in

Homer's *Iliad* during the Trojan War (1190 B.C.). Plague is a term derived from the Greek *plege* meaning blow and Latin *plungere* meaning lament, which refers to the devastating pestilence observed during any epidemic disease especially those marked by fever. Plague has produced at least three great pandemics and multiple epidemics throughout history. The first pandemic, known as the Justinian plague in 543–542 B.C., was believed to have started in Egypt spreading into the Middle East and Mediterranean killing 100 million people over 60 years. On some days, it was believed that up to 10,000 people died in Constantinople during this pandemic. The second pandemic, known as the Black Plague as a result of the gangrenous extremities and extensive subcutaneous hemorrhage of its victims, began in 1347 when traders from central Asia introduced plague infested rats into the ports of Sicily. As the plague moved through Europe, it hit London and quickly multiplied. At that time, the number of rats found within cities was high, sanitation was poor, and knowledge of treatments for diseases unknown. The violent cough produced from the disease was believed to be the main suspect in its quick spread. By 1350 the Black Death had killed approximately one-fourth of Europe's population and was responsible for the death of one-third of the world's population (about 20 to 30 million people). The third pandemic began in China in the 1860s spreading to Hong Kong in the 1890s. Plague was spread by rats transported on ships to Africa, Asia, California, and South America. Plague made its way to the western United States through the port of San Francisco around 1900. The 1906 San Francisco earthquake caused infected rats to abandon the city and mingle with native populations of rodents. The infection of native rodent populations caused the pathogen to gradually spread eastward to the Great Plains. Epidemics appeared sporadically in the United States including one in Los Angeles between 1924 and 1925 that continued until World War II. The insecticide DDT stopped its spread; however, it can still be found in ground squirrels and wild rodent burrows.

Plague is caused by the rod-shaped bacterium *Yersinia pestis* and is a disease that presents itself in three forms: bubonic, pneumonic, and septicemic. *Yersinia* is named in honor of Swiss-born (but French naturalized) Alexander Yersin, who successfully isolated the bacteria in 1894 during the Chinese pandemic of the 1860s. Yersin originally named the bacterium *Pasteurella pestis* in honor of Louis Pasteur, his French patron, but in 1970 it was named in Yersin's honor. The term bubonic refers to the buboes (from the Greek *boubon*, which refers to the groin) that are painful swollen lymph nodes that occur in the groin area as an early sign of disease.

Yersiniosis

Yersiniosis is caused by *Yersinia enterocolitica*, an enteric pathogen that was first described in 1939 by Schleifstein and Coleman. Both *Ye. enterocolitica* and *Ye. pseudotuberculosis* have species names that correlate with the disease that they cause in people and genera named after Alexander Yersin who worked with Émile Roux, the discoverer of diphteria exotoxin. Yersin met Dr. Roux after he had cut himself while performing an autopsy on a patient who had died of rabies. Yersin was saved by an injection of a serum given to him by Roux and in 1888 was hired as his assistant. Yersin was also known for his work with Robert Koch on tuberculosis. The first case of *Ye. enterocolitica* in the United States was in 1976 when contaminated chocolate milk was consumed by school children in Oneida County, New York. This prompted the FDA to establish a laboratory to study *Ye. enterocolitica* and *Ye. pseudotuberculosis*. Other United States outbreaks include an outbreak of *Ye. enterocolitica* in King County, Washington caused by ingestion of tofu in 1982 and an outbreak in Arkansas, Tennessee, and Mississippi associated with consumption of pasteurized milk in the summer of 1982.

Ye. pseudotuberculosis is found worldwide with many cases reported in Europe and Japan in the early 1990s (most cases occur in the winter season because of the bacterium's enhanced growth characteristics in cold temperatures). In November 1998, four laboratory-confirmed cases of *Ye. pseudotuberculosis* were reported in British Columbia possibly as a result of contaminated homogenized milk; in 1991, children consuming untreated drinking water in Japan led to clinical disease; and the bacterium has been isolated in Czechoslovakian well water. In the 1980s, outbreaks in Finland and Japan were responsible for most of the sporadic cases reported in the literature.

Causative Agent

The genus *Yersinia* contains 11 named species but only three that are human pathogens (*Ye. pseudotuberculosis*, *Ye. enterocolitica*, and *Ye. pestis*); other species are not associated with significant human disease. These three species are normally pathogens of animals.

- Plague is caused by *Ye. pestis*, a gram-negative, facultatively anaerobic, nonendospore forming, plump, pleomorphic rod that is a member of the Enterobacteriaceae family of bacteria (Figure 3-55). It is motile when isolated from the environment, but is becomes nonmotile in a mammalian host. When stained with special stains such as Wayson or Wright-Giemsa stain, *Ye. pestis* stains heavily at both ends (bipolar staining). *Ye. pestis* ferments glucose and is oxidase negative and catalase positive. It grows well on most nonselective standard laboratory media (blood, chocolate, and tryptic soy agars) producing pinpoint, gray-white, nonhemolytic colonies at 24 hours and after 48 to 72 hours, the colonies are gray-white to slightly yellow opaque raised with irregular fried egg appearance. It also grows on MacConkey agar appearing as small nonlactose-fermenting colonies (clear). *Ye. pestis* makes a lipopolysaccharide

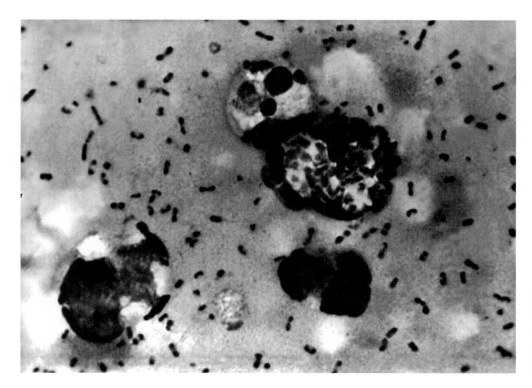

Figure 3-55 *Yersinia pestis* bacteria seen on a slide prepared from lymph node aspirate.

(Courtesy of CDC/Margaret Parsons, Dr. Karl F. Meyer)

endotoxin, coagulase, and a fibrinolysin, which are the principal factors in the pathogenesis of this disease. *Ye. pestis* contains virulence plasmids that code for adhesions, secretion systems, outer membrane protein plasminogen activator, and capsular protein. The adhesions allow the bacterium to attach to cells and after attachment the secretion system is used to inject proteins that trigger cell death in macrophages and neutrophils. The outer membrane protein plasminogen activator is a protease that interferes with blood coagulation and complement activation. The capsular protein plasmid is believed to enhance resistance to phagocytosis. All virulent strains produce exotoxin composed of two protein components, toxin A and toxin B.

> *Ye. enterocolitica* can grow at refrigerator temperatures.

- *Ye. enterocolitica* and *Ye. pseudotuberculosis* are similar to *Ye. pestis* except that *Ye. enterocolitica* produces bull's-eye colonies when grown on agar. *Ye. enterocolitica* should be cultivated using cold enrichment. *Ye. enterocolitica* grows best in environments with a pH of 5.0 to 9.0; therefore, the disease is seen more frequently in people who take antacids. *Ye. enterocolitica* requires iron to survive and is seen more commonly in people who have iron overload diseases (such as hemochromatosis) and in children following accidental iron overdose.

Epizootiology and Public Health Significance

Plague

Plague occurs worldwide and can affect people of any age and during any season of the year; however, the elderly and immunocompromised are more likely to contract the disease. There are approximately 10 cases of plague reported annually in the United States typically from the states of Arizona, California, Colorado, New Mexico, and Utah. Most cases of plague reported outside of the United States are from developing countries in Africa and Asia. Approximately 300 cases worldwide are reported to the WHO annually; however, between 1990 and 1995 approximately 13,000 cases of plague were reported to the WHO (the following countries reported more than 100 cases of plague: China, Congo, India, Madagascar, Mozambique, Myanmar, Peru, Tanzania, Uganda, Vietnam, and Zimbabwe). In the United States an average of 13 cases (between 1 and 40) has been reported annually for many years. Mortality rates are between 50% and 90% in untreated cases and about 15% when diagnosed and treated.

Yersiniosis

Yersiniosis occurs worldwide and appears to be increasing in frequency. People at greater risk of contracting these bacteria are immunocompromised people and people who work in the meat handling industry such as butchers. There are 50 known serovars of *Ye. enterocolitica* and specific serovars are common in certain areas (serogroups 0:3 and 0:9 found in Europe and 0:8 found in the United States). Infections with *Ye. enterocolitica* can occur throughout the year but are seen more frequently in late fall and winter. Strains of pathogenic *Ye. enterocolitica* can be found in the intestinal tracts of a variety of healthy animals including pigs, dogs, and cats as well as in animal feed.

 Ye. enterocolitica infections are infrequent causes of diarrhea and abdominal pain in the United States (about 1 case per 100,000 people). The CDC monitors *Ye. enterocolitica* infections through its foodborne disease surveillance network (FoodNet) and conducts investigations of outbreaks to control their spread. The FDA inspects imported foods and milk pasteurization plants and promotes consumer education of proper food preparation techniques. The USDA monitors food animal health and quality of slaughtered and processed foods, whereas the EPA regulates and monitors the drinking water supplies.

There are 14 serogroups of *Ye. pseudotuberculosis* with serogroup 0:1 causing 70% of human disease. Particular serovars are found in certain areas (0:4 and 0:6 are seen more in Japan, whereas 0:1 is seen in Europe). Infections with *Ye. pseudotuberculosis* can occur throughout the year but are seen more frequently in the late fall, winter, and spring as a result of this bacterium's preferred growth characteristics in cold temperatures. *Ye. pseudotuberculosis* infections occur in animals such as birds, Guinea pigs, and rabbits. Distribution of *Ye. pseudotuberculosis* is worldwide. There have not been any reported foodborne outbreaks caused by *Ye. pseudotuberculosis* in the United States, but human infections transmitted via contaminated water and foods have been reported in Japan.

> Humans are accidental hosts in the natural cycle of plague.

Transmission

Plague

People can contract plague through contact with wild animals (sylvatic plague), domestic or semidomestic animals (urban plague), or infected humans (pneumonic plague). *Ye. pestis* is harbored in more than 200 different mammalian animal reservoir hosts. Infected animals typically do not show clinical signs. These hosts spread the bacterium to other mammals (called amplifying hosts) that become infected and experience massive die-offs during epidemics. Amplifying hosts include the brown rat, black rat, ground squirrel, wood rat, chipmunk, prairie dog, and rabbit, and the mammal most important in this process depends on the area of the world (Figure 3-56).

Figure 3-56 Prairie dogs (genus *Cynomys*) can harbor fleas infected with *Yersinia pestis*, the bacterium that causes plague. Prairie dogs typically do not show clinical signs of plague. (Courtesy of CDC)

The principle vectors in the transmission of *Ye. pestis* from reservoir hosts to amplifying hosts to people are fleas. After an uninfected flea ingests a blood meal from an animal with *Ye. pestis*, bacteria multiply in the flea's gut. In time the increase in bacterial numbers as a result of multiplication and accumulation of clotted blood will block the flea's esophagus. The flea is unable to feed properly making the fleas extremely hungry. In an attempt to gain nourishment the flea jumps from animal to animal regurgitating infectious material into the bite wounds it inflicts in these animals.

The flea prefers its natural host and will remain with that species and transmit the bacterium within that population. If the natural host is not available the fleas will seek out other species including humans. Plague can be transmitted to humans when an infected flea attempts to take a blood meal and regurgitates bacteria into the human's bloodstream. The classical vector for transmission is the oriental rat flea, *Xenopsylla cheopis*, but over 1500 flea species can transmit *Ye. pestis* (see Figure 6-81).

Humans can also be infected with *Ye. pestis* by handling infected animals (including skins and meat) and by inhalation. Ticks and human lice have also been identified as possible vectors. In the United States the role of domestic cats in the transmission of *Ye. pestis* is important. Domestic cats become infected by eating infected prey and may transmit the bacterium to humans through aerosols or bite and scratch wounds. From 1977 to 1998 nearly 300 people became infected with *Ye. pestis* with 7.7% occurring from contact with infected cats.

Yersiniosis

Yersiniosis is transmitted by consumption of contaminated food. In the United States reported contaminated sources include secondarily contaminated pasteurized milk, chocolate milk, tofu, beans, and home-slaughtered pork and preparation of chitterlings.

Pathogenesis

Plague

After a flea ingests a blood meal from an animal infected with *Ye. pestis*, bacterial coagulase causes the blood to clot. *Ye. pestis* multiply in the blood clot and as the flea bites an animal it inoculates thousands of bacteria through a host's skin. The bacteria migrate to the regional lymph nodes, are phagocytosed by the segmented neutrophils and monocytes, and multiply in them. Most bacteria are killed by the segmented neutrophils, but the ones taken up by macrophages are not killed because the macrophages provide a protected environment for the bacterium to synthesize their virulence factors. The bacteria then kill the macrophage and bacteria are released into the blood where they resist phagocytosis. *Ye. pestis* quickly spread to lymph nodes (usually in the groin, axilla, or neck) and become hot, hemorrhagic, swollen, and tender (black buboes for which the disease is named). Organisms may then spread to distant organs.

The manifestations of infection with *Ye. pestis* lead to bubonic, septicemic, or pneumonic plague. In bubonic plague, *Ye. pestis* multiplies in the flea bite, enters the lymph, and is filtered by the regional lymph nodes causing inflammation and swelling of the lymph node (bubo). Bubonic plague can progress to septicemic plague if bacteria move from the lymphatic into the circulatory system. Release of virulence factors from *Ye. pestis* causes circulatory problems that may degenerate into necrosis and gangrene (which often appeared black leading to the term "black death"). Pneumonic plague occurs when the bacterium becomes lodged in the lungs either through the bacterium's presence in the blood or via inhalation. Pneumonic plague develops rapidly with people developing fever, lethargy, and pulmonary disease within a day of infection.

Yersiniosis

The incubation period for *Ye. enterocolitica* is between 3 and 10 days; for *Ye. pseudotuberculosis* it is 7 and 21 days. The disease process is similar for both organisms; however, each bacterium has a predilection for a certain disease course and individual it infects.

After ingestion of *Ye. enterocolitica*, it adheres to and penetrates the mucosal surface of the terminal ileum where the bacterium reproduces in Peyer's patches. Nonspecific ileocolitis occurs and bacteria may spread to the ileocecal lymph nodes leading to bacteremia or to abscess development and pain in the right lower quadrant that mimics the pain of appendicitis. Plasmid-encoded proteins of the outer membrane represent the major factor in the pathogenicity of *Ye. enterocolitica* by contributing to adherence and invasion of the organism.

After ingestion of *Ye. pseudotuberculosis*, it also adheres to and penetrates the mucosal surface of the ileum where the bacterium spreads to the mesenteric lymph nodes leading to the more characteristic pseudoappendicitis associated with this species of *Yersinia*. Fecal excretion of the organism can occur for several weeks

Septicemic and bubonic plague is transmitted with direct contact with a flea, whereas pneumonic plague is transmitted through airborne droplets of infected saliva.

Ye. pestis can remain viable in fleas for up to three years in animal habitats.

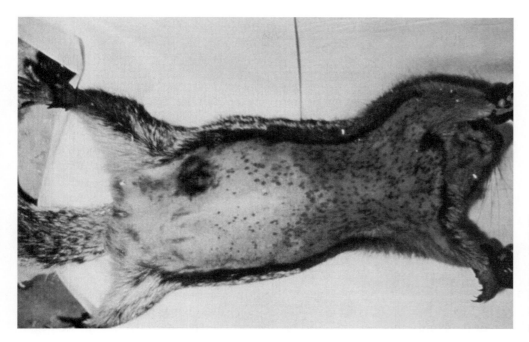

Figure 3-57 This squirrel has a petechial rash (small, pinpoint, flat lesions of the skin and mucous membranes that are associated with hemorrhages beneath the skin surface) as a result of infection with *Yersinia pestis*. (Courtesy of CDC)

after the appearance of clinical signs but often does not result in secondary person-to-person cases or clinical relapses.

Clinical Signs in Animals

Plague

Ye. pestis can infect hundreds of species of animals who serve as reservoirs of plague without showing clinical disease; however, plague is fundamentally an infection of rats, mice, and voles (Figure 3-57). Natural infections have also been reported in dogs, cats, prairie dogs, camels, elephants, buffalo, and deer. Cats and dogs are now recognized as sources of infection for humans with hundreds of cases of domestic cats reported with plague since it was first recognized less than 40 years ago. In the United States cats have been the source of infection for approximately 5% of recent cases of plague in man. Clinical signs in cats and dogs include acute illness within 24 to 48 hours of infections typically resulting is high fever (as high as 106°F). Approximately 50% of naturally-infected cats contract the bubonic form. In endemic areas, plague should be suspected in cats (including wild cats such as bobcats) that have clinical signs of fever, pneumonia, and lymphadenitis.

Yersiniosis

Ye. enterocolitica causes gastroenteritis, ileitis, and mesenteric adenitis in animals. It has been isolated from animals such as pigs, chinchillas, hares, deer, rabbits, dogs, horses, mink, birds, goats, and cattle as well as from water and milk.

    Ye. pseudotuberculosis produces less severe intestinal inflammation than *Ye. enterocolitica*. *Ye. pseudotuberculosis* infection in animals causes respiratory disease such as that resembling tuberculosis in rabbits, cats, chinchillas, rodents, Guinea pigs, and turkeys; orchitis in rams; and abortion in goats. It can produce pyogranulomatous disease in cats.

> The major animal reservoir for *Ye. enterocolitica* is pigs where it is most often found on the tonsils.

Clinical Signs in Humans

Plague

In humans plague can present itself in three ways:

- Bubonic plague in which people develop buboes (enlarged, tender lymph

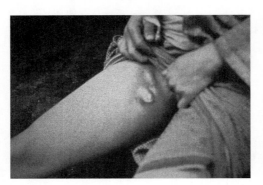

nodes especially in the groin area) (Figure 3-58), sudden high fever, weakness, and chills. Buboes are unilateral, oval, extremely tender lymph nodes and can range in size from 2 to 10 centimeters. Enlargement of the buboes leads to rupture and discharge of malodorous pus. Vesicles may be observed at the site of the infected flea bite and with advanced disease pustules or papules may be observed in areas of the skin drained by the involved lymph nodes. A papular rash of the hands and feet may also be observed. Hepatomegaly and splenomegaly often occur. The incubation period usually lasts 2 to 6 days.

- Septicemic plague in which people develop fever, chills, abdominal pain, shock, and bleeding into the skin and other organs. Buboes are not observed with septicemic plague and this form of plague is associated with a high mortality rate. Infected people have a toxic appearance and may present with hypothermia, tachycardia, tachypnea, and hypotension. Bleeding into the skin (purpura) may lead to necrosis and gangrene.

- Pneumonic plague is a highly contagious form of plague that is transmitted by aerosol droplets. People develop an abrupt onset of fever and chills, accompanied by cough, chest pain, dyspnea, purulent sputum, or hemoptysis (coughing up blood). Buboes may or may not appear in pneumonic plague.

Yersiniosis

Ye. enterocolitica and *Ye. pseudotuberculosis* are pathogens causing enteric yersiniosis and are acquired by consumption of food or water contaminated by infected animal feces. *Ye. enterocolitica* is a common cause of intestinal tract inflammation that can last weeks or months. Involvement of the terminal portion of the small intestine can result in painful inflammation of the mesenteric lymph nodes, mimicking the pain of appendicitis. Infections with *Ye. enterocolitica* and *Ye. pseudotuberculosis* can cause a variety of symptoms depending on the age of the people infected. Yersiniosis occurs most frequently in young children who develop fever, abdominal pain, and diarrhea (often bloody) that presents 4 to 7 days after exposure and may last 1 to 3 weeks. In older children and adults, abdominal pain (usually right sided, which is often confused with appendicitis) and fever are the predominant signs. Complications such as skin rash and joint pain may be seen in a small number of cases.

Diagnosis in Animals

Plague

Diagnosis can be confirmed by special stains (Wayson stains show light-blue bacilli with dark-blue polar bodies), media culture (pinpoint colonies on blood agar at

24 hours and cauliflower appearance at 48 hours), or by a fluorescent antibody test (FA) of lymph node aspirate or biopsy (for F1 antigen). *Ye. pestis* bacteria express a unique diagnostic envelop glycoprotein called the Fraction 1 (F1) antigen or capsular antigen during growth at greater than 33°C. Serum and blood can be taken for antibody detection by agglutination. Fleas or pooled flea material can be examined by FA if samples are pre-incubated at 37°C for 24 hours to encourage F1 antigen expression. Both animal and flea material may be tested by PCR to determine if plague DNA is present in the specimens.

Yersiniosis

Intestinal *Yersinia* spp. can be isolated with cefsulodin-Irgasan-novobiocin (CIN) agar using cold enrichment. MacConkey and *Salmonella-Shigella* agar may also be used for identification. The Widal agglutination reaction is used to detect antibodies against the species of *Yersinia* that cause zoonotic disease. An ELISA test and an immunoblotting assay have also been developed for these bacteria.

Diagnosis in Humans

Plague

Diagnosis of *Ye. pestis* in people is similar to animals using blood cultures and lymph node specimens. Testing for plague is limited to reference laboratories in plague-endemic states and the CDC. In addition to the tests describe above, a direct immunoflurorescence test of fluid or cultures may help in rapid diagnosis and a passive hemagglutination test with a fourfold or greater increase in titer indicates disease in humans.

Yersiniosis

The diagnosis for *Ye. enterocolitica* and *Ye. pseudotuberculosis* in humans is the same as for animals using stool specimens. Stool culture is usually positive within 2 weeks of onset of clinical signs. Many laboratories do not test for these organisms so laboratory notification is important. These bacteria can also be recovered from lymph nodes, joint fluid, urine, bile, and blood. Agglutination titers rise 1 week after onset of illness and elevated levels can be found for years after infection.

Treatment in Animals

Plague in cats can be rapidly lethal; therefore, therapy should be started immediately with antibiotics such as streptomycin and tetracycline in combination and should be continued for more than five days after the temperature returns to normal to avoid relapse.

Yersiniosis is typically self-limiting in animals. It can be treated with antibiotics and supportive therapy such as fluid administration if an animal becomes dehydrated.

Treatment in Humans

Plague

As soon as plague is suspected in a person, he/she should be isolated and local and state health departments notified. Antibiotic therapy is begun as soon as possible after collecting laboratory samples and includes antibiotics such as streptomycin or gentamicin.

All suspected human plague cases must be reported to local and state health departments and the diagnosis confirmed by the CDC. International Health Regulations require the CDC to report all U.S. plague cases to the WHO.

Yersiniosis

In uncomplicated cases of intestinal yersiniosis, clinical signs of disease such as diarrhea usually resolve without treatment. Antibiotics are only needed for septicemic, chronic, and recurrent cases of yersiniosis with ciprofloxacin and doxycycline being effective against these bacteria. Rehydration therapy may be needed in severely dehydrated people.

Management and Control in Animals

Attempts to eliminate wild rodent plague are costly and unproductive.

Management and Control in Humans

Epidemic plague is best prevented by controlling rat populations in urban and rural areas. Close surveillance of human plague cases associated with plague in rodents and the use of effective insecticide for controlling rodent fleas when outbreaks occur are two things that could limit the spread of disease. Some other measures taken to control plague are:

- Prophylactic antibiotic therapy as recommended by the CDC for people who have been exposed to bites of potentially infected rodent fleas during a plague outbreak or people who have handled an animal known to be infected with *Ye. pestis*.
- Plague vaccine is recommended for field workers in endemic areas for plague and for scientists and laboratory personnel who routinely work with *Ye. pestis* (it is no longer commercially available in the United States). Vaccination is of limited use and is not mandatory for entry into any country and is ineffective against the pneumonic form of plague.
- Environmental sanitation including removing food sources used by rodents, rodent-proofing homes, buildings, or warehouses, and applying chemicals to kill fleas and rodents.
- Treatment of pets such as dogs and cats for fleas and proper flea control.
- Educating the public on ways to prevent exposure to the plague.

Ways to avoid contracting *Ye. enterocolitica* and *Ye. pseudotuberculosis* include:

- Not eating raw or undercooked pork.
- Consuming only pasteurized milk or milk products.
- Washing hands before eating and preparing food, after contact with animals, and after handling raw meat.
- Preventing cross-contamination in the kitchen by using separate cutting boards for meat and other foods and carefully cleaning all cutting boards, surfaces, and utensils after preparing raw meat.
- Disposing of animal feces in a sanitary manner.

Summary

Plague

Plague is a bacterial disease caused by *Ye. pestis*, a gram-negative, facultatively anaerobic, nonendospore forming, plump, pleomorphic rod that is a member of the Enterobacteriaceae family of bacteria. Plague presents itself in three forms in people: bubonic, pneumonic, and septicemic. Transmission is via a flea bite in most cases; however, pneumonic plague is transmitted by aerosol transmission. Most animals that are infected with *Ye. pestis* are asymptomatic except for cats and dogs that may have high fever and respiratory signs. Diagnosis can be confirmed by special stains (Wayson stains), media culture, or by a fluorescent antibody test of lymph node aspirate or biopsy (for F1 antigen). In addition to the animal tests listed, people can be diagnosed via a direct immunoflurorescence test of fluid or cultures and a passive hemagglutination test with a fourfold or greater increase in titer indicates disease in humans. Early antibiotic treatment is needed to eliminate the bacterium and clinical signs. Mortality can be high if antibiotic treatment is delayed. Epidemic plague is best prevented by controlling rat populations in urban and rural areas. Endemic areas of sylvatic plague exist in the western United States and several other areas of the world. Throughout history, outbreaks of human plague involving rats in densely populated agricultural or urban areas have occurred when rats commingled with infected sylvatic rodents.

Yersiniosis

Yersiniosis is caused by *Ye. enterocolitica* and *Ye. pseudotuberculosis*, gram-negative, facultatively anaerobic, nonendospore forming, plump, pleomorphic rods that are acquired by consumption of food or water contaminated by infected animal feces. Yersiniosis produces clinical signs of enteric disease. *Ye. enterocolitica* is a common cause of intestinal tract inflammation that can last weeks or months and the organism's involvement in the terminal portion of the small intestine can result in painful inflammation of the mesenteric lymph nodes, mimicking the pain of appendicitis. *Ye. pseudotuberculosis* produces less severe intestinal inflammation than *Ye. enterocolitica*. *Ye. enterocolitica* causes gastroenteritis, ileitis, and mesenteric adenitis in animals such as pigs, chinchillas, hares, deer, rabbits, dogs, horses, mink, birds, goats, and cattle. *Ye. pseudotuberculosis* infection in animals causes respiratory disease in rabbits, cats, chinchillas, rodents, Guinea pigs, and turkeys; orchitis in rams; abortion in goats, and pyogranulomatous disease in cats. Intestinal *Yersinia* spp. can be isolated with cefsulodin-Irgasan-novobiocin (CIN) agar using cold enrichment. The Widal agglutination reaction is used to detect antibodies against these two species of *Yersinia* and an ELISA test and an immunoblotting assay have been developed for these bacteria. Most cases of yersiniosis resolve on their own; however, antibiotic treatment may be needed in the young, old, and immunocompromised. Ways to avoid contracting *Ye. enterocolitica* and *Ye. pseudotuberculosis* include proper food handling.

RAT-BITE FEVER

Overview

Rat-bite fever is a disease caused by two different bacteria, *Streptobacillus moniliformis* and *Spirillum minus*, and may actually be two zoonotic diseases rather than one. Once thought to be confined to Asia, rat-bite fever (RBF) is now confirmed to occur worldwide. RBF was once thought to be a disease of rat-infested parts of inner cities, but it is now known to occur in a variety of places. RBF caused by *Stre. moniliformis* is also known as Haverhill Fever and is a systemic infection acquired through a rodent bite or scratch or ingestion of food or water contaminated with rat feces. The name Haverhill Fever comes from the 1926 outbreak of this organism in

Haverhill, MA, attributed to the consumption of rat-feces contaminated milk. This form of RBF is also known as streptobacillary fever and epidemic arthritic erythema. The second cause of RBF is *Sp. minus* (also known as *Sp. minor*). This form of RBF is also known as Sodoku (Japanese for rat poisoning) or spirillary fever and is transmitted by the bite of a rodent. Both forms of RBF cause acute febrile disease in humans; however, their epidemiology and clinical presentations vary.

Causative Agent

Both *Stre. moniliformis* and *Sp. minus* are present in the oral and respiratory tracts of rodents, particularly mice and rats. *Stre. moniliformis* is a gram-negative, nonmotile, nonendospore forming, pleomorphic rod that is a facultative anaerobic bacterium. Microscopically it may show branching filaments, bead-like chains, and fusiform swelling that taper toward each end of the bacillus. The virulence factors for *Stre. moniliformis* are unknown, but the bacterium develops L forms (bacteria without cell walls) spontaneously that may allow for its persistence in some body sites.

 Sp. minus is a gram-negative, aerobic, short, thick, tightly coiled, spiral-shaped bacterium with two to three coils and bipolar flagella. Little is known about the virulence factors of *Sp. minus*, but its pathogenesis is believed to be similar to *Stre. moniliformis*.

Epizootiology and Public Health Significance

One rat can produce 20 to 50 fecal droppings and excrete 14 mL of urine per day.

Rodents are the principle reservoir of both *Stre. moniliformis* and *Sp. minus*. Both bacteria are oral and respiratory normal flora of asymptomatic rodents, with *Stre. moniliformis* found in 10% to 100% of healthy laboratory rats and in 50% to 100% of wild rats. *Sp. minus* is mainly found in rats, but may also be found in weasels, squirrels, ferrets, mice, pigs, dogs, and cats. *Stre. moniliformis* is the most common cause of RBF in the United States, whereas *Sp. minus* is most commonly found in Asia.

 RBF is rare in the United States and accurate data about the rate of RBF is not reliable because it is not a reportable disease. RBF is also rare in other countries.

Transmission

RBF is mainly transmitted through rat bites, although contact with urine or oral or conjunctival secretions from an infected animal can also transmit these bacteria. The main source of infection is rats, but animals such as squirrels, weasels, and gerbils may transmit the disease. *Sp. minus* is not found in rat saliva, but is found in blood and occasionally the conjunctiva. If there is damage to the oral mucosa of rats, *Sp. minus* may be found in the animal's saliva. *Sp. minus* is typically transmitted by the bite of an infected rat, rodent, or animal that has ingested an infected rodent. *Sp. minus* is not typically transmitted by contaminated food or water. *Stre. moniliformis* is typically transmitted by contaminated food and water.

Pathogenesis

The pathogenesis of infection caused by *Stre. moniliformis* is unknown. This bacterium can produce L forms (a bacterial form without cell walls) that may help it survive in a lesion. The pathogenesis of infection caused by *Sp. minus* involves a bite wound that may produce edema and ulceration. There is minimal local inflammation of the bite wound and the wound heals quickly producing little lymph node inflammation. Approximately 1 to 3 days following the bite, bacteremia may be present and the wound may reappear and be pyogenic (pus-like).

Clinical Signs in Animals

Both *Stre. moniliformis* and *Sp. minus* are normal oral and respiratory flora of rodents and can serve as opportunistic pathogens under some circumstances. As a secondary invader, *Stre. moniliformis* causes pneumonia in rats and septicemia, septic arthritis, hepatitis, and lymphadenitis in mice. As a secondary invader, *Sp. minus* causes septicemia, ocular infections, and pneumonia in rats.

Clinical Signs in Humans

Clinical signs of RBF in humans depend on the bacterium involved. Infections as a result of *Stre. moniliformis* have an incubation period of 3 to 5 days where there is an abrupt onset of irregularly relapsing fever (101°F to 104°F). Other signs include chills, regional lymph node enlargement, vomiting, headache, lethargy, joint pain, asymmetric polyarthritis, and muscle pain. In greater than 90% of cases within 2 to 4 days there are dark red rashes on the extremities, particularly on the palms and soles and over joints such as the elbows and wrists. Sore throat with severe dysphagia (difficulty swallowing), diffuse redness of the pharynx, and painful laryngitis (with cough, hoarseness, and voice change) may also be seen. Complications include endocarditis, abscess formation in organs, pneumonia, hepatitis, and meningitis. Most cases resolve spontaneously within 2 weeks; however, 13% of untreated cases are fatal.

Infections as a result of *Sp. minus* have an incubation period of 2 to 3 weeks. The initial wound quickly heals but soon becomes painful, edematous, and ulcerated. Regional lymph nodes enlarge and the person develops fever, chills, headache, and lethargy. The fever may last for 3 to 5 days and then recur for 4 to 8 weeks, losing their intensity as they recur. Red rashes may develop around the bite. Headache, muscle pain, diarrhea, vomiting, joint pain, and central nervous system signs may occur. Complications include endocarditis, hepatitis, and meningitis. Arthritis is rarely seen with *Sp. minus* infections.

Diagnosis in Animals

Both *Stre. moniliformis* and *Sp. minus* can be visualized in affected tissue using Giemsa stain. *Stre. moniliformis* can be visualized using Gram stain as well. Diagnosis in animals is typically not performed as a result of the nature of the bacterium, but culture of *Stre. moniliformis* is possible from lesions such as joints, lymph nodes, or other affected area. *Stre. moniliformis* is typically diagnosed by blood culture because the bacterium has strict growth requirements and is slow growing. It can be cultured on blood agar when it is incubated in very moist conditions with 5% to 10% carbon dioxide and 20% horse serum. Colonies are embedded in the agar and may appear to look like a fried egg with a dark center and lacy edges. Growth in brain-heart infusion broth may show puffball growth in 2 to 3 days. Gram-stained organisms show pleomorphic rods that may be club-shaped. Dienes stain, acridine orange, and Giemsa stain may be necessary for visualization of the organism. Biochemical tests are also available for identification of this organism (indole negative, catalase negative, and nitrate negative). Specialized serologic tests are only available at national reference laboratories.

Sp. minus cannot be cultured on synthetic media. Direct visualization of the spirochetes in clinical specimens using Giemsa stain or darkfield microscopy is used to help identify this bacterium. Diagnosis is definitively made by injection of lesion material or blood into experimental mice or Guinea pigs and recovery of the organism 1 to 3 weeks after inoculation.

Diagnosis in Humans

RBF in humans is diagnosed by clinical symptoms and tests based upon which bacterium is involved. Culture of *Stre. moniliformis* is the same as described for animals. *Sp. minus* is diagnosed by darkfield microscopy of blood smears or tissue/ exudates. Giemsa and Wright stains are the most common stains used for visual identification of the thick, spiral bacterium with 2 to 3 coils and polar flagella. No serologic test is available. Mice or Guinea pigs can be injected SQ or IP with material containing the organism and then recovering the organism 1 to 3 weeks after the inoculation.

Treatment in Animals

Treatment of RBF in animals is not documented.

Treatment in Humans

Following a rat or rodent bite, the wound should be thoroughly cleaned. Antibiotic treatment with penicillin IV for 5 to 7 days is recommended by the CDC. Erythromycin, chloramphenicol, clindamycin, tetracyclines, and cephalosporins have also been used to treat RBF. Tetanus prophylaxis may be administered if needed.

Management and Control in Animals

There is no documentation of attempts to control the disease in animals.

Management and Control in Humans

Prevention of RBF includes rodent control especially in overcrowded, unsanitary conditions. Animal handlers, laboratory works, and sanitation and sewer workers need to take extra precautions to avoid rodent bites. Wild rodents should not be touched and pets should not be allowed to ingest rodents.

Prompt cleaning of wounds with antiseptic solution can minimize the risk of RBF with rodent bites. No vaccines are currently available for RBF.

Summary

Rat-bite fever is a disease caused by two different bacteria, *Stre. moniliformis* and *Sp. minus*. *Stre. moniliformis* is a gram-negative, nonmotile, nonendospore forming, pleomorphic rod that is a facultative anaerobic bacterium. *Sp. minus* is a gram-negative, aerobic, short, thick, tightly coiled, spiral-shaped bacterium with 2 to 3 coils and bipolar flagella. Both bacteria are present as oral and respiratory normal flora of asymptomatic rodents, particularly mice and rats. RBF is mainly transmitted through rat bites, although contact with urine or oral or conjunctival secretions from an infected animal can also transmit these bacteria. *Stre. moniliformis* causes pneumonia in rats (as a secondary invader) and septicemia, septic arthritis, hepatitis, and lymphadenitis in mice. *Sp. minus* causes septicemia, ocular infections, and pneumonia in rats. Human infections as a result of *Stre. moniliformis* have an incubation period of 3 to 5 days presenting with an abrupt onset of irregularly relapsing fever (101°F to 104°F). Dark red rashes may occur in 2 to 4 days on the extremities and arthritis may develop. Most cases resolve spontaneously. Human infections as a result of *Sp. minus* have an incubation period of 2 to 3 weeks. Regional lymph nodes enlarge and the person develops fever, chills, headache, and lethargy. The fever may last for 3 to 5 days and then recur for 4 to 8 weeks, losing their intensity as they recur. Red rashes may develop around the bite. Arthritis is rarely seen with *Sp. minus* infections. Diagnosis is based on clinical signs, culture, and staining methods (vary depending on the bacterium involved). Antibiotics are used to treat RBF in humans.

SALMONELLOSIS

Overview

Salmonellosis, also called *Salmonella* food poisoning, *Salmonella* enterocolitis, or enteric paratyphosis, is an infectious disease caused by the microbe *Salmonella*, a gram-negative, rod-shaped bacterium that lives in the gastrointestinal tracts of warm-blooded animals (such as mammals and birds), cold-blooded animals (such as reptiles), and insects. Some species of *Salmonella* are ubiquitous, whereas others are specifically adapted to a particular host. *Salmonella* derives its name from the American veterinary pathologist Daniel E. Salmon, who together with Theobald Smith (who also worked on anaphylaxis) first described the *Salmonella* bacterium in 1880 and cultured it from porcine intestine in 1884. In humans the *Salmonella* bacillus was first observed by Eberth in spleen sections and mesenteric lymph nodes from a patient who died from typhoid fever in the 1880s. Robert Koch successfully cultured *Salmonella* in 1881.

In animals salmonellosis presents itself as one of three syndromes: septicemia, acute enteritis, or chronic enteritis. In humans, *Salmonella* bacteria are the cause of two diseases: enteric fever (also known as typhoid fever), which occurs as a result of bacterial invasion of the bloodstream and acute gastroenteritis (nontyphoid salmonellosis), which results from foodborne infection.

The most virulent form of *Salmonella* is *S. typhi*, the cause of the severe systemic disease typhoid fever. Typhoid fever derives its names from its resemblance of clinical signs to the rickettsial disease typhus (other than clinical signs the two diseases are very different). Typhoid fever was described in 1659 by Thomas Willis; however, a very similar disease was known for a long time. Hippocrates described a fever that probably was typhoid fever and Antonius Musa, a Romanian physician, treated Emperor Augustus with cold baths when he became ill with typhoid fever. Historically, typhoid fever affected all classes of people with both royalty and common people susceptible to the disease. Prince Albert, the Prince Consort, died of typhoid fever in 1861, and it has been calculated that about 50,000 cases per year occurred in England alone about this time. Typhoid fever mainly affected the poor who many times lived in crowded and unsanitary conditions. In 1875, water supplies and sanitary conditions in England were

Figure 3-59 Unsanitary conditions may lead to the transmission of *Salmonella* bacteria. This privy was located in an area easily accessible to humans, animals, and the water supply. Conditions such as this helped the transmission of typhoid fever in the early 1900s.

(Courtesy of CDC/Minnesota Dept. of Health, R.N. Barr Library; Librarians Melissa Rethlefsen and Marie Jones)

improved by a Public Health Act, which caused the cases of typhoid fever to decline by more than half in the next decade (Figure 3-59).

The most famous case of typhoid fever in the United States is that of a middle-aged Irish cook named Mary Mallon who was commonly called Typhoid Mary. Mary Mallon immigrated to New York to work as a cook and over a 7-year period was responsible for seven of the eight families she worked for to be stricken by typhoid fever. Dr. George Soper, a physician trying to trace the source of the infection, examined water, milk, food, and the environment to determine the origin of the typhoid bacillus, but was unsuccessful until he began to consider the possibility that the cook was a carrier of the disease. When confronted with the fact that she may be the source of the disease, Mary Mallon ran away. As a result of the threat to public health, police captured and imprisoned her. Tests for the pathogen were positive and Mary could either agree to have her gallbladder removed (to stop the shedding of bacteria) or stop working as a cook. Mary refused both choices and remained imprisoned for 3 years after which time she agreed to stop cooking. When she was released, Mary returned to being a cook, shedding bacilli, and causing more disease and death. She was rearrested and spent the rest of her life as a ward of the hospital continuing to deny that she was the source of typhoid fever.

Nontyphoid salmonellosis causes less severe disease than typhoid fever and is caused by many serotypes of *S. enterica* all of which are zoonotic. Nontyphoid salmonellosis is more prevalent than typhoid fever and its primary reservoirs are cattle, rodents, poultry, and reptiles. *Salmonella* are normal intestinal flora in many vertebrates and because meat and milk may be contaminated during slaughter, collection, and processing outbreaks of nontyphoid salmonellosis may occur as a result of poorly cooked meat and unpasteurized dairy products. Foods contaminated by rodent feces, eggs contaminated with the bacillus during egg formation, and pet reptiles are additional sources of *Salmonella* bacteria. The sale of small pet turtles has been illegal in the United States since 1975 as a result of the risk of children placing the turtles in their mouths and being infected with *Salmonella*.

In 1985 an epidemic of nontyphoid salmonellosis occurred in a six-state area of the Midwest United States as a result of contaminated milk originating from the same dairy. Salmonellosis continues to be a problem with outbreaks occurring from time to time.

Causative Agent

Salmonella bacteria are motile, gram-negative, rod-shaped bacteria of the family Enterobacteriaceae, commonly known as enteric bacteria. *Salmonella* nomenclature has been confusing because the original naming system was not based on DNA, but rather names were assigned based on clinical considerations (*S. typhi* causes typhoid fever) or species affected (*S. choleraesuis* is found in swine, *S. abortusovis* is found in sheep). Names were also derived from the geographical origin of the first isolated strain of the newly discovered serovars (*S. london*, *S. saint paul*, *S. heidelberg*, *S. panama*, *S. stanleyville*). The nomenclature for the genus *Salmonella* has evolved from the initial one serotype/one species concept, which was based on serologic identification of O (somatic) and H (flagellar) antigens because this system would result in over 2,400 species of *Salmonella*. In 1973 *Salmonella* taxonomy was revamped based on DNA hybridization and resulted in virtually all *Salmonella* belonging to a single species (with the exception of *S. bongori*) and these species are further characterized by the use of serovar identification. By 1986, it was recommended that only two species of *Salmonella* be recognized: *S. enterica* and *S. bongori*. This naming system, however, did not address the significance of the highly virulent, causative organism of typhoid fever. Bacteriologists have currently agreed upon the naming system using *S. enterica*, *S. typhi*, *S. choleraesuis*, and *S. bongori* (which is not of clinical significance). *S. enterica* accounts for more than 99.5% of the *Salmonella* strains isolated from humans and other warm-blooded animals. The preferred terminology used in medical practice is *S. enterica* serovar Typhimurium, which is usually shortened to *S.* Typhimurium (the serovar name is capitalized). Presently, more than 2,400 *Salmonella* serovars exist with *S. enterica* serovar Typhimurium being the predominant serovar. *S. enterica* serovar Typhimurium is also known as Salmonella DT 104 as a result of its phage type (bacteriophages or phages are viruses that infect bacteria and bacteria may be differentiated into phage types based on their susceptibility to infection and lysis by these phages). Phages may determine pathogenicity.

All Enterobacteriaceae, including the genus *Salmonella*, have three kinds of major antigens: somatic or cell wall (O), surface or envelope (Vi, which represents virulence), and flagellar (H).

- Somatic or cell wall antigens are heat stable, alcohol resistant, and are used for serologic identification. O factors with the same number are closely related, but not always antigenically identical.
- Surface or envelope antigens may be found in some *Salmonella* serovars. Surface antigens in *Salmonella* spp. may mask O antigens, and bacteria will not agglutinate with O antisera. One specific surface antigen is well known: the Vi antigen, and it is found in only three *Salmonella* serovars (Typhi, Paratyphi C, and Dublin).
- Flagellar antigens are heat-labile proteins and when mixing *Salmonella* cells with flagella-specific antisera, a characteristic pattern of agglutination is observed (Figure 3-60).

Salmonella bacteria are found in the intestinal tract of humans and animals with the primary reservoirs being livestock, poultry, and reptiles. *Salmonella* serovars can be found predominantly in one particular host, can be ubiquitous, or

> The FDA in 1975 made it illegal to sell viable turtle eggs and live turtles with a shell length less than 4 inches (unless they were sold to educational and scientific institutions).

> A serovar is a strain of bacteria that is differentiated by serologic methods.

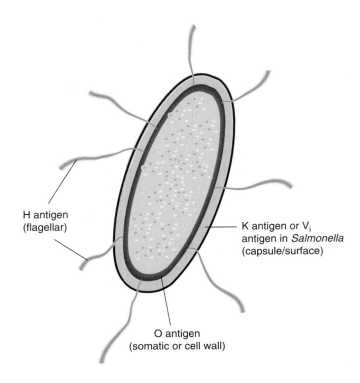

H antigen
(flagellar)

K antigen or V$_i$
antigen in *Salmonella*
(capsule/surface)

O antigen
(somatic or cell wall)

Figure 3-60 Antigenic structures found in gram-negative enteric rods.

can have an unknown habitat. Typhi and Paratyphi A are strictly human serovars that may cause severe diseases associated with bacteremia (bacteria in the blood). Salmonellosis in these cases is transmitted through fecal contamination of water or food. Gallinarum (birds), Abortusovis (sheep), and Typhisuis (swine) are *Salmonella* serovars found in animals. Ubiquitous (nonhost-adapted) *Salmonella* serovars (e.g., Typhimurium) cause very diverse clinical disease, from asymptomatic infection to typhoid-like syndromes in young or highly susceptible animals such as mice. In human adults, ubiquitous *Salmonella* organisms are mostly responsible for foodborne toxic infections.

Epizootiology and Public Health Significance

Salmonellosis is usually associated with ingestion of contaminated meat, dairy products, or eggs; however, plants can also be infected with *Salmonella* bacteria. In 1995, 15 outbreaks of *Salmonella* were as a result of ingestion of raw sprouts.

All *Salmonella* bacteria except *S. typhi* occur worldwide in a variety of animals, including calves, pigs, poultry, wild birds, rodents, and pets such as dogs and cats (*S. typhi* is only found in the blood and intestinal tract of humans). *Salmonella* bacteria can survive for months in wet, warm areas in which animals typically are housed. Clinical disease is commonly found after a period of stress such as food deprivation, transportation, drought, overcrowding, recent parturition, and drug administration. The frequency at which *Salmonella* is found in these animals varies with the species and geographic location. In calves, salmonellosis is typically endemic on a particular farm that may have sporadic outbreaks. Adult cattle typically have subclinical infection with occasional herd outbreaks. Horses can be *Salmonella* carriers. Mice may be a reservoir of infection in both human and animal salmonellosis. Pets can also harbor *Salmonella*; 10% of the United States canine population and up to 27% of cats worldwide have been cultured positive for *Salmonella*. Approximately 94% of reptiles harbor *Salmonella*.

Increased global trade has resulted in increasing amounts of food contamination with *Salmonella*. *Salmonella* can also be found in sewage and contaminated surface waters. The incidence of disease (especially typhoid fever) decreases when the level of development of controlled water sewage systems increases in a country. Where these hygienic conditions are missing, the probability of fecal contamination of water and food remains high and so is the incidence of disease. The incidence of nontyphoid salmonellosis in the United States is greatest among children younger than 5 years of age (approximately 62 cases per 100,000 people), with the highest numbers occurring in children younger than 1 year of age. Infants and people older than 60 years tend to have more severe infections. In adults about 14 cases occur per 100,000 people; however, most cases are unreported, and the true incidence of this disease may be much higher. Outbreaks of nontyphoid salmonellosis are common among people who are institutionalized and in nursing homes. In the United States from 1985 to 1991 the mortality rate from nontyphoid salmonellosis was 0.4% with fatality rates up to 70 times higher in nursing homes and hospitals. In many countries, the incidence of salmonellosis has markedly increased. Over 1 million people develop nontyphoid *Salmonella* infection annually in the United States. Two-thirds of these patients are younger than 20 years of age with the highest incidence occurring from July through October.

Typhoid fever caused by *S. typhi* rarely occurs in the United States. Cases of typhoid fever occur each year (0.2 per 100,000 people), and these are increasingly associated with travel to developing countries (currently 72% of cases). Typhoid fever is endemic in many developing countries of the Indian subcontinent, South and Central America, and Africa. Sporadic outbreaks of typhoid fever have also occurred in Eastern Europe. An estimated 12 to 33 million cases of typhoid fever occur globally each year. Mortality rates from typhoid fever are low in the United States (<1%), but mortality rates of 10% to 30% have been reported in some parts of Asia and Africa.

Transmission

Salmonella bacteria live in the intestinal tracts of several animals and are found in the natural environment (water, soil, and plants) as a result of human or animal excretion. Humans and animals can excrete *Salmonella* either when clinically diseased or after having had salmonellosis, if they remain carriers. *Salmonella* organisms do not multiply significantly in the environment out of the intestinal tracts of animals, but they can survive several weeks in water and several years in soil if conditions of temperature, humidity, and pH are favorable.

Salmonella bacteria are usually transmitted to humans and animals by the fecal-oral route (ingestion of foods contaminated with feces, unpasteurized milk, food contaminated by the unwashed hands of an infected food handler, or animal feed contaminated with *Salmonella*-contaminated animal by-products). The most common sources of *Salmonella* are beef, poultry, and eggs. In the United States in 2000 the consumption of egg shell fragments contaminated with *S. enterica* serovar Enteritidis was responsible for approximately 182,000 cases of enteritis. Zoonotic transmission of *Salmonella* can occur through exposure to amphibians and reptiles that harbor *Salmonella* as part of their skin normal flora. Approximately 74,000 *Salmonella* infections occur annually in the United States as a result of handling amphibians and reptiles. Veterinary workers, farmers, and animal dealers can be infected by contact with infected animal fluids, membranes, or aborted fetuses. Transmission of nontyphoidal Salmonella is summarized in Figure 3-61.

Pathogenesis

The clinical manifestations of salmonellosis vary with the serovar, the infectious dose, and the health status of the host. Although the infectious dose varies among

Most human food-poisoning *Salmonella* bacteria do not cause clinical illness in animals.

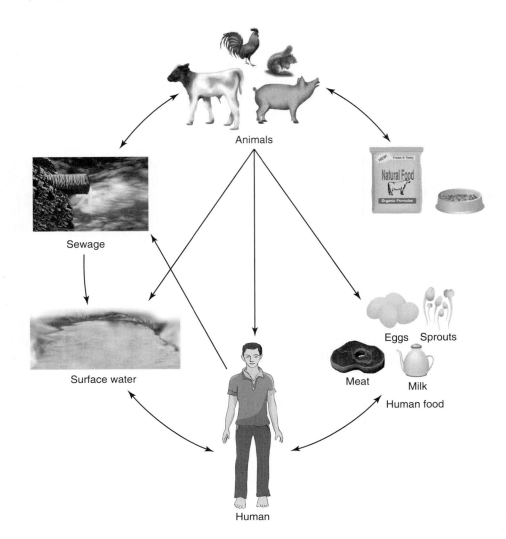

Figure 3-61 Transmission of nontyphoidal *Salmonella* bacteria.

In the 1970s, stringent procedures for cleaning and inspecting eggs were implemented in the United States and have made salmonellosis caused by fecal contamination of egg shells extremely rare. However, eggs can be contaminated by *Salmonella* in ways other than via the shell. *S. enterica* serovar Enteritidis infects the ovaries of healthy appearing hens and contaminates the eggs before the shells are formed.

serovars, a large inoculum is thought to be necessary to overcome stomach acidity and to compete with normal intestinal flora (an oral dose of at least 10^5 *S. typhi* bacteria are needed to cause typhoid fever in 50% of humans and at least 10^9 *S. enterica* serovar Typhimurium bacteria are needed to cause clinical disease in people). Large infectious doses are associated with higher rates of illness and shorter incubation periods; however, lower infectious doses may cause infection if these bacteria are ingested with foods that move through the stomach rapidly (liquids) or that raise gastric pH (cheese, milk).

After ingestion, *Salmonella* bacteria attach by fimbriae or pili to cells lining the intestinal lumen. The bacteria are then engulfed by endocytosis and transported within phagocytes to the mucosa of the intestinal wall where they are released. Once in the intestinal mucosa, *Salmonella* bacteria cause an influx of macrophages (typhoidal strains) or neutrophils (nontyphoidal strains). Nontyphoid *Salmonella* generally cause a localized response, whereas *S. typhi* and other especially virulent strains invade deeper tissues via the lymphatics and capillaries, activating an immune response. When *S. typhi* penetrate human intestinal mucosa they do not cause lesions and are stopped in the mesenteric lymph nodes where bacteria multiply. From the mesenteric lymph nodes, viable bacteria and endotoxin may be released into the bloodstream resulting in septicemia. This endotoxin causes water secretion into the intestinal lumen resulting in diarrhea. Bacteria and endotoxin can then spread via the blood to a variety of organs (such as the brain, meninges, reproductive tract, extremities, urinary bladder, liver, and gallbladder) causing inflammation.

Clinical Signs in Animals

Clinical manifestations of salmonellosis range from none (subclinical carrier) to acute, severe diarrhea. Clinical infection may cause septicemia, acute enteritis, and chronic enteritis. Septicemia tends to occur in young and aged animals, acute enteritis tends to occur in adult animals, and chronic enteritis tends to occur following an acute episode. Chronic and temporary carrier states can also occur. Species specific clinical signs include:

- *Cattle.* The clinical form of salmonellosis varies with the age of the animal. Septicemia is seen in newborn calves resulting in lethargy, fever, and death in 24 to 48 hours. Nervous signs and pneumonia may occur in calves. In adult cattle, subclinical infection is common with sporadic outbreaks occurring as a result of external stressors such as feed/water deprivation, long transportation, calving, or crowding in feedlot animals. Acute enteritis is the common clinical form seen in adult cattle and calves older than 1 week of age. The first signs observed may be fever and abortion, followed several days later by diarrhea (Figure 3-62). Other clinical signs of acute enteritis include severe watery diarrhea and tenesmus (straining to defecate). In herd outbreaks, several hours may pass before the onset of diarrhea, during which time the fever may resolve. The consistency of feces varies considerably and may have a putrid odor and contain mucus, casts, pieces of mucous membrane, and blood clots. Milk production may decline in dairy cows. Chronic enteritis is also seen in adult cattle. Persistent, small amounts of diarrhea with or without mucus or blood, emaciation, and intermittent fever may be seen with chronic enteritis.

- *Sheep.* Septicemia is seen in newborn lambs producing fever, lethargy, and death in 24 to 48 hours. Subacute enteritis may occur in sheep on endemic farms producing

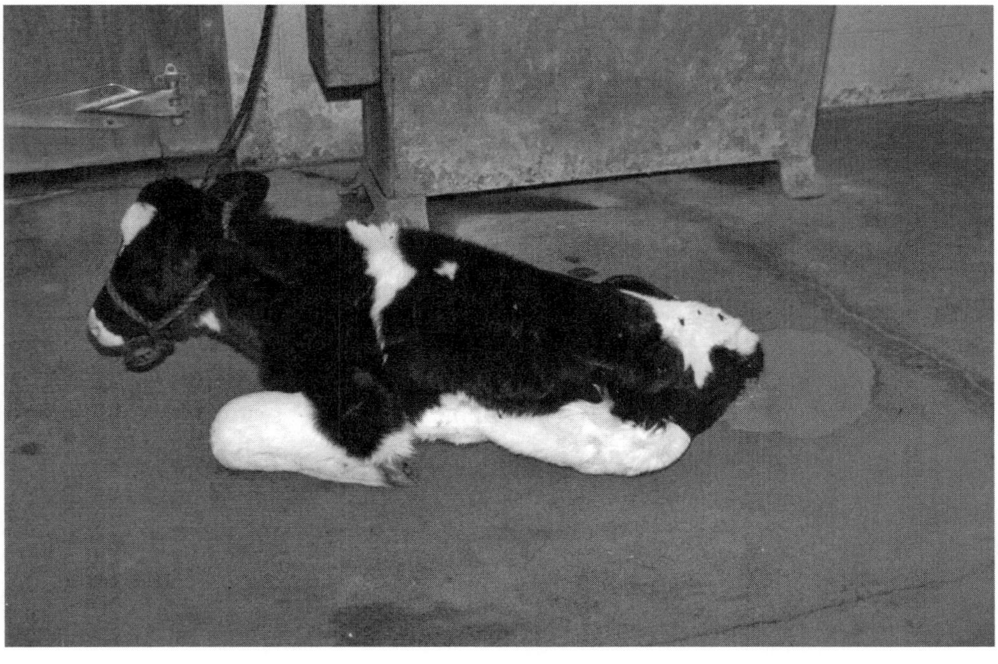

Milk can be contaminated with *Salmonella* bacteria in a variety of ways: contamination of milk with feces during the milking process, use of contaminated water or dirty equipment during the processing of milk, or excretion of bacteria in the milk during the febrile stage of clinical disease in cows (*Salmonella* is adapted to and can colonize the bovine mammary gland).

Figure 3-62   Bovine with diarrhea from *Salmonella.*

(Courtesy of Dr. Ramos-Vara, College of Veterinary Medicine, Michigan State University)

signs that include mild fever, soft feces, anorexia, and dehydration. There may be a high incidence of abortion in ewes, some ewe death after abortion, and a high mortality rate in lambs under a few weeks of age as a result of enteritis.

- *Horses.* Septicemia is seen in newborn foals producing fever and death in 24 to 48 hours. Subacute enteritis may occur in adult horses on endemic farms producing mild fever and soft feces. Abdominal pain is common and severe in horses. Affected horses are severely dehydrated and may die within 24 hours after the onset of diarrhea and mortality may reach 100%. A marked leukopenia (low white blood cell count) with neutropenia (low neutrophil count) is characteristic of the acute disease in horses. Many horses become carriers. Mares may be inapparent shedders and may shed bacteria at parturition and infect the newborn foal. Salmonellosis in hospitalized horses is a major problem for equine clinics and stud farms.

- *Swine.* Septicemia is rare in pigs but may be seen in newborn piglets and outbreaks may occur in pigs up to 6 months of age. The onset of illness is acute producing fever and death in 24 to 48 hours. In pigs, a dark red to purple discoloration of the skin is common, most noticeably seen on the ears and ventral abdomen. Nervous signs and pneumonia may occur in pigs. Chronic enteritis is a common form of salmonellosis in pigs, producing persistent diarrhea, emaciation, and intermittent fever. In growing pigs, rectal stricture may form if the terminal part of the rectum is infected (Figure 3-63).

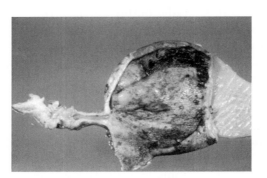

Figure 3-63 Marked narrowing of a pig rectum because of inflammation caused by *Salmonella* Typhimurium.

(Courtesy of Leland S. Shapiro)

Salmonella-infected pigs are anorectic and lose weight; however, the abdomen becomes grossly distended because of rectal stricture. Most outbreaks of salmonellosis in pigs can be traced to a purchased, infected pig; therefore, purchasing pigs from *Salmonella*-free herds and using an all-in/all-out policy in finishing units is desired to minimize the risk of exposure to *Salmonella.*

- *Cats and dogs.* Many dogs and cats are carriers of *Salmonella* and do not produce clinical signs. When clinical disease occurs it produces acute diarrhea with septicemia and occurs in puppies and kittens or adults stressed with concurrent conditions such as hospitalization, co-infection or a debilitating condition in adults, or exposure to large numbers of the bacteria in puppies and kittens. Other manifestations of salmonellosis include pneumonia, abortion, and conjunctivitis (mainly in affected cats).

- *Wildlife.* Carnivores such as foxes can be infected with *Salmonella* as a result of contaminated feed, producing clinical signs of acute enteritis. Several rodents (Guinea pigs, hamsters, rats, and mice), rabbits, and wild birds are susceptible and act as a source of infection on farms where the disease is endemic.

The development of chronic and temporary carrier states in animals is also possible. Chronic carrier states occur in animals that have recovered from clinical illness and appear normal; however, they excrete organisms for several weeks or months. Temporary carrier states occur when bacteria that have been continually acquired from the environment via contaminated feed or fecal contamination are shed. Temporary carriers are typically found in slaughter animals and animals coming from stockyards or kennels.

A latent form of salmonellosis has also been described. Animals with latent infection will appear healthy, but will have *Salmonella* bacilli isolated from the mesenteric lymph nodes at slaughter. Latent infection is most likely a result of repeated ingestion of contaminated feed.

Clinical Signs in Humans

Salmonella infections in humans produce three distinct syndromes: gastroenteritis, focal disease, or typhoid (enteric) fever. Infection with nontyphoid *Salmonella* manifests itself as gastroenteritis and causes nausea, vomiting, and diarrhea (usually bloodless) that occurs 6 to 48 hours after ingestion of contaminated food or drink. The diarrhea is self-limiting and typically resolves in 3 to 7 days. Occasionally nontyphoid *Salmonella* may spread to the cardiovascular, neurologic, and musculoskeletal systems causing inflammation. Focal disease produces variable lesions and is defined by the site of infection (cardiovascular, genitourinary, pulmonary, and neurologic systems). Focal disease is rare and is typically caused by *S. enterica* serovar Typhimurium.

Infection with *S. typhi* is associated with acute abdominal pain, either constipation or recurrent diarrhea, high fever, chills, and headache. The fever usually resolves within 48 hours. Typhoid fever may present with both gastrointestinal and skin signs (raised red macules called rose spots may be seen on the chest and abdomen) (Figure 3-64).

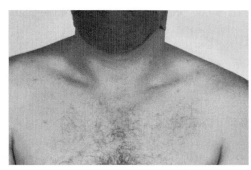

Figure 3-64 Rose spots in person with typhoid fever.
(Courtesy of CDC/AFIP, Charles Farmer)

Diagnosis in Animals

The lesions associated with salmonellosis are those of a septicemia or a necrotizing enteritis, or both. Lesions are typically present in the terminal ileum and the large intestine, whereas the stomach and proximal small intestine are usually not affected (Figure 3-65). On biopsy, lesions vary from shortening of intestinal villi with loss of the epithelium to complete loss of intestinal architecture. The mucosal layer of the intestine may be hyperemic, thickened, and is often covered with yellow or grey exudates. Microscopically there is edema, necrosis, and a neutrophilic reaction in the intestine. Blood vessels in this region may contain thrombi.

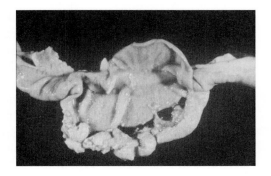

Figure 3-65 Jejunum from a calf demonstrating the classic plug of tissue (cast) seen in intestines infected with *Salmonella* Typhimurium.
(Courtesy of Leland S. Shapiro)

Lesions in the liver include small foci of necrosis and paratyphoid nodules containing small clusters of histiocytes or macrophages. Numerous leukocytes may be in the liver sinusoids. Hyperplasia is present in lymph nodes and the spleen, causing these tissues to enlarge. Lesions in other organs are less consistent.

Diagnosis of salmonellosis in animals depends on clinical signs and laboratory examination of feces, affected tissues, feed, water, and possible sources of infection such as rodent feces. *Salmonella* are gram-negative, lactose-negative rods that produce hydrogen sulfide. Culture techniques should involve suppression of fecal normal flora *E. coli* and should include several daily fecal cultures to isolate the organism. A variety of lowly selective media have been developed for culturing of *Salmonella* bacteria including MacConkey agar, eosin-methylene blue (EMB) agar, and bromocresol purple lactose agar. Intermediately selective agars used to help identify *Salmonella* are *Salmonella-Shigella* (SS) agar, bismuth sulfite agar, Hektoen enteric (HE) agar, and xylose-lysine-deoxycholate (XLD) agar. Highly selective media such as selenite agar with brilliant green can also be used to identify *Salmonella*. Blood cultures in septicemic animals may be rewarding but are expensive. Serologic testing is difficult to interpret. Individual isolates can then be distinguished by serotyping, bacteriophage typing, and genotyping.

Diagnosis in Humans

Diagnosis of salmonellosis in humans is similar to that in animals. Freshly passed stool is the preferred specimen for culturing of nontyphoidal *Salmonella* species. Various selective media are routinely used, including MacConkey agar, EMB agar, HE agar, SS agar, and XLD agar. *Salmonella* species may be differentiated by conventional biochemical assays, phage typing, and serologic typing; specialized laboratories may use plasmid analysis, gel electrophoresis, and PCR to identify *Salmonella* bacteria.

For diagnosis of typhoid fever, *S. typhi* may be isolated from stool, urine, or biopsy of the rose spot rash and from blood, bone marrow, or gastrointestinal secretions using the media described above. Agglutinating antibodies can be measured to diagnose tphoid fever. The white blood cell count of people with typhoid fever is usually within reference ranges, with approximately one fourth of patients being leukopenic, neutropenic, or anemic.

Treatment in Animals

Early treatment using parenterally administered, broad-spectrum antibiotics is essential for septicemic salmonellosis. Trimethoprim-sulfonamide combinations, ampicillin, fluoroquinolones, or third-generation cephalosporins are effective antibiotics used to treat *Salmonella* bacteria. Treatment should be continued daily for up to 6 days. Antibiotic treatment can lead to lysis of bacteria resulting in endotoxin release. Fluid therapy to correct dehydration and acid-base and electrolyte imbalance is necessary to correct these problems. Nonsteroidal anti-inflammatory drugs (NSAIDs) may be used to reduce the effects of endotoxemia. Corticosteroids are not recommended because of their immunosuppressive effects.

The use of antibiotics to treat enteric forms of salmonellosis is questionable. Oral antibiotics may alter the intestinal normal flora and prolong shedding of the bacteria. The development of antibiotic-resistant strains of *Salmonella* is also a concern when treating enteric salmonellosis. Fluid therapy is essential to correct electrolyte imbalance and dehydration. NSAIDs may be used to reduce the effects of endotoxemia. Animals should have stool samples cultured to assess the effectiveness of treatment. Carriers of *Salmonella* are not treated with antibiotics, but are treated with an attenuated *Salmonella* vaccine.

Treatment in Humans

Salmonella gastroenteritis is usually a self-limiting disease in people. Supportive therapy including oral fluid and electrolyte replacement as recommended by WHO is essential. Antibiotics do not appear to shorten the duration of clinical signs and actually prolong the duration of convalescence; therefore, they are not routinely used to treat uncomplicated nontyphoidal *Salmonella* gastroenteritis. Antibiotic sensitivity testing and treatment is indicated in infants, small children, immuno-compromised people, and severe cases.

Treatment of typhoid fever historically consisted of oral antibiotics such as ampicillin, trimethoprim-sulfamethoxazole, or chloramphenicol; however, emerging drug resistance has limited the usefulness of these antibiotics. Fluoroquinolone, macrolide, and third-generation cephalosporin antibiotics are now used to treat typhoid fever. *Salmonella* excretion by asymptomatic carriers is potentially dangerous following antibiotic treatment. Typhoid fever may occasionally be complicated by intestinal perforation or hemorrhage necessitating surgical intervention. Patients with *S. typhi* residing in the gallbladder as the reservoir may require a cholecystectomy.

Management and Control in Animals

Control of *Salmonella* in animals includes preventing introduction of infected animals into a herd or facility and limitation of spread within a herd or facility. Carrier animals and contaminated feedstuffs are major source of *Salmonella* in livestock facilities. Ways to prevent carrier animals include:

- Purchasing animals directly from disease-free farms or facilities
- Isolating animals for 1 week or more while their health status is checked
- Purchasing *Salmonella*-free supplies from reputable sources.

  Limiting the spread of *Salmonella* within a herd or facility includes the following:
- Identifying and culling or isolating and aggressively treating carrier animals (treated animals should be rechecked several times before they are considered noncarriers)
- Restricting the movement of animals from one area of the facility to another to limit any infection to the smallest group
- Avoiding random mixing of animals
- Cleaning and disinfecting contaminated buildings
- Protecting feed and water supplies from fecal contamination
- Disposing of contaminated material properly
- Educating workers about how to handle infected animals and using proper hygiene for both themselves and the animals
- Culturing equipment such as drains and milk filters to monitor the *Salmonella* status of a herd
- Using vaccines in certain situations such as outbreaks in pregnant cattle (to protect adults and calves)
- Minimizing stresses on animals
- Properly disposing of fetal membranes and other tissues that may be infected and come in contact with other animals
- Using antibiotics prophylactically in feed or water supplies; however, this is widely believed to lead to antibiotic resistant strains of *Salmonella* and is typically not used.

Management and Control in Humans

To prevent nontyphoid salmonellosis, proper food handling and storage are essential (reducing fecal contamination of food and water was previously covered in the *E. coli* section). Since foods of animal origin may be contaminated with *Salmonella* bacilli, people should not eat raw or undercooked eggs, poultry, or meat and should wash their hands after handling eggs, poultry, and meat. Raw eggs may be found in foods such as homemade sauces and salad dressings, homemade ice cream, homemade mayonnaise, cookie dough, and frostings. Poultry and meat, including hamburgers, should be well-cooked until they are no longer pink in the middle and reach their recommended cooking temperatures (internal temperature of 165°F). People also should not consume raw or unpasteurized milk or other dairy products. Meat, eggs, and dairy products should be stored at temperatures less than 40°F and frozen meat should be thawed at refrigerator temperatures and served soon after cooking. Plants grown with organic fertilizer or in developing countries (40% of U.S. produce is grown in foreign countries) may be contaminated with *Salmonella*; therefore, produce (including already chopped lettuce mixtures) should be thoroughly washed before consuming.

Avoiding cross-contamination of foods is another important step in controlling the spread of *Salmonella*. Ways to avoid cross-contamination include:

- Keeping uncooked meats and eggs separate from produce, cooked foods, and ready-to-eat foods
- Washing hands, cutting boards, counters, knives, and other utensils thoroughly after handling uncooked foods
- Washing hands before handling any food and between handling different food items
- Restricting people who have salmonellosis from preparing food or pouring water for others until they are no longer considered to be carrying the *Salmonella* bacterium.

To prevent nontyphoid salmonellosis from animal contact, people should:

- Thoroughly wash hands after contact with animal feces and after handling reptiles
- Wear gloves when handling a reptile or its feces. Thorough hand washing after handling a reptile is imperative. Reptiles (including turtles) are not appropriate pets for small children and are not considered good pets in households with an infant
- Not clean cages, water or food bowls, or enrichment items of reptiles in the kitchen or bathroom sink. Reptiles should not be allowed to bathe in the bathtub.

There is no human vaccine to prevent nontyphoid salmonellosis. There is a vaccine for *S. typhi* that is not routinely recommended for people in the United States, but is recommended for people traveling to some foreign countries.

Summary

Salmonellosis is an enteric disease caused by the gram-negative bacillus *Salmonella*. There are over 2,400 serovars of *Salmonella* that are distinguished serologically on the basis of surface antigens O (lipopolysaccharide), H (flagellar), and Vi (surface). Typhoid fever is a potentially fatal enteric disease caused by *S. typhi* and is strictly a human pathogen transmitted by the fecal-oral route. All other species of *Salmonella* occur worldwide in the intestinal tract of a variety of animals. Nontyphoid salmonellosis is caused mainly by the serovars *S. enterica* serovar Enteritidis (mainly from eggs) and *S. enterica* serovar Typhimurium (mainly from meat and dairy products). Nontyphoid salmonellosis occurs when people or animals ingest large numbers of *Salmonella* bacilli that invade the epithelium of the small intestine and begin to replicate. These bacteria typically stay in the intestinal tract causing diarrhea, abdominal cramps, fever, and vomiting. Diagnosis of *Salmonella* infection in animals and humans includes bacterial culture and serologic tests. Nontyphoid salmonellosis in both animals and people is usually self-limiting except in the young, old, and immunocompromised. Antibiotic treatment in both animals and people is questionable since treatment can produce carriers of *Salmonella* as well as antibiotic-resistant strains. Control of *Salmonella* in animals includes preventing introduction of infected animals into a herd or facility and limitation of spread within a herd or facility. Carrier animals and contaminated feedstuffs are major source of *Salmonella* in livestock facilities. To prevent nontyphoid salmonellosis in people, proper food handling and storage are essential. There is a vaccine for *S. typhi* that is not routinely recommended for people in the United States, but is recommended for people traveling to some foreign countries.

SHIGELLOSIS

Overview

Shigellosis, also known as bacillary dysentery, was first described in the 4th century B.C. and is caused by several serovars of *Shigella* bacteria. *Shigella* is named for Kiyoshi Shiga, a Japanese scientist who in 1897 identified this bacterium as the cause of an outbreak of diarrhea that killed more than 22,000 people in a 6-month period in Japan. Shigellosis was once known as asylum dysentery because of the large numbers of outbreaks that occurred in mental institutions. Outbreaks continue to occur sporadically especially when people are in confined areas such as cruise ships, hotels, and institutions. A 1994 outbreak on the Viking Serenade cruise ship occurred during a round trip voyage from San Pedro, California, to Ensenada, Mexico. Thirty-seven percent (586) of 1589 passengers and 4% (24) of 594 crew members were infected during the cruise, with one case being fatal. Each year 1.1 million people die from *Shigella* infection and 580,000 cases of shigellosis are reported among travelers from industrialized countries. A total of 69% of all episodes and 61% of all deaths attributable to shigellosis involve children younger than 5 years of age. Because *Shigella* bacteria only replicate in the gastrointestinal tracts of primates, humans and higher level primates (chimpanzees and gorillas) are the only animals that acquire shigellosis.

Causative Agent

Shigella spp. are gram-negative bacilli that are facultative aerobes. *Shigella* spp. are very infectious pathogens that produce acute disease with as few as 10 bacteria. Virulence factors for *Shigella* include the production of toxin and its ability to adhere to and invade mucosal cells. There are several different species of *Shigella* bacteria with many serotypes of each species:

- *Sh. sonnei*, also known as group D, causes over two-thirds of the shigellosis in the United States. *Sh. sonnei* and *Sh. boydii* are usually associated with mild illness of short duration that produces watery or bloody stool.
- *Sh. flexneri*, or group B, causes most of the remaining cases of shigellosis in the United States. *Sh. flexneri* is a principal cause of endemic shigellosis in many developing countries and generally produces more severe, longer lasting bloody diarrhea.
- *Sh. dysenteriae*, or group A, causes the most severe cases of dysentery and is responsible for deadly epidemics in other parts of the world, particularly the eastern hemisphere. *Sh. dysenteriae*, particularly type 1, causes the most severe cases of diarrhea with the highest death rates.

- *Sh. boydii*, or group C, causes less than 1% of shigellosis cases in the United States and along with *Sh. dysenteriae* are responsible for international cases of shigellosis.

Epizootiology and Public Health Significance

The source of shigellosis is typically other humans, who are sick or carriers of this bacterium. Apparently healthy nonhuman primates can also be carriers of these organisms.

Shigellosis has a worldwide distribution and has been seen with higher frequency in Africa, Guinea, Sierra Leone, and India. Changes in the worldwide distribution of *Shigella* spp. have occurred in the last part of the 20th century. In industrialized regions, *Sh. sonnei* has become the principle source of shigellosis, whereas *Sh. flexneri* remains the leading cause of shigellosis in the developing world. The annual number of *Shigella* cases worldwide was estimated in 1997 to be 164 million, of which 163 million were in developing countries (with 1.1 million deaths) and 1 million in industrialized countries. About 70% of all cases of shigellosis and about 6% of all deaths attributable to shigellosis involve children younger than 5 years of age. Increases in cases of shigellosis can be seen during times of war and natural disasters such as floods and earthquakes.

Approximately 300,000 cases of shigellosis occur annually in the United States; however, the number of cases is probably underreported because of the self-limiting nature of mild cases. Shigellosis is common and recurrent in settings where hygiene is poor and can sometimes sweep through entire communities. Shigellosis is more commonly seen in summer than in winter. Children, especially toddlers younger than 5 years of age, are more likely to get shigellosis. Antibiotic-resistant strains of *Shigella* are seen with increasing frequency worldwide and outbreaks of antibiotic-resistant shigellosis, especially in the developing world, are a cause for concern. In the developing world, shigellosis is endemic in some communities.

Transmission

Flies, fingers, feces, and food (the 4 Fs) are usual vehicles for transmission of *Shigella*.

Shigella organisms only multiply in the gastrointestinal tract of primates. The organisms are primarily transmitted via the fecal-oral route, which can include fecally contaminated water and foods. Food may become contaminated by infected food handlers who forget to wash their hands with soap after using the bathroom, by contaminated vegetables that are harvested from a field with sewage on it, and by flies that bred in infected feces and then contaminate food. *Shigella* infections can also be acquired by drinking or swimming in contaminated water if contaminated sewage runs into the water or if someone with shigellosis swims in it (*Shigella* can survive in water and infect other swimmers who swallow the water or even get their lips wet). Playing, bathing, and washing clothes in contaminated water are significant sources of *Shigella* transmission.

Shigella can survive for long periods of time in contaminated water and on fomites. *Shigella* bacteria are present in stool from infected individuals for 1 to 2 weeks after the person becomes ill. People who live in crowded conditions with marginal cleanliness are likely to contract this bacterium. Children are more likely to contract *Shigella* as a result of poor hygiene habits, especially toddlers who are not fully toilet-trained. *Shigella* can also be sexually transmitted by the oral-fecal route.

Pathogenesis

Shigella bacteria can persist in foods for up to one month.

Shigella spp. can be highly virulent and colonize the intestine using their pili to adhere to receptors on cells of the colon. Once contaminated food or water is ingested, *Shigella* survive the acidity of the stomach, pass through the small intestine, and adhere to portions of the small and large intestine. The adhesion of this bacterium to mucosa cells of the colon stimulates phagocytosis. Once phagocytized, the bacterium lyses the phagocyte to enter its cytoplasm. *Shigella* organisms multiply rapidly in the phagocyte, produce toxins, and in about 6 hours lyse the phagocyte releasing many bacteria that infect neighboring epithelial cells. The organisms continue to kill intestinal epithelial cells by spreading from cell to cell. These events cause a great deal of inflammation and produce classic patchy areas called microabscesses in the colon. *Shigella* bacteria can stay within the intestinal epithelium causing blood samples of people or animals infected with this organism to test negative for these bacteria.

Some strains of *Shigella* produce toxins. *Sh. dysenteriae* produces Shiga toxin, which causes severe abdominal cramps and frequent, bloody stools. Other strains of *Shigella* produce shiga-like toxins. These toxins damage blood vessels in the intestinal wall and cause intense intestinal inflammation. Bleeding and inflammation cause stool to contain streaks of blood and to contain strings of mucus. The toxin can also produce watery diarrhea and can also spread from the intestines to the nervous system, resulting in nervous system signs such as convulsions.

Clinical Signs in Animals

Shigellosis in animals and people has similar presentations. Animals affected by *Shigella* bacteria are typically nonhuman primates in captivity where the disease can cause high mortality. Monkeys and other nonhuman primates are not believed to be the origin of the infection, but become infected by contact with infected humans. The infection can spread rapidly in nonhuman primate colonies as a result of the fact that these animals defecate on the cage floor, throw their food on the floor, and throw feces at other animals. It is not believed that monkeys can harbor the bacteria in their natural habitat. The incubation period for shigellosis is less than 4 days. The disease presents with fever and abdominal pain, followed by diarrhea and dehydration that typically lasts 1 to 3 days (Figure 3-66). A

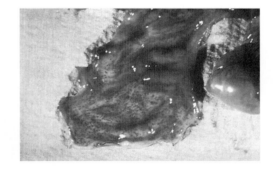

Figure 3-66 Intestine of a monkey with *Shigella* lesions. *Shigella* causes inflammation of the intestine and may produce petechial hemorrhages. (Courtesy of CDC)

second phase of shigellosis lasts for several weeks with the major disease sign of tenesmus. Stool may contain blood, mucus, and pus.

Shigella spp. have been isolated from horses, bats, and rattlesnakes, but they pose little threat of disease spread for this bacterium.

Clinical Signs in Humans

Shigellosis presents within 1 to 2 days after exposure as diarrhea (usually bloody), cramping, and fever (Figure 3-67). Most cases are minor and resolve in 5 to 7 days except in the young, the elderly, the pregnant, and those that are immunocompromised (in these groups of people the diarrhea

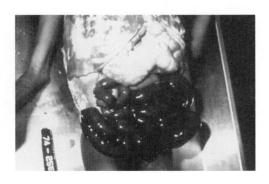

Figure 3-67 Intestinal necrosis in a human with *Shigella*. (Courtesy of CDC/Dr. Eugene Gangaroa)

can be so severe that the patients need to be hospitalized as a result of dehydration). A severe infection with high fever may cause seizures in children younger than 2 years of age. Some people infected with *Shigella* may have no symptoms at all, but may become carriers and pass the *Shigella* bacteria to others.

People with diarrhea caused by *Shigella* usually recover completely, although it may take several months for their stool to become normal. About 3% of people infected with *Sh. flexneri* will develop Reiter's syndrome (joint pain, eye irritation, and painful urination) that can last for months or years and can lead to chronic arthritis. Reiter's syndrome is caused by a reaction to *Shigella* bacteria. Bacteremia from *Shigella* organisms in malnourished children has a mortality rate of 20% as a result of renal failure, hemolysis, thrombocytopenia, gastrointestinal hemorrhage, and shock. People with shigellosis may develop hemolytic uremic syndrome (HUS), which is characterized by acute hemolysis, renal failure, uremia, and disseminated intravascular coagulation (DIC), which has a mortality rate greater than 50%.

Once someone has contracted and recovered from shigellosis, they are unlikely to become infected with that specific type of *Shigella* bacteria for several years. They can however become infected with other species of *Shigella*.

Diagnosis in Animals

Diagnosis in animals consists of bacteriologic or serologic identification. Bacteriologic identification involves the Gram stain and cellular morphology. *Shigella* spp. are gram-negative bacilli that are facultative aerobes, nonmotile, nonencapsulated, and nonendospore forming rods. Growth on MAC (MacConkey), XLD (xylose-lysine-dextrose), HE (Hektoen enteric), and SS (*Salmonella-Shigella*) agars produces lactose negative, nonspreading colonies (Figure 3-68). It should be noted that SS agar inhibits *Sh. dysenteriae* growth. A variety of biochemical tests, including triple sugar iron inoculation, urease testing, and Simmon's citrate inoculation, are needed to confirm *Shigella* spp. because of antigenic cross-over with serologic test methods.

Figure 3-68 *Shigella* bacteria growing on HE agar.
(Courtesy of CDC)

Serologic testing can be done to identify *Shigella* antigen or toxin. A variety of commercially available agglutination tests are available for subtyping of this bacterium. PCR testing is also available.

Diagnosis in Humans

Diagnosis of shigellosis in humans revolves around culturing of microorganism from stool and mucus (as described above), detection of antigens or antibodies to the organism via serology, immunofluorescent detection of microorganism in stool samples, and PCR testing. Colonic tissue biopsies can be taken to identify *Shigella* bacteria. Microabscesses may be seen histologically.

Treatment in Animals

Nonhuman primates have been treated with fluoroquinolone antibiotics until they are no longer shedding the organism. Symptomatic treatment consisting of fluid therapy may be indicated in animals with dehydration.

Treatment in Humans

People with mild cases of shigellosis usually recover without treatment. In more severe cases, shigellosis can be treated with antibiotics such as ampicillin, trimethoprim-sulfamethoxazole, nalidixic acid, or ciprofloxacin, although multiple antibiotic-resistant strains are common. Antidiarrheal agents are likely to make the illness worse by keeping the bacterium in the intestinal tract for a longer time and should be avoided. Dehydration can be treated with fluid therapy.

Management and Control in Animals

Control of shigellosis in animals involves containment of disease within a laboratory or zoological setting. In laboratory or zoo settings control consists of isolation and treatment of sick or carrier animals, sterilization of cages, disinfection of the facility, prompt removal of animal waste, control of insects, and prevention of animal overcrowding. Animal caretakers should be educated in proper hygiene procedures to avoid spreading the disease to animals or from animal to animal.

Management and Control in Humans

In the United States, the CDC monitors the frequency of *Shigella* infections and assists local and state health departments to investigating outbreaks. The FDA helps control diseases like shigellosis by inspecting foods and promoting better food preparation techniques in restaurants and food processing plants. The EPA helps control waterborne spread of shigellosis by regulating and monitoring the safety of the drinking water.

Control of shigellosis in humans includes a variety of hygiene procedures. Some control methods include:

- Frequent and careful handwashing with soap. Supervised hand washing of all children in day care centers and in homes with children who are not completely toilet-trained is recommended.
- Isolation of people with *Shigella* infection especially young children.
- Proper supervision of the preparation of foods, in particular salads (potato, tuna, shrimp, macaroni, and chicken), raw vegetables, milk and dairy products, and poultry. This includes not allowing people with shigellosis to prepare food or to pour water for others.

- Using clean water that has not been contaminated with human waste for washing food and personal hygiene. Drink only treated or boiled water.
- Providing potable water for use outdoors and providing enough bathrooms near swimming areas to help keep the water from becoming contaminated with fecal material.
- Controlling flies to prevent this mode of transmission.
- Notifying the public health department about cases of shigellosis. Isolates of *Shigella* should be sent to the city, county, or state public health laboratory so the specific type can be determined.

A live vaccine that provides protection for 6 to 12 months can be given to prevent clinical disease (live vaccines have been developed for *Sh. flexneri* and *Sh. sonnei*). The vaccine is given orally in three or four doses. Additional research is attempting to produce vaccines for the other species of *Shigella*.

Summary

| Cases of shigellosis must be reported to the local health department. |

Shigella is a common cause of intestinal illness of developing countries of the world and is often seen in primates, but rarely in other animals. *Shigella* infections may also be seen in laboratory and zoological settings that house nonhuman primates. *Shigella* is a gram-negative rod that causes diarrhea, abdominal cramping, and fever in primates. *Shigella* is classified as follows: *Sh. dysenteriae* (group A), the one originally described by Shiga; *Sh. flexneri* (group B); *Sh. boydii* (group C); and *Sh. sonnei* (group D). *Shigella* spp. are very infectious pathogens that produce acute disease with as few as 10 bacteria. Virulence factors for *Shigella* include the production of toxin and its ability to adhere to and invade mucosal cells. Shigellosis is usually self-limiting, but can cause dehydration in the young, old, pregnant, and immunocompromised individuals. *Shigella* is usually spread through fecal-oral contamination, including contaminated food and water. Animals other than nonhuman primates play an insignificant role in the epidemiology of shigellosis. Humans and nonhuman primates with shigellosis present with fever and abdominal pain, followed by diarrhea and dehydration that typically lasts 1 to 3 days. A second phase of shigellosis lasts for several weeks with the major disease sign of tenesmus. Stool may contain blood, mucus, and pus. Diagnosis of shigellosis consists of bacteriologic or serologic identification. Serologic testing is done to identify *Shigella* antigen or toxin. Shigellosis has been treated in animals and humans with antibiotics and fluid therapy if the patient is dehydrated. People with mild cases of shigellosis usually recover without treatment. Control of shigellosis in animals involves containment of disease within a laboratory or zoological setting. Control of shigellosis in humans includes a variety of hygiene procedures. In the United States, the CDC monitors the frequency of *Shigella* infections and assists local and state health departments in investigating outbreaks.

STAPHYLOCOCCAL INFECTIONS

Overview

Staphylococcal infections have been present since recorded history. It is believed that the boils Job scraped with broken pottery in the Bible were caused by *Staphylococcus* bacteria. Both Robert Koch (in 1878) and Louis Pasteur (in 1880) observed *Staphylococcus* bacteria. In 1884, Sir Alexander Ogston of the

Royal Aberdeen Infirmary named the organism *Staphylococcus* (*staphyle* is Greek for *bunch of grapes* and *kokkos* is Greek for kernal or berry) as a result of its appearance under the microscope. Also in 1884, Rosenback grew *Staphylococcus* on agar for the first time. From 1884 to1929, the role of *Staphylococcus* bacteria was linked to foodborne human illness from contaminated cheese, meat from ill cattle, milk from cows with mastitis, and contaminated baked goods. *St. aureus* were the bacteria Sir Alexander Fleming destroyed by the mold *Penicillium notatumin* in 1928.

There are about thirty species of *Staphylococcus* with most being completely harmless as normal flora of the skin and mucous membranes of humans and animals. They can also be found in small amounts in soil. *Staphylococcus* can cause a wide variety of diseases either through toxin production or invasion (food poisoning as a result of the heat-resistant enterotoxins the bacterium produces), wound infections (as a result of its ability to survive on dry surfaces), shock syndromes (when toxins are released into the bloodstream), and infections in immunocompromised individuals. In recent years other *Staphylococcus* spp. such as *St. lugdunensis*, *St. schleiferi*, and *St. caprae* have been implicated in human infections.

A concern with *Staphylococcus* bacteria is the development of antibiotic resistance. One main concern is methicillin-resistant *St. aureus* (MRSA). Before penicillin was discovered, a *St. aureus* infection was often fatal. *St. aureus* infection could start on the skin and spread to internal organs causing a slow and painful death. When penicillin was first used in 1941 it was effective against *St. aureus*, but it was only within a few years that doctors began to see signs of resistance. During these years, the bacterium acquired a plasmid containing a gene for the enzyme penicillinase, which destroys the penicillin molecule. By the mid-1950s, penicillin was no longer effective against *St. aureus*. In 1959, a new drug called methicillin was discovered. This new drug was a synthetic penicillin that the penicillinase enzyme could not recognize. By 1961, the first report of methicillin-resistant *St. aureus* was reported in England. *St. aureus* bacteria acquired resistance to this antibiotic by a gene called mecA that encodes for an altered penicillin binding protein in the bacterial cell wall that also blocks the binding of cephalosporins. By the 1980s, MRSA was endemic in U.S. hospitals and currently only one antibiotic, vancomycin, is fully effective against all MRSA strains. MRSA is now emerging as a disease that can be spread from person to person in the community (community-acquired (CA) MRSA) and is seen in animals that may spread the bacterium from animal to person.

Causative Agent

Staphylococcus bacteria are found on the skin of most mammals. *Staphylococcus* bacteria are small, nonmotile, salt-tolerant, gram-positive, facultatively anaerobic spherical bacteria that are typically clustered. The cluster arrangement is a result of two characteristics of cell division: the division in successively different planes and the fact that daughter cells remain attached to one another (Figure 3-69). *Staphylococcus* bacteria are also tolerant of drying, radiation, and heat (up to 60°C for 30 minutes), which explains their survival on environmental surfaces in addition to skin. There are two species commonly associated with staphylococcal disease in humans: *St. aureus* and *St. epidermidis*. Species of *Staphylococcus* that are of interest in humans and animals include:

> Because *Staphylococcus* bacteria are salt-tolerant and capable of growing in up to 10% NaCl, they can tolerate the salt deposited on human skin by sweat glands.

- *St. aureus.* *St. aureus* is the main pathogen in humans, but has also been recovered from dogs and cats. Most strains of *St. aureus* have preferred-host

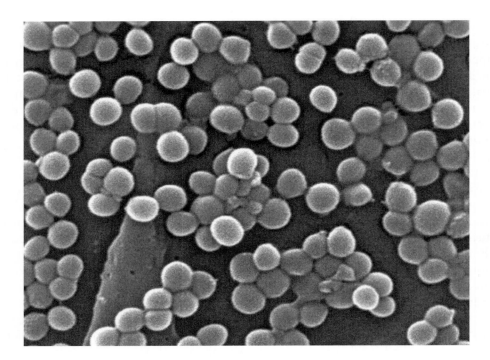

Figure 3-69   Scanning electron micrograph demonstrating the spherical clumping arrangement of *Staphylococcus* bacteria.
(Courtesy of CDC/Janice Carr/Jeff Hageman, M.H.S.)

species; however, they may opportunistically infect other species under some circumstances. Cases of reverse zoonoses are rare and typically involve pets of healthcare workers who are exposed to *St. aureus* through direct contact with the owner. *St. aureus* is also a common cause of mastitis in cattle and is spread cow-to-cow via milking equipment, udder cleaning cloths, and human hands. The spread of *St. aureus* from bovine udder to human is rare.

- *St. epidermidis.* This species of *Staphylococcus* is part of the normal flora of human skin, but can be an opportunistic pathogen in immunocompromised people especially via the use of intravenous catheters and other medical equipment.

- *St. intermedius.* *St. intermedius* is similar to *St. aureus* and many strains of this species have been previously identified as *St. aureus*. It is normal flora of canine skin; however, it can cause opportunistic disease. Canine pyoderma is almost exclusively caused by *St. intermedius*. *St. intermedius* was first reported in 1976 and is almost exclusively pathogenic in dogs; however, it has been recovered from numerous species including humans.

- *St. schleiferi.* This organism was first identified in 1988 (named after German microbiologist Carl Heinz Schleiferi) and the first infection in humans was reported in 1989. *St. schleiferi* has been reported in cases of canine pyoderma, canine otitis in Japan, and in numerous case studies in Europe. This organism has been found in infected wounds, osteomyelitis, catheter-related infections, pleural effusion, urinary tract infections, sepsis, and corneal ulcers. Humans can carry this bacterium in the preaxillary area. In 2002 the first cases of *St. schleiferi* skin infections were reported in the United States. *St. schleiferi* is believed to be normal flora of canine skin. Cases of zoonoses and reverse zoonoses have not been reported; however, both humans and dogs have the ability to serve as reservoir species for this organism.

- *St. haemolyticus, St. saprophyticus, St. simulans, St. cohnii, St. warneri,* and *St. lugdunensis* can also cause infections in humans, but do not appear to be zoonotic.

Staphylococcus bacteria were traditionally divided into two groups based on their ability to clot blood plasma (the coagulase reaction). The coagulase-positive staphylococci constitute the more pathogenic species *St. aureus* and *St. intermedius*. The coagulase-negative staphylococci include *St. epidermidis* and *St. schleiferi* as well as other species of *Staphylococcus*. There is no direct evidence that coagulase is a virulence factor and some isolates of *St. aureus* are coagulase negative; therefore, this distinction is more historical than clinically significant.

Epizootiology and Public Health Significance

Staphylococcus spp. have worldwide distribution; *St. epidermidis* is ubiquitous on human skin, whereas *St. aureus* is typically found only on moist skin folds. Both species are found in the upper respiratory, gastrointestinal, and urogenital tracts of humans. In the United States *St. aureus* occurs in about 80% of people at some point in their lives. Most people are only colonized intermittently and about 20% to 30% of people are colonized persistently. People with higher rates of colonization include healthcare workers, diabetics, and patients on dialysis. Adults tend to colonize *St. aureus* in the anterior nares with other sites being the axilla, rectum, and perineum. Cases of untreated *St. aureus* bacteremia have an 80% mortality rate; infections as a result of coagulase-negative *Staphylococci* have a low mortality rate.

> Since 1999, there have been published reports of MRSA being isolated from dogs, cats, rabbits, horses, dairy cows, and chickens.

In dogs, pyoderma caused primarily by *St. intermedius* is the most common skin disease of dogs. Approximately 80% of dogs with atopy have a secondary bacterial infection at the time of diagnosis.

Until recently, extremely resistant *Staphylococcus* bacteria were limited to a few species of *St. aureus* and had been primarily a human problem. Multiple drug-resistant strains of *St. aureus* have been identified in dogs and *St. schleiferi* (which has the ability to develop multiple drug resistance) has been reported to cause canine pyoderma. Until recently, MRSA has been primarily a nosocomial infection with 5% to 10% of U.S. hospital patients having this infection in 2002. Many reports of community-acquired MRSA have been documented making the presence of this bacterium in the population at large a potential problem. Between October 1, 2000, and November 16, 2002, MRSA was isolated from 82 horses and 29 humans in one Canadian study. The infected horses were clinically ill from other diseases for which they required prolonged hospitalization (MRSA was the cause of death in only one horse). The strain in this study was the same in 95% of the horses and 93% of the humans, which demonstrates the zoonotic potential of this bacterium.

> In 2006, the COC reported that 2.3 million people in the United States carry MRSA asymptomatically in their noses.

Transmission

Staphylococcal infections can be transmitted through direct contact between people and through fomites such as contaminated bedding, clothing, and medical equipment. Staphylococcal infections can be transmitted from animals to people through skin lesions that occur when working with tissue or bones from infected animals, contact with animals with skin infections or those carrying *Staphylococcus*, and from bites and scratches.

> Humans are the principal reservoir of MRSA; however, in some circumstances other animals can be colonized by MRSA, causing disease in animals.

Pathogenesis

Staphylococcal infections usually begin when the bacterium enters the skin through wounds, follicles, or skin glands. In order to initiate infection the pathogen must gain access to the host and attach to host cells or tissues. The pathogenicity of *Staphylococcus* results from structures that allow it to avoid phagocytosis, the production of enzymes, and the production of toxins. *Staphylococcus* bacteria are able to produce many cell surface–associated and extracellular proteins that are potential virulence factors that cause tissue damage.

St. aureus bacteria are coated with a protein called protein A that interferes with humoral immunity by binding to IgG antibodies (which are antibodies that enhance phagocytosis). Protein A also inhibits the complement cascade. The outer surface of *St. aureus* and *St. intermedius* also contain bound coagulase, an enzyme that changes blood fibrinogen into fibrin molecules that help form blood clots around bacteria. These fibrin clots are believed to hide bacteria from phagocytic cells. Several species of *Staphylococcus* produce a slime layer or surface mucoid substance that is believed to help the adhesiveness of the organism.

Staphylococcus bacteria also produce a number of enzymes that contribute to their pathogenicity. These include:

- Cell-free coagulase that triggers blood clotting (done by *St. aureus* and *St. intermedius*);
- Hyaluranidase, which breaks down hyaluronic acid, damaging connective tissue;
- Kinases, which convert plasminogen to plasmin, a substance that digests fibrin;
- Staphylokinase, which activates proteolytic activity that dissolves clots allowing bacteria to free themselves to spread to new locations;
- Lipases that digest lipids allowing the bacterium to grow on skin surfaces and oil glands;
- Streptolysin, which repels and damages phagocytes;
- Hemolysins that destroy cells by lysing them; and
- B-lactamase, which breaks down penicillin.

Some *Staphylococcus* bacteria also produce toxins such as:

- Cytolytic toxins (such as alpha, beta, gamma, and delta) that disrupt the plasma membranes of many cells including leukocytes. Leukocidin is another cytolytic toxin that disrupts neutrophil membranes and destroys the neutrophil;
- Exfoliative toxins that affect the epidermis causing the skin to separate, blister, and slough;
- Toxic-shock-syndrome toxins that induce a severe systemic inflammatory reaction; and
- Enterotoxins (there are five named A through E) that stimulate intestinal muscle contractions and vomiting associated with staphylococcal food poisoning.

In contrast to *St. aureus*, little is known about mechanisms of pathogenesis of *St. epidermidis* infections. *St. epidermidis* produces catalase and lipase, but does not have nearly the number of virulence factors of *St. aureus* and *St. intermedius*.

St. aureus becomes resistant to antibiotics by acquiring genetic sequences from other bacteria. Methods used by bacteria to transfer genetic material include:

- *Plasmids.* Circular double-stranded DNA molecules that are separate from the bacterium's chromosomal DNA and can pass antibiotic-resistance information from one bacterium to the next;
- *Transposons.* Genes that can jump from one DNA molecule to another in a cell;
- *Chromosomal cassettes.* Small chromosome regions that confer a specific function, such as drug resistance;

- *Bacteriophages*. Viruses that infect bacteria and cause them to be more virulent;
- *Pathogenicity islands*. Large chromosomal regions in pathogenic bacteria that encode virulence genes that are often acquired from several, unrelated organisms.

Clinical Signs in Animals

Staphylococcal infections in animals vary with the species. Some examples of the animal species and the resulting infections include:

- *Dogs*. *Staphylococcus* pyoderma caused by *St. intermedius* (and occasionally *St. schleiferi*) is the most common skin disease of dogs and is almost always caused by changes in the normal skin integrity associated with a primary dermatosis such as atopy (hypersensitivity reaction in animals involving pruritus with secondary dermatitis) or endocrine disease. Lesions seen in canine superficial pyoderma include complete alopecia, follicular pustules, epidermal collarettes, and serous crusts (Figure 3-70). Lesions are usually associated with the trunk, head, and proximal extremities. Lesions seen in canine deep pyoderma include erythema, swelling, ulcerations, hemorrhagic crusts, alopecia, and draining tracts with purulent discharge. Lesions are usually associated with the bridge of the muzzle, chin, elbows, hocks, interdigital areas, and lateral stifles.

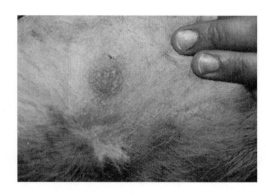

Figure 3-70 *Staphylococcus* infection in a dog.
(Courtesy of Michael Richards, DVM)

- *Cats*. Feline superficial pyoderma is usually caused by *St. intermedius* and usually presents with alopecia, papules, and focal crusting. Deep pyodermas in cats present with alopecia, ulcerations, hemorrhagic crusts, and draining tracts. These conditions are rare and often indicate other systemic disease, such as feline immunodeficiency virus or feline leukemia virus.
- *Birds*. Infections in birds are usually caused by *St. aureus* and most often manifest as a synovitis (lameness is the most common clinical presentation). The proximal tibiotarsus and proximal femur are most commonly affected; however, the proximal tarsometatarsus, distal femur, and tibiotarsus are also involved when infection is extensive. Navel and yolk sac infections may also be seen with staphylococcal infections in birds. Chicks with navel infection have navel areas that are dark and wet; infected yolk sacs are retained longer than normal and are abnormal in color, consistency, and odor.
- *Pigs*. Exudative epidermitis (greasy pig disease) is an acute, generalized dermatitis caused by *St. hyicus* (*hyos*) that occurs in 5- to 60-day-old pigs and is characterized by sudden onset, with morbidity of 10% to 90% and mortality of 5% to 90%. It has been reported from most swine-producing areas of the world. *St. hyicus* is unable to penetrate intact skin; therefore, abrasions on the feet and legs or lacerations on the body precede infection. Older pigs tend not to be infected, but *St. hyicus* may be recovered from the skin of older pigs, the vagina of sows, and the preputial diverticulum of boars. These animals can serve as a source of contamination for naive herds. Clinical signs include lethargy, reddening of the skin in one or more piglets in the litter, anorexia, and transient fever. The infected skin thickens and

reddish brown spots appear. The lesions produce a moist, greasy exudate of sebum and serum that becomes crusty. The feet are nearly always involved with erosions at the coronary band and heel. Mortality is low except in the very young, but recovery is slow and growth is retarded. *St. hyicus* is not considered zoonotic.

- *Cattle. St. aureus* causes both acute and chronic mastitis that is easily transmitted at milking time and colonizes the teat canal but not the skin. In herds in which staphylococcal mastitis is a problem, greater than 50% of the cows may have chronic, subclinical infections. Infections lasting more than a few months often are difficult to treat as a result of the development of a tissue barrier between the antibiotic and the organism. Coagulase-negative staphylococci can also cause subclinical and clinical mastitis and are commonly found in the udder of heifers.

- *Horses.* MRSA infection has been reported in hospitalized horses as well as the transmission of this bacterium between infected horses and veterinary personnel. Most horses with MRSA acquire it during prolonged hospitalization, usually at referral institutions.

Clinical Signs in Humans

Figure 3-71   A cutaneous abscess on the knee caused by methicillin-resistant *Staphylococcus aureus* bacteria (MRSA). Staph infections, including MRSA, occur most frequently among persons in hospitals and healthcare facilities or who have weakened immune systems; however, *Staph* infections may be acquired by otherwise healthy individuals.
(Courtesy of CDC/Bruno Coignard, M.D.; Jeff Hageman, M.H.S.)

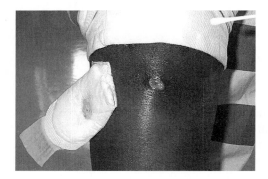

Most *Staphylococcus* infections are limited to the body surfaces such as boils, furuncles, styes, impetigo, and other superficial skin infections in humans (Figure 3-71). A wound or surgical opening allows *Staphylococci* to colonize the blood and internal organs, causing conditions such as toxic shock syndrome, osteomyelitis, endocarditis, meningitis, and pneumonia.

St. aureus and *St. epidermidis* are common causes of infections associated with indwelling devices such as intravenous catheters, joint prostheses, cardiovascular devices, and artificial heart valves.

Diagnosis in Animals

Diagnosis of staphylococcal infections involves demonstration of gram-positive bacteria in a cluster of spheres from pus, blood, or other fluid samples. Bacterial culture on blood agar will grow all species of *Staphylococcus*; *St. aureus* will grow as medium to large, smooth, creamy-yellow colonies that are usually beta-hemolytic; *St. epidermidis* will grow as small to medium, translucent, gray-white colonies that are usually nonhemolytic; *St. intermedius* will grow as large, glossy, translucent colonies; and *St. schleiferi* will grow as medium to large, glossy, unpigmented colonies. Selective media such as mannitol salt agar and deoxyribonuclease (DNase) agar can differentiate the species of *Staphylococcus*. All species of *Staphylococcus* are catalase positive (the catalase test is important in distinguishing *Streptococcus* [catalase negative] from *Staphylococcus* [catalase positive]). *St. aureus* and *St. intermedius* are coagulase positive. Commercial identification methods that agglutinate latex particles and contain multiple wells for multiple tests are also available.

Diagnosis in Humans

Diagnosis in people is similar to animals.

Treatment in Animals

Staphylococcal infections may be difficult to treat as a result of the development of antibiotic resistance. Proper antibiotic selection based on culture and sensitivity, appropriate dose and duration of treatment, and controlling underlying disease are all critical in ensuring successful treatment. Amoxicillin with clavulanic acid and cephalexin are commonly used antibiotics for Staphylococcal infections; clindamycin, sulfa antibiotics, and erythromycin are also effective. Antibiotics are typically given for a minimum of 21 days. Controlling pruritus with corticosteroids and the use of shampoos to reduce superficial colonization of bacteria are essential in treating these infections.

Treatment in Humans

Treatment of staphylococcal infections in humans involves the use of proper antibiotics, which can be determined via culture and sensitivity. Resistant strains such as MRSA have made sensitivity testing vital in the treatment of these infections. Early treatment is critical and prolonged treatment may be needed for severe or complicated cases.

Management and Control in Animals

The best way to prevent dogs from contracting staphylococcal infections is to eliminate underlying skin disease and prevent damage to intact skin. Proper milking procedures and disinfection of equipment can prevent staphylococcal mastitis in cows. There are not any vaccines to prevent staphylococcal infections in animals.

Management and Control in Humans

The best way to prevent staphylococcal infections in people is by using hygienic precautions when in contact with animals and by avoiding bites and scratches. At risk persons should wear gloves when working with animals and animal tissues. Gloves should be worn when examining any animal with open wounds. There is no human vaccine for staphylococcal infections.

The main mode of transmission of MRSA is human to human through hands, especially those of healthcare workers. Human-to-human transmission of MRSA can be limited by following proper hygiene procedures, isolation of infected patients, and prompt treatment of wounds. Antibiotic-resistant strains of *Staphylococcus* bacteria can be avoided by judicious use of antibiotics.

Summary

Staphylococcal infections typically involve the skin and mucous membranes of mammals. There are about thirty species of *Staphylococcus* bacteria with most being completely harmless as normal flora of the skin and mucous membranes. *Staphylococcus* bacteria can cause a wide variety of diseases either through toxin production or invasion (food poisoning as a result of the heat-resistant enterotoxins the bacterium produces), wound infections (as a result of its ability to survive on dry surfaces), shock syndromes (when toxins are released into the bloodstream), and infections in immunocompromised individuals. *Staphylococcus* bacteria are small, nonmotile, salt-tolerant, gram-positive, facultatively anaerobic spherical bacteria that are typically clustered. Species of *Staphylococcus* that are of interest in humans and animals include *St. aureus* (the main pathogen in humans, but has also been recovered from dogs and cats), *St. epidermidis* (this species of *Staphylococcus* is part of the normal flora of human skin, but can be an opportunistic pathogen in immunocompromised people), *St. intermedius* (normal flora of canine skin; however, it can cause opportunistic disease), and *St. schleiferi* (reported in cases of canine pyoderma, canine otitis in Japan, and in numerous case studies in Europe). *Staphylococcus* spp. have worldwide distribution. Staphylococcal infections can be transmitted through direct contact between people and through fomites such as contaminated bedding, clothing, and medical equipment. Staphylococcal infections can be transmitted from animals to people through skin lesions that occur when working with tissue or bones from infected animals, contact with animals with skin infections or carrying staphylococci, and from bites and scratches. Staphylococcal infections in animals vary with the species and include pyoderma in dogs, synovitis in birds, greasy pig disease in pigs (caused by a nonzoonotic species), mastitis in cows, and nasal colonization in ill horses. In humans, most staphylococcal infections are limited to the body surfaces such as boils, furuncles, styes, impetigo, and other superficial skin infections. A wound or surgical opening allows *Staphylococcus* to colonize the blood and internal organs, causing conditions such as toxic shock syndrome, osteomyelitis, endocarditis, meningitis, and pneumonia. Diagnosis of staphylococcal infections involves staining, culture, and the use of commercial identification methods. Staphylococcal infections may be difficult to treat as a result of the development of antibiotic resistance. Proper antibiotic selection based on culture and sensitivity, appropriate dose and duration of treatment, and controlling underlying disease are all critical in ensuring successful treatment. The best way to prevent dogs from contracting staphylococcal infections is to eliminate underlying skin disease and prevent damage to intact skin. The best way to prevent staphylococcal infections in people is by using hygienic precautions when in contact with animals, and bites and scratches should be avoided.

STREPTOCOCCAL INFECTIONS

Overview

Bacteria in the *Streptococcus* genus are a diverse group that are all gram-positive spheres arranged in pairs or chains. The name *Streptococcus* was given to this group of bacteria in 1874 by Theodor Billroth, a Viennese surgeon, because of the bacterium's spherical chain morphology (*streptos* is Greek for twisted chain and *kokkos* is Greek for kernel or berry). There are numerous species of *Streptococcus* bacteria that cause a variety of diseases including strep throat (*Str. pyogenes*), scarlet fever (*Str. pyogenes*), and pneumococcal pneumonia (*Str. pneumoniae*) in humans and strangles in horses (*Str. equi*), pneumonia in pigs (*Str. suis*), and mastitis in cattle (*Str. agalactiae*). In humans *Streptococcus* bacteria has caused many epidemics throughout history. The first recorded epidemic of scarlet fever, a sequela of strep throat, occurred in Sicily in 1543 and deadly epidemics occurred across the United States from the 1830s until the 1880s. Pneumococcal pneumonia caused by *Str. pneumoniae* is the most common cause of bacterial pneumonia worldwide. Necrotizing fasciitis, a severe skin and tissue infection caused by a strain of *Str. pyogenes* that emits a flesh-destroying toxin, has been reported in humans since 1994. In animals *Streptococcus* infections can cause mastitis, polyarthritis, genital infections, gastric disorders, meningitis, septicemia, pneumonia, and abscesses. Of the numerous species of *Streptococcus*, only a few of them are zoonotic. Because of better understanding of gene sequencing techniques and relationships between members of the *Streptococcus* genus, nomenclature and taxonomy of this bacterial group has changed and their role in disease transmission and pathogenesis has been clarified.

Causative Agent

Streptococcus bacteria are gram-positive spheres that grow in chains and may appear ovoid under the microscope when grown in broth culture. They are facultative anaerobes and are catalase negative. Species of *Streptococcus* are differentiated from each other using many techniques based on antibody reactions, hemolysis patterns, cell arrangement, and biochemical tests. Historically, species of *Streptococcus* have been classified into Lancefield groups, based on a method developed by Rebecca Lancefield in the 1930s. Lancefield grouping differentiates these bacteria based on a cell wall carbohydrate called component C. The antigenic parts of component C are amino sugars and a precipitation test is used to identify which Lancefield group a species of *Streptococcus* belongs. There are twenty Lancefield serogroups that are designated by capital letters such as group A *Streptococcus*, group B *Streptococcus*, and include groups A through H and K through V. There are some streptococcal species that have no Lancefield group antigens and some with newly identified antigens. The use of Lancefield groups is diminishing, but they will be referred to in the literature and in clinical practice. Another method of distinguishing strains of *Streptococci* is based on their reaction on blood agar. Some strains of *Streptococcus* produce extracellular enzymes that produce zones of hemolysis (damage to the red blood cells in the blood agar) and the type of hemolysis they produce or do not produce can help in their identification. Some *Streptococcus* species produce alpha-hemolysis (partial lysis of the red blood cells to produce a greenish discoloration around the colony), beta-hemolysis (complete lysis of the red blood cells producing complete clearing around the colony), or gamma reaction (no change in the red blood cells leaving the agar unchanged and are referred to as nonhemolytic). Currently *Streptococci* are subdivided into three large groups: pyogenic (where most pathogens reside), oral, and other.

Streptococcal spp. have been isolated from a variety of animals and are found widely distributed worldwide in nature and as commensals. There are more than 37 species of *Streptococcus* with only a few being pathogenic and even less being zoonotic. Strains of *Streptococcus* that cause disease are summarized in Table 3-8. Strains of *Streptococcus* that cause zoonotic disease include:

- *Str. equi* (group C) is a beta-hemolytic bacterium. *Str. equi* infections contracted from animals are sporadic in humans and are more frequently associated with *Str. equi* subspecies *zooepidemicus* (formerly known as *Str. zooepidemicus*) than *Str. equi* subspecies *equi*. *Str. equi* subspecies *equi* occurs almost exclusively in equine causing strangles primarily in young animals without prior infection or immunization and is found in the nasopharynx, upper and lower respiratory tract, and the genital mucous membranes of healthy equine and cattle. *Str. equi* subspecies *zooepidemicus* is an opportunistic pathogen found in a large spectrum of animal hosts. *Str. equi* subspecies *zooepidemicus* causes pneumonia, wound infections, endometritis, arthritis, and mastitis in animals. Animals can be carriers of this bacterium and horses are the animals most commonly affected by this organism.

- *Str. suis* (groups R, S, and T) is a nonhemolytic pathogenic and commensal bacteria in swine causing epizootic outbreaks of meningitis, septicemia, and arthritis especially in young, growing pigs. There are at least 35 serotypes of *Str. suis* with type 2 being the most frequently isolated and the one predominantly isolated

Classification of streptococcal species has changed over time; in the 1980s some species moved to new genera (*Lactococcus* and *Enterococcus*) and more new genera were established (*Abiotrophia*, *Granulicatella*, *Dolosicoccus*, *Facklamia*, *Globicatella*, and *Ignavigranum*).

Table 3-8   *Streptococcus* spp. and Their Features

Species	Lancefield Group	Habitat	Disease
Str. pyogenes*	A	Human throat	Strep throat, scarlet fever, impetigo, erysipelas, necrotizing fasciitis, streptococcal toxic shock syndrome
Str. agalactiae	B	Vaginal tract of some women GI and urinary tracts of people Udder of cow	Neonatal septicemia, meningitis, pneumonia, and death in humans; mastitis in cattle
Str. dysgalactiae subspecies *equisimilis*#	A, C, G, and L	Swine, dogs, cats, poultry, and cattle	Various infections in animals; pharyngitis and endocarditis in humans
Str. dysgalactiae subspecies *dysgalactiae*	A, C, G, and L	Variety of animals	Skin and respiratory infections (dog and cats), mastitis (cattle), septicemia, arthritis, and meningitis (pigs), and nasopharygeal infections (poultry)
Str. equi subspecies *equi*	C	Various mammals (mainly horses)	Strangles in horses
Str. equi subspecies *zooepidemicus*#	C	Large spectrum of animal species	Wound infections, endometritis, arthritis, and mastitis in animals; bacteremia, meningitis, and streptococcal toxic shock syndrome in humans
Str. suis#	R, S, and T	Swine, ruminants, bison, humans	Meningitis, septicemia, and arthritis in pigs; meningitis and hearing loss in humans
Str. canis#	G	Dogs, cats, humans	Otitis media, septicemia, lymphadenopathy, polyarthritis, reproductive tract infections, and mastitis in dogs; wound infections and bacteremia in humans
Str. iniae#	No Lancefield grouping	Freshwater fish, dolphins, humans	Abscesses and skin infections in fish; cellulitis, meningitis, and endocarditis in humans
Str. porcinus~	E, P, U, V, none, and new	Swine, humans	Isolated from swine and humans, but incidence of infection unknown
Str. bovis group~	D	Normal flora of humans and animals	Endocarditis, urinary tract infections, osteomyelitis, and sepsis in humans
Str. pneumoniae~	No Lancefield grouping	Horses, Guinea pigs, rats, humans	Respiratory disease in horses, Guinea pigs, and rats; number one cause of bacterial pneumoniae in humans
viridans *Streptococcus* group~	No Lancefield grouping	Commensal in many animals	May cause endocarditis and other infections in humans

* Can cause reverse zoonosis in dogs, udders of cows
# zoonotic
~ zoonotic potential uncertain

from humans. *Str. suis* causes disease more frequently in swine housed in facilities with high animal densities and poor ventilation and is usually introduced into herds by carrier swine with bacteria in their noses or tonsils (there is also evidence of sow-to-pig transmission). The disease in swine is manifested by meningitis, septicemia, paralysis, convulsion, cutaneous erythema, and fever. In humans meningitis and hearing loss (as a result of cranial nerve VIII involvement) are common signs of *Str. suis* infection.

- *Str. canis* (Group G) is a beta-hemolytic opportunistic pathogen and is the most important species of *Streptococcus* in dogs and cats. *Str. canis* causes otitis media, septicemia in neonates, lymphadenopathy in cervical lymph nodes, polyarthritis, and reproductive tract infections with abortion and mastitis in dogs. In humans *Str. canis* has been cultured from wound infections and people with bacteremia.

- *Str. iniae* (no Lancefield grouping) is a beta-hemolytic bacteria found in freshwater fish and dolphins causing abscesses in these animals. There are both virulent and commensal strains of *Str. iniae*. *Str. iniae* has been reported to cause cellulitis and systemic infections (meningitis and endocarditis) in humans. Most human infections have been associated with contact with tilapia fish.

- *Str. dysgalactiae* (groups A, C, G, and L) is a bacterium that is a pathogen and commensal of animals. *Str. dysgalactiae* subspecies *equisimilis* is a beta-hemolytic streptococcal species that occurs worldwide particularly in dogs and cats, cattle, pigs, and poultry. *Str. dysgalactiae* subspecies *dysgalactiae* is a nonhemolytic species that is nonzoonotic and causes a variety of animal infections including mastitis in cattle.

- *Str. porcinus* (group E, P, U, V, none or new) is a beta-hemolytic bacterium that has been isolated from swine and has been isolated from the urogenital tract of women. The zoonotic significance of *Str. porcinus* is unknown.

- *Str. bovis* group (group D) are nonhemolytic normal flora bacteria of both humans and animals and include the bacteria *Str. bovis*, *Str. equines*, *Str. gallolyticus*, *Str. infantarius*, *Str. pasteurianus*, and *Str. lutetiensis*. Bacteria in this group are found in humans with endocarditis, urinary tract infections, osteomyelitis, and sepsis and their zoonotic significance is uncertain.

- *Str. pneumoniae* (no Lancefield grouping) are alpha-hemolytic human pathogens that cause pneumonia, otitis media, and meningitis. *Str. pneumoniae* causes respiratory disease in horses and is a commensal or respiratory pathogen in other animal species including Guinea pigs and rats. There is some evidence of reverse zoonosis with this particular species of *Streptococcus*. The zoonotic significance of *Str. pneumoniae* is uncertain.

- viridans *Streptococcus* group (no Lancefield grouping) are a diverse group of approximately 26 species of bacteria that are alpha-hemolytic. They are commensals of the mouth, gastrointestinal tract, and vagina of healthy humans and can be found in animals. Some species may be zoonotic causing endocarditis and other infections in people (especially neutropenic people). Some names given to bacteria in this group include *Str. mitis* and *Str. sanguis*.

- *Str. agalactiae* (group B *Streptococcus*) are beta-hemolytic (subtle) and cause mastitis in cattle and neonatal sepsis and meningitis in people; however, human infections are not zoonotic as *Str. agalactiae* can be part of the normal flora of human genital and gastrointestinal tracts.

- *Str. pyogenes* (group A *Streptococcus*) are beta-hemolytic (large zone) and cause strep throat, scarlet fever, streptococcal toxic shock syndrome, and necrotizing fasciitis in people and may be isolated from animals such as dogs, cattle, ducks, and monkeys that serve as occasional reservoirs (chain of infection probably is human to animal to human).

> *Str. pyogenes* bacteria are adapted to humans and have no natural animal reservoir. *Str. pyogenes* can be transmitted to animals (reverse zoonosis) and animals can retransmit the infection to humans. Examples of reverse zoonosis of *Str. pyogenes* include humans infecting the bovine udder causing contamination of raw milk that caused human outbreaks and family dogs that carry the bacterium after family members had strep throat.

Epizootiology and Public Health Significance

Streptococcus spp. are found worldwide in nature and as commensals in animals. *Str. equi* subspecies *zooepidemicus* has the largest spectrum of animal hosts with the presence of animal carriers and diseased animals. *Str. suis* infections are being diagnosed more frequently in humans; however, human cases have not been seen in the United States. Most human *Str. iniae* infections have occurred in North America. The other species of *Streptococcus* that cause human disease are found in a variety of animals including dogs, cats, poultry, and fish.

Human infections with *Str. equi* subspecies *zooepidemicus* are rare and occur sporadically usually associated with consumption of unpasteurized dairy products. Mortality rates for group C bacteremia (to which *Str. equi* subspecies *zooepidemicus* belongs) is about 20% to 30% and for group C meningitis is about 57%. Human infections with *Str. suis* are rare with fewer than 100 human cases reported to date. Virulent strains of *Str. suis* are found in pigs in the United States; however, no human cases have been seen in the United States. The mortality rate for *Str. suis* meningitis is about 7% with common sequelae of deafness and vertigo. Human infection with *Str. canis* is rare with less than ten documented cases. *Str. iniae* infections in humans have been seen in people handling live or freshly killed fish and is mainly seen in the elderly.

Str. pyogenes is a common human pathogen that accounts for more than 10 million infections, with approximately 9,000 cases occurring annually in the United States. The mortality rates are 10% to 15% for invasive disease, 20% to 25% for necrotizing fasciitis, and about 45% for streptococcal toxic shock syndrome.

Transmission

Streptococcus bacteria are often normal flora of animals and humans and can be transmitted in a variety of ways. *Str. equi* subspecies *zooepidemicius* is spread to humans by direct contact with animals excreting the pathogen in large amounts (contact with purulent discharges or bites), indirect transmission by fomites, aerosols, or by ingestion (consumption of raw milk and dairy products). *Str. suis* is transmitted to humans by direct contact with pigs and pork usually through nonintact skin or via the conjunctiva, indirect transmission by fomites, ingestion of undercooked pork from infected pigs, or aerosol. Carriers transmit the infection to other pigs by close contact (mainly between weaned pigs). *Str. canis* is found on the skin and mucosa of dogs and other animals and transmission seems to be via close contact (colonization of open wounds or through bites). *Str. iniae* transmission appears to be mainly through contact of traumatized skin with live or freshly killed fish or contaminated instruments. *Str. dysgalactia* subspecies *equisimilis* and *Str. porcinus* can be transmitted to people via contact with infected animals or via bites or scratch wounds from infected animals.

Pathogenesis

Streptococcal infections can result in a variety of diseases and their development depends on several factors such as animal species, bacterial species, body system affected, and portal of bacterial entry. Pyogenic *Streptococcus* bacteria produce pus as part of their pathogenesis. When pyogenic *Streptococcus* bacteria enter tissue they cause an inflammatory response including vasodilation and invasion of neutrophils. As a result of chemotaxis (chemical signaling by bacteria) neutrophils move toward bacteria and phagocytose them. Following phagocytosis, bacteria may be digested, but some are resistant to enzymes in the neutrophil and will multiply in the neutrophil. Some *Streptococcus* bacteria will produce toxins that kill the neutrophils and the enzymes released upon death of the neutrophil will cause liquefaction of dead tissue. This liquefied mass becomes thick pus as a result of the large amount of protein from the nuclei of dead cells.

Virulence factors vary among species of *Streptococcus* and include the following:

- *Protein M.* Some strains have a membrane protein called protein M that inactivates complement and is antiphagocytic.
- *Hyaluronic acid capsule.* Hyaluronic acid is a natural substance in the body and its presence in a capsule makes phagocytic leukocytes (WBCs) ignore the presence of bacteria with this virulence factor.
- *Streptokinases.* There are two streptokinase enzymes that break down blood clots allowing bacteria to spread rapidly throughout infected tissue.
- *Streptolysins.* There are two different, membrane-bound proteins called streptolysins, which lyse erythrocytes (RBCs), leukocytes, and platelets. These proteins interfere with the oxygen-carrying capacity of blood, immunity, and blood clotting. After phagocytosis, some *Streptococcus* bacteria will release streptolysins into the cytoplasm of the phagocyte, causing the lysosomes to release their contents causing death to the phagocytes and release of bacteria.
- *Enzymes.* Some species of *Streptococcus* produce proteases that break down proteins, others produce deoxyribonucleases (DNases) that reduce the viscosity of fluid containing DNA, whereas others produce hyaluronidases that promote the rapid spread of infection.
- *Protein adhesin.* Some *Streptococcus* bacteria produce protein adhesin, a protein that causes binding of the cells to epithelial cells allowing bacteria a place to enter the cells.

Clinical Signs in Animals

Streptococcal infection can cause a variety of diseases in animals as listed in Table 3-8 (see p. 198.) The incubation period of each disease varies with the organism causing the disease and can vary from several hours to days. *Streptococcus* bacteria that cause disease in animals include:

- *Str. equi* subspecies *equi* causes strangles in horses and is considered nonzoonotic. *Str. equi* subspecies *equi* occurs almost exclusively in equine and is found in the nasopharynx, upper and lower respiratory tract, and the genital mucous membranes of healthy equine and cattle. *Str. equi* subspecies *equi* is transmitted through purulent discharges from one animal to another. Affected horses are infectious for at least 4 weeks following onset of disease and carrier animals can exist. The incubation period in horses is 3 to 6 days and signs of strangles include anorexia, fever, inflammation of the upper respiratory mucosa and lymph nodes, followed by mucopurulent nasal discharge and abscessed lymph nodes in the neck region (Figure 3-72). The normal course of strangles is 10 to 14 days. Morbidity can be close to 100% in a naive population although mortality is low. "Bastard strangles" is a term used to describe strangles that has abscesses in other lymph nodes of the body as a result of the animal's inability to confine the disease to the upper respiratory tract. *Str. equi* subspecies *zooepidemicus* is a commensal and opportunistic bacterium in horses. Other animal species may be carriers of *Str. equi* subspecies *zooepidemicus*. This bacterium causes a wide variety of infections in animals including secondary bacterial infection following viral respiratory infection that mimics strangles (horses), metritis and placentitis leading to abortion (horses), cervical lymphadenitis, pneumonia, and septicemia (Guinea pigs), polyarthritis, bronchopneumonia,

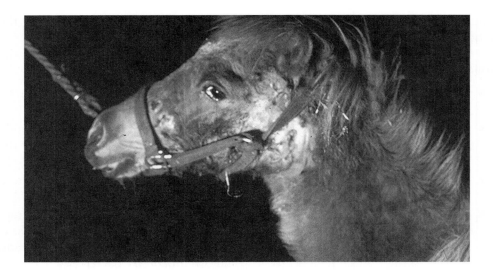

Figure 3-72   Horse with strangles.
(Courtesy of Drs. Tom Swerczek and R.C. Giles, University of Kentucky)

diarrhea, endocarditis, and meningitis (monkeys), and mastitis (cattle and goats). Morbidity and mortality rates can be very high with outbreaks of septicemia.

- *Str. suis* mainly affects pigs, but has been isolated from cattle, sheep, goats, and bison. Although *Str. suis* can be carried in swine without producing signs of disease, some virulent strains can cause meningitis, arthritis, or subclinical disease in swine (especially suckling pigs). Acute meningitis in growing pigs can have high mortality rates.

- *Str. canis* is mainly found in dogs, but can also be found in cats, cattle, rats, mink, mice, rabbits, and foxes. *Str. canis* has been isolated from dogs with skin infections, reproductive tract infections, mastitis, pneumonia, and septicemica and in cats with arthritis, wound infections, cervical lymphadenitis, pneumonia, and septicemia.

- *Str. iniae* has been found in freshwater dolphins and wild and farmed fish (tilapia fish can be carriers of this organism). Most fish do not show clinical signs of disease but some fish have exhibited signs of meningoencephalitis and panophthalmitis.

- *Str. dysgalactiae* subspecies *equisimilis* produces a variety of clinical diseases in animals including dogs and cats (causing skin infections and respiratory infections), cattle (causing mastitis), pigs (septicemia, arthritis, and meningitis), and poultry (nasopharyngeal infections).

Clinical Signs in Humans

Streptococcal infections in humans vary with the species of bacteria causing the disease. The incubation periods can range from less than 24 hours to approximately 2 to 3 days. *Streptococcus* that cause disease in people from potential animal sources include:

- *Str. equi* subspecies *zooepidemicus* has been isolated from people with pneumonia, arthritis, meningitis, septicemia, glomerulonephritis, and streptococcal toxic shock syndrome.

- *Str. suis* has mainly been linked to meningitis in people that results in some degree of hearing loss in approximately 50% of infected individuals.

- *Str. canis* rarely causes human disease but may cause septicemia, meningitis, and peritonitis in people.
- *Str. iniae* can rarely cause cellulites (Figure 3-73), arthritis, endocarditis, meningitis, and osteomyelitis in people.
- *Str. dysgalactiae* subspecies *equisimilis* causes human infections mainly from direct contact with rodents and can cause skin infections, septicemia, endocarditis, and thrombophlebitis.

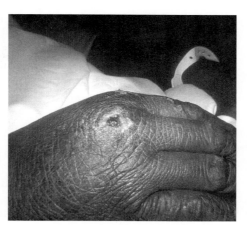

Figure 3-73 A systemically disseminated streptococcal infection in a person. The initial portal of entry was a hand wound.
(Courtesy of CDC/Dr. Thomas F. Sellers/ Emory University)

- *Str. bovis* group bacteria are found in humans with endocarditis, urinary tract infections, osteomyelitis, and sepsis.
- *Str. porcinus* has been isolated from the urogenital tract of women.
- *Str. pneumoniae* causes pneumonia, meningitis, otitis media, and sinusitis in people. Pneumococcal disease is highest in children and the elderly.
- viridians *Streptococcus* group can cause purulent abdominal infections, dental caries, and endocarditis in people, but typically does not cause human disease.
- *Str. pyogenes* is a reverse zoonotic agent that can cause pharyngitis (strep throat), abscesses, pneumonia, septicemia, necrotizing fasciitis (flesh-eating bacteria), rheumatic fever, glomerulonephritis, and scarlet fever.

Diagnosis in Animals

Samples obtained from biopsies, smears, or aspirates from wounds, pharyngeal secretions, blood, CSF, or other sites that demonstrate gram-positive cocci in pairs or chains can lead to a presumptive diagnosis of streptococcal infection. Gross lesions produced vary with the species of *Streptococcus* causing the infection. *Str. equi* subspecies *zooepidemicus* typically produce mucus with respiratory infections and abortions with placentitis (in horses the placenta is edematous with brown fibronecrotic exudates and the fetus may be severely necrotic). Lesions in pigs with *Str. suis* include patchy erythema of the skin, enlarged lymph nodes, and thickened joint capsules. Fish with *Str. iniae* infection can show exudative meningitis and diffuse visceral hemorrhages.

Diagnosis of *Streptococcus* is made via culture and identification of the bacteria based on their hemolysis pattern on blood agar, colony morphology, biochemical reactions, and serology to detect antigens. *Streptococcus* spp. grow on blood agar plate producing pinpoint colonies with varying hemolytic patterns that are species dependent. *Streptococcus* spp. do not grow on MacConkey agar, but will grow on gram-positive selective agar such as CNA (Columbia agar with colistin and nalidixic acid) and PEA (phenylethyl alcohol agar). Various species of *Streptococcus* can be identified using biochemical tests such as hippurate and CAMP tests (positive for *Str. agalactiae*), PYR reaction and susceptibility to bacitracin (positive for *Str. pyogenes*), and bile solubility and susceptibility to Optochin (positive and inhibition of growth for *Str. pneumoniae*).

Lancefield group identification can be done using the capillary precipitation test, but other serologic methods can also be used. ELISA tests can also be used to rapidly identify some species of *Streptococcus*.

Diagnosis in Humans

Diagnosis in humans is similar to that described for animals. Additional serology is sometimes performed to diagnose human streptococcal disease and may include tests like the antistreptolysin O (ASO) titer to determine poststreptococcal sequealea, antihyaluronidaste titer, and anti-DNase B.

Treatment in Animals

Treatment of *Streptococcus* bacteria in animals typically involves penicillin in which resistance to this antibiotic are uncommon. Other antibiotics that have been used to treat streptococcal infections include tetracyclines and sulfa antibiotics. Carrier animals may or may not be treated.

Treatment in Humans

Streptococcus infections in humans are typically treated with a variety of antibiotics such as penicillin, amoxicillin, ampicillin, cephalosporins, vancomycin, and clindamycin. In cases of shock (usually seen with streptococcal toxic shock syndrome and necrotizing fasciitis), supportive treatment such as IV fluid administration, IV immunoglobulin treatment, and surgical debridement may be warranted. Dialysis may be necessary in cases of glomerulonephritis.

Management and Control in Animals

It is difficult to prevent some *Streptococcus* bacterial infections because these bacteria are part of the normal flora in some species. Poor husbandry and stress predispose animals to streptococcal infections so good hygiene practices can reduce the risk of these infections. Good hygiene practices during milking can .reduce the risk of foodborne spread of *Streptococcus* spp. All in/all out management of swine can be helpful in reducing *Str. suis* in a herd. Conditions such as adding sick animals to a healthy colony or naïve animals should also be avoided. If outbreaks occur in animal colonies (such as with Guinea pigs) depopulation of the facility may be warranted. There are no commercial vaccines for *Str. equi* subspecies *zooepidemicus*; however, autogenous vaccines for have been used in Guinea pigs. Killed vaccines are available for *Str. suis*.

Management and Control in Humans

Most *Streptococcal* infections are transmitted through wounds and abrasions, so the use of protective clothing and gloves are valuable in controlling the spread of this bacterium from animals to people. To prevent foodborne infections the consumption of raw milk and unpasteurized milk products should be avoided. Good hygiene practices when caring for horses with respiratory disease can help decrease the spread of *Str. equi* subspecies *zooepidemicus* infections.

Summary

Streptococcus spp. are gram-positive bacteria that occur in pairs or chains. Many species are pathogenic for humans and animals and a few are zoonotic. Clinical identification of *Streptococcus* bacteria is based on their hemolytic reactions on blood agar and Lancefield grouping. Zoonotic species of *Streptococcus* include

Str. equi subspecies *zooepidemicus* (formerly known as *Streptococcus zooepidemicus*), *Str. suis*, *Str. canis*, *Str. iniae*, and *Str. dysgalactiae* subspecies *equisimilis*. *Str. porcinus*, *Str. bovis* group, *Str. pneumoniae*, and viridans *Streptococcus* group have uncertain zoonotic potential. *Streptococcus* bacteria are often normal flora of animals and humans and can be transmitted in a variety of ways including direct and indirect contact. Streptococcal diseases in animals range from respiratory infection to septicemia to meningitis and range from meningitis to pharyngitis to pneumonia to endocarditis in people. Diagnosis of streptococcal infections includes Gram stain reaction and morphology, growth and hemolysis pattern on culture, biochemical tests, and antigenic testing. Treatment of streptococcal infection includes antibiotics and in cases of shock supportive care. Streptococcal infections can be controlled with proper hygiene practices for animals and people. Avoidance of ingestion of unpasteurized dairy products can also limit the risk of acquiring streptococcal infections.

TUBERCULOSIS

Overview

Tuberculosis, also known as consumption, is so named because of the characteristic tiny nodule (tubercle) resulting from infection by *Mycobacterium* spp. Tuberculosis (TB) has been present in human and animal populations since antiquity (Egyptian mummies from 2400 B.C. show pathologic lesions of TB) and was known as phthisis (Greek for wasting or decay) or consumption because of the way the disease consumed the infected individual. In 460 B.C. Hippocrates named phthisis as the most prevalent disease of the day killing nearly everyone it infected. The lesions produced in patients with TB were first called tubercles in 1689 by Dr. Richard Morton, a London physician. Treatment, diagnosis, and prevention of TB have changed with a variety of scientific discoveries. Treatment of TB changed when Hermann Brehmer in 1854 advised that a TB patient be moved to a healthier climate, which resulted in curing this patient's disease and the beginning of sanatorium construction for TB patients. In 1882, Robert Koch discovered a staining technique that allowed him to visualize *My. tuberculosis* and gave hope that a cure would be forthcoming. Wilhelm Konrad von Roentgen's 1895 discovery of radiation allowed physicians to monitor the progression of TB through chest x-rays. French bacteriologist Calmette, together with Guerin, used special culture media to lower the virulence of *My. bovis*, creating the basis for the BCG (Bacille Calmette Guerin) vaccine for people still in use today. Antibiotic treatment for TB became available in the middle of World War II (1943) when streptomycin was found to be effective at treating TB infections. Despite all of the efforts aimed at understanding and treating TB, it remains the leading killer of people among all infectious diseases. Famous people who contracted and died of TB include King Tutankhamen, John Keats, Frédéric Chopin, Edgar Allan Poe, Eleanor Roosevelt, and Vivien Leigh. Nelson Mandela contracted TB, but did not die from the disease.

Causative Agent

Mycobacterium spp. are gram-positive (though weakly staining), acid-fast bacilli that are nonendospore forming, obligate aerobes that have an unusual waxy cell wall that affects many of its properties, including its slow growth (doubling time is 20 hours compared with 20 minutes with other bacteria) and staining ability (Figure 3-74). This waxy cell wall allows it to survive prolonged periods of drying

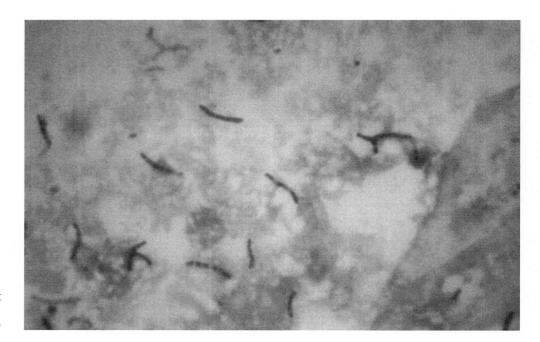

Figure 3-74 *Mycobacterium* spp. appear as red rods using acid-fast staining methods (and are referred to as acid-fast bacilli)
(Courtesy of CDC/Dr. George P. Kubica)

and also affects its acid-fast staining property. The waxy cell wall also contains mycolic acid components that protect the bacterium from lytic enzymes and oxidants within phagocytes. One mycolic acid is called cord factor, which causes the bacterium to clump into cordlike masses that adds to the organism's virulence (strains of *Mycobacterium* without cord factor are avirulent). Cord factor also inhibits migration of neutrophils and is toxic to mammalian cells. There are three main types of tubercle bacilli recognized: human (*My. tuberculosis*), bovine (*My. bovis*), and avian (*My. avium*) with the two mammalian species (tuberculosis and bovis) being more closely related. *My. tuberculosis* along with *My. bovis*, *My. africanum*, and *My. microti* all cause the disease known as tuberculosis and are members of the tuberculosis complex; however, *My. tuberculosis* is usually pathogenic for primates (including humans), whereas *My. bovis* is pathogenic for animals. The types of *Mycobacterium* differ in their characteristics in culture media and pathogenicity and may produce infection in host species other than their own.

- *My. tuberculosis* infects humans and other nonhuman primates. Humans are the ultimate reservoir for *My. tuberculosis*. This species will occasionally infect dogs and parrots, but is mainly found and spread in primates. Cattle infected with *My. tuberculosis* can cause false-positive skin test results for *My. bovis*, but are not clinically sick. Pigs may be infected with *My. tuberculosis* by eating table scraps from the household of an infected person (leads to granulomas in the gastrointestinal tract and regional lymph nodes). Dogs infected with *My. tuberculosis* can develop granulomas in many parts of the body including the pharynx where it can be transmitted to other animals and people. Birds can develop cutaneous granulomas from *My. tuberculosis*.

- *My. bovis* can produce disease in most warm-blooded vertebrates including humans. This organism is the principle agent of zoonotic tuberculosis.

- *My. avium* is the only species found in birds, but can also infect swine, cattle, sheep, mink, dogs, cats, and cold-blooded animals. *My. avium* is found in soil and water (perhaps the real reservoir in zoonotic transmission) and can cause disease in humans and both domestic and wild animals. *My. avium complex* (MAC) consists of 2 species: *My. avium* and *My. intracellulare* that cause disseminated infection mainly in people with immune system compromise secondary to AIDS and immunosuppressive chemotherapy.

- *My. marinum, My. fortuitum, My. platypolcitis, My. scrofulaceum, My. kansasii,* and *My. intracellulare* have been reported in a variety of animals and may be present in soil and water.
- *My. africanum* is a cause of human tuberculosis in tropical Africa and its retention as a distinct species is probably not warranted.
- *My. microti* causes tuberculosis in voles, wood mice, shrews, llamas, badgers, ferrets, cattle, pigs, and cats.

Epizootiology and Public Health Significance

TB affects nearly all species of vertebrate animals and at one time was one of the major diseases of humans and domestic animals. *My. tuberculosis* has been reported in some animals, but infections are often traced to infected humans who expose susceptible animals to infection. *My. bovis* is found in susceptible animals such as ungulates, carnivores, primates (including humans), lagomorphs, and rodents. Several wildlife reservoir hosts have also been identified including white-tailed deer, opossums, and badgers.

WHO estimates that 3 million people die from TB annually and 8 million people are infected with the disease (95% of cases are in developing countries). The registered number of new TB cases worldwide correlates with economic conditions with the highest incidence of TB found in Africa, Asia, and Latin America (countries with the lowest gross national product). In industrialized countries, the incidence of TB was in decline until the mid-1980s when cases of TB began increasing. This rise in tuberculosis cases was attributed to high rates of immigration from TB-ravaged countries and HIV infection. The resurgence of TB can also be attributed to multiple drug-resistant strains of *Mycobacterium*; causing physicians to use antibiotic combination therapy to ensure that the bacilli are killed during the treatment.

Transmission

Animals contract *Mycobacterium* infections primarily via the respiratory and gastrointestinal tracts. Respiratory transmission (inhalation of contaminated aerosols or fomites) is the most efficient type of TB transmission. Animals or humans with active TB produce respiratory droplets that contain one to three organisms and these TB-containing droplets can remain suspended in air for several hours. Infection occurs when an infected animal coughs and inhalation of the contaminated droplet reaches the alveoli of the lung. Indirect transmission via contaminated feed, pasture, or water (contaminated by mucus secretions, feces, urine, or milk with infective organisms) is another important route of transmission. Other less frequent routes of transmission include bites, vertical transmission (which may actually result from aerosol transmission from parent to offspring in close quarters), and direct contact from infected animals grooming healthy animals. The route of infection determines the disease manifestations of TB. Aerosol spread leads to involvement of the lungs and thoracic lymph nodes; ingestion of contaminated food and water results in involvement of the lymph tissues associated with the intestinal tract.

Humans contract TB by the respiratory tract (aerosol transmission of droplets or inhalation of dust), gastrointestinal tract (consumption of uncooked meat or unpasteurized milk from TB-infected animals), or integumentary system (direct

> Overcrowding, poor sanitation, malnutrition, and lack of disease knowledge have all contributed to problems with treating and preventing the spread of TB.

injury to skin and mucous membranes). One of the main reasons people pushed for pasteurization of milk was to eliminate TB organisms in milk and dairy products. Transmission of *Mycobacterium* is summarized in Figures 3-75A and 3-75B.

Pathogenesis

My. bovis can survive for about 4 days on fomites and several months on feces or in animal carcasses; however, its survival in the environment is reduced by drying, exposure to sunlight, and high temperature.

Tubercular infection typically begins with inhalation of infective respiratory droplets containing the tubercle bacilli. The inoculum of bacteria is usually no more than three organisms, which are engulfed by alveolar macrophages, but not killed. The bacilli continue to multiply slowly inside and outside host cells. From the alveolar macrophages the organisms may be carried from the primary focus to the regional lymph nodes and spread via lymph and blood to multiple organs (most of these disseminated lesions heal). Lymphatic drainage from the primary focus leads to the formation of caseous (dry and crumbly) lesions in an adjacent lymph node. The primary focus and the lymph node lesion are called the primary complex. The primary complex seldom heals.

Several weeks later, cell-mediated immunity is activated and T lymphocytes stimulated by mycobacterial antigens begin to proliferate and secrete lymphokines. The immune response also stimulates tuberculin hypersensitivity; a form of type IV delayed hypersensitivity, which causes the reactive tuberculin test used in diagnosis of TB.

When bacteria localize in an area, they stimulate the formation of tumor-like masses called tubercles. The tubercle is an inflammatory lesion produced when

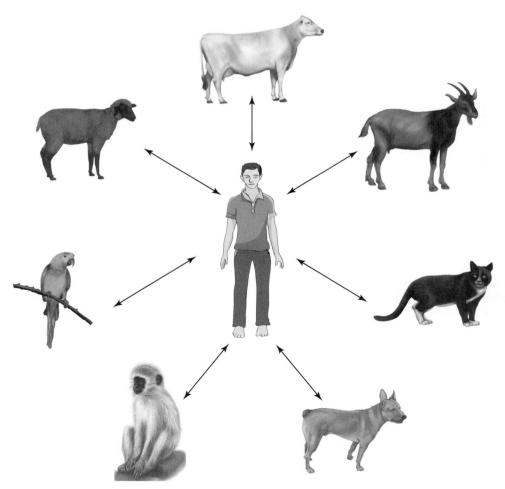

Figure 3-75A Cycle of *Mycobacterium tuberculosis* infection.

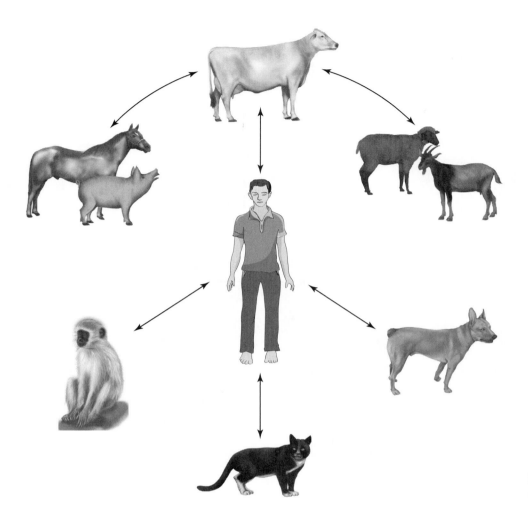

Figure 3-75B Cycle of *Mycobacterium bovis* infection.

the activated macrophages engulf the bacteria and isolate them. The tubercle contains a few phagocytized mycobacteria surrounded by many activated macrophages and lymphocytes. These viable bacteria can continue to multiply causing the tubercles to enlarge. As the tubercles increase in size, the host cells often die within the tubercle and the dead tissue looks caseous (cheesy). A few bacteria can survive in the caseous center and remain viable for many years. The caseous masses have a tendency to undergo mineralization (calcification), liquefaction, or may become enclosed by dense, fibrous connective tissue and the disease becomes inactive. Reactivation of the infection can occur months or years later if the immune response is weakened. Bacteria can escape from the primary site and travel via the lymphatic and vascular systems to other organs where they establish other tubercles. If the vascular system contains numerous organisms from a local lesion, many tubercles develop in major organs such as the lungs causing a rapidly fatal disease (this acute, generalized infection is called miliary tuberculosis, named after the small lesions that form that resemble millet seeds). If small numbers of organisms enter the circulation from a primary complex a few isolated lesions develop in other organs, become encapsulated, remain small for long periods of time, and do not cause clinical signs of disease. The time it takes to develop clinical TB depends upon the interaction of the host's immune response and proliferation of bacteria in macrophages. Tubercule formation is illustrated in Figure 3-76.

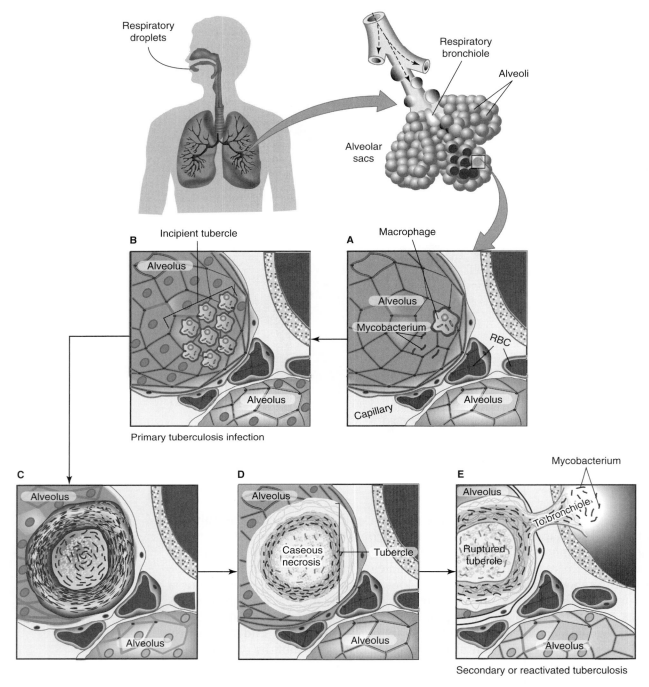

Figure 3-76   Although *Mycobacterium tuberculosis* can infect any organ, approximately 85% of infections remain in the lung. There are five stages of primary infection with *Mycobacterium tuberculosis* as shown in this figure. (A) *Mycobacterium* infects the respiratory tract via inhalation of droplets. A respiratory droplet is about 5 μm in diameter and carries one to three bacilli. The minimum infectious dose is about 10 bacilli. (B) Alveolar macrophages phagocytize the pathogens but are unable to digest them as a result of the bacilli's waxy cell walls. (C) Bacteria freely replicate within macrophages, gradually killing the macrophages. Bacteria are released from dead cells and are phagocytized by other macrophages. This stage of infection is typically asymptomatic or is associated with mild fever. (D) Infected macrophages present antigen to T lymphocytes, which in turn produce lymphokines that attract and activate more macrophages. The macrophages surround the infection site and tightly pack around the organisms forming a tubercle. (E) Other body cells deposit collagen further walling off the tubercle. Infected cells in the center of the tubercle die, release bacilli, and produce caseous (cheeselike) necrosis due to the presence of protein and fat released from dying cells. Occasionally the center will liquefy and then fill with air forming a tuberculous cavity. *My. tuberculosis* may remain dormant for decades within macrophages and in the centers of tubercles. In time if the immune system kills all of the bacteria, the body deposits calcium around the tubercles (which are now called Ghon complexes).

Clinical Signs in Animals

Clinical signs of TB depend on the extent and location of the lesions. General signs of TB include weakness, anorexia, dyspnea, emaciation, and low-grade fever. When lungs are extensively involved there may be an intermittent, hacking cough. Species specific signs include:

- *Cattle.* Acute TB lesions are usually found in the thorax and occasionally lymph nodes causing the classical signs of wasting, weakness, anorexia, and dyspnea. In advanced stages of TB lesions may be found in many organs including the udder, uterus, kidneys, and meninges (Figure 3-77). Skeletal muscles are seldom affected.

- *Swine.* Swine can be infected with *My. tuberculosis*, *My. bovis*, and *My. avium*, which are usually contracted by ingestion. The clinical signs of TB in swine are associated with the gastrointestinal tract and regional lymph nodes. Swine infected with *My. bovis* typically have rapidly progressive disease with caseation of lesions.

- *Sheep and goats.* TB is rare in sheep and goats. When infected with *My. bovis* the disease in sheep and goats is similar to that of cattle; when infected with *My. avium* the disease causes disseminated lesions.

- *Dogs.* TB lesions in dogs usually resemble neoplasms (especially carcinomas). Lesions are typically found in the lungs, pleura, and liver. The lesions appear differently depending on the area involved varying from ones with depressed centers and hemorrhagic edges (liver) to ones containing liquid.

- *Cats.* Cats are resistant to *My. tuberculosis* infection, but may become infected with *My. bovis* by ingestions of contaminated milk. TB lesions in cats are found primarily in the gastrointestinal tract; however, respiratory infection and contaminated wounds can also be infection sources.

- *Nonhuman primates.* Monkeys and apes are susceptible to *My. bovis*, *My. tuberculosis*, and *My. avium*. Signs of TB in nonhuman primates include

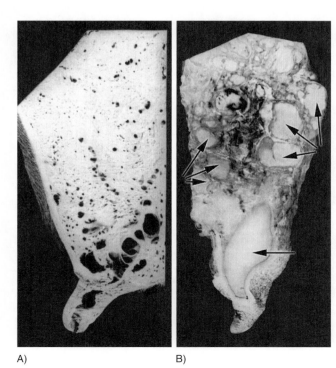

A)   B)

Figure 3-77   Tuberculosis lesions in a cow udder that may lead to shedding of the bacterium into milk. (A) Healthy udder. (B) Tuberculous udder.

(Courtesy of CDC/Minnesota Dept. of Health, R.N. Barr Library; Librarians Melissa Rethlefsen and Marie Jones)

behavioral changes, anorexia, lethargy, and sudden death while appearing to be in good condition. Depending on the organ system involved other signs may include diarrhea, skin ulceration, splenomegaly, and hepatomegaly.

- *Poultry*. The digestive tract is the entry site of *My. avium* with predominant lesions found in the liver, spleen, intestine, and bone marrow. Poultry TB is common on small rural farms where poultry has been housed for many years in contaminated housing. TB is less common in commercial poultry operations as a result of rapid turnover of animals.

- *Equine*. Horses, donkeys, and mules are less susceptible to TB than bovine and the lesions seen in these animals usually involve the gastrointestinal tract. Tubercles in equine are rarely caseous or calcified.

Clinical Signs in Humans

> The principle sign of TB is chronic wasting or emaciation despite adequate nutrition and care.

TB in people is a multifaceted disease. Primary TB typically is asymptomatic and may only be recognized by the development of a positive skin test. People who develop clinical signs of TB are typically homeless or alcoholic people, immuno-suppressed people, the elderly, nursing home residents, and people suffering from such diseases as diabetes mellitus and lymphoma. The incubation period is from 4 to 6 weeks. Clinical signs of TB in humans depend on the organ system involved. Signs of pulmonary TB include cough, production of sputum, and hemoptysis (coughing up blood) (Figure 3-78). Regional lymph nodes are affected but heal and develop calcifications. As pulmonary TB progresses people may develop a subfebrile temperature, night sweats, lymphadenopathy, fatigue, anorexia, and erythema of the extremities. The clinical signs of cutaneous TB are highly variable and may show ulcerated, nodular, or hemorrhagic skin lesions, which spread

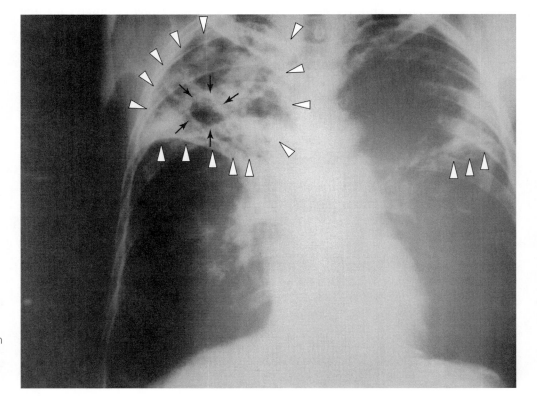

Figure 3-78 Thoracic X-ray of a patient with advanced bilateral tuberculosis. Black arrows outline caving formation and the white triangles outline bilateral pulmonary infiltrates.
(Courtesy of CDC)

superficially or extend to deeper skin layers. Generalized or extrapulmonary TB may affect almost any organ system, but typically affects the lymph nodes, pleura, genitourinary tract, skeletal system, meninges, and peritoneum. Clinical signs of extrapulmonary TB include anorexia, weight loss, fatigue, high and sustained fever, night sweats, and chills; however, the patient may be asymptomatic for years. Bone and joint TB mainly affects the hip and stifle regions with clinical signs of pain, swelling, and decreased range of motion. Destruction of vertebrae with damage to the spinal cord is a sequela of spinal TB (also known as Pott's disease).

> Laryngeal TB is highly infectious because it can be spread to other people by talking.

Diagnosis in Animals

Pathologic findings for TB include the tubercle, which microscopically consists of a cluster of macrophages surrounding bacteria. Since the macrophages engulf the bacteria yet do not inhibit their growth, the lesion can be of varying sizes (Figure 3-79). As the bacteria multiply, the nearby cells undergo necrosis, the center may become caseous, and in time the center of the tubercle may calcify (except with *My. avium* infections). Simple tubercles are approximately between 1 mm and 2 cm in diameter, but large tubercles may be formed when tubercles coalesce with other tubercles. Tubercles in which bacteria are eventually killed may contain fibrous scar tissue. If this does not occur secondary infection with other organisms may occur followed by liquefaction and necrosis of the tubercle producing cavitation.

Grossly, tubercles either in organs or on a surface are firm, pale nodules that contain yellowish, caseous, necrotic centers that are dry (versus pus filled as in an abscess). If calcification has occurred the tubercle has a white gritty appearance. If the tubercle breaks into a blood vessel or large numbers of bacilli are released into the bloodstream, many small tubercles (2 to 3 mm in size) that are the same age and size are seen. These lesions are seen in miliary tuberculosis, so named because the lesions look like a scattering of millet seeds.

Clinical diagnosis of TB can only be done when the disease has reached an advanced state, at which time animals are typically shedding organisms and are an infection source for other animals. Antemortem identification of TB is

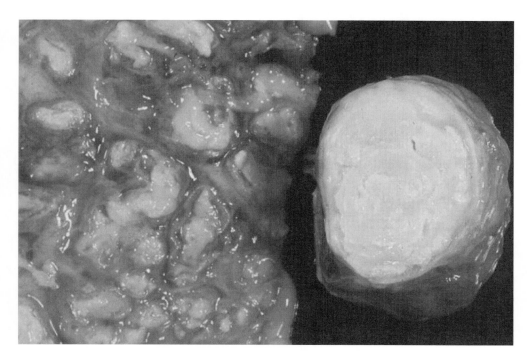

Figure 3-79 Lung lesions of elk with tuberculosis. The lung contains multiple areas of caseous necrosis surrounded by thin pale fibrous tissue capsules (tubercles).

(Courtesy of Dr. G. Wobeser, Canadian Cooperative Wildlife Health Centre)

critical in controlling this disease. The tuberculin skin test is used in animals and is based on the premise that animals infected with *Mycobacterium* bacteria are allergic to the proteins in the tuberculin and develop delayed-type hypersensitivity reactions when exposed to this protein. The tuberculin is placed intradermally in the deep layers of the skin and in infected animals will elicit a local reaction characterized by inflammation and swelling. The USDA has accepted the tuberculin test for identification of *My. bovis* in cattle, bison, goats, and captive cervids.

The purified protein derivative (PPD) tuberculin for veterinary use is made from *My. bovis* and can identify infections caused by *My. bovis* and *My. tuberculosis*. A *My. avium* PPD tuberculin is used when testing for avian tuberculosis because animals infected with *My. avium* react less to tuberculin made from *My. bovis*. Most countries use PPD tuberculin at a dose of 0.1 mL (0.1 mg of protein) containing 5,000 tuberculin units in mammals and 0.05 mL containing 2,500 tuberculin units in chickens. The results from PPD tuberculin tests are used to classify animals as negative for infection (response to the test is negative), suspected to have infection (response to the test is unclear), or reactor (response is positive). In the United States, two specific skin tests are used:

- The caudal fold skin test is used for large mammals such as cattle, sheep, goats, and bison and tuberculin is typically injected in one of the folds at the tail base (Figure 3-80A). Swine are injected in the skin behind the ear or vulva. Chickens are injected in the skin of the wattle. The injection sites are observed and palpated 72 hours after injection for cattle, sheep, and goats and 48 hours after infection for swine and chickens. The caudal fold tuberculin test used to be done every three years on cattle; now it is used in cases of transport or sale or during slaughter-suspect cases.

- The comparative cervical tuberculin test is conducted only by regulatory state or federal veterinarians and is performed by injecting *My. avium* and *My. bovis* PPD tuberculins into separate sites in the skin of the neck. Both injection sites are then observed and palpated and the difference in size of the two responses allows for the differentiation of infection by *My. bovis* versus *My. avium*, *My. avium* subspecies *paratuberculosis*, or transient saprophytic mycobacteria found in the environment. The comparative cervical tuberculin test has greatly reduced the incidence of reactors without gross lesions (decreased the number of false positives). This test is not used in herds where *My. bovis* has been diagnosed.

Nonhuman primates should be tested with tuberculins prepared for veterinary use as the tuberculins prepared for human use are not of sufficient potency to elicit the delayed-type hypersensitivity response in these species. Eyelids are used as injection sites as they can be viewed from a distance (used in zoos).

Animals that test positive or suspect are removed from the farm and examined postmortem for confirmation of mycobacterial infection. Microscopy with acid-fast stains can detect acid-fast bacilli in tissue samples. Bacterial culture using Löwenstein-Jensen agar is still required to confirm a diagnosis of TB; however, this can take 6 to 8 weeks. PCR techniques have been developed for the diagnosis of *My. tuberculosis* and *My. bovis*. DNA fingerprinting techniques have also been developed and are helpful in identifying potential sources of infection or relatedness of strains.

Diagnosis in Humans

Clinical symptoms, especially in high risk individuals, along with laboratory testing are used to diagnose TB in humans. Chest radiographs are not pathognomonic since any pattern of infiltrate can be seen. Samples of sputum, bronchial secretions, gastric secretions, CSF, body fluids, and biopsies are used for microscopy, culture, and molecular methods of detected *Mycobacterium* infections. A tissue sample obtained surgically or at necropsy show acid-fast bacilli within tubercles and is sufficient to establish a diagnosis of TB. Final confirmation is based on bacterial culture. Microscopy with acid fast stains can detect acid-fast bacilli; however, culture is mandatory for identification and can take several weeks. Culture media for

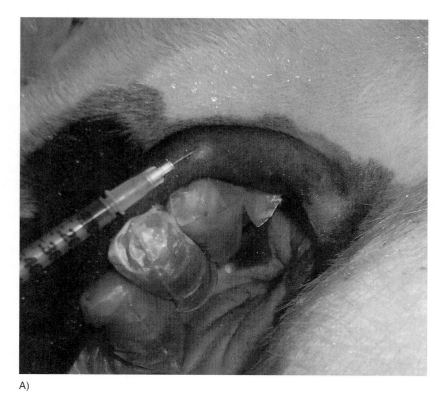

A)

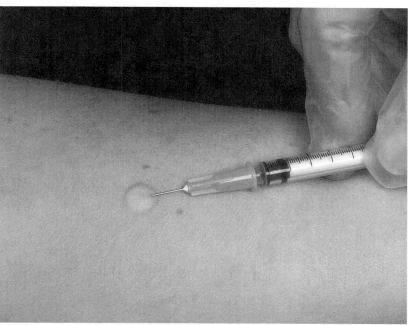

B)

Figure 3-80 (A) Caudal fold test in a bovine.
(Courtesy of James Meronek, DVM)

(B) PPD skin test involves intradermal injection of tuberculin.
(Courtesy of Greg Knobloch)

Mycobacterium spp. include Löwenstein-Jensen agar and Middlebrook 7H10 and 7H11 agar. Automated systems may also be used. ELISA testing is also available; however, results should be verified by culture and histopathology. PCR can detect less than 10 bacteria, but should not replace culture.

Tuberculin skin testing is also used in humans and indicates past exposure to *My. tuberculosis* and some cross-reactivity with other mycobacteria. The preferred skin test for *My. tuberculosis* infection is the intradermal or Mendel-Mantoux method (Figure 3-80B). Five tuberculin units of PPD in 0.1 mL of solution is injected intradermally into the volar surface of the forearm. Tests are read after 48 to 72 hours and the transverse diameter (in mm) of induration (hardened mass) is recorded. Three cutoff levels are used for defining a positive test (≥ 5, ≥ 10, and ≥ 15). The cutoff used depends on the health status of the individual (immunosuppressed people with chest lesions are read at the low range, whereas people with no known risk are read at the high range). In some individuals (those without a known prior-positive skin test, have not had a tuberculin skin test within a 1-year period, and have an initial negative skin test) it is recommended to have a two-step PPD test to detect individuals with past TB infection who now have diminished skin test reactivity. The two-step PPD test involves performing the first skin test (which turns out negative) and repeating the skin test 7 days after the initial one. The second test is read in 48 to 72 hours after its placement. If the second test shows a positive response, it may demonstrate a boosted response to an old infection (in time the antibody levels from a previous infection become too low to detect with the one-step test). Boosting can last up to 2 years; therefore, the two-step test reduces the likelihood that repeated skin testing might falsely indicate new infection. People with positive skin tests then have chest radiographs taken as well as sputum samples for culture.

Treatment in Animals

The antituberculosis drug, isonicotinic acid hydrazide (INH) or isoniazid, can treat bovine TB; however, the disadvantages to treatment in animals include emergence of drug-resistant strains, excretion of antibiotic in milk and meat, and relapse when therapy is discontinued making treatment of TB impractical.

Treatment in Humans

People are treated with a variety of antibiotics; however, isoniazid has the highest antibacterial activity for *Mycobacterium* bacteria and inhibits the development of resistance making it the treatment of choice. Treatment typically lasts 9 months with isoniazid (other drugs are given for a shorter amount of time). Isoniazid is hepatotoxic and liver enzymes are monitored during treatment. Some strains of *My. tuberculosis* are extremely drug-resistant (XDR) making treatment difficult.

Management and Control in Animals

In 1917, the Cooperative State-Federal Tuberculosis Eradication Program, administered by the USDA Animal and Plant Health Inspection Service (APHIS), state animal health agencies, and United States livestock producers, was initiated because TB caused more losses among U.S. farm animals in the early part of the 1900s than all other infectious diseases combined. This program has nearly eradicated bovine TB from the U.S. livestock population and has reduced human disease. Initially, all cattle herds were systematically tested, and all reactors were slaughtered. Premises were cleaned and disinfected after infected cattle were removed. As a result of this program, the reactor rate in U.S. cattle was reduced from 5% to approximately 0.02%.

Control programs for TB currently revolve around four components:

- *Prevention.* TB prevention focuses on reducing exposure to the pathogen and reducing the likelihood that an exposed animal will become infected after exposure. Cattle are mainly infected by infected cattle that reside on the farm or are introduced into the herd. If livestock in a herd are tested for TB and are negative, maintaining a closed herd is the best way to prevent TB on a farm. Basic hygiene practices and biosecurity practices (routine testing and quarantine of imported animals, manure management, and maintenance of feed and water) have reduced the risks of *My. bovis* spread on cattle farms. Population control in wild reservoir animals has also been necessary as a result of their interaction with cattle. The BCG vaccine does not completely prevent infection in cattle or other animals and as a result of the fact that vaccinated animals will test positive on the tuberculin skin test, the vaccine has not been used in the United States.

- *Treatment*. The antituberculosis drug INH has made treatment practical for TB; however, the treatment of cattle in the United States and many other countries is forbidden as a result of the shedding of organisms to other animals during treatment and the effects on the eradication program of keeping animals that will test positive for TB in the herd.

- *Eradication*. In domestic livestock herds, culling of reactors and after a waiting period (12 months in the United States) allowing livestock to repopulate the site. The USDA also allows regulatory agencies to develop herd-specific test-and-slaughter programs for individual livestock operators and managers of wildlife farms (such as deer, elk, and buffalo). Eradication of *My. bovis* from wildlife reservoirs is more difficult and has included such strategies as trapping and removal programs, directed hunts to reduce animal numbers, restrictions on feeding and baiting animals, or provision incentives to hunters to increase the number of animals harvested during a hunting season.

- *Surveillance*. Surveillance of TB involves antemortem testing and slaughter surveillance of livestock and captive animal species.

Management and Control in Humans

There are a variety of ways to prevent the spread of TB. One important step to preventing the spread of TB is to isolate and treat all disease carriers until they are no longer an infective risk. The live, attenuated BCG vaccine is used in humans in many foreign countries (it is considered effective in about 80% of vaccinates) and may be recommended in high risk groups. BCG vaccine protects children from miliary TB and TB meningitis. The concern with human vaccination is that vaccinated individuals will have a positive PPD TB test, making surveillance procedures ineffective. If traveling to a country where TB is a concern, people should get vaccinated and avoid people with persistent coughs. Improvement of socioeconomic conditions that can lead to persistent levels of TB within a group (such as prisons, homeless shelters, nursing homes, AIDS victims, and immigrants living in crowded conditions) can greatly reduce TB mortality. Laboratories that may handle *Mycobacterium* organisms (including performing cultures, identification, and susceptibility testing) are designated as Biosafety level 3 facilities and specific precautions are taken.

A 1994 outbreak of bovine TB infection in the wild white-tailed deer population spread into nearby cattle prompting testing and surveillance of a variety of wildlife.

Summary

Tuberculosis is a chronic disease in humans and animals caused by *My. tuberculosis*, *My. bovis*, and *My. avium*. Zoonotic TB has been greatly reduced by control and eradication programs in cattle. Animals contract *My. bovis* infections primarily via the respiratory and gastrointestinal tracts, whereas humans contract TB by the respiratory tract, gastrointestinal tract, or integumentary system. Clinical signs of TB in humans and animals depend on the organ system involved. General signs of TB in animals include weakness, anorexia, dyspnea, emaciation, and low-grade fever. When lungs are extensively involved there may be an intermittent, hacking cough. Clinical presentation of TB in humans includes cough, production of sputum, and hemoptysis (coughing up blood); cutaneous TB lesions are typically ulcerated. Generalized signs of TB include anorexia, weight loss, fatigue, fever, and chills and the patient may be asymptomatic for years. PPD testing is used in both animals and humans to diagnose and control TB. Animals are not treated for TB and are culled from the herd. People are treated with INH for extended periods of time (9 months). Drug-resistant strains of *My. tuberculosis* have made treatment more difficult. Control programs for TB revolve around prevention, treatment, eradication, and surveillance.

VIBRIOSIS AND CHOLERA

Overview

In the past, vibriosis was a term used to describe diseases caused by the curved bacterium *Campylobacter*. *Campylobacter* bacteria are sexually transmitted in animals and the diseases they cause are now called campylobacteriosis.

The disease hog cholera is caused by an RNA virus and is not associated with the diseases caused by *Vibrio* bacteria.

Vibriosis and cholera, diseases primarily of the gastrointestinal tract, are caused by bacteria in the *Vibrio* genus. Vibrio comes from the Latin *vibrare* meaning to quiver and this genus of bacteria moves in an undulating pattern much like quivering. Vibriosis is most commonly caused by *Vi. parahemolyticus* (which causes diarrhea) or *Vi. vulnificus* (which causes skin infections and/or septicemia). The diarrhea-causing *Vi. parahemolyticus* is a relatively harmless infection, but *Vi. vulnificus* infection can lead to septicemia and death. In 1951, Japanese bacteriologists discovered *Vi. parahemolyticus* as a cause of illness among people who eat fish from contaminated waters.

Cholera, also known as Asiatic Cholera, is a severe bacterial infection of the gastrointestinal tract caused by *Vi. cholerae*. Cholera was a rare disease confined to India before the 1800s. It became one of the world's first epidemics as the disease broke out in Calcutta in 1817 from contaminated water. The spread of cholera was aided by the Industrial Revolution and the accompanying growth of urban tenements and slums. In the United States, the Industrial Revolution caused immigrants to move into crowded living quarters, which helped cholera spread to a large degree and provided an ongoing source of new infections. Cholera and its effects helped develop the United States infrastructure because most municipal water mains and sewer systems were built in the late 1800s to help contain the disease. Public health agencies were also formed and funded during this time as well as building codes and ordinances.

Dr. John Snow began investigating the spread of cholera during the London epidemics of 1848 and 1849, developing his theory that cholera was waterborne and taken into the body orally. Dr. Snow proved that water used from a common pump had the cholera organisms, whereas private water sources did not. The actual discovery of the comma-shaped bacillus of cholera was made by the German bacteriologist Robert Koch in 1883. Through microscopic examination, he concluded that feces may contain cholera bacteria for a period of time after the actual onset of the disease. All cases of cholera today are part of a pandemic strain (*Vi. cholerae* 01, biotype El Tor) that started in Indonesia in 1961.

Causative Agent

Vibrio infections are considered foodborne diseases because the majority of these infections are associated with the consumption of contaminated food. Infections caused by *Vibrio* species are classified into two groups: *Vi. cholerae* and non-cholera *Vibrio* (which are further classified as halophilic [need sodium chloride for growth] *Vibrio* species and nonhalophilic [do not need sodium chloride for growth] *Vibrio* species). *Vibrio* spp. are oxidase positive, gram-negative curved bacteria with a single flagellum (Figure 3-81). *Vibrio* spp. can produce multiple

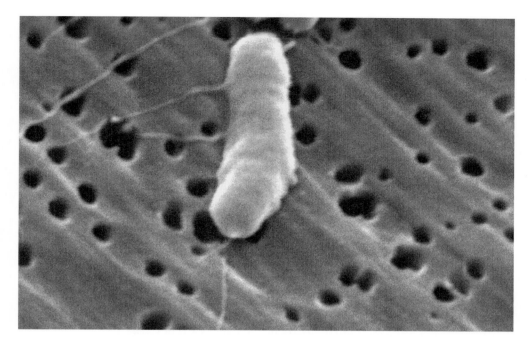

Figure 3-81   Microscopic appearance of *Vibrio cholerae* demonstrating its flagella (scanning electron micrograph). (Courtesy of CDC/Janice Carr)

exotoxins and enzymes that are associated with extensive tissue damage that play a major role in disease development.

Cholera transmission and onset of epidemics is influenced by the season and climate. Cold, acidic, dry environments inhibit the survival of *Vibrio,* whereas warm, monsoon, alkaline, and saline conditions favor their survival.

- *Vi. parahemolyticus* and *Vi. vulnificus* are halophilic, noncholera *Vibrio* found in salt water. Both organisms live in areas where the temperature exceeds 18°C (in the United States it is found along the coasts that border the Gulf of Mexico, New England, and the northern Pacific coast). Infection with either of these two bacteria primarily occurs through eating contaminated raw seafood such as raw oysters. Both bacteria are ingested by filter-feeding mollusks such as oysters, mussels, clams, and scallops; filter-feeding concentrates bacteria in these mollusks providing a large inoculum of bacteria for the person ingesting them. *Vi. parahemolyticus*, which has a rapid doubling time of 9 minutes under ideal conditions, causes acute watery diarrhea with abdominal cramps and fever approximately 24 hours after eating shellfish. *Vi. vulnificus* infection occurs following ingestion of contaminated fish or contact of a wound with seawater or contaminated seafood. *Vi. vulnificus* causes cellulitis and septicemia. Cellulitis can result in severe tissue destruction leading to amputation in some cases. Infection with *Vi. vulnificus* can cause death in up to 50% of immunocompromised people who contract it.
- *Vi. cholerae* is primarily found in humans; however, it can survive in freshwater and seawater. In epidemic areas cholera bacteria are found in contaminated water systems. In areas where public sanitation is adequate, the majority of cholera cases are a result of ingestion of infected shellfish. *Vi. cholerae* has also been isolated from birds and herbivores, and has experimentally caused disease in dogs, Guinea pigs, and rabbits.

Epizootiology and Public Health Significance

Vibriosis

In the United States, vibriosis is not commonly recognized as a cause of illness, partly because of proper sanitation systems and because clinical laboratories rarely use the culture techniques necessary to identify this organism. In addition, not all states require that *Vi. parahemolyticus* infections be reported to the state health department (CDC collaborates with the Gulf Coast states of Alabama, Florida, Louisiana, and Texas to monitor the number of cases of *Vibrio* infection in these regions and the Foodborne Diseases Active Surveillance Network (Food Net) tracks *Vi. parahemolyticus* in regions outside the Gulf Coast). Most cases of vibriosis in the United States occur in coastal states between June and October. Between 1988 and 1991, there were only 21 reported cases of *Vi. parahemolyticus* infection in the United States. Between 1988 and 1995, there were over 300 reports of *Vi. vulnificus* infection in the United States (although it is believed to be underreported). In 1997, the incidence of diagnosed *Vi. parahemolyticus* infection in the United States was 0.25 per 100,000 people. In Asia, *Vi. parahemolyticus* is a common cause of foodborne disease.

Since 1988, the CDC has maintained a voluntary surveillance system for culture-confirmed *Vibrio* infections in Alabama, Florida, Louisiana, Mississippi, and Texas. The estimated illnesses caused by all noncholera *Vibrio* species in the United States are 8,000 per year resulting in approximately 60 deaths. *Vi. parahemolyticus* is the most common noncholera *Vibrio* species reported; however, *Vi. vulnificus* is associated with 94% of reported deaths. Because clinical laboratories do not routinely use the selective medium thiosulfate-citrate-bile salts-sucrose (TCBS) for stool culture, many cases of *Vibrio*-associated gastroenteritis are not reported.

In recent years in the United States, the prevalence rate of infections caused by noncholera *Vibrio* species appears to be increasing. The combination of increased water temperature where shellfish have been harvested and increased salt levels in these bodies of water may have contributed to the increased contamination rate of shellfish. Worldwide, areas such as Japan, Taiwan, China, Hong Kong, and Korea frequently report noncholera *Vibrio* infections, which may be related to the high prevalence of hepatitis B resulting in liver disease.

Cholera

> Cholera was generally thought to be the scourge of the depraved, poor masses because these groups tended to be affected the worst.

In the United States, cholera was prevalent in the 1800s but has been rare in industrialized nations for the last 100 years. In industrialized nations cholera has been virtually eliminated through the use of modern sewage and water treatment systems. Improved transportation, however, results in more people from the United States traveling to Latin America, Africa, or Asia where epidemic cholera occurs. U.S. travelers to areas with epidemic cholera may be exposed to the cholera bacterium or may bring contaminated seafood back to the United States. Between 1995 and 1999 there were 53 confirmed cases of cholera in the United States, including one death, caused by the *Vi. cholerae* 01, biotype El Tor strain. Most of these infections were acquired in other countries; however, four came from eating Gulf Coast seafood.

Cholera is still common today in some parts of the world, including the Indian subcontinent and sub-Saharan Africa (Figure 3-82). In January 1991, epidemic cholera appeared in South America (Peru) and quickly spread to 21 countries infecting 700,000 people and killing 6,000.

Transmission

Vibriosis

Halophilic *Vibrio* (*Vi. parahemolyticus* and *Vi. vulnificus*) inhabit coastal waters and associate with zooplankton. Halophilic marine organisms such as crustaceans and shellfish contract *Vibrio* infections primarily by ingesting these zooplankton. Ecologically, *Vibrio* organisms degrade chitin and recycle the breakdown products.

Figure 3-82   Cholera is caused by *Vibrio cholerae* and is typically found in contaminated water.
(Courtesy of CDC/Dr. Jack Weissman)

In temperate zones, *Vibrio* bacteria survive over winter by settling into the ocean sediment and are resuspended during the warmer seasons. As they are resuspended, they are incorporated into the food chain where they eventually grow on fish, shellfish, and other edible seafood. People contract the disease by ingesting contaminated seafood or through direct contact of an open wound with contaminated seawater. There is no evidence of person-to-person transmission.

Cholera

Cholera can be contracted by drinking water or eating food contaminated with *Vi. cholerae*. When epidemics occur, the source of the contamination is usually the feces or vomitus of an infected person. The disease can spread rapidly in areas with inadequate treatment of sewage and drinking water. Cholera is often spread by someone who is infected with the bacterium who prepares food for others or by sharing a drinking cup with a healthy individual.

    Vi. cholerae can also live in the environment in brackish rivers and coastal waters. Raw shellfish have been a source of cholera, and people in the United States have contracted cholera after eating raw or undercooked shellfish from the Gulf of Mexico. The disease is not spread directly from one person to another.

Pathogenesis

Once *Vibrio* organisms are ingested with food or water, they enter the stomach where some of the organisms die. Those bacteria that survive the extreme acidity of the stomach penetrate the mucus layer of the duodenum and jejunum using their flagella, adhering to the outside of the epithelium but not penetrating the mucosal cells. The bacteria begin replicating and release toxin that disrupts the normal physiology of the intestinal cells by altering the levels of chemical messengers resulting in removal of anions from the cell through the cell membrane. Depending upon the species of *Vibrio* and the toxin it produces, the epithelial cells become damaged and secrete large amounts of electrolytes (chloride and

Vibrio bacteria cause disease in humans by production of toxins and enzymes rather than invasion of epithelial cells (the intestinal lining may appear normal in affected individuals).

bicarbonate ions in particular) into the intestinal lumen. As the cell secretes large amounts of chloride and bicarbonate ions into the intestinal lumen, water follows resulting in massive amounts of fluid loss through the intestines (cholera) or fluid loss that is self-limiting (vibriosis). The chemicals produced by the bacteria can also damage epithelial cells resulting in ulceration of the affected areas. As a result, profuse diarrhea occurs producing stools that resemble water flecked with small particles of mucus known as the rice-water stools of cholera. The loss of water as a result of cholera infection is extensive (up to one liter per hour in humans).

Vi. vulnificus causes cellulitis and septicemia following ingestion of shellfish or exposure to sea water. *Vi. vulnificus* can get through the intestinal wall and into the bloodstream and produces multiple chemicals, such as proteases, hemolysins, and cytolysins, which are associated with its virulence.

Clinical Signs in Animals

Shellfish and other halophilic marine animals do not show clinical signs of disease when infected with *Vibrio* bacteria.

Clinical Signs in Humans

Vibriosis

Vi. parahemolyticus causes watery diarrhea often with abdominal cramping, nausea, vomiting, headache, fever, and chills. Symptoms typically occur within 24 hours of ingestion and the disease is usually self-limiting lasting for about 3 days (ranges from 2 to 10 days). Severe disease is rare and occurs more commonly in immunocompromised people. *Vi. parahemolyticus* can also cause a skin infection producing large, fluid-filled blisters on the arms or legs when an open wound is exposed to warm, contaminated seawater.

Vi. vulnificus can cause disease in people who ingest contaminated seafood or water or have an open wound that is exposed to contaminated seawater. Ingestion of *Vi. vulnificus* can cause vomiting, diarrhea, and abdominal pain in healthy people; however, in immunocompromised people (especially those with chronic liver disease) *Vi. vulnificus* can result in septicemia, causing a severe and life-threatening illness characterized by fever and chills, decreased blood pressure, and blistering skin lesions. Septicemia from *Vi. vulnificus* is fatal about 50% of the time. *Vi. vulnificus* can also cause a skin infection when open wounds are exposed to warm seawater resulting in skin breakdown and ulceration.

Cholera

Clinical signs of cholera in humans appear 1 to 7 days following consumption of contaminated food or water. Abdominal bloating and cramps that quickly progresses to produce large quantities of very watery stool are classic signs of cholera. The stool has little odor and is often referred to as rice-water stool because of its appearance (very watery, light colored, and laced with tiny flecks of mucus but not blood). Typically the person is afebrile and occasionally will be vomiting. People infected with cholera may experience intense thirst, extreme weakness, sunken eyes, decreased urination, dry, wrinkled skin, rapid heart rate, lowered blood pressure, weakened pulse, sleepiness, unconsciousness, seizures, or kidney failure. Death is caused by the dehydration.

Diagnosis in Animals

Diagnosis in animals is not performed other than for surveillance.

Diagnosis in Humans

Vibriosis can be diagnosed via Gram stain and culture from samples of stool, blood, or blister fluid. Cholera can be diagnosed via gram stain and stool culture. For isolation of *Vibrio* pathogens the use of a selective medium that has thiosulfate, citrate, bile salts, and sucrose (TCBS agar) is recommended (Figure 3-83). These organisms are all oxidase positive and vary in their fermentation properties (lactose versus sucrose) and ability to grow in varying concentrations of sodium chloride (*Vi. cholerae* cannot grow in salt, *Vi. parahemolyticus* and *Vi. vulnificus* cannot grow without salt). Suspicion of *Vi. parahemolyticus* infection is warranted if a patient has watery diarrhea and has eaten raw or undercooked seafood, especially oysters, or when a wound infection occurs after exposure to contaminated seawater. *Vi. vulnificus* skin lesions or septicemia may be suspected in immunocompromised people with skin blisters or ulcerations or signs of blood infection such as fever or shock. Suspicion of *Vi. cholerae* is warranted if a patient has returned from an endemic country or coastal region where the water or food may be contaminated.

Treatment in Animals

Shellfish and halophilic marine animals are typically not treated. In cultured, aquarium fish and confined fish skin, fins, and tail ulcerations may be treated with sulfamerazine (there is a withdrawl time for fish marketed for human consumption).

Treatment in Humans

Treatment in humans is typically symptomatic including fluids either by mouth or intravenously. Treatment is not necessary in most cases of *Vi. parahemolyticus* infection and there is no evidence that antibiotic treatment decreases the severity or the length of the illness. Patients should drink plenty of liquids to replace fluids lost through diarrhea.

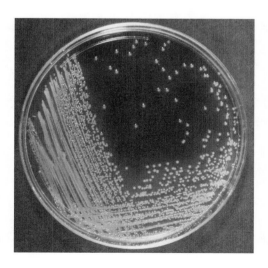

Figure 3-83 *Vibrio cholerae, Vibrio parahemolyticus,* and *Vibrio vulnificus* produce good growth on thiosulfate citrate bile salts-sucrose (TCBS) agar. (Courtesy of CDC)

Table 3-9   Less Common Bacterial Zoonoses

Bacterium	Bacterial Description	Predominant Signs in People	Transmission	Animal Source	Geographic Distribution	Diagnosis	Treatment
Actinobacillus lignieressi, A. equuli, A. suis	Gram-negative rod	Wounds with abscess formation are seen on hands and forearms; septicemia may occur	Animal bites; direct contact with preexisting wounds during exposure to animals	Normal oropharyngeal bacteria of horses, cattle, sheep, and pigs	Worldwide	Bacterial culture on blood agar and biochemical tests	B-lactam antibiotics or fluoroquinolones
Aeromonas spp. (*A. caviae, A. hydrophila, A. sobia, A. jandaei, A. schubertii, A. veronii*)	Gram-negative rod	Diarrhea; gastroenteritis, skin wounds. Healthy people can contract this bacterium; however, children are more prone to this illness than any other age group. Those with weak immune systems develop septicemia	Ingestion of contaminated food or water; skin or mucous membrane exposure to contaminated water; skin trauma by fish fin or fish hook; seafood	Fresh aquatic environments; various warm- and cold-blooded animals	Worldwide	Bacterial culture on blood and MacConkey agars and biochemical tests (string test differentiates it from *Vibrio* spp.)	Diarrhea may be self-limiting; cephalosporins, quinolones, and aminoglycoside antibiotics
Arcanobacterium pyogenes (formerly *Corynebacterium pyogenes* and *Actinomyces pyogenes*)	Gram-positive curved rod	Septicemia, endocarditis, meningitis, arthritis, pneumonia, and abscesses on extremities	Direct transmission from animals to humans not proven; fly vector possible	Normal bacteria of cattle, sheep, and pig mucous membranes	Worldwide	Bacterial culture on blood, chocolate, Columbia colistin-nalidixic acid (CNA) agar and biochemical tests	Usually susceptible to penicillin, erythromycin, and clindamycin
Bordetella bronchiseptica	Gram-negative rod	Pertussis-like disease in immunecompromised people	Aerosols or close contact with infected animals	Normal bacteria of horses, pigs, dogs, cats, rabbits, and Guinea pigs (has caused respiratory disease in this animals as well)	Worldwide	Bacterial culture on blood and MacConkey agar and biochemical tests	Aminoglycoside and quinolone antibiotics; commonly resistant to ampicillin and cephalosporins
Corynebacterium pseudotuberculosis	Gram-positive rod	Necrotixing lymphadentitis of mandibular, axillar, or inguinal lymph nodes	Consumption of raw milk; close contact with infected animals	Normal bacteria of sheep, goats, horses, and cattle (may cause caseous lymphadenitis in sheep)	Worldwide; more frequent in Australia and New Zealand	Bacterial culture on blood and commercial identification (API) agars and biochemical tests (identification is complex in this genus of bacteria)	Vancomycin; resistance to penicillins and cephalosporins has been seen

Organism	Morphology	Disease	Transmission	Source	Distribution	Diagnosis	Treatment
Corynebacterium ulcerans	Gram-positive rod	Diphtheria-like sore throat	Consumption of raw milk; close contact with animals especially in summer	Normal bacteria of cattle (can cause mastitis in cattle)	United Kingdom	Bacterial culture on blood and commercial identification (API) agars and biochemical tests (identification is complex in this genus of bacteria)	Vancomycin; resistance to penicillins and cephalosporins has been seen
Dermatophilus congolensis	Gram-positive branching filaments	Eczema-like lesion, pustules, or furuncles on the hands and forearms	Contact with infected animals or animal products; fly vectors	Source is soil that causes suppurative skin disease in wild and domestic animals (most frequently cattle, sheep, and horses)	Worldwide, but more prevalent in humid, tropical, and subtropical regions such as Africa, Australia, New Zealand, and India	Gram stain, bacterial culture on routine agars such as blood and Sabouraud dextrose, and biochemical tests	Topical antibiotic treatment is usually sufficient
Edwardsiella tarda	Gram-negative rod	Wide variety of infections in immune-compromised people (usually in normally sterile sites such as the lungs, urinary tract, and blood)	Uncertain; most likely close contact with contaminated water or infected animal	Gastrointestinal tract of cold-blooded animals (reptiles)	Worldwide	Gram stain, bacterial culture on blood, MacConkey, and HE or XLD agar, and biochemical tests	Susceptible to many antibiotics
Helicobacter spp.	Gram-negative spirillum	Gastric ulcers and gastritis (*H. pylori*); enteritis and sepsis (*H. cinaedi* and *H. fennelliae*); other species in animals include *H. heilmananii*, *H. canis*, *H. felis*, and *H. suis*	Unknown other than that the bacterium is obtained orally	Humans and hamsters (*H. pylori*); other species have been found in pigs, cattle, dogs, and cats	Worldwide	Stained tissue biopsy and bacterial culture on selective agar such as Skirrow's and modified Thayer-Martin agar may require 1 week; serology for *H. pylori*	Resistance to antibiotics is a problem; quinolones recommended
Rhodococcus equi	Gram-positive, slightly acid-fast rod	Pulmonary infections in immune compromised people	Contact with infected animals may be source; most common route is through contaminated soil	Source is soil that causes pneumonia in foals and sporadic infections in cattle, sheep, and pigs	Worldwide	Gram and acid-fast stain, culture on routine agars such as blood and Sabouraud dextrose, and biochemical tests	Antibiotic combinations such as erythromycin and rifampin

Vi. vulnificus infections are treated with antibiotics such as tetracycline or doxycycline plus ceftazidime. One out of five patients with vibriosis requires hospitalization.

Cholera is treated with oral rehydration therapy (ORT), which was developed in the 1960s that is cheaper, easier, and more effective than IV fluid replacement. Patients are treated with an oral rehydration solution, a prepackaged mixture of sugar and salts to be mixed with water and drunk in large amounts. This solution is used throughout the world to treat diarrhea. With prompt rehydration therapy, fewer than 1% of cholera patients die. Antibiotics, typically tetracycline, are given to prevent the bacterium from replicating in the intestinal lumen.

Management and Control in Animals

Vi. vulnificus and *Vi. parahemolyticus* are bacteria naturally present in marine environments and elimination of them from natural water systems is impossible. Contamination with *Vibrio* bacteria does not clinically affect shellfish and does not change the look, smell, or taste of the seafood. Management of these organisms is based on lowering the risk of disease is people. In the United States oysters can only be harvested legally from waters free from fecal contamination; however, even legally harvested oysters can be contaminated. Voluntary reporting of *Vi. vulnificus* infections to CDC and to regional offices of the FDA help state officials with tracking of shellfish and sampling of harvest waters to discover possible sources of infection and to close oyster beds when problems are identified.

Management and Control in Humans

Vibriosis can be prevented by avoiding raw or undercooked shellfish, keeping raw shellfish and its juices away from cooked foods (cross-contamination), and avoiding contact of wounded skin with seawater or raw seafood. When an outbreak is traced to an oyster bed, health officials recommend closing the oyster bed until conditions are less favorable for *Vi. parahemolyticus*. There is no national surveillance system for *Vi. vulnificus*, but the CDC collaborates with the states of Alabama, Florida, Louisiana, Texas, and Mississippi to monitor the number of cases of *Vi. vulnificus* infection in the Gulf Coast region.

The risk for cholera is very low in the United States; however, travelers visiting areas with epidemic cholera should only drink water that has been boiled or treated with chlorine or iodine (including tea, coffee, and ice), eat only foods that have been thoroughly cooked and are still hot, eat only fruit that you have peeled yourself, and eat only cooked vegetables. Seafood from these areas should not be brought to the United States. Cholera is a reportable disease in the United States.

The only licensed cholera vaccine in the United States has been discontinued because of the brief and incomplete immunity it provided. Cholera vaccination is not required for entry or exit in any country. Two recently developed vaccines for cholera are licensed and available in other countries and appear to provide better immunity and fewer side-effects than the previously available vaccine.

U.S. and international public health authorities continue to enhance surveillance for cholera, investigate cholera outbreaks, and design and implement preventive measures. The EPA works with water and sewage treatment operators in the United States to prevent contamination of water. The FDA tests imported and domestic shellfish for *Vibrio* organisms and monitors the safety of U.S. shellfish beds.

Summary

Vibriosis and cholera are diseases caused by *Vi. parahemolyticus*, *Vi. vulnificus*, and *Vi. cholerae*. Cholera, caused by *Vi. cholerae*, is typically spread person to person, but may be found in contaminated water and aquatic animals. *Vi. cholerae* produces diarrhea with severe electrolyte imbalances and extreme levels of fluid loss. Cholera is rarely seen in the United States. Cholera is diagnosed via gram stain and stool culture. Cholera is treated with oral rehydration therapy. Vibriosis is typically spread from consumption of contaminated water and shellfish or through skin wounds from contaminated water. *Vi. parahemolyticus* typically produces a self-limiting gastroenteritis; whereas *Vi. vulnificus* produces skin lesions and septicemia, which is fatal in about 50% of affected people. Vibriosis can be seen in higher amounts in coastal states as a result of the consumption of undercooked shellfish or contact with contaminated water. In humans, vibriosis can be diagnosed via culture from samples of stool, blood, or blister fluid. *Vi. parahemolyticus* infections in people are treated with fluid therapy. *Vi. vulnificus* infections are treated with antibiotics such as tetracycline or doxycycline plus ceftazidime. Vibriosis can be prevented by avoiding raw or undercooked shellfish, keeping raw shellfish and its juices away from cooked foods (cross-contamination), and avoiding contact of wounded skin with seawater or raw seafood. The risk for cholera is very low in the United States; however, travelers visiting areas with epidemic cholera should only drink water that has been boiled or treated with chlorine or iodine, eat only foods that have been thoroughly cooked and are still hot, eat only fruit that you have peeled yourself, eat only cooked vegetables, and should not bring seafood from these areas to the United States.

For a list of less common bacterial zoonoses, see Table 3-9 (p. 224).

Review Questions

Multiple Choice

1. In regard to the taxonomic scheme, what is true?
   a. Kingdoms are the smallest and most general taxon in the Linnaean system of classification.
   b. Binomial nomenclature uses kingdom, phylum, class, order, family, genus, and species in its naming system.
   c. Writing scientific names involves capitalizing the genus name and using a lower case letter to begin the species name.
   d. There are seven levels in the Linnaean system arranged in descending ranks with species being the most diverse.

2. Prokaryotic cells
   a. usually have a single, circular chromosome located in the nucleoid region of the cell.
   b. have chromosomes that are not surrounded by a nuclear membrane.
   c. have few organelles in comparison to eukaryotic cells.
   d. all of the above.

3. Bacteria are identified by
   a. their growth on culture.
   b. their Gram stain reaction and morphology (shape).
   c. biochemical tests.
   d. all of the above.

4. The principle of the Gram stain is based upon the fact that
   a. bacteria have different shapes that the stain makes easier to see.
   b. some bacteria grow differently on the various types of agars.
   c. bacteria may be grouped together in special arrangements.
   d. some bacteria have a different cell wall structure than others.

5. Ways that bacteria can cause infection include
   a. the presence of flagella that allow bacteria to attach to certain body sites.
   b. the presence of a capsule that inhibits chemical identification of the bacterium by phagocytes.
   c. the presence of fimbria that help bacteria reach a body site where they can survive and multiply.
   d. the presence of normal flora that help foreign bacteria colonize an area of the body.

6. What are true regarding anthrax?
   a. Anthrax is primarily a disease of herbivores such as cattle, sheep, goats, and horses.
   b. Humans are susceptible to anthrax in industrial (animal-based industry) and nonindustrial (contaminated milk and meat) settings.
   c. The pulmonary form is called woolsorter's disease, which is contracted by inhaling endospores during handling of infected animal hair or products.
   d. All of the above are true.

7. This organism is most commonly associated with foodborne illness following ingestion of contaminated dairy products. Small numbers of this organism are significant because they can survive and multiply at refrigerator temperatures.
   a. *Listeria monocytogenes*
   b. *Erysipelothrix rhusiopathiae*
   c. *Bacillus anthracis*
   d. *Pasteurella multocida*

8. *Erysipelothrix* infections in humans characteristically produce
   a. CNS pathology.
   b. lesions at the point of entry into the host.
   c. formation of abscesses in visceral organs.
   d. urinary tract pathology.

9. A person develops a nonspecific disease several weeks after receiving a gift of Mexican goat cheese. A gram-negative coccobacilli is isolated from the person's blood culture. Based on this information, what is the most likely organism causing this person's symptoms?
   a. *Bacillus*
   b. *Streptococcus*
   c. *Pasteurella*
   d. *Brucella*

10. A picnic of potato salad, chicken, and hamburgers was consumed during a family reunion on a hot summer day. Within 24 hours ten family members had diarrhea, fever, and abdominal pain. Symptoms lasted for 3 days. Laboratory analysis of stool samples revealed that the organism was gram-negative bacteria and H_2S producers that grew on selenite agar with brilliant green. What is the most likely organism?
    a. *Shigella dysenteriae*
    b. *Salmonella typhimurium*
    c. *Listeria monocytogenes*
    d. *Staphylococcus aureus*

11. A miner in the southwestern United States who develops the symptoms of bubonic plaque most likely had contact with
    a. rodents.
    b. arthropods.
    c. wild dogs.
    d. dead birds.

12. What bacteria is most likely to cause watery diarrhea?
    a. Enterohemorrhagic *E. coli*
    b. Enteropathogenic *E. coli*
    c. Enteroinvasive *E. coli*
    d. *Shigella sonnei*

13. What bacteria causes strangles in horses and pneumonia in people?
    a. *Pasteurella multocida*
    b. *Streptococcus equi*
    c. *Burkholderia pseudomallei*
    d. *Staphylococcus aureus*

14. Spontaneous abortion in cattle is related to infections with
    a. *Bacillus abortus.*
    b. *Staphylococcus aureus.*
    c. *Brucella abortus.*
    d. *Streptococcus abortus.*

15. A person was admitted to the hospital with a diagnosis of appendicitis. During surgery, the appendix appeared normal and an enlarged lymph node was removed and cultured. Small, gram-negative rods were isolated from the room temperature agar plate. What is most likely to have been involved with this scenario?
    a. *Shigella sonnei*
    b. *Salmonella enteritidis*
    c. *Escherichia coli*
    d. *Yersinia enterocolitica*

16. The bubonic plague is transmitted by the bite of an infected
    a. rat.
    b. flea.
    c. raccoon.
    d. mouse.

17. What bacterium is highly communicable because of its low infective dose?
    a. *Staphylococcus aureus*
    b. *Streptococcus pyogenes*
    c. *Salmonella typhi*
    d. *Shigella* spp.

18. A recent immigrant from Africa was seen by his physician because he had developed nodules, abscesses, and ulcers in his oral mucous membranes. A mucopurulent, blood-tinged discharge was seen from these mucous membrane lesions. It was revealed upon staining of the discharge that bacteria were present (gram-negative rods). A definitive diagnosis was done by observing the Straus reaction after inoculating Guinea pigs with some of the material. What organism and disease does this person have?
    a. *Burkholderia pseudomallei*; melioidosis
    b. *Streptococcus pyogenes*, necrotizing fasciitis
    c. *Staphylococcus aureus*, toxic shock syndrome
    d. *Burkholderia mallei*; glanders

19. What bacterium causes wound infections and septicemia in immunocompromised people who have eaten raw or improperly cooked seafood?
    a. *Vibrio vulnificus*
    b. *Salmonella enteritidis*
    c. *Yersinia pestis*
    d. *Camplyobacter jejuni*

20. Travelers to areas of the world where cholera is endemic
    a. have no means of prevention.
    b. should avoid tick bites.
    c. can be immunized against cholera.
    d. should wear protective clothing.

21. The botulism toxin induces paralysis in the body by
    a. degrading the cell membrane of leukocytes.
    b. preventing the release of acetylcholine at the synapse.
    c. causing pseudomembranous colitis.
    d. altering the normal flora balance of the gastrointestinal tract.

22. What bacterium causes pneumonia in people, air sacculitis in birds, and eye infections in cats?
    a. *Chlamydophila psittaci*
    b. *Campylobacter jejuni*
    c. *Clostridium tetani*
    d. *Clostridium perfringens*

23. The most common transmission route of *Mycobacterium tuberculosis* in people is
    a. ingestion.
    b. inhalation.
    c. cutaneous.
    d. hematologic.

24. What is false in regards to *Mycobacterium* spp.?
    a. acid-fast
    b. obligate anaerobe
    c. slow replication time
    d. high cell wall lipid content

25. What bacteria cause rat-bite fever?
    a. *Staphylococcus aureus* and *Streptococcus pyogenes*
    b. *Streptobacillus moniliformis* and *Spirillum minus*
    c. *Mycobacterium tuberculosis* and *Mycobacterium leprae*
    d. *Capnocytophaga* spp. and *Bergeyella zoohelcum*

26. The foods most often associated with *Clostridium botulinum* food poisoning are
    a. poultry and eggs.
    b. improperly home canned goods.
    c. fish and shellfish.
    d. poorly cooked pork products.

27. The foods most often associated with *Salmonella* food poisoning are
    a. poultry and eggs.
    b. improperly home canned goods.
    c. fish and shellfish.
    d. poorly cooked pork products.

28. Rice-water stools are characteristic of
    a. typhoid fever.
    b. dysentery.
    c. plague.
    d. cholera.

29. A person is admitted to the hospital with fever, chills, and night sweats. Blood cultures were obtained and on the third week grew a gram-negative rod. This patient worked in a pig processing plant. The cultured organism is most likely
    a. *Pasteurella multocida.*
    b. *Brucella suis.*
    c. *Bacillus anthracis.*
    d. *Capnocytophaga* spp.

30. What bacteria can be found in animal bite wounds and can cause snuffles in rabbits?
    a. *Pasteurella multocida*
    b. *Brucella suis*
    c. *Bacillus anthracis*
    d. *Capnocytophaga* spp.

Matching

31. _____ *Clostridium tetani*

32. _____ *Clostridium perfringens*

33. _____ *Clostridium difficile*

34. _____ Enterotoxigenic *E. coli*

35. _____ Enteropathogenic *E. coli*

36. _____ *Capnocytophaga* spp.

37. _____ *Vibrio parahemolyticus*

38. _____ Enterohemorrhagic *E. coli*

39. _____ Enteroinvasive *E. coli*

40. _____ *Brucella abortus*

41. _____ *Streptococcus pyogenes*

42. _____ *Burkholderia pseudomallei*

43. _____ *Burkholderia mallei*

44. _____ *Mycobacterium leprae*

45. _____ *Bartonella henselae*

46. _____ *Chlamydophila psittaci*

47. _____ *Campylobacter jejuni*

48. _____ *Mycobacterium tuberculosis*

49. _____ *Bacillus anthracis*

50. _____ *Staphylococcus aureus*

A. antibiotic-associated pseudomembranous colitis

B. traveler's diarrhea

C. hemolytic uremic syndrome

D. pathogen that causes sustained muscle contraction

E. infantile diarrhea

F. pathogen that causes gas gangrene

G. shigellosis-like diarrhea

H. pyoderma in dogs; boils in people

I. woolsorter's disease

J. tuberculosis

K. shellfish-associated gastroenteritis

L. cat-scratch disease

M. ornithosis

N. #1 cause of human gastroenteritis in United States

O. tuberculoid and lepromatous leprosy

P. glomerulonephritis and necrotizing fasciitis

Q. pathogen found in dog and cat bite wounds

R. melioidosis

S. glanders

T. undulant fever in people; Bang's disease in animals

Case Studies

51. A previously healthy 9-year-old girl was awakened during the night by excruciating abdominal pain, nausea, and copious watery diarrhea every 30 to 60 minutes. She was brought into the acute care clinic where she developed bright red bloody diarrhea. She was sent to the emergency room and admitted to the hospital. She had not had contact with any other people with diarrhea and no history of recent travel outside of the United States. Her symptoms appeared 24 hours after eating a hamburger at a local fast-food restaurant. Laboratory work revealed a leukocytosis on the CBC (complete blood count) and a normal UA (urinanalysis). Routine stool cultures were set up in the laboratory.

    a. What bacteria could cause foodborne gastrointestinal illness?

    b. Is the consumption of hamburger significant in this case history?

    c. Because this patient had bloody diarrhea, the stool sample should be cultured on what additional agar?

    d. What additional condition should be of a concern in this patient?

52. A 52-year-old male developed severe vomiting and diarrhea during a flight from Bangkok to New York. On landing he was examined at the airport clinic, but left against medical advice. Approximately 12 hours later he went into shock and was brought into an emergency room by family members. This patient spent the next 3 days in the intensive care unit of the hospital. Upon getting a thorough history from him, he remembered that he ate fried fish in Bangkok and sushi on the airplane.

    a. Based on this person's history, what bacterium could be making him sick?

    b. Why did this person go into shock?

    c. What is this bacterium's virulence factor?

    d. What is the mechanism of action of this virulence factor?

53. A 22-year-old college student went to his physician complaining of severe abdominal pain and diarrhea. This episode had begun approximately 4 days ago as a mild stomachache that has now developed into intermittent abdominal pain in the lower right quadrant. On physical examination the physician determined the student had abdominal tenderness, normal vital signs, and was afebrile. The student had never traveled outside of the United States and had not consumed raw seafood. Laboratory results revealed a normal CBC and chemistry panel. Stool culture on *Salmonella-Shigella* and Hektoen enteric agar were negative. An additional culture was done using Skirrow's medium incubated at 42°C, which produced slow-growing, pinpoint colonies.

    a. Based on this person's history and laboratory results, what bacterial infection do you think he has?

    b. What are some possible sources of this infection?

    c. Is this type of enteritis common in the United States?

54. A 65-year-old man presented to the emergency room with localized chest pain. Past medical history was unremarkable; however, this person's social history included being homeless, a heavy smoker, and a moderate drinker (two to three glasses of wine or beer per day). On physical examination he had poor dental health, clear lungs on auscultation, and pain on deep inspiration.

    a. Based on this person's history what might be a cause of his illness?

    b. What organism causes this disease?

    c. What specimens should be sampled and what test should be run?

d. How is the disease transmitted?

e. Is this infection easy to treat?

55. A 10-year-old boy presented to his physician with an infected dog bite wound. Material from the wound was Gram stained revealing gram-negative rods that exhibited bipolar staining. Culture on blood agar revealed convex, smooth, gray, nonhemolytic colonies. Biochemical tests indicated that this organism was indole positive, urea negative, catalase positive, and ferments mannitol, sucrose, and maltose.

a. What is the most likely cause of this boy's wound contamination?

b. What were the dog's clinical signs of this infection?

c. What else should the physician ask about this case?

References

Abedon, S. 2004. An Introduction to Taxonomy. http://www.mansfield.ohio-state.edu/~sabedon/black09.htm (accessed February 10, 2004).

Alcamo, I. E. 1998. *Schaum's Outlines: Microbiology*. New York: McGraw-Hill.

Altekruse, S. F., N. J. Stern, P. I. Fields, and D. L. Swerdlow. 1999. Campylobacter jejuni: an emerging foodborne pathogen. *Emerging Infectious Diseases* 5(1):28–35.

Badshah, C. 2005. Cholera. http://www.nlm.nih.gov/medlineplus/ency/article/000303.htm (accessed November 30, 2005).

Barrali, R. Jr. 2006. *Salmonella* infection. http://www.emedicine.com/emerg/topic515.htm (accessed April 15, 2006).

Bauman, R. 2004. *Microbiology*. San Francisco, CA: Pearson Benjamin Cummings.

Biddle, W. 2002. *A Field Guide to Germs*. New York: Anchor Books.

Bronfin, D. 2003. *Yersinia enterocolitica* infection. http://www.emedicine.com/ped/topic2465.htm (accessed May 20, 2003).

Bruning-Fann, C. 1998. History of bovine tuberculosis. http://www.michigan.gov/emergingdiseases/0,1607,7-186-25804_26354-74838--,00.html (accessed May 14, 2005).

Burgess, B. A., P. S. Morley, and D. R. Hyatt. 2004. *Salmonella enterica* in a teaching hospital. *JAVMA* 225(9):1344–8.

Burton, R., and P. Engelkirk. 2004. *Microbiology for the Health Sciences*, 7th ed. Philadelphia, PA: Lippincott, Williams & Wilkins.

Center for Food Security and Public Health. 2005. Listeriosis. http://www.cfsph.iastate.edu/Factsheets/pdfs/listeriosis.pdf (accessed May 1, 2005).

Centers for Disease Control and Prevention. 1998. Rat-bite fever—New Mexico, 1996. *Morbidity and Mortality Weekly, Vol 48: No 5*. http://www.cdc.gov/mmwr/preview/mmwrhtml/00051368.htm (accessed December 15, 2004) pp. 89–91.

Centers for Disease Control and Prevention. 2003. Anthrax. http://www.bt.cdc.gov/agent/anthrax/needtoknow.asp (accessed October 25, 2005).

Centers for Disease Control and Prevention. 2004. General information on *Clostridium difficile* infections. http://www.cdc.gov/ncidod/dhqp/id_CdiffFAQ_general.html (accessed May 20, 2005).

Centers for Disease Control and Prevention. 2005. *Campylobacter* infections. http://www.cdc.gov/ncidod/dbmd/diseaseinfo/campylobacter_g.htm (accessed October 6, 2005).

Centers for Disease Control and Prevention. 2005. Leptospirosis. http://www.cdc.gov/ncidod/dbmd/diseaseinfo/leptospirosis_g.htm (accessed November 12, 2005).

Centers for Disease Control and Prevention. 2005. Listeriosis. http://www.cdc.gov/ncidod/dbmd/diseaseinfo/listeriosis_g.htm (accessed February 2006).

Centers for Disease Control and Prevention. 2005. *Streptococcus pneumoniae* disease. http://www.cdc.gov/ncidod/dbmd/diseaseinfo/streppneum_t.htm (accessed November 11, 2005).

Centers for Disease Control and Prevention. 2005. Yersiniosis. http://www.cdc.gov/ncidod/dbmd/diseaseinfo/yersinia_g.htm (accessed March 30, 2005).

Centers for Disease Control and Prevention. 2006. Staphylococcal food poisoning. http://www.cdc.gov/ncidod/dbmd/diseaseinfo/staphylococcus_food_g.htm (accessed May 29, 2006).

Centers for Disease Control and Prevention Division of Bacterial and Mycotic Diseases. 2005. Melioidosis. http://www.cdc.gov/ncidod/dbmd/diseaseinfo/melioidosis_g.htm (accessed October 12, 2005).

Chamberlain, N. 2003. Rat-bite fever. http://www.kcom.edu/faculty/chamberlain/website/lectures/lecture/ratfever.htm (accessed November 27, 2006).

Chamberlain, N. R. 2004. Plague. http://www.kcom.edu/faculty/chamberlain/website/lectures/lecture/plague.htm (accessed August 3, 2004).

Cliver, D., and M. Hajmeer. 2006. Veterinary Food Safety. http://www.vetmed.ucdavis.edu/PHR/VMD413/2006/41306SYL2.pdf (accessed February 13, 2006).

Cowan, M., and K. Talaro. 2006. *Microbiology: A Systems Approach*. New York: McGraw-Hill.

Cranmer, H., and M. Martinez. 2007. CBRNE-Anthrax Infection http://www.emedicine.com/emerg/topic864.htm (accessed February 13, 2007).

Cruz, L., E. Fletcher, A. Jones, and N. Prakash. 2004. The history of tuberculosis. http://efletch.myweb.uga.edu/history.htm (accessed May 14, 2005).

Cuthill, S. 2003. *Escherichia coli* infections. http://www.emedicine.com/ped/topic2696.htm (accessed October 16, 2005).

Ettinger, S. J., and E. C. Feldman. 2004. *Textbook of Veterinary Internal Medicine*, 6th ed. Philadelphia, PA: W. B. Saunders.

Factsheets on Chemical and Biological Warfare Agents. 1999. Glanders: essential data. http://www.cbwinfo.com/Biological/Pathogens/BMa.html (accessed August 22, 2005).

Factsheets on Chemical and Biological Warfare Agents. 1999. Melioidosis: essential data. http://www.cbwinfo.com/Biological/Pathogens/BP.html (accessed August 22, 2005).

Factsheets on Chemical and Biological Warfare Agents. 2005. Brucellosis. http://www.cbwinfo.com/Biological/Pathogens/BM.shtml (accessed November 11, 2005).

Forbes B., D. Sahm, and A. Weissfeld. 2002. *Bailey & Scott's Diagnostic Microbiology*, 11th ed. St. Louis, MO: Mosby.

Ghorayeb, Z. N., and M. Matta-Muallem. 2005. Erysipeloid. http://www.emedicine.com/derm/topic602.htm (accessed November 8, 2005).

Harmes, J. 2005. Tuberculosis: death's captain. http://www.bact.wisc.edu (accessed May 15, 2005).

Haubrich, W. S. 1997. *Medical Meanings: A Glossary of Word Origins*. Washington, DC: American College of Physicians.

Herchline, T. 2005. Staphylococcal infections. http://www.emedicine.com/med/topic2166.htm (accessed July 6, 2005).

Hnilica, K., and E. May. n.d. *Staphylococcus pyoderma*: an emerging crisis. http://www.utskinvet.org/pdf/staphylococcus_an_emerging_crisis.pdf (accessed July 20, 2005).

Ingraham, J., and C. Ingraham. 2004. *Introduction to Microbiology: A Case History Approach*, 3rd ed. Pacific Grove, CA: Thomson Brooks Cole.

Institute for International Cooperation in Animal Biologics, Iowa State University. 2005. Streptococcosis. http://www.cfsph.iastate.edu/Factsheets/pdfs/streptococcosis.pdf (accessed June 15, 2005).

Jakowski, R., and G. Kaufman. 2004. *Avian Bacterial, Mycoplasmal and Chlamydial Diseases*. Tufts Open Course Ware. http://www.myoops.org/twocw/tufts/courses/5/content/215761.htm (accessed November 12, 2005).

Jani, A. 2003. Pseudotuberculosis Yersinia. http://www.emedicine.com/med/topic1947.htm (accessed July 28, 2003).

Johnstom, W., M. Eidson, K. Smith, and M. Stobierski. 2000. Compendium of measures to control *Chlamydia psittaci* infection among humans (psittacosis) and pet birds (avian chlamydiosis), 2000. *Morbidity and Mortality Weekly Review* 49(RR14):1–17.

Joint FAO/WHO Expert Committee on Brucellosis. 1986. Sixth Report, Technical Report Series 740. Geneva: World Health Organization.

Jones, T. C., R. D. Hunt, and N. W. King. 1997. *Veterinary Pathology*, 6th ed. Baltimore, MD: Williams & Wilkins.

Kaneene, J., and C. Thoen. 2004. Zoonosis update: tuberculosis. *Journal of the American Veterinary Medical Association* 224(5):685–91.

Kotloff, K. L., et al. 1999. Global burden of *Shigella* infections: implications for vaccine development and implementation of control strategies. *Bull World Health Organ* 77(8):651–66.

Krauss, H., A. Weber, M. Appel, B. Enders, H. D. Isnebergy, H. G. Schiefer, W. Slenczka, A. von Graeventiz, and H. Zahner. 2003. *Zoonoses Infectious Diseases Transmissible From Animals to Humans*, 3rd ed. Washington, DC: ASM Press.

Kumamoto, K. S., D. J. Vukich. 1998. Clinical infections of *Vibrio vulnificus*: a case report and review of the literature. *Journal of Emergency Medicine* 16(1):61–66.

Lappin, M. R. 2004. Zoonotic diseases: What you can catch at work Parts I and II. *The NAVTA Journal* Fall:33–45.

Lundstrom, K., and G. Allen. 2005. Cat Scratch Disease. http://www.emedicine.com/ent/topic523.htm (accessed April 3, 2005).

Maloney, G., and W. Fraser. 2006. Brucellosis. http://www.emedicine.com/emerg/topic883.htm (accessed May 10, 2006).

Martin, B. Institutional Animal Care and Use Committee. 1996. Zoonotic diseases. http://research.ucsb.edu (accessed February 23, 2005).

Medline Plus. 2005. Brucellosis. http://www.nlm.nih.gov/medlineplus/ency/article/000597.htm (accessed November 11, 2005).

Medline Plus. 2005. *E. coli* infections. http://www.nlm.nih.gov/medlineplus/ecoliinfections.html (accessed January 12, 2006).

Medline Plus. 2005. Leprosy. http://www.nlm.nih.gov/medlineplus/ency/article/001347.htm (accessed October 30, 2005).

Medline Plus. 2005. Salmonellosis. http://www.nlm.nih.gov/medlineplus/salmonellainfections.html (accessed September 5, 2005).

Medline Plus. 2005. Tuberculosis. http://www.nlm.nih.gov/medlineplus//tuberculosis.html (accessed May 10, 2005).

Medline Plus. 2006. Gas gangrene. http://www.nlm.nih.gov/medlineplus/ency/article/000620.htm (accessed February 17, 2006).

Medline Plus. 2006. Rat-bite fever. http://www.nlm.nih.gov/medlineplus/ency/article/001348.htm (accessed November 27, 2006).

Merck Veterinary Manual. 2006. Clostridia-associated enterocolitis infections. http://www.merckvetmanual.com/mvm/index.jsp?cfile=htm/bc/22205.htm&word=clostridial%2cinfection (accessed February 13, 2006).

Merck Veterinary Manual. 2006. Staphylococcus: introduction. http://www.merckvetmanual.com/mvm/index.jsp?cfile=htm/bc/204200.htm&word=staphylococcal%2cinfection (accessed February 13, 2006).

Merck Veterinary Manual. 2006. Vibriosis. http://www.merckvetmanual.com/mvm/servlet/CVMHighLight?file=htm/bc/tzns01.htm&word=vibrio (accessed July 26, 2006).

Minnaganti, V., and B. Cunha. 2003. Plague. http://www.emedicine.com/MED/topic3381.htm (accessed January 22, 2003).

Mylonakis, E., C. Go, and B. Cunha. 2006. *Escherichia coli* infections. http://www.emedicine.com/med/topic734.htm (accessed May 26, 2006).

National Center for Animal Health Programs. 2006. OIE Listed Diseases. http://www.aphis.usda.gov/vs/ncie/oie/rtf_files/animal-dis-noti_dec03.rtf (accessed March 1, 2006).

National Tuberculosis Center. 1996. Brief history of tuberculosis. http://www.umdnj.edu/~ntbcweb/history.htm (accessed May 10, 2005).

Netahlo-Barrett, L. n.d. The bubonic plague (*Yersinia pestis*): The black death. http://militaryhistoryonline.com (accessed April 10, 2005).

Oba, Y., and G. Salzman. 2005. Chlamydial pneumonias. http://www.emedicine.com/med/topic341.htm (accessed March 8, 2005).

Office of Communications and Public Liaison, Anthrax, National Institutes of Health, United States Department of Health and Human Services. 2002. The Jordan Report 20th Anniversay Accelerated Development of Vaccines, 2002. http://www.niaid.nih.gov/dmid/vaccines/jordan20/jordan20_2002.pdf (accessed May 15, 2002) pp. 3–261.

Olson, C. 2004. *Campylobacter*. http://www.vetmed.wisc.edu/pbs/zoonoses/GIk9fel/campylobacter.html (accessed April 14, 2005).

Olson, C. 2004. Streptococcal zoonoses. http://www.vetmed.wisc.edu/pbs/zoonoses/Streptococcus/strpindex.html (accessed January 4, 2006).

Olson, C. 2004. Zoonosis Tutorial, Tuberculosis. zoonoseshttp://www.vetmed.wisc.edu/pbs/zoonoses/Streptococcus/strpindex.html (accessed May 11, 2005).

Olson, C., and G. Mejicano. Partners in Agriculture Health. n.d. *Erysipelothrix rhusiopathiae* infection. Module VI. http://worh.org/new_orh_docs/farmershealth/pdf_etc/fhpdf/zoon.pdf (accessed February 23, 2005).

Pelzer, K. D. 1995. *Salmonellosis: Zoonosis Updates*, 2nd ed. Schaumburg, IL: AVMA.

Perry, D., and J. Fetherston. 1997. *Yersinia pestis*: etiologic agent of plaque. *Clinical Microbiology Reviews* 10(1):35–66.

Prescott, L. M., D. A. Klein, and J. P. Harley. 2002. *Microbiology*, 5th ed. New York: McGraw-Hill.

Rao, P., S. Riccardi, and D. Birrer. 2004. The history of salmonellosis. www.columbia.edu/cu/biology/courses/g4158/presentations/2004/Salmonella.ppt (accessed September 14, 2005).

Reboli, A., and W. E. Farrar. 1989. *Erysipelotrhix rhusiopathiae*: an occupational pathogen. *Clinical Microbiology Reviews* 2 (4):354–9.

Rega, P., and D. Batts. 2005. CBRNE: Glanders and Melioidosis. http://www.emedicine.com/emerg/topic884.htm (accessed April 6, 2005).

Regnery, R., and J. Tappero. 1995. Unraveling mysteries associated with cat-scratch disease, bacillary angiomatosis, and related syndromes. *Emerging Infectious Diseases* 1(1):16–21.

Rollins, D. M., and S. W. Joseph. 2000. *Campylobacter*. http://www.clfs.umd.edu/classroom/bsci424/PathogenDescriptions/Campylobacter.htm (accessed May 24, 2005).

Sack, D. A. 1998. Cholera and related illnesses caused by *Vibrio* species and *Aeromonas*. In *Infectious Diseases*, edited by S. L. Gorbach, J. G. Bartlett, and N. R. Blacklow. Philadelphia, PA: W. B. Saunders Co.

Schwartzman, W. A. 2000. Cat scratch disease and other bartonella infections. *Current Treatment Options in Infectious Diseases* 2:155–62.

Shimeld, L. A. 1999. *Essentials of Diagnostic Microbiology*. Albany, NY: Delmar Publishers.

Short, B. 2002. Pasteurellosis: an important emerging infectious disease—a military problem? *ADF Health* 3(1):13–21.

Smith, B. P. 2001. *Large Animal Internal Medicine*, 3rd ed. St. Louis, MO: Mosby.

Smith, K., K. Bradley, M. Stobierski, and L. Tengelsen. 2005. Compendium of measures to control *Chlamydophila psittaci* (formerly *Chlamydia psittaci*) infection among humans (psittacosis) and pet birds. http://www.avma.org/pubhlth/psittacosis.asp (accessed November 12, 2005).

Stump, J. L. 2004. Animal bites. http://www.emedicine.com/emerg/topic60.htm (accessed October 12, 2004).

Subekti, D., B. Oyofo, P. Tjaniadi, A. Corwin, W. Larasati, M. Putri, C. Simanjuntak, N. Punjabi, J. Taslim, B. Setiawan, A. Agung, G. S. Djelantik, L. Sriwati, A. Sumardiati, E. Putra, J. Campbell, and M. Lesmana. 2001. *Shigella* spp. surveillance in Indonesia: the emergence or reemergence of *S. dysenteriae*. *Emerging Infectious Diseases* 7(1):137–40.

Taplitz, R. A. 2004. Managing bite wounds. *Postgraduate Medicine Online*. http://www.postgradmed.com (accessed June 23, 2005).

Todar, K. 2003. *Listeria monocytogenes* and Listeriosis. In Todar's Online Textbook of Bacteriology. http://textbookofbacteriology.net/Listeria.html (accessed May 1, 2005).

Todar, K. 2005. *Bacillus anthracis* and anthrax. In Todar's Online Textbook of Bacteriology. http://textbookofbacteriology.net (accessed April 14, 2005).

Todar, K. 2005. The pathogenic Clostridia. In Todar's Online Textbook of Bacteriology. http://textbookofbacteriology.net/clostridia.html (accessed May 21, 2005).

USDA APHIS. 2005. *Brucellosis*, National Center for Animal Health Programs. http://www.aphis.usda.gov/vs/nahps/brucellosis/ (accessed July 30, 2005).

Wassenaar, T. 2005. Virtual Museum of Bacteria. http://www.bacteriamuseum.org (accessed April 15, 2005).

Weese, J. 2002. A review of equine zoonotic diseases: risks in veterinary medicine. *AAEP Proceedings* 48:362–9.

Wong, J., and W. Reenstra-Buras. 2006. Gas gangrene. http://www.emedicine.com/emerg/topic211.htm (accessed January 31, 2006).

World Health Organization. 2000. Cholera. http://www.who.int/mediacentre/factsheets/fs107/en/ (accessed April 25, 2005).

World Health Organization. 2001. Water-related Diseases: Leptospirosis. http://www.who.int/water_sanitation_health/diseases/leptospirosis/en/ (accessed November 12, 2005).

World Health Organization. 2005. Brucellosis. http://www.who.int/zoonoses/diseases/brucellosis/en/ (accessed November 11, 2005).

World Health Organization. 2005. Leprosy. http://www.who.int/mediacentre/factsheets/fs101/en/ (accessed October 30, 2005).

World Health Organization. 2005. Salmonellosis. http://www.who.int/topics/salmonella/en/ (accessed September 14, 2005).

World Health Organization. 2005. Shigellosis. http://www.who.int/vaccine_research/diseases/shigella/en/ (accessed September 17, 2005).

World Health Organization. 2005. Tuberculosis. http://www.who.int/mediacentre/factsheets/fs104/en/index.html (accessed May 10, 2005).

World Health Organization. n.d. Water sanitation: Introduction. http://www.who.int/water_sanitation_health/bathing/recreadischap1.pdf (accessed January 12, 2006).

Young, E. J., and M. J. Corbel, eds. 1989. *Brucellosis: Clinical and Laboratory Aspects*. Boca Raton, FL: CRC Press.

Zach, T. 2006. Listeria infection. http://www.emedicine.com/PED/topic1319.htm (accessed February 27, 2006).

Zapor, M., and D. Dooley. 2005. Salmonellosis. http://www.emedicine.com/med/topic2058.htm (accessed August 2, 2005).

4

Tick-Borne Bacterial Zoonoses

Objectives

After completing this chapter, the learner should be able to

- Briefly describe the history of ticks and tick-borne diseases
- Describe basic tick biology
- Identify unique characteristics of ticks
- Differentiate between soft and hard ticks
- Describe methods of controlling ticks
- Describe protective measures veterinary professionals can take to prevent contracting tick-borne zoonoses
- Describe the causative agent of specific tick-borne bacterial zoonoses
- Identify the geographic distribution of tick-borne bacterial zoonoses
- Describe the transmission, clinical signs, and diagnostic procedures for tick-borne bacterial zoonoses
- Describe methods of controlling tick-borne bacterial zoonoses

Key Terms

capitulum	hypostome	nymph	transovarial transmission
chelicera	idiosoma	pedipalps	transstadial transmission
festoons	larva	questing	vertical transmission
horizontal transmission	molting	scutum	

OVERVIEW

Ticks transmit a wide array of pathogens including bacteria, viruses, and parasites. Ticks typically transmit disease via a bite; however, direct contact, ingestion, aerosol exposure, transfusion, and maternal transmission can also occur. Ticks have been around in the same form for about 200 million years making them one of the oldest and most successful groups of arthropods. Greeks described ticks as pests and ticks are typically regarded in literature with disgust.

The discovery that ticks transmit disease preceded the discovery of mosquito-borne (malaria and yellow fever) and flea-borne (plague) diseases. In 1893, Smith and Kilbourne established ticks as the vector of *Babesia bigemina*, the protozoal agent that causes Texas cattle fever. In 1903, J. E. Dutton proved that ticks were vectors of human disease when he discovered the cause of endemic relapsing fever and its vector, the soft-tick *Ornithodoros moubata*. In a short period of time, *Dermacentor* ticks were discovered as the vectors of Rocky Mountain spotted fever (RMSF) (H. T. Ricketts) and tularemia (R. R. Parker and E. Francis). An extensive list of tick-borne diseases now exists including viral encephalitides, hemorrhagic fevers, Colorado tick fever, Q fever, ehrlichiosis, anaplasmosis, and babesiosis.

Mosquitoes are the only arthropods that transmit more disease than ticks.

TICKS

Causative Agent

Ticks are primitive, obligate, blood-sucking parasites that prey on every class of vertebrate (including mammals, birds, reptiles, and amphibians) in all parts of the world. Ticks are members of the Kingdom *Animalia*, phylum *Arthropoda*, subphylum *Chelicerata*, class *Arachnida*, and order *Acarina*. Ticks that transmit disease to animals and humans can be divided into two groups: the hard ticks (Ixodidae) (Figure 4-1) and the soft ticks (Argasidae). Table 4-1 characterizes these ticks. A third group of ticks, Nuttallielloidea, contains only the African species, *Nuttalliella namaqua*, and do not cause clinical disease. In contrast to other blood-sucking arthropods, ticks have long life cycles, consume large volumes of blood (up to 4 to 5 mL per tick), and produce large numbers of eggs.

Epizootiology and Public Health Significance

Ticks are found in nearly every country throughout the world; however, only a few species are known to cause disease in humans and animals. Globally, tick-borne pathogens account for more than 100,000 human illnesses annually. Because humans are atypical hosts, tick-borne diseases are usually acute and severe. Examples of tick-borne diseases of humans include Lyme borreliosis, Crimean-Congo hemorrhagic fever (CCHF), ehrlichiosis, anaplasmosis, Q fever, babesiosis, tularemia, RMSF, Colorado tick fever, Russian spring summer encephalitis (TBE), and African tick typhus.

In animals, ticks can affect livestock production, cause irritation, and produce disease. Livestock production can be affected through blood loss (some ticks consume large amounts of blood that cause anemia); wounds that may cause pain, ulcerate, and lead to secondary infection (which can lower weight gain or milk

Figure 4-1 *Rhipicephalus sanguineus* is an example of a hard tick.

(Courtesy of CDC/James Gathany; William Nicholson)

Table 4-1   Features of Hard and Soft Ticks

Common Name	Family Name	Unique Features	Examples
Hard tick	Ixodidae	■ Have a hard plate called a scutum covering the dorsal body surface. ■ Mouthparts protrude and are visible from above.	There are seven genera that transmit disease: ■ *Ixodes* ■ *Dermacentor* ■ *Amblyomma* ■ *Haemaphysalis* ■ *Hyalomma* ■ *Rhipicephalus* ■ *Boophilus**
Soft tick	Argasidae	■ Lack a scutum. ■ Have a cuticle that is soft and leathery.	There is only one genus that transmits disease to humans: ■ *Ornithodoros* ■ *Otobius** ■ *Argas**

*do not transmit disease to humans

production); and initiation of disease (including paralysis or infection). Wildlife may show signs of tick-borne disease or they may be asymptomatic and serve as a reservoir host for some diseases. Examples of tick-borne diseases of animals include Lyme disease, tick paralysis, tularemia, RMSF, hemobartonellosis, canine ehrlichiosis, Q fever, and cytauxzoonosis.

Identifying Ticks

All ticks have a general morphology consisting of two body regions: the **capitulum** (the cranial portion containing mouthparts) and the **idiosoma** (the caudal portion containing most internal organs and bearing the legs). Tick identification is based on characteristics of this morphologic scheme such as its shape, size, mouth parts, color, dorsal shield (**scutum**), and **festoons** (caudal abdominal markings). Figure 4-2 illustrates the parts of ticks.

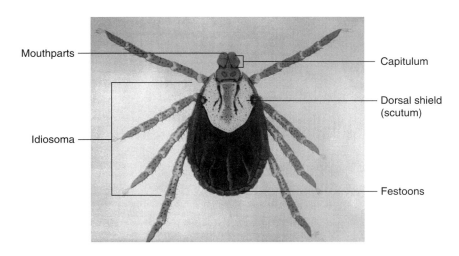

Figure 4-2 Parts of a hard tick (*Dermacentor viriabilis*).

(Courtesy of CDC/Andrew J. Brooks)

- Mouthparts are found on the capitulum. Mouth parts of adult soft ticks are located on the ventral surface and are not visible dorsally. For adult hard ticks, mouth parts are visible dorsally and are either short (*Dermacentor*), long (*Ixodes*) or the longest (*Amblyomma*).

- The scutum is the dorsal plate or shield covering the cranial part of the body in female hard ticks and the entire dorsal surface in male hard ticks (the full scutum of male hard ticks limits the increase in size with engorgement). Soft ticks have a leathery or wrinkled appearance because they lack a scutum.

- Festoons are rectangular grooves seen on the caudal edge of some hard ticks and appear as a string of pearls. *Dermacentor*, *Haemaphyalis*, *Rhipicephalus*, *Anocentor*, and *Amblyomma* have festoons present, whereas *Boophilus*, *Hyalomma*, and *Ixodes* ticks do not. Soft ticks such as *Ornithodoros* do not have festoons.

- In adult, nonengorged ticks, there are three major body shapes: teardrop, oval, and rounded. Both *Dermacentor* and *Ixodes* hard ticks have a teardrop shape that tapers at the mouthparts. *Amblyomma* hard ticks are more rounded. *Ornithodoros* soft ticks have a characteristic full oval body shape.

- Size is commonly used to identify ticks, but is not as specific as other characteristics. In general, **larvae** and **nymphs** are significantly smaller than adult ticks and engorged adult ticks are significantly larger than nonengorged adults. Male ticks are often somewhat smaller than female ticks. Relative size of ticks from one species to another can also aide in tick identification (adult *Dermacentor* and *Ornithodoros* are approximately the size of a sesame seed; adult *Ixodes* and adult male *Amblyomma* ticks are approximately one-half the size of a sesame seed; adult female *Amblyomma* ticks are larger than a sesame seed).

Tick Life Cycles

All ticks undergo four basic stages: egg, larva, nymph, and adult. These stages may take 6 weeks to 3 years to complete depending on the species of tick and environmental conditions, such as temperature (ticks in tropical regions have short life cycles, whereas ticks in cold climates have long life cycles) and host availability (some hard ticks can go several months without feeding). All stages except the egg require a blood meal from a host for transition to the next stage of development. Following emergence from the egg, ticks molt from one form to the next. Each stage between molts is called an instar and many times each instar will have a somewhat different body conformation. Ixodidae (hard ticks) have only one nymphal instar, whereas Argasidae (soft ticks) may have as many as five. Ticks are paurometabolous meaning that the immature stages resemble smaller forms of the adults.

Hard Ticks

Stages of Hard Ticks

Ixodidae (hard) ticks have three distinct life stages: larvae, nymphs, and adults. Only one blood meal is taken during each of the three life stages. After feeding the tick then drops off the host.

- Larvae emerge from the egg and have three pairs of legs (six legs total). Larvae obtain a blood meal from a vertebrate host and molt to the nymphal stage.
- Nymphs are the outcome of larvae molting and have four pairs of legs (eight legs total). Nymphs feed and molt to the next and final stage (adult).

- Adults are the outcome of nymph molting and also have four pairs of legs (eight legs total). After feeding, the adult female hard ticks lay one batch of thousands of eggs and then die.

Appearance of Hard Ticks

Hard ticks have mouthparts that are visible dorsally. In addition to the mouthparts, the capitulum contain **chelicera** (appendages on each side of the mouth) that pierce, tear, or grasp host tissues and protect the **hypostome; pedipalps** (a pair of jointed appendages) that aid feeding; a hypostome (ventral fusion of the pedipalps) that helps anchor the tick to the host; and a rostrum that extends dorsally over the mouth. When ticks find their host, the pedipalps grasp the skin, the chelicerae cut through the skin, and the hypostome enters the host's skin while feeding and its backward directed projections prevent easy removal of the attached tick.

Biology of Hard Ticks

Copulation in ticks occurs almost always on the host. In females, a blood meal is usually necessary for egg production. An engorged female tick will drop to the ground to deposit eggs in soil. From the many eggs (hundreds to tens of thousands) that are laid by female hard ticks emerge six-legged larvae.

Hard ticks seek hosts by a behavior called **questing**, a process by which ticks crawl up a piece of grass or perch on leaf edges with their front legs extended. Stimuli such as movement, odor, sweat, color, size, carbon dioxide, or heat can initiate questing. Questing occurs when ticks use extended front legs to climb on to a potential host that brushes against them. Upon finding a host, the tick feeds and then molts to an eight-legged nymph. When feeding the cuticle (outer surface) of hard ticks grows to accommodate the blood ingested. Female hard ticks can increase their size to a greater degree because their scutum only covers the cranial part of their bodies.

Molting in hard ticks can occur by a variety of different themes. Some ticks feed on only one host and molt through all three life stages on the same host. These ticks are called one-host ticks and an example is *Boophilus* spp. When a one-host tick becomes an adult, the female will drop off the host after feeding to lay her batch of eggs. Other ticks feed on two hosts during their lives and are called two-host ticks such as *Rhipicephalus* spp. Two-host ticks feed and stay on the first host during the larva and nymph life stages, then the nymph drops off and molts to an adult, and the adult attaches to a different host for its final blood meal. The adult female then drops off after feeding to lay eggs. Two-host ticks are adapted to feeding on a wide-range of hosts. Other ticks (incluing many hard ticks) feed on three hosts and are on a different host during each life stage. These ticks are called three-host ticks and examples include *Dermacentor* spp. and *Ixodes* spp. Three-host ticks drop off and reattach to a new host during each life stage. In all of the various types of host ticks, the engorged adult stage is terminal with the female dying after laying one batch of eggs and the male dying after he has reproduced. Figure 4-3 illustrates the types of tick host cycles.

Hard Tick Genera

There are three subfamilies of hard ticks within the Ixodidae family: Ixodinae (consisting of the single genus *Ixodes*), Amblyominae (containing the genera *Amblyomma*, *Haemaphysalis*, *Aponomma*, and *Dermacentor*), and Rhipicephalineae (containing the genera *Rhipicephalus*, *Anocentor*, *Hyalomma*, *Boophilus*, and *Margaropus*).

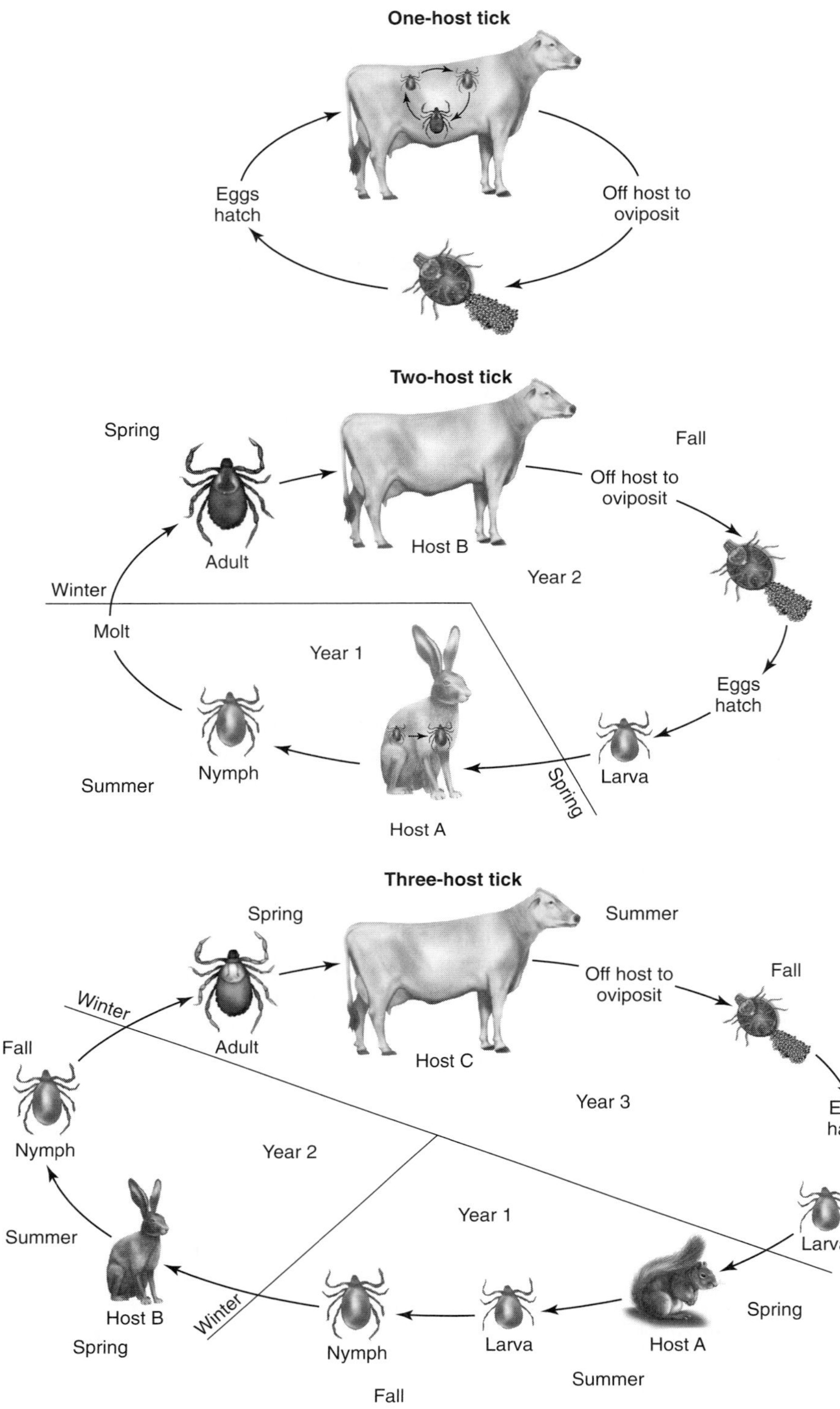

One-host tick

Eggs hatch

Off host to oviposit

Two-host tick

Spring

Fall

Off host to oviposit

Adult

Host B

Year 2

Winter

Molt

Year 1

Summer

Nymph

Host A

Spring

Larva

Eggs hatch

Figure 4-3   In one-host ticks, the larva, nymph, and adult stages all feed on the same host. Molting also takes place on this host (Host A). The only stage off the host is when the blood-engorged female tick drops to the ground to lay eggs (oviposit). An example of a one-host tick is *Boophilus* spp.

In two-host ticks, larvae that have completed feeding remain on the host and molt to a nymph on the same host (Host A). The blood-engorged female tick drops to the ground, molts, and the adult feeds on another host (Host B). An example of a two-host tick is *Hyalomma* spp.

In three-host ticks, the larvae, nymph, and adult stages each have its own host. These hosts may be the same or different animal species. Molting occurs on the ground. Examples of three-host ticks are *Ixodes*, *Dermacenter*, and *Amblyomma* spp.

Three-host tick

Spring

Summer

Winter

Off host to oviposit

Fall

Adult

Host C

Year 3

Fall

Eggs hatch

Nymph

Year 2

Year 1

Larva

Summer

Host B

Winter

Nymph

Larva

Host A

Spring

Spring

Summer

Fall

Ixodes ticks

There are over 200 species of *Ixodes* ticks worldwide and approximately 40 species are found in North America. All *Ixodes* ticks are three-host ticks that parasitize small mammals and are easily overlooked because of their small size. Examples include *Ixodes scapularis*, the blacklegged tick that is the primary vector of Lyme disease in the eastern and southcentral United States as well as a vector of *Ehrlichia* spp.; *I. pacificus*, the vector of Lyme disease and ehrlichosis in the western United States; and *I. ricinus*, *I. persulcatus*, and *I. pavlovskyi*, which transmit tick-borne encephalitis in Europe and Asia.

Ticks once known as *I. dammini*, occurring in the northern and eastern United States, are now considered the same species as *I. scapularis*.

Amblyominae ticks

Types of Amblyominae ticks include *Amblyomma* spp., *Haemaphysalis* spp., *Dermacentor* spp., and *Aponomma* spp.

- *Amblyomma* spp. are restricted to tropical areas, appear to be one-host ticks, and contain approximately 100 species. The species of concern in the United States is *A. americanum*, the lone star tick, which is the vector for RMSF and tularemia.

- *Dermacentor* spp. are some of the most medically important ticks containing about 30 species with at least seven found in the United States. *Dermacentor andersoni*, the Rocky Mountain wood tick, is found in the western United States (west of the Rockies) and is the vector for tick paralysis, Powassan encephalitis virus, Colorado tick fever virus, tularemia, RMSF, and anaplasmosis. *De. variabilis*, the American dog tick, is found in the eastern United States (east of the Rockies) and isolated areas in the Pacific Northwest and is the principle vector of RMSF in the central and eastern United States. *De. occidentalis*, the Pacific Coast tick, is a vector of Colorado tick fever virus, RMSF, tularemia, and anaplasmosis. *De. occidentalis* also transmits chlamydial infection in cattle. *De. albipictus*, the horse tick or winter tick, is found in the northern United States and Canada and feeds on elk, moose, horses, and deer causing winter tick infestations.

- *Haemaphysalis* spp. are small ticks with about 150 species known worldwide. There are two species found in North America: *Haemaphysalis leporispalustris*, the rabbit tick which occasionally feeds on domestic animals and can serve as a vector of tularemia and RMSF in wildlife and *H. cordeilis*, the bird tick which is common on game fowl and may transmit diseases among these birds.

- *Aponomma* spp. parasitize reptiles, particularly in Asia and Africa. This genus of tick is not of medical or economic significance.

Rhipicephalinae ticks

Types of Rhipicephalinae ticks include *Rhipicephalus* spp., *Anocentor* spp., *Hyalomma* spp., *Boophilus* spp., and *Margaropus* spp. *Rhipicephalus* ticks contain approximately 60 species in areas of limited rainfall, show very little host specificity, and are typically two- or three-host ticks.

- *Rhipicephalus sanguineus*, the brown dog or kennel tick, is found in most of North America and is a three-host tick with all three stages feeding mainly on dogs. *R. sanguineus* transmits *Rickettsia rickettsii* (RSMF), *Babesia canis*, *Rickettsia canis*, *Borrelia theileri*, and *Rickettsia connorii* (in Europe and Africa).

- *R. appendiculatus*, the brown ear tick, transmits *Theileria parva*, a protozoan that causes East Coast Fever in cattle.

- *Anocentor* ticks have only one species, *Anocentor nitens* (the tropical horse tick), found in South America and parts of the southern United States. This tick transmits *Babesia caballi*, a protozoan blood parasite of horses.

- *Hyalomma* species are found in desert conditions where hosts are few and environmental conditions are extreme. *Hyalomma* ticks transmit CCHF, Dugbe virus, and occasionally West Nile virus, *Babesia canis* (transmitted by *Hy. marginatum*), Thogoto virus, swine poxvirus, and *Theileria annulata* (transmitted by *H. anotolicum*),

- *Boophilus* ticks are one-host ticks and consist of three species: *Boo. annulatus*, the American cattle tick which was once widespread in the southern United States and transmitted the protozoan *Babesia bigemina*, the causative agent of Texas cattle fever (also known as red-water disease); *Boo. microplus*, which affects cattle in Mexico, Africa, and Australia; and *Boo. decoloratus*, the blue tick that attacks cattle in Africa.

- *Margaropus* ticks contain four species that are rarely seen.

Soft Ticks

Stages of Soft Ticks

The life cycle of Argasidae (soft) ticks consists of three life stages: larvae, nymphs, and adults. Soft ticks feed several times during each life stage, females lay multiple small batches of eggs between blood meals throughout their lives, and most soft ticks have a multihost life cycle involving more than three hosts.

- Six-legged larvae emerge from the egg, take a blood meal from a host, and molt to the first nymphal stage. Depending on environmental conditions such as temperature and humidity, larvae will hatch from the eggs in anywhere from 2 weeks to several months.

- Eight-legged nymphs are the outcome of larval molting. Many soft ticks go through multiple nymphal stages, gradually increasing in size until the final molt to the adult stage. Each nymphal instar feeds separately, drops, and molts to the next stage.

- Eight-legged adults are the outcome of nymphal molting. The time to completion of the entire life cycle is generally much longer than that of hard ticks and can last over several years because they can survive for many years without a blood meal. Adults do not molt.

Appearance of Soft Ticks

Like hard ticks, soft ticks have two body regions: the capitulum and the idiosoma. Unlike hard ticks, soft ticks have mouthparts that are not readily visible dorsally because their capitulum in the nymph and adult stages lies in a depression (called a camerostome) that has a dorsal wall (called a hood) that extends over the capitulum (Figure 4-4). In soft ticks, the idiosoma has a sac-like shape. Soft ticks do not

Figure 4-4 Comparison of soft tick structures to hard tick structures.

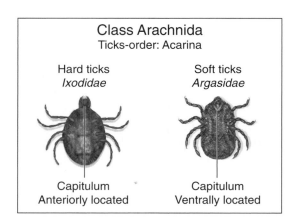

have a scutum (protective covering) on the dorsum of the body, and the exoskeleton is leathery and rough in texture.

Biology of Soft Ticks

Some soft ticks seek hosts by questing; however, the majority of soft ticks are nest parasites. Most soft ticks feed repeatedly and rest off the host between meals. Soft ticks usually feed on one type of animal during their lifetime and when not on the host they tend to stay near the habitats of the host animal (examples include soil, burrows, nests, and crevices). Soft ticks feed for short periods of time on their hosts (ranging from several minutes to days, depending on life stage, host type, and tick species). The nymph and adult stages feed rapidly (usually 30 to 60 minutes), whereas the larvae may feed rapidly or may require days to fully engorge. The feeding behavior of many soft ticks resembles that of fleas because the ticks reside in the host's environment and feed rapidly when the host returns. The cuticle (outer surface) of soft ticks expands, but does not grow to accommodate the large volume of blood ingested (which may be 5 to 10 times their unfed body weight). Adult females lay eggs in these hiding places several times between feedings. Some adult female soft ticks will feed several times and lay 20 to 50 eggs after each meal compared to adult female hard ticks that feed only once and lay one large batch of eggs, often containing as many as 10,000 or more eggs.

> Ticks feed only on the blood of vertebrates.

> Soft ticks feed rapidly and leave their host after engorging. Hard ticks secrete a cement-like substance that helps secure them to the host for longer feeding times. This substance dissolves after feeding is complete.

Soft Tick Genera

There is one family of soft ticks (Argasidae) that contains five genera: *Argas*, *Ornithodoros*, *Otobius*, *Nothoaspis*, and *Antricola*.

Argas ticks

Argas ticks parasitize birds and bats and may occasionally bite humans without transmitting disease. Examples in this group include *Argas persicus* (the fowl tick), *Ar. reflexus* (the pigeon tick found in the Near and Middle East, Europe, and Russia), and *Ar. cooleyi* (found in cliff swallows in the United States).

Ornithodoros ticks

There are over 100 species of *Ornithodoros* ticks that parasitize mammals.

- *Ornithodoros hermsi* is found in the Rocky Mountain and Pacific Coast states of the United States and is the vector of *Borrelia recurrentis*, the causative agent of relapsing fever.
- *O. cariaeceus* is found from Mexico to the Oregon coast and produces a painful bite in people.
- *O. moubata* is found in Africa and is primarily a parasite of warthogs.

Otobius ticks

Otobius ticks are called spinose ear ticks because the nymph stage has a spiny covering and they typically feed in the folds of the external auditory canal.

- *Otobius megnini* is found in the warmer parts of the United States, feeding mainly on cattle. Heavy infestations can effect production in livestock. *Ot. megnini* can also infest horses, sheep, goats, antelope, mule deer, dogs, and humans. This tick can be the vector for Q-fever, tularemia, Colorado tick fever, and RMSF.
- *Ot. lagophilus* is found in western North America and the larvae and nymphs feed on the faces of rabbits.

Nothoaspis and *Antricola* **ticks**

Ticks in these genera are found in cave-dwelling bats in North and Central America and are of limited medical and economic importance.

Transmission

Tick bites may cause local reaction, possible infection, or tick paralysis. The outcome of a tick bite depends on the species of tick, life stage of the tick, presence or absence of an infectious agent, and host factors such as immune status, age, and duration of tick attachment. Two important factors that affect the outcome of a tick bite are attachment time and tick stage. Infectious agents can be transmitted to the host via infected saliva only after there is sufficient attachment time to the host. An engorged tick signifies longer attachment time and increased risk of disease transmission. *I. scapularis* usually requires undisturbed attachment and feeding for approximately 24 to 52 hours before the spirochete *Borrelia burgdorferi* can be transmitted and cause Lyme disease. *Ornithodoros* spp. requires less than one hour of attachment time, usually at night, to transmit the *Borrelia* spirochete causing relapsing fever. *De. andersoni* usually requires 5 to 7 days of attachment time before tick paralysis occurs. The stage of tick development (larva versus nymph versus adult) also affects disease transmission. *I. scapularis* nymphs are more likely to transmit Lyme disease than adult ticks. Table 4-2 summarizes different diseases transmitted from ticks.

Types of Transmission

The cycle of tick-borne disease agents require a vertebrate host that develops a level of infection so that the infectious agent can be passed on to a feeding tick, a tick that acquires the agent and is able to pass the agent to another host, and adequate numbers of vertebrate hosts that are susceptible to tick-borne infection. The transmission of tick-borne agents can follow several different mechanisms including

- *horizontal transmission*: transfer of the agent from tick to susceptible host usually through the saliva of the tick when it is feeding, but may occur by inoculation or inhalation of aerosolized agents from dried tick feces,
- *vertical transmission*: transfer of the agent from one tick generation to another tick generation,
- *transovarial transmission*: transfer of the agent from the female to her eggs so that hatched larvae are infected (this is a type of vertical transmission),
- *transstadial transmission*: transfer of the agent from one tick life stage (instar) through molting to the next instar.

Management and Control

Tick surveillence involves the use of chemical agents and management techniques to control both the parasitic and nonparasitic stages of the tick.

Management techniques to control ticks include

- keeping grass and weeds cut short to discourage tick infestation (by allowing sunlight and heat to penetrate an area that is detrimental to tick survival) and tick transfer to a host,
- controlling moisture because lack of moisture is detrimental to ticks,
- using predators to reduce tick populations,

Table 4-2   Different Types of Diseases Transmitted From Ticks

Tick	Diseases	Region Found (U.S.)	Hosts
Dermacentor andersoni (Wood tick)*	Rocky Mountain spotted fever Tularemia Q fever Tick paralysis Cytauxzoonosis Colorado tick fever virus Powassan encephalitis virus	Western states south to Arizona and New Mexico	Dogs Horses Livestock Mammals Humans
Dermacentor variablilis (Dog tick)*	Rocky Mountain spotted fever Tularemia Q fever Tick paralysis Ehrlichiosis Cytauxzoonosis	Eastern two-thirds of the United States	Dogs Cats Mammals Humans
Ixodes scapularis (Blacklegged tick)* (sometimes called the deer tick)	Lyme disease Ehrlichiosis Babesiosis Tick-borne encephalitis Tick paralysis	Most of the United States, especially the Northeast, upper midwest, and Northern California	Mammals Birds Humans
Amblyomma americanum (Lone star tick)*	Rocky Mountain spotted fever Tularemia Q fever Ehrlichiosis Tick paralysis	East of central Texas to the Atlantic coast, north to Iowa	Livestock Dogs Deer Birds Humans
Rhipicephalus sanguineus (Brown dog tick)*	Tick paralysis Babesiosis Ehrlichiosis Rocky Mountain spotted fever Hepatozoonosis Haemobartonellosis	Most of the United States	Dogs
Ornithodoros hermsii (Relapsing fever tick)†	Relapsing fever	Idaho, Oregon, Washington, California, Nevada, and Colorado	Chipmunks Squirrels Humans
Ornithodoros tunicatae (Relapsing fever tick)†	Relapsing fever	Southwest United States	Chipmunks Squirrels Humans
Ornithodoros cariaceus (Pajaroello tick)†	Painful tick bite reaction	California	Mammals Humans

*Diseases transmitted by hard ticks.

†Diseases transmitted by soft ticks.

1. Use fine-tipped tweezers and protect bare hands with a tissue or gloves to avoid contact with tick fluids.

2. Grab the tick close to the skin. Do not twist or jerk the tick, as this may cause the mouthparts to break off and remain in the skin.

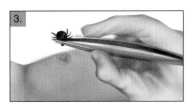

3. Gently pull straight up until all parts of the tick are removed.

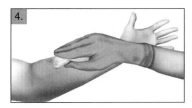

4. After removing the tick, wash hands with soap and water (or waterless alcohol-based hand rubs when soap is not available). Clean the tick bite with an antiseptic such as iodine scrub, rubbing alcohol, or water containing detergents.

Figure 4-5 Proper removal of a tick.

- selective grazing of livestock to remove the favorable habitat for ticks,
- manually removing of ticks when present in low numbers (making sure mouthparts are not broken or left embedding in the host) (Figure 4-5),
- rotating pasture to control the nonparasitic stages of ticks, and
- potentially using animal breeds that are resistant to some tick species.

Chemical control of ticks includes

- appropriate use of pesticides or acaricides on animals and their environments keeping in mind that some species of ticks have become resistant to the recommended levels of some insecticides. Chemicals that have historically been used to control ticks include organophosphates, carbamates, pyrethrins, synthetic pyrethroids, formamidines, fipronil, and selamectin.
- appropriate application of chemicals which may include solutions, emulsions, or suspensions that are applied as dips, impregnated tags, sprays, spot-ons, washes, pour-ons, or dusting.

Summary

Ticks transmit a variety of pathogens and typically transmit disease through a bite. Ticks are obligate, blood-sucking parasites that attack all vertebrates in all parts of the world. Ticks that cause disease to animals and humans are either hard ticks (Ixodidae) or soft ticks (Argasidae). Ticks have a general morphology consisting of two body regions: the capitulum (cranial portion) and the idiosoma (caudal portion). The capitulum has the mouthparts and the idiosoma contains most internal organs and bears the legs. All ticks undergo four basic stages: egg, larva, nymph, and adult. There are three subfamilies of hard ticks: Ixodinae (consisting of the single genus *Ixodes*), Amblyominae (containing the genera *Amblyomma*, *Haemaphysalis*, *Aponomma*, and *Dermacentor*), and Rhipicephalineae (containing the genera *Rhipicephalus*, *Anocentor*, *Hyalomma*, *Boophilus*, and *Margaropus*). There is one family of soft ticks that contains five genera: *Argas*, *Ornithodoros*, *Otobius*, *Nothoaspis, and Antricola*. Tick-borne disease can be transmitted horizontally, vertically, transovarially, and transstadially. Tick control can be accomplished through management and chemical means.

LYME DISEASE

Overview

Lyme disease, also known as Lyme borreliosis, borreliosis, Bannworth's syndrome, tick-borne meningopolyneuritis, erythema chronicum migrans (EM), and Steere's disease, was first discovered in 1975, but many manifestations of this disease had been previously described. In Europe, a red, slowly expanding rash associated with tick bites known as erythema chronicum migrans or EM was first described at the beginning of the 20th century. The signs of neurological disease and the association of Lyme disease with *Ixodes* ticks (commonly referred to as blacklegged ticks or deer ticks) were recognized by the mid-1930s and were attributed to a disease called tick-borne meningoencephalitis. In the 1940s, a similar tick-borne illness was described that began with EM and developed into multisystem illness. In the late 1940s, spirochete-like structures were observed in skin lesions.

In the United States, Lyme disease was not recognized until 1975, when a geographic cluster of childhood arthritis cases occurred in Lyme, Connecticut. The epidemic of arthritis in these children in Connecticut along with seasonal onset of signs and onsets of arthritis within the same families in different years prompted Dr. Allen Steere from Yale to investigate this disease phenomenon. It was determined that the patients in the United States had what was known as EM, which in turn led to the recognition that Lyme arthritis was one manifestation of the same tick-borne condition known in Europe.

Lyme disease is caused by the spirochete bacteria *Borrelia burgdorferi* and is the most common tick-borne disease in the United States. In 1982, Willy Burgdorfer, a Montana bacteriologist, identified a spirochete in the midgut of adult *Ixodes scapularis* ticks as the cause of Lyme disease. In 1984 spirochetes were cultured from the blood of patients with EM and from cerebrospinal fluid (CSF) of a patient with meningoencephalitis and EM. Different strains of the bacteria are now recognized, which explains why the clinical manifestations of Lyme disease are different in the United States and Europe.

The emergence of Lyme disease in the United States is likely a result of the explosion of deer and tick populations with the reforestation of the northeast, as well as the increased contact between ticks and humans as people moved into deer habitats. By the early 1800s, many parts of the eastern United States had been cleared of trees for use in lumber products or for development of farmland. The virgin forest was replaced with fields or low brush, and many animals such as deer and their predators normally found in these regions had disappeared. By the

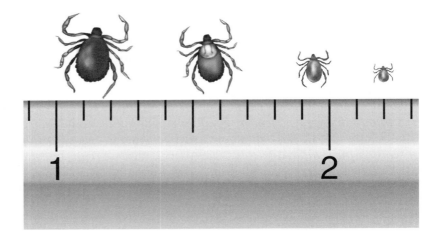

Figure 4-6 Various stages of *I. scapularis* ticks. From left to right adult female, adult male, nymph, and larva on a centimeter scale.

mid-1800s, agriculture moved westward and abandoned farmland in the eastern United States. In time trees and brush that were different than the original forest began to grow in this region. Deer returned to the eastern United States by the early 1900s, but their predators did not, and deer populations reached record levels by the end of the 1900s. Deer are the preferred hosts of the adult *I. scapularis*; humans and other animals are accidental hosts.

Bo. burgdorferi is found in white-footed mice and does not cause disease signs in affected mice. White-footed mice are also the preferred feeding site of the larva and nymph stages of *I. scapularis* (Figure 4-6), the blacklegged or deer tick, which may contain *Bo. burgdorferi* bacteria. These juvenile stages of the *I. scapularis* consume mouse blood and become infected with the spirochete. The nymph stage of the *I. scapularis* will then feed on almost any type of vertebrate, which may include humans and other animals, thus spreading the spirochete. Lyme disease is now endemic in some areas of the United States including parts of the east coast and Midwestern states such as Wisconsin and Minnesota where deer and human interaction is common.

> The use of the common term deer tick has fallen out of favor when describing *I. scapularis* because many types of ticks feed on deer making it an imprecise term.

Causative Agent

Bacteria belonging to the genus *Borrelia* are spirochetes composed of 3 to 10 loose coils and are vigorously motile by means of flagella (Figure 4-7). *Borrelia* bacteria

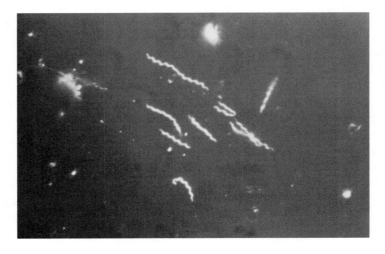

Figure 4-7 Darkfield microscope image of *Borrelia* spirochetes, which are approximately about 10 to 25 μm in length.
(Courtesy of CDC)

are longer and more loosely coiled than the other spirochetes and stain well with Giemsa stain. *Bo. burgdorferi* is the longest and narrowest species of *Borrelia* and it has fewer flagella. *Bo. burgdorferi* spirochetes are fastidious, microaerophilic bacteria that grow best at 33°C to 34°C in broth called Barbour-Stoenner-Kelly medium. This bacterium grows slowly, with a doubling time of 12 to 24 hours.

A variety of *Borrelia* spp. cause disease including:

- *Bo. burgdorferi* causes disease in the United States and Europe and has been subdivided into multiple species including three that cause human disease. *Bo. burgdorferi sensu lato* (broad sense) refers to the group of related *Borrelia* organisms as a whole, whereas *Bo. burgdorferi sensu stricto* (strict sense) is the specific strain of bacteria associated with classical Lyme disease. Antigenic variation among *sensu stricto* isolates has been documented making vaccine development challenging. *Bo. burgdorferi* has lipoproteins on its surface termed outer surface proteins (Osp). The North American *Bo. burgdorferi* has two major Osp: Osp A and Osp B. When the surface structure of the spirochete changes (termed host-adaptation), the production of Osp A and Osp B is turned off and the production of another outer surface protein, Osp C, is often turned on.
- *Bo. garinii* and *Bo. afzelii* are genospecies associated primarily with neurologic Lyme disease and chronic skin disease in Europe and Asia.
- *Bo. lonestari* is the cause of southern tick-associated rash syndrome in the United States. There are other nonpathogenic strains of *Borrelia* found in the United States, Europe, and Asia.

Epizootiology and Public Health Significance

Lyme disease is the most common tick-borne disease reported in the United States. Cases of Lyme disease have been increasing in most parts of the country since the 1980s. In 1982 there were 497 cases of Lyme disease reported in 11 states. From 1993 to 1997, an average of 12,451 cases of human cases of Lyme disease were reported to the Centers for Disease Control and Prevention (CDC) annually; 17,730 human cases were reported in 2000; and 23,763 human cases were reported to the CDC in 2002. A national surveillance program was initiated in 1982 and more than 157,000 cases have been reported to health authorities in the United States since that time. The overall incidence rate of reported cases in the United States is approximately 7 per 100,000 people (it is believed to be underreported).

Lyme disease has been reported in 49 states and the District of Columbia (Figure 4-8). It has a distinctive geographic concentration in the northeastern, upper Midwest, and Pacific coastal regions of the United States. It is a reportable disease throughout the United States, but the actual incidence is likely to be much higher in endemic areas where not all cases are reported or early disease may have been treated without antibody testing. Approximately 90% of cases were reported in the Northeast (the states between Maryland and Maine), 8% from the upper Midwest (mainly Wisconsin and Minnesota), and 2% from the Pacific coast (northern California and Oregon in particular). Lyme disease occurs in geographically limited areas and the incidence in endemic areas may reach 1% to 3% per year. People of all ages and both genders are equally susceptible, although the highest rates are in children aged 0 to 14 years, and in persons 30 years of age and older. Although Lyme disease has been reported from 49 states and the District of Columbia, the significant risk of infection is found in only about 100 counties

Ticks must be attached to a host for 24 to 52 hours before transmission of *Bo. burgdorferi* can occur because time is required for bacteria to migrate from the midgut of the tick to its salivary glands. During this time the surface structure of the spirochete changes in response to cues from the host and the blood meal.

1 dot placed randomly within county of residence for each reported case

Figure 4-8   Geographic location of Lyme disease in the United States in 2005.
(Courtesy of CDC)

in 12 states located along the northeastern and mid-Atlantic seaboard and in the upper Midwest region, and in a few counties in northern California. In the Northeast and Midwest, the white-footed mouse is the major reservoir host and *I. scapularis* is the primary tick vector. In the Pacific coast region, the dusky-footed wood rat is the major reservoir host and *I. pacificus* is the major tick vector. Lyme disease is also seen in Europe, Soviet Union, China, Japan, Southeast Asia, South Africa, Australia, and Canada.

People who live or work in residential areas surrounded by tick-infested woods or overgrown brush are the most at-risk group of contracting Lyme disease. People who work or play in their yard, participate in recreational activities such as hiking, camping, fishing and hunting, or engage in outdoor occupations, such as landscaping, forestry, and wildlife and parks management in endemic areas may also be at greater risk of contracting Lyme disease. Most cases of Lyme disease occur in the late spring and summer when the tiny nymphs are most active and human outdoor activity is greatest.

Transmission

For Lyme disease to exist in an area, three closely related elements must be present: *Bo. burgdorferi* bacteria, *Ixodes* ticks, and mammals such as mice and deer that provide a blood meal for the ticks through their various life stages. Ticks, small rodents, and other vertebrate animals all serve as natural reservoirs (bacteria can live and grow within these hosts without causing them disease) for the spirochete. *Bo. burgdorferi* is transmitted by the Ixodidae (*I. scapularis* or blacklegged

Areas in the United States with *I. scapularis* report the highest incidence of Lyme disease cases. Principal vectors of Lyme disease are *I. scapularis* in the northeast and upper-Midwest states and *I. pacificus* along the west coast of the United States. Although *I. scapularis* is widely distributed in the southern United States, it is not an established vector of Lyme disease in that area.

ticks) family of ticks known as the *I. ricinus* complex. This complex consists of 14 closely related tick species that are nearly identical in their appearance. The important vectors of human Lyme disease consist of four ticks that bite humans: *I. scapularis* (in northeastern and north-central United States), *I. pacificus* (in the western United States), *I. ricinus* (the sheep tick in Europe), and *I. persulcatus* (in Asia). Ticks of the *I. ricinus* complex have larva, nymph, and adult stages and feed once during each of the three stages of their usual two-year cycle; therefore, these ticks require 2 years to complete their life cycle and must feed on three independent hosts during this cycle (Figure 4-9). The 2-year cycle includes:

- Adult female ticks lay eggs on the ground in early spring.
- By summer, eggs hatch into larvae and feed only once. Larvae feed on mice, other small mammals, deer, and birds in the late summer and early fall. The tick larvae first become infected by feeding on these animals especially rodents (white-footed mice are the preferred reservoir hosts) which are hosts for *Bo. burgdorferi*. These hosts do not develop immune responses to the bacteria nor do they develop organ damage and they allow *Bo. burgdorferi* to replicate to a sufficient level to be infectious for subsequently feeding ticks.
- Larvae molt into nymphs by fall, and they are dormant (inactive) until the next spring. This period of inactivity is sometimes referred to as overwinter. In

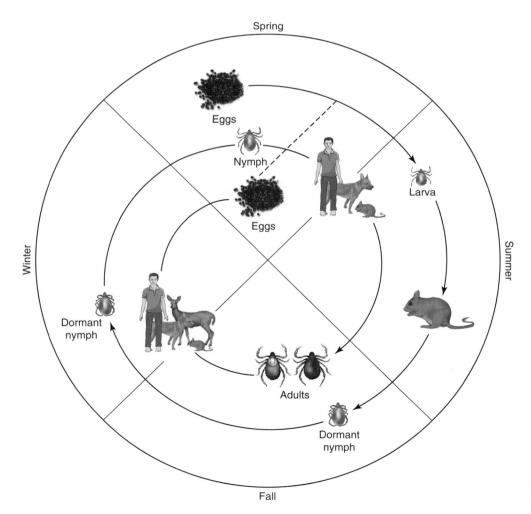

Figure 4-9 The 2-year life cycle of the *Ixodes* tick.

their larva and nymph stages, *Ixodes* spp. are smaller than a pinhead and their attachment to humans and domestic animals is often inapparent.

- In late spring and early summer, nymphs feed only once on rodents (white-footed mice are the main reservoir host), small mammals, birds, and humans.
- Nymphs molt into adults in the fall.
- In the fall and early spring, adult ticks feed and mate on large mammals (especially deer which are the preferred host) and bite humans. The adult female ticks will then drop off these animals and begin laying eggs in spring.

The nymph stage is responsible for disease transmission to humans; *Bo. burgdorferi* is not directly transmitted from animals to humans. The nymphs and adults obtain their blood meals by feeding on larger mammals. Deer are the preferred feeding hosts; other mammals such as dogs, horses, cattle, and people are accidental hosts of these ticks. Deer are not involved in maintaining *Bo. burgdorferi*; their role is to maintain the ticks. Ticks only crawl (they do not fly or jump) and can attach to any part of an animal's body but often crawl to the more hidden areas such as the groin or axilla (armpit) to feed. There is minimal transovarial transmission of *Bo. burgdorferi* in ticks, so each tick must be infected by feeding on an infected host. *Bo. burgdorferi* can be transmitted transstadially from larvae to nymph to adult. There is some evidence of in utero transmission of *Bo. burgdorferi* in dogs, which is important because there have been cases of adverse fetal outcomes in women who were infected with *Bo. burgdorferi* during pregnancy (neonatal deaths).

Ixodes ticks are found in temperate regions with high humidity at ground level. In the eastern United States, ticks are associated with the deciduous forest and habitat containing leaf litter that provides a moist cover from wind, snow, and other elements. In the upper Midwestern states, these ticks are generally found in heavily wooded areas often surrounded by land cleared for agriculture. On the Pacific Coast, the habitats are more diverse and ticks have been found in forest, north coastal scrub, high brush, and open grassland areas. Coastal tick populations prefer areas of high rainfall, but ticks are also found at inland locations.

> Risk of infection with *Borrelia* bacteria depends on the density of ticks, their feeding habits, and their animal hosts.

A tick must be attached to an animal for 24 to 52 hours to transmit the bacterium because of the life cycle of *Bo. burgdorferi* in ticks (peak transmission occurs between 48 and 52 hours). In previously infected ticks (nymphs and adults), only small numbers of bacteria are present until the tick feeds. Once the tick begins feeding, bacteria multiply in the midgut of the tick. The bacteria then migrate to the tick's salivary glands after 2 to 3 days. From the salivary gland they can be injected into the animal by the tick when it finishes its feeding. Without this multiplication of bacteria in the midgut, ticks are rarely able to pass on enough organisms to cause infection. *Bo. burgdorferi* bacteria persist in infected *I. scapularis* ticks from stage to stage (transstadially).

Pathogenesis

> By harboring infected ticks, domestic animals may increase the chances for human exposure to Lyme disease.

Following tick transmission of *Bo. burgdorferi*, the spirochete replicates in the skin at the site of the bite. After a few days to weeks, the spirochete is disseminated via the blood to a variety of sites such as joints, blood vessels, and connective tissues of the heart by binding to a number of host-cell receptors. Once in organs the spirochete can elicit a vigorous inflammatory reaction that causes carditis, arthritis, vasculitis, and dermatitis. Within a few weeks, the immune system can clear the organism from infected organs and blood causing disease signs to resolve. Despite this intense immune response, some spirochetes can survive for years in some organ systems causing recurrent bouts of disease.

Once *Bo. burgdorferi* bacteria gets into a host, one of the following events can occur:

- Animals clear the infection, do not develop clinical signs, and become seropositive
- *Bo. burgdorferi* disseminates throughout the body and produces clinical signs related to the invasion of bacteria into a particular organ system
- *Bo. burgdorferi* causes an immune response leading to clinical signs in various organs, without evidence of bacterial invasion

Clinical Signs in Animals

Lyme disease in animals is seen throughout the active tick season, but is most often diagnosed in late spring and fall when nymphs are more active, rather than during summer. Clinical signs in animals vary with the species involved:

- *Dogs.* In the initial or acute stage of Lyme disease dogs typically present with an abrupt onset of fever, anorexia, lethargy, and lameness that usually occurs in one or two joints and may be shifting (Figure 4-10). In the secondary or chronic stage of Lyme disease dogs may present with chronic, intermittent arthritis, heart conditions (such as pericarditis, endocarditis, and conduction disturbances),

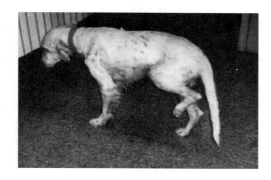

Figure 4-10 Joint swelling and arthritis in a dog with Lyme disease.
(Courtesy of Lyme Disease Foundation, www.lyme.org)

neurologic conditions (such as behavioral changes and seizures), and renal conditions (such as glomerulonephritis and protein-losing nephropathies). Rheumatoid arthritis may occur secondary to *Bo. burgdorferi* infection.
- *Horses.* In horses Lyme disease causes arthropathy, uveitis, and encephalitis.
- *Cattle.* In cattle Lyme disease causes arthropathy, cutaneous lesions, and multisystemic disease.
- *Cats.* Cats do not produce consistent clinical signs and pathologic lesions when naturally-exposed to *Bo. burgdorferi*.

Clinical Signs in Humans

In humans Lyme disease represents itself in three stages as follows:

- Early local disease (previously called stage 1 or localized): the most notable lesion in this stage is an erythema migrans (EM) (previously referred to as erythema "chronicum" migrans or ECM) which is also known as the bull's-eye rash (Figure 4-11). The rash is an expanding, red skin lesion that develops at the site of a tick bite. Other clinical signs in this stage include fever, lethargy, joint and muscle pain, and headaches.

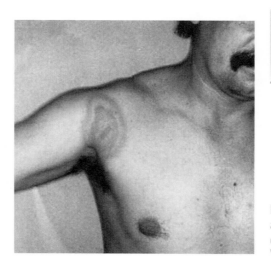

In humans Lyme disease is a multistage, multisystem disease that is categorized into early localized, early disseminated, and late disseminated stages.

Figure 4-11 Bull's-eye rash associated with Lyme disease.
(Courtesy of Lyme Disease Foundation, www.lyme.org)

- Early disseminated disease (previously called stage 2 or disseminated): this stage occurs weeks to months after initial infection producing clinical signs such as multiple smaller EM-like lesions on various parts of the body (apparently as a result of spread of the bacterium through the skin without additional tick bites). Other clinical signs in this stage include fatigue, neurologic disease (especially meningitis and facial nerve paralysis), myocarditis, and arthropathy without joint effusion.

- Late disseminated disease (previously called stage 3 or persistent): this stage occurs months to years after initial infection and is characterized by chronic arthritis and/or encephalopathy (sleep disturbances, fatigue, personality changes).

Diagnosis in Animals

> People contract Lyme disease from ticks in the nymph stage 85% of the time (spring to summer) and the adult stage 15% of the time (fall).

Pathology associated with Lyme disease is a result of the host's immune response to the spirochetes. Skin lesions on histological examination consist of perivascular infiltrates of lymphocytes and macrophages. In disseminated infection, all infected tissues show an infiltration of lymphocytes, macrophages, and plasma cells. Vasculitis may be seen at multiple sites. *Bo. burgdorferi* can be visualized in tissue sections stained with Warthin-Strarry silver stain and in blood and CSF stained with acridine orange or Giemsa stain.

Bo. burgdorferi is difficult to culture, but may be cultured from ticks or vertebrate hosts using BSK (Barbour-Stoenner-Kelly) broth that has been incubated at 33°C for 6 weeks or longer. Weekly samples are taken from the broth and examined by darkfield microscopy for the presence of spirochetes. BSK agar with 1.3% agarose produces slow-growing colonies.

Serologic tests for Lyme dis-ease include immunofluorescent assay (IFA), enzyme-linked immunoabsorbant assay (ELISA), and Western blot tests that detect antibody production to *Bo. burgdorferi*. ELISA tests serve as rapid screening methods because they are quick, reproducible, and less expensive; however, false positive rates are high making confirmatory tests necessary (Figure 4-12). A combination of ELISA screening and immunoblot confirmation is recommended for the diagnosis of Lyme disease. Some other problems exist with serologic testing of this spirochete. Studies have shown that the rate of positive serology tests in enzootic areas is higher than the prevalence of clinical disease. Addtionally, IgM cannot be used as an indicator of recent infection because IgM titers remain elevated in dogs for long periods of time after infection. Antibody titers also tend to remain elevated after antibiotic treatment making tracking of the progress of treatment difficult. Polymerase chain reaction (PCR)-based diagnostic testing is becoming more widely available and has been performed on urine and serum samples.

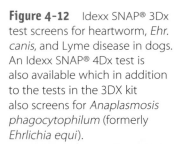

Figure 4-12   Idexx SNAP® 3Dx test screens for heartworm, *Ehr. canis*, and Lyme disease in dogs. An Idexx SNAP® 4Dx test is also available which in addition to the tests in the 3DX kit also screens for *Anaplasmosis phagocytophilum* (formerly *Ehrlichia equi*).

(Courtesy of Dr. Eli Larson, Pewaukee Veterinary Service)

Diagnosis in Humans

The diagnosis of Lyme disease in people is based primarily on clinical findings and known exposure with serologic testing providing supportive diagnostic information.

The most widely used tests for identifying Lyme disease in people are antibody detection tests. The current recommendation from the CDC is for a two-step testing process. In the first step, patients with symptoms consistent with Lyme disease are tested with an ELISA or an IFA test. The second step is to confirm positive tests with the more specific Western blot (WB) test (Figure 4-13). Patients with early disseminated or late disseminated disease usually have strong serologic reactivity and demonstrate banding patterns diagnostic for *Bo. burgdorferi*. Because antibodies may persist for months or years following successfully treated or untreated infection (they are believed to be re-exposed to new borrelial antigens), positive antibody test results alone cannot be used as an indicator of active disease.

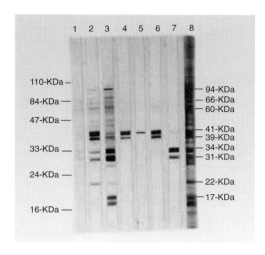

Figure 4-13 An example of a Western blot test used to detect Lyme disease in people.
(Courtesy of Lyme Disease Foundation)

Patients that are not treated continue to produce IgM antibodies long after the initial infection, thus patients may have both IgM and IgG antibodies at the same time making the correlation between antibody type and length of exposure invalid. Antibody testing in patients with erythema migrans is not indicated because the rash may develop before antibodies are produced and the test results could be misinterpreted.

PCR has been used to amplify genomic DNA of *Bo. burgdorferi* in skin, blood, CSF, and synovial fluid, but PCR has not been standardized for the diagnosis of Lyme disease.

Treatment in Animals

Treatment of *Bo. burgdorferi* infection in dogs consists of doxycycline or amoxicillin (in young animals in which tetracycline products are contra-indicated or in dogs that are not eating). Both treatments are done for 30 days to completely clear all of the *Bo. burgdorferi* organisms from the body (*Bo. burgdorferi* can persist in skin, CNS, and joint and connective tissues).

Prophylactic treatment with antibiotics in animals that are bitten by an *I. scapularis* tick is controversial. In enzootic areas, treating all dogs that are found to have been bitten by an *I. scapularis* tick is questionable, because the tick must be attached to the animal for greater than 24 hours to transmit the spirochete.

Animals showing signs of lameness may be treated concurrently with nonsteroidal anti-inflammatory drugs; corticosteroids should not be used because they decrease both the humoral and cell-mediated immune response to the spirochete. Animals with carditis may exhibit bradycardia (abnormally slow heart rate) making medical treatment with cardiac drugs and pacemakers necessary. Lyme nephritis may require fluid therapy and supportive medical therapy in addition to antibiotics. Seizures observed in animals with neuroborreliosis should be treated with anticonvulsants.

Treatment in Humans

Lyme disease in human adults is treated with oral doxycycline for 30 days because, with long-term exposure, accurately pinpointing the date of infection is not always possible. A 30-day course of antibiotics also reduces the number of patients who relapse after the shorter courses of antibiotics. Children and pregnant women are treated with amoxicillin or cephalosporin antibiotics as a result of the side effects of doxycycline use during growth.

Management and Control in Animals

In enzootic regions, regular examination of animals for ticks should be performed daily since transmission of Bo. burgdorferi requires a minimum 24- to 52-hour feeding time. Tick collars, spot-on products, sprays, baths, and dips should be used to eliminate ticks from animals or prevent their attachment. Environmental control, such as cutting brush and mowing grass more frequently, may alter the environment ticks thrive in and minimize their ability to parasitize small mammals that serve as the reservoir for the spirochete and a host for immature ticks.

Vaccination is also recommended for dogs that are at increased risk of contracting Lyme disease. Both a whole-cell, killed bacterin and a recombinant vaccine containing only the Osp A (outer-surface protein) from Bo. burgdorferi are available. The whole-cell bacterin induces antibodies against a wide variety of spirochete antigens including Osp A and Osp B. These antibodies are directed against antigens on the spirochete in the tick midgut; other antibodies may also be present in this vaccine to act against spirochetes present in the tick salivary glands and in the host. The recombinant vaccine is only effective against Osp A and must kill all spirochetes within the tick before the down regulation of Osp A occurs. If the antibodies are ingested by ticks during the early stages of feeding and then inactivate Bo. burgdorferi within the gut of the tick, transmission to the host is blocked. Because there is substantial genetic and antigenic variation among Bo. burgdorferi isolates multivalent vaccines may be needed to provide protection against a wide range of geographically distinct strains. In North America and Europe, at least five seroprotection groups of Bo. burgdorferi have been identified.

Management and Control in Humans

In humans, just like animals, deterrence and prevention of Lyme disease is emphasized. Several recommendations to reduce the risk of tick bite include:

Larva and nymph stages of *Ixodes* ticks attach to birds and may be the primary means by which infected ticks are spread from one area to another (thus potentially spreading disease from one area to another).

- Avoidance of ticks and tick-infested areas.
- Alteration of the tick habitat. Back yard patios, decks, and grassy areas that are mowed regularly are less likely to have ticks present. Areas around ornamental plantings and gardens are more hospitable for mice and ticks. The highest concentration of ticks is found in wooded areas.
- If tick-infested areas cannot be avoided, application of repellents containing DEET (N, N-diethylmetatoluamide) or permethrin will deter ticks. Permethrin can only be applied on clothing and DEET may cause side effects with frequent, long-term use (especially in children).
- Performance of tick checks after potential exposure to ticks. The groin, axilla, and hairline should be inspected particularly well.
- The Lyme disease vaccine for humans is no longer available (it was introduced to the market in 1998 and was withdrawn from the market in 2002).

- Some physicians advocate the use of prophylactic antibiotics if an *I. scapularis* tick bite is observed. If the attached tick is removed quickly, no other treatment should be necessary. If an engorged nymphal stage of the *I. scapularis* is found, a single dose of doxycycline may be given if the tick is found within 72 hours of the bite.

Summary

Lyme disease is an important infectious disease in North America, Europe, and Asia. The causative agent of Lyme disease is the spirochete *Bo. burgdorferi* that has now been subdivided into multiple species. This spirochete exists in nature in enzootic cycles involving ticks of the *I. ricinus* complex and a wide range of animal hosts. A tick must be attached to an animal for 24 to 52 hours to transmit the bacterium because of the life cycle of *Bo. burgdorferi* in ticks (peak transmission occurs between 48-52 hours). Lyme disease in animals is seen throughout the active tick season, but is most often diagnosed in late spring and fall when nymphs are more active, rather than during summer. In dogs, the initial or acute stage of Lyme disease typically presents with an abrupt onset of fever, anorexia, lethargy, and lameness that usually occurs in one or two joints and may be shifting. In the secondary or chronic stage dogs may present with chronic, intermittent arthritis, heart conditions, neurologic conditions, and renal conditions. In horses Lyme disease causes arthropathy, uveitis, and encephalitis, whereas in cattle Lyme disease causes arthropathy, cutaneous lesions, and multisystemic disease. Cats do not produce consistent clinical signs and pathologic lesions when naturally-exposed to *Bo. burgdorferi*. In humans, Lyme disease represents itself in three stages. The first stage (early local disease) produces erythema migrans (bull's-eye rash); the second stage (early disseminated disease) presents with fatigue, neurologic disease, myocarditis, and arthropathy; the final stage (late disseminated disease) is characterized by chronic arthritis, and/or encephalopathy.

Bo. burgdorferi is difficult to culture, but may be cultured from ticks or vertebrate hosts using BSK (Barbour-Stoenner-Kelly) broth that has been incubated at 33°C for 6 weeks or longer. Serologic tests for Lyme disease include IFA, ELISA, and Western blot tests that detect antibody production to *Bo. burgdorferi*. Treatment of *Bo. burgdorferi* infection consists of doxycycline or amoxicillin. In enzootic regions, regular examination of animals for ticks; preventative products such as tick collars, spot-on products, sprays, baths, and dips; environmental control; and vaccination in dogs are important in prevention of Lyme disease.

RELAPSING FEVER

Overview

Relapsing fever, more accurately known as either tick-borne relapsing fever (TBRF) or louse-borne relapsing fever (LBRF), is an arthropod-borne infection caused by several species of spirochetes in the genus *Borrelia*. TBRF is usually zoonotic and is endemic on most continents, each with its own species of *Borrelia* (for example, *Bo. hermsii* in North America and *Bo. duttonii* in Africa). LBRT is found in developing countries (Africa, China, and Peru mainly) and is caused by *Bo. recurrentis* and is spread from person to person by human lice.

In 1843 in Edinburgh, Scotland, the name relapsing fever was given to this disease by Craigie and Henderson, who described a human epidemic fever, with a characteristic pattern of acute fever followed by remission and relapse. Relapsing fever is believed to have been described by Hippocrates in the 4th century B.C. The German physician Otto Obermeier first discovered highly motile thread-like spirochetes as the cause of an epidemic of relapsing fever in Berlin in 1868. Gregor Munch first suggested that relapsing fever was transmitted by the bite of arthropods such as lice, fleas and bugs in 1878. The theory of lice being one of the vectors of relapsing fever was confirmed by the French microbiologists Sergent and Foley in 1910.

TBRF is transmitted by the soft tick of the genus *Ornithodoros* and occurs in Africa, Spain, Saudi Arabia, Asia, and certain areas in the western United States and Canada (it is endemic in the higher elevations and coniferous forests of the western United States and southern British Columbia). From 1903 to 1905, British physicians Joseph Dutton and John Todd in the Congo, and independently Ross and Milne in Uganda, identified the spirochete responsible for "human tick disease." Both Dutton and Todd contracted the disease during their work. Dutton died of the disease and the causative agent of East African relapsing fever (*Bo. duttonii*) is named after him. In 1904, German microbiologist Robert Koch was called to East Africa to investigate fever in cattle and learned that Europeans traveling into the interior regions of Africa had been suffering recurrent fever as well. Koch unknowingly confirmed the vector role of the soft tick *Ornithodoros* and demonstrated that spirochetes were transmitted to eggs (transovarial transmission) from infected female ticks. In the United States, TBRF is caused by the soft tick *Ornithodoros hermsii* (a rodent tick) and was first reported in North America from gold miners near Denver in 1915.

The causative agent of LBRF (*Bo. recurrentis*) was first observed in the blood of patients during an outbreak of the disease in Berlin, Germany, in 1868; however, it was not until 1907 that the human body louse (*Pediculus humanus*) was determined to transmit the spirochete. *Bo. recurrentis* has no wild animal reservoir and is transmitted solely among humans by the human body louse. In the 19th century, outbreaks of LBRF occurred in the British Isles, Europe, and the United States. More recently, LBRF has been recorded only in northeastern and central Africa, especially Ethiopia, Somalia, and Sudan, where infestations of human body lice are prevalent. Some authorities believe the Yellow Plague of Europe starting around A.D. 550, which halved the European population, was a result of LBRF.

Causative Agent

Relapsing fever is caused by many different species of the spirochete *Borrelia*. These spirochetes are morphologically indistinguishable from each other and are motile, slightly staining gram-negative, and have between 3 and 10 loose coils. *Borrelia* spp. are able to avoid immune destruction by undergoing antigenic variation (periodically changing surface antigens to avoid recognition by antibodies). Antigenic variation usually results from gene conversions or gene rearrangements in the deoxyribonucleic acid (DNA) of the bacterium. These variants express a unique variable major protein (Vmp), which occurs in two classes: the variable large proteins (Vlps) and the variable small proteins (Vsps). In each phase of bacteremia a population of one serotype predominates. After *Borrelia* spp. invade the body, bacteria multiply in tissues and cause a fever until the onset of an

Soft ticks feed for short periods of time (15 to 20 minutes), so infection occurs in minutes, in contrast to hard ticks (such as those that cause Lyme disease) that feed for days, where infection requires several hours.

immune response. Bacteria levels drop because of antibody mediated phagocytosis and the fever resolves. In time an antigenically distinct mutated *Borrelia* spp. appears in the infected individual, multiplies, and reappears in the blood causing another febrile attack. The immune system is stimulated and responds to the new antigenic variant, but the cycle of antigenic variation continues producing a new set of antibodies in the host.

- TBRF, also known as endemic relapsing fever, may be caused by over 20 *Borrelia* species. The nomenclature used when naming these species of *Borrelia* often mimics the name of the tick vector (e.g., *Bo. turicatae* infecting *Or. turicatae*, *Bo. hermsii* in *Or. hermsii*). In the mountains of California, Utah, Arizona, New Mexico, Colorado, Oregon, and Washington, infections are usually caused by *Bo. hermsi*. In Africa, the main species are *Bo. duttonii* and *Bo. crocidura*. *Borrelia* spirochetes in ticks affect the salivary glands and transmission occurs when the tick feeds on new hosts. Small mammals such as rodents, chipmunks, tree squirrels, bats, and rabbits are hosts for the spirochete, whereas ticks serve as the reservoir and are able to pass the bacterium transovarially. Lizards have also been known to be hosts for *Borrelia* bacteria.

- LBRF, also known as epidemic relapsing fever, is the more severe form of relapsing fever and is caused by the spirochete *Bo. recurrentis*. LBRF is transmitted by body lice (*Pediculus humanus corporis*), and to a limited extent the head louse (*Pediculus humanus capitis*). Humans are most likely the only reservoir with lice being unable to transmit it transovarially to its progeny. *Bo. recurrentis* does not invade the salivary or genital glands of the louse; therefore, transmission is not from the bite or the saliva of the louse but rather by inoculation of hemolymph from a crushed louse through conjunctiva, broken skin, or scratched through intact skin.

Epizootiology and Public Health Significance

TBRF or endemic relapsing fever has worldwide distribution and is endemic in most of Africa. Few cases of TBRF are reported in the United States; those reported are typically seen in the late spring and summer in the western mountainous states and the high deserts and plains of the Southwest, south into Texas, and northwest into Washington. Cases may appear in clusters where people are in contact with rodents on which the ticks feed (such as infested camp sites). The mortality rate of people with TBRF who are treated is less than 1%. A poor prognosis is given to people with signs of severe jaundice, severe change in mental status, and severe bleeding. The number of TBRF cases peak in the summer months. TBRF has been reported in 15 states: Arizona, California, Colorado, Idaho, Kansas, Montana, Nevada, New Mexico, Ohio, Oklahoma, Oregon, Texas, Utah, Washington, and Wyoming. TBRF was removed from the list of nationally notifiable conditions in 1987; however, 11 states require TBRF to be reported to their State Health Departments (Arizona, California, Colorado, Idaho, Nevada, New Mexico, Oregon, Texas, Utah, Washington, and Wyoming).

LBRF or epidemic relapsing fever is found in areas of overcrowding, war, and poverty. Currently LBRF is found in Africa, China, and Peru. Epidemic relapsing fever has not been reported in the United States since 1906. Mortality rates from LBRF vary from 30% to 70% in untreated patients with the mortality rate decreasing to about 5% with treatment.

Transmission

TBRF

In the United States, TBRF results from infection by *Bo. hermsii* and *Bo. turicatae*. The disease is transmitted to humans principally by the bites of the infected

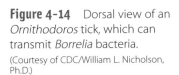

Cabins and camping areas are attractive nesting sites for rodents potentially infected with *Borrelia*, and may be areas where cluster outbreaks occur.

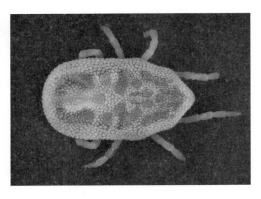

ticks *Or. hermsii* and *Or. turicata* (possibly also by *Or. parkeri* and *Or. rudis*) (Figure 4-14). These ticks inhabit the burrows and nests of rodents in southern British Columbia, Washington, Idaho, Oregon, California, Nevada, Colorado, and the northern regions of Arizona and New Mexico. The reservoir hosts are typically rodents (squirrels, mice, and chipmunks) and the natural infection cycle of *Borrelia* occurs without producing apparent disease in these animals. Other animals that may serve as reservoir hosts are pigs, goats, sheep, rabbits, bats, opossums, armadillos, foxes, cats, and dogs.

Figure 4-14 Dorsal view of an *Ornithodoros* tick, which can transmit *Borrelia* bacteria.
(Courtesy of CDC/William L. Nicholson, Ph.D.)

Humans are incidental hosts of TBRF when bitten by an infected tick. *Ornithodoros* ticks are night feeders and their bites often go unnoticed. Direct transmission from human to human is rare.

LBRF

LBRF is transmitted from person to person by the body louse (*Pediculus humanus corporis*) and less commonly the head louse (*Pediculus humanus capitis*) (Figure 4-15). Lice only become infected with *Bo. recurrentis* through a blood meal;

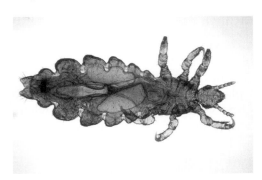

there is not transovarial transmission. Infection in humans occurs when the lice are crushed and their infected hemolymph invades the human through nonintact surfaces. Infection does not occur from the louse bite. Transmission may also occur from person to person via infected blood by needle-stick injury, blood transfusion, conjunctiva, and broken skin. There are no animal reservoirs for LBRF.

Figure 4-15 Dorsal view of the body louse, *Pediculus humanus corporis*, the vector of *Borrelia recurrentis*, the agent of relapsing fever.
(Courtesy of CDC/Dr. Dennis D. Juranek)

Pathogenesis

Regardless of transmission, once the spirochete gets into the body a spirochetemia develops in about 3 to 18 days (average is 7 to 8 days). Spirochete levels in blood may reach 500,000 per milliliter; however, clinical signs do not develop during the incubation period. The spirochete then invades the endothelium of many body systems producing fever, headache, fatigue, a low-grade disseminated intravascular coagulation (DIC), and thrombocytopenia (low numbers of clotting cells). As antibodies are produced, the spirochetes are cleared from the blood (many times undetectable in 24 to 48 hours). At this time the person becomes susceptible to other serotypes. The waxing and waning relapses occur as a result of shifting of outer surface proteins of the *Borrelia* bacterium that allows a new clone to avoid antibody destruction. The person will clinically improve until the new clone multiplies in sufficient quantities to cause another relapse. In time, recovery is because of the development of antibodies to all or most serotypes during the course of infection or to a rise of cross-protective antibodies. TBRF tends to produce more relapses (average of three) compared to LBRF (often just one).

Clinical Signs in Animals

Animals such as rodents (mice, rats, hamsters, and chipmunks), rabbits, and domestic animals (pigs, horses, and cattle) can serve as reservoirs for TBRF, but do not show clinical signs of disease.

Ornithodoros ticks can live for 10 to 20 years and can survive without a blood meal for several years.

Clinical Signs in Humans

After an incubation of 2 to 12 days (usually 7 to 8 days), affected people develop acute high fever and chills. A 2- to 3-inch itchy black scab may develop at the site of the tick bite, but typically the bite goes unnoticed (Figure 4-16). Other signs include headache, muscle and joint pain, rapid pulse, weakness, anorexia, and weight loss. In TBRF, multiple episodes of fever occur that may last up to 3 to 4 days in this form of relapsing fever and the person may remain free of fever for up to 2 weeks prior to a relapse. In LBRF, the initial episode usually lasts 5 to 7 days and is

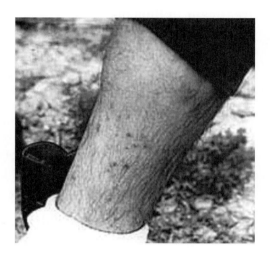

Figure 4-16 Scabs at the site of tick bites may be one of the signs of TBRF in people. (Courtesy of Lyme Disease Foundation, www.lyme.org)

usually followed by a single, milder episode. In both forms, the fever may end in "crisis," which consists of shaking chills, followed by intense sweating, decreased temperature, and hypotension (low blood pressure) that results in death in up to 10% of individuals.

After several cycles of fever, some people may develop central nervous system signs such as seizures, stupor, and coma. The *Borrelia* spirochete may also invade heart and liver tissues, causing myocarditis (inflammation of the heart muscle) and hepatitis (resulting in jaundice). Diffuse bleeding and pneumonia are other complications of this illness.

Diagnosis in Animals

Relapsing fever is only diagnosed in animals for experimental purposes and would be similar to those methods used in humans.

Diagnosis in Humans

Diagnosis of relapsing fever is confirmed by microscopy of a blood smear stained with Giemsa or Wright stain taken from a patient during a febrile episode (70% of people will show spirochetes in their blood during the febrile period). Multiple thick and thin smears should be examined. The spirochete is not found in the blood between relapses. Actively motile spirochetes may be seen in an unstained drop of blood when viewed using phase-contrast or darkfield microscopy. Culture of *Borrelia* should be done on BSK II medium in a microaerophilic atmosphere at 30°C to 35°C for 6 weeks. PCR, IFA, and ELISA tests have been developed for diagnosing relapsing fevers but are not used routinely. The IFA and ELISA tests have approximately a 10% false-positive rate. Injection of blood from infected humans into laboratory animals followed by examination of the laboratory animal's blood sometimes is useful in diagnosing this disease (xenodiagnosis).

Treatment in Animals

Animals are not treated for relapsing fever.

Treatment in Humans

Both forms of relapsing fever can be treated with tetracycline for 10 days. The development of a Jarisch-Herxheimer reaction (apprehension, diaphoresis (sweating), fever, tachycardia, and tachypnea especially with LBRF) can occur with tetracycline treatment; therefore, treatment should be started in a hospital setting. Erythromycin can be used for the treatment of pregnant women and children. Antipyretics are used to reduce fever. Poor prognostic signs include severe jaundice, severe change in mental status, severe bleeding, and prolonged QT interval on electrocardiogram (ECG). Tick or louse removal is imperative.

Management and Control in Animals

Tick control in pet animals and livestock may control the spread of TBRF in an area. LBRF is not found in animals.

Management and Control in Humans

Tick and rodent control is essential for preventing TBRF. Ways to control ticks include:

- Wearing proper clothing.
- Using tick repellent with DEET (use with caution on children and do not apply to infants younger than the age of 2).
- Changing and washing all bedding before use, especially in endemic areas.
- Removing existing ticks promptly and properly by grasping them close to the skin with tweezers. Avoid crushing the tick's body.

  Ways to control rodents include:

- Checking sleeping areas for evidence of rodents when camping (finding holes in the floor or walls, shredded material from mattresses, and rodent droppings on counters or in cupboards) and avoiding sleeping on the floor or on a bed that touches the wall.
- Checking external doors and windows to make sure they close with a tight seal.
- Keeping all food and garbage in tightly sealed containers and cleaning up any leftover or spilled food.

- Not feeding squirrels, chipmunks, and other rodents around dwellings.
- Eliminating woodpiles in or near the house. Store firewood outside, away from walls and stack wood on pallets, or raised a few inches off the ground.
- Eliminating existing rodents in your home via extermination, poison bait, or the use of spring-loaded mousetraps. Wear gloves and spray with bleach and water solution before and after handling dead mice.

Delousing and improved hygiene are needed for LBRF.

- Treating the person and other family members with head lice with a pediculicide to kill the lice.
- Machine wash using hot water (130°F) all clothing and bedding used by the infested person in the 2-day period prior to when treatment is started.
- Storing all clothing, stuffed animals, comforters, and other items that cannot be washed or dry cleaned into a sealed plastic bag for 2 weeks.
- Soaking combs and brushes for 1 hour in rubbing alcohol or wash with soap and hot (130°F) water.
- Vacuuming the floor and furniture, including the car interior.
- Avoiding head-to-head contact.
- Not sharing clothing (such as hats, scarves, coats, sports uniforms, or hair ribbons) or combs, brushes, or towels.
- Avoiding things that have recently been in contact with an infested person.

Summary

Relapsing fever is a disease characterized by relapsing or recurring episodes of fever, often accompanied by headache, muscle and joint aches, and nausea. There are two forms of relapsing fever: tick-borne relapsing fever (TBRF) and louse-borne relapsing fever (LBRF). TBRF is caused by several species of *Borrelia* spirochetes (mainly *Bo. hermsii* in the United States) and is transmitted to humans through the bite of infected soft ticks of the genus *Ornithodoros*. Most cases of TBRF occur in the summer months in mountainous areas of the western United States. LBRF is caused by *Bo. recurrentis* that is transmitted from human to human by the body louse (*Pediculus humanus corporis*). LBRF is rare in the United States. The relapsing nature of these diseases is associated with the development of antigenic variants. When the immune response develops to a predominant antigenic strain, variant strains multiply and cause a recurring infection.

Animals do not show clinical signs of disease. Clinical illness in people is marked by a febrile episode that lasts about 3 to 6 days, resolves, and then recurs 7 to 10 days later. Other signs include headache, muscle and joint pain, and lethargy. TBRF tends to have more relapses than LBRF. *Borrelia* organisms can be seen on microscopic examinations of thick and thin blood smears stained with Giemsa or Wright stain and darkfield or phase-contrast wet mounts. PCR, IFA, and ELISA tests are also available. Both tetracycline and erythromycin have been effective treatment in people; the disease is not treated in animals as a result of the lack of disease signs. Prevention of relapsing fever consists of avoiding likely tick and rodent habitats (TBRF) or eliminating lice infestations (LBRF). Cases of relapsing fever should be reported to local and/or state health departments.

TULAREMIA

Overview

Tularemia, also known as deerfly fever or rabbit fever in the United States, yato-byo in Japan, and lemming fever in Norway, is an infectious disease caused by the gram-positive bacterium *Francisella tularensis*. Tularemia was first described as a disease entity in Japan in 1837, but was not isolated until 1911 where it was cultivated by McCoy and Chapin from ground squirrels with plague-like illness in Tulare County, California. McCoy and Chapin named the organism *Bacterium tularense*. In 1914 the first link of this agent to human disease was documented when Wherry and Lamb isolated *Bacterium tularense* from an Ohio meat cutter. Several years later, Dr. Edward Francis, an epidemiologist of the U.S. Public Health Service, isolated the bacterium from the blood of patients with an ulcer-type illness following deerfly bites in Utah. In 1959 the organism was renamed *Fr. tularensis* because of the major contributions to the understanding of tularemia by Dr. Francis whose last name was designated for the genus name and the rural county in central California where it was first identified was designated for the species name. *Fr. tularensis* is found worldwide in more than 100 species of wild animals (mainly small mammals such as rabbits, rodents, squirrels, muskrats, beavers, and hares), birds, and insects and is found in nature from 20 degrees north of the equator to the Arctic Circle.

There are four major strains of *Francisella*, which differ in both virulence and geographical range. *Fr. tularensis*, found primarily in North America, has two strains. The bacterium can survive for weeks at low temperatures in water, moist soil, hay, straw, or decaying animal carcasses. Its ability to infect whole populations was seen during outbreaks of waterborne disease in Europe and the Soviet Union in the 1930s and 1940s and in outbreaks associated with deerfly bites in the United States in 1919, 1936, 1937, and 1973.

Causative Agent

Fr. tularensis is one of the most infective agents known with the infective dose as low as ten bacteria. Typically, 90% to 100% of those exposed to the bacterium develop disease. Even though tularemia is highly infectious and has many routes of transmission, it is not contagious between people.

Fr. tularensis is a small, pleomorphic, nonspore-forming, faintly staining, gram-negative bacillus (some report it as a coccobacillus) that can be readily aerosolized in laboratory settings. Cultivation of *Fr. tularensis* is difficult because it requires strict aerobic conditions and enriched media containing cysteine and cystine for isolation. Media used to cultivate *Fr. tularensis* include glucose cystein agar, cystine-heart agar, Thayer-Martin agar, chocolate agar, and tryptic soy or Mueller-Hinton agar supplemented with IsoVitaleX. *Fr. tularensis* may require 2 to 4 days for maximal colony growth and are weakly catalase positive. Colony morphology is transparent and mucoid.

Fr. tularensis is an intracellular pathogen that has a capsule and is not known to produce toxins. *Fr. tularensis* produces β-lactamase and can survive freezing, but is sensitive to water chlorination and heat and can be killed by cooking at 56°C for at least 10 minutes. *Fr. tularensis* is a Biosafety level 3 pathogen that requires laboratory workers to wear gloves/goggles and work under a biological safety cabinet when handling the bacterium.

There are four subspecies of *Fr. tularensis:*

- *Fr. tularensis* subspecies *tularensis* or type A, which is the most virulent form, is found in lagomorphs (rabbits and their relatives), and is common in North America.

- *Fr. tularemia* subspecies *holoarctica* or type B, is found in rodents (rats, mice, and their relatives) in Europe and Asia.
- *Fr. tularemia* subspecies *mediasiatia* is found in Central Asia and is not known to cause human disease.
- *Fr. tularemia* subspecies *novicida* is found in North America and is genetically similar to type A, but has lower virulence.

Epizootiology and Public Health Significance

Fr. tularensis is found throughout the United States, Europe, Russia, parts of the Middle East, central Asia, China, Japan, and the northern coast of Africa. *Fr. tularensis* is found in many different species of animals, but the most important animals in maintaining its cycle in nature and transmission to people are rodents such as voles, beavers, muskrats, and mice and lagomorphs such as hares and rabbits. Different rodents and lagomorphs vary in their resistance to tularemia; cottontail rabbits are highly susceptible to the disease and are usually killed by the infection. In North America, tularemia infections are most common in snowshoe hares, black-tailed jackrabbits, and eastern and desert cottontails. Carnivores are susceptible to the disease, but require high bacterial doses to become infected and rarely exhibit signs of disease. Infections in birds, fish, amphibians, and reptiles are relatively rare. Arthropods such as hard ticks, tabanid flies, mosquitoes, and fleas are important vectors of infection among vertebrates.

Approximately 200 cases of human tularemia are reported annually in the United States, mainly in rural areas of the southcentral and western states. The highest prevalence of cases is seen in Arkansas, Illinois, Missouri, Texas, Oklahoma, Utah, Virginia, and Tennessee. Most cases in the United States are associated with the bites of infected ticks, mosquitoes, and biting flies or with the handling of infected rodents, rabbits, or hares. Many cases are likely unreported or misdiagnosed. The frequency of tularemia in the United States has decreased markedly over the last 50 years, and has shifted from winter disease (usually from rabbits) to summer disease (more likely from ticks). Untreated cases of tularemia have a mortality rate of 5% to 15%; the use of appropriate antibiotics reduces this rate to about 1%.

The number of international cases of tularemia is unknown because it is not reportable.

Transmission

Fr. tularensis prefers cooler weather and in nature it is found north of the tropics. Natural sources of transmission include the bites of blood sucking ticks and insects or from water contaminated by infected animals (ticks are the most important vectors of *Fr. tularensis*, transferring the bacterium between rabbits, hares, and rodents). Ticks are biological vectors and maintenance hosts of *Fr. tularensis* and the bacterium can be introduced into mammals by both tick saliva and feces. Transmission in ticks is transstadially and transovarially. Deerflies also transmit *Fr. tularensis* and have been responsible for many outbreaks in the United States.

Another common source of human infection is contaminated wild meat (usually rabbit meat). Direct contact with excretions, blood, or organs of infected wild animals can be the source of *Fr. tularensis* infection in people. The bacterium enters the body through small skin lesions as well as through mucous membranes during contact with infected animals, especially during skinning or eviscerating of

> *Fr. tularensis* is classified as a category A bioterrorism agent as a result of its high infectivity, ease of dissemination, and potential to cause disease. Anticipated mechanisms for dissemination include contamination of food or water and aerosolization.

> Wild rabbits should not be kept as pets, since they may carry *Fr. tularensis*. Domestically raised rabbits are typically free of this bacterium.

carcasses. Contact with contaminated environmental sources such as water, soil, or vegetation are also possible sources of infection. Inhaling airborne bacteria (which can be generated during transportation of infected bedding material or from handling contaminated animal skins) can also cause tularemia.

Pathogenesis

The dog tick, *De. variabilis*, is considered the principle vector of *Fr. tularensis* in the United States.

The pathophysiology of tularemia depends on the route of infection, strain of bacteria, and host response. As few as 10 to 50 type A *Fr. tularensis* bacteria can cause disease if directly inoculated into the skin or inhaled (higher numbers are needed when the organism is ingested). After an incubation period usually between 2 to 10 days (varies from hours to weeks), acute disease develops. If the portal of entry is through skin a red papule appears, which enlarges, becomes purulent, and ulcerates. As bacteria spread through the blood and lymphatics, a primary complex arises in which regional lymph nodes enlarge, fill with pus, and ulcerate. Occasionally, there is lymph node enlargement without a primary lesion. Bacteria can spread to other organs of the body, particularly the lungs, pleura, spleen, liver, and kidneys where granulomas may develop. Widespread involvement of lymph nodes and viscera typically results in death. It is believed that *Fr. tularensis* is an obligate intracellular parasite that replicates within macrophages.

In rabbits and ground squirrels, multiple chalky lesions of variable size scattered through the liver, spleen, and lymph nodes. Caseous necrosis is surrounded by a zone of lymphocytes, with a few neutrophils and macrophages. Early lesions may have a purulent center that becomes necrotic with time.

Clinical Signs in Animals

Clinical signs of tularemia in animals vary with the species infected. Species infected with *Fr. tularensis* include:

- *Rabbits, hares, and rodents.* Naturally-infected rabbits, hares, and rodents are most often found dead without any clinical signs. Experimentally infected rabbits, hares, and rodents show signs of weakness, fever, ulcers, regional lymphadenopathy, and abscesses. Death usually occurs in these animals in 8 to 14 days.
- *Sheep.* Clinical signs of tularemia in sheep include high fever, rigid gait, diarrhea, polyuria, weight loss, tachycardia, tachypnea, and dyspnea. Affected sheep may isolate themselves from the remainder of the flock. Death is most common in young animals within hours or days of developing clinical signs. Pregnant ewes may abort. Tularemia in sheep is typically a seasonal disease, coinciding with tick infestations.
- *Domestic cats.* Cats infected with *Fr. tularensis* may show signs that include fever, lymphadenopathy, abscesses, oral or lingual ulceration, gastroenteritis, hepatomegaly, splenomegaly, icterus, anorexia, weight loss, pneumonia, and sepsis.
- *Dogs.* Clinical signs of tularemia in dogs may be inapparent or mild and are related to the route of transmission. Clinical signs include fever, mucopurulent oculonasal discharge, pustules at inoculation sites, lymphadenopathy, and anorexia.
- *Cattle.* Natural infection in cattle occurs, but clinical disease is not observed.

■ *Horses*. Reports of tularemia in horses are limited; however, fever, dyspnea, incoordination, and depression have been described following extensive tick infestation.

Clinical Signs in Humans

The incubation period for tularemia in people is typically 3 to 5 days, but may range from 1 to 14 days. Fever, chills, lethargy, myalgia (muscle pain), and vomiting are followed by more specific signs of disease that depend on route of entry. All forms of tularemia can progress to pleuropneumonia, meningitis, sepsis, shock, and death.

Occasionally tularemia in humans is divided into 2 categories: the external form which includes the ulceroglandular form (in which local or regional signs predominate) and the internal form which includes the more lethal typhoidal form (in which systemic signs dominate the clinical picture). More commonly tularemia is divided into six forms that reflect the mode of transmission and include:

■ *Ulceroglandular*. Ulceroglandular tularemia is the most common form (75% to 85% of reported cases) with the bacterium entering the body through a scratch, abrasion, or tick or insect bite. After introduction of as few as 10 organisms into the body bacteria spread via the lymphatic system. At the site of entry, usually the fingers or hands in cases associated with exposure to rabbits, hares, or rodents, an ulcer develops that progresses to necrosis and lymphadenopathy (Figure 4-17). Lymph nodes may suppurate and ulcerate.

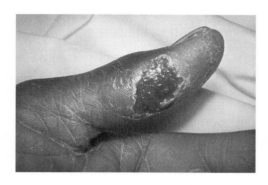

Figure 4-17 Skin ulcer on the thumb of a person with tularemia.
(Courtesy of Emory University/Dr. Sellers)

■ *Glandular*. The glandular form of tularemia is rare. There is no development of an ulcer and the organism is believed to have entered the lymphatic system and/or bloodstream through clinically undetectable abrasions. Signs of glandular tularemia are similar to ulceroglandular tularemia with the exception of a skin ulcer.

■ *Oculoglandular*. The oculoglandular form of tularemia is rare (approximately 1% of cases) and results from contamination of the conjunctiva from either a splash of infected blood into the mucous membrane or rubbing the eyes after contact with infectious materials such as blood from an infected rabbit carcass. Ulcerated papules, which are usually located on the lower eyelid, are seen as well as lymphadenopathy.

■ *Oropharyngeal*. Oropharyngeal tularemia is rare and is contracted through ingestion of *Fr. tularensis* (usually undercooked rabbit meat containing the bacterium) and results in acute pharyngeotonsillitis, which may be exudative or membranous, with cervical lymphadenopathy.

■ *Pulmonary*. Pulmonary tularemia represents about 30% of tularemia cases and develops after inhalation of aerosolized bacteria. Pneumonia in one or both lungs is the typical clinical sign. Pulmonary tularemia is most frequently observed in laboratory workers. Pneumonia also occurs in 10% to 15% of patients with ulceroglandular tularemia and in one-half of those patients with typhoidal tularemia.

■ *Typhoidal.* Typhoidal tularemia, also known as septicemic tularemia, occurs in about 10% to 15% of cases and results from ingestion of contaminated food or water. Clinical signs include fever, weight loss, gastroenteritis, and sepsis. Mortality rates in untreated cases can range from 40% to 60%. This form is more severe than the others and often includes pneumonia. Ingestion may be the mode of transmission; however, in most cases, the portal of entry may be unknown.

Diagnosis in Animals

Definitive diagnosis of *Fr. tularensis* is through bacterial culture from clinical specimens such as blood, exudates, or biopsy. Cultivation of this bacterium is difficult as a result of its fastidious nature and the fact that is grows very slowly in culture. Cultures should be considered dangerous and many laboratories may not attempt to isolate this bacterium because of the potential to infect laboratory personnel.

ELISA, hemagglutination, microagglutination, and tube agglutination are all serologic methods used to identify agglutinating antibodies to *Fr. tularensis*. Tularemia is generally a postmortem diagnosis in wild animals.

Diagnosis in Humans

For humans, a presumptive diagnosis is based on a patient's history, clinical signs, and a history of exposure. Definitive diagnosis is through bacterial culture of lesions, pus, biopsy, sputum, or blood. In routine laboratories, serology is the preferred method of identification because of the risk of infection to personnel. Agglutinating antibodies can be expected in the second week of the disease with titers of $>1:40$ being suspicious. In nonendemic areas, a single convalescent titer of 1:160 or greater is considered diagnostic. In endemic areas, acute and convalescent titers are required and a fourfold change of titer between samples obtained 2 to 4 weeks apart is considered to be diagnostic. ELISA tests for detection of IgM, IgA, and IgG antibodies are also available as well as lymphocyte stimulation tests. PCR tests are being evaluated for the diagnosis of tularemia that directly measure DNA from the organisms.

Treatment in Animals

Streptomycin and tetracycline are the antibiotics of choice for treating wild and domestic animals. Early treatment should prevent death.

Treatment in Humans

The treatment of tularemia in humans consists of antibiotics such as streptomycin, tetracyclines (especially doxycycline), gentamicin, fluoroquinolones, and chloramphenicol. Tularemia typically responds well to antibiotics. Fluid therapy and antipyretics may be helpful in treating clinical signs.

Management and Control in Animals

Control of tularemia between animals is difficult. Tick control is an important part of tularemia prevention as are rapid diagnosis and treatment of the disease.

Management and Control in Humans

Preventing tularemia in people involves avoiding the various routes of exposure. People who hunt, trap, butcher, skin, or handle wild animals are most at risk and should thoroughly cook wild game before consumption; wear gloves when handling wild game, their skins, and carcasses; and properly dispose of or disinfect equipment used in the diagnosis, care, or collection of animals suspected or known to be infected with *Fr. tularensis*. In endemic areas, handling of dead and diseased animals should be avoided. Washing of hands after handling any wild animal is recommended. Tick and insect bites should be controlled by the use of insect repellent containing DEET for skin application and insect repellant containing permethrin for clothing.

Contact with untreated water should be avoided when contamination with *Fr. tularensis* is suspected. Healthcare professionals working with animal and human tularemia patients should wear personal protective clothing including gowns, gloves, and face masks. Diagnostic laboratories should be notified if tularemia is suspected when specimens are submitted. Biosafety level 2 precautions are recommended for diagnostic work on suspect material and biosafety level 3 precautions are required for culture. Tularemia is a reportable disease in the United States, but not internationally.

A live attenuated vaccine to protect against typhoidal tularemia administered by scarification (the method used for smallpox vaccines) is no longer available under its Investigational New Drug protocol.

Summary

Tularemia is a disease caused by *Fr. tularensis*, a small, pleomorphic, nonspore-forming, faintly staining, gram-negative bacillus. This bacterium is transmitted by tick and insect bites and by contact with infected animals. There are four subspecies of *Fr. tularensis*: *Fr. tularensis* subspecies *tularensis* or type A, which is the most virulent form, is found in lagomorphs (rabbits and their relatives), and is common in North America; *Fr. tularemia* subspecies *holoarctica* or type B, is found in rodents (rats, mice, and their relatives) in Europe and Asia; *Fr. tularemia* subspecies *mediasiatia* is found in Central Asia and is not known to cause human disease; and *Fr. tularemia* subspecies *novicida* is found in North America and is genetically similar to type A, but has lower virulence. Clinical signs of tularemia in animals vary with the species infected and include rabbits, hares, and rodents, sheep, domestic cats, and dogs (rarely cattle and horses). Most animals exhibit signs of high fever, lethargy, anorexia, gait abnormalities, and altered respiratory rates. In humans the incubation period is typically 3 to 5 days and clinical signs include fever, chills, lethargy, myalgia (muscle pain), and vomiting. In humans, tularemia is divided into six forms that reflects the mode of transmission and include ulceroglandular, glandular, oculoglandular, oropharyngeal, pulmonary, and typhoidal. *Fr. tularensis* is diagnosed via bacterial culture (it is difficult to culture requiring strict aerobic conditions and enriched media) and serology. Treatment of tularemia involves the early use of antibiotics in both humans and animals. Prevention of tularemia involves tick control, rapid diagnosis and treatment of the disease, using proper hygiene procedures, and using personal protection when handling suspect tissues. Tularemia is a reportable disease in the United States.

RICKETTSIAL INFECTIONS

Overview

Rickettsioses are infections caused by *Rickettsiae* in which animals are the hosts and arthropods are the vectors. *Rickettsiae* are named after Dr. Howard

Ricketts (1871–1910), who was the scientist that first identified them and described the transmission of one rickettsial species via its tick vector. *Rickettsiae* are bacteria with atypical morphology, physiology, and behavior. In general they are small, gram-negative, pleomorphic (cocci or small bacilli) bacteria that are obligate intracellular parasites of eukaryotic cells. Some species that originally were classified in the Rickettsiaceae family have been reassigned to different families. *Rickettsia, Orientia, Ehrlichia, Anaplasma,* and *Neorickettsia* are all small obligate intracellular bacteria which were once thought to be part of the same family; however, they are currently considered to be distinct unrelated bacteria. There are several genera in the Rickettsiaceae family including *Rickettsia* and *Orientia* and several genera in the Anaplasmataceae family including *Anaplasma, Neorickettsia,* and *Ehrlichia.* Typically these bacteria are not transmissible directly from person to person (other than by blood transfusion or organ transplantation); transmission typically occurs via an infected arthropod vector or through exposure to an infected animal reservoir host.

All rickettsioses are zoonoses with the exceptions of epidemic typhus and trench fever in which humans are the hosts and lice are the vectors. Rickettsioses may cause relatively mild disease (rickettsialpox, cat scratch disease, and African tick-bite fever) or they may cause severe disease (epidemic typhus, RMSF, and Oroya fever). They can also vary in duration from those that are self-limiting to chronic or those that recur.

Causative Agent

For years *Rickettsiae* were believed to be related to viruses because they are very small and can only multiply in living host cells.

Rickettsiaceae are small bacteria in the family Rickettsiaceae and are fastidious, nonmotile, aerobic, gram-negative, obligate intracellular bacteria that survive only briefly outside the host. They multiply by binary fission intracellularly in host cells. *Rickettsiae* are among the smallest cells, ranging in size from 0.3 to 0.6 µm wide and from 0.8 to 2.0 µm long. The nutritional requirements of *Rickettsiae* are based on the host cell because of their inability to metabolize a precursor to the energy producing molecule ATP (adenosine triphosphate). Most *Rickettsiae* have life cycles that depend on an exchange between blood-sucking arthropod and vertebrate host. Humans are often accidental (dead end) hosts. The mechanism by which *Rickettsiae* cause disease is not clearly understood; however, most infections target the endothelial lining of small blood vessels.

Epizootiology and Public Health Significance

The taxonomy of this group of organisms has recently changed and will likely change again.

Table 4-3 summarizes the distribution of rickettsiosis, which varies with the disease.

Rickettsial zoonoses may occur sporadically or endemically. Rarely do they occur in epidemics (such as Q fever as a result of its aerosol transmission). The incidence of rickettsial disease varies, but in general is uncommon in humans (but when they occur can cause serious disease).

Transmission

Rickettsioses are transmitted by arthropod vectors that vary with the disease. These arthropod vectors feed on the blood or tissue fluid of the vertebrate host. The rickettsial bacterium is transmitted in a variety of ways including through direct inoculation with arthropod saliva, direct inoculation into the skin lesion as they feed, release of the rickettsiae onto the skin or into a wound via the smashing of the vector or the arthropod defecating into the area, or other means (such as fomites, animal products, and food).

Pathogenesis

A common target in most rickettsioses is the endothelial lining of small blood vessels such as venules and capillaries. *Rickettsiae* recognize, enter, and multiply in endothelial cells causing necrosis. In an effort to repair the necrotic endothelium, the host responds by proliferating endothelial cells that eventually block the vascular lumen (center of the blood vessel). Pathologic changes such as vasculitis, perivascular infiltration by inflammatory cells, increased vascular permeability resulting in fluid leakage, and thrombosis are common conditions seen with rickettsioses. Specific organ involvement would depend on the blood vessels affected. For example, intravascular clotting of blood cells in vessels supplying blood to the brain results in changes in mentation and other neurologic signs that may occur with some diseases. Target organs of rickettsial diseases include skin, lung tissue, heart, brain, gastrointestinal tract, pancreas, liver, coagulation system, and kidney.

Clinical Signs in Animals

Typically there are not clinical signs of rickettsial infection in animals and animals serve as reservoir hosts of the bacterium.

Clinical Signs in Humans

Clinical signs of rickettsial illnesses vary in humans, but typical early nonspecific signs include fever, headache, and lethargy. Rashes and eschars (black scabs) may be associated with some rickettsioses.

Diagnosis in Animals

Diagnosis of rickettsioses is typically not done in animals.

Diagnosis in Humans

Rickettsioses are diagnosed based on clinical signs, history of arthropod exposure, development of specific acute and convalescent antibody levels reactive for a specific pathogen or antigenic group, a positive result for a serologic test method, such as PCR, IFA, or ELISA test, or isolation of the rickettsial bacterium.

Treatment in Animals

Animals are not routinely treated for rickettsioses.

Treatment in Humans

Treatment of rickettsioses is similar and includes antibiotics (most often doxycycline, tetracycline, or chloramphenicol) and supportive care (antipyretics, analgesics, and fluid therapy). Treatment is initiated based on clinical signs and arthropod exposure prior to obtaining test results.

Table 4-3   Some Rickettsioses and Their Properties

Antigenic Group	Disease	Bacterium	Predominant Signs	Vector or Acquisition Mechanism	Animal Reservoir	Geographic Distribution
Typhus fevers	Epidemic typhus, Sylvatic typhus	*Rickettsia prowazekii*	Headache, chills, fever, prostration, confusion, photophobia, vomiting, rash (generally starting on trunk)	Human body louse, squirrel flea and louse	Humans, flying squirrels (United States)	Cool mountainous regions of Africa, Asia, and Central and South America
	Murine typhus	*Ri. typhi*	As above, generally less severe	Rat flea	Rats, mice	Worldwide
Spotted fevers	Rocky Mountain spotted fever	*Ri. rickettsii*	Headache, fever, abdominal pain, rash (generally starting on extremities)	Tick	Rodents	United States, Mexico, Central and South America
	Mediterranean spotted fever	*Ri. conorii*	Fever, eschar, regional adenopathy, rash on extremities	Tick	Rodents	Africa, India, Europe, Middle East, Mediterranean
	African tick-bite fever	*Ri. africae*	Fever, eschar(s), regional adenopathy, rash subtle or absent	Tick	Rodents	Sub-Saharan Africa
	North Asian tick typhus	*Ri. sibirica*	As above	Tick	Rodents	Russia, China, Mongolia
	Oriental spotted fever	*Ri. japonica*	As above	Tick	Rodents	Japan
	Rickettsialpox	*Ri. akari*	Fever, eschar, adenopathy, disseminated vesicular rash	Mite	House mice	Russia, South Africa, Korea
	Tick-borne disease	*Ri. slovaca*	Necrosis erythema, lymphadenopathy	Tick	Lagomorphs, rodents	Europe

Genus	Disease	Species	Symptoms	Vector	Reservoir	Distribution
	Aneruptive fever	*Ri. helvetica*	Fever, headache, myalgia	Tick	Rodents	Old World
	Cat flea rickettsiosis	*Ri. felis*	As murine typhus, generally less severe	Cat and dog flea	Domestic cats, opossums	Europe, South America
	Queensland tick typhus	*Ri. australis*	Fever, eschar, regional adenopathy, rash on extremities	Tick	Rodents	Australia, Tasmania
	Flinders Island spotted fever, Thai tick typhus	*Ri. honei*	As above but milder; eschar and adenopathy are rare	Tick	Not defined	Australia, Thailand
Orientia	Scrub typhus	*Orientia tsutsugamushi*	Fever, headache, sweating, conjunctival injection, adenopathy, eschar, rash (starting on trunk), respiratory distress	Mite	Rodents	Indian subcontinent, Central, Eastern, and Southeast Asia and Australia
Ehrlichia	Ehrlichiosis	*Ehr. chaffeensis*	Fever, headache, nausea, occasionally rash	Tick	Various large and small mammals, including deer and rodents	Worldwide
Anaplasma	Anaplasmosis	*Anaplasma phagocytophilum*	Fever, headache, nausea, occasionally rash	Tick	Small mammals and rodents	Europe, Asia, Africa
Neorickettsia	Sennetsu fever	*Neorickettsia sennetsu*	Fever, chills, headache, sore throat, insomnia	Fish, fluke	Fish	Japan, Malaysia

Management and Control in Animals

The best way to prevent domestic animals from contracting rickettsioses is to limit their exposure to arthropods, particularly ticks. Inspection for ticks and tick removal and the use of topical agents or tick collars are effective methods of tick control. Vaccines that protect against rickettsiosis in the United States are not available (other than *Neorickettsia risticii*, the agent of equine monocytic ehrlichiosis, more commonly known as Potomac horse fever).

Management and Control in Humans

The best way to prevent humans from contracting rickettsioses is to limit their exposure to arthropods as previously described. There are no commercially licensed vaccines for rickettsioses available in the United States. Vaccinations to prevent rickettsial infections are not required by any country as a condition for entry.

Summary

Rickettsioses are infections caused by Rickettsiae in which animals are the hosts and arthropods are the vectors. *Rickettsiae* are bacteria with atypical morphology, physiology, and behavior and are typically small, gram-negative, pleomorphic bacteria that are obligate intracellular parasites of eukaryotic cells. There are two families of bacteria that cause rickettsioses: Rickettsiaceae (*Rickettsia* and *Orientia*) and Anaplasmataceae (*Anaplasma*, *Neorickettsia*, and *Ehrlichia*). Typically these bacteria are transmitted via an infected arthropod vector or through exposure to an infected animal reservoir host.

All rickettsioses are zoonoses with the exceptions of epidemic typhus and trench fever in which humans are the hosts and lice are the vectors. Rickettsioses may range from those that cause relatively mild disease to those that cause severe disease. Typically there are not clinical signs of rickettsial infection in animals and animals serve as reservoir hosts of the bacterium. Clinical signs of rickettsial illnesses vary in humans, but typical early nonspecific signs include fever, headache, and lethargy. Rashes and eschars may be associated with some rickettsioses. Rickettsioses are diagnosed based on clinical signs, history of arthropod exposure, development of specific acute and convalescent antibody levels reactive for a specific pathogen or antigenic group, a positive result for a serologic test method such as IFA or ELISA testing, isolation of the rickettsial bacterium, or PCR testing. Treatment of rickettsiosis includes antibiotics and supportive care. The best way to prevent domestic animals and humans from contracting rickettsioses is to limit their exposure to arthropods—particularly ticks.

EHRLICHIOSIS/ANAPLASMOSIS

Overview

Ehrlichioses and Anaplasmoses are tick-borne diseases caused by small, intracellular bacteria belonging to the Rickettsiaceae family and originally belonging to the *Ehrlichia* genus. Ehrlichiosis was first described in Algerian (Africa) dogs in 1935 (caused by *Ehr. canis*). The next outbreak of canine ehrlichiosis was in military guard dogs stationed in Vietnam during the 1960s in which a large number of dogs became ill and died because of hemorrhagic complications of the disease. Ehrlichiosis in humans was first described in 1954 in Japan and was called Sennetsu fever. Sennetsu fever, caused by *Ehr. sennetsu*, occurs in limited areas of the Far East (primarily Japan) and is extremely rare. Sennetsu fever was the only form of ehrlichiosis known to afflict humans for many years until 1986 when a Detroit man became sick after being exposed to ticks in rural Arkansas. From that time on, cases of human ehrlichiosis have been diagnosed in the United States annually primarily in the southeastern and southcentral states. Originally human cases of ehrlichiosis were attributed to *Ehr. canis*, but in 1990, the CDC isolated a new species of *Ehrlichia* from the blood of a U.S. Army reservist training at Fort Chaffee, Arkansas. This new species of *Ehrlichia* was named *Ehr. chaffeensis*. In 1956, American C. B. Philip gave the name Ehrlichieae to this family of bacteria, after Paul Ehrlich, who initially described a disease associated with small, gram-negative bacteria known to infect cattle, sheep, goats, and dogs. *Ehr. ewingii* was named in 1992 after Sidney Ewing, a veterinary pathologist and investigator of ehrlichioses, who found the organism in neutrophils of a febrile dog.

In 2001, the taxonomy of this group changed with some species of *Ehrlichia* being reclassified into the genera *Anaplasma* or *Neorickettsia*. All of these organisms were placed in the Anaplasmataceae family. *Anaplasma phagocytophilum* now contains the bacteria known as *Ehr. equi* and *Ehr. phagocytophila*.

Ehrlichiosis is a term historically used to describe three tick-borne diseases caused by intracellular bacteria of the genus *Ehrlichia*. Ehrlichiosis was originally named according to the host species and type of white blood cell most often infected. Human monocytic ehrlichiosis (HME) was described in 1986 and is caused by *Ehr. chaffeensis*; human granulocytic anaplasmosis (HGA), which was formerly known as human granulocytic ehrlichiosis (HGE) was described in 1993 and is caused by *Ana. phagocytophilum* (formerly *Ehr. phagocytophilis*); and *Ehr. ewingii* ehrlichiosis caused by *Ehr. ewingii* was first described in St. Louis, MO, in 1999. Canine monocytic ehrlichiosis is caused by *Ehr. canis* and occasionally *Ehr. chaffeensis*; canine granulocytic ehrlichiosis is caused by *Ana. phagocytophilum* (formerly *Ehr. phagocytophilis*) and *Ehr. ewingii*; equine granulocytic ehrlichiosis is caused by *Ana. phagocytophilum*; and equine monocytic ehrlichiosis/Potomac horse fever is caused by *Neorickettsia risticii* (formerly *Ehr. risticii*). Zoonotic species include *Ehr. chaffeensis*, *Ehr. ewingii*, *Ana. phagocytophilum*, and *Neorickettsia sennetsu* (*Ehr. canis* may also be zoonotic, but this is not confirmed).

Causative Agent

The family Anaplasmataceae now contains four genera: *Ehrlichia*, *Anaplasma*, *Neorickettsia*, and *Wolbachia* (found only in arthropods). The zoonotic species of this group of organisms include *Ehr. chaffeensis*, *Ehr. ewingii*, *Ana. phagocytophilum*, and *N. sennetsu*. *Ehr. canis* may be zoonotic. The organisms currently believed not to be zoonotic are *Ehr. bovis* (causes bovine petechial fever in cattle in the Middle East and Africa), *Ehr. muris* (found in rodents in Japan and does not cause disease), *Ehr. ondiri* (found in cattle and wild ruminants in Africa), *Ehr. ovina* (found in sheep in the Middle East), *Ehr. ruminantium* (causes heartwater disease in ruminants), *Ana. platys* (causes cyclic canine thrombocytopenia in the United States, Taiwan, Greece, and Israel), and *N. risticii* (causes Potomac horse fever/equine monocytic ehrlichiosis in the United States). The bacteria in this group of organisms are small, pleomorphic, nonmotile, gram-negative, obligate, intracellular bacilli (they parasitize leukocytes). These bacteria survive only briefly outside the host (reservoir or vector) and only multiply intracellularly (Figure 4-18).

> *Ana. phagocytophilum* contains the organism formerly known as *Ehr. equi* and *Ehr. phagocytophila*.

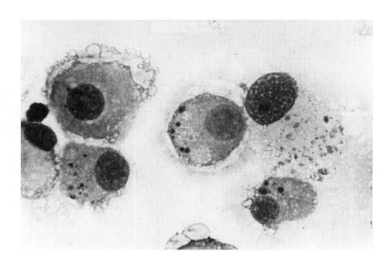

Figure 4-18 *Ehrlichia* organisms within leukocytes. (Courtesy of Lyme Disease Foundation, www.lyme.org)

Within a cell, small elementary bodies develop into larger initial bodies and eventually into intracytoplasmic inclusion bodies called morulae (which are diagnostic in blood smears). These organisms do not grow well on routine culture media and are typically cultured in embryonated eggs and in tissue culture.

Epizootiology and Public Health Significance

Ehr. chaffeensis, *Ehr. canis*, and *Ana. phagocytophilum* have worldwide distribution. In the United States, *Ehr. chaffeensis* occurs in more than 30 states (particularly Missouri, Tennessee, Oklahoma, Texas, Arkansas, Virginia, and Georgia). In the United States *Ana. phagocytophilum* is endemic is Wisconsin, Minnesota, Connecticut, and Massachusetts. *Ehr. ewingii* has only been found in the southeastern and southcentral United States. *N. sennetsu* has been reported mainly in Japan, but is probably found in other parts of Asia as well. In some cases, particular diseases are not reported from the bacterium's entire geographic range (for example *Ana. phagocytophilum* is found worldwide, but only causes tick-borne fever in ruminants in Europe, India, and South Africa).

Approximately 1,200 cases of ehrlichiosis/anaplasmosis were reported in the United States from 1986 to 1997. Human monocytic ehrlichiosis is most commonly seen in the Southeast and Midwest United States; human granulocytic anaplasmosis is most commonly seen in the Northeast and upper Midwest United States. Most cases of ehrlichiosis occur from April to September. Approximately half of all people with human monocytic ehrlichiosis require hospitalization. About 2% to 3% of human monocytic ehrlichiosis cases, 7% of human granulocytic anaplasmosis cases, and 5% to 10% of *Ehr. chaffeensis* infections are fatal. Infection with *Ehr. ewingii* is rare and most people recover without complications. Sennetsu fever is usually a mild illness and is not a significant cause of disease in the United States.

Transmission

Ehrlichiosis is a tick-borne disease and is transmitted by ticks in the family Ixodidae. The following bacteria are transmitted by the following hard ticks (Figure 4-19):

- *Ehr. chaffeensis* is transmitted mainly by *Amblyomma americanum* (Lone Star tick), but has also been seen in *De. variabilis* (American dog tick).

Transovarial transmission is not believed to occur with members of the *Ehrlichia* genus; ticks appear to become infected as larvae or nymphs.

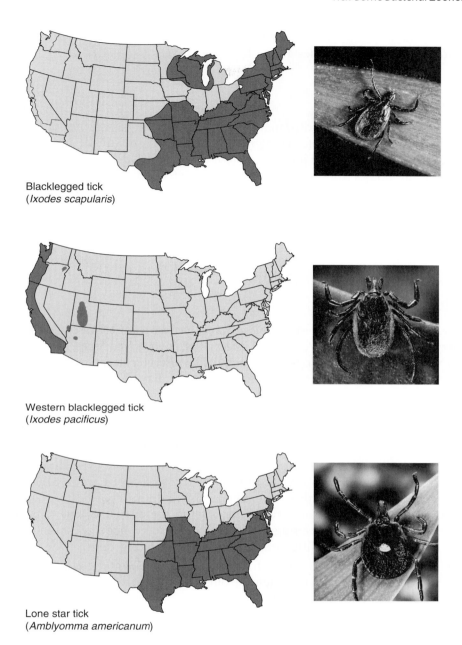

Blacklegged tick
(*Ixodes scapularis*)

Western blacklegged tick
(*Ixodes pacificus*)

Lone star tick
(*Amblyomma americanum*)

Figure 4-19   The distribution and identification of some tick vectors in the United States. (Courtesy of CDC)

- *Ana. phagocytophila* is transmitted by *I. scapularis* (the black-legged tick) in the eastern United States, *I. pacificus* (western United States), *and I. ricinus* (Europe).
- *Ehr. canis* is transmitted mainly by *Ri. sanguineus* (brown dog-tick; refer to Figure 4-1) and also by *De. variabilis* (American dog tick; refer to Figure 4-2).
- *Ehr. ewingii* is transmitted by *Amb. americanum*.
- *N. sennetsu* has an unknown vector but may be transmitted by consumption of raw fish infested with helminths or by a tick.

Most cases of ehrlichiosis/anaplasmosis are acquired during the months of highest tick activity, which is typically April to October with maximum activity occurring in June and July.

Transmission can also occur by blood transfusion. Viable bacteria have been found in refrigerated samples at 4°C for up to a week.

Pathogenesis

Bacteria that cause ehrlichiosis/anaplasmosis bind to the cell surface of leukocytes (a different leukocyte is preferred by different bacteria) and invade and live in these cells ultimately altering the immune system of the infected animal/person, thereby lessening the body's ability to fight secondary infections. These bacteria live and reproduce in the cytoplasm and are most frequently found clustered together as aggregates of many organisms. These clusters are berry-like in appearance and are called morulae. Little is known about how the infection spreads from the initial tick bite site, what cells or tissues are involved, what causes illness, and how tissue damage occurs.

The ticks that transmit *Bo. burgdorferi* and *Bab. microti* also transmit ehrlichiosis.

Clinical Signs in Animals

The variety of different bacteria in this group can cause a variety of clinical signs in a variety of animals. *Ehr. chaffeensis* can infect dogs, coyotes, red foxes, deer, goats, and lemurs (causing disease in dogs and lemurs). The reservoir hosts for *Ehr. chaffeensis* are deer. *Ehr. ewingii* causes disease in dogs. Dogs may also be the reservoir host. *Ehr. canis* infects dogs, wolves, and jackals; these animals are also the reservoir hosts. *Ana. phagocytophilum* causes disease in dogs, horses, llamas, cats, cattle, sheep, goats, and nonhuman primates. Reservoir hosts are deer, elk, and rodents.

Ehrlichia infections have been reportable to the CDC since 1998.

Diseases in animals caused by bacteria in this group include:

- *Canine monocytic ehrlichiosis.* Most cases of canine monocytic ehrlichiosis are caused by *Ehr. canis* (however, *Ehr. chaffeensis* infections are possible and are clinically indistinguishable from *Ehr. canis*) and are reported throughout the year (the tick vector can survive indoors and the disease course is prolonged in dogs). There are three stages of this disease: acute, subclinical, and chronic, which are difficult to differentiate in naturally-infected dogs.
  - Acute disease typically lasts for one to four weeks and can display a wide variety of clinical signs ranging from mild to severe with nonspecific signs such as fever, lethargy, anorexia, lymphadenopathy, splenomegaly, and weight loss. Bleeding disorders such as anemia and petechial hemorrhages may be seen. Ocular lesions such as anterior uveitis, oculonasal discharge, corneal opacity, and subretinal hemorrhages may occur. Other signs that may be seen include vomiting, diarrhea, lameness, neurologic signs (ataxia, seizures, and vestibular dysfunction), coughing, and dyspnea.
  - Subclinical disease occurs when dogs recover from the acute phase and remain infected for months or years. Dogs with subclinical disease can remain infected without showing clinical signs, may clear the infection, or may progress to the chronic phase.
  - Chronic disease presents as chronic weight loss, weakness, fever, peripheral edema, bleeding disorders (pale mucous membranes, petechial hemorrhages, hematuria, and melena), ocular lesions (anterior uveitis, retinal disease, and blindness), arthritis, renal failure, and pneumonia. Neurologic signs such as ataxia, hyperesthesia, head tremors, paresis, and seizures may be seen. Death may occur as a consequence of hemorrhages or secondary infections. Canine monocytic ehrlichiosis is difficult to cure once it has reached the chronic stage.

- *Canine granulocytic anaplasmosis (formerly canine granulocytic ehrlichiosis)*. This disease is caused by the organisms *Ana. phagocytophilum* and *Ehr. ewingii* and resembles monocytic ehrlichiosis. The most commonly seen sign with canine granulocytic ehrlichiosis is polyarthritis (it is uncommon with monocytic ehrlichiosis). Moderate to severe anemia has also been seen with this disease process.

- *Equine granulocytic ehrlichiosis*. Equine granulocytic ehrlichiosis is caused by the organism *Ana. phagocytophilum* and the disease varies from a mild infection with fever to severe disease. Clinical signs are more severe in older animals and include fever, anorexia, ataxia, jaundice, petechial hemorrhages, and peripheral edema (mainly hind limb). Equine granulocytic ehrlichiosis is most commonly seen in California, with sporadic cases occurring in other states. Most cases are seen in late fall, winter, and spring. Illness is more severe in older horses and animals are immune for at least two years following recovery.

- *Equine monocytic ehrlichiosis (Potomac horse fever)*. Equine monocytic ehrlichiosis is caused by *N. risticii* and is a serious illness of horses first described in the area around the Potomac River in Maryland in 1979 (it is now recognized throughout the United States and other countries). After the organism is ingested, it multiplies in the intestinal tract, where it can cause marked inflammation. Clinical signs include fever, depression, poor appetite, and diarrhea. Some horses will develop laminitis and pregnant mares can abort.

- *Tick-borne fever*. Tick-borne fever is seen in domestic and wild ruminants especially sheep and cattle and is caused by the organism *Ana. phagocytophilum*. Most ruminants will recover from this disease within 2 weeks, but relapses may occur. Tick-borne fever usually occurs in the spring and early summer when dairy cattle are turned out onto pasture. Impaired immunity seen with this disease will make the animals more susceptible to concurrent infections with some infections persisting for up to 2 years following clinical recovery. After one or two bouts of tick-borne fever sheep and cattle can develop immunity that can last for several months, but will decrease rapidly once the animal is removed from an endemic region. Death is rare.

  - *Sheep*. Tick-borne fever in sheep is mainly seen in young lambs born in tick-infested areas and in newly introduced older sheep. The main clinical sign is sudden fever that lasts for 4 to 10 days; other signs include weight loss, lethargy, coughing, tachypnea, and tachycardia. Pregnant ewes introduced onto infected pastures during the last stages of pregnancy can abort. Abortion is usually seen 2 to 8 days after fever onset. Infected rams may develop reduced sperm quality.

  - *Cattle*. Tick-borne fever is usually seen in dairy cattle turned out onto pasture with animals displaying a variety of clinical signs and severity of signs. Anorexia, decreased milk production, dyspnea, coughing, abortions, and reduced semen quality can be seen with tick-borne fever; abortions resulting in reduced milk yield and respiratory disease are the most common clinical findings in cattle.

- *Sennetsu fever*. Dogs have been experimentally infected with *N. sennetsu* developing fever as the only clinical sign. Mice have been experimentally infected with the same organism causing diarrhea, weakness, lymphadenopathy, and death.

- *Ehrlichiosis in cats*. Documented cases of ehrlichiosis in cats are rare. Cats infected with *Ana. phagocytophilum* have had clinical signs of fever, anorexia, lethargy, dehydration, and tachycardia.

■ *Ehrlichiosis in nonhuman primates.* *Ehr. chaffeensis* has caused naturally-occurring infection in captive ring-tailed and red ruffed lemurs. Clinical signs included anorexia, fever, lethargy, and lymphadenopathy.

Clinical Signs in Humans

Human infections with Anaplasmataceae organisms have been reported since the 1950s (*N. sennetsu*); however, most cases of infection were found in the 1980s (*Ehr. chaffeensis*, *Ehr. ewingii*, and *Ana. phagocytopyhilum*). *Ehr. canis* has been isolated from only one asymptomatic person.

■ Ehrlichiosis in humans consists of the clinically similar diseases human monocytic ehrlichiosis (HME) (Figure 4-20), which affects monocytic cells and is caused by *Ehr. chaffeensis*; human granulocytic anaplasmosis (HGA) (Figure 4-21), which affects neutrophils and is caused by *Ana. phagocytophilum*; and *Ehr. ewingii* ehrlichiosis caused by *Ehr. ewingii*. Human disease caused by *Ehr. ewingii* has only been reported in a few immunocompromised patients. Ehrlichiosis in people resembles RMSF (with or without the rash) with clinical signs beginning approximately 7 to 10 days after infection. The rash occurs in 20% to 40% of cases particularly in children and tends to be spotted in nature and is less prominent than that seen in RMSF. The rash can involve the trunk, legs, arms, and face, but usually spares the hands and feet. In people, symptoms vary greatly in severity, ranging from mild infection where no medical attention is needed, to a severe, life-threatening condition. Early symptoms are nonspecific and include high fever, headache, chills, and muscle pain (which mimics the early symptoms of many other tick-borne diseases). Other signs include vomiting, diarrhea, abdominal pain, anorexia, photophobia, conjunctivitis, joint pain, coughing, and mental confusion. Severe symptoms are seen in immunocompromised people and include fever, renal failure, opportunistic infections, hemorrhages, multisystem organ failure, cardiomegaly, seizures,

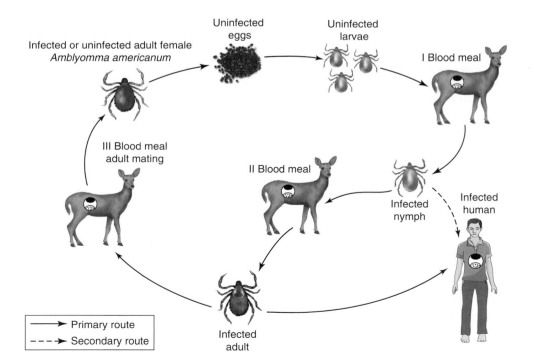

Figure 4-20 Life cycle of *Ehrlichia chaffeensis*, the causative agent of human monocytic ehrlichiosis (HME).

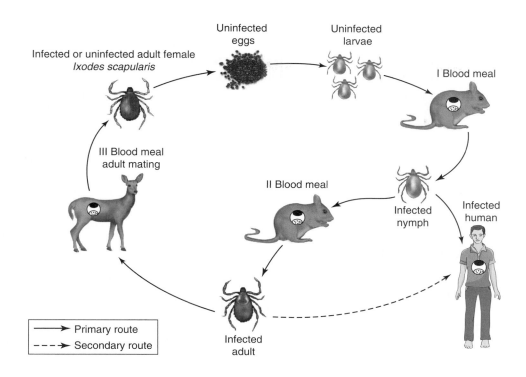

Figure 4-21　Life cycle of *Anaplasma phagocytophilum*, the causative agent of human granulocytic anaplasmosis (HGA).

and coma. Complications are more commonly seen in human granulocytic anaplasmosis.

- Sennetsu fever (caused by *N. sennetsu*) in people is a mild infection resembling mononucleosis. Clinical signs include fever, lethargy, anorexia, lymphadenopathy, hepatosplenomegaly, chills, headache, and backache. Circulating mononuclear cells and atypical lymphocytes are often increased. People with Sennetsu fever rarely have a rash and deaths from this disease have not been reported.

- *Ehr. canis* may be rarely zoonotic and its zoonotic potential needs to be confirmed.

Diagnosis in Animals

Ehrlichiosis/anaplasmosis in animals can be suspected based on clinical signs and gross pathology. Dogs with canine ehrlichiosis develop nonspecific gross lesions that include splenomegaly, lymphadenopathy, and congested, discolored lungs. Animals may be emaciated, have pale mucous membranes, and hemorrhages of the gastrointestinal tract, heart, urinary bladder, lungs, and eyes. Lymph nodes may be enlarged and discolored. Ascites and edema of the legs may also be seen. Horses with equine granulocytic ehrlichiosis may have petechial hemorrhages, subcutaneous edema, edema of the legs, and interstitial pneumonia. Sheep and cattle with tick-borne fever may abort. Complete blood count (CBC) abnormalities include thrombocytopenia (most common), anemia, and leukopenia. Diagnosis can also be supported by a response to treatment.

　　Laboratory tests for ehrlichiosis include serology or detection of the organism. Co-infection and cross-reactions may make diagnosis of this disease difficult. Bacterial culture is not used because these organisms can be difficult to culture and can take up to 30 days to grow. Detection of the organism is done

Unlike *Rickettsiae*, *Ehrlichiae* do not cause vasculitis, but cause multisystem diseases and can be found in many organs by lymphohistolytic (lymphocytes and macrophages) infiltrates (such as gastrointestinal, kidney, hearts, bone marrow, liver, spleen, meninges, and CNS).

by finding morulae in peripheral blood smears or impression smears from fresh tissues stained with Giemsa or by immunofluorescence. The morulae are typically seen in monocytes or granulocytes. Detection of organisms is more useful in cases of equine granulocytic ehrlichiosis than in cases of canine ehrlichiosis. Serologic tests include indirect immunofluroescent antibody tests (equine granulocytic ehrlichiosis, canine ehrlichiosis/anaplasmosis, and tick-borne fever), ELISA tests (canine monocytic ehrlichiosis and canine granulocytic anaplasmosis), and immunoblotting techniques such as Western blotting (research use). Disease is usually confirmed by the presence of rising antibody titers; in dogs, a single positive titer is evidence of exposure. PCR assays that detect antigens in blood are available for equine ehrlichiosis and may become available for canine ehrlichiosis/anaplasmosis.

Diagnosis in Humans

Diagnosis in people is based on history, clinical signs, and abnormalities on blood work (CBC and serum chemistry panels). Bacterial culture is difficult and time-consuming (*Ana. phagocytophilum* and *Ehr. chaffeensis* have been isolated from the blood of acutely ill people using various cell lines). Detection of the organism can be done by finding morulae in neutrophils or mononuclear cells. Disease confirmation is done through serologic testing which consists of indirect immunofluorescence assay (human monocytic ehrlichiosis or human granulocytic anaplasmosis and Sennetsu fever). The current case definition by the CDC for human ehrlichiosis/anaplasmosis is a fourfold rise or fall in antibody titer. ELISA tests are being developed for ehrlichiosis. PCR testing is available; immunohistochemistry and in situ hybridization has been done on tissue samples such as the spleen and lymph nodes.

Treatment in Animals

Ehrlichiosis/anaplasmosis is treated with tetracycline antibiotics. Early treatment of equine granulocytic ehrlichiosis and tick-borne fever are usually effective. Early treatment is critical for canine ehrlichiosis and uncomplicated cases respond well. Treatment of the chronic severe form in dogs is difficult and may require combination therapies (glucocorticoids, chemotherapy drugs such as vincristine, and hematopoietic growth factors).

Treatment in Humans

Treatment of ehrlichiosis/anaplasmosis in humans involves the use of tetracycline antibiotics; the current drug of choice is doxycycline. Early treatment is critical and prolonged treatment may be needed for severe or complicated cases.

Management and Control in Animals

The best way to prevent dogs from contracting ehrlichiosis is to limit their tick exposure. Dogs should be inspected daily for ticks and any ticks that are found should be removed quickly and safely with a gloved hand. Topical agents (such as fiprinol or permethrin) and tick collars containing amitraz are effective methods of tick control.

Vaccines are not available for canine ehrlichiosis, equine granulocytic ehrlichiosis, or tick-borne fever. There is a vaccine for equine monocytic ehrlichiosis (Potomac horse fever). Prophylactic antibiotics are sometimes used to prevent tick-borne fever in ruminants.

Management and Control in Humans

The best way to prevent ehrlichiosis or anaplasmosis in people also includes tick control. Strategies to reduce ticks include area-wide application of acaricides (chemicals that will kill ticks and mites), application of tick repellent with DEET, and control of tick habitats. Prompt removal of ticks is also essential. Tick control has been covered in the tick biology section and should be referred to. There is no vaccine for ehrlichiosis.

Summary

Ehrlichiosis and anaplasmosis are tick-borne diseases that are transmitted by ticks in the family Ixodidae. The various types of ehrlichioses/anaplasmosis are named according to their host species and white blood cell type infected. Human monocytic ehrlichiosis is caused by *Ehr. chaffeensis*; human granulocytic anaplasmosis is caused by *Ana. phagocytophilum*; and *Ehr. ewingii* ehrlichiosis is caused by *Ehr. ewingii*. Canine monocytic ehrlichiosis is caused by *Ehr. canis* and occasionally *Ehr. chaffeensis*; canine granulocytic ehrlichiosis is caused by *Ana. phagocytophilum* and *Ehr. ewingii*; equine granulocytic ehrlichiosis is caused by *Ana. phagocytophilum*; equine monocytic ehrlichiosis/Potomac horse fever is caused by *N. risticii*; tick-borne fever is caused by *Ana. phagocytophilum*; Sennetsu fever is caused by *N. sennetsu*; ehrlichiosis in cats is caused by *Ana. phagocytophilum*, and ehrlichiosis in nonhuman primates is caused by *Ehr. chaffeensis*. Zoonotic species include *Ehr. chaffeensis*, *Ehr. ewingii*, *Ana. phagocytophilum*, and *N. sennetsu* (*Ehr. canis* has been zoonotic, but this is not confirmed). The bacteria in this group of organisms are small, pleomorphic, nonmotile, gram-negative, obligate, intracellular bacilli (they parasitize leukocytes). These bacteria survive only briefly outside the host (reservoir or vector) and only multiply intracellularly. *Ehr. chaffeensis*, *Ehr. canis*, and *Ana. phagocytophilum* have worldwide distribution. *Ehr. ewingii* has only been found in the southeastern and southcentral United States. *N. sennetsu* has been reported mainly in Japan, but is probably found in other parts of Asia as well. Laboratory tests for ehrlichiosis or anaplasmosis in animals and people include serology or detection of the organism. Bacterial culture is not used because these organisms can be difficult to culture and can take up to 30 days to grow. Detection of the organism is done by finding morulae in peripheral blood smears or impression smears from fresh tissues stained with Giemsa or by immunofluorescence. Serologic tests in animals include indirect immunofluroescent antibody tests (equine granulocytic ehrlichiosis, canine ehrlichiosis/anaplasmosis, and tick-borne fever), ELISA tests (canine monocytic ehrlichiosis and canine granulocytic anaplasmosis), immunoblotting techniques such as Western blotting (research use). PCR assays that detect antigens in blood are available for some types of ehrlichiosis. Human monocytic ehrlichiosis and human granulocytic anaplasmosis are diagnosed by a fourfold rise or fall in antibody titer via immunofluorescence assay. Ehrlichiosis/anaplasmosis are treated with tetracycline antibiotics in both animals and people. The best way to prevent contracting ehrlichiosis/anaplasmosis is to limit tick exposure.

> Ticks that are removed from animals and people should be kept frozen in a plastic bag for identification in case of illness.

Q FEVER

Overview

Q fever, also known as query fever, was first described in 1935 in Australia by Dr. Edward Derrick who was investigating abattoir fever in a group of 800 Brisbane slaughterhouse workers who had symptoms of fever, headache, shivers, and sweats. He called the disease query fever because its causative agent was unknown. In 1936, Drs. Burnet and Freeman successfully identified rickettsial bacteria as the infectious agent of Q fever based on agglutination of infected animal tissues with convalescent sera obtained from Q fever patients. Dr. Derrick named the organism *Rickettsia burnetti* in honor of Dr. Burnet. In the United States around the same time period, scientists at the Rocky Mountain Laboratory in Montana were conducting research on the Nine Mile agent, a microbe transmitted by ticks. Dr. Herald Rae Cox is credited with identifying the "nine mile agent" as a rickettsial bacterium. In the United States, Dr. Cox called the organism *Ri. diaporica* in recognition of its ability to pass through filters used in those times to distinguish between bacteria (impermeable) and viruses (permeable). The bacterium has since been reclassified placing it genus on its own called *Coxiella*, within the family of Legionellaceae. The bacterium's name, *Coxiella burnetii*, honors the contributions of both Dr. Burnet and Dr. Cox.

Q fever has been known as abattoir fever (because of the epidemic among slaughterhouse workers), Balkan grippe (because of the epidemic among soldiers in the Balkans), and goat boat fever (because the disease commonly occurred among boat crews transporting infected goats). Allied forces experienced Q fever outbreaks in Italy and other Mediterranean countries during World War II. Explosive outbreaks of Q fever occurred in slaughterhouses in Texas and Chicago in the 1940s and the disease is still recognized as an occupational hazard among slaughterhouse workers.

Causative Agent

Cox. burnetii can survive for months and even years in dust or soil.

Cox. burnetii is a small, aerobic, gram-negative coccobacillus that is an obligate intracellular parasite in eukaryotic cells; however, unlike other rickettsial bacteria it multiplies in the acidic environment of phagosomes. *Cox. burnetii* forms an internal, stable, resistant infective body (sometimes called a spore) that is similar in structure and function to an endospore of some gram-positive bacilli. This infective body allows the bacterium to survive harsh environmental conditions. *Cox. burnetii* exists in two antigenic states: phase I (the virulent form which is also known as the smooth phase) and phase II (the avirulent form which is also known as the rough phase). The different states relate to its cell coating (changes in lipopolysaccharides) and antigenic, pathogenic, and immunogenic properties. Phase I bacteria possess a full complement of lipopolysaccharides, whereas phase II bacteria posses a simpler structure. The phase I bacterium is the form isolated from animals and is highly infectious; the phase II bacterium is isolated in cultured cell lines and is not infectious.

Epizootiology and Public Health Significance

Cox. burnetii does not replicate in bacteriologic culture media.

Cox. burnetii is distributed worldwide except in New Zealand. It has been found in various wild and domestic mammals, arthropods, and birds. Domestic cattle, sheep, goats, dogs, and cats are susceptible to infection, and the disease is found in most areas where these animals are kept. Both Ixodidae (hard) and Argasidae (soft) ticks can be reservoirs of the organism with greater than 40 species of ticks serving as natural reservoirs that remain infected throughout life and can transmit the bacterium transovarially. Infected animals shed this bacterium in urine, feces, reproductive tissues/fluid, and milk.

In 1999, Q fever became a notifiable disease in many U.S. states but reporting is not required in many other countries. In the United States Q fever is a reportable disease in all states except Delaware, Iowa, Oklahoma, Vermont, and West Virginia. In 2001, 26 cases of Q fever were reported to the CDC and in 2002, 61 cases of Q fever were reported to the CDC (0.05 per 100,000 people). The incidence of Q fever worldwide varies in frequency and presentation from country to country.

The mortality rate with acute Q fever is reportedly as high as 2.4%. People at greatest risk for infection are veterinarians, farmers, sheep and dairy workers, and laboratory workers who work with this organism.

Transmission

Cox. burnetii is transmitted via inhalation, direct or indirect contact with infected animals, or direct or indirect contact with their dried excretions. People typically contract Q fever by inhaling contaminated droplets of the highly infectious phase I spore forms excreted by infected animals. Consumption of raw milk has also been associated with infection. Infected ruminants can shed *Cox. burnetii* in their milk and amniotic fluid, and animals or humans can be infected by inhaling aerosols from the amniotic fluid or from unpasteurized infected milk. Pregnancy stimulates the replication of bacteria in reproductive and mammary gland tissues of many mammals. The amniotic fluid of infected animals carries high numbers of bacteria and is particularly dangerous. Person-to-person transmission is extremely rare. As a result of its inhalational route of transmission, it can be used as a biological agent and is classified as a category B agent.

> The most important transmission route from domestic ruminants to humans is through airborne transmission of particles from reproductive fluids.

Transmission occurs among wild and domestic animals by the bites of ticks (humans are not typically infected by tick bites, although it may be possible). Animal-to-animal transmission can also be through airborne particles or direct contact and ingestion of reproductive tissues/fluids or milk.

Pathogenesis

Once inside the body, *Cox. burnetii* is phagocytized by host cells and replicates within vacuoles. The incubation period varies from 9 to 40 days (average 18 to 21 days) during which time bacteria proliferate in the lungs, are engulfed by macrophages, and are transported to the lymph nodes. From the lymph nodes bacteria are carried to the bloodstream where they reach many areas of the body.

> As few as ten *Cox. burnetii* can initiate infection.

Clinical Signs in Animals

Goats, sheep, and cattle are the primary domestic reservoirs of *Cox. burnetii* (Figure 4-22). Inapparent infection is typical in these animals since clinical signs of infection rarely develop in infected livestock. If the infected animal is pregnant, abortion sometimes results. Occasionally an abortion storm (series of abortions) occurs when Q fever passes through a previously uninfected flock or herd. If a flock or herd is infected, most animals in the group will be infected.

Figure 4-22   Sheep, goats, and cattle may have inapparent disease from *Coxiella burnetii* infection or may show clinical signs such as abortion.
(Courtesy of Getty Images)

Clinical Signs in Humans

In people, Q fever can present as an inapparent, acute, or chronic disease.

- Inapparent Q fever is seen in about half of the people infected with *Cox. burnetii* and these people do not show any clinical signs.
- Acute Q fever is a generalized disease that presents like influenza. The incubation period is 2 to 4 weeks (average is 20 days). Clinical signs include sudden fever, chills, lethargy, muscle and joint pain, headache, and photophobia. This flu-like syndrome is usually self-limiting and lasts up to three weeks. Pneumonia can occur in about one third of people. Hepatitis can also occur alone or concurrently with pneumonia. Less common features of acute Q fever include rashes, meningitis, myocarditis, and pericarditis.
- Chronic Q fever develops in individuals who have been infected for over 6 months without effective treatment. Its main feature is endocarditis and/or chronic hepatitis. Clinical signs include prolonged fever, night sweats, chills, fatigue, and shortness of breath.

Diagnosis in Animals

Necropsy lesions in animals with Q fever are nonspecific and of little value in diagnosing the disease. *Cox. burnetii* cannot be cultured using current bacteriologic media. Serologic testing is the diagnostic tool of choice with complement fixation, IFA, and ELISA tests available.

Diagnosis in Humans

In people, Q fever is confirmed by serology using IFA (most widely used). Immunohistochemical staining methods and PCR tests are also available. Because *Cox. burnetii* exists in two antigenic phases, assessing both phase I and phase II levels are important in diagnosis; therefore, baseline and 3- to 4-week samples are taken for analysis. In acute Q fever, the antibody level to phase II is usually higher than phase I and is generally first detectable during the second week of illness. A fourfold rise in complement-fixing antibodies against phase II antigen confirms acute Q fever. In acute Q fever, patients will have IgG antibodies to phase II and IgM antibodies to phases I and II. In chronic Q fever, high levels of antibody to phase I in combination with constant or falling levels of phase II antibodies are found. Antibodies to both phase I and II antigens have been shown to last for months or years after initial infection. Increased IgG and IgA antibodies to phase I are often indicative of chronic Q fever.

Treatment in Animals

In animals, tetracycline antibiotic treatment is effective for treating Q fever. Separation of pregnant animals and burning or burying infective reproductive tissues/fluids can reduce the spread of bacteria. Resistance to physical and chemical agents makes ridding the environment of *Cox. burnetii* difficult. Recommended disinfectants include a formulation of two quaternary ammonium compounds, 70% ethanol, and 1:10 bleach solution.

> Antibodies to phase I antigens generally take longer to appear and indicate continued exposure to *Cox. burnetii*.

Treatment in Humans

Antibiotics are used to treat both acute and chronic Q fever. The most common treatment is doxycycline for acute Q fever and combinations of doxycycline plus an additional antibiotic such as fluoroquinolone, rifampin, or trimethoprim-sulfamethoxazole for chronic Q fever. Chronic Q fever antibiotic treatment is recommended for 3 years. Disinfection of contaminated areas is also important.

Management and Control in Animals

There is not a vaccine to protect animals from acquiring *Cox. burnetii*. Proper hygiene, especially around birthing animals, is important in preventing animal-to-animal spread of disease.

Management and Control in Humans

A formalin inactivated phase I whole cell vaccine is licensed in Australia and Eastern Europe (a single dose provides greater than 95% protection against naturally occurring Q fever within 3 weeks and lasts for at least 5 years). A live attenuated vaccine (Strain M44) has been used in the former USSR. In the United States, a noncommercial inactivated vaccine is available for at risk laboratory personnel through the U.S. Army Medical Research Institute. Standard precautions are recommended for health care workers taking care of patients with suspicion or diagnosis of Q fever.

CDC recommendations for preventing Q fever include:

- Disposing appropriately of the placenta, birth products, fetal membranes, and aborted fetuses at facilities housing sheep and goats.
- Restricting access to barns and laboratories used in housing potentially infected animals.
- Using only pasteurized milk and milk products.
- Using appropriate procedures for bagging, autoclaving, and washing of laboratory clothing.
- Vaccinating (where possible) individuals engaged in research involving pregnant sheep or live *Cox. burnetii*.
- Quarantining imported animals.
- Ensuring that sheep holding facilities are located away from populated areas. Animals should be routinely tested for antibodies to *Cox. burnetii*, and measures should be implemented to prevent airflow to other occupied areas.

- Counseling persons at highest risk for developing chronic Q fever, especially persons with pre-existing cardiac valvular disease or individuals with vascular grafts.

Summary

Q fever is an infection caused by the bacterium *Cox. burnetii*, a small, aerobic, gram-negative coccobacillus that is an obligate intracellular parasite in eukaryotic cells. *Cox. burnetii* forms an internal, stable, resistant infective body that allows the bacterium to survive harsh environmental conditions. *Cox. burnetii* exists in two antigenic states: the highly infectious phase I (the virulent form which is also known as the smooth phase) and the noninfectious phase II (the avirulent form which is also known as the rough phase). *Cox. burnetii* does not replicate in bacteriologic culture media.

Cattle, sheep, and goats are the primary reservoirs of *Cox. burnetii*. Clinical disease is rare in these animals, although abortion in pregnant ruminants may occur. Organisms are excreted in milk, urine, and feces of infected animal; and high numbers of bacteria are shed in amniotic fluids and the placenta. Infection of humans usually occurs by inhalation of these organisms from air that contains airborne particles contaminated by infected animals. Ingestion of contaminated milk is a less common mode of transmission. In people Q fever can cause flu-like signs, pneumona, and hepatitis in its early form, and endocarditis in its chronic form. Q fever is diagnosed by serology and treated with antibiotics such as doxycycline. In the United States, Q fever outbreaks have resulted mainly from occupational exposure involving veterinarians, slaughterhouse workers, sheep and dairy workers, livestock farmers, and laboratory workers. Prevention and control efforts should be directed primarily toward these groups and environments.

ROCKY MOUNTAIN SPOTTED FEVER

Overview

Rocky Mountain spotted fever (RMSF), originally known as black measles because of its characteristic rash and referred to as tick typhus outside the United States, is the most important rickettsiosis in the western hemisphere. RMSF is one of the spotted fevers, a large group of arthropod-borne infections caused by closely related *Rickettsiae* bacteria. *Rickettsiae* are small, gram-negative, pleomorphic (cocci or small bacilli) bacteria that are obligate intracellular parasites of eukaryotic cells. Rickettsioses are rickettsial infections in which mammals are the hosts and arthropods are the vectors.

RMSF was first recognized in 1896 in Idaho as a frequently fatal disease affecting hundreds of people in the Snake River Valley area. Outbreaks of RMSF spread rapidly and by the early 1900s, its geographic distribution in the United States went as far north as Washington and Montana and as far south as California, Arizona, and New Mexico. In response to its rapid spread and severity of clinical signs, the Rocky Mountain Laboratory was established in Hamilton, Montana (it is currently part of the National Institute of Allergy and Infectious Diseases, National Institutes of Health).

RMSF is caused by the bacterium *Rickettsia rickettsii*, named after Dr. Howard T. Ricketts, the first person to identify the infectious organism causing this disease in blood smears of infected animals and humans. Ricketts and his researchers also identified the cycle of infection involving ticks and mammals with humans considered accidental hosts and not a critical component in the natural transmission cycle of *Ri. rickettsii*. In 1910 after completing his work on RMSF, Ricketts died of typhus, a different rickettsial disease.

In the 1930s, it became clear that RMSF occurred in many areas of the United States other than the Rocky Mountain region. This vast distribution is a result of the ticks that serve as vector and reservoir of the disease: *Dermacentor variabilis* (commonly known as the American dog tick) in the eastern United States and *Dermacentor andersoni* (commonly known as the Rocky Mountain wood tick) in the western United States. The majority of RMSF cases are currently concentrated in the southeast and eastern seaboard regions of the United States as well as southern Canada, Central America, Mexico, and parts of South America.

Causative Agent

RMSF is caused by *Ri. rickettsii*, a small bacterium in the family Rickettsiaceae (which consists of genera: *Rickettsia* and *Orientia*). *Rickettsiae* are fastidious, nonmotile, aerobic, gram-negative, obligate intracellular bacteria that survive only briefly outside the host. They only multiply intracellularly in the cytoplasm of host cells. *Rickettsiae* have cell walls consisting of small amounts of peptidoglycan (making them seem as though they lack a cell wall) and an outer lipopolysaccharide membrane that has little endotoxin activity. The bacterium is surrounded by a loosely organized slime layer causing them not to react well to Gram stain; hence they stain a pale pink. This slime layer is believed to play a role in transmission. Tick feeding results in growth of the slime layer (called reactivation) and is believed to attribute to the bacterium's virulence. To better visualize these bacteria Giemsa stain is routinely used. *Rickettsiae* consist of three groups of bacteria: the spotted fever group, the typhus group, and the scrub typhus group. More than 20 species of *Rickettsia* are known and not all of them cause disease.

> Over 90% of RMSF infections occur between April and September when increased numbers of adult and nymphal *Dermacentor* ticks are seen. Infection can occur during winter in warmer regions such as Central and South America.

Epizootiology and Public Health Significance

The distribution of *Rickettsiae* is limited to the geographic region of their arthropod hosts. RMSF was originally found in the western United States; however, in the last 100 years there has been a shift to the eastern United States with the occurrence of disease highest in the south-Atlantic region (Delaware, Maryland, Washington D.C., Virginia, West Virginia, North Carolina, South Carolina, Georgia, and Florida). Infection also occurs in other parts of the United States, such as the Pacific region (Washington, Oregon, and California) and west southcentral region (Arkansas, Louisiana, Oklahoma, and Texas). The states with the highest incidences of RMSF are North Carolina and Oklahoma.

Ri. rickettsii infection has also been reported in Argentina, Brazil (called São Paulo fever and fiebre maculosa), Colombia (called Tobia fever), Costa Rica, Mexico (called fiebre manchada), and Panama. RMSF does not exist in Europe, Africa, or Asia.

RMSF has been a reportable disease in the United States since the 1920s. Approximately 600 to 800 cases of RMSF are reported annually in the United States, although many cases go unreported. RMSF is highest among males, Caucasians, and children with two-thirds of the cases occurring in children younger than 15 years of age (peak age being 5 to 9 years old). People with frequent exposure to dogs and who reside near wooded areas or areas with high grass are at increased risk of infection. Seasonal outbreaks parallel tick activity with 90% of cases reported from April 1 to September 30 (peaks seen in May and June). Human RMSF mortality rates are approximately 4%, with death usually occurring 8 days after onset of symptoms.

Transmission

In general, about 1% to 3% of the tick population carries *Ri. rickettsii* making the risk of exposure low even in areas where the majority of human cases are reported.

Ri. rickettsii bacteria typically infect and are transmitted by Ixodidae (hard) ticks. *Ri. rickettsii* is most frequently transmitted to a vertebrate host through saliva while the tick feeds. It usually takes several hours (between 6 and 10) of attachment and feeding before *Ri. rickettsii* is transmitted to the host. After an immature tick develops into the next stage, *Ri. rickettsii* may be transmitted to a second host during the feeding process. This bacterium may also be transmitted to a vertebrae host through contact with infected tick hemolymph or excrement when engorged ticks are crushed.

There are two major Ixodidae tick vectors of *Ri. rickettsii* in the United States: *Dermacentor variabilis* and *De. andersoni*. *De. variabilis* is found east of the Rocky Mountains and in limited areas on the Pacific Coast (Figure 4-23 and Figure 4-2). Dogs and medium-sized mammals are the preferred hosts of adult *De. variabilis* ticks. *De. variabilis* also feeds on other large mammals (including humans) and is the tick most commonly responsible for transmitting *Ri. rickettsii* to humans. *De. andersoni* is found in the Rocky Mountain states and in southwestern Canada. Adult ticks feed primarily on large animals, whereas larvae and nymphs feed on small rodents. The life cycle of this tick may require up to 2 to 3 years for completion.

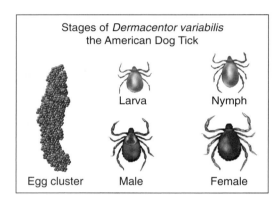

Figure 4-23 The growth stages of the "American dog tick," *Dermacentor variabilis*, from eggs to adult insects.

The ticks *Rhipicephalus sanguineus* (in Mexico) (Figure 4-1) and *Amblyomma cajennense* (in Central and South America) have been shown to be naturally and experimentally infected with *Ri. rickettsii*. Although these species play only a minor role in the transmission of *Ri. rickettsii* in the United States, they are important vectors of RMSF in Central and South America.

Ticks become infected with *Ri. rickettsii* by two methods. One way ticks acquire the bacterium is by feeding on infective small mammals reservoirs such as chipmunks and squirrels. Dogs and humans may also serve as reservoirs for RMSF; however, they are incidental hosts and are the only reservoirs that show clinical signs. The larva and nymph forms of *Dermacentor* ticks feed on small mammals, whereas the adult ticks feed on larger mammalian hosts. Larger mammals rarely achieve the level of organisms in blood necessary to transmit disease to a feeding tick; therefore, it is the larva and nymph stages that are frequently infected with *Ri. rickettsii* during feeding on small mammals.

The second way ticks can become infected with *Ri. rickettsii* is via other ticks. Transstadial spread of RMSF occurs through the transfer of bodily fluids or spermatozoa during mating from one tick to another. Transovarial spread of RMSF occurs from the pregnant female tick to her eggs. Transovarial infection is the primary means by which *Ri. rickettsii* is spread in nature.

Pathogenesis

Ri. rickettsii enter the skin typically through a tick bite and undergoes a 3- to 14-day (usually 7 day) incubation period in which the organism replicates. Following the incubation period, bacteria spread via lymphatics to the bloodstream and attach to endothelial cells of venules and capillaries and begin replicating. This bacterial replication causes vasculitis and increased vascular permeability. Fluid moves to the interstitial spaces leading to edema (typically in the extremities including the scrotum, prepuce, and ears), hemorrhage, hypovolemia, shock, and vascular collapse. Severity of the vasculitis can be directly correlated to the infective dose. Vascular endothelial damage contributes to development of petechiae and ecchymotic hemorrhages as a result of the destruction of platelets in response to vasculitis. Petechial hemorrhages are often seen on exposed mucosal surfaces in the dog. Organ damage, secondary to vascular collapse, is common in the brain, skin, heart, and kidneys. Vascular leakage also triggers activation of the animal's platelets and coagulation system. In skin, vascular injury initially appears as erythematous (red) macules that are usually 1 to 5 mm in diameter and progress to the classic petechial rash of RMSF. Because of damage to the vascular system RMSF is a multisystem disease.

> Once infected, a tick can carry *Ri. rickettsii* for life.

Clinical Signs in Animals

Ri. rickettsii causes disease in dogs. RMSF, also known as tick fever in dogs, is usually seen in dogs younger than 3 years old with a recent history of exposure to ticks or their habitat. RMSF is usually reported in dogs between the months of March and October when there is an increased prevalence of ticks in the environment. Early signs may include fever (up to 105°F), anorexia, lymphadenopathy, polyarthritis, coughing or dyspnea, abdominal pain, and edema of the face or extremities. In severe cases petechial hemorrhages may be seen on the mucous membranes. Neurologic signs, such as altered mental states, vestibular dysfunction, and hyperesthesia, are commonly seen with RMSF. Focal retinal hemorrhage is usually seen in the early stages of this disease.

> Dogs are susceptible to RMSF and serve as excellent sentinels of the disease.

Clinical Signs in Humans

RMSF in humans typically presents with three classic signs: fever, rash, and history of tick bite. Initial clincial signs may include fever, nausea, vomiting, severe headache, muscle pain, and lack of appetite. The rash appears 2 to 5 days after the onset of fever and is often very subtle appearing as small, flat, pink, nonitchy macules (spots) on the wrists, forearms, and ankles (Figure 4-24). These macules turn pale when pressure is applied and eventually become raised on the skin. Younger people usually develop the rash earlier than older people. Clinical signs that appear later in the disease include rash, abdominal pain, joint pain, and diarrhea. The classic red, petechial rash of RMSF is usually not seen until the sixth day or later after disease onset and occurs in only 40% to 60% of patients. The rash usually involves the palms or soles.

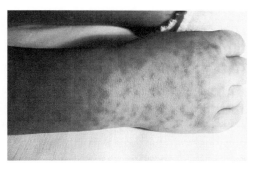

Figure 4-24 The characteristic spotted rash of Rocky Mountain spotted fever. (Courtesy of CDC)

Diagnosis in Animals

Laboratory abnormalities seen in dogs with RMSF include thrombocytopenia (platelet counts ranging from 23,000 to 220,000/µl); a moderate leukocytosis with a mild left shift (a mild leucopenia occurs early in infection); normocytic, normochromic anemia; azotemia, elevated glucose, cholesterol, alkaline phosphatase (ALP) and alanine aminotransferase (ALT); and hyponatremia, hypocalcemia, and hypoalbuminemia (secondary to vasculitis). Definitive diagnosis of RMSF is through IFA serologic testing along with clinical signs. IFA can be used to detect either IgG or IgM antibodies. Blood samples taken early (known as the acute sample) and late (known as the convalescent sample) in the disease are the preferred specimens for evaluation. IgG antibody titers to *Ri. rickettsii* that increase or decrease fourfold are considered diagnostic for RMSF but may not be clinically useful because IgG antibody concentration does not increase until 2 to 3 weeks postinfection. A single high IgG titer (>1024) is suggestive of exposure within the last tick season, whereas a single positive IgM titer (>8) indicates a more recent exposure. Positive IgG titers may persist for 3 to 10 months following infection; however, positive IgM titers normally decrease after 4 weeks. Cross-reactivity to other spotted fever *Rickettsiae* exist and may affect test interpretation.

Diagnosis in Humans

Laboratory abnormalities suggestive of RMSF in humans may include abnormal white blood cell counts, thrombocytopenia, hyponatremia, or elevated liver enzyme levels. Serologic assays are most frequently used for confirming cases of RMSF, which include IFA, ELISA, latex agglutination, and immunoassays. In humans increased IgM titers appear by the end of the first week of illness and diagnostic levels of IgG antibody do not appear until 7 to 10 days after the onset of illness. The most rapid and specific diagnostic assays are PCR tests, which can detect DNA present in as few as 5 to 10 bacteria in a sample. PCR testing is done on fresh skin biopsies or fixed or unfixed tissues samples. Diagnosis can be confirmed by isolation of *Ri. rickettsii* from clinical samples such as whole blood and biopsies. Isolation may require several weeks and samples should be shipped unfrozen or frozen and on dry ice to the CDC. Immunostaining is another method used to identify *Ri. rickettsii* from a skin biopsy of the rash from an infected person prior to therapy, but may not always detect the bacterium as a result of its focally distributed lesions.

Treatment in Animals

In dogs, antibiotic treatment is initiated immediately after samples are taken for laboratory testing to reduce the disease signs. Tetracycle or doxycycline is the treatment of choice with chloramphenicol recommended in pregnant bitches and puppies younger than 6 months of age to avoid dental staining in growing fetuses/puppies. Supportive care should be considered along with antibiotic administration; however, fluid therapy should be administered conservatively as a result of the vasculitis and potential for pulmonary and cerebral edema. Mild ocular lesions should resolve with systemic antibiotic therapy and the use of topical corticosteroids may help conditions such as uveitis. Dogs that have recovered from RMSF have protective immunity to further reinfection.

Treatment in Humans

Treatment of RMSF in humans involves the use of antibiotics such as doxycycline for at least 3 days after fever subsides. Tetracycline and chloramphenicol are alternative antibiotics used to treat RMSF; however, they are associated with side effects that limit their use.

Management and Control in Animals

The best way to prevent dogs from contracting RMSF is to limit their tick exposure particularly between the months of March through October. Dogs should be inspected daily for ticks and any ticks that are found should be removed quickly and safely with a gloved hand. Topical agents (such as fiprinol or permethrin) and tick collars containing amitraz are effective methods of tick control. There is not a vaccine for protection against *Ri. rickettsii*.

Management and Control in Humans

The best way to prevent RMSF in people also includes tick control. Strategies to reduce ticks include area-wide application of acaricides (chemicals that will kill ticks and mites), application of tick repellent with DEET, and control of tick habitats. Prompt removal of ticks is also essential. Tick control has been covered in a previous section and should be referred to.

Summary

RMSF is a clinical disease of humans and dogs (with small mammals occasionally infected) that is caused by *Ri. rickettsii*. *Ri. rickettsii* is a small, gram-negative bacterium that is spread to humans and dogs by the Ixodidae ticks *De. andersoni* and *De. variabilis*. Clinical signs in dogs include fever, anorexia, lymphadenopathy, polyarthritis, coughing or dyspnea, abdominal pain, edema of the face or extremities, petechial hemorrhages, neurologic signs, and retinal hemorrhage. Clinical signs in people include fever, headache, and muscle pain, followed by development of rash. The disease can be difficult to diagnose in the early stages, and without prompt and appropriate treatment it can be fatal. RMSF is a seasonal disease and occurs throughout the United States during the months of April through September (peak tick times are March to October). Most of the cases occur in the south-Atlantic region of the United States (Delaware, Maryland, Washington D.C., Virginia, West Virginia, North Carolina, South Carolina, Georgia, and Florida) and the highest incidence rates have been found in North Carolina and Oklahoma. RMSF is diagnosed based on clinical signs and serologic testing such as IFA. Treatment involves the use of antibiotics such as doxycycline. Once a person or dog clears the infection it is believed that they have long lasting immunity to *Ri. rickettsii*. The disease is prevented by controlling ticks.

Review Questions

Multiple Choice

1. The tick-borne disease that manifests initially as erythema migrans and later as chronic arthritis is
   a. Rocky Mountain spotted fever.
   b. Lyme disease.
   c. relapsing fever.
   d. ehrlichiosis.

2. What is not a characteristic of the Rickettsiae
   a. obligate intracellular organisms.
   b. transmitted by arthropods.
   c. gram-negative, pleomorphic bacilli.
   d. multiply extracellularly.

3. What disease can be transmitted by aerosol inhalation?
   a. Q fever
   b. tularemia
   c. Rocky Mountain spotted fever
   d. Lyme disease

4. A 19-year-old female is admitted to a local hospital with fever, chills, headache, and a rash on her palms and soles. The woman states that she has recently been bitten by a tick. The physician has ruled out babesiosis and Lyme disease based on laboratory tests. The probable cause of her symptoms is infection with
   a. *Coxiella burnetii.*
   b. *Rickettsia rickettsii.*
   c. *Ehrlichia canis.*
   d. *Borrelia burgdorferi.*

5. What is false regarding ticks?
   a. They have long life cycles.
   b. They consume large volumes of blood.
   c. They produce large numbers of eggs.
   d. They have three body regions: capitulum, idiosoma, and scutum.

6. All ticks undergo which basic stages?
   a. egg, larva, nymph, adult
   b. egg, nymph, adult
   c. egg, larva, adult
   d. egg, larva, nymph, instar

7. The process by which ticks crawl up a piece of grass or perch on leaf edges with their front legs extended is called
   a. perching.
   b. questing.
   c. trolling.
   d. engorging.

8. Transfer of an infectious agent from one tick life stage through molting to the next stage is called
   a. horizontal transmission.
   b. vertical transmission.
   c. transovarial transmission.
   d. transstadial transmission.

9. A common target in most rickettsioses is the
   a. liver.
   b. nervous system.
   c. endothelial lining of small blood vessels.
   d. lymphatic channels.

10. Bacteria that cause ehrlichiosis bind to and are named for the type of cell they infect. These cells are
    a. erythrocytes.
    b. thrombocytes.
    c. leukocytes.
    d. monocytes.

Matching

11. _____ Rocky Mountain spotted fever   A. Argasidae

12. _____ Q fever   B. *Francisella tularensis*

13. _____ Lyme disease   C. *Rickettsia rickettsii*

14. _____ TBRF (endemic relapsing fever)   D. *Anaplasma phagocytophilum*

15. _____ LBRF (epidemic relapsing fever)   E. *Ehrlichia chaffeensis*

16. _____ Tularemia   F. *Coxiella burnetii*

17. _____ HME   G. *Borrelia recurrentis*

18. _____ HGA   H. Ixodidae

19. _____ hard tick   I. *Borrelia hermsi*

20. _____ soft tick   J. *Borrelia burgdorferi*

Case Studies

21. A 41-year-old man was admitted to the hospital complaining of severe headache, moderate fever, chest pain, and a productive cough. Swollen lymph nodes and a tender, enlarged liver were noted on the examination. This man is a professional furrier and trapper and had recently returned from an excursion on which he had trapped and skinned approximately 50 rabbits. Routine sputum and blood cultures were collected and revealed very faintly staining gram-negative bacilli on Gram stain and no growth on routine bacteriological media (blood agar and MacConkey) after 72 hours. After 6 days, growth was observed on chocolate agar plates.
    a. Given this person's history and symptoms, what disease might he have?
    b. What organism causes this disease?
    c. Why did the organism grow on chocolate agar (what chemical does this organism need for growth)?
    d. What special precautions need to be taken when handling this organism?

22. Almost 2 weeks after returning from a camping trip in the Grand Canyon, a 50-year-old man developed fever, chills, headache, muscle pain, and profuse sweating. These symptoms typically lasted for 2 days. Over the next 2 weeks he experienced three febrile relapses and was hospitalized. Physical examination and laboratory tests did not conclusively lead to a diagnosis. While in the hospital the patient had a fourth episode of fever during which time a

peripheral blood sample was taken and examined. Spirochetes were observed and although the patient did not remember a tick bite, he was treated with tetracycline and recovered.

    a. What disease did this patient most likely have?

    b. What is the causative agent of this disease?

    c. What is the vector of endemic relapsing fever?

    d. What is the vector of epidemic relapsing fever?

23. A 3-year-old male Coonhound presented to the clinic with an acute fever (T=104°F), anorexia, and lameness. Physical examination reveals a swollen left rear hock.

    a. What questions would you want to ask this owner when taking the animal's history?

    b. What test would you recommend for this dog?

    c. If this test comes back positive, what would be used to treat this dog?

    d. What preventative measures could this owner take to prevent this disease?

References

Adams, D. R., B. E. Anderson, C. T. Ammirati, and K. F. Helm. 2003. Identification and diseases of common U.S. ticks. *The Internet Journal of Dermatology.* http://www.ispub.com/ostia/index.php?xmlFilePath=journals/ijd/vol2n1/tick.xml (accessed April 15, 2004).

AVMA. 2003. Tularemia facts. http://www.avma.org (accessed June 27, 2003).

Barbour, A. 2005. Relapsing Fever: Tick-borne Diseases of humans. In *Tick-Borne Diseases of Humans*, edited by J. Goodman, D. Dennis, and D. Sonenshine, pp. 268–91. Washington, DC: ASM Press.

Biddle, W. 2002. *A Field Guide to Germs.* New York, NY: Anchor Books.

Burgdorfer, W. 1993. How the discovery of *Borrelia burgdorferi* came about. *Clinical Dermatology* 11(3):335–8.

Centers for Disease Control and Prevention. 2000. Human ehrlichiosis in the United States. http://www.cdc.gov/Ncidod/dvrd/ehrlichia/Index.htm (accessed May 20, 2005).

Centers for Disease Control and Prevention. 2005. Tularemia. http://www.cdc.gov/ncidod/dvbid/tularemia.htm (accessed May 1, 2005).

Centers for Disease Control and Prevention. 2006. Rickettsial infections. http://www2.ncid.cdc.gov/travel/yb/utils/ybGet.asp?section=dis&obj=rickettsial.htm (accessed January 5, 2006).

Centers for Disease Control and Prevention. 2005. Rocky Mountain spotted fever. http://www.cdc.gov/ncidod/dvrd/rmsf/index.htm (accessed May 20, 2005).

Centers for Disease Control and Prevention. 2004. Relapsing fever http://www.cdc.gov/ncidod/dvbid/RelapsingFever/index.htm (accessed March 15, 2005).

Centers for Disease Control and Prevention. 2003. Q fever. http://www.cdc.gov/ncidod/dvrd/qfever/ (accessed May 20, 2005).

Center for Food Security and Public Health, Iowa State University. 2005. Ehrlichiosis. http://www.cfsph.iastate.edu (accessed May 1, 2005).

Chomel, B., D. Behymer, and H. Riemann. *Coxiella burnetii* infection (Q fever). In *Zoonosis Updates*, 2nd ed. Schaumburg, IL: American Veterinary Medical Association.

Dawson, J., S. Ewing, W. Davidson, J. Childs, S. Little, and S. Standaert. 2005. Human monocytic ehrlichiosis. In *Tick-Borne Diseases of Humans*, edited by

J. Goodman, D. Dennis, and D. Sonenshine, pp. 239–57. Washington, DC: ASM Press.

Edlow, J. 2005. Tick-borne diseases: relapsing fever. http://www.emedicine.com/EMERG/topic588.htm (accessed May 16, 2005).

Edlow, J. 2005. Tick-borne diseases: tularemia. http://www.emedicine.com (accessed April 25, 2005).

Forbes, B., D. Sahm, A. Weissfeld. 2002. *Bailey & Scott's Diagnostic Microbiology*, 11th ed. St. Louis, MO: Mosby.

Goodman, J. 2005. Human Granulocytic Anaplasmosis (Ehrichiosis). In *Tick-Borne Diseases of Humans*, edited by J. Goodman, D. Dennis, and D. Sonenshine, pp. 218–38. Washington, DC: ASM Press.

Hayes, E. 2005. Tularemia. In *Tick-Borne Diseases of Humans*, edited by J. Goodman, D. Dennis, and D. Sonenshine, pp. 207–17. Washington, DC: ASM Press.

Ingraham, J., and C. Ingraham. 2004. *Introduction to Microbiology: A Case History Approach*, 3rd ed. Belmont, CA: Thomson Brooks/Cole.

Institutional Animal Care and Use Committee. 1996. Zoonotic diseases. http://research.ucsb.edu/connect/pro/disease.html (accessed April 25, 2005).

Jones, T. C., R. D. Hunt, and N. W. King. 1997. *Veterinary Pathology*, 6th ed. Philadelphia, PA: Williams & Wilkins.

Krauss, H., A. Weber, M. Appel, B. Enders, H. D. Isenberg, H. G. Schiefer, W. Slenczka, A. von Graeventiz, and H. Zahner. 2003. Rickettsioses. In *Zoonoses: Infectious Diseases Transmissible from Animals to Humans*, 3rd ed., pp. 221–33. Washington, DC: ASM Press.

Levy, S. 2002. Use of a C_6 ELISA test to evaluate the efficacy of a whole-cell bacterin for the prevention of naturally transmitted canine *Borrelia burgdorferi* infection. *Veterinary Therapeutics* 3(4):420–4.

Levy, S., S. Barthold, D. Dombach, and T. Wasmoen. 1993. Canine lyme borreliosis. *The Compendium* 15(6):833–47.

Levy, S., K. Clark, and L. Glickman. 2005. Infection rates in dogs vaccinated and not vaccinated with an OspA *Borrelia burgdorferi* vaccine in a Lyme disease-endemic area of Connecticut. *International Journal Applied Research Veterinary Medicine* 3(1):1–5.

Macaluso, K. and A. Azad. 2005a. Rocky Mountain spotted fever and other spotted fever group riskettsioses. In *Tick-Borne Diseases of Humans*, edited by J. Goodman, D. Dennis, and D. Sonenshine, pp. 258–67. Washington, DC: ASM Press.

Macaluso, K., and A. Azad. 2005b. Rocky Mountain spotted fever and other spotted fever group riskettsioses. In *Tick-Borne Diseases of Humans*, edited by J. Goodman, D. Dennis, and D. Sonenshine, pp. 292–301. Washington, DC: ASM Press.

Merck Veterinary Manual. 2006. *Ixodes* spp. http://www.merckvetmanual.com/mvm/index.jsp?cfile=htm/bc/72112.htm&word=tick (accessed April 26, 2005).

Merck Veterinary Manual. 2006. Tickborne fever. http://www.merckvetmanual.com/mvm/index.jsp?cfile=htm/bc/56400.htm&word=tick (accessed March 1, 2006).

Merck Veterinary Manual. 2006. Tick Control. http://www.merckvetmanual.com/mvm/index.jsp?cfile=htm/bc/72120.htm&word=tick (accessed March 1, 2006).

National Center for Animal Health Programs. 2002. Heartwater. http://www.aphis.usda.gov/lpa/pubs/fsheet_faq_notice/fs_ahheartw.html (accessed May 28, 2005).

National Institutes of Health. 2005. Lyme disease. http://www.nlm.nih.gov/medlineplus/lymedisease.html (accessed April 17, 2005).

Nochimson, G. 2004. CBRNE-Q fever. http://www.emedicine.com/emerg/topic492.htm (accessed June 28, 2004).

Nochimson, G. 2005. Tick-borne diseases: ehrlichiosis. http://www.emedicine.com/EMERG/topic159.htm (accessed November 30, 2005).

Office of Environmental Health and Safety, Washington State Department of Health. 2005. Ticks and tick-borne diseases. http://www.doh.wa.gov/ehp/ts/Zoo/WATickDiseases.htm (accessed October 7, 2005).

Olson, C. 2004. Zoonosis diseases tutorial: Ehrlichiosis. http://www.vetmed.wisc.edu/pbs/zoonoses/Ehrlichia/ehrlindex.html (accessed February 18, 2005).

Paddock, C., A. Liddell, and G. Storch. 2005. Other causes of tick-borne ehrlichioses, including *Ehrlichia ewingii*. In *Tick-Borne Diseases of Humans*, edited by J. Goodman, D. Dennis, and D. Sonenshine, pp. 258–67. Washington, DC: ASM Press.

Porcella, S., S. J. Raffel, M. E. Scrumpf, M. E. Schriefer, D. T. Dennis, and T. G. Schwan. 2000. Serodiagnosis of louse-borne relapsing fever with glycerophosphodiester phosphodiesterase (GlpQ) from *Borrelia recurrentis*. *Journal of Clinical Microbiology* 38(10):3561–71.

Roberts, L., and J. Janovy. 2005. Parasitic Arachnids: Subclass Acari, Ticks and Mites. In *Foundations of Parasitology*, 7th ed., pp. 637–46. New York, NY: McGraw-Hill.

Schwan, T. G., P. F. Policastro, Z. Miller, R. L. Thompson, T. Damrow, and J. E. Keirans. 2003. Tick-borne relapsing fever caused by *Borrelia hermsii*. *Emerging Infectious Disease* 9(9):1151–4.

Stiles, J. 2000. Rocky Mountain spotted fever. *Veterinary Clinics of North America Small Animal Practice* 30:1144–8.

Synder, R., and M. Spevak. 2006. Rocky Mountain spotted fever. http://www.emedicine.com/med/topic2043.htm (accessed April 6, 2006).

Thompson, H., D. Dennis, and G. Dasch. 2005. Q fever. In *Tick-Borne Diseases of Humans*, edited by J. Goodman, D. Dennis, and D. Sonenshine, pp. 328–43. Washington, DC: ASM Press.

Tidy, C. 2004. Ehrlichiosis. http://www.patient.co.uk/showdoc/40000062/ (accessed November 24, 2004).

U.S. Army Center for Health Promotion and Preventative Medicine, Entomological Sciences Program. 1994. Ehrlichiosis. http://chppm-www.apgea.army.mil/ento/erlichio.htm (accessed April 26, 2005).

VetMedTeam. 2005. Ticks from A to Z. MerialEdu, a continuing education workbook for veterinarians and clinic staff. http://www.vetmedteam.com (accessed July 15, 2005).

Vredovoe, L. 2005. Background information on the biology of ticks. http://entomology.ucdavis.edu/faculty/rbkimsey/tickbio.html (accessed June 20, 2005).

Warner, R. D., W. W. Marsh 2002. Rocky Mountain spotted fever. *Journal of Veterinary Internal Medicine* 10:1413–17.

Chapter 5

Fungal Zoonoses

Objectives

After completing this chapter, the learner should be able to

- Describe properties unique to fungi
- Identify the appearance of fungi microscopically and by culture methods
- Identify the different types of fungal groups and describe the properties of these groups
- Identify the appearance of fungi microscopically and by growth patterns
- Briefly describe the history of specific fungi zoonoses
- Describe the causative agent of specific fungal zoonoses
- Identify the geographic distribution of specific fungal zoonoses
- Describe the transmission, clinical signs, and diagnostic procedures of specific fungal zoonoses
- Describe methods of controlling fungal zoonoses
- Describe protective measures professionals can take to prevent transmission of fungal zoonoses

Key Terms

anthropophilic	geophilic	nonseptate	septate
budding	heterotrophic	opportunistic mycoses	superficial mycoses
dimorphic fungi	hyphae	opportunistic pathogen	systemic mycoses
ectothrix	macroconidia		true pathogen
endothrix	microconidia	pseudohyphae	yeast
fermentation	mold	saprophytes	zoophilic
fungi	mycelium		

OVERVIEW

The Kingdom Fungi includes some of the most important organisms, both in terms of their ecological and economic roles, found on earth. **Fungi** perform a variety of different functions, such as breaking down dead organic material, providing the roots of plants with essential nutrients, providing the medical communities with numerous drugs (such as penicillin and other antibiotics), providing foods such as mushrooms, truffles and morels, and supplying the bubbles in bread, champagne, and beer. Macroscopic fungi include mushrooms and puffballs, whereas microscopic fungi include molds and yeasts. About 70,000 species of fungi have been described; however, the medical community focuses on molds and yeast because they can produce a number of diseases including ringworm, yeast infections, and several systemic diseases. Fungi are eukaryotic; therefore, fungi are more chemically and genetically similar to animals than other organisms making them more challenging to treat. Fungi are nonmotile and **heterotrophic** (cannot make their own food).

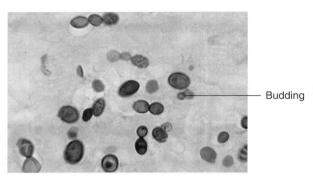

Figure 5-1 Yeast cells undergoing replication by budding.
(Courtesy of CDC/Dr. Libero Ajello)

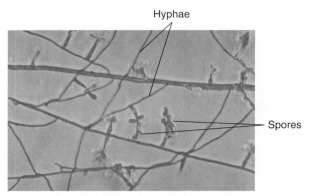

Figure 5-2 A photomicrograph of mold showing hyphae and spores.
(Courtesy of CDC/Dr. Libero Ajello)

TYPES OF FUNGI

Fungi are classified into a variety of categories based on the way in which the fungus reproduces and the presence or absence of branching filaments called **hyphae.** Some fungi exist only in a yeast form; others exist primarily as hyphae; a few are dimorphic.

- **Yeasts** are unicellular fungi with round to oval cells that reproduce asexually by giving off small cells called buds (**budding**) (Figure 5-1). Some yeast may form **pseudohyphae** (a string of yeast cells that have not separated after each new bud is formed).
- **Molds** or filamentous fungi are multicellular and form long, tubular chains of cells called hyphae (Figure 5-2).
- **Dimorphic fungi** can exist as yeast or mold depending upon the growth conditions.

WHAT DO FUNGI LOOK LIKE TO THE NAKED EYE?

Fungi are routinely grown on Sabouraud Dextrose Agar (SDA), which contains starch and has an acidic pH favored by fungi, but there are other specialized agars for fungal growth as well. When growing fungi, their appearance on culture media varies depending upon whether they are a yeast or mold. Yeast grow in colonies on agar much like typical bacterial colonies (Figure 5-3). Depending upon the yeast the colonies may appear mucoid (especially yeast cells that have capsules). Some commercial agars allow particular yeast to grow particular colors, aiding in their identification.

Molds grow as fuzzy colonies. Some molds grow particular colors or may be one color growing on the surface of the plate and another color when viewing the agar from the underside (called the reverse growth) (Figure 5-4).

WHAT DO FUNGI LOOK LIKE UNDER THE MICROSCOPE?

Molds

Molds are eukaryotic and multicellular organisms. Another feature of mold is the presence of chitin in their cell walls (not cellulose as in plants). Chitin is a long

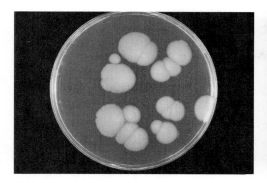

Figure 5-3 Yeast colonies on Sabouraud Dextrose Agar (SDA).
(Courtesy of CDC/Dr. William Kaplan)

Figure 5-4 Mold colonies on SDA.
(Courtesy of CDC/Dr. Lucille K. George)

carbohydrate polymer that also occurs in the exoskeletons of insects, spiders, and other arthropods. Chitin adds rigidity and structural support to the thin cells of the fungus.

Molds are composed of filaments called hyphae; their cells are long and thread-like and connected end-to-end. Hyphae twist and tangle together to form a fuzzy mass called **mycelium**, a term which is applied to the mass of hyphae. Hyphae may be **nonseptate** (consists of one long, continuous cell not divided into individual compartments) or **septate** (consists of cross walls that divide the hyphae into individual compartments) (Figure 5-5). Nonseptate hyphae allow the cytoplasm and organelles to move freely from one region to the next and each hypha may contain many nuclei. Septate hyphae may have solid partitions with no communication between compartments or they may have small pores in the cross walls to allow some degree of communication between compartments.

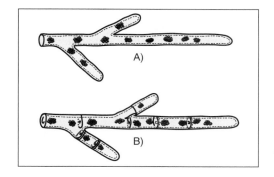

Figure 5-5 Types of hyphae;
A) nonseptate hyphae,
B) septate hyphae.

Hyphae can be categorized by their function. Vegetative hyphae are responsible for the fuzzy appearance of mold. Vegetative hyphae also penetrate into agar or substrate to digest and absorb nutrients. Vegetative hyphae give rise to aerial hyphae. Aerial hyphae, also known as reproductive hyphae, form the reproductive spores of the fungus.

Yeast

Microscopically, yeast cells are larger than bacteria and are round or oval in shape (Figure 5-6). Budding may be seen in yeast that are reproducing asexually; however, some yeast produce sexual spores. Nuclei may be seen in yeast cells because they are eukaryotic. Some yeast are encapsulated.

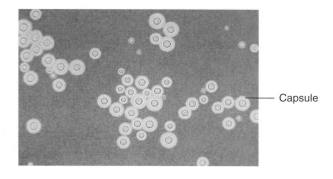

— Capsule

Figure 5-6 Photomicrograph of yeast cells with capsule using India Ink technique.
(Courtesy of CDC/Dr. Leanor Haley)

HOW DO FUNGI REPRODUCE?

Molds

Molds have a complex reproductive cycle that may be either sexual or asexual. In either case, the important feature of mold reproduction is a single cell known as a spore. In contrast to bacterial endospores (commonly called spores) that are used as a resistance mechanism, fungal spores are used for reproduction. A single spore is capable of regenerating the entire mycelium of a fungal mold.

Asexual spores are produced by mitosis from a single parent cell. Asexual reproduction in mold occurs at the tip of aerial hyphae. Daughter cells develop into spores that move with wind currents and land at distant locations. The spore then germinates and new cells emerge to form filaments and reproduce the mold. Asexual reproduction occurs without a union of sexual cells, and asexually produced spores are generally very numerous. All the spores are genetically identical.

A mold can produce a variety of asexual spores. An arthrospore is a spore formed by fragmentation of the tip of the hyphae, whereas a blastospore is produced as an outgrowth along a septate hypha. Conidiospores are unprotected spores formed by mitosis at the tips of hyphae, and sporangiospores are spores produced within a sac called a sporangium.

Although mold spends much of its time in the asexual phase of its life cycle, some types of mold produce sexual spores. Sexual spores are formed through a process of fusion of parental nuclei, followed by meiosis. Most molds that cause disease in animals and humans do not reproduce sexually.

Yeast

Budding is a type of asexual reproduction that occurs in yeast cells. The yeast cell undergoes mitosis and one daughter nucleus is isolated to one part of the cell. This nucleus is in a small amount of cytoplasm and becomes divided from the parent cell by the formation of a new wall. This cell is called a bud and it increases in size and eventually separates from the parent cell. In some species such as *Candida albicans*, a series of buds remain attached to one another and to the parent cell to form long filaments called pseudohyphae.

HOW DO FUNGI OBTAIN NUTRIENTS?

Fungi are not able to ingest their food nor can they manufacture their own food. Instead, fungi feed by absorption of nutrients from the surrounding environment. Molds accomplish this by growing their vegetative hyphae through and within the substrate on which they are feeding. Many yeast are facultative anaerobes (organisms that use oxygen to grow when it is available but can also grow without oxygen) that obtain energy from **fermentation** (energy obtained through anaerobic degradation of substrates into simpler metabolites). Fermentation in yeast typically involves their use of sugar as an energy source.

Some fungi are **saprophytes** (feeding on dead or decaying material) and some fungi are parasitic (feeding on living organisms without killing them). Ringworm is a disease caused by parasitic fungi.

HOW DO FUNGI CAUSE INFECTION?

Most fungi are free-living and do not require a host to complete their life cycles. Most human and animal infection occurs through accidental contact from an environmental source such as soil, water, or dust. Humans and animals are generally tolerant of fungi except for two main kinds of pathogens: **true pathogens** (those that can infect healthy people and animals) and **opportunistic pathogens** (those organisms present in low numbers in people or animals that cause disease when the host environment is altered).

Fungal diseases are described based on the location of the infection they cause. Categories of fungal diseases include **superficial mycoses** (those located on the outer surface of hair, nails, and skin known as dermatophytoses caused by such dermatophytes as *Trichophyton* and *Microsporum*, the fungi that cause ringworm and those located subcutaneously such as *Fonsecaea*, the fungi that causes chromblastomycosis), **systemic mycoses** (those that invade body tissues and tend to be dimorphic fungi such as *Blastomyces*), and **opportunistic mycoses** (those fungi such as *Candida* that are normally harmless, but can cause disease in a compromised/altered host).

DERMATOPHYTOSES

Overview

Dermatophytoses are infections caused by dermatophytes (which literally mean plants that live on skin) but are actually fungi that only grow on skin, hair, and nails. Dermatophytoses are commonly referred to as ringworm as a result of its classic circular, scaly patches (ring) that resemble a worm lying below the skin surface (worm). The fungi feed on dead skin and hair cells causing a round, red lesion with a ring of scale around the edges and normal skin in the center. The characteristic ring is typically seen in humans; ringworm in animals frequently appears as a dry, grey, scaly patch that looks like many other skin lesions.

The Greeks referred to dermatophytoses as herpes (to creep around) and the Romans thought the disease resembled the larval stage of the worm *Tinea* (which is Latin for worm). The actual cause of ringworm was not discovered until the 1800s; however, people understood that it was transmitted from person-to-person because infected individuals were isolated from uninfected people. In 1841, a Hungarian Jew living in Paris named David Gruby demonstrated for the first time that an infection of the scalp called favus was caused by a fungus. *Favus* is Latin for honeycomb and this disease is characterized by thick yellow honeycomb-like crusts over the hair follicles. The following year Gruby described *Trichophyton ectothrix*, another fungi, found at the roots of a man's beard. Shortly afterward he discovered *Oidium albicans* (*Monilia albicans*, which is now known as *Candida albicans*), the cause of thrush in infants. In 1843, Gruby described another fungus, *Microsporum audouini*, which he named in honor of Jean Victor Audouin, a French entomologist. In humans, *Microsporum* causes a form of tinea (ringworm) that is also called microsporia or Gruby's disease. Gruby's research in these areas had mostly been ignored during his time, possibly a result of strong anti-Semitic feelings. It was only slowly realized that in the short period of his scientific activity Gruby had made very original and important contributions to science and is now known as the father of medical microbiology.

Table 5-1 Dermatophytes and the Diseases They Cause

Fungus	Disease Name	Target Tissue	Transmission
Microsporum	▪ Tinea capitis (ringworm of scalp) ▪ Tinea corporis (ringworm of body)	▪ Scalp hair ▪ Skin, not nails	*Microsporum canis* is spread through contact with dogs, cats, horses, pigs, sheep, rabbits, hamsters, rats, and zoo animals *Microsporum canis* is the most common dermatophyte seen in humans
Trichophyton	▪ Tinea capitis (ringworm of scalp) ▪ Tinea corporis (ringworm of body) ▪ Tinea barbae (ringworm of the beard) ▪ Tinea unguium (ringworm of the nail; also known as onychomycosis) ▪ Tinea pedis (athlete's foot)	▪ Scalp hair ▪ Skin, not nails ▪ Beard ▪ Nails ▪ Feet	*Trichophyton mentagrophytes, T. verrucosum, T. equinum, T. quinckeanum,* and *T. erinacei* are the only zoonotic species of the 26 *Trichophyton spp.* *Trichophyton rubrum* is the most common dermatophyte in the United States, but is not found in animals
Epidermophyton	▪ Tinea cruris (ringworm of the groin or jock itch) ▪ Tinea unguium (ringworm of the nail)	▪ Groin area ▪ Nail	*Epidermophyton* is not zoonotic; it is only transmitted by person-to-person contact

The spores of *M. canis* cannot penetrate healthy, intact skin. A scratch or fleabite, for example, is needed to start the infection. The fungus invades the hair shafts and top layer of the skin.

Table 5-1 summarizes tinea infections, which are named for the area they infect. Tinea capitis, ringworm of the scalp, was epidemic in North America during World War II. It was believed to be spread by children leaving dermatophytes on movie theater seats followed by another child picking up the fungus from these seats. Some doctors prescribed x-ray treatments to kill the fungus because they believed the rays made the infected hairs fall out (these people later died of radiation toxicity). During the 1940s, U.S. military personnel who were fighting in the South Pacific during World War II contracted ringworm and other fungi in the humid tropics. This led to intensive study of these organisms by the U.S. government. At one time ringworm was a common disease especially in poor children, where it was believed that poor children were not cleaned as often with soap and water as were affluent children. It is now believed that dietary deficiencies and unsanitary conditions both play a key role in the pathogenesis of this disease.

Causative Agent

Dermatophytes are a group of fungi that invade the dead keratin of skin, hair, and nails. Dermatophytes use keratin as a nutrient source; therefore, they only colonize dead layers of skin, hair, and nails. Dermatophytes have the ability to penetrate all layers of skin, but generally stay to the nonliving keratin layer. These fungi do not penetrate beyond the stratum corneum layer as a result of antifungal activity of serum and body fluids and perhaps the decreased tolerance for temperature over 35°C. Three genera of dermatophytes infect humans (*Epidermophyton, Microsporum,* and *Trichophyton*) and only two infect animals (*Microsporum* and *Trichophyton*). Some sources only consider fungi dermatophytes if they cause ringworm in both animals and humans; therefore, only certain species of *Microsporum* and *Trichophyton* would be considered dermatophytes (See Table 5-2).

Dermatophytes are classified based on their natural habitats. These categories are **anthropophilic** (those associated with humans only and are transmitted by close, direct human contact or through contaminated fomites), **zoophilic** (those

Table 5-2   Summary of Veterinary Dermatophytes

Organism	Principle Hosts	Fluorescence	Spore Location	Culture Appearance	Macroconidia	Microconidia
Microsporum canis	Dog, cat, humans, monkey, horse	Positive (50%)	Ectothrix	White to buff; reverse yellow to orange	Many; spindle shaped	Few; attached at the base
Microsporum gypseum	Dog, cat, horse	Negative	Ectothrix	Buff; reverse is orange brown to yellow	Many; elliptical shaped	Few; attached at the base; club-shaped
Microsporum nanum	Swine	Negative	Ectothrix	White to buff; reverse is red	Many; oblong to elliptical shaped	Club-shaped
Trichophyton mentagrophytes	Many animal species	Negative	Ectothrix	Granular, light buff to tan; reverse is variable from red to yellow	Spindle or club-shaped; 5–6 septa (divisions within each macroconidia)	Abundant; ovoid or club-shaped
Trichophyton verrucosum	Cattle, sheep	Negative	Ectothrix	Deeply folded; white to brilliant yellow	Rare; long and thin walled	Abundant; ovoid or club-shaped
Trichophyton gallinae	Fowl	Negative	Ectothrix	Radial folds, white to pale rose; reverse red	Infrequent; club-shaped	Single; club-shaped
Trichophyton equinum	Horse	Negative	Ectothrix	White, cottony, yellow; reverse yellow to red-brown	Rare; club-shaped	Many; spherical

Source: A Concise Review of Veterinary Virology by G. R. Carter, D. L. Wise, and E. F. Flores

associated with animals and are transmitted by close contact with animals or contaminated animal products), and **geophilic** (soil fungi transmitted to humans through direct exposure to soil or to dusty animals).

Dermatophytes are aerobic, nonfastidious organisms that require moisture for growth. Dermatophytes are so closely related and morphologically similar that they are difficult to differentiate; however, different species have unique **macroconidia** (large, multinuclear asexual spore), **microconidia** (small, single-celled asexual spore), and hyphae (filaments). *Trichophyton* spp. produce thin-walled, smooth macroconidia, and numerous microconidia (Figure 5-7); *Microsporum* spp. produce thick-walled, rough macroconidia, and fewer microconidia (Figure 5-8); and

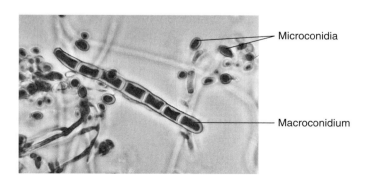

Microconidia

Macroconidium

Figure 5-7   Macroconidium and microconidia of *Trichophyton rubrum*.
(Courtesy of CDC/Dr. Libero Ajello)

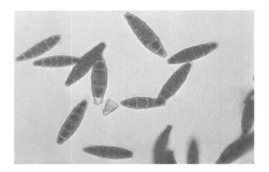

Figure 5-8 Spindle-shaped macroconidia of *Microsporum gypseum.*
(Courtesy of CDC/Dr. Lucille K. Georg)

Epidermophyton infects skin and nails; *Microsporum* infects skin and hair; *Trichophyton* infects skin, hair, and nails.

Humid environments or moist skin provides a favorable environment for establishing a fungal infection.

Epidermophyton spp. produce smooth, ovoid, clustered macroconidia, and no microconidia.

Dermatophytes are keratinophilic and keratinolytic, meaning they digest keratin and utilize it as a nutrient source. These fungi produce keratinases and collageneases that provide nutrients to the fungi by digesting host tissues. Some dermatophytes produce extracellular products that induce inflammation at the infection site.

Epizootiology and Public Health Significance

Microsporum infections occur worldwide with cats being an important transmission source of the most common species *M. canis*. *Trichophyton* infections occur worldwide with a variety of animal transmission sources.

Dermatophyte infections in humans occur worldwide with tinea pedis being the most common type (including in the United States). Tinea capitis is one of the most common infections in children, accounting for up to 92% of dermatophytoses in children younger than 10 years of age. Tinea capitis is rare in adults, but may be found in elderly people. Tinea capitis is widespread in some urban areas in North America, Central America, and South America and is common in parts of Africa and India. In Southeast Asia, the rate of infection has been reported to have decreased dramatically as a result of improved general sanitary conditions and personal hygiene. Onychomycosis (fungal infection of the nail) is a common problem, especially in adults. Onychomycosis accounts for approximately 30% of all cutaneous fungal infections and the prevalence of onychomycosis in the United States is approximately 3% in males and 1.4% in females.

Transmission

Dermatophytes can be transmitted in a variety of ways. One way is through direct contact with an infected person or animal. Other ways include indirect contact with contaminated fomites (such as soil, clothing, shower floors, riding equipment, straw, towels, and bedding), and through air-borne spores. Transmission by lice, fleas, flies, mites, or spiders has not been ruled out.

Pathogenesis

The pathogenicity of dermatophytes depends on its ability to produce enzymes, such as proteinases, collagenase, and keratinases. Dermatophytes invade keratinized tissue and are not able to penetrate the dead cornified layer of the epidermis. Dermatophyte infection may result in a hypersensitivity state with the nature of the lesion depending on the host immune response. Although no living tissue is infected, the fungi can provoke cell-mediated immunity that can damage living tissues. Blister-like lesions may appear on various body parts as a result of hematogenous spread of the fungus or its products.

Initially, the skin is infected with a fungal spore and the dermatophyte begins to grow in its filamentous form within the stratum corneum. The hyphae extend downward, but growth is restricted to the superficial layers because certain nutrients (mainly iron) are not available below the stratum corneum. Lateral expansion of the hyphae within the stratum corneum continues for 10 to 35 days during which time the infected skin appears normal or only very slightly inflamed. After 2 to 3 weeks, the advancing border of the infection may become inflamed producing the typical ringworm lesions.

During fungal colonization the host begins to mount an immune response by becoming sensitized to soluble fungal antigens. The first response is cell-mediated immunity, which is characterized by an intense inflammatory process and produces much of the pathology of these diseases. This inflammation produces clinical signs ranging from erythema (redness) and edema of the dermis and epidermis to the formulation of vesicles and pustules. As skin becomes damaged, oozing and weeping of tissue fluid occurs and hair follicles become inflamed and potentially infected with secondary microbes. The second part of the disease pathogenesis is the host-parasite interaction phase. Several host defense mechanisms such as antibodies or lymphocytes do not appear to minimize fungal growth. One host substance that affects dermatophyte growth and spread is transferin, a serum protein that binds and transports iron. Transferin will diffuse into the stratum corneum and bind iron making it unavailable for fungal utilization.

If the cell-mediated response to the dermatophyte antigen does not develop or is suppressed, the skin will not become sufficiently inflamed to reject the fungus. Therefore, dermatophyte infections can occur in two ways: an acute or inflammatory type of infection, which is associated with cell-mediated immunity to the fungus (which generally heals spontaneously or responds well to treatment) and a chronic or noninflammatory type of infection (which is associated with an inability to mount a cell-mediated response to the fungus at the site of infection resulting in relapses and poor response to treatment).

Hair infection with dermatophytes can occur in two ways. One way is **endothrix** in which fungal growth and spore formation are confined mainly within the hair shaft without formation of external spores. Endothrix infections begin by penetration of the hair, and the organism then grows up the interior main axis of the hair. Examples of fungi that cause endothrix infections are *Trichophyton tonsurans* and *T. violaceum*. The second way is **ectothrix** in which fungal growth occurs within the hair shaft but also produces spores on the outside of the hair. Ectothrix infections begin as in endothrix, but they then extend back out through the hair cuticle (the outer wall of the hair) and form a mass of spores both within and around the hair shaft. Examples of fungi that cause ectothrix infections include *M. canis*, *M. audouinii*, *M. gypseum*, *T. verrucosum*, and *T. mentagrophytes*.

> Dermatophytes do not produce irritants or toxins to cause disease.

Clinical Signs in Animals

Ringworm affects a wide variety of animals with a number of different fungal species causing skin lesions. In dogs, approximately 70% of cases are caused by *M. canis*, 20% by *M. gypseum*, and 10% by *T. mentagrophytes*. Lesions in dogs include alopecia, scaly patches, broken hairs, folliculitis, and pustules. Generalized ringworm in adult dogs is uncommon and usually results from immunodeficiency conditions such as hyperadrenocorticism.

In cats, 98% of ringworm cases are caused by *M. canis*. Lesions in cats are quite variable, with kittens being most frequently affected. Lesions in cats include focal alopecia, scaling, and crusting (mostly around the ears and face or on the extremities). Some cats have no clinical signs but can still serve as a source of infection to other cats or people. Ringworm in cats can also cause feline miliary dermatitis (lesions resemble millet seeds) and these cats are pruritic. Cats with generalized dermatophytosis may develop cutaneous ulcerated nodules, known as dermatophyte granulomas or pseudomycetomas. Ringworm in cats is more frequently seen in indoor-outdoor cats, those in poor health, older cats that may not be able to groom themselves as well, kittens who may not be able to groom their face and ears as well, or those with a compromised immune system.

Dermatophytosis in cattle is usually caused by *T. verrucosum*, but *T. mentagrophytes*, *T. equinum*, *M. gypseum*, *M. nanum*, *M. canis*, and other fungi have occasionally been isolated from cattle. Dermatophytosis in cattle occurs

most frequently in calves with the classic clinical sign of periocular lesions (that are not pruritic). Cows and heifers tend to develop lesions on the chest and limbs, while bulls develop lesions in the dewlap and intermaxillary skin regions. Lesions in cattle include scaling patches of hair loss with gray-white crusts or thick crusts with discharge. Ringworm as a herd health problem is more common in the winter and is more commonly seen in temperate climates (Figure 5-9).

Figure 5-9 Holstein heifer with ringworm.

In horses, ringworm is most commonly caused by *T. equinum* and the disease signs can vary from subclinical or mild disease (crusty lesions and alopecia) to severe lesions such as folliculitis.

In Guinea pigs, ringworm is most often caused by *Tricho. mentagrophytes* and less frequently by *Microsporum* spp. Lesions are more often seen in younger or stressed Guinea pigs, with patchy alopecia (usually starting at the head), crusts, and flakey skin lesions with reddened margins. Facial lesions may occur around the eyes, nose, and ears, but the disease can spread over the back.

In pigs, ringworm is usually caused by *M. nanum*, which produces lesions that appear as rings of inflammation or brown discoloration. Lesions are not pruritic in adult pigs.

In sheep, ringworm is common in lambs (especially show lambs) but is otherwise uncommon in production flocks of sheep. Lesions in lambs are most often seen on the head, but widespread lesions under the wool may be apparent in sheared lambs. The fungi that cause ringworm in sheep include *M. canis*, *M. gypseum*, and *T. verrucosum*. Ringworm is uncommon in goats.

Clinical Signs in Humans

The incubation period of dermatophytoses is uncertain as a result of the insidious onset of infection, but is typically believed to vary from several days to several weeks. Ringworm infections in people are divided into categories based on the body part infected and includes:

■ Tinea capitis, dermatophytosis of the scalp, appears as single or multiple round or ovoid lesions of varying size over the scalp. Hairs will break off

---

Long-haired cats may suffer more from ringworm because their hair protects spores from mechanical removal through grooming.

Sometimes one person in a family or one animal in a herd will get ringworm and it will not spread to others, even though other ringworm infections are highly contagious.

2 to 4 centimeters from the skin and the hair stumps lose their shiny appearance and appear gray. Inflammation is typically absent and if present redness and swelling may occur around lesion margins. The hair of head, eyebrows, and eyelashes may also be involved with tinea capitis. Tinea capitis commonly affects children, mostly in late childhood or adolescence (Figure 5-10).

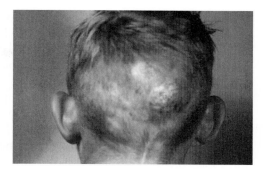

Figure 5-10 Child with tinea capitis (ringworm of the head). (Courtesy of CDC)

- Tinea corporis, dermatophytosis of the body, can present on any area of the body (Figure 5-11). Zoophilic dermatophytes commonly affect exposed areas like the face, neck, and arms, whereas anthropophilic dermatophytes typically affect occluded areas of the skin or areas of trauma. The clinical appearance of tinea corporis is quite variable and can present as circular lesions with active, erythematous (red), spreading borders with central clearing, as an erythematous, scaly rash, or as nodular granulomas.

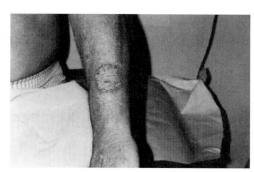

Figure 5-11 Person with tinea corporis, which commonly appears on the extremities. (Courtesy of Lyme Foundation, www.lyme.org)

- Tinea barbae (or barber's itch), dermatophytosis of the bearded area of the face and neck, presents with swelling, crusting, often pruritus, and occasional hair breakage.

- Tinea cruris (or jock itch), dermatophytosis of the groin, tends to produce reddish-brown lesions that extend from the folds of the groin down onto one or both thighs. Tinea cruris may or may not be symmetrical.

- Tinea pedis (or athlete's foot), dermatophytosis of the feet, causes scaling and inflammation in the toe webs (especially the one between the fourth and fifth digits) or thickening or scaling of the skin on the heels and soles.

- Tinea unguium (or onychomycosis), dermatophytosis of the nails, causes the fingernails and, more often, the toenails to appear yellow, thick, and crumbly.

Diagnosis in Animals

Dermatophytoses in animals are diagnosed with Wood's light examination, direct examination of hairs, and fungal culture. A Wood's light emits 253.7 nm of ultraviolet light, which causes the tryptophan metabolites produced by some dermatophytes to emit fluorescent light that appears bright apple green. Only 50% of *M. canis* fluoresce. The light must be used for several minutes before fluorescence will appear and absence of fluorescence does not mean the animal does not have a dermatophyte. Electric Wood's lights are preferred over battery-powered models. There are many false-positive and -negative results with Wood's light examination.

Direct examination of hairs involves examining hair under a microscope using a clearing agent such as potassium hydroxide (KOH). Superficial scraping from the spreading border of the lesion is recommended. The presence of fungal spores indicates an ectothrix fungal infection; however, interpretation is difficult as a result of confusion with melanin granules or saprophytic fungi.

Fungal culture is the definitive test for dermatophytes. Samples are collected by combing a toothbrush over the suspect area of the skin and then gently stabbing the bristles into a fungal culture plate (SDA). The culture plate is examined daily for up to 21 days. Pathogens produce pale or white growth. Cultures are then examined under a microscope to identify the fungus. Dermatophyte test media (DTM) is used to detect dermatophytes via a color change in the media if the fungus is present. Dermatophytes metabolize nutrients in the media to produce alkaline by-products causing the phenol red acid-base indicator in the media to turn red. Dermatophytes produce alkaline by-products as soon as colonies are visible on the media. Dermatophytes will produce the color change within 7 days and should be checked daily. Saprophytic fungi metabolize the nutrients in the media to alkaline byproducts after several weeks (2 to 3 weeks typically). Any growth is then examined under a microscope to diagnose the fungus. It may take up to 3 weeks for macroconidia to form to enable identification of fungal species.

Diagnosis in Humans

Diagnosis of dermatophytoses is the same for humans as in animals. Occasionally a biopsy is needed diagnosis. Biopsies are stained with periodic acid-Schiff (PAS), silver methenamine, or other fungal stains.

Treatment in Animals

Dermatophytoses in dogs and short-haired cats are usually self-limiting, but treatment usually speeds recovery. Some advocate total body hair clipping (especially in cats) as a treatment adjunct for ringworm. Whole-body topical therapy is controversial and is typically done with enilconazole (a rinse not currently available in North America), miconazole shampoo, or lime sulfur dip. Local lesions can be effectively treated with topical miconazole or clotrimazole. Systemic treatment is recommended for chronic or severe cases and for ringworm cases in long-haired breeds of cats. The microsized formulation of griseofulvin can be used in dogs and in cats. Alternative treatments include terbinafine or itraconazole, but neither of these drugs is approved for use in domestic animals. Systemic and topical treatments for dermatophytosis should be continued for 2 to 4 weeks past clinical cure or until a negative toothbrush culture is obtained. The efficacy of lufenuron in treating ringworm in dogs and cats was promising, but is now dropping out of favor with dermatologists. Thorough and repeated vacuuming and scrubbing of surfaces daily is necessary to prevent contamination of the home. Bedding and blankets should be washed daily in hot water and a 1:10 bleach solution. Cleaning heating ducts and vents, installing air filters, dusting with electrostatic dust clothes, and using HEPA filters have been recommended in households with ringworm-positive animals.

Many topical treatments have been reported to be successful in cattle, but their claims of efficacy are difficult to substantiate. Valuable individual animals should still be treated to limit both progression of existing lesions and spread to others in the herd. Thick crusts should be manually removed gently with a brush and the infective material burned or disinfected with hypochlorite solution. Effective agents include washes or sprays of 4% lime sulfur, 0.5% sodium hypochlorite (1:10 household bleach), 0.5% chlorhexidine, 1% povidone-iodine, natamycin, and enilconazole. Individual lesions can be treated with miconazole or clotrimazole lotions. In sheep treatment is application of sodium hypochlorite solutions or enilconazole rinses (where available). Laboratory animals such as guinea pigs are treated for 5 to 6 weeks with oral griseofulvin. Isolated skin lesions may be treated effectively with topical griseofulvin, tolnaftate, or butenafine creams applied daily for 7 to 10 days.

> Dermatophytosis will spontaneously resolve in most healthy cats within 60 to 100 days, but treatment is recommended because the disease is highly contagious.

Treatment in Humans

Treatment of dermatophytoses in humans includes terbinafine, ketoconazole, or griseofulvin. Animals should be treated at the same time as people to avoid reinfection.

Management and Control in Animals

A killed fungal cell wall vaccine is approved for treatment and prevention of *M. canis* in cats. The vaccine helps decrease the time needed for improvement of clinical signs, but does not affect cure time. The vaccine also reduces the severity, but not the frequency, of re-exposed kitten infections.

An attenuated fungal vaccine for cattle that prevents development of severe clinical lesions and has reduced the incidence of zoonotic disease in animal care workers is used in some European countries. Vaccinated animals shed fungal spores for a period after vaccination. No live vaccine is available in North America.

Some rugs used in animals are treated with chemicals such as tolnaftate can help prevent the fungus from persisting within the rug.

Management and Control in Humans

There are a variety of recommendations for limiting the potential for acquiring dermatophytoses including:

- Keep skin, hair, and nails clean and dry. Fungal colonization is favored by moisture, so keeping skin and hair clean and dry helps minimize infection.
- Do not walk barefoot outside or on unwashed floors.
- Never lend or exchange hair brushes, combs, or clips.
- Practice good hygiene.
- Reduce stress levels and seek treatment of immunocompromising diseases.
- Wear gloves and practice proper hygiene procedures when handling animals.
- Treat infective animals.
- People who have ringworm and are involved in contact sports should be restricted from practices and events until cleared by a physician.

Summary

Dermatophytes are a group of fungi that invade the dead keratin of skin, hair, and nails causing dermatophytoses (commonly known as ringworm). Three genera of dermatophytes infect humans (*Epidermophyton*, *Microsporum*, and *Trichophyton*) and only two infect animals (*Microsporum* and *Trichophyton*). Categories of dermatophytes include anthropophilic, zoophilic, and geophilic. Dermatophytes are aerobic, nonfastidious organisms that require moisture for growth. *Microsporum* infections occur worldwide with cats being an important transmission source of the most common species *M. canis*. *Trichophyton* infections occur worldwide with a variety of animal transmission sources. Ringworm affects a wide variety of animals with a number of different fungal species causing skin lesions. In dogs, approximately 70% of cases are caused by *M. canis*, 20% by *M. gypseum*, and 10% by *T. mentagrophytes*. In cats, 98% of ringworm cases are caused by *M. canis*. Lesions in cats are quite variable, with kittens being most frequently affected. Dermatophytosis in cattle is usually caused by *T. verrucosum*, but *T. mentagrophytes*, *T. equinum*, *M. gypseum*, *M. nanum*, *M. canis*, and other fungi. In horses, ringworm is most commonly caused by *T. equinum*. In guinea pigs, ringworm is most often caused by *T. mentagrophytes* and less frequently by *Microsporum* spp. In pigs, ringworm is usually caused by *M. nanum*. The fungi that cause ringworm in sheep include *M. canis*, *M. gypseum*, and *T. verrucosum*.

Ringworm infections in people are divided into categories based on the body part infected and include tinea capitis, tinea corporis, tinea barbae, tinea cruris, tinea pedis, and tinea unguium. Dermatophytoses are diagnosed with Wood's light examination, direct examination of hairs, and fungal culture. Treatment of dermatophytoses includes oral and topical antifungal treatment (although the disease may be self-limiting, it is treated to prevent spread). Cleaning of the area and any equipment is essential to controlling these fungi. A killed fungal cell wall vaccine is approved for treatment and prevention of *M. canis* in cats. There is no vaccine in people. The disease can be prevented with proper hygiene practices.

SPOROTRICHOSIS

Overview

Rose gardener's disease is a common name of sporotrichosis because people tend to become infected by thorn pricks of rosebushes.

Sporotrichosis, also called rose gardener's disease, is a subcutaneous fungal disease caused by *Sporothrix schenckii*, a dimorphic fungus that occurs in nature and is associated with soil, wood, and vegetation (such as rosebushes, barberry bushes, sphagnum moss, and other mulches). The first case of sporotrichosis was identified in 1898 by Benjamin Robinson Scheck at the Johns Hopkins Hospital in Baltimore. He described the fungus as related to *Sporotricha*, because it resembled a species of the plant *Sporotrichum*. The second report of the disease occurred in 1900 and involved a boy who had developed a lesion on a finger that had been hit with a hammer. Hektoen and Perkins isolated the fungus and named the organism *Sporotrichum schenckii*. The name *Spo. schenckii* was used in a few of the early reports; however, the name *Spo. schenckii* was used for about 50 years. During the early 1900s in France, sporotrichosis was a common disease and scientists such as Beurmann, Ramond, and Gougerot described the fungus and its clinical signs. The use of potassium iodide to treat sporotrichosis was suggested by Sabouraud to Beurmann and Gougerot in 1903 and this treatment is still used today. The incidence of the sporotrichosis in France declined after the 1920s. In 1908 in Brazil, Splendore described the asteroid bodies seen around *Spo. schenckii* that is used in the histologic diagnosis of sporotrichosis. In 1927, Pijper and Pullinger reported a sporotrichosis outbreak involving 14 gold mine workers in Witwatersrand, South Africa. Between 1941 and 1944 nearly 3,000 workers from these mines were infected (the origin of infection was traced to the mine timbers). The outbreak was brought under control by treating the timbers with fungicides and using potassium iodide as therapy for the affected miners. In 1988 an outbreak of sporotrichosis in the United States affected 84 people in 15 states who handled conifer seedlings that were packed in Pennsylvania with sphagnum moss harvested in Wisconsin. From 1998 to 2001, 178 culture-proven cases have been identified and treated, predominantly among women at a median age of 39 years and predominantly among those with infected cats.

Causative Agent

Sporotrichosis is a cutaneous or extracutaneous infection caused by *Spo. schenckii*, a rapidly growing fungus. *Spo. schenckii* exists as a saprophytic mold on vegetative matter in humid climates worldwide. It is a dimorphic fungus that grows in the yeast form in the body and in culture at 37°C and grows as a filamentous mold exhibiting mycelial forms at 25°C (Figure 5-12). Microscopically, *Spo. schenckii* appears as small (2 to 5 micrometers), round to oval to cigar-shaped yeast cells or fine, branching septate hyphae with either ovoid microconidia borne in clusters

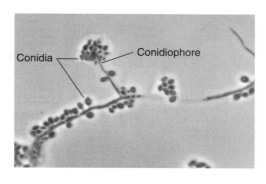

Conidia — Conidiophore

Figure 5-12 A conidiophore and conidia of the fungus *Sporothrix schenckii.* (Courtesy of CDC/Dr. Libero Ajello)

from the ends of conidiophores (having a flowerette arrangement) or are brown, oval or triangular, thick-walled, and directly attached to the sides of the hyphae. If stained with periodic acid-Schiff (PAS) stain, an amorphous pink material may be seen around the yeast cells. *Spo. schenckii* grows readily on brain heart infusion agar, blood agar, and Sabouraud dextrose agar (SDA).

Epizootiology and Public Health Significance

Spo. schenckii is found worldwide and sporotrichosis occurs worldwide especially in areas in which the humidity is between 92% and 100% and there is temperature between 80°F and 85°F. Sporotrichosis occurs mainly in moist tropical and subtropical areas such as Brazil, Columbia, and Mexico; although outbreaks from infective sphagnum moss harvested in Wisconsin have been reported. In the United States, sporotrichosis is most commonly found in coastal regions and river valleys.

The incidence of sporotrichosis is unknown. The mold itself is endemic to the Missouri and the Mississippi River Valleys. Sporotrichosis is the most common subcutaneous mycosis in South America and it is rare in Europe. Sporotrichosis is associated with minimal morbidity, unless the fungus infects patients who are immunologically compromised.

> *Spo. schenckii* survives in the environment and becomes pathogenic in animals as a result of its dimorphic qualities.

Transmission

Any compromise of the skin barrier with fungal contact could potentially cause infection. Cutaneous infection often results from a puncture wound involving thorns or other plant matter (less common infection sources have been squirrel bites and trauma induced by liposuction). Sporotrichosis has been transmitted by cat bites and occasionally dog and squirrel bites. Direct skin contact with the ulcerated and draining lesions or exudates also transmits the fungus. Small wounds created by splinters, thorns, and insect bites serve as portals of fungal entry. Person-to-person transmission is rare. Pulmonary sporotrichosis is rare and is caused by inhalation of the fungus into the lungs. Sporotrichosis usually occurs sporadically.

> Proteases are possible virulence factors of *Spo. schenckii.*

Pathogenesis

Sporotrichosis infections are classified as cutaneous (more common) or extracutaneous. Cutaneous infections are subclassified into fixed cutaneous and lymphocutaneous forms. Fixed cutaneous sporotrichosis infections occur at the site of inoculation and remain confined entirely to the skin. Fixed cutaneous lesions are painless, nodular lesions that form around the site of inoculation and with time produce pus that eventually drains because of ulceration of the lesion. Lymphocutaneous disease results from the fungi entering the lymphatic system near the primary lesion site with satellite lesions developing along the path of the lymphatic vessels (sporotrichoid spread) resulting in lymphadenopathy. The lymphocutaneous form is restricted to the subcutaneous tissues and does not enter the blood. Extracutaneous, or disseminated sporotrichosis, can present as pyelonephritis, orchitis, osteomyelitis, septic arthritis (monoarthritis), mastitis, synovitis, or meningitis. Extracutaneous sporotrichosis usually occurs in immuno-compromised animals and people. Pulmonary involvement is rare.

Infection in animals and humans is a result of the fungus' ability to change phases from an organism that survives on living or decaying plant material to a yeast phase upon entering the skin. The fungus is typically acquired by traumatic implantation into the skin and the fungus converts to its yeast phase causing local or systemic infection. Clinically a local pustule or an ulcer with nodules develops along the draining lymphatics. As the fungus grows, it is recognized by the immune system and an inflammatory response occurs. Clinical signs relate to the location and degree of inflammation. Primary organ systems involved in sporotrichosis include the skin and the lungs. Unless the fungus is inhaled or acquired by an immunocompromised person or animal, it remains cutaneous. In a host who is immunocompromised, disseminated infection can occur from skin involvement or from primary pulmonary infection.

Clinical Signs in Animals

Sporotrichosis is a sporadic chronic granulomatous disease of various domestic and laboratory animals. Sporotrichosis has been reported in dogs, cats, horses, cows, camels, dolphins, goats, mules, birds, pigs, rats, and armadillos. In animals, sporotrichosis occurs in three forms: cutaneous, lymphocutaneous, and disseminated, with the lymphocutaneous form being most common. Small nodules (1 to 3 centimeters in diameter) develop at the inoculation site. The cutaneous form remains localized to the inoculation site, although lesions may be multicentric. As infection ascends along the lymphatic vessels, new nodules develop producing the lymphocutaneous form of the disease. Lesions typically ulcerate and discharge a serohemorrhagic fluid. Chronic illness may result in fever and lethargy. Disseminated sporotrichosis is rare and may involve the bone, lungs, liver, spleen, testes, gastrointestinal tract, or central nervous system (CNS).

Sporotrichosis is more commonly seen in cats and tends to occur in sexually intact male cats with initial lesions occurring more frequently on the distal extremities, head, or base of the tail. Lesions appear as small, draining, puncture wounds that look like bacterial abscesses or cellulitis from cat bite wounds. Lesions will ulcerate and drain potentially exposing muscle and bone. Cats spread the disease by licking and grooming affected sites resulting in secondary infection sites and disseminated disease. Cats with disseminated disease are febrile and anorexic. Cats produce the greatest zoonotic potential, and transmission from cat to human has been reported without evidence of trauma. Cats shed a large number of organisms from wounds and in the feces. In contrast, transmission from other infected species requires inoculation of traumatized skin.

Clinical Signs in Humans

In people there are a few clinical manifestations of sporotrichosis including:

- Fixed cutaneous sporotrichosis in which primary lesions develop at the inoculation site and begin as a painless nodule that enlarges and ulcerates releasing a serous or purulent fluid (Figure 5-13). This form is the most common.

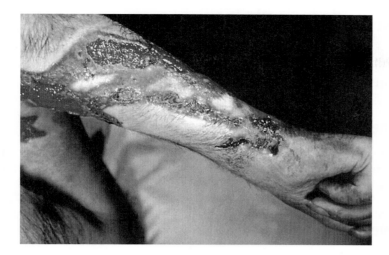

Figure 5-13 Sporotrichosis caused by the fungus *Spororthrix schenckii* on a person's arm.
(Courtesy of CDC/Dr. Lucille K. Georg)

- Lymphocutaneous sporotrichosis in which primary lesions develop at the inoculation site, but secondary lesions also appear along the lymphatic channels. Lesions are most common on the nose, mouth, pharynx, larynx, and trachea.
- Pulmonary sporotrichosis in which conidia are inhaled or there is hematogenous spread of fungi produce clinical signs such as cough, sputum production, fever, and weight loss.
- Osteoarticular sporotrichosis in which people have cutaneous lesions that spread to joints and bones. People typically have one enlarged joint (usually the knee).

Diagnosis in Animals

Diagnosis can be made by culture or microscopic examination of the exudate or of biopsy specimens. In tissues and exudate, the tissue is stained using PAS, GMS (Grocott's methenamine silver), or Gram stain. Tissue biopsies will demonstrate few to many organisms that are cigar-shaped, single cells within macrophages. The fungal cells are pleomorphic and small and ping-pong paddle buds may be present. In species other than cats, *Sporothrix* fungi numbers are low so diagnosis usually requires culturing the organism on primary isolation media such as SDA and brain heart infusion agar. In cultures incubated at 25°C colonies are white and soft initially then become wrinkled and coarsely matted with the color becoming tan to dark brown or black and the colonies leathery. At 25°C, true mycelia are produced, with fine, branching, septate hyphae bearing pear-shaped conidia on slender conidiophores. Cultures incubated at 37°C grow rapidly (3 to 5 days) and produce small, moist, soft cream to white-colored yeast-like colonies. A fluorescent antibody technique has been used to identify the yeast-like cells in tissues.

Diagnosis in Humans

Diagnosis in people is similar to animals except there is a sporotrichin test available that is an intradermal skin test.

Treatment in Animals

Few treatments have been critically evaluated in animals; however, itraconazole is considered the treatment of choice. Treatment is continued 3 to 4 weeks beyond apparent clinical cure.

Treatment in Humans

Itraconazole is the treatment of choice in people. Oral potassium iodide, terbinafine, and local heat therapy have also been used. The systemic form of sporotrichosis is treated with amphotericin B.

Management and Control in Animals

Sporotrichosis in animals can be prevented by limiting their exposure to potentially infective animals (such as outdoor cats) and by proper cleaning and examination of wounds.

Management and Control in Humans

People can limit their exposure to *Spo. schenckii* by practicing strict hygiene when handling animals with suspected or diagnosed sporotrichosis. Wearing gloves and long sleeves when handling pine seedlings, rose bushes, hay bales, and other plants that can cause minor skin breaks is recommended. Prudent use of pine seedling packing materials especially sphagnum moss is also recommended.

Summary

Sporotrichosis is a cutaneous or extracutaneous infection caused by *Spo. schenckii*, a rapidly growing fungus. It is a dimorphic fungus found worldwide most commonly seen in areas in which the humidity is between 92% and 100% and there is temperature between 80°F and 85°F. Sporotrichosis occurs mainly in moist tropical and subtropical areas such as Brazil, Columbia, and Mexico; although outbreaks from infective sphagnum moss harvested in Wisconsin have been reported. In the United States, sporotrichosis is most commonly found in coastal regions and river valleys. Sporotrichosis has been transmitted by animal bites (cats, dogs, and squirrels) and direct skin contact with the ulcerated lesions or exudates. Small wounds created by splinters, thorns, and insect bites serve as portals of fungal entry. Sporotrichosis is a sporadic chronic granulomatous disease of various domestic and laboratory animals with cats shedding a large number of organisms from wounds. In contrast, transmission from other infected species requires inoculation of traumatized skin. In animals, sporotrichosis occurs in three forms, cutaneous, lymphocutaneous, and disseminated, with the lymphocutaneous form being most common. Sporotrichosis in cats tends to occur in sexually intact male cats with initial lesions occurring more frequently on the distal extremities, head, or base of the tail. In people there are a few clinical manifestations of sporotrichosis including fixed cutaneous sporotrichosis, lymphocutaneous sporotrichosis, pulmonary sporotrichosis, and osteoarticular sporotrichosis. Diagnosis can be made by culture or microscopic examination of the exudate or of biopsy specimens. Itraconazole is considered the treatment of choice in animals and people. Sporotrichosis in animals can be prevented by limiting their exposure to potentially infective animals (such as outdoor cats) and by proper cleaning and examination of wounds. People can limit their exposure to *Spo. schenckii* by practicing strict hygiene when handling animals with suspected or diagnosed sporotrichosis.

   321

Table 5-3   Fungal Zoonoses Acquired Through Animal-Contaminated Soil

Fungus	Disease	Infectious Form	Predominant Signs in People	Clinical Signs in Animals	Transmission	Animal Source	Geographic Distribution	Pathology	Diagnosis	Treatment
Cryptococcus neoformans (serotypes A, B, C, and D) (See Figure 5-14)	Cryptococcosis (older names include torulosis and European blastomycosis)	Yeast: spherical to ovoid in shape with small, constricted buds and a large capsule (that contributes to its virulence)	Meningitis (in 2/3 of people), disseminated disease in immunocompromised people, respiratory disease (cough, fever, and lung nodules); skin infection	■ Dogs and cats are most commonly infected: CNS signs, eyes and orbit, skin, & nasal cavity; some cases show dissemination to the lungs, kidney, and joints ■ Cattle: sporadic cases of mastitis ■ Horses: paranasal sinus infection, CNS, abortion	Contact with contaminated soil (usually respiratory inhalation)	Feces of birds (especially pigeons), bird nests, and soil contaminated with bird excreta	Worldwide with the primary ecological niche revolving around birds	Prevalent in urban areas where pigeons congregate; proliferated in the high nitrogen content of bird feces; as masses of yeast cells dry they are scattered in the air and dust	Identification of yeast with thick capsule in India ink wet mounts; yeast stain gram positive; isolation on Sabouraud dextrose agar yields wrinkled, whitish granular colonies in about 7 days when incubated at 25° C; colonies will become mucoid and cream to brown as they age; commercial identification systems such as API-20C AUX Yeast System are also available; cryptococcal latex agglutination antigen assay	Amphotericin B; amphotericin B and flucytosine combination therapy; ammonia foot baths are recommended for people entering a facility
Blastomyces dermatitidis (See Figure 5-15)	North American blastomycosis, Gilchrist's disease, Chicago disease	Dimorphic fungi; conidia are the likely infectious form with the organism growing and causing disease as a yeast in the body; the cell wall structure is responsible for its virulence	Produces a chronic suppurative and granulomatous infection; begins as an upper respiratory infection and may spread to the lungs, bones, soft tissue, and skin	■ Dogs: produces granulomatous nodules in lungs and on skin. Skin lesions and disseminated blastomycosis result from hematogenous spread from the original respiratory lesions. Disseminated disease can spread to the bone, eyes, brain, and genitalia. ■ Has been described in horses, cats, dolphins, ferrets, and sea loins but is rare	Inhalation; infection via wound contamination is rare	Unknown; found in soil and wood	Primarily the northcentral and southeastern U.S. extending from Canada to the Mississippi, Ohio, and Missouri River Valleys, Mexico, and Central America; has been reported in Africa	After inhalation of only 10-100 conidia disease can begin in the respiratory system when the conidia convert to yeast and multiply. As these organisms multiply an inflammatory response begins and alveolar granulomas are formed.	Direct examination of microscopic smears showing large, spherical, thick-walled yeast cells about 8-15 μm in diameter, typically with a single bud connected to the parent cell by a broad base; culture at 37° C on enriched media produces waxy colonies that are whitish and turn gray to brown with age; mold form has septate hyphae with single, pyriform conidia; identification can also be done with an exoantigen test or nucleic acid probe	Amphotericin B; ketaconazole Disseminated disease does not respond well to treatment

(Continued)

Table 5-3 (Continued)

Fungus	Disease	Infectious Form	Predominant Signs in People	Clinical Signs in Animals	Transmission	Animal Source	Geographic Distribution	Pathology	Diagnosis	Treatment
Histoplasma capsulatum (See Figure 5-16)	Histoplasmosis, Ohio Valley fever, Darling's disease, reticuloendotheliosis	Dimorphic fungi; conidia are the infectious form; virulence factors include its cell wall composition, intracellular growth, and thermotolerance	Can produce pulmonary, systemic, or cutaneous lesions	■ Dogs and cats: ulceration of the intestinal tract, hepatomegaly, splenomegaly, lymphadenopathy; necrosis and tubercle-like lesions in the lungs, liver, kidneys, and spleen ■ Other animals infected include cattle, nonhuman primates, horses, sheep, swine, and various wild animals	Inhalation (and less commonly ingestion) of soil contaminated with bat and bird feces (especially starlings and pigeons)	Bat and bird feces; soil	All continents except Australia; especially common in the eastern and central regions of the U.S. (the Ohio River Valley) and the midwest	A benign form occurs after inhaling a small amount of conidia into the deep recesses of the lung establishing a primary fungal infection with mild signs (cough, aches, and pains) that may become severe in some people (fever, night sweats, and weight loss); chronic disease occurs in immunocompromised people in which the fungus disseminates within the macrophages leading to systemic disease	Direct microscopic examination often fails to reveal this fungi but if found will appear intracellularly within mononuclear cells as small, round to oval yeast cells that are 2–5 μm in diameter; fungal culture reveals a slow-growing mold at 25° C taking 2–4 weeks to grow appearing as white, fluffy mold that turns brown to buff with age; yeast grows at 37° C as wrinkled, moist, yeastlike colonies that are soft and cream, tan, or pink; microscopically hyphae are small with spherical or pyriform, smooth-walled macroconidia that become roughened or tuberculate with age; exoantigen test (complement fixation and immunodiffusion) and nucleic acid probes are available	Amphotericin B, ketoconazole or miconazole; surgery may be needed to remove masses
Coccidioides immitis (See Figure 5-17)	Coccidioidomycosis, Valley Fever, San Joaquin Fever, California Disease	Dimorphic fungi; arthroconidia are the infectious form; virulence factors include extracellular proteinases	Asymptomatic in 60% of infected people; self-limiting respiratory infection; disseminated disease can affect visceral organs, meninges, bone, skin, lymph nodes, and subcutaneous tissue	■ Cattle: resembles tuberculosis with nodular lesions in the bronchial and mediastinal lymph nodes (occasionally lungs) ■ Dogs: formation of nodules in lung, brain, liver, spleen, bones, and kidney ■ Horses: disseminated infection and abortion	Inhalation	Soil (dispersal is aided by windstorms, dust storms, drainoff water, and burrowing animals)	Desert southwestern portion of the U.S., semiarid regions of Mexico and Central and South America	This fungus favors a habitat with high carbon and salt content and a semiarid, relatively hot climate. Infection follows a cyclic pattern with a period of dormancy in winter and spring, followed by growth in summer and fall. Arthrospores are inhaled and converted to spherules in the lung where they swell, sporulate, burst, and release spores. Chronic pulmonary disease manifests itself with nodular growths called fungomas and cavity formation in the lungs. Spores will disseminate in immunocompromised people/ animals in which multisystem organ involvement may occur.	Direct microscopic examination demonstrates nonbudding, thick-walled spherules that are 20–200 μm in diameter and contain granular material or numerous small nonbudding endospores; fungal culture is a biohazard to laboratory workers and needs to occur inside a biosafety cabinet; culture shows delicate, cobweb-like growth at 25° C with most isolates appearing fluffy white; small septate hyphae with barrel-shaped arthroconidia that stain darkly with lactophenol cotton blue; exoantigen test and nucleic acid probes are available	Amphotericin B

Organism	Disease	Characteristics	Disease features	Animal reservoir	Transmission	Source	Distribution	Pathogenesis	Diagnosis	Treatment
Paracoccidioides brasiliensis (See Figure 5-18)	Paracoccidioidomycosis, paracoccidioidal granulomas, South American blastomycosis	Dimorphic fungi with conidia as infectious form; virulence factors include estrogen-binding proteins and its cell wall components	Most infections are self-limiting; if systemic disease occurs the lungs, skin and mucous membranes, and lymphatic organs are most frequently involved	Nine banded armadillo may have lung granulomas; typically asymptomatic	Inhalation and trauma	Soil and plants	Cool, humid soil of tropical and semitropical regions of South and Central America	Paracoccidiodomycosis begins as a primary pulmonary infection that is often asymptomatic. It can disseminate to produce ulcerative lesions of the mucous membranes with a serpiginous (snakelike) active border and a crusted surface. Cervical lymph node involvement is common.	Direct microscopic examination shows multiple "captain's wheel" buds, a thin cell wall, and a narrow base; fungal culture on Sabouraud dextrose agar at 25°C shows that the colony has a dense, white mycelium; culture at 37°C produces slow growing white-tan, thick colonies; immunodiffusion tests are also available	Amphotericin B, ketoconzaole, (sulfa drugs may also be effective)
Pneumocystis carinii	Pneumocystosis, parasitic pneumonia, interstitial plasma cell pneumonia	Atypical fungus that is currently classified as a fungi based on RNA analysis. Has flexible-walled trophozoite form (predominant) and firm-walled cystic form (infectious form). Originally thought to be a trypanosome	Many infections are asymptomatic and rarely caused disease until the advent of HIV. Now seen mainly in cancer patients with T-cell deficiency caused by chemotherapy and HIV-infected people producing moderate fever, intense dyspnea, unproductive cough, tachypnea, and cyanosis.	Common in mice and rats that are asymptomatic. May also be found in zoo animals, pigs, rabbits, sheep, goats, and dogs. Some believe the infection is not zoonotic.	Inhalation of dust particles or aerosolized respiratory secretions.	Dust contaminated with cysts or respiratory secretions	Worldwide; most common in an immuno-compromised host	Following inhalation of the cyst, the fungus attaches to type I pneumocytes where it exists and replicates extracellularly. Replication of the organism fills the alveolar spaces impairing oxygen diffusion and producing hypoxia.	Direct examination of bronchoalveolar lavage fluid or sputum using Giemsa stain, calcofluor white stain, or methenamine silver stain. Cysts are spherical to concave and 4–7 µm in diameter and may contain intracystic bodies. Trophozoites are difficult to see and are pleomorphic. Monoclonal antibody tests and PCR tests are also available.	Co-trimoxazole or dapsone plus trimethoprim or clindamycin plus primaquine. Human-to-human transmission is possible making patient isolation important.

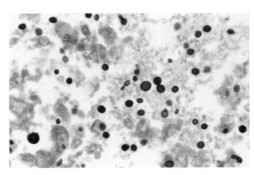

Figure 5-14 *Cryptococcus neoformans* yeast show narrow-base budding and characteristic variation in size.

(Courtesy of CDC/Dr. Edwin P. Ewing, Jr.)

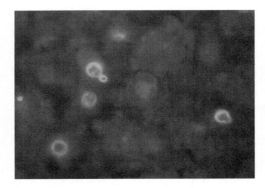

Figure 5-15 *Blastomyces dermatitidis* identification using direct FA stain.

(Courtesy of CDC/Dr. William Kaplan)

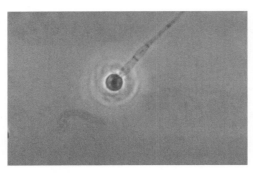

Figure 5-16 A conidiophore of the fungus *Histoplasma capsulatum*.

(Courtesy of CDC/Dr. Libero Ajello)

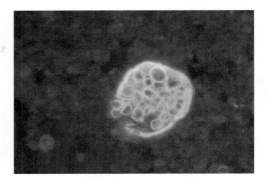

Figure 5-17 Spherule of *Coccidioides immitis* with endospores.

(Courtesy of CDC/Mercy Hospital Toledo, OH/Brian J. Harrington)

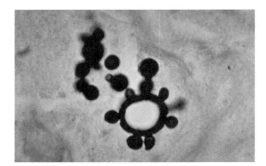

Figure 5-18 Budding cells of *Paracoccidioides brasiliensis* using methenamine silver stain.

(Courtesy of CDC/Dr. Lucille K. Georg)

Review Questions

Multiple Choice

1. Fungi of medical importance are
   a. yeasts and molds.
   b. yeasts and endospores.
   c. molds and puffballs.
   d. molds and sphirochetes.

2. Yeast cells look differently than bacteria under the microscope because
   a. they are stained with different stains that allow visualization of the bacterium's nucleus.
   b. yeast are larger than bacteria, are round or oval in shape, contain nuclei, and may be seen budding.
   c. they contain filaments called hyphae that bacteria do not possess.
   d. they form pseudohyphae, are filamentous in shape, and feed by growing into the surrounding environment.

3. Most fungal diseases are spread through
   a. ingestion of the fungal spores.
   b. absorption through the skin and mucous membranes.
   c. accidental contact from an environmental source.
   d. vertical transmission.

4. The function of 10% potassium hydroxide in the direct examination of a fungal specimen is
   a. to permanently fix fungal elements.
   b. to digest and dissolve cellular debris.
   c. to eradicate bacterial contaminants.
   d. to stain existing hyphae.

5. The toxicity of antifungal drug therapy is attributed to
   a. the virulence of fungal organisms.
   b. the similarities between mammalian and fungal cells.
   c. the poor health of the patient.
   d. the resistance of fungal organisms.

6. A dimorphic fungus is one that
   a. changes its form on different media types.
   b. changes its form as a result of different temperatures.
   c. starts to grow as a mold, but becomes more yeast-like.
   d. exhibits yeast and mold forms at the same time.

7. Yeast reproduce by (and can be identified by) a process called
   a. budding.
   b. binary fission.
   c. binary fusion.
   d. meiosis.

8. The India ink stain is combined with cerebrospinal fluid (CSF) to identify
   a. *Coccidioides immitis.*
   b. *Cryptococcus neoformans.*
   c. *Candida albicans.*
   d. *Histoplasma capsulatum.*

9. A human patient with subcutaneous ulcerated lesions on the arm is seen by a physician. The lesions appear to follow the distribution of the lymphatic drainage, and regional lymph nodes are enlarged. The patient states that she

has several indoor-outdoor cats at home. Microscopic examination of the subcutaneous aspirate shows cigar-shaped yeast cells. What disease does this patient most likely have?
a. ringworm
b. histoplasmosis
c. coccidiodes
d. sporothrichosis

10. Bats serve as a source of infection for
a. *Coccidioides immitis.*
b. *Histoplasma capsulatum.*
c. *Paracoccidioides brasiliensis.*
d. *Blastomyces dermatitidis.*

Matching

11. _____ dermatophyte	A. dimorphic fungi that produces lymphocutaneous infection
12. _____ *Microsporum canis*	B. captain's wheel buds seen in South and Central America
13. _____ *Trichophyton*	C. fungus that invades dead hair, skin, or nails
14. _____ *Microsporum*	D. 50% of this dermatophyte fluoresces
15. _____ *Cryptococcus neoformans*	E. produce thick-walled, rough macroconidia, and fewer microconidia
16. _____ *Sporothrix schenckii*	F. broad-based budding yeast seen in southeastern and northcentral United States
17. _____ *Coccidioides immitis*	G. large, tuberculate macroconidia in fungal form seen in Ohio River Valley
18. _____ *Blastomyces dermatitis*	H. spherule at 37°C seen in San Joaquin Valley
19. _____ *Paracoccidioides brasiliensis*	I. yeast with thick capsule seen in urban areas with pigeons
20. _____ *Histoplasma capsulatum*	J. produce thin-walled, smooth macroconidia, and numerous microconidia

Case Studies

21. A 3-year-old child is seen by her pediatrician for alopecia and inflammation of the scalp. Broken hair shafts fluoresced under a Wood's lamp, and skin and hair samples were collected for fungal culture. The colonies grew quickly on SDA and appeared fluffy and white with a yellow reverse side. Microscopic examination revealed numerous spindle-shaped, thick-walled, multicellular macroconidia.
a. What disease does this child most like have?
b. What is the most likely etiologic agent?
c. How may this child have acquired this infection?
d. What is the recommended treatment and guidelines for this child?

22. A 65-year-old male presents to his physician with complaints of prolonged fever, cough, and shortness of breath. His illness began as a mild flu-like episode 1 week ago, just after returning from a visit to his sister in Ohio. His dog has also developed a cough following their trip. The physician ordered

a chest radiograph of the patient which showed nodular infiltrates and enlarged lymph nodes. A fungal culture of the patient's sputum produced slow-growing white mold colonies, with small microconidia and large tuberculate macroconidia.

a. Based on this person's history and laboratory results, what disease do you think he might have?

b. What organism causes this disease?

c. What tests can be done to confirm the diagnosis?

d. Describe the tissue phase of this organism.

e. Since this patient's dog is also sick, is the dog the source of the infection?

23. A 4-year-old intact male Labrador retriever was presented to his veterinarian with nystagmus, ataxia, and circling. During his examination the dog had a grand mal seizure. Diagnostic tests including complete blood count (CBC), chemistry panel, and urinalysis (UA) were normal. As the dog worsened he was sent to a referral center where a cerebrospinal fluid (CSF) tap was done. A Gram stain of the CSF was negative for bacteria, but revealed large purple ovals that were budding.

a. What type of infection might this dog have?

b. Based on the answer for a, what is the most likely genus and species?

c. What is a quick test that should be done next?

d. How could this dog have acquired this infection?

e. Is there anything else of concern in this dog?

References

Alcamo, I. E. 1998. *Schaum's Outlines: Microbiology*. New York: McGraw-Hill.

Baugh, W., and B. Graham. 2005. Sporotrichosis. http://www.emedicine.com/PED/topic2144.htm (accessed June 29, 2005).

Bauman, R. 2004. *Microbiology*. San Francisco, CA: Pearson Benjamin Cummings.

Biddle, W. 2002. *A Field Guide to Germs*. New York: Anchor Books, pp. 157–8.

Burton R., and Engelkirk P. 2004. *Microbiology for the Health Sciences*, 8th ed. Philadelphia, PA: Lippincott, Williams & Wilkins, pp. 79–85.

Carter, G., M. Chengappa, and A. Roberts. 1995. *Essentials of Veterinary Microbiology*, 5th ed., Baltimore, MD: Williams & Wilkins, pp. 245–72.

Centers for Disease Control and Prevention. 2005. Ringworm and animals. http://www.cdc.gov/healthypets/diseases/ringworm.htm (accessed May 20, 2005).

Centers for Disease Control and Prevention. 2006. Sporotrichosis. http://www.cdc.gov/ncidod/dbmd/diseaseinfo/sporotrichosis_g.htm (accessed February 20, 2006).

Cowan M., and K. Talaro. 2006. *Microbiology: A Systems Approach*. New York: McGraw-Hill.

Doctor Fungus. 2006. Tinea barbae. http://www.doctorfungus.org/mycoses/human/other/tinea_barbae.htm (accessed February 19, 2006).

Doctor Fungus. 2006. Tinea capitis and tinea favosa. http://www.doctorfungus.org/mycoses/human/other/tinea_capitis_favosa.htm (accessed February 19, 2006).

Doctor Fungus. 2006. Tinea corporis, tinea cruris, and tinea pedis. http://www.doctorfungus.org/mycoses/human/other/tineacorporis_cruris_pedis.htm (accessed February 19, 2006).

Dunstan, R., K. Reimann, and R. Langham. 2003. Feline sporotrichosis. http://www.avma.org/reference/zoonosis/znsporotrichosis.asp (accessed February 20, 2006).

Forbes, B., D. Sahm, and A. Weissfeld. 2002. *Bailey & Scott's Diagnostic Microbiology*, 11th ed. St. Louis, MO: Mosby, pp. 747–54 and 759–68.

Kao, G. 2005. Tinea capitis. http://www.emedicine.com/DERM/topic420.htm (accessed June 22, 2005).

Krauss, H., A. Weber, M. Appel, B. Enders, H. Isenberg, H. G. Schiefer, W. Slenczka, A. von Graeventiz, and H. Zahner. 2003. *Zoonoses Infectious Diseases Transmissible From Animals to Humans*, 3rd ed. Washington, DC: American Society of Microbiology.

Merck Manual. 2006. Dermatophytoses. http://www.merckvetmanual.com/mvm/index.jsp?cfile5htm/bc/71300.htm&word5dermatophytes (accessed February 13, 2006).

Merck Manual. 2006. Sporotrichosis. http://www.merckvetmanual.com/mvm/index.jsp?cfile5htm/bc/51117.htm&word5sporotrichosis (accessed February 19, 2006).

Miller, S., and P. Vogel. 2005. Sporotrichosis. http://www.emedicine.com/derm/topic400.htm (accessed July 22, 2005).

Narqadzay, J., and N. Rubeiz. 2006. Tinea. http://www.emedicine.com/emerg/topic592.htm (accessed October 12, 2006).

Prescott, L. M., D. A. Klein, and J. P. Harley. 2002. *Microbiology*, 5th edition. New York: McGraw-Hill.

Shimeld, L. A. 1999. *Essentials of Diagnostic Microbiology*. Albany, NY: Delmar Publishers.

Simpanya, M. 2000. Dermatophytes: their taxonomy, ecology and pathogenicity. *Revista Iberoamericana de Micologia Apdo* http://www.dermatophytes.reviberoammicol.com/p001012.pdf (accessed January 31, 2006).

Weese, J. 2002. A review of equine zoonotic diseases: risks in veterinary medicine. *AAEP Proceedings* 48:362–9.

Welch, R. 2003. Zoonosis update: sporotrichosis. *JAVMA* 223(8):1123–6.

Wong, G. 2005. Ringworm fungal infections. http://www.earthtym.net/ref-ringworm.htm (accessed February 18, 2006).

Parasitic Zoonoses

Objectives

After completing this chapter, the learner should be able to

- Describe properties unique to the different types of parasites
- Describe the different types of medically important parasites
- Identify the visual appearance of parasites either microscopically or to the naked eye
- Describe the roles arthropods play in disease
- Differentiate direct versus indirect parasitic life cycles based upon the types of hosts involved
- Briefly describe the history of specific parasitic zoonoses
- Describe the causative agent of specific parasitic zoonoses
- Identify the geographic distribution of specific parasitic zoonoses
- Describe the transmission, clinical signs, and diagnostic procedures of specific parasitic zoonoses
- Describe methods of controlling parasitic zoonoses
- Describe protective measures professionals can take to prevent transmission of parasitic zoonoses

Key Terms

amastigote	definitive host	hemoflagellete	nematode	protozoa
arthropod	dioecious	hermaphrodite	oocyst	reservoir host
autoinfection	direct life cycle	host	ovoviviparous	rostellum
cestode	ectoparasite	indirect life cycle	parasite	sporozoite
coenurosis	endoparasite	intermediate	parthenogenic	trematode
cyst	epimastigote	host	platyhelminthes	trophozoite
cysticercosis	helminth	metamorphosis	promastigote	trypomastigote

OVERVIEW

Parasitology is the study of unicellular and multicellular organisms that parasitize people and animals. **Parasites** are organisms that derive nourishment and protection from other living organisms. Parasites are organisms that live in association with, and at the expense of, other organisms. The organism that provides the nutrition is called the host, and the organism obtaining the nutrition is the parasite. Parasites found on the exterior of the host are called **ectoparasites** (such as lice or fleas), whereas those found inside the host are called **endoparasites** (such as worms, amebae, and malaria protozoa) (Figure 6-1A and B). The effect a parasite has on a host ranges from causing minor harm (allowing the host to live

Figure 6-1 A and B
(A) A louse is an example of an ectoparasite.
(Courtesy of CDC/Dr. Dennis D. Juranek)

(B) *Ancylostoma caninum*, the hookworm of dogs, is an example of an endoparasite.
(Courtesy of CDC)

(A)

(B)

and reproduce normally and complete its normal life cycle) to completely interfering with reproduction or causing the premature death of the host.

Parasites range in size from tiny, single-celled organisms to worms visible to the naked eye. Several parasites have emerged as significant causes of illness especially foodborne and waterborne disease. These organisms live and reproduce within the tissues and organs of infected human and animal hosts, and are often excreted in feces.

Types of Parasitic Relationships

A **host** is a source of nourishment and housing for parasites. There are many different types of hosts involved in parasitic life cycles. Examples of different types of hosts include:

- The host in which a parasite reaches sexual maturity and reproduces sexually is referred to as the **definitive host**. In parasites that only reproduce asexually the host in which the parasite spends the majority of its life cycle is also referred to as the definitive host. The definitive host of *Toxoplasma gondii* is the cat because sexual reproduction occurs in the epithelial tissue of the feline intestine.
- **Intermediate hosts** harbor an immature stage of a parasite and are required for parasite development. A parasitic life cycle may include one or more intermediate hosts. Intermediate hosts are sometimes called alternate hosts. Arthropods such as mosquitoes, ticks, and flies are common intermediate hosts for parasitic disease. Intermediate hosts for *Toxoplasma gondii* are humans, wild animals, and domestic animals other than cats.
- **Reservoir hosts** serve as a source of the parasite that can be transmitted to humans from animals, even if the animal is the normal host of the parasite. For example, beavers can be reservoir hosts of *Giardia duodenalis*.

Types of Parasites

Parasites can be described as obligate (meaning the parasite cannot live apart from its host) or facultative (meaning the organism may be parasitic on another organism but is also capable of living independently). Most parasites are obligate parasites even if they have free-living stages outside the host. Facultative parasites are not normally parasitic, but can become parasitic when they are accidentally eaten or enter a wound or other body orifice. An example of a facultative parasite is *Naegleria fowleri*, a free-living ameba that causes primary amoebic meningoencephalitis.

Parasites are classified into different groups based upon their cell type, size, locomotion, and a variety of other factors. General categories of medically significant parasites include protozoa, helminths, and arthropods.

- **Protozoa** are unicellular, eukaryotic organisms that cannot make their own food and are usually found in moist, if not fully aquatic, environments. Protozoa are classified by the way they move: some move by tube-like structures called pseudopodia that project and withdraw to cause movement (such as *Entamoeba histolytica*), some move by flagella which causes whip-like movement (such as *G. duodenalis*), some move by cilia that cause waves of movement (such as *Balantidium coli*), whereas others have no organs for motion in the adult form (such as *T. gondii*).

- Another type of parasite is **helminths** commonly known as worms. Helminths are multicellular, eukaryotic organisms that cannot make their own food and are usually found in animals. **Platyhelminthes**, commonly called "flatworms," and **nematodes**, commonly called "roundworms," are the two medically important groups of helminths. Adult flatworms have flattened bodies and consist of two groups: **cestodes** ("tapeworms") and **trematodes** ("flukes"). Roundworms have cylindrically shaped bodies usually tapered at both ends.

- A third type of parasite is **arthropods**. Arthropods are multicellular organisms with jointed appendages and an exoskeleton. They may function as vectors of disease (such as the *Anopheles* mosquito that helps transmits *Plasmodium* organisms that cause malaria) or may cause disease directly (such as lice which are blood-sucking insects that live on skin).

Protozoa

The Kingdom Protista contains organisms called protozoa whose single unifying trait is that they exist as a single cell. There are over 50,000 species of protozoa; however, only about one-fifth are parasitic, whereas the majority are free-living. Parasitic protozoa are small, eukaryotic, and have high rates of reproduction.

One way protozoa are classified into phyla is according to their method of movement. There are seven phyla of protozoa; however, only four are parasitic in humans and animals. The four phyla of medical concern are Sarcomastigophora (motile by means of pseudopods such as ameba or by flagella such as *Trichomonas vaginalis*), Apicomplexa (nonmotile when mature such as *Cryptosporidium parvum*), Cilophora (motile by means of cilia such as *Bal. coli*), and Microspora (lack locomotion extensions on their body such as *Enterocytozoon* spp.).

> Structurally, a protozoan is equivalent to a single eukaryotic cell.

- Sarcomastigophora consist of the amebae and flagellates. Amebae are motile by cytoplasmic extensions called pseudopods and most amebae have **cyst** and **trophozoite** stages in their life cycle. The cyst stage is the infective stage, whereas the active stage that feeds and moves is called the trophozoite (commonly called a troph). In most species of ameba the trophozoite is not infectious because it is unable to survive the harsh conditions of the stomach. An example of a pathogenic ameba is *En. histolytica*; examples of ameba that are nonpathogenic in healthy people and animals include *En. coli*, *Endolimax nana*, and *Iodamoeba butschlii* (Figure 6-2).

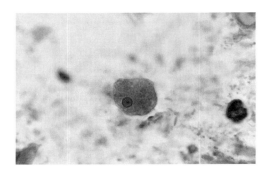

Figure 6-2 A photo micrograph of *Entamoeba histolytica* cyst that has a halo appearance around its outer ectoplasm.
(Courtesy of CDC)

Flagellates are also members of the phylum Sarcomastigophora and are motile by one or more long, thin cytoplasmic extensions called flagella. Flagellates are surrounded by a flexible cell covering called a pellicle that gives them more definite shape than ameba. Flagellates also have a cyst stage that is infective and an active trophozoite stage. Some flagellates have a deoxyribonucleic acid (DNA) containing structure in the mitochondrion called a kinetoplast. Two examples of pathogenic flagellates are *G. duodenalis* and *Tr. vaginalis*. A subcategory of the flagellate group is the hemoflagellates which infect blood. **Hemoflagellates** are transmitted by the bite of a blood-sucking arthropod and are found in the blood smears of infected individuals. Examples of hemoflagellates are *Trypanosoma* and *Leishmania* spp. (Figure 6-3).

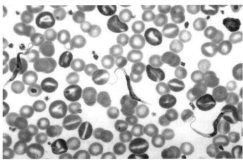

Figure 6-3 A photomicrograph of the hemoflagellate *Trypanosoma brucei.*
(Courtesy of CDC/Dr. Myron G. Schultz)

- Apicomplexa is the largest taxon of parasitic protozoa with about 4,000 known species. Protozoa in the group Apicomplexa have an apical complex, which is a unique arrangement of microtubules, vacuoles, and fibrils at one end of the cell. Apicomplexans infect both invertebrates and vertebrates; they may be relatively benign or may cause serious illnesses. Examples of Apicomplexa parasites include species of *Plasmodium* (Figure 6-4), the protozoa that cause malaria in humans and other animals, *Eimeria* spp., which cause coccidiosis, *Cry. parvum*, and *T. gondii*.

  Apicomplexans have varied and complex life cycles involving two different hosts (usually a mammal and a mosquito). Their life cycles have both asexual and sexual phases that involve the formation of a thick-walled **oocyst** after fertilization, followed by meiosis to produce infective spores. Apicomplexans are transmitted to new hosts in various ways; some, like the malaria parasite, are transmitted by infected mosquitoes, whereas others may be transmitted in the feces of an infected host (*Cry. parvum*) or when a predator eats infected prey (*T. gondii*).

- Ciliophora contain the ciliates, which are free-living protozoal

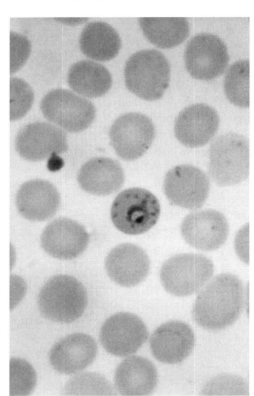

Figure 6-4 A photomicrograph of a thin blood smear showing the trophozoite stage of *Plasmodium falciparum.*
(Courtesy of CDC/Steven Glenn, Laboratory and Consultation Division)

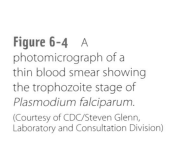

forms. Relatively few are parasitic, and only one species, *Bal. coli*, is known to cause human disease (Figure 6-5). Some ciliates cause diseases in fish; others are parasites or commensals on various invertebrates; others live in great numbers in the digestive tracts of many hoofed mammals, where they help breakdown cellulose.

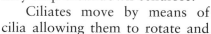

Figure 6-5 A photomicrograph of the ciliate *Balantidium coli* (cyst). (Courtesy of CDC)

Ciliates move by means of cilia allowing them to rotate and move forward and backward. Many ciliates have one or more macronuclei and micronuclei. The micronucleus functions in mitosis and meiosis, whereas the macronucleus maintains routine cell function. Their life cycle consists of a trophozoite and a cyst. Ciliates reproduce asexually by division: the micronucleus undergoes mitosis, whereas in most ciliates the macronucleus simply pinches apart into two.

- Members of Microspora are tiny, obligate intracellular parasites that infect many vertebrates and invertebrates. Transmission is by direct contact or by an intermediate host. As a result of their small size (1 to 4 μm), electron microscopy is needed to recognize and identify these organisms. Five genera of microsporidians have been reported in immunocompromised humans: *Encephalitozoon*, *Pleistophora*, *Nosema*, *Microsporidium*, and *Enterocytozoon*.

> Protozoa enter the cyst stage when moisture and food supplies dwindle; the cyst stage is the tougher, dormant form, whereas the trophozoite stage is the motile, active form.

Helminths

Helminths are macrocytic worms in the Kingdom Animalia. The typical helminth life cycle includes three stages:

- *Egg.* Adults produce eggs either by production of eggs and sperm in the same worm (**hermaphroditic**) or in separate male and female worms (**dioecious**). Fertilized eggs are usually released in the environment with a protective shell and food source to aid their development into larvae. Most eggs are vulnerable to environmental extremes such as heat, cold, and drying; therefore, certain helminths lay hundreds of thousands of eggs in an attempt to successfully complete its life cycle.

- *Larva.* Larvae emerge from eggs and are sometimes called the juvenile form. The larva stage is found on the intermediate host. There may be more than one intermediate host as is the case with *Diphyllobothrium latum*, the fish tapeworm, which has a freshwater crustacean and a freshwater fish as intermediate hosts.

- *Adult.* As the larvae mature they become adults. The adult stage is found in the definitive host. There may be more than one definitive host as is the case with *Dipylidium caninum*, the dog tapeworm, which can have a definitive host of a dog, cat, or human.

Helminth infections are typically acquired by ingestion of the larval stage (some exceptions are larval injection into the body via an insect bite or entry into the body by skin penetration). Helminths that are parasitic in animals belong to two phyla: Platyhelminthes and Nematoda.

Figure 6-6A A photograph of monkey colon with an intestinal fluke infection.
(Courtesy of CDC/Dr. Roger Broderson)

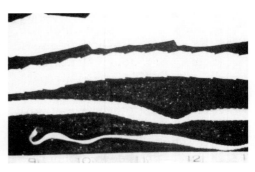

Figure 6-6B A photograph of the adult tapeworm, *Taenia saginata.*
(Courtesy of CDC)

Platyhelminthes

Platyhelminthes are the flatworms and have no body cavity other than the gastrointestinal tract, which lacks an anus. The same pharyngeal opening in flatworms takes in food and expels waste. Because of the lack of any other body cavity, in larger flatworms the digestive system is often highly branched in order to transport food to all parts of the body. Flatworms consist of two medically significant groups. The Trematoda, or flukes, are all parasitic, and have complex life cycles specialized for parasitism in animal tissues. The Cestoda, or tapeworms, are intestinal parasites in vertebrates. Trematodes, named from the Greek *trema* meaning hole, which refers to the muscular sucker containing a mouth (hole) at the anterior end of the fluke, are unsegmented, are flattened dorsoventrally, have leaf-shaped or elongated bodies, and have suckers or hooks, which attach the organism to the host allowing them to draw fluids from host tissue. Trematodes are hermaphroditic (contain both types of sex organs) and can cross-fertilize. Trematodes have indirect life cycles and infection occurs by ingestion or penetration of infective larvae (Figure 6-6A). As seen in Table 6-1, trematodes are named according to the tissue they inhabit in the definitive host (for example, *Fasciola hepatica* is the liver fluke).

Table 6-1   **Classification of Trematodes According to Their Habitat**

Blood flukes	■ *Schistosoma haematobium* ■ *Schistosoma mansoni* ■ *Schistosoma japonicum* ■ *Schistosoma mekongi*
Liver flukes	■ *Fasciola hepatica* ■ *Clonorchis sinensis* ■ *Opisthorchis felineus* ■ *Opisthorchis viverrini*
Lung flukes	■ *Paragonimus westermani*
Intestinal flukes	■ *Fasciolopsis buski* ■ *Heterophyes heterophyes*

Cestodes, named from the Greek *kestos* meaning embroidered belt as a result of their appearance, have long, ribbon-like, segmented bodies (Figure 6-6B). Adult cestodes inhabit the intestines of animals and they do not have their own digestive system. Cestodes obtain nutrients by absorption through their external covering. A specialized attachment organ called the scolex is located at the cranial end and has hooks, suckers, or both. The segments of the tapeworm are called proglottids and contain male and female sex organs. Tapeworms cross-fertilize or self-fertilize and eggs are produced in the proglottids. As the tapeworm grows, proglottids are pushed to the caudal end where they eventually break free to be passed in the host's feces or the proglottids disintegrate releasing ova. Most species of cestodes have two hosts. The adult worm lives in the definitive host and the larva stage lives in the intermediate host. The intermediate host must ingest the egg for the life cycle to be completed. For humans to get infected the larva stage must be ingested.

> Trematodes and cestodes are hermaphroditic (monoecious), whereas nematodes are dioecious (have separate sexes).

Nematoda

Nematoda is a diverse animal phylum containing more commonly known parasites such as hookworms (*Necator americanus* and *Ancylostoma duodenale* in people, *An. caninum* in dogs, and *An. braziliense* in cats), pinworms (*Enterobius vermicularis*), Guinea worms (genus *Dracunculus*), and intestinal roundworms (genus *Ascaris*). Nematodes are long, cylindrical worms whose bodies are covered by a tough external covering and are tapered at both ends (Figure 6-7).

Nematodes have a complete digestive system with both a mouth and an anus. Longitudinal muscle fibers allow the roundworms to move with a side-to-side motion. Nematodes are dioecious (the sexes are separate), undergo sexual reproduction, and males are typically smaller than females. The females are extremely prolific, producing hundreds or thousands of eggs that must embryonate (incubate) outside of the host. After this embryonation period, they are infectious to another indi-

Figure 6-7 An example of a nematode (*Strongyloides*) demonstrating its cylindrical appearance.
(Courtesy of CDC/Dr. Mae Melvin)

vidual. Some nematodes require an intermediate host. Nematodes are transmitted by ingestion of eggs or infective larvae or by penetration of the skin by larvae. The pathogenicity of the worms is often determined by the worm burden (the number of adult worms living in the intestine). The adults will mate in the intestine and the females will lay eggs that will pass out with the feces. Most ova are not infective immediately after passing out of the body in the feces (if most ova were infective immediately humans would have great numbers of worms in their bodies).

> Pinworm infections in people are often blamed on the family's pet dog or cat; however, *Enterobius vermicularis* (the pinworm) does not parasitize dogs or cats. Pinworms parasitize such animals as mice, rats, monkeys, rabbits, and horses.

Arthropods

Arthropods are found in the Kingdom Animalia and make up over 80% of all species on Earth. There are over 750,000 species of insects and the vast majority of arthropods are either beneficial or harmless to us.

Arthropods are involved in virtually every type of parasitic relationship serving as both definitive and intermediate hosts for protozoa, trematodes, nematodes, and other arthropods, functioning as vectors to transmit infective stages of parasites to vertebrates, and serving as parasites in their own right (such as fleas, ticks, and some crustaceans). Arthropod life cycles are as varied as their

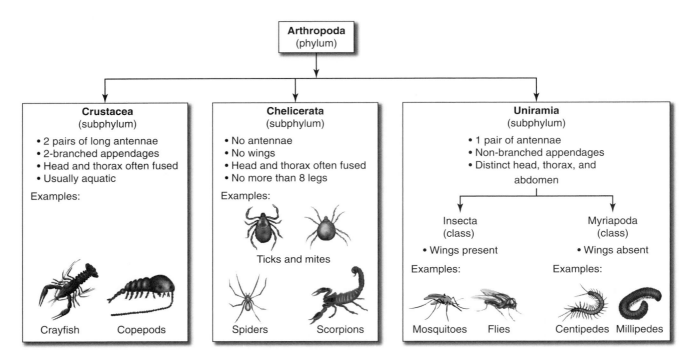

Figure 6-8 The Arthropoda phylum.

structures, which can provide a series of challenges in attempting to control parasitic infestations.

The Arthropoda phylum can be divided into three subphyla, each of which contains medically significant organisms (Figures 6-8 and 6-9). The subphylum Crustacea are arthropods with two pairs of antennae, two-branched appendages, and are usually aquatic. Examples in Crustacea are crayfish and copepods. The subphylum Uniramia are arthropods that have one pair of antennae, non-branched appendages, a distinct head, thorax, and abdomen, and in the class Insecta never have more than six legs. Examples in *Uniramia* include organisms in the class Insecta (insects) and the class Myriapoda (centipedes and millipedes). The subphylum Chelicerata are arthropods that have no antennae, no wings, a typically fused head and thorax, no more than eight legs, and possess specialized feeding appendages.

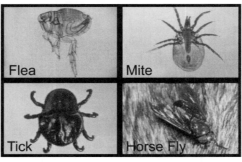

Figure 6-9 Examples of four arthropods that are often the source of vector-borne disease.

Examples in Chelicerata are in the order Acarina (ticks and mites), the order Araneae (spiders), and the order Scorpiones (scorpions).

Two key evolutionary adaptations are important in understanding arthropods. One of these adaptations is the development of jointed appendages. Jointed appendages allow arthropods to move quickly. The other important adaptation of arthropods is the rigid carbohydrate exoskeleton. The exoskeleton not only provides protection but also serves as an anchor for muscles that attach to the inner surface of this exoskeleton. The arthropod body can only grow so much before it expands into its exoskeleton, thus limiting the size arthropods can attain. The vast majority of arthropods are less than 1 centimeter in length. One problem

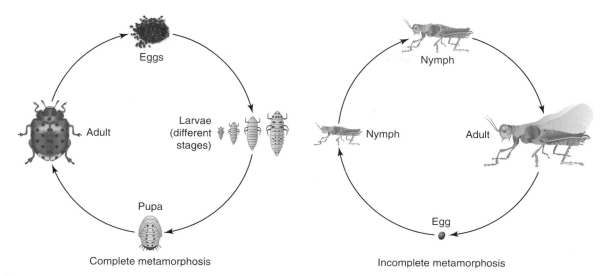

Figure 6-10 Comparison of complete (left) and incomplete (right) metamorphosis.

that arthropods encounter with this exoskeleton is that the animal is enclosed in a rigid, nonexpandable covering. The solution to this problem is a series of molts (changes) that all arthropods go through during their development. The larval stages between each molt are referred to as instars. The length of time between molts depends on the species involved. The number of instars also varies with species, the season or annual cycle, and sometimes with the species' nutritional state. The process of molting is controlled by hormones.

Arthropods differ from one another in their postembryonic development. In most species embryos develop into larvae. A larva is a life cycle state that is structurally distinct from the adult, normally occupies an ecological niche separate from the adult, is sexually immature, and must undergo a structural reorganization (**metamorphosis**) before becoming an adult. Metamorphosis is the change in size, form, and function that takes place as an immature arthropod reaches adulthood. In insects, two types of metamorphoses exist: complete and incomplete (also known as gradual) (Figure 6-10).

- In complete metamorphosis, the immature stages bear little resemblance to the adult, they are often found in different environments, and there is a pre-adult "resting" stage called the pupa. The stages in this type of metamorphosis are egg, larva, pupa, and adult (ELPA). If there is more than one larva stage, they are referred to as L_1, L_2, and so on and each larval stage is referred to as an "instar." An example of an insect that undergoes complete metamorphosis is the flea (insect order *Siphonaptera*).

- In incomplete or gradual metamorphosis, the immature stages look very similar to the adults (just miniature versions of them) and are usually found in the same environment. The stages in this type of metamorphosis are egg, nymph, and adult (ENA) and there are several nymphal instars. An example of an insect that undergoes incomplete metamorphosis is the bed bug (insect order *Hemiptera*).

In some arthropods, the terms complete and incomplete metamorphosis are not used to describe development of these organisms. The terms egg, larva, nymph, and adult (ELNA) are used to describe the developmental stages found in arachnids (spiders). Larva of arachnids have six legs (three pairs), whereas nymph and adult stages have eight legs (four pairs). In other cases (such as ticks), there's only one instar per stage (for example, one larva stage, which then molts into a nymph, which then molts into an adult). In still other cases (such as mites), there may be two or more nymph instars.

How Are Parasites Transmitted?

Three elements are necessary for the transmission of parasitic diseases: a source of infection; a mode of transmission; and a susceptible host. The source of infection may be the cyst or less often the trophozoite of a protozoan. Eggs or infective larvae are the source of infection for helminths.

Parasites may be transmitted from host to host through consumption of contaminated food and water, by injection of the parasite into the host by an arthropod vector, by transplacental transmission, or by putting anything into a mouth that has touched the stool (feces) of an infected person or animal. Insects and arachnids can transmit infectious agents to humans and animals either mechanically or biologically. Mechanical transmission involves the physical transfer of an infectious agent from an infected host to a noninfected host by the insect or arachnid. For example a horse with equine infectious anemia (EIA), a viral infection of horses, is fed upon by a blood-feeding horse fly. The blood on the fly's mouthparts is transferred to another horse when the fly feeds again. This is the same outcome as if blood from one horse has been transferred to another by a needle.

Biological transmission uses the insect or arachnid as a necessary part of the transmission cycle. Depending upon the organism, the infectious agent multiplies and/or matures within the insect/arachnid, and when the insect/arachnid feeds on an uninfected host, the agent is transferred. This may seem similar to the arthropod being an intermediate host; however, the term intermediate host is only used when the arthropod transmits parasitic organisms (not viruses, rickettsiae, or bacteria). An example of biological transmission is *Ixodes scapularis*, the vector for Lyme disease. Uninfected larval ticks acquire the infection by feeding on rodents (usually white-footed mice), which have *Borrelia burgdorferi* bacteria circulating in their blood. The organism multiples in the larval tick and persists as the tick molts to the nymph stage, and when the nymph tick feeds on an uninfected human or animal, the bacteria are transmitted to that host.

How Do Parasites Reproduce?

Parasites reproduce sexually or asexually or a combination of both. Asexual reproduction in parasites can be accomplished by several methods including mitotic fission and schizogony. Simple mitotic fission is common in protozoan parasites and occurs by production of two daughter cells. *En. histolytica* and *Tr. vaginalis* reproduce by mitotic fission. Many protozoa have increased their reproductive potential with a method known as schizogony. Schizogony (multiple fission) occurs when the nucleus undergoes several mitotic divisions after which the cytoplasm is divided among daughter cells. *Plasmodium* is a protozoan parasite that reproduces by schizogony.

Sexual reproduction in parasitic organisms is accomplished in several ways. Protozoa reproduce by conjugation and the production of gametocytes. Helminths reproduce by cross-fertilization and self-fertilization. Many cestodes and trematodes are hermaphroditic and can self-fertilize.

Some organisms undergo both sexual and asexual reproduction. *Bal. coli*, a ciliate, reproduces asexually by fission and sexually by conjugation. Sexes are separate in most arthropods and reproduction is generally sexual.

What Are Parasitic Life Cycles?

Parasitic life cycles are complex and have exact requirements. There are two basic types of parasitic life cycles: direct (simple) and indirect (complex). **Direct life cycles** involve only one definitive host, whereas **indirect life cycles** involve a definitive host and one or more intermediate hosts. Both types of life cycles are seen in protozoa and helminths.

Direct life cycles involve a single host where the parasite often spends most of its life (usually as an adult) and where it reproduces. In direct life cycles the parasites are transmitted from one host to another by direct contact or by ingestion of a form of the parasite (usually a cyst or egg) (Figure 6-11).

Indirect life cycles are more complicated than direct life cycles as a result of the need for one or more intermediate hosts and typically further development of the parasite that will only occur in a specific environment (Figure 6-12).

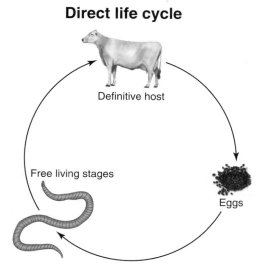

Direct life cycle

Definitive host

Free living stages

Eggs

Figure 6-11 A generic example of a direct life cycle.

How Do Parasites Cause Disease?

Most human parasites cause disease by going through three general stages. The first stage is transmission of the parasite to the animal/human host from a source such as water, soil, food, or other animal. The microbe then invades and multiplies

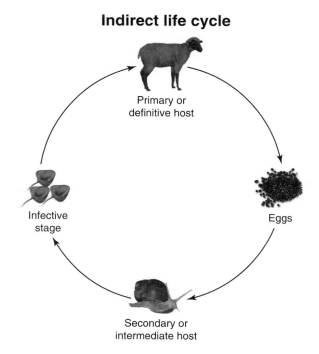

Indirect life cycle

Primary or definitive host

Infective stage

Eggs

Secondary or intermediate host

Figure 6-12 A generic example of an indirect life cycle.

in the host, producing more parasites that can infect other suitable hosts. The final stage is when the parasite leaves the host in large numbers by a specific means and, in order to survive, must find and enter a new host. There are variations on this scheme that can include the parasite invading more than one host species (using an intermediate host), the parasite being spread by means of vectors, and the parasite being spread through body fluids.

Arthropods are unique in that they may cause disease by three different categories: direct effect, miscellaneous effects, and transmission of infectious diseases.

- Direct effect is the physical irritation caused by bites or stings of the arthropod. Examples of arthropods that cause disease by direct effect are mites, ticks, spiders, scorpions, lice, bedbuds, kissing bugs, fleas, bees, flies, mosquitoes, and beetles.

- Miscellaneous effects include tissue invasion and psychological effects caused by the arthropod. Examples of arthropods that cause disease by miscellaneous effects are scabies (invasion of the skin), demodetic mange (invasion of hair follicles), screwworm larvae (invasion of the brain), and spinal paralysis (continued biting by Ixodidae ticks). Some of these arthropods also cause direct effects.

- Transmission of infectious diseases is the arthropod serving as a means of moving infectious organisms from one individual to another. Examples of arthropods serving as means of infectious disease transmission include mosquitoes (malaria, encephalitis, dengue fever, yellow fever), tsetse flies (trypanosomiasis), black flies (onchocerciasis), fleas (plague), ticks (Lyme disease, ehrlichiosis), and kissing bugs (Chagas' disease).

How Are Parasitic Diseases Diagnosed?

A variety of techniques are used to detect and identify parasitic infections.

- *Stool specimens.* The examination of fecal samples for parasitic organisms begins with examination of unpreserved fecal samples. Protozoan trophozoites and oocysts are found in diarrhea, whereas cyst forms are found in formed stool. A direct wet film (commonly called a fecal smear or direct smear) may be made from a small portion of the fecal sample and a drop of saline on a clean slide. Specimens that cannot be promptly examined need to be preserved in a fixative such as polyvinyl alcohol (PVA).

- *Blood specimens.* The examination of blood samples usually involves the preparation of both thick and thin blood smears. Thin preps must be thin enough to avoid obscuring the parasites with blood cells (ideally one cell layer thick). Thick preps will contain more organisms per field; however, this technique can cause some distortion. Buffy coat preps (using the layer between the red blood cells and serum after centrifugation) concentrate the white blood cells and are especially useful for detecting *Leishmania* and trypanosomes.

- *Sputum specimens.* Pulmonary infection with parasites can be detected by examination of sputum (lower respiratory tract secretion). The best sample for examination is induced, free of saliva, and collected early in the morning. Fresh sputum is examined by wet mount or is fixed with PVA.

- *Urine and vaginal specimens.* Wet mounts are used for examination of urine, vaginal secretions, or prostate exudate.

- *Cerebrospinal fluid (CSF) specimens.* CSF is best examined promptly (within 20 minutes of collection) and is often centrifuged to concentrate parasitic organisms.

- *Serologic tests*. Serologic tests are used when conventional methods of parasite identification fail to yield results. Many test kits are available; however, cross-reaction with other antigens/antibodies is possible, which limits their usefulness. Some examples of serologic tests include complement fixation (Chagas' disease, *Leishmania*, and *Paragonimus*), agglutination tests (Chagas' disease), and Western blot tests (schistosomiasis).

PROTOZOAN ZOONOSES

Amebiasis

Overview

Amebiasis, also called amebic dysentery or invasive amebiasis, is caused by *Entamoeba histolytica*, an anaerobic parasitic protozoan that predominantly infects humans and other primates. Amebiasis is the third leading parasitic cause of death worldwide, surpassed only by malaria and schistosomiasis. In 1875 Fedor Aleksandrovich Lošch, in St. Petersburg, Russia, first described amebiasis in a young farmer suffering from chronic dysentery who was admitted to his clinic two years earlier. He described the organisms as round, pear-shaped, or irregular forms that were in a state of almost continuous motion. After finding large numbers of ameba in the patient's feces, Losch named the organism *Amoeba coli* and documented its pathogenicity by injecting amebae from his patient into a dog's rectum. The dog had the same colonic ulcers that the patient had on his autopsy. In 1903 German bacteriologist Fritz Schaudinn subsequently renamed the organism *En. histolytica* because of its ability to kill host cells and named another nonpathogenic species *En. coli*. Schaudinn was the first to differentiate between *En. histolytica*, the cause of amebic dysentery, and its harmless counterpart, *En. coli*. In 1886 in Egypt, Kartulis proved that amebae caused intestinal and hepatic lesions in patients with diarrhea. In 1890, Sir William Osler reported the first North American case of amebiasis in stool and abscess fluid from a physician who had previously lived in Panama. In 1891, Councilman and Lafleur, at Johns Hopkins University Hospital, differentiated between bacillary and amebic dysentery. Walker and Sellards, in the Philippines, described the pathogenic role of amebae in extensive studies in 1913 and in 1925, Dobell further described the organism's life cycle. By the start of the 20th century it was clear to some scientists that many people shedding *En. histolytica* cysts were asymptomatic and in 1925 Emile Brumpt suggested that there were two separate but indistinguishable types of ameba. It is now believed that *En. histolytica* and *En. dispar* are two of eight species of *Entamoeba* that sometimes inhabit the human colon. *En. histolytica* is an invasive organism capable of causing life-threatening intestinal and extra-intestinal disease whereas *En. dispar* appears not to be invasive.

Causative Agent

Entamebae in the order Amoebida are anaerobic protozoa that lack mitochondria and have a nucleus with a distinct endosome (nucleolus-type organelle). *En. histolytica* is the main pathogen of this group and is mainly a disease in humans; however, it can occur in nonhuman primates, dogs, cats, and rats. *En. histolytica* exists in two forms: a large (20 to 40 µm), motile, feeding, reproducing trophozoite form and a smaller (10 to 15 µm), nonfeeding, nonmotile, environmentally resistant cyst form, which is the form responsible for the transmission of infection. The trophozoites of *En. histolytica* inhabit the crypts of the large intestine producing the lesions of amebic colitis.

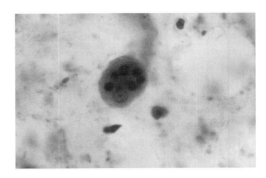

Figure 6-13   Trophozoite
of *Entamoeba histolytica.*
(Courtesy of CDC)

The trophozoite form of *En. histolytica* has a clear ectoplasm and a somewhat granular endoplasm containing several vacuoles (Figure 6-13). The trophozoite has a single 3- to 5-μm spherical nucleus with fine peripheral chromatin and a central endosome. Food vacuoles and ingested RBCs may be present in the endoplasm of the trophozoite.

The fully mature cyst form of *En. histolytica* has 1 to 4 nuclei that are morphologically similar to the nuclei of the trophozoite (Figure 6-2). Its wall is thin and transparent. Immature cysts may have iodine-stainable glycogen clumps and rod-like structures called chromatoid bodies with smooth rounded edges in their endoplasm. Cysts can live for up to 30 days in water; however, they are rapidly killed by drying and temperatures below 5°C and above 40°C. Cysts are resistant to chlorine levels commonly used for water purification.

En. histolytica has been divided into two species based on isozyme (enzymes that catalyze the same reaction, but are coded on different gene loci) analysis. These species are *En. histolytica* (pathogen) and *En. dispar* (noninvasive species). *En. dispar* is a nonpathogenic protozoan morphologically identical to *En. histolytica*. *En. dispar* may produce intestinal lesions experimentally and is often found in captive primates.

Morphologically distinct organisms that are nonpathogenic include:

- *En. coli*, a commensal in the intestinal tract of humans and pigs, in its trophozoite form has a large eccentric endosome, irregular peripheral chromatin clumping along the nuclear membrane, and endoplasm that may contain ingested bacteria, but not erythrocytes. The cyst form has up to eight nuclei and if chromatoid bars are present they have pointed rather than rounded ends.

- *En. gingivalis*, found between teeth and in tonsillar crypts in primates, was the first ameba of humans to be described and is nonpathogenic. This species has no cyst form (therefore transmission must be direct via kissing or droplet spray) and it is the only species to ingest leukocytes, which can be seen in permanently stained smears.

- *En. hartmanni* is a small, nonpathogenic ameba in humans that for many years was considered to be an *En. histolytica* form. It is now known that these different sized amebae are two species. The life cycle, morphology, and overall appearance of *En. hartmanni* is identical to *En. histolytica* except for its smaller size (12 to 15 μm for the trophozoite and 5 to 9 μm for the cyst).

- *En. polecki* is typically a parasite of pigs, monkeys, and rarely humans where it is nonpathogenic. *En. polecki* cysts are rare and have just one nucleus.

- *Endolimax nana* lives in the human large intestine as a commensal. The trophozoite is tiny (6 to 15 μm) with a small nucleus and large centrally or eccentrically located endosome. Large glycogen vacuoles and food vacuoles may be present. The mature cyst is 5 to 14 μm and contains four nuclei.

- *Iodamoeba butschlii* is found in humans, nonhuman primates, and pigs worldwide and is nonpathogenic. Trophozoites are 9 to 14 µm long and move by short, blunt pseudopodia. The nucleus is relatively large and contains a large endosome. Food vacuoles may contain bacteria and yeast. The cyst almost always has one nucleus and a large glycogen vacuole that stains deeply with iodine (hence its name).

Epizootiology and Public Health Significance

En. histolytica is found worldwide with approximately 10% of the world population being infected with *Entamoeba* spp. with the Indian subcontinent, southern and western Africa, the Far East, South America, and Central America having the highest number of infected people. In endemic areas, as many as 25% of people may be carrying antibodies to *En. histolytica* as a result of prior infections. This parasite occurs in all climatic zones, but is predominantly found in the tropics and subtropics such as Asia, Africa, and Central and South America.

Approximately 500 million people are infected (50 million new cases per year) with what appears to be *En. histolytica* worldwide with up to 100,000 deaths occuring annually (mostly from liver abscesses). The number of infected people increases as urban migration and low econonic status of some developing countries produce unhygenic conditions. Infection is more common in warmer areas because the cysts can survive longer in warm, moist conditions; however, the parasite has been found in arctic areas as well.

The prevalence rate of amebiasis in the United States is about 4% with *En. dispar* infection 10 times more common than *En. histolytica* infection. Approximately 10% of *En. histolytica* infections cause invasive disease. In the United States, amebiasis is more prevalent in immigrants from endemic areas, in people of lower socioeconomic status, in institutionalized people, and people who live in communal situations.

Transmission

Fecal-oral transmission of cysts involves oral ingestion of cyst-contaminated food or water. Amebae can be directly transmitted by anal sexual contact. Flies and cockroaches can serve as mechanical vectors. Dogs, monkeys, and rodents are animal reservoirs for some amebic species. Laboratory monkeys should be considered a potential source of *En. histolytica* protozoa and handled properly to avoid zoonotic transmission of the parasite.

> In theory, the ingestion of one viable cyst can cause an infection.

Pathogenesis

The incubation period for *En. histolytica* varies from 2 days to 4 months. After ingestion of fecally contaminated food or water, cysts pass through the stomach unharmed and are not active in the acidic environment of the stomach. From the stomach they travel to the small intestine, where excystation occurs (transformation from cyst to trophozoite). After excystation, the parasite may lead a commensal existence on the mucosal surface and in the crypts of the colon. In asymptomatic infections the ameba lives by eating and digesting bacteria and food particles in the gut. The protective layer of mucus that lines the intestine prevents the parasite from coming into contact with the intestine itself. Disease occurs only when amebae come in contact with the cells lining the intestine where trophozoites live and multiply in the crypts of the large intestine (typically the cecum and proximal colon initially then spreading the entire length of the colon). *En. histolytica* trophozoites adhere to colonic mucins, epithelial cells, and leukocytes and may invade the colonic epithelium to produce the flask-shaped ulcerative lesions

> The adverse colonic environment signals the trophozoite form to change to its cystic form, which is better adapted to survive in harsh conditions of the colon. Trophozoites passed in stool are not able to encyst.

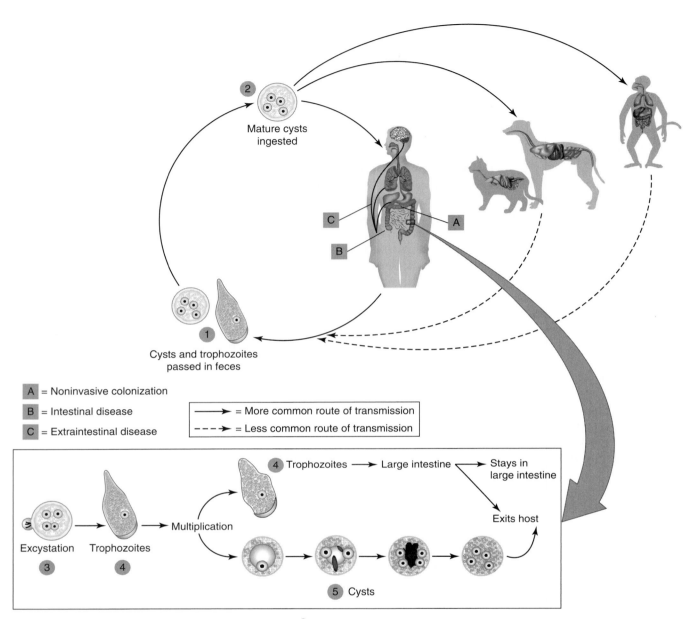

Figure 6-14 Cysts and trophozoites are passed in feces❶ with cysts typically found in formed stool and trophozoites typically found in diarrheal stool. Infection by *Entamoeba histolytica* occurs by ingestion of mature cysts❷ in fecally contaminated food, water, or hands. Excystation❸ occurs in the small intestine and trophozoites❹ are released, which migrate to the large intestine. The trophozoites multiply and produce cysts❺, and both stages are passed in the feces❶. In many cases, the trophozoites remain confined to the intestinal lumen🄰 producing a noninvasive infection resulting in asymptomatic carriers who pass cysts in their stool. In some patients the trophozoites invade the intestinal mucosa🄱 producing intestinal disease. Extraintestinal sites such as the liver, brain, and lungs can be infected via the bloodstream🄲 producing extraintestinal disease.

typical of intestinal amebiasis. *En. histolytica* is unique among amebae in that they hydrolyze host tissue causing the intestinal cells to leak ions such as Na⁺, K⁺, and Ca⁺⁺ from their cytoplasm. A number of hemolysins, an extracellular cysteine kinase, and protein kinase activators increase the cytolytic activity of *En. histolytica* (Figure 6-14).

En. histolytica also feeds on the destroyed cells by phagocytosis and is often seen with red blood cells inside. A granulomatous mass (ameboma) may form in the wall of the colon that can cause obstruction and is sometimes confused with cancer.

Spread of amebiasis to the liver occurs via the portal blood with trophozoites ascending the portal veins to produce liver abscesses filled with acellular protein-aceous debris. The trophozoites of *En. histolytica* lyse the hepatocytes and the neutrophils. Metastatic infection first involves the liver then may spread from the liver to the lung, brain, or other viscera.

In most people (about 90%), the trophozoites re-encyst and produce asymp-tomatic infection that typically resolves within 12 months. In a few people (about 10%), the parasite causes symptomatic amebiasis.

> The prevalence of *En. histolytica* has declined in the United States over the past several decades, but amebiasis is still important in many tropical areas, particularly in times of disasters.

Clinical Signs in Animals

Humans are the natural host for *En. histolytica* and they are the usual source of infection for domestic animals. In animals, amebiasis is an acute or chronic coli-tis characterized by persistent diarrhea or dysentery. It is common in nonhuman primates (rhesus monkeys and chimpanzees), occasionally seen in dogs and cats, and rarely seen in other mammals. Although several species of amebae are found in mammals, the only known pathogen is *En. histolytica*. The parasite lives in the lumen of the large intestine and cecum and may produce no obvious clinical signs (especially in cats) or it may invade the intestinal mucosa producing mild to severe, ulcerative, hemorrhagic colitis (acute and chronic disease is seen sporadically in dogs). In acute disease, dysentery may develop, which may be fatal or resolve spontaneously. In chronic disease, weight loss, anorexia, tenesmus, and chronic diarrhea or dysentery may develop. Amebae may invade perianal skin, genitalia, liver, brain, lungs, kidneys, and other organs producing clinical signs associated with the organ system involved.

Clinical Signs in Humans

Ameba infections in humans can last for years and may present with no symptoms (as many as 90% of cases), gastrointestinal signs ranging from mild diarrhea to dysentery with blood and mucus, or systemic (extraintestinal) disease. Clinical signs of these presentations include:

> *En. invadens* is found in reptiles and is morphologically identical to *En. histolytica*, but it is not transmissible to mammals.

- Asymptomatic infections are common following ingestion of *En. histolytica*, are typically self-limiting, and may be recurrent.
- Amebiasis may cause gastrointestinal signs including acute amebic colitis (with a gradual onset of a 1- to 2-week history of abdominal pain, diarrhea, and tenesmus), fulminant amebic colitis (a rare complication of amebic dysentery that presents in children under two years of age with severe bloody diar-rhea, severe abdominal pain, and high fever with intestinal perforation), or chronic amebic colitis (a disease that is clinically similar to inflam-matory bowel disease with recur-rent episodes of bloody diarrhea and vague abdominal discomfort). Occasionally a granulomatous mass called an ameboma forms in the intestinal wall and may cause bowel obstruction (Figure 6-15).
- Extraintestinal amebiasis occurs when the organism spreads from the gastrointestinal tract to other

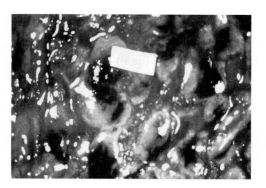

Figure 6-15 Gross pathology of intestinal lesions caused by *Entamoeba histolytica*. (Courtesy of CDC)

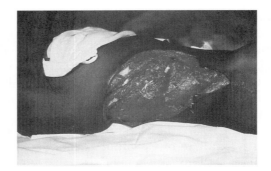

Figure 6-16 Tissue necrosis because of invasive extraintestinal amebiasis. (Courtesy of CDC)

organs in the body. Secondary lesions can be found in nearly every organ of the body, but the liver and lungs are most commonly affected. Hepatic amebiasis results when the trophozoites enter the venules and travel to the liver. As the trophozoites enter the portal capillaries and sinusoids, they begin to form liver abscesses that may remain pinpoint in size or grow to the size of a grapefruit and rupture. Signs of hepatic amebiasis appear as an abrupt onset of high fever; a cough and dull, aching, constant abdominal pain in the right cranial abdomen usually lasting fewer than 10 days. Pulmonary amebiasis typically develops in 20- to 40-year old people by metastasis from a hepatic lesion (usually a ruptured liver abscess) with clinical signs of cough, pleural pain, and dyspnea. Other extraintestinal infections that occur secondary to liver abscess rupture include amebic peritonitis, amebic pericarditis, cerebral amebiasis, and genitourinary ulcers (Figure 6-16).

Diagnosis in Animals

> Cysts of *En. histolytica* can remain viable and infective in moist, cool environments for about 12 days and in water they can live up to 30 days.

Definitive diagnosis of amebiasis is accomplished by finding *En histolytica* trophozoites or cysts in feces. In intestinal infections, repeated examinations may be necessary because parasites may be passed periodically in the feces. Trophozoites are best seen in direct smears of diarrheic stool examined either as a wet mount or stained smear (iodine, trichrome, iron hematoxyline, or periodic acid-Schiff stain). Stained biopsy sections of affected colonic tissue is also used to diagnosis amebiasis. Colonoscopy or biopsy of ulcerations is more effective than fecal examination in diagnosing amebic colitis. Cysts may be recovered and identified from formed stool using zinc sulfate flotations or in fixed and stained preparations (iodine, trichrome, or iron hematoxyline); however, *En. histolytica* cysts are seldom excreted by some species such as dogs or cats. Animals with extraintestinal amebiasis typically have no concurrent intestinal infection and would need to be diagnosed via immunologic tests. An enzyme-linked immunoabsorbant assay (ELISA)-based antigen test used in humans may aid in diagnosis of other mammals. Immunostaining techniques may also be used.

Diagnosis in Humans

> Feces should be examined promptly because trophozoites die quickly once outside the body.

Asymptomatic human infections are usually diagnosed by finding cysts in formed stool via various flotation or sedimentation procedures with stains used to help visualize the cysts. A minimum of three stools should be examined because cysts are not shed constantly. It is important to distinguish the *En. histolytica* cyst (up to four nuclei and a centrally located endosome) from the cysts of nonpathogenic ameba such as *En. coli* (up to eight nuclei and an off-centered endosome). In symptomatic infections, the trophozoite can often be seen in fresh feces. *En. dispar* cannot be distinguished from *En. histolytica* using microscopy. Polymerase chain reaction (PCR) and monoclonal antibody methods can distinguish between *En. histolytica* and *En. dispar*.

Serologic tests such as ELISA tests, indirect hemagglutination antibody, and immunodiffusion are available and most individuals whether they have symptoms or not will test positive for the presence of antibodies. Antibody levels are much higher in people with liver abscesses. Serologic tests become positive about 2 weeks after infection. More recently developed tests include detection of the presence of ameba proteins in the feces and the detection of ameba DNA in feces (these tests are not widely used as a result of their expense).

Treatment in Animals

Limited information is available on treatment in animals; however, metronidazole or furazolidone have been suggested as acceptable therapies. Animals, especially dogs, may continue to shed trophozoites after therapy.

Treatment in Humans

En. histolytica infections occur in both the intestine and in tissue resulting in the use of two different drugs to clear the infections. Metronidazole is used to destroy amebae that have invaded tissue because it is rapidly absorbed into the bloodstream and transported to the infection site. Metronidazole's rapid absorption limits its time in the intestine; therefore, other drugs such as paromomycin are used for treating intestinal infections. Ornidazole and tinidazole are used to treat amebic liver abscesses.

Management and Control in Animals

Humans are the natural hosts for *En. histolytica* and are the usual source of infection for animals. Proper management of food and water to prevent human disease will also decrease the disease in animals.

Management and Control in Humans

There are many ways to limit the spread of *En. histolytica* because its transmission relies on contaminated food and water. Proper disposal of human and animal waste, eliminating vectors such as flies and cockroaches, proper water sanitation including proper plumbing techniques, and limiting the use of human waste for fertilizer can limit the spread of ameba. Humans are the most important reservoir of *En. histolytica*; however, other animals may be the source of infection. Proper hygiene after handling animals is also important in limiting the spread of this disease.

Summary

Amebiasis is caused by *En. histolytica*, an anaerobic parasitic protozoan that predominantly infects humans and other primates. *En. histolytica* exists in two forms: a large (20 to 40 μm), motile, feeding, reproducing trophozoite form and a smaller (10 to 15 μm), nonfeeding, nonmotile, environmentally resistant cyst form responsible for the transmission of infection. *En. histolytica* is mainly a disease in humans; however, it can occur in nonhuman primates, dogs, cats, and other mammals such as rats. *En. dispar* is a nonpathogenic protozoon morphologically identical to *En. histolytica* and needs to be differentiated from the pathogenic form. Morphologically distinct organisms that are nonpathogenic include

En. coli (a commensal in the intestinal tract of humans and pigs), *En. gingivalis* (found between teeth and in tonsillar crypts of primates), *En. hartmanni* (small, nonpathogenic ameba in humans), *En. polecki* (a parasite of pigs, monkeys, and rarely humans), *End. nana* (found in the human large intestine as a commensal), and *Io. butschlii* (a nonpathogen found in humans, nonhuman primates, and pigs). *En. histolytica* is found worldwide with approximately 10% of the world population being infected with *Entamoeba* spp. Fecal-oral transmission of cysts involves ingestion of cyst-contaminated food or water. Amebae can be directly transmitted by anal sexual contact. Flies and cockroaches can serve as mechanical vectors. Humans are the natural host for *En. histolytica* and they are the usual source of infection for domestic animals. In animals, amebiasis is an acute or chronic colitis characterized by persistent diarrhea or dysentery. In acute disease, dysentery may develop, which may be fatal or resolve spontaneously. In chronic disease, weight loss, anorexia, tenesmus, and chronic diarrhea or dysentery may develop. Ameba infections in humans can last for years and may present with no symptoms (as many as 90% of cases), gastrointestinal signs ranging from mild diarrhea to dysentery with blood and mucus, or systemic (extraintestinal) disease. Definitive diagnosis of amebiasis is accomplished by finding *En. histolytica* trophozoites or cysts in feces. Trophozoites are best seen in direct smears of diarrheic stool examined either as a wet mount or stained smear and may be recovered and identified from formed stool using zinc sulfate flotations or in fixed and stained preparations. An ELISA-based antigen test and immunostaining techniques may also be used to diagnose amebiasis. *En. histolytica* infections are usually treated with metronidazole and paromomycin. Ways to limit the spread of *En. histolytica* include proper disposal of human and animal waste, eliminating vectors such as flies and cockroaches, proper water sanitation including proper plumbing techniques, and limiting the use of human waste for fertilizer. Humans are the most important reservoir of *En. histolytica*; however, other animals may be the source of infection making proper hygiene after handling animals important in limiting the spread of this disease.

Babesiosis

Overview

Babesiosis, also known as piroplasmosis and Nantucket fever in people, is caused by protozoan red blood cell parasites of the genus *Babesia*. Although the first human case of babesiosis was reported in 1957 in a Yugoslavian cattle farmer and the first U.S. human case was reported from Nantucket, Massachusetts in 1969, *Babesia* organisms were known as pathogens for some time. *Babesia* is named after the Romanian bacteriologist Victor Babeș, who was the first person to identify the red blood cell protozoan in 1888 when he noticed intraerythrocytic protozoa in cattle with febrile hemoglobinuria. *Babesia* parasites were the first protozoa linked to a disease by Theobald Smith and Frank Kilbourne as the causative agent of red-water fever (also known as Texas cattle fever and hemoglobinuria) in cattle. The discovery of the protozoan that causes Texas cattle fever (currently named *Bab. bigemina*) in 1889 and the demonstration (1889–1893) of its transmission by the cattle tick (*Boophilus annulatus*) was instrumental in understanding the role of vectors in disease transmission. Cattle with red-water fever develop an acute, febrile, hemolytic disease characterized by the destruction of erythrocytes and the development of hemoglobinuria.

Babesiosis affects a wide range of domestic and wild animals and occasionally humans with its major economic impact on the cattle industry. All *Babesia* organisms are tick-transmitted intraerythrocytic parasites that are collectively called piroplasms as a result of their pear-shaped figures within red blood cells (*pirum* is Latin for pear and *plasma* is Greek for something formed). Most piroplasms are transmitted by hard ticks and are capable of infecting a wide variety of vertebrate hosts that serve as reservoirs of the parasite.

Babesia species are informally grouped by size: small *Babesia* species have trophozoites of 1 to 2.5 μm and include *Bab. gibsoni, Bab. microti,* and *Bab. rodhaini*; large *Babesia* species have trophozoites of 2.5 to 5 μm and include *Bab. bovis, Bab. caballi,* and *Bab. canis.*

Causative Agent

Babesia spp. are intracellular protozoal organisms that parasitize erythrocytes ultimately causing anemia in the host (Figure 6-17). Many different species exist with varying host specificity. *Babesia* parasites belong to the Apicomplexa phylum and have two host life cycles involving an invertebrate host (a tick) and a vertebrate host (mammal). In the mammalian host the organisms reproduce asexually in the host's red blood cells. The infective stage (found in the tick) is the **sporozoite**, which is about 2-μm long and pyriform, spherical, or ovoid in shape. Sporozoites are injected into the host by the tick bite, enter red blood cells, and become trophozoites (the form seen diagnostically in blood smears).

Although more than 100 species have been reported, only a few have been identified as causing human infections. *Bab. microti* and *Bab. divergens* have been identified in most human cases, but different species have also been identified in humans. The various species of *Babesia* include:

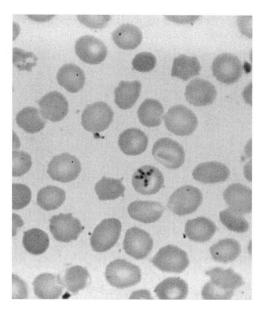

Figure 6-17 The tetrad configuration of *Babesia* trophozoites.

(Courtesy of CDC/Steven Glenn, Laboratory & Consultation Division)

- *Bab. microti*. Human babesiosis in the United States was rare until 1969 when cases involving *Bab. microti* were identified. *Bab. microti* was described in white-footed mice in the 1930s, but was not identified as the primary reservoir until 1976. *Bab. microti* has been found in meadow voles and other rodents that can also infect pets. It is transmitted from rodents to humans by *I. scapularis*, the same tick that transmits Lyme disease and human granulocytic anaplasmosis.

- *Bab. divergens*. This species of *Babesia* usually infects cattle mainly in Europe, but there have been rare human infections caused by this parasite. It is small like *Bab. bovis* and is transmitted by *I. ricinus*.

- *Bab. bigemina*. The vector for *Bab. bigemina* is a hard tick (*Boophilus annulatus, Boo. microplus,* and *Boo. decoloratus*) and this parasite infects a variety of ruminants (cattle, deer, and water buffalo). The parasite often occurs in pairs (hence the name bigemina) in the animal's red blood cells. In cattle *Bab. bigemina* can cause massive destruction of the red blood cells resulting in hemoglobinuria (the urine appears red), and the disease can kill cattle within a week. This parasite causes more severe disease in adult cattle than in calves.

- *Bab. bovis*. This species of *Babesia* is transmitted by *Boo. annulatus* and *Boo. microplus* ticks and infects cattle in Europe, Russia, and Africa. Both *Bab. bigemina* and *Bab. bovis* are widespread in tropical and subtropical areas.

- *Bab. canis* and *Bab. gibsoni*. These two organisms commonly infect dogs and are transmitted by ixodid tick vectors (*Rhipicephalus sanguineus* and *Dermacentor variabilis*). *Bab. canis* is the large babesia found in the southern United States and South Africa, whereas *Bab. gibsoni* is the small babesia found in Asia and northern Africa. These organisms are found throughout Asia, Africa, Europe, the Middle East, and North America, with *Bab. canis* being more prevalent (*Bab. gibsoni* infection is increasing in frequency).
- *Bab. equi* and *Bab. caballi* cause anemia, icterus, hemoglobinuria, and death in horses. Equine babesiosis caused by the large *Bab. caballi* is widely distributed throughout the tropics and subtropics. *Bab. equi* is the small babesia and has a wide geographic distribution. *Bab. caballi* and *Bab. equi* are transmitted by ticks of the genera *Dermacentor*, *Hyalomma*, and *Rhipicephalus* and are passed transovarially from one tick generation to the next. *Bab. equi* was reclassified as *Theileria equi* in 1998.
- Other species of *Babesia* that can infect animals include *Bab. felis* (cats), *Bab. motasi* (large babesia of sheep), *Bab. ovis* (sheep and goats), *Bab. trautmanni* (large babesia of swine), and *Bab. perroncitol* (small babesia of swine).

Epizootiology and Public Health Significance

The *Babesia* species found in humans include *Bab. microti, Bab. divergens, Bab. bovis, Bab. equi, Bab. gibsoni,* and *Bab. canis.*

Babesiosis is a relatively rare, underreported disease. Since 1969, 30 cases have been reported in Europe and more than 300 cases have been documented in the United States. Babesiosis occurs mainly in coastal areas in the northeastern United States, especially the offshore islands of New York and Massachusetts. U.S. cases have also been reported in Wisconsin, California, and Georgia.

There has been an increase in babesiosis over the past 30 years potentially as a result of the restocking of the deer population, curtailment of hunting, and an increase in outdoor recreational activities. The mortality of babesioisis in the United States is low with most cases being asymptomatic that improve without treatment. Babesiosis in Europe is a more devastating disease and is often fatal.

Transmission

Ticks are responsible for transmission of most cases of babesiosis. Transmission to the host via saliva contaminated with sporozoites occurs when the tick feeds on animal hosts or humans. *Babesia* spp. can also be transmitted from contaminated blood transfusions, mechanical transmission by insects or during surgical procedures, and intrauterine infection; however, these methods are extremely rare.

Pathogenesis

Ticks must feed for 2 to 3 days for transmission of *Babesia* to occur.

When an infected tick feeds on a vertebrate host, the *Babesia* parasite enters the host in the sporozoite form. Sporozoites immediately enter red blood cells where they become trophozoites. In the red blood cells, the trophozoites invade the host's red blood cells and multiply asexually by binary fission producing merozoites. The merozoite form of *Babesia* destroys red blood cells and causes anemia. Uninfected ticks ingest the infected vertebrate's blood when feeding, and the merozoite-containing red blood cells enter the tick's midgut. After multiplying in the tick's midgut (gamogony), *Babesia* sporozoites migrate to the salivary glands. Sporozoites enter the animal's or human's blood stream during the tick bite and become intraerythrocytic (Figure 6-18).

The clinical manifestations of babesiosis are a result of the presence of the asexual reproductive stage in host red blood cells and their subsequent lysis. A critical host defense against this infection is the spleen because the spleen traps

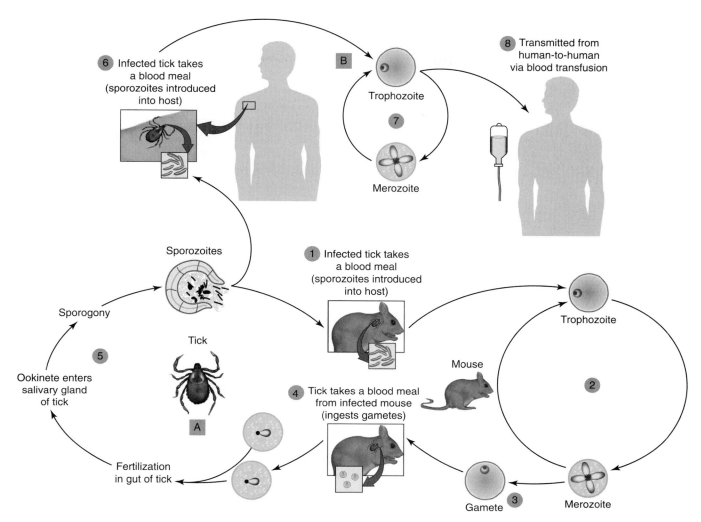

Figure 6-18 The *Babesia microti* life cycle involves two hosts, which includes a rodent (primarily the white-footed mouse, *Peromyscus leucopus*). During a blood meal, a *Babesia*-infected tick introduces sporozoites into the mouse host❶. Sporozoites enter erythrocytes and undergo asexual reproduction❷ to trophozoites. Trophozoites multiply asexually to form merozoites. In the blood, some parasites differentiate into male and female gametes❸. The definitive host is a tick (such as the deer tick, *Ixodes scapularis*). Once ingested by an appropriate tick❹, gametes unite and undergo a sporogonic cycle resulting in sporozoites❺. Transovarial transmission has been documented for large *Babesia* spp. but not for the small babesiae, such as *Babesia microti*🅐.

Humans enter the cycle when bitten by infected ticks. During a blood meal, a *Babesia*-infected tick introduces sporozoites into the human host❻. Sporozoites enter erythrocytes🅑 and undergo asexual replication❼ to trophozoites. Trophozoites multiply asexually to form merozoites. Merozoites destroy erythrocytes causing anemia. Human-to-human transmission is known to occur through blood transfusions❽.

Deer are the hosts upon which the adult ticks feed and are indirectly part of the *Babesia* cycle as they influence the tick population. When deer populations increase, the tick population also increases, thus heightening the potential for transmission of *Babesia* protozoa.

infected erythrocytes, which are then ingested by macrophages. People or animals without a spleen are at higher risk of complications from babesiosis.

Clinical Signs in Animals

Babesiosis in animals varies with the species involved; however, clinical signs are related to the parasitism of red blood cells by *Babesia* organisms.

- *Cattle.* Texas cattle fever, red-water fever, or hemoglobinuria are synonyms for the disease caused by *Bab. bigemina* in the United States (*Bab. bovi* and *Bab. divergens* can also cause disease in cattle). The incubation period is 8 to 15 days in cattle. Acute disease typically occurs over a period of 1 week with the first clinical sign being fever (frequently 105.8°F (41°C) or higher). Other clinical signs include anorexia, tachypnea, muscle tremors, anemia, jaundice, and weight loss with hemoglobinemia and hemoglobinuria occurring in the final stages. The high fever can cause late-term pregnant cows to abort and bulls may become temporarily infertile. Virulent strains of *Bab. bovis* can cause hypotensive shock, nonspecific inflammation, and coagulation disturbances. Cattle that recover from the acute disease remain infected without showing clinical signs for a number of years (*Bab. bovis*) and for a few months (*Bab. bigemina*). Lesions include an enlarged and friable spleen; a swollen liver; enlarged gallbladder; congested, dark-colored kidneys; and generalized anemia and jaundice. The urine is often red (hence the hemoglobinuria name).
- *Horses.* Equine babesiosis is caused by *Th. equi* or *Bab. caballi* with the disease prevalent in Africa, Europe, Asia, South and Central America, and the southern United States. Clinical signs include fever, anemia, jaundice, hemoglobinuria, and death.
- *Dogs. Bab. canis* has been reported in dogs from most regions and consists of subspecies *Bab. canis canis, Bab. canis vogeli,* and *Bab. canis rossi.* Clinical signs of *Bab. canis* infection vary from a mild, transient illness to acutely fatal disease. *Bab. gibsoni* causes a chronic disease with progressive, severe anemia as the main sign.

Clinical Signs in Humans

The incubation period of babesiosis is from 1 to 4 weeks with clinical signs related to red blood cell damage caused by the parasite. Clinical signs include fever, hemolytic anemia, and hemoglobinuria. Capillary blockage may occur as red blood cell fragments are formed resulting in systemic signs involving the liver, spleen, kidney, and central nervous system (CNS). Red blood cell fragments are removed from circulation by the spleen. Red blood cell destruction results in hemolytic anemia. People who develop babesiois often have a history of travel to an endemic area between the months of May and September. This is the period during which the *Ixodes* tick is in its infectious nymph stage.

Diagnosis in Animals

Babesiosis in animals is diagnosed by identifying intraerythrocytic merozoites using both Giemsa-stained thick and thin blood smears. Serologic tests such as indirect immunofluorescent antibody and ELISA tests can confirm asymptomatic animals. PCR assays that can detect low levels of organisms are available but are not in routine use.

Diagnosis in Humans

Diagnosis of babesiosis in humans is similar to that described for animals. In people immunofluorescent antibody titers of greater than 1:64 is positive and a fourfold rise in titer or a single titer greater than 1:256 is suggestive of acute infection. ELISA tests are also available. PCR tests are highly specific and can be used to confirm diagnosis. Inoculation of hamsters with the person's blood and subsequent antibody analysis of the animal's blood is used to confirm diagnosis when peripheral blood smear and laboratory tests are equivocal.

Treatment in Animals

A variety of drugs have been used to treat babesiosis, but only diminazene aceturate and imidocarb dipropionate are currently in use. These drugs are not available in all endemic countries, or their use may be restricted. Supportive treatment may include the use of anti-inflammatory drugs, antioxidants, and corticosteroids. Blood transfusions may be needed in extremely anemic animals.

Treatment in Humans

Treatment of babesiosis in people has historically been with clindamycin and quinine, but atovaquone and azithromycin have also been used. Supportive care may be needed in people with more severe disease.

Management and Control in Animals

Tick control is the best way to control babesiosis. A live, attenuated vaccine has been used successfully in cattle in a number of countries (such as Argentina, Australia, Brazil, Israel, South Africa, and Uruguay) with a single inoculation providing immunity for the commercial life of the animal. If *Bab. bigemina* is found in cattle, both the protozoan and the tick must be reported to state and federal authorities. A canine vaccine for protection against *Babesia canis* infection is available in Europe.

Management and Control in Humans

Tick control is the best way to prevent babesiosis in people. No vaccine is available to protect humans against babesiosis.

Summary

Babesiosis is caused by a variety of *Babesia* spp., protozoal parasites that infect erythrocytes ultimately causing anemia in the host. Babesia parasites have two host life cycles involving an invertebrate host (a tick) and a vertebrate host (mammal). The various species of Babesia include *Bab. microti* (rodents), *Bab. divergens* (cattle), *Bab. bigemina* (cattle), *Bab. bovis* (cattle), *Bab. canis* (dogs), *Bab. gibsoni* (dogs), *Th. (Bab.) equi* (horses), *Bab. caballi* (horses), *Bab. felis* (cats), *Bab. motasi* (large babesia of sheep), *Bab. ovis* (sheep and goats), *Bab. trautmanni* (large babesia of swine), and *Bab. perroncitol* (small babesia of swine). The *Babesia* species found in humans include *Bab. microti*, *Bab. divergens*, *Bab. bovis*, *Bab. equi*, *Bab. gibsoni*, and *Bab. canis* although *Bab. microti* is most commonly seen in humans. Babesiosis is a relatively rare, underreported disease found mainly in coastal areas in the northeastern United States. Ticks are responsible for transmission of most cases of babesiosis. *Babesia* spp. can also rarely be transmitted from contaminated blood transfusions, mechanical transmission by insects, or during surgical procedures, and intrauterine infection. The clinical manifestations of babesiosis are a result of the presence of the asexual reproductive stage in host red blood cells and their subsequent lysis. Fever, anemia, jaundice, and hemoglobinuria are seen clinically in a variety of animal species and humans. Babesiosis in animals and humans is diagnosed by identifying intraerythrocytic merozoites using both Giemsa-stained thick and thin blood smears. Serologic tests such as indirect immunofluorescent antibody and ELISA tests can confirm asymptomatic animals. PCR assays that can detect low levels of organisms are also available. A variety of drugs have been used to treat animals with babesiosis, but only diminazene aceturate and imidocarb dipropionate are currently in use. Treatment of babesiosis in people is with clindamycin, quinine, atovaquone, or azithromycin. Tick control is the best way to control babesiosis. A live, attenuated vaccine is available for animals in other countries; no vaccine is available for people.

Cryptosporidiosis

Overview

Cryptosporidiosis is a disease caused by the Apicomplexa protozoan *Cryptosporidium* that was once thought to only infect animals, but in 1976 human cases were documented in the United States in people who handled calves with scours. *Cryptosporidium* is derived from the Latin term *crypta* meaning hidden and the Greek term *spora* meaning seed. *Cryptosporidium* parasites were first described by Tyzzer in 1907 (*Cry. muris* from mice). In 1912 Tyzzer described *Cry. parvum* and from the time of discovery until 1976 only fifteen reports describing *Cryptosporidium* infection in eight species of animals were published. In 1987 a *Cryptosporidium* outbreak causing acute diarrhea in Georgia affected 13,000 people and the pathogen was shown to be transmitted in water systems affecting healthy people. From 1989 to 1997 nearly 1,000 reports of human and domestic animal infection were reported. Animal species included in these reports were calves, lambs, Guinea pigs, turkeys, chickens, and pet and zoo animals. Cryptosporidiosis in people causes self-limiting diarrhea in healthy individuals and severe prolonged diarrhea in immunocompromised people. The source of most endemic human infections is human-to-human fecal-oral transmission, but it also may include animal-to-person transmission and waterborne transmission. In 1993 a waterborne cryptosporidiosis outbreak occurred in Milwaukee, Wisconsin, where an estimated 403,000 people became ill, including the hospitalization of 4,400 people. In the summer of 2005, numerous reports of gastrointestinal upset in people attending a water park at Seneca Lake State Park, New York, was found to have two water storage tanks infected with *Cryptosporidium*. In recent years, other species of *Cryptosporidium* have been found in humans including *Cry. hominis* (morphologically indistinguishable from *Cry. parvum*), *Cry. meleagridis* (occurring in turkeys), *Cry. baileyi* (occurring in birds), *Cry. canis* (occurring in dogs), and *Cry. felis* (occurring in cats).

Causative Agent

Cry. parvum protozoa are in the Apicomplexa phylum (these members have a complex of special intracellular structures located at the apex of its infective stage). *Cry. parvum* is an obligate intracellular parasite that is a member of the class Conoidasida in the subclass Coccidiasina, which are commonly referred to as coccidia. When the parasite is isolated from different hosts it is given different names; however, it is currently believed that the form infecting humans is the same species that causes disease in young calves. The infective stage of the organism is the oocyst, which is round, between 2 to 6 µm in diameter, is highly refractile, and contains four slender sporozoites (the motile infectious stage developing asexually) (Figure 6-19). Only oocysts are seen in feces and can live a long time in water (including seawater).

Cry. parvum is widespread in mammals and is an important pathogen of calves and lambs.

Cryptosporidium spp. infects many herd animals such as cows, goats, sheep, deer and elk.

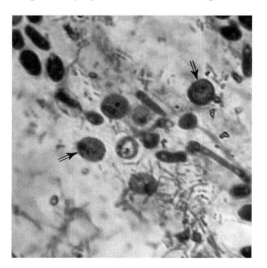

Figure 6-19 Direct fecal smear stained to demonstrate *Cryptosporidium* protozoa. (Courtesy of CDC)

Epizootiology and Public Health Significance

Cry. parvum is found worldwide with the highest prevalence in developing countries with inadequate sanitation. This parasite is carried asymptomatically in 30% of people living in developing nations. Most endemic outbreaks are caused by contaminated water or food. Waterborne outbreaks tend to be associated with improper water filtration techniques because the cysts are chlorine-resistant and chlorine levels in treated water may not be high enough to kill the oocyst.

The prevalence of cryptosporidiosis in North America is about 2% of the immunocompetent population and 10% to 15% of the immunocompromised population. Serologic studies indicate that 30% of the U.S. population has been exposed to cryptosporidiosis. The rate of cryptosporidiosis is higher in developing countries, ranging from 12% to 48% of people with diarrhea. In developing nations in Asia and Africa, 5% to 10% of immunocompetent people have *Cry. parvum* oocysts in their stool. Serologic studies in South America show that more than 60% have been exposed to cryptosporidiosis. The mortality rate of cryptosporidiosis in immunocompromised people is higher than 50%.

The prevalence of cryptosporidiosis in animals is variable among species. Cattle are the main source of zoonotic transmission. One national study reported that approximately 60% of cattle and 25% of calves are infected with *Cryptosporidium* protozoa. Incidence of cryptosporidiosis in other animals is not known.

Cryptosporidium has a spore phase (oocyst) and can survive for long periods outside a host.

Transmission

Cryptosporidium organisms are typically transmitted by ingestion of thick-walled oocysts and are frequently associated with traveling, contact with livestock, person-to-person (including anal sex), animal-to-animal, indirectly via fomites, from contamination in the environment, or by fecal contamination of the feed or water supply. Airborne transmission has been reported mainly in children and immunocompromised people. The sporulated thick-walled oocyst containing four infectious sporozoites is excreted in feces and is the infectious agent for a susceptible host. Another type of oocyst that is thin-walled and not very resistant can result in autoinfections and may be the cause of persistent infections in people and animals.

Pathogenesis

Cry. parvum is highly infectious (requiring only 10 to 100 oocysts to cause disease) with oocysts that are immediately infectious. Once in the intestine the parasite becomes intracellular but extracytoplasmic, which may contribute to the difficulty in treating this parasite. The mechanism by which *Cry. parvum* causes diarrhea is not completely understood, but may be a combination of secretory and malabsorptive alterations. After invasion of the small intestinal cells by the parasite, the epithelial cells release cytokines. These cytokines activate phagocytes and recruit new leukocytes. These cells release soluble factors resulting in intestinal secretion of chloride and water, whereas the parasites themselves cause damage to the intestinal villi. The result is nutrient malabsorption and osmotic diarrhea.

The life cycle of *Cry. parvum* begins with ingestion of the sporulated oocyst. Each oocyst contains four infective sporozoites, which exit from a suture located along one side of the oocyst when exposed to the acid environment of the stomach. Sporozoites attach to and penetrate individual epithelial cells typically in the ileum (or epithelium of the respiratory tract if the oocyst is inhaled). Multiple fission (schizogony) occurs; resulting in the formation of eight merozoites within the meront (referred to as type I meronts). The type I meronts rupture, releasing free merozoites. Once these merozoites penetrate new cells, they undergo merogony

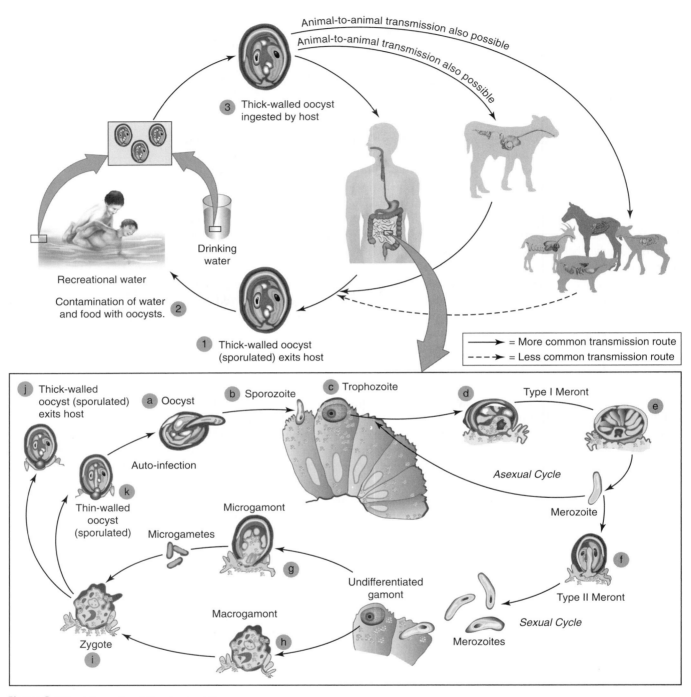

Figure 6-20 Life cycle of *Cryptosporidium*.

Sporulated oocysts, containing four sporozoites, are excreted by the infected host through feces and possibly other routes such as respiratory secretions❶. Transmission of *Cryptosporidium* occurs mainly through contact with contaminated water (occasionally contaminated food sources may transmit the protozoan). Zoonotic transmission of *Cryptosporidium parvum* occurs through exposure to infected animals or exposure to water contaminated by feces of infected animals❷. Following ingestion (and possibly inhalation) by a suitable host❸, excystation❶ occurs. The sporozoites are released and parasitize epithelial cells(❶, ❶) of the gastrointestinal tract or other tissues such as the respiratory tract. In these cells, the parasites undergo asexual multiplication (schizogony or merogony)(❶, ❶, ❶) and then sexual multiplication (gametogony) producing microgamonts (male)❶ and macrogamonts (female)❶. Upon fertilization of the macrogamonts by the microgametes(❶), oocysts(❶, ❶) develop that sporulate in the infected host. Two different types of oocysts are produced, the thick-walled (commonly excreted from the host)❶, and the thin-walled oocyst❶ (primarily involved in autoinfection). Oocysts are infective upon excretion, thus permitting direct and immediate fecal-oral transmission.

to form additional meronts. Type I merozoites are triggered into forming a second type of meront (type II meront), which contains only four merozoites. Upon release, the Type II merozoites form the sexual stages. Some type II merozoites enter cells, enlarge, and form macrogametes. Others undergo multiple fission inside cells, forming microgametocytes containing 16 nonflagellated microgametes. Microgametes rupture from the microgametocyte and penetrate macrogametes forming a zygote (fusion of gametes). A resistant oocyst wall is then formed around the zygote, meiosis occurs, and four sporozoites are formed (sporogony). These oocysts are passed in the feces and into the environment. Approximately 20% of the oocysts produced in the gastrointestinal tract fail to form a thick oocyst wall producing only a series of membranes surrounding the developing sporozoites. These thin-walled oocysts contain sporozoites that can excyst while still within the gastrointestinal tract and infect new cells of the same individual (termed autoinfection) (Figure 6-20).

Infected animals shed large numbers of oocysts that are resistant to harsh environmental and chemical conditions, including chlorine levels usually used in water treatment. Although disease can present in any animal, in general the infection is limited to the jejunum and typically without clinical signs in immunocompetent people and adult animals. In immunocompromised people and neonatal animals the entire gastrointestinal tract and respiratory tract may be involved producing clinical signs.

Clinical Signs in Animals

Cryptosporidiosis in animals is caused by *Cry. parvum* and occurs primarily in neonatal calves but also causes disease in lambs, kids, foals, and piglets. Seropositivity in dogs and cats suggests that cryptosporidiosis is common in these species, but produces a self-limiting disease and typically occurs in animals younger than 6 months of age.

> *Cryptosporidium* parasites live in or under the membrane of epithelial cells lining the digestive and respiratory systems.

- *Calves.* Cryptosporidial infection is common in 1- to 3-week-old dairy calves (cryptosporidiosis is a common cause of diarrhea in calves 5 to 15 days of age). Calves usually have a mild to moderate diarrhea producing feces that are yellow or pale, watery, and contain mucus. The persistent diarrhea may last for several days regardless of treatment resulting in marked weight loss and emaciation. In most cases, cryptosporidial diarrhea is self-limiting after several days. Oocysts are usually passed in the feces of calves for 3 to 12 days.

- *Lambs and kids.* *Cry. parvum* is a common gastrointestinal infection in young lambs and goats (there are high fatality rates in lambs 4 to 10 days of age and in goat kids 5 to 21 days of age).

- *Pigs.* Cryptosporidial infection in pigs typically does not produce clinical signs and is seen over a wide range of ages (1 week of age through market age).

- *Foals.* Cryptosporidial infection in foals is less prevalent than in ruminants, is seen at an older age (5 to 8 weeks of age), and is usually subclinical in immunocompetent foals.

- *Wildlife.* Cryptosporidiosis can also be seen in young deer, especially in artificially reared orphans.

Clinical Signs in Humans

Cryptosporidiosis in humans can be classified as intestinal or extraintestinal.

> Diarrhea caused by *Cry. parvum* can occur by itself, but more commonly is associated with mixed infections.

- Intestinal cryptosporidiosis in people can be asymptomatic or cause severe watery diarrhea with abdominal cramps. Asymptomatic infections occur in normal and immunocompromised people. Acute diarrhea in immunocompetent people is a predominant clinical manifestation of *Cry. parvum* infection particularly in underdeveloped countries, medical personnel, travelers, and institutionalized people. In immunocompromised people symptoms range from asymptomatic infection to severe life-threatening diarrhea, dehydration, and chronic malabsorption syndromes.

- Extraintestinal disease usually involves infection in the bile duct (causing cholangitis and pancreatitis) or the respiratory tract (causing coughing and a low-grade fever).

Diagnosis in Animals

Diagnosis of cryptosporidiosis in animals is based on detection of oocysts by examination of fecal smears using Ziehl-Neelson stains, by fecal flotation, or by serologic test methods. In fecal samples the oocysts are small (4 to 5 μm in diameter) and are best seen using phase-contrast microscopy.

Small intestinal histological samples of calves will have villous atrophy (villi are shorter than normal and crypts are hyperplasic) with large numbers of the parasite embedded in the microvilli. In mild infections, only a few parasites are present with only minor histological changes.

> Spreading cattle manure on frozen farm fields may cause *Cryptosporidium* oocysts to enter the waterways as the ground thaws.

Diagnosis in Humans

Round, sporulated oocysts (4 to 5 μm) can be seen microscopically in stool, duodenal secretions, or bile. The oocysts can be found using Sheather's sugar or zinc sulfate fecal flotation methods, by direct examination using phase-contrast microscopy, or by staining with acid-fast, Giemsa, or Ziehl-Neelson stains. A monoclonal antibody-based immunofluorescent kit used to stain the organisms and an ELISA test is available commercially to diagnosis cryptosporidiosis. Diagnosis has also been made by staining the trophozoites in intestinal biopsies; however, many people whose stools tested positive for *Cry. parvum* can test negative by duodenal biopsy. Pulmonary cryptosporidiosis is diagnosed by biopsy and staining.

Treatment in Animals

In the United States, there is no currently licensed treatment for *Cry. parvum* infection in food animals. Calves are treated with oral and parenteral fluids and electrolytes, with cow's whole milk given in small quantities several times daily to minimize weight loss. Hyperimmune bovine colostrum can reduce the severity of diarrhea and the period of oocyst shedding in calves. Halofuginone (in infected lambs and calves) and paromomycin sulfate (in infected goat kids) have been used experimentally for the treatment of cryptosporidiosis. In dogs the disease is usually self-limiting but has been treated with paromomycin, tylosin, and azithoromycin.

Treatment in Humans

Nitazoxanide is an antiprotozoal drug approved in 2002 for the treatment of human cryptosporidiosis. Intestinal cryptosporidiosis is self-limiting in most healthy people with watery diarrhea lasting 2 to 4 days. Immunocompromised people may have the disease for life, with the severe watery diarrhea contributing to death in some people. Disease involving the respiratory system may also be fatal.

Management and Control in Animals

Cryptosporidiosis is difficult to control completely and the goal may be reducing the number of oocysts ingested, which will reduce the severity of infection. Oocysts are resistant to most disinfectants and may survive for several months in cool and moist conditions. Oocysts are sensitive to drying and can be destroyed by ammonia, hydrogen peroxide, chlorine dioxide, formalin, freeze-drying, and exposure to temperatures less than 32°F (0°C) or greater than 149°F (65°C). Some ways to manage this disease include the following:

- Parturition should occur in a clean environment,
- Neonates should receive adequate amounts of colostrum,
- Animal-to-animal contact should be limited for the first weeks of life,
- Strict hygiene protocols should be enforced when feeding young animals,
- Animals with diarrhea should be isolated from healthy animals during the course of the diarrhea and for several days after recovery,
- Mechanical transmission of infection should be avoided,
- Animal housing should be vacated and cleaned on a regular basis,
- An all-in/all-out management system is optimal for preventing disease transmission,
- Rats, mice, and flies should be controlled and should not have access to feeding and feed storage areas.

Management and Control in Humans

People should avoid direct contact with human or animal feces and garden soil. Drinking water should be boiled or filtered to prevent water-borne disease. All produce should be throughly washed because of the potential use of manure fertilization used in some countries. Swimming in public baths or water potentially contaminated by animal feces should be avoided. Immunocompromised people should limit their access to young animals.

Summary

Cryptosporidiosis is caused by the coccidian protozoan, *Cry. parvum*, an obligate intracellular parasite. *Cry. parvum* is found worldwide with the highest prevalence in developing countries with inadequate sanitation. The infective stage of the organism is the oocyst, which is round, between 2 to 6 μm in diameter, and contains four slender sporozoites. Only oocysts are seen in feces and can live a long time in water (including seawater). *Cryptosporidium* organisms are typically transmitted by ingestion of thick-walled oocysts and are frequently associated with traveling, contact with livestock, person-to-person, animal-to-animal, indirectly via fomites, from contamination in the environment, or by fecal contamination of the feed or water supply. Waterborne outbreaks tend to be associated with improper water filtration techniques because the cysts are chlorine-resistant and chlorine levels in treated water will not necessarily kill the cyst.

Cry. parvum is highly infectious (requiring only 10 to 100 oocysts to cause disease) with oocysts that are immediately infectious. Infected animals shed large numbers of oocysts that are resistant to harsh environmental and chemical conditions, including chlorine levels usually used in water treatment. Cryptosporidiosis in animals is caused mainly by *Cry. parvum* and occurs primarily in neonatal calves but also causes disease in lambs, kids, foals, and piglets (dogs and cats are seropositive indicating that they have been exposed to the protozoan). Calves usually have a mild to moderate diarrhea producing feces that are yellow or pale, watery, and contain mucus. Diarrhea is the main clinical sign in other affected animals. Cryptosporidiosis in humans can be classified as intestinal or extraintestinal. Intestinal cryptosporidiosis in people can be asymptomatic or cause severe watery diarrhea with abdominal cramps. Extraintestinal disease usually involves infection in the bile duct or the respiratory tract. Diagnosis of cryptosporidiosis in animals is based on detection of oocysts by examination of fecal smears using Ziehl-Neelson stains, by fecal flotation, or by serologic test methods. Diagnosis in humans is done using fecal flotation methods such as with Sheather's sugar or zinc sulfate solution, by direct examination using phase-contrast microscopy, or by staining with acid-fast, Giemsa, or Ziehl-Neelson stains. A monoclonal antibody-based immunofluorescent kit used to stain the organisms and an ELISA test is available commercially to diagnosis cryptosporidiosis. In the United States, there is no currently licensed treatment for *Cry. parvum* infection in food animals. Treatment in animals is supportive. Nitazoxanide is an antiprotozoal drug approved in 2002 for the treatment of human cryptosporidiosis. Intestinal cryptosporidiosis is self-limiting in most healthy people with watery diarrhea lasting 2 to 4 days. Cryptosporidiosis is difficult to control completely and the goal may be reducing the number of oocysts ingested, which will reduce the severity of infection. Oocysts are resistant to most disinfectants and can survive for several months in cool and moist conditions. Proper management techniques and avoiding direct contact with contaminated sources is recommended to reduce the incidence of cryptosporidiosis.

Giardiasis

Overview

Giardiasis is the most common nonbacterial cause of diarrhea in North America.

Giardiasis, also called lambliasis, backpacker disease, beaver fever, and runner's diarrhea, is caused by the flagellated protozoan *Giardia duodenalis* (also known as *G. intestinalis or G. lamblia*). *G. duodenalis* was first discovered in 1681 by Dutch microbiologist Antony van Leeuwenhoek, who found the organism in his own stool. This protozoan was initially named *Cercomonas intestinalis* by Lambl in 1859 and renamed *G. lamblia* by Stiles in 1915, in honor of French biologist Alfred M. Giard of Paris and Dr. F. Lambl of Prague. Originally each species of *Giardia* was given a different name when it was found in a new host such as *G. canis* (dogs), *G. bovis* (cattle), etc., but currently only five species are recognized and each species can infect more than one host. It appears that some *Giardia* spp. are infective to a variety of mammals, whereas others are more species specific.

Giardiasis is known as beaver fever because people who drank water downstream from a beaver dam often became sick.

Giardiasis is a chronic, intestinal protozoal infection that is seen worldwide in most domestic and wild mammals, many birds, and people. *G. duodenalis* is the most commonly diagnosed intestinal parasite in the United States producing symptoms ranging from none to severe, chronic diarrhea. Wildlife such as elk, deer, beaver, and muskrats can carry *G. duodenalis* and can contaminate streams or lakes. People and animals can become infected by drinking water in mountain streams that is presumed to be clean and uncontaminated. Infection is common in dogs and cats, occasional in ruminants, and rare in horses and pigs. *Giardia* spp. have been found in fecal samples from pet and shelter dogs and cats, with higher rates of infection in younger animals.

Causative Agent

There have been more than 40 species of *Giardia* described but only five are currently considered valid: *G. duodenalis* (which is the same as G. *intestinalis* or G. *lamblia*), *G. muris* from mammals, *G. ardeae* and *G. psittaci* from birds, and *G. agilis* from amphibians. Members of the genus *Giardia* are found in the Hexamitidae family whose members are recognized because they have two equal nuclei lying side by side. *Giardia* spp. lack mitochondria and are believed to be relatively primitive organisms. The parasite occurs in two morphologically distinct forms: trophozoites, the motile, vegetative form that is pear-shaped with eight

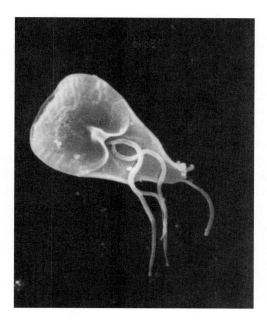

Figure 6-21 Scanning electron micrograph of flagellated *Giardia duodenalis* protozoan (trophozoite form). (Courtesy of CDC/Janice Carr)

flagella and is about 10 to 20 μm long (Figure 6-21); and cysts, the oval-shaped, thin-walled form that contains 4 nuclei (2 nuclei are seen in newly formed cysts) and is 8 to 14 μm long. The trophozoite has what is referred to as falling leaf motility that may be difficult to see in mucus. The cyst often has the cytoplasm pulled away from the cyst wall and there may be a halo effect around the outside of the cyst wall as a result of shrinkage caused by dehydrating agents.

Epizootiology and Public Health Significance

G. duodenalis is found worldwide with higher percentages found in tropical areas than in temperate climates. WHO estimates the incidence of symptomatic giardiasis in Asia, Africa, and Latin America to be about 200 million, with about 500,000 new infections per year. *G. duodenalis* is the most commonly isolated intestinal parasite throughout the world with infection rates of 20% to 40% in developing countries (especially in children). In the United States prevalence varies from 3% to 13%. *G. duodenalis* is the most common protozoan organism in humans in the United States. *G. duodenalis* is endemic in some childcare facilities and can be detected in approximately 20% of asymptomatic children.

The prevalence of *Giardia* in animals is not known as a result of the lack of clinical signs in some animals and the absence of surveillance testing. It is believed that between 30% to 50% of cattle may carry *Giardia*.

Transmission

G. duodenalis is easily transmitted from human to human as a result of the number of cysts excreted daily in the feces of infected people. Zoonotic transmission occurs when animal reservoirs such as beavers, dogs, cats, sheep, and livestock contaminate water with feces containing infective cysts. Beavers are a particularly significant source of human infection. Indirect transmission via food contaminated by infected workers has been documented as well as contaminated fomites in daycare facilities. Human and animal fertilizers can contaminate vegetables and fruits. Flies are possible carriers of cysts.

Giardia spp. can be transmitted from humans to dogs and rodents and vice versa. Mammals such as dogs, cattle, cats, primates, sheep, pigs, and rodents are direct or indirect sources of human infection.

G. canis and *G. cati* are morphologically indistinguishable from *G. duodenalis*.

Pathogenesis

> *Giardia* cysts can survive in cold water for several months.

G. duodenalis lives in the duodenum, jejunum, and proximal ileum. After ingestion of cyst-contaminated water or food (infective dose varies from 10 to 100 cysts) the trophozoite excysts in the small intestine, rapidly multiplies, and attaches to the cell surface of small intestinal villi by means of a disk on its posterior or ventral surface. A protein on the trophozoite lining (lectin) recognizes specific receptors on the intestinal cell and causes tight attachment between the parasite and the villi, resulting in mucosal damage and mechanical obstruction (in the small intestine and in diarrheic stools, only the trophozoites are found). As feces enter the colon, they begin to dehydrate and the parasites become encysted. Mature infective cysts pass in formed feces and complete the cycle. Cyst shedding may be continuous over several days and weeks and is often intermittent. Overcrowding and high humidity favor the survival of cysts (Figure 6-22).

> The pathogenesis of diarrhea in giardiasis is related to the number of organisms ingested, the specific strain ingested, non–antibody protective factors in the gastrointestinal tract, and the immune response of the host.

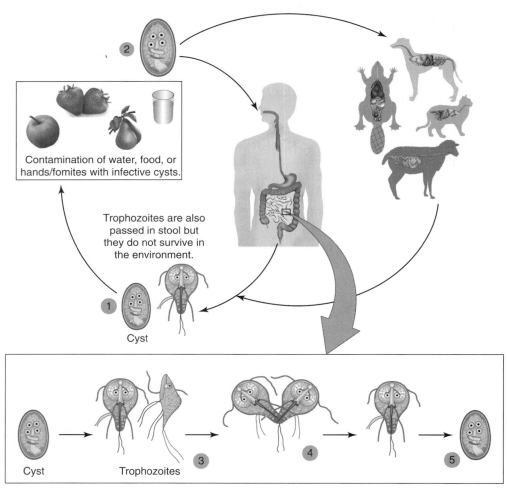

Figure 6-22   Cysts are resistant forms and are responsible for transmission of giardiasis. Both cysts and trophozoites can be found in the feces (diagnostic stages)❶. The cysts are hardy and can survive several months in cold water. Infection occurs by the ingestion of cysts in contaminated water, food, or by the fecal-oral route (hands or fomites)❷. In the small intestine, excystation releases trophozoites (each cyst produces two trophozoites)❸. Trophozoites multiply by longitudinal binary fission remaining in the lumen of the proximal small intestine where they can be free or attached to the mucosa by a ventral sucking disk❹. Encystation occurs as the parasites transit toward the colon. The cyst is the stage found most commonly in nondiarrheal feces❺. Because the cysts are infectious when passed in the stool or shortly afterward, person-to-person transmission is possible.

Clinical Signs in Animals

Giardia infections in dogs and cats may produce inapparent signs or may produce weight loss, chronic diarrhea, or steatorrhea that may be continual or intermittent (particularly in puppies and kittens). Feces are typically soft, pale in color, malodorous, and contain mucus. Blood is typically not present in feces and vomiting only occurs occasionally. Gastrointestinal disease caused by *Giardia* has also been reported in calves. Guinea pigs infected with *Giardia* are asymptomatic. Mucoid diarrhea is seen in *Giardia*-infected rodents. *Giardia*-infected horses have chronic diarrhea. Birds infected with *Giardia* have voluminous, pea-soup consistency feces.

Clinical Signs in Humans

People with giardiasis can be asymptomatic carriers (about 3% to 7% of U.S. population), can develop acute infectious diarrhea (short-lasting acute diarrhea (with or without low-grade fever), nausea, abdominal distension, mucoid or fatty stools, and anorexia), or chronic diarrhea (intermittent, loose, foul-smelling stools, abdominal distension, belching, flatulence, nausea, and anorexia). The incubation period lasts from 7 to 21 days. Illness typically lasts 1 to 2 weeks, but in chronic cases may last for months to years.

Diagnosis in Animals

Trophozoites are occasionally seen in saline smears of loose or watery feces. Cysts are best detected in feces concentrated by the zinc sulfate flotation. Staining cysts with iodine can help with identification (Figure 6-23). Several fecal examinations should be performed if giardiasis is suspected because of the intermittent shedding of this organism. An ELISA test that detects *Giardia* antigen in the feces of dogs and cats is available. Gross intestinal lesions are rarely seen, although microscopically villous atrophy may be present. PCR tests have been developed to detect parasite DNA from both *Giardia* trophozoites and cysts. These assays are extremely sensitive and specific and are able to detect low levels of active or inactive parasite.

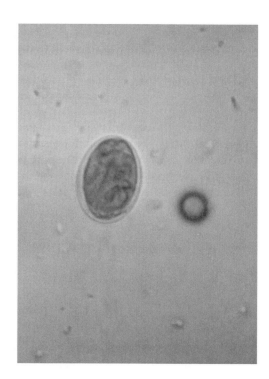

Figure 6-23 Cyst of *Giardia duodenalis* using an iodine staining method.
(Courtesy of CDC)

Diagnosis in Humans

Giardiasis is most commonly diagnosed in humans using an ELISA test that detects *G. duodenalis* antigen in stool (sensitivity is 92% to 98% and specificity is 87% to 100%). Fecal specimens should be examined within 1 hour after being collected or should be preserved in polyvinyl alcohol (PVA) or 10% formalin. Fecal examination for both trophozoites and cysts can be done with direct wet saline preparations with/without iodine; trichrome stained slides and formalin-ethyl-acetate concentration methods are additional tests that can be done. Duodenum or proximal jejunum aspiration may be performed in people suspected of having

Intermittent excretion of the parasite, administering contrast media for radiographs, improper specimen collection, and inadequate training of the observer contribute to the difficulty in identifying the organism.

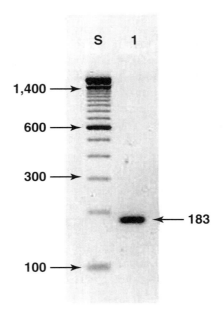

Figure 6-24 PCR diagnostic test for detection of *Giardia* DNA. Lane S is the molecular base pair standard (100-bp ladder). The black arrows show the size of the standard bands. Lane 1 is a positive fecal test for *Giardia* whose size is 183 bp.

giardiasis who have negative fecal examinations. Small intestinal biopsies and touch preparations that are stained with Giemsa stain can also be used to identify *G. duodenalis*. PCR tests for *Giarida* are also available (Figure 6-24).

Treatment in Animals

Fenbendazole is used to treat giardiasis in dogs and cats. A combination product of praziquantel, pyrantel pamoate, and febantel has been shown to decrease cyst excretion in infected dogs. Occasionally metronidazole is used to treat giardiasis in dogs. Giardiasis in calves may be treated with albendazole or fenbendazole. A killed vaccine is available for dogs and cats and reduces clinical signs and cysts shed into the environment.

Treatment in Humans

G. duodenalis is treated with metronidazole, furazolidone, paromonmycin, or tinidazole in people.

Management and Control in Animals

Giardia cysts are immediately infective when passed in the feces serving as a source of infection and reinfection for animals (particularly those in crowded, confined conditions). Promptly removing feces from cages, runs, and yards limits environmental exposure. Cysts can be inactivated by most quaternary ammonium compounds, household bleach (1:32 or 1:16 dilution), steam, and boiling water. Solutions should be left on the surface for 5 to 20 minutes before rinsing to increase their effectiveness. Grass areas should be considered contaminated for at least one month after infected dogs have left the premises. Cysts contaminating the hair of dogs and cats may be a source of reinfection making shampooing and rinsing of the animals important in limiting disease spread. Vaccination is available for dogs and cats to aide in disease prevention by decreasing or preventing cyst shedding. Fenbendazole deworming in calves in also effective against *Giardia*.

Management and Control in Humans

Giardiasis can be controlled in people by protecting food from contaminated cysts and filtering or boiling drinking water. In endemic areas, human and animal feces should not be used for fertilizer. Eating undercooked food or unpeeled fruit should be avoided. While swimming in ponds or streams, swallowing water should be avoided. A vaccine for *Giardia* is not available for people.

Summary

Giardiasis is one of the most common intestinal parasitic diseases in the world. Giardiasis is a chronic, intestinal protozoal infection caused by *G. duodenalis* and is seen worldwide in most domestic and wild mammals, many birds, and people. The disease is most prevalent in developing countries, usually associated with poor sanitary conditions, poor water quality control, and overcrowding. *G. duodenalis* is also a major cause of waterborne outbreaks of diarrhea in the United States, primarily in mountainous areas where water supplies may be contaminated with feces from humans or wildlife. *G. duodenalis* has two stages: trophozoites, the active form of the parasite inside the body, and cysts, the resting and infective stage that enables the parasite to survive outside the body. Infection typically begins from ingestion of *Giardia* cysts. *Giardia* infections in dogs and cats may produce inapparent signs or may produce weight loss, chronic diarrhea, or steatorrhea that may be continual or intermittent (particularly in puppies and kittens). Feces are typically soft, pale in color, malodorous, and contain mucus. Gastrointestinal disease caused by *Giardia* has also been reported in calves. The symptoms of giardiasis in people include development of a sudden explosive, watery, foul-smelling diarrhea that produces excessive gas and abdominal pain. Giardiasis can last for months or even years if left untreated. Giardiasis is diagnosed by observing trophozoites in saline smears of loose or watery feces or cysts in formed feces concentrated by the zinc sulfate flotation. Staining cysts with iodine can help with identification. Several fecal examinations should be performed if giardiasis is suspected because of its intermittent shedding. An ELISA test that detects *Giardia* antigen in the feces of dogs and cats is available. Giardiasis is most commonly diagnosed in humans using an ELISA test that detects *G. duodenalis* antigen in stool. Additional tests include PCR tests and fecal examination for both trophozoites and cysts using direct wet saline preparations with/without iodine, trichrome stained slides, and formalin-ethyl-acetate concentration methods. Duodenum or proximal jejunum aspiration and small intestinal biopsies can be taken to identify *G. duodenalis*.

Fenbendazole is the drug most commonly used to treat giardiasis in dogs and cats. Giardiasis in calves may be treated with albendazole or fenbendazole. A killed vaccine is available for dogs and cats and reduces clinical signs and cysts shed into the environment. *G. duodenalis* is treated with metronidazole, furazolidone, paromonmycin, or tinidazole in people. Promptly removing feces from cages, runs, and yards limits environmental exposure in an attempt to control *Giardia* outbreaks in a given area. Giardiasis can be controlled in people by protecting food from contaminated cysts and filtering or boiling drinking water. In endemic areas, human and animal feces should not be used for fertilizer. Eating undercooked food or unpeeled fruit should be avoided. A vaccine is not available for people.

Leishmaniasis

Overview

Leishmaniasis is a vector-transmitted protozoal complex of diseases caused by a variety of *Leishmania* species producing a spectrum of clinical diseases ranging from cutaneous ulcerations to systemic infections. The genus *Leishmania* is unique among parasites in that its species are divided based on pathology and disease signs rather than form and structure of the parasite. Traditionally divided between Old World and New World parasites, more than 20 pathogenic species have been identified. Common Old World hosts are domestic and feral dogs, rodents, foxes, jackals, wolves, dogs, and hyraxes (African rock-rabbits). Common New World hosts include sloths, anteaters, opossums, and rodents. Leishmaniasis is prevalent throughout the tropical and sub-tropical regions of Africa, Asia, the Mediterranean, Southern Europe (Old World) and South and Central America (New World). The protozoa are transmitted to mammals via the bite of the female sand fly (*Phlebotomus* spp. in the Old World and *Lutzomyia* spp. in the New World). There are three forms of leishmaniasis: the most serious, visceral leishmaniasis, affects the internal organs (causing fever, anemia, splenomegaly, and discoloration of the skin), the cutaneous form (causing deep, disfiguring sores at the site of the bite), and the mucocutaneous form (causing flat, ulcerated plaques, degeneration of nasal cartilage, and secondary bacterial infection).

Visceral leishmaniasis was described as the cause of an epidemic, which occurred in the Garo hills of Assam in Saudi Arabia as far back as 1870 (it was named kala-azar, which is derived from the Hindi terms *kala* meaning black and *azar* meaning sickness). A similar disease that occurred in 1885 was described by Cunningham that was later determined to be caused by a parasite later named *Le. tropica*. In 1900 an Irish soldier developed kala-azar after a stay in Dum Dum, near Calcutta, India. When he died in England, the Scottish physician Dr. William Leishman performed his autopsy and discovered small particles within the macrophages of the spleen as the cause of Dum Dum fever (another name for visceral leishmaniasis). The Irish physician Dr. Charles Donovan investigated splenic aspirates from kala-azar patients and confirmed Leishman's discovery. The small particles within the macrophages were called Leishman-Donovan bodies. In 1908 Nicolle reported that mammals could act as reservoir hosts for the *Leishmania* parasite. Swaminath in 1942 proved that the *Leishmania* parasite could be transmitted by the *Phlebotomus* sand flies.

The first recognized epidemic of visceral leishmaniasis appears to have occurred in 1824, in Bangladesh. In 1875 kala-azar was noted in Assam, India (another named for visceral leishmaniasis is Assam fever) and over the next 25 years 25% of the population died of the disease. Between 1918 and 1923 more than 200,000 people died of kala-azar in Assam and in the Brahmaputra valley. This was followed by another epidemic in 1944.

Cutaneous leishmaniasis may have been described as early as 650 B.C. and possibility earlier in the Tigris/Euphrates basin. The typical skin lesions from this disease were described in 900 B.C. and have been referred to by a variety of names such as Balkan sore in the Balkans, Delhi boil in India, Baghdad boil in Iraq, and saldana in Afghanistan. Cutaneous leishmaniasis caused by *Le. infantum* may have occurred in Crete in the 18th century. In 1929, cutaneous leishmaniasis was described in Jordan by Adler and Theodor and in 1970, Jordan was recognized as an endemic region for cutaneous leishmaniasis. Recent conflicts in Iraq, Kuwait, and Afghanistan have led to greater than 600 cases of cutaneous leishmaniasis and 4 cases of visceral leishmaniasis in American soldiers in 2004.

Mucocutaneous leishmaniasis was first described in 1913 by Bates and is a sequela of new world cutaneous leishmaniasis and results from direct extension or metastasis from blood or lymph to the nasal or oral mucosa. Most cases of mucocutaneous leishmaniasis are from metastasis rather than extension from a cutaneous lesion.The progression of ulcers is slow and steady and secondary infection plays a significant role in the size and persistence of ulcers. Ninety percent of mucocutaneous leishmaniasis cases occur in Boliva, Brazil, and Peru. Illustrations of skin lesions suggestive of mucocutaneous leishmaniasis are seen on pre-Inca earthenware indicating that the disease was in existence in Peru and Ecuador in the 1st century A.D. Text from the 15th- to 16th-century Inca period and the Spanish conquest mention the risk of cutaneous ulcers in seasonal farmers, which was described as "white leprosy" (espundia).

The transmission of *Leishmania* took longer to understand with the first breakthrough occurring in 1904 by Leonard Rodgers. He put some spleen tissue from an infected patient into a flask with simple culture medium and observed elongated organisms with flagellum instead of the spherical form found in the body. This discovery implied there were different stages of the parasite's life cycle. Irishman John Sinton discovered that the distribution area of kala-azar in India coincided with a map of the distribution of *Phlebotomus argentipes* (the silverfoot sand fly). Twenty-five years later Knowles in Calcutta demonstrated *Leishmania* parasites in *Phlebotomus* flies. In 1940 it was shown that if the *Phlebotomus* fly ingests infected blood and then takes its next blood meal, it regurgitates the parasites into the new host.

Today leishmaniasis is found worldwide and is believed to have been transported to the Western world through the African slave trade. Air travel has also generated the spread of *Leishmania* parasites. It is believed that approximately 15 million people in 67 countries are infected with leishmaniasis.

Causative Agent

Leishmania organisms are hemoflagelletes with the **amastigote** form known as the Leishman-Donovan (L-D) body. The amastigote (aflagellated, developmental stage) form is an intracellular parasite in cells of the reticuloendothelial system (Figure 6-25). They are oval, approximately 1.5 to 5 μm, and contain a nucleus and a kinetoplast (DNA containing portion of the mitochondrion). *Leishmania* organisms exist in the amastigote form in humans and animals and the promastigote form in the invertebrate host (arthropod). The **promastigote** form is similar to the amastigote form except it possesses a prominent flagellum. Special surface membrane binding site molecules allow the promastigote form to be taken up by macrophages in the host. In the macrophage, the flagellum is lost producing the spherical, amastigote form. Although all *Leishmania* spp. have similar morphology, they differ clinically. Many strains of *Leishmania* make up each main group: *Le. donovani* complex causes visceral leishmaniasis, *Le. tropica* complex causes cutaneous leishmaniasis of the Old World, and *Le. braziliensis* complex causes mucocutaneous leishmaniasis in the New World. Other complexes include *Le. mexicana* complex and *Le. peruviana*.

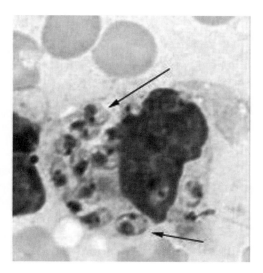

Figure 6-25 *Leishmania* amastigotes in macrophage from skin. Arrows are pointing to amastigotes in the macrophage cytoplasm. (Courtesy of CDC/NCID/DPD)

Epizootiology and Public Health Significance

Visceral leishmaniasis is caused by protozoan parasites in the *Le. donovani* complex: *Le. donovani donovani* (India, Southwest Asia, China, Nepal, and East Africa),

Le. donovani infantum (Mediterranean, Balkan, Middle East, North and East Africa, Central and Southwest Asia, and China), and *Le. donovani chagasi* (Central and South America). *Le. donovani infantum* and *Le. donovani chagasi* are believed to be genetically identical and may be referred to as the same species. The Old World species (*Le. donovani donovani* and *Le. donovani infantum*) are transmitted by the sand fly vector *Phlebotomus* spp. and the New World species (*Le. donovani chagasi*) is transmitted by the sand fly vector *Lutzomyia* spp. Humans, wild animals, and domestic animals are the reservoir hosts for the Old World species and wild and domesticated dogs are the reservoir hosts for the New World species.

Old world cutaneous leishmaniasis (also known as oriental sore, Jericho boil, Aleppo boil, and Delhi boil) is caused by protozoan parasites in the *Le. tropica* complex: *Le. tropica minor* (Eastern Mediterranean, East Africa, Southwest Asia to India) and *Le. tropica major* (Central and Southwest Asia, Middle East, and North, East, and South Africa) and a separate species *Le. aethiopica* (Ethiopia and Kenya). All are transmitted by the sand fly vector *Phlebotomus* spp. (Figure 6-74).

New World mucocutaneous leishmaniasis (also known as espundia and uta) is caused by *Le. braziliensis* (Central and South America) and other distinct species such as *Le. peruviana* (Peru, Bolivia, and Ecuador), *Le. amazonensis* (Venezuela and the Amazonian basin), *Le. tropica mexicana* (commonly called *Le. mexicana*) (Mexico and Central and South America) and *Le. venezuelensis* (Dominican Republic and Venezuela). New World mucocutaneous leishmaniasis is vectored by *Lutzomyia* spp.

It is estimated that approximately 12 million people are currently infected with leishmaniasis and another 367 million are at risk of acquiring the disease in 88 countries. Approximately 1.5 million new cases of cutaneous leishmaniasis and 500,000 cases of visceral leishmaniasis occur worldwide annually (mucocutaneous leishmaniasis is less common). It is estimated that leishmaniasis is responsible for more than 80,000 worldwide deaths annually. Leishmaniasis is occasionally reported in areas in the United States that border Mexico. Most of the cases found in the United States are acquired elsewhere through travel.

Transmission

Leishmaniasis is most commonly transmitted by the bite of sand fly vectors (*Phlebotomus* spp. for Old World species and *Lutzomyia* spp. for New World species). Only female sand flies obtain a blood meal and some sand fly species prefer humans, whereas others prefer animals. Female sand flies take blood only from vertebrates and move at dawn and dusk in search of food. Sand fly activity is high under conditions of calm weather and high humidity. Person-to-person, congenital, and bloodborne transmission of visceral leishmaniasis is possible.

Sand flies do not fly long distances and complete their life cycles in less than 1 kilometer area.

Pathogenesis

The sand fly (intermediate host and vector) bites a human or reservoir host (typically a canine) and either ingests infected macrophages containing amastigotes (aflagellated developmental stage) or free amastigotes. Amastigotes pass to the midgut of the sand fly, where they transform into extracellular promastigotes that attach to the gut. Over the next 4 to 5 days, promastigotes multiply in the gut and migrate toward the pharynx. In the pharyngeal area the organism replicates, causing an obstruction in the sand fly's esophagus. The feeding sand fly clears its esophagus by expelling *Leishmaniae* into the skin of the host, from where they pass into the blood and tissues. Promastigotes are phagocytosed by macrophages of the reticuloendothelial system (liver, spleen, and bone marrow) of the newly infected host, where they lose their flagella and become amastigotes that begin to replicate. As replication continues the infected cells

Leishmaniae spend part of their life cycle in the gut of the sand fly, but their life cycle is completed in a vertebrate host.

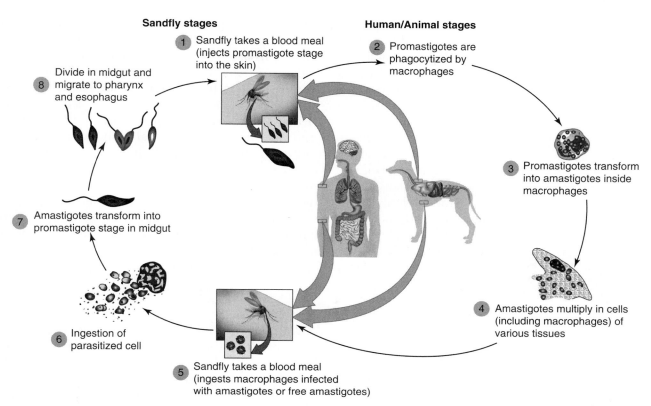

Figure 6-26   Life cycle of *Leishmania* spp. The sand fly bites a human or reservoir host (typically a canine)❺ and either ingests infected macrophages containing amastigotes or free amastigotes❻. Amastigotes pass to the midgut of the sand fly, where they transform into extracellular promastigotes that attach to the gut❼. Promastigotes multiply in the gut and migrate toward the pharynx and esophagus❽. The feeding sand fly clears its esophagus by expelling *Leishmaniae* into the skin of the host❶, from where they pass into the blood and tissues and are phagocytosed by macrophages❷. Promastigotes lose their flagella❸ and become amastigotes that begin to replicate❹. As replication continues the infected cells become engorged with parasites and the infected macrophages rupture spreading the organism to other macrophages that carry the organism throughout the body to be ingested by another biting sand fly.

become engorged with parasites and the infected macrophages rupture spreading the organism to other macrophages. Macrophages carry the organism throughout the body and may be ingested by another biting sand fly. The pathophysiology and clinical signs associated with leishmaniasis are related to the extensive damage to the immune system that results when the parasites destroy large numbers of phagocytic cells (Figure 6-26).

Clinical Signs in Animals

Leishmaniasis can occur in dogs, foxes, jackals, marsupials, sloths, anteaters, hyraxes, and certain rodents. Cats, horses, and nonhuman primates occasionally contract leishmaniasis. The reservoir host varies with the *Leishmania* spp. and the geographic region: rodents, marsupials, sloths, anteaters, and dogs are reservoir hosts for cutaneous leishmaniasis in the Americas; dogs, foxes, jackals, gerbils, hyraxes, and rodents are reservoir hosts for cutaneous leishmaniasis in the Old World. Dogs and wild canines are the reservoir hosts for visceral leishmaniasis in the Americas; whereas dogs, wild canines, and wild rodents are reservoir hosts for visceral leishmaniais in the Old World.

Leishmaniasis may produce a wide variety of clinical signs in animals or may be asymptomatic. Visceral and cutaneous signs are found simultaneously

> Temperature is an important factor in the localization of leishmanial lesions: species causing visceral leishmaniasis are able to grow at core body temperatures, whereas those species responsible for cutaneous leishmaniasis grow best at lower temperatures.

in dogs and include cutaneous signs (alopecia with severe crusting commonly found around the eyes, ears, head, and feet; excessive claw growth) and visceral signs (fever, lymphadenopathy, weight loss, anorexia, anemia, lameness, renal failure, chronic diarrhea, splenomegaly, epistaxis (nose bleeds) or ocular lesions). The incubation period is variable, ranging from 3 months to several years. The causative parasites in dogs are *Le. donovani infantum*, *Le. donovani donovani*, *Le. tropica*, and *Le. major* (Mediterranean area and Middle East) and *Le. donovani chagasi*, *Le. braziliensis*, and *Le. mexicana* (Central and South America). The disease in dogs is prevalent in these areas and endemic in foxhounds in North America. Dogs serve as a reservoir of the parasite for human disease in areas where there is a sand fly vector.

Horses, donkeys, and other equine species infected with *Leishmania* protozoa may develop nodules, scabs, or ulcers near the earflap. Cats are rarely infected and may develop crusted cutaneous lesions near the lips, nose, eyelids, or pinnae. Rodents and other wild animals typically are asymptomatic reservoirs of the protozoan, but may develop alopecia or cutaneous ulcers.

Clinical Signs in Humans

Visceral leishmaniasis (kala-azar; Dum Dum Fever) presents with the small, inapparent lesion at the site of an infected sand fly bite. *Le. donovani* complex organisms disseminate from the skin via the bloodstream to the lymph nodes, spleen, liver, and bone marrow. Clinical signs develop over a period of 2 weeks to 1 year. Clinical signs of disseminated disease include persistent, irregular, or remittent fever, hepatosplenomegaly secondary to production of phagocytic blood cells (weight loss and abdominal pain), pancytopenia (lethargy), darkening of the skin (hence the name black fever), diarrhea, and lymphadenopathy. People may die of hemorrhage (secondary to infiltration of the hematopoietic system), severe anemia, secondary bacterial infections of mucous membranes, and secondary bacterial pneumonia. Emaciation and death occur within 1 to 2 years in 80% to 90% of untreated symptomatic patients. One to 2 years after apparent cure, some people develop nodular cutaneous lesions full of parasites, which can last for years.

Cutaneous leishmaniasis (oriental or tropical sore, Delhi boil, Baghdad boil, Aleppo boil, Uta ulcer, Chiclero ulcer, Bauru ulcer, forest yaws, Pian bois, Panama leishmaniasis) presents with a sharply demarcated skin lesion at the bite site 1 to 4 weeks following a sand fly bite (Figure 6-27). Multiple lesions may occur after multiple bites, accidental autoinoculation by scratching, or metastatic spread. The initial lesion at the fly bite is a small (up to 2 cm), red papule that enlarges, ulcerates centrally producing either a moist center with a seropurulent exudate or a dry crusted scab. In about 3 to 6 months lesions will often develop a raised, hyperpigmented border (the border is where the intracellular parasites are concentrated). The ulcers are painless and cause no systemic symptoms unless they become secondarily infected. Skin ulcers typically heal spontaneously over several months and leave a depressed scar. People with limited cell-mediated immunity may develop diffuse cutaneous leishmaniasis, an uncommon form characterized by widespread nodular skin lesions.

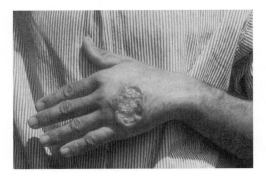

Figure 6-27 Cutaneous leishmaniasis lesion.
(Courtesy of CDC/Dr. D.S. Martin)

Mucocutaneous leishmaniasis (Espundia) presents with a primary cutaneous ulcer 2 to 3 months after the sand fly bites, which lasts for 6 to 15 months. The lesions closely resemble lesions of simple cutaneous leishmaniasis; however, mucocutaneous leishmaniasis can metastasize to mucocutaneous areas such as nasopharyngeal tissues. Mucosal lesions may progress to involve the entire nasal mucosa and the hard and soft palates causing deformity, ulceration, and erosion of the nasal septum, lips, and palate.

> In some endemic areas, people deliberately inoculate their children with material from sores of sick people on an unexposed body part to avoid development of a disfiguring scar in the future.

Diagnosis in Animals

Canine leishmaniasis is diagnosed by identification of the parasite in Giemsa-stained bone marrow, spleen, or lymph node smears (amastigotes appear as oval basophilic bodies in the cytoplasm of macrophages), culture (Novy-MacNeil-Nicole agar, brain-heart infusion agar, Evan's modified Tobie's agar, or Schneider's *Drosophila* agar), or serology (indirect immunofluorescence, direct agglutination, immunoblotting, and ELISA) (Figure 6-28).

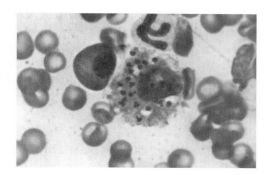

Figure 6-28 *Leishmania* protozoa from a Giemsa-stained bone marrow aspirate in a dog. (Courtesy of CDC/Francis W. Chandler)

Diagnosis in Humans

Visceral leishmaniasis in people is diagnosed by observing the parasite (amastigotes in tissue) on stained Giemsa smears or by observing cultures grown on Novy-McNeal-Nicole agar or liquid media with fetal calf serum (such as Schneider's *Drosophila* medium). Culture growth takes 1 to 3 weeks. Serologic testing using indirect fluorescent antibody test, ELISA test, or Western blot is also available.

Cutaneous leishmaniasis in people is diagnosed based on the appearance of the lesion. Punch biopsy specimens obtained at the border of the lesion can demonstrate the organism. Culture is difficult as the lesion becomes older. The leishmanin skin test, also known as the Montenegro skin test, which is negative during active infection, becomes positive a few weeks to 2 years after cure and remains positive for life.

Mucocutaneous leishmaniasis in people is diagnosed by culture of the organism. Biopsy may show a nonspecific granulomatous reaction and if Giemsa-stained may show the organisms. The leishmanin skin test is positive after 2 to 3 months of infection (skin tests are not approved for use in the United States). As a result of the scarcity of organisms, the diagnosis often is based on epidemiology.

Treatment in Animals

Dogs with leishmaniasis are treated with pentavalent antimony derivatives (N-methylglucamine antimoniate and sodium stibogluconate). Amphotericin B has also been effective. Relapses after treatment are common.

Treatment in Humans

Visceral leishmaniasis in people is treated with pentavalent antimony and liposomal formulations of amphotericin B. Miltefosine is an effective oral drug for visceral leishmaniasis but is not currently available in the United States. Transfusions and antibiotics may be needed to treat complications from this form. Cutaneous leishmaniasis typically resolves on its own; however, large and disfiguring lesions may be treated with pentavalent antimony or amphotericin B. Mucocutaneous leishmaniasis is treated with pentavalent antimony or amphotericin B.

Management and Control in Animals

In endemic areas, rapid treatment of infected dogs, control of stray and homeless dogs, and controlling the insect vectors are effective ways to control leishmaniasis. Currently there is no vaccine against canine leishmaniasis.

Mangement and Control in Humans

Controlling the sand fly population, treating infected and seropositive dogs, and controlling wild reservoirs are all ways to control human leishmaniasis. In endemic areas, vaccination with parasite-containing material has been practiced for centuries (however, it can also spread other diseases such as syphilis or HIV). Exposing the buttock of children to sand flies to induce a leishmanial lesion at an inconspicuous site has also been practiced.

Summary

Leishmaniasis is caused by a complex of *Leishmania* organisms, hemoflagellete protozoal parasites transmitted by sandfly vectors (*Phlebotomus* spp. and *Lutzomyia* spp.). Many strains of *Leishmania* make up three main groups: *Le. donovani* causes visceral leishmaniasis, *Le. tropica* causes cutaneous leishmaniasis of the Old World, and *Le. braziliensis* causes the majority of mucocutaneous leishmaniasis in the New World. In addition to vector transmission, person-to-person, congenital, and blood-borne transmission of visceral leishmaniasis is possible. The pathophysiology and clinical signs associated with leishmaniasis are related to the extensive damage to the immune system that results when the parasites destroy large numbers of phagocytic cells. Leishmaniasis may produce a wide variety of clinical signs in animals or may be asymptomatic. Visceral and cutaneous signs are found simultaneously in dogs and include cutaneous signs (alopecia with severe crusting commonly found around the eyes, ears, head, and feet; excessive claw growth) and visceral signs (fever, lymphadenopathy, weight loss, anorexia, anemia, lameness, renal failure, chronic diarrhea, splenomegaly, epistaxis (nose bleeds) or ocular lesions). Other animals may be asymptomatic reservoirs of the protozoan or have mild cutaneous lesions such as alopecia or cutaneous ulcers. Visceral leishmaniasis in people presents with the small, inapparent lesion at the site of an infected sand fly bite. Clinical signs of disseminated disease include persistent, irregular, or remittent fever, hepatosplenomegaly (weight loss and abdominal pain), pancytopenia (lethargy), darkening of the skin, diarrhea, and lymphadenopathy. Cutaneous leishmaniasis presents with a sharply demarcated skin lesion at the bite site 1 to 4 weeks following a sand fly bite. Mucocutaneous leishmaniasis presents with a primary cutaneous ulcer 2 to 3 months after the sand fly bites, which lasts for 6 to 15 months. The lesions closely resemble lesions of simple cutaneous leishmaniasis; however, mucocutaneous leishmaniasis can metastasize to nasopharyngeal tissues. Canine leishmaniasis is diagnosed by identification of the parasite in bone marrow or lymph node smears (amastigotes appear as oval basophilic bodies in the cytoplasm of macrophages) or serology (indirect immunofluorescence and ELISA). Visceral leishmaniasis in people is diagnosed by observing amastigotes on stained Giemsa smears or by culture growth. Serologic testing using indirect fluorescent antibody test, ELISA test, or Western blot is available. Cutaneous leishmaniasis in people is diagnosed based on the appearance of the lesion. Punch biopsy specimens obtained at the border of the lesion can demonstrate the organism. The leishmanin skin test, also known as the Montenegro skin test, may also be used for diagnosis. Mucocutaneous leishmaniasis in people is diagnosed by culture of the organism. Dogs with leishmaniasis are treated with pentavalent antimony derivatives or amphotericin B. Visceral leishmaniasis in people is treated with pentavalent antimony and liposomal formulations of amphotericin B. Cutaneous leishmaniasis typically resolves on its own. Mucocutaneous leishmaniasis is treated with pentavalent antimony or amphotericin B. In endemic areas, rapid treatment of infected dogs, control of stray and homeless dogs, and controlling the insect vectors are effective ways to control leishmaniasis. Currently there is no vaccine against canine or human leishmaniasis. Historically, exposing people to the parasite in an inconspicuous site has been used to prevent the development of disfiguring lesions in conspicuous sites.

Toxoplasmosis

Overview

Toxoplasmosis, also called litterbox disease, is caused by the coccidian protozoa *Toxoplasma gondii*. *T. gondii* is an obligate intracellular protozoan that was first described in 1908 by Nicolle and Manceaux who observed the parasites in the blood, spleen, and liver of the rodent *Ctenodactylus gondii* while doing research in North Africa and by Splendore who was doing research in Brazil on rabbits. The parasite was named *Toxoplasma* (because of the crescent shape of the tachyzoite with the name derived from the Greek terms *toxo* meaning arc or bow and *plasma* meaning form) *gondii* (after the rodent) in 1909. In 1923, Janku reported parasitic cysts in the retina of an infant who had hydrocephalus, seizures, and unilateral microphthalmia. From 1937 to 1939, Wolf, Cowan, and Paige studied a syndrome of severe congenital infection in infants with toxoplasmosis. Since its discovery the parasite has been found in almost every country and in many species of carnivores, insectivores, rodents, pigs, herbivores, primates, mammals, and birds.

This parasite is found worldwide with more than 60 million people in the United States potentially infected with the *Toxoplasma* parasite. Very few infected people have symptoms because an immunocompetent person's immune system usually prevents the parasite from causing illness. However, pregnant women, infants that were infected in utero, and immunocompromised people could develop serious health problems with toxoplasmosis.

> The definitive host of *T. gondii* is the domestic cat and other Felidae, in which the sexual cycle takes place in the epithelial cells of the intestine.

Causative Agent

T. gondii is an obligate intracellular parasite of many tissues including muscle and intestinal epithelium (Figure 6-29). *Toxoplasma* belongs to the phylum Apicomplexa, which consists of unicellular, intracellular parasites that have a characteristic cell structure at their apical end. *T. gondii* is the only representative of its genus. *T. gondii* undergoes schizogony in all nucleated cells of almost all animals

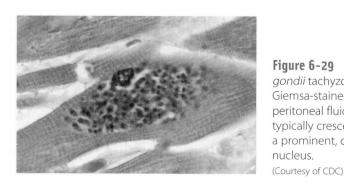

Figure 6-29 *Toxoplasma gondii* tachyzoites from a Giemsa-stained smear of peritoneal fluid. Tachyzoites are typically crescent-shaped with a prominent, centrally placed nucleus.

(Courtesy of CDC)

and birds. There are three forms of *T. gondii*: oocysts, tachyzoites, and tissue cysts (containing bradyzoites).

- *Oocysts. Toxoplasma* oocysts are 10 to 13 μm by 6 to 11 μm in size. Noninfectious, unsporulated oocysts may be excreted in feces of cats for up to 2 weeks (large numbers of oocysts may be shed during this period). Once these oocysts are outside the cat's body, sporogony occurs (up to 21 days), resulting in development of infectious oocysts. Infectious oocysts contain sporozoites. Only members of the cat family shed oocysts. Cats become infected either by ingesting oocysts from fecal contamination or by ingesting tissue cysts (containing bradyzoites) present in the flesh of eaten animals. Once the oocyst is ingested digestive enzymes release the sporozoites, which invade either the feline small intestinal epithelium (enteroepithelial cycle, which only occurs in cats) or penetrate through the mucosa where the protozoan develops and is spread to other organs.
- *Tachyzoites.* Tachyzoites (actively proliferating trophozoites; *tachy* means fast) are crescent-shaped infectious trophozoites observed during the acute stage of infection and may be found in any organ but occur most commonly in the brain, skeletal muscle, and heart muscle. Intracellular replication continues until host cells lyse or a tissue cyst is formed. In an immunocompromised person, tachyzoite replication results in development of focal necrosis, such as necrotizing encephalitis, pneumonitis, or myocarditis. Tachyzoites are less resistant forms of *Toxoplasma*; therefore, they are less important infection sources.
- *Tissue cysts.* Tissue cysts are found most commonly in the brain, skeletal muscle, and cardiac muscle but can occur in any organ. These cysts contain infectious slower replicating bradyzoites (slowly growing trophozoites; *brady* means slow). Tissue cysts usually are observed during the chronic or latent stage of infection causing minor inflammation. These cysts may be involved in disease transmission if present in tissues ingested by carnivores. After ingestion, cysts are broken down by digestive enzymes, which release the bradyzoites, allowing them to invade the gastrointestinal tract, where they spread via blood and lymphatics.

Epizootiology and Public Health Signficance

T. gondii is found worldwide with nearly one-third of all adults in the United States and Europe seropositive for antibodies to *Toxoplasma*, which means they have been exposed to the parasite. Approximately 3% to 70% of healthy adults in the United States have been infected with *T. gondii* with the incidence of the infection varying with the population group and geographic location studied. *T. gondii* infection affects more than 3,500 newborns in the United States annually. *T. gondii* seropositivity rates among HIV-infected patients vary from 10% to 45%. In foreign countries such as France, the seropositivity prevalence rate is as high as 75% by the fourth decade of a person's life. As many as 90% of adults in Paris are seropositive for *Toxoplasma* as a result of the fact that they eat raw meat. In West African and Latin American countries seroprevalence can be as high as 75%. Overall the prevalence of seropositivity for *T. gondii* increases with age.

The prevalence of *Toxoplasma* in cats ranges from 10% to 80% (average is 30%); in pigs the prevalence is less than 1% (in the 1970s the range was 10% to 55%); in ruminants the prevalence is between 20% to 90%; in horses the prevalence is very low; and in wild ruminants the prevalence may be up to 60%. Dogs are less commonly infected with *Toxoplasma* and are not significant sources of infection.

In warm, moist soil, *Toxoplasma* oocysts can remain viable for more than 1 year. *Toxoplasma* oocysts can survive freezing, extreme heat, and dehydration for several months. Oocysts may be carried long distances in wind and water.

Transmission

Toxoplasma protozoa may be transmitted by:

- Ingestion of infectious (sporulated) oocysts from dirt in which cats have defecated, from contaminated litter, or by ingestion of infective oocysts in food or water contaminated with feline feces. Cats usually do not exhibit signs of illness while passing oocysts making it is difficult to determine when a particular cat's feces may be infectious to people or other mammals. Most adult cats will not pass oocysts again after recovering from an initial exposure to *Toxoplasma*.

- Consumption of undercooked or raw meat from animals with tissue cysts containing bradyzoites or contaminated milk or milk products. Pigs, sheep, goats, and poultry are sources of meat commonly infected with *Toxoplasma* tissue cysts. Toxoplasma tissue cysts found in meat can be killed by cooking at 160°F (67°C) or higher or freezing for 24 hours in a household freezer. Raw, unpasteurized milk (especially goat's milk) can lead to toxoplasmosis. Transmammary transmission from infected lactating dams to offspring is also possible (and is a common transmission route for kittens).

- Transplacentally from pregnant mother to unborn offspring when the mother becomes infected with *T. gondii* tachyzoites during pregnancy. Transplacental transmission of *T. gondii* occurs in any mammal and is the only form of human-to-human transmission of toxoplasmosis. Rate of transplacental transmission has been reported to be 55% for untreated mothers and 25% for treated mothers.

- Transplantation of infected organs, through blood transfusion, and through laboratory accidents.

These routes of transmission are illustrated in Figure 6-30.

> Felines, predominantly kittens, are the sole animals to shed *Toxoplasma* oocysts.

> *Toxoplasma* oocysts in feces become infectious after 24 hours. Prompt removal of cat feces from a litterbox may eliminate the risk of contracting toxoplasmosis from pet cats. Because most cats do not leave feces on their fur for 1 day, it is unlikely that humans become infected from direct contact with cats themselves.

> Cats can shed up to 1,000,000 oocysts per gram of feces.

Pathogenesis

T. gondii has two distinct life cycles: the sexual cycle (occurring only in cats, the definitive host) and the asexual cycle (involving other mammals (including humans) and birds, the intermediate hosts).

- *Sexual phase in the definitive host.* The sexual cycle begins in the cat gastrointestinal tract where gametocytes (sex cells) develop from ingested bradyzoites (slow growing form found in tissue cysts). The microgamete (male gametocyte) fertilizes the macrogamete (female gametocyte) in the intestinal cells to form zygotes (fusion of gametes). Zygotes become encapsulated within a rigid wall forming oocysts. Oocysts are expelled into the intestinal lumen after rupture of the intestinal cell and are excreted unsporulated in feces. Within the oocyst the zygote sporulates by dividing into two sporocysts. Each sporocyst contains four sporozoites.

- *Asexual phase in the definitive host.* Bradyzoites that do not become gametocytes penetrate the intestinal epithelial cells of cats and multiply as tachyzoites (motile infective stage). Tachyzoites are disseminated throughout the body in a few days and encyst in tissues.

- *Asexual phase in the intermediate host.* Tachyzoites are disseminated throughout the body of the intermediate host in macrophages, lymphocytes, and plasma. Tachyzoites continue to divide in host cells until the host cell is filled with parasites. When the tachyzoites cannot be contained within the host cell, the host cell ruptures. Tachyzoites are released and are usually cleared by the host's immune response, although some manage to infect new host cells. Some

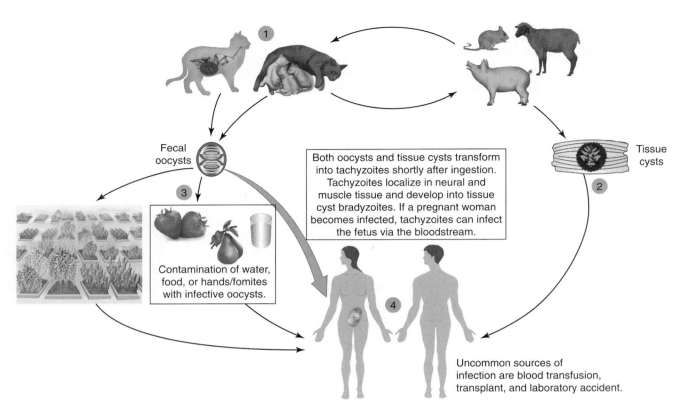

Fecal oocysts

Tissue cysts

Both oocysts and tissue cysts transform into tachyzoites shortly after ingestion. Tachyzoites localize in neural and muscle tissue and develop into tissue cyst bradyzoites. If a pregnant woman becomes infected, tachyzoites can infect the fetus via the bloodstream.

Contamination of water, food, or hands/fomites with infective oocysts.

Uncommon sources of infection are blood transfusion, transplant, and laboratory accident.

Figure 6-30   Life cycle of *Toxoplasma gondii*. Felines are the only known definitive hosts for the sexual stages of *Toxoplasma gondii* and are the main reservoirs of infection. Cats become infected with *Toxoplasma gondii* by ingesting tissue cysts or oocysts❶ (or transplacentally and transmammary). After ingestion of tissue cysts or oocysts, viable organisms are released and invade epithelial cells of the small intestine where they undergo an asexual followed by a sexual cycle forming oocysts, which are excreted. The unsporulated oocyst takes 1 to 5 days after excretion to sporulate (become infective). Human infection may be acquired in several ways: A) ingestion of undercooked infected meat containing *Toxoplasma* cysts❷; B) ingestion of the oocyst from fecally contaminated hands or food❸; C) organ transplantation or blood transfusion; D) transplacental transmission❹; E) accidental inoculation of tachyzoites. *Toxoplasma* parasites then form tissue cysts, most commonly in skeletal muscle, myocardium, and brain.

time after the tachyzoites enter new host cells they transform to bradyzoites in tissue cysts. Intact tissue cysts probably do not cause inflammation and may persist for the life of the animal or human.

Clinical disease and pathology from toxoplasmosis is a result of the spread of *T. gondii* to other parts of the body. As organisms are transported via the lymphatics to the bloodstream they are spread throughout the body. Tachyzoites replicate producing necrotic focal areas that are surrounded by a cellular reaction and in an immunocompetent animal or human tachyzoites are cleared from tissues. In immunocompromised animals or humans the acute infection progresses and may cause potentially lethal consequences such as pneumonitis, myocarditis, or necrotizing encephalitis. Tissue cysts form as early as 7 days after infection and remain in the host its entire life producing little or no inflammatory response.

When a mother becomes infected with *T. gondii* during gestation, the parasite may be disseminated via the bloodstream to the placenta. When this occurs, infection may be transmitted to the fetus transplacentally or during vaginal delivery. Mothers who acquire the infection in the first trimester and are not treated have a fetal infection rate of approximately 17% resulting in severe disease in the infant. Mothers who acquire the infection in the third trimester and are not treated have a fetal infection rate of approximately 65% resulting in mild or inapparent disease in infants. These different rates of transmission are most likely related to placental blood flow, the virulence and amount of *T. gondii* acquired, and the immune status of the mother. Almost all congenitally infected people have signs of infection by adolescence if they are not treated in the newborn period.

Clinical Signs in Animals

Clinical signs associated with toxoplasmosis are a result of tissue damage by the tachyzoite. The presentation of clinical signs depends on the number of tachyzoites released, the ability of the host's immune system to limit tachyzoite spread, and the organs infiltrated and damaged by the tachyzoites. Adult immunocompetent animals have the ability to control tachyzoite spread resulting in subclinical illness. Young animals, particularly puppies, kittens, and piglets, cannot contain the spread of tachyzoites and may show signs of pneumonia (cough and dyspnea), myocarditis (fever and lethargy), hepatic necrosis (jaundice), meningoencephalomyelitis (seizures and death), chorioretinitis (visual impairment), and lymphadenopathy. *T. gondii* may also cause abortion and stillbirth in sheep, goats, and pigs.

Clinical Signs in Humans

Toxoplasmosis in people and its clinical presentation depends on the type of infections and status of the host. Clinical presentations include:

- *Acute infection in immunocompetent adults.* Toxoplasmosis in this group of people is typically asymptomatic (80% to 90%). If symptoms are present they are localized (mainly cervical) or generalized lymphadenopathy producing signs such as low grade fever, headache, lethargy, and muscle pain. Most symptoms will resolve in a few weeks.

- *Infection in immunocompromised adults.* Most cases of toxoplasmosis in immunocompromised adults are typically from reactivation of a latent infection and are rarely a result of new infection. Clinical signs include CNS abnormalities (personality changes, apathy, ataxia, vision problems, and seizures), dyspnea, and diarrhea.

- *Ocular toxoplasmosis.* Chorioretinitis in people is caused by infection acquired in utero with symptoms including blurred vision, photophobia, and loss of central vision.

- *Toxoplasmosis in pregnancy and congenital toxoplasmosis.* Toxoplasmosis in pregnancy is a result of primary infection with *T. gondii* during gestation when helper T lymphocytes have decreased. The risk of fetal infection depends on the time of material infection: if the infection is acquired more than 6 months prior to conception the fetuses are not affected. During the first trimester, the risk of fetal infection is 15%, but the fetal disease is more severe. During the third trimester, the risk of fetal infection is 65%, but the fetal disease is less severe or asymptomatic. Toxoplasmosis in pregnancy may cause abortion, stillbirth, or preterm delivery. Congenital infection can be manifested immediately postpartum or many years later. Signs in newborns include microcephalus, hydrocephalus, chorioretinitis, seizures, anemia, and jaundice.

Diagnosis in Animals

Toxoplasmosis is mainly diagnosed via serology. Clinical signs of toxoplasmosis are nonspecific, and many times animals are asymptomatic, making them of limited use in making a definite diagnosis. Indirect hemagglutination assay, indirect fluorescent antibody assay, latex agglutination test, or ELISA serologic tests are available for diagnosing the disease antemortem. Postmortem diagnosis includes the observation of tachyzoites in tissue impression smears or tachyzoites and/or bradyzoites in tissue sections. Examination of cat feces for oocysts has been used to diagnosis toxoplasmosis; however, the examination should be repeated three times at weekly intervals before infections can be excluded (oocysts are detectable for only 1 to 2 weeks).

Diagnosis in Humans

Toxoplasmosis in humans is diagnosed via serology such as immunofluorescent antibody, ELISA, Sabin-Feldman dye, and complement fixation tests. Specific IgM antibodies appear during the first 2 weeks of illness, peak within 4 to 8 weeks, and then become undetectable within several months. IgG antibody levels increase more slowly, peak in 1 to 2 months, and may remain high and stable for months to years. Detection of IgM antibodies or a fourfold rise in one of the IgG tests usually indicates acute toxoplasmosis. Past exposure, which provides resistance to reinfection, typically produces a positive IgG and a negative IgM test. Detection of specific IgM antibody in neonatal disease suggests congenital infection because maternal IgG crosses the placenta but IgM does not. *T. gondii* may be isolated during the acute phase of disease by inoculating mice or tissue cultures with biopsy materials or body fluids. Tests for the rapid detection of *Toxoplasma* antigens or PCR-amplified DNA in blood, CSF, or amniotic fluid are under investigation. PCR analysis of amniotic fluid is the most sensitive method used to diagnose toxoplasmosis in utero.

Treatment in Animals

For animals treatment is seldom given. Clindamycin is the treatment of choice for cats and is prescribed for 14 to 21 days.

Treatment in Humans

Most immunocompetent people do not require treatment unless disseminated disease is present or severe symptoms persist. Treatment is indicated for acute toxoplasmosis of newborns, pregnant women, and immunocompromised people. The most effective regimen is a combination of pyrimethamine and sulfadiazine because of their synergistic effect. These drugs are effective if given in the acute stage of the disease when there is active replication of the parasite; they will not eradicate infection. People with ocular toxoplasmosis should also receive corticosteroids to limit damage to ocular organs. Treatment in pregnant women may need to be adjusted depending on their stage of gestation.

Management and Control in Animals

Prevention of toxoplasmosis in cats includes not allowing them to hunt rodents and birds, keeping them indoors, and feeding them only cooked meat or processed food from commercial sources. There is currently no vaccine for toxoplasmosis in cats.

Management and Control in Humans

In people toxoplasmosis can be prevented by

- Thoroughly washing hands with soap and water when handling meat,
- Thoroughly washing all cutting boards, sink tops, knives, and other materials after preparing meat on them with soap and water (kills all stages of the parasite),
- Cooking meat throughout to 160°F (67°C) or by cooling to −13°C,

- Refraining from tasting food until it is properly prepared,
- Using gamma irradiation of meat in areas where traditional food processing techniques do not include heating or freezing,
- Testing women who are planning on becoming pregnant for *Toxoplasma*,
- Testing immunocompromised people for toxoplasmosis,
- Having pregnant women avoid contact with cat litter, soil, and raw meat,
- Feeding pet cats only dry, canned, or cooked food,
- Cleaning the cat litter box daily, preferably not by a pregnant woman,
- Wearing gloves while gardening, and
- Washing vegetables thoroughly before eating because they may have been contaminated with cat feces.

Currently there is no vaccine to prevent toxoplasmosis in humans.

Summary

Toxoplasmosis is caused by *T. gondii*, an obligate intracellular parasite of many tissues including muscle and intestinal epithelium. There are three forms of *T. gondii*: oocysts, tachyzoites, and tissue cysts (containing bradyzoites). *T. gondii* is found worldwide with nearly one-third of all adults in the United States and Europe seropositive for antibodies to *Toxoplasma*. The prevalence of *Toxoplasma* in cats ranges from 10% to 80%; in pigs the prevalence is less than 1%; in ruminants the prevalence is between 20% to 90%; in horses the prevalence is very low; and in wild ruminants the prevalence is up to 60%. Toxoplasmosis may be transmitted by ingestion of infectious (sporulated) oocysts from dirt in which cats have defecated, from contaminated litter, by ingestion of infective oocysts in food or water contaminated with feline feces; by consumption of undercooked or raw meat from animals with tissue cysts; directly from pregnant mother to unborn child when the mother becomes infected with *T. gondii* tachyzoites during pregnancy; by transplantation of infected organs; through blood transfusion; and through laboratory accidents. *T. gondii* has two distinct life cycles: the sexual cycle (occurring only in cats, the definitive host) and the asexual cycle (involving other mammals (including humans) and birds, the intermediate hosts). Clinical disease and pathology from toxoplasmosis is a result of the spread of *T. gondii* to other parts of the body. Clinical signs associated with toxoplasmosis result from tissue damage by the tachyzoite. Young animals, particularly puppies, kittens, and piglets, cannot contain the spread of tachyzoites and may show signs of pneumonia (cough and dyspnea), myocarditis (fever and lethargy), hepatic necrosis (jaundice), meningoencephalomyelitis (seizures and death), chorioretinitis (visual impairment), and lymphadenopathy. *T. gondii* may also cause abortion and stillbirth in sheep, goats, and pigs. Toxoplasmosis in people and its clinical presentation depends on the type of infections and status of the host. Clinical presentations include acute infection in immunocompetent adults, infection in immunocompromised adults, ocular toxoplasmosis, and toxoplasmosis in pregnancy and congenital toxoplasmosis. Toxoplasmosis is mainly diagnosed via serology. Indirect hemagglutination assay, indirect fluorescent antibody assay, latex agglutination test, or ELISA serologic tests are available for diagnosing the disease antemortem in animals. Animals other than cats are rarely treated for toxoplasmosis. Clindamycin is the treatment of choice for cats and is prescribed for 14 to 21 days. Most immunocompetent people do not require treatment unless disseminated disease is present or severe symptoms persist. Treatment is indicated for acute toxoplasmosis of newborns, pregnant women, and immunocompromised people and consists of a combination of pyrimethamine and sulfadiazine. To prevent toxoplasmosis in cats they should not be allowed to hunt rodents and birds, they should be kept indoors, and they should only be fed cooked meat or processed food from commercial sources. There is currently no vaccine for toxoplasmosis in cats. In people toxoplasmosis can be prevented by a variety of ways including thorough hand and equipment washing, cooking meat properly, being tested prior to pregnancy, and avoiding contact with things that may be contaminated by cat feces. Currently there is no vaccine to prevent toxoplasmosis in humans.

Trypanosomiasis

Overview

Trypanosomiasis is a general term applied to diseases caused by the flagellated protozoan parasite *Trypanosoma*. The name Trypanosoma is derived from the Greek *trypanon* meaning borer and *soma* meaning body, which reflects the invasiveness of infection. There are two main forms of trypanosomiasis, African and American.

American trypanosomiasis (found mainly in South America), also known as Chagas' disease, is a disease causing chronic cardiomyopathy, megaesophagus, or megacolon. The disease was named for the Brazilian physician Carlos R.J. Chagas, who was the first to describe the flagellated protozoan *Trypanosoma cruzi* as the pathogen causing the disease in 1909. Chagas discovered the parasite in the hindgut of the vector insect (bugs in the Reduviidae family). Chagas sent a number of Reduviid bugs to the Oswaldo Cruz Institute, where the bugs fed on marmosets and Guinea pigs. These animals had trypanosomes in the blood within one month. Chagas believed that the parasites replicated in the lungs and named the organism *Schizotrypanum cruzi* (named after Oswaldo Cruz, a Brazilian epidemiologist). By 1916, Chagas linked the organism to the disease that would eventually bear his name, but because he also linked the organism to diseases such as goiter and cretinism his research was dismissed by some. Not until the 1930s was it shown that Chagas' disease was transmitted by the feces of the reduviid bug (also known as the assassin bug, benchuca, vinchuca, kissing bug, chipo, and barbeiro). *Try. cruzi* infections are found in small mammals (sylvatic cycle) and human disease results from transmission of the protozoan by some vector species (domestic cycle). Chagas' disease is almost exclusively found in rural areas, where the reduviid bug can breed and feed on the natural reservoirs (opossums and armadillos) of *Try. cruzi*. Other infected humans, domestic animals (cats, dogs, Guinea pigs), and wild animals (rodents, monkeys, ground squirrels) could also serve as important parasite reservoirs.

African trypanosomiasis, also known as sleeping sickness or *maladie du sommeil*, is a chronic disease producing generalized lymphadenopathy and often fatal meningoencephalitis. African trypanosomiasis is endemic in certain regions of sub-Saharan Africa (affecting 36 countries and 60 million people) and has been described by Arabian historians as occuring in Africa since the 14th century. African trypanosomiasis is caused by *Try. brucei gambiense* and *Try. brucei rhodesiense* and is transmitted by the tsetse fly. The causative agent and the vector were identified in 1902 by Sir David Bruce. Differentiation between the two protozoa was not made until 1910. There have been three severe epidemics in Africa over the last century: one between 1896 and 1906 (mostly in Uganda and the Congo Basin), one in 1920 (in several African countries), and one that began in 1970 and is still in progress. The 1920 epidemic was contained when mobile health teams began screening millions of people at risk, nearly eradicating the disease between 1960 and 1965. At that time, disease screening and surveillance were relaxed allowing the disease to reappear in its endemic form over the last thirty years.

Causative Agent

Chagas' Disease

Try. cruzi, in the Trypanosomatidae family, is a large flagellate (17 to 20 µm × 2 µm) with an undulating membrane (Figure 6-31). It is further classified in a special section called Stercoraria because it is the only human trypanosome to be transmitted by the feces of its invertebrate vector (others are transmitted by saliva). Four forms (Table 6-2) can be identified during the *Try. cruzi* life cycle:

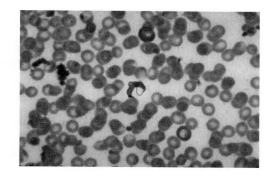

Figure 6-31   *Trypanosoma cruzi* in a Giemsa-stained blood smear.
(Courtesy of CDC/Dr. Mae Melvin)

- **Trypomastigote** (*trypanon* is Greek for auger, describing the flagellum of this form). The trypomastigote is the infective flagellate form found in the blood of the mammalian host (blood trypomastigote) and in the terminal part of the digestive and urinary tracts of vectors (metacyclic trypomastigote, the infective form for the vertebrate host). The trypomastigote does not replicate and has a moderately long flagellum that runs alongside the parasite's body. The flagellum extends beyond the body aiding motility. The trypomastigote also has a narrow undulating membrane and kinetoplast (an organelle associated with the large mitochondrion containing the extranuclear DNA) that is subterminal.

- **Epimastigote** (*epi* is Greek for upon, describing the flagellum on the anterior end). The epimastigote is the replicating form of the parasite in the gut of the insect vector. It has a free (unattached) flagellum at the anterior part of the parasite and an undulating membrane.

- **Promastigote** (*pro* is Greek for before, describing the flagellum before development of the undulating membrane). The promastigote is found in the vector's stomach and is unable to replicate.

- **Amastigote** (*a* is Greek for without, describing the lack of flagellum in this form). The amastigote is the intracellular replicating form of the parasite in the vertebrate host. It is spherical and nonmotile (aflagellate).

Table 6-2   Cellular and Infective Stages of the Hemoflagellates

Genus/Species	Amastigote	Promastigote	Epimastigote	Trypomastigote
Trypanosoma cruzi	Intracellular	Occurs	In gut of reduviid bug	In feces of reduviid bug; transferred to humans
Trypanosoma brucei	Does not occur	Does not occur	In salivary gland of tsetse fly	In mouthparts of tsetse fly
Leishmania	Intracellular	In sand fly gut	Does not occur	Does not occur

The insect vectors of Chagas' disease belong to the Hemiptera order, Reduviidae family, and Triatominae subfamily (kissing bugs). The three most important vector species of the human Chagas' disease are *Triatoma infestans* (south of the Amazonian basin), *Rhodnius prolixus* (Central America, Venezuela, and Columbia), and *Tri. dimidiate* (Columbia, Ecuador, and along the Pacific coast from Mexico to north of Peru).

African Trypanosomiasis

Try. brucei, in the Trypanosomatidae family, is a large flagellate (15 to 20 μm × 3 μm) with an undulating membrane. It is further classified in a special section called Salivaria because it is the human trypanosome transmitted by the saliva of its invertebrate vector. *Try. brucei* tends to be pleomorphic in the vertebrate host and has a small kinetoplast near the posterior end. Two forms can be identified during the *Try. brucei* life cycle:

- *Trypomastigote.* The trypomastigote is the infective flagellate form found in the blood of the mammalian host (blood trypomastigote) and in the biting mouthparts of vectors (metacyclic trypomastigote, the only form infective to a vertebrate host).
- *Epimastigote.* The epimastigote is the replicating form of the parasite in the salivary gland of the insect vector.

The two variants of African trypanosomiasis are the Gambian (West African) strain, caused by the subspecies *Try. brucei gambiense* and the Rhodesian (East African) strain, caused by the subspecies *Try. brucei rhodesiense*. These two subspecies of *Trypanosoma* are isolated based on different ecological niches of the tsetse fly. In the West African form, the fly inhabits a niche in dense vegetation along rivers and forests, whereas in the East African form, the fly inhabits a niche of savanna woodland and lakefront thickets. The insect vectors of *Try. brucei gambiense* are *Glossina palpalis* and *Gl. tachinoides*, whereas the insect vectors of *Try. brucei rhodesiense* are *Gl. morsitans*, *Gl. pallidipes*, and *Gl. swynnertoni*.

Other trypanosomes that affect animals are described later in this chapter.

Epizootiology and Public Health Significance

Chagas' Disease

Chagas' disease causes economic loss and human disease mainly in Mexico, Central America, and South America (30% of deaths are from Brazil). There are a few cases of Chagas' disease in the United States (Texas and California). Chagas' disease currently affects 16 to 18 million people, killing approximately 20,000 people annually. Higher percentages of Chagas' disease are found in poorer areas of Central and South America. Even though human cases in the United States are rare, increased travel has increased the risk of the United States becoming a potential area where Chagas' disease could be found. *Try. cruzi* has been found infecting wild opossums and raccoons as far north as North Carolina and is occasionally reported in dogs in the southern half of the United States. Infection is frequently found in immigrants from Mexico and Central and South America (approximately 100,000 to 675,000 people).

African Trypanosomiasis

African trypanosomiasis has geographic isolation based on the subspecies involved. The Gambian form is found in central and western Africa causing a chronic condition that can extend in a passive phase for months or years before symptoms emerge. The Rhodesian form has a much more limited range being found in southern and eastern Africa and is the acute form of the disease.

Try. cruzi trypomastigotes do not replicate in blood; replication only occurs when the parasites enter another cell or are ingested by another vector. *Try. brucei* trypomastigotes replicate in blood.

A single tsetse fly lives up to 3 months and is capable of delivering 50,000 cells when it bites the host (only 500 cells are needed to initiate infection).

When transfer of an agent occurs via saliva the parasite has undergone development in the middle or anterior intestinal portions of the arthropod host and is termed anterior station. When transfer of an agent occurs via feces the parasite has undergone development in the hindgut or posterior midgut of the arthropod host and is termed posterior station. *Try. cruzi* replicates in the gut of the reduviid bug and is said to undergo posterior station development. *Try. brucei* replicates in saliva of the tsetse fly and is said to undergo anterior station development.

African trypanosomiasis is confined to areas in Africa between latitudes 15°N and 20°S (north of South Africa to South of Algeria, Libya, and Egypt). Prevalence of disease varies from country to country with major outbreaks occurring in Angola, the Democratic Republic of the Congo, and Sudan in 2005. In 1986 WHO estimated that 70 million people lived in areas where disease transmission could take place and in 1998 40,000 cases of African trypanosomiasis were reported. Current case estimates are 50,000 to 70,000 (with an estimated 300,000 to 500,000 cases undiagnosed). All U.S. cases of African trypanosomiasis are found in immigrants from or travelers to Africa.

Trypanosoma infection occurs in wild and domestic animal species (especially cattle) who serve as reservoirs of disease. The disease in cattle is a major economic factor for agriculture in this area of Africa.

Transmission

Chagas' Disease

Try. cruzi is mainly transmitted by the feces of the reduviid bug (80% of human infections) (Figure 6-32). Reservoir hosts include humans, dogs, cats, opossums, mice, rats, guinea pigs, and rabbits. Pigs, goats, cattle, and horses do not produce a high enough level of parasitemia for disease transmission and do not play a role in disease transmission. People can transmit the metacyclic trypomastigotes by inoculating the oral or ocular mucosa by rubbing or scratching these areas. Direct skin penetration of the parasite is possible, but difficult. Contaminated blood transfusions (mainly in urban centers) are responsible for 5% to 20% of the human cases of Chagas' disease. Congenital transmission may also occur by transplacental infection of the fetus or infections of infants via breast milk. Ingestion of food contaminated by feces of infected reduviid bugs is another source of transmission.

African Trypanosomiasis

Try. brucei parasites are transmitted by tsetse flies (*Glossina* spp.). The main reservoir for *Try. brucei gambiense* is humans, but it can also be found in pigs and other animals. The main reservoirs of *Try. brucei rhodesiense* are wild game animals and cattle. African trypanosomiasis is also transmitted transplacentally (causing abortion and perinatal death) and by laboratory accidents.

Pathogenesis

Chagas' Disease

Adult and nymph stages of the reduviid bug acquire *Try. cruzi* organisms when trypomastigotes are taken from infected mammals when the arthropod takes a blood meal. In the intestinal tract of insects, trypomastigotes transform into promastigotes and epimastigotes. Some epimastigotes migrate and bind to the terminal part of the intestine (becoming metacyclic trypomastigotes). The infective forms of the parasite (metacyclic trypomastigotes) are then passed with the insect's feces and urine at the end of the blood meal. The complete cycle of *Try. cruzi* within the vector takes 2 to 4 weeks. The infected reduviid bug defecates on the new host who becomes infected by scratching the bite, which allows trypomastigotes to enter through the wound or through intact mucous membranes (such as conjunctiva). Once in the mammalian host, the metacyclic trypomastigote must invade a nucleated cell to complete its life cycle. The trypomastigote fuses with lysosomes in the host cell and begins to differentiate into an amastigote. In time the amastigote is released in the cytoplasm of the host cells. When the host cell is full of parasites, amastigotes begin to differentiate into trypomastigotes. The motility

Figure 6-32 The reduviid bug, the vector of Chagas' disease.
(Courtesy of CDC/World Health Organization, Geneva, Switzerland)

Tsetse flies are restricted to Africa from about latitude 15°N to 29°S. Tsetse fly species inhabit relatively distinct environments—*Gl. morsitans* usually is found in savanna country, *Gl. pallidipes* prefers areas around rivers and lakes, and other species live in high forest areas.

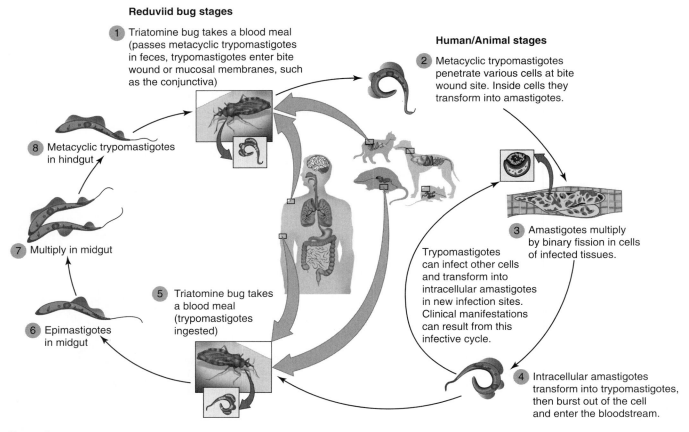

Reduviid bug stages

1. Triatomine bug takes a blood meal (passes metacyclic trypomastigotes in feces, trypomastigotes enter bite wound or mucosal membranes, such as the conjunctiva)

8. Metacyclic trypomastigotes in hindgut

7. Multiply in midgut

6. Epimastigotes in midgut

5. Triatomine bug takes a blood meal (trypomastigotes ingested)

Human/Animal stages

2. Metacyclic trypomastigotes penetrate various cells at bite wound site. Inside cells they transform into amastigotes.

3. Amastigotes multiply by binary fission in cells of infected tissues.

Trypomastigotes can infect other cells and transform into intracellular amastigotes in new infection sites. Clinical manifestations can result from this infective cycle.

4. Intracellular amastigotes transform into trypomastigotes, then burst out of the cell and enter the bloodstream.

Figure 6-33 Life cycle of *Trypanosoma cruzi.*

An infected reduviid bug takes a blood meal and releases trypomastigotes in its feces near the site of the bite wound. Trypomastigotes enter the host through the wound or through intact mucosal membranes (conjunctiva)❶. Inside the host, the trypomastigotes invade cells, where they differentiate into intracellular amastigotes❷. The amastigotes multiply❸ and differentiate into trypomastigotes, and then are released into the circulation as bloodstream trypomastigotes❹. Trypomastigotes infect cells from a variety of tissues and transform into intracellular amastigotes in new infection sites. The bloodstream trypomastigotes do not replicate (different from the African trypanosomes). Replication resumes only when the parasites enter another cell or are ingested by another vector. The reduviid bug becomes infected by feeding on human or animal blood that contains circulating parasites❺. The ingested trypomastigotes transform into epimastigotes in the vector's midgut❻. The protozoa multiply and differentiate in the midgut❼ and differentiate into infective metacyclic trypomastigotes in the hindgut❽. *Trypanosoma cruzi* can also be transmitted through blood transfusions, organ transplantation, transplacentally, and in laboratory accidents.

of trypomastigotes ruptures the cell, releasing free parasites that can invade other cells or be taken up by the vector when it takes a blood meal (Figure 6-33).

Try. cruzi causes the development of organ lesions by inducing an inflammatory response, cellular lesions, and fibrosis frequently seen in the heart, esophagus, and colon. The inflammatory response results from the rupture of infected cells releasing trypomastigotes and cellular debris. Inflammation is intense in the acute phase, but it is less intense in the chronic phase. The cellular lesions mainly affect the muscle cells and the nervous cells. Fibrosis appears slowly and gradually as part of the healing process and typically involves the heart. Significant destruction of the conduction system, cardiac muscle cells, and parasympathetic cardiac nerves frequently produce arrhythmias. Hypertrophy of muscle cells and fibrosis predispose the animal or person to cardiac dilatation and failure. Gastrointestinal lesions are irregularly dispersed and mainly affect the esophagus and the colon producing megaesophagus or megacolon.

African Trypanosomiasis

The cycle begins when the tsetse fly ingests the trypomastigote form of the parasite from an infected reservoir host such as wild animals (antelopes, pigs, lions, or hyenas), domestic animals (cattle or goats), or humans. In the posterior section of the intestinal tract of insects, trypomastigotes replicate and migrate anteriorly into the foregut. After approximately 20 days, they migrate farther forward to the esophagus, pharynx, and salivary gland of the vector. Once in the salivary glands, trypomastigotes transform into epimastigotes and attach to host cells or remain free in the lumen. After several asexual generations, they transform into metacyclic trypomastigotes (infective stage for the vertebrate host). As the tsetse fly feeds it inoculates the new host where trypanosomes undergo a series of divisions at the bite site producing a sore called a primary chancre. Once the parasite is within the vertebrate host, it replicates as trypomastigotes in blood and lymph. In chronic cases, many parasites will invade the CNS, replicate, and enter intracellular brain spaces. The onset, severity, and duration of the disease vary with the two forms: the Rhodesian form is acute and has brain involvement in three to four weeks with death in a few months, whereas the Gambian form has a longer incubation period, is chronic, and may not affect the brain for several years (Figure 6-34).

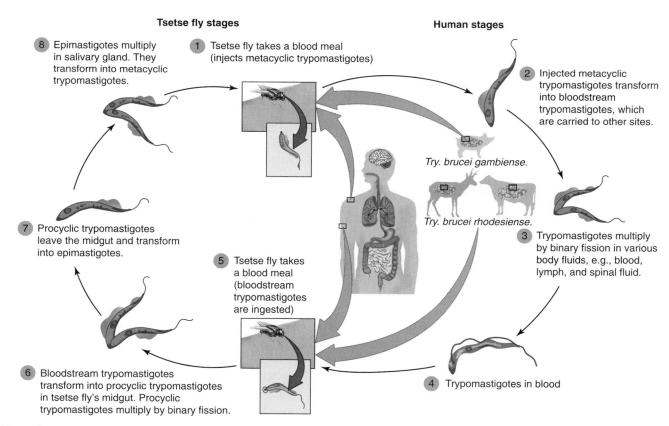

Figure 6-34 Life cycle of *Trypanosoma brucei.*

An infected tsetse fly injects metacyclic trypomastigotes into skin tissue when it takes a blood meal. The trypomastigotes enter the lymphatic system and pass into the bloodstream❶. Inside the host, they transform into bloodstream trypomastigotes❷, are carried to other sites throughout the body, reach other body fluids (e.g., lymph, spinal fluid), and continue replicating❸. The tsetse fly becomes infected with bloodstream trypomastigotes when taking a blood meal on an infected mammalian host❹,❺. In the fly's midgut, the protozoa transform into procyclic trypomastigotes, multiply❻, leave the midgut, and transform into epimastigotes❼. The epimastigotes reach the tsetse fly's salivary glands and continue replicating❽. The cycle in the fly takes approximately 3 weeks. Humans are the main reservoir for *Trypanosoma brucei gambiense,* but this species can also be found in animals. Wild game animals are the main reservoir of *Trypanosoma brucei rhodesiense.*

Trypanosomes live in blood, lymph nodes, the spleen, and cerebrospinal fluid. They do not invade or live in cells, but inhabit connective tissue spaces within a variety of organs. The pathophysiology of trypanosomiasis is directed at lymphatics and areas surrounding blood vessels. In cases of acute disease mortality is likely as a result of disruption of normal physiological processes by overstimulation of the immune system producing large amounts of immunoglobulins. Lysis of host cells, especially erythrocytes, also contributes to the disease process.

Clinical Signs in Animals

Chagas' Disease

The entire life cycle of *Try. brucei* is represented by extracellular stages.

Domestic animals may become infected with *Try. cruzi* without showing clinical signs yet having the ability to introduce the trypanosome into dwellings where reduviid bugs are present. Infection with *Try. cruzi* is usually asymptomatic in many animals except puppies and kittens. The only clinical sign in dogs may be acute death as a result of myocarditis.

African Trypanosomiasis

African trypanosomiasis is also called tsetse-transmitted trypanosomiasis or zulu in animals. Animals infected with trypanosomes mount a vigorous immune response and these immune complexes cause inflammation that produce the signs and lesions of the disease. Severity of disease varies with the species and age of the animal infected and the species of trypanosome involved. Some animals, such as horses, mules, donkeys, some ruminants, and dogs suffer acute disease and death. After a typical 1- to 4-week incubation period, the primary clinical signs are intermittent fever, anemia, and weight loss. Within a few days the animals become emaciated, uncoordinated, and paralyzed. Blindness is common in dogs. Cattle usually have a chronic course with high mortality. Ruminants may gradually recover from the disease; however, stress results in relapse. Necropsy lesions may include petechiation of serosal membranes (especially the peritoneal cavity), swollen lymph nodes and spleen, and anemia.

Trypanosomes that affect animals are summarized in Table 6-3.

Clinical Signs in Humans

Chagas' Disease

Chagas' disease in humans occurs in two stages:

- *Acute.* The acute stage starts shortly after infection in which a localized skin nodule called a chagoma appears at the inoculation site. If the inoculation site is the conjunctiva, unilateral periorbital edema and conjunctivitis may occur producing a lesion called Romaña's sign. The acute stage is usually asymptomatic, but may manifest itself with fever, anorexia, lymphadenopathy, mild hepatosplenomegaly, and myocarditis. Approximately 10% to 20% of acute cases resolve in 2 to 3 months only to reappear several years later. After a symptomatic or asymptomatic acute phase, infected people enter a latent phase that may last years, decades, or for life. Clinical signs of chronic Chagas' disease develop in 20% to 40% of infected people.
- *Chronic.* The chronic stage may not develop for years or decades after the initial infection and affects the nervous system, gastrointestinal system, and cardiovascular system. Clinical signs include dementia, heart damage, mega-colon, megaesophagus, and weight loss. Difficulty swallowing may be the first

Table 6-3   Trypansomes That Affect Animals

Trypanosome Species	Animals Affected	Geographic Distribution	Pathogenicity
Try. congolense	Cattle, sheep, goats, dogs, pigs, camels, horses, most wild animals	Africa (transmitted by tsetse fly)	Pathogen; clinical signs include intermittent fever, anemia, and weight loss
Try. vivax	Cattle, sheep, goats, camels, horses, various wild animals	Africa, Central and South America, West Indies (transmitted by tsetse fly)	Pathogen; clinical signs include intermittent fever, anemia, and weight loss
Try. brucei brucei	All domestic and various wild animals; most severe in dogs, horses, cats	Africa (transmitted by tsetse fly)	Pathogen; clinical signs include intermittent fever, anemia, and weight loss
Try. simiae	Domestic and wild pigs, camels	Africa (transmitted by tsetse fly)	Pathogen; clinical signs include intermittent fever, anemia, and weight loss
Try. melophagium	Sheep	Worldwide (transmitted by sheep ked)	Nonpathogen
Try. theileri	Cattle	Worldwide (transmitted by tabanid flies)	Nonpathogen
Try. equiperdum	Horses (disease is called dourine)	Mediterranean coast of Africa and in the Middle East, southern Africa, and South America	Pathogen; clinical signs of venereal disease develop over weeks or months. Signs include mucopurulent discharge from the urethra in stallions and from the vagina in mares, edema of the genitalia, plaques 2 to 10 cm in diameter appear on the skin, emaciation. Mortality in untreated cases is 50% to 70%.
Try. evansi	Mainly camels and horses, but all domestic animals are susceptible (disease is called surra).	North Africa, the Middle East, Asia, the Far East, and Central and South America (transmitted by biting flies)	Pathogen of camels, horses, and dogs; clinical signs include intermittent fever, anemia, and weight loss. Surra can be fatal, particularly in camels, horses, and dogs. In other species it is nonpathogenic.
Try. cruzi	Puppies and kittens; opossums, armadillos, rodents, and wild carnivores are reservoirs	Central and South America and localized areas of the southern United States	Pathogen of young dogs and cats; other domestic animals act as reservoir hosts. Clinical signs include myocarditis or acute death

Cattle, sheep, and goats are infected, in order of importance, by *Trypanosoma congolense, Trypanosoma vivax,* and *Trypanosoma brucei brucei* . In pigs, *Trypanosoma simiae* is the most important. In dogs and cats, *Trypanosoma cruzi* and *Trypanosoma brucei* are the most important.

observable sign. If left untreated the disase may be fatal usually as a result of cardiomyopathy.

African Trypanosomiasis

In African trypanosomiasis, the first sign of disease is a primary, localized inflammatory lesion at the inoculation site where intensive localized replication of the parasites occurs. The parasites are spread from the inoculation site via lymphatics and blood to regional lymph nodes. Symptoms begin with fever, headaches, and joint pain. If untreated, the disease slowly spreads to produce anemia, endocrine problems, cardiovascular problems, edema, lymphadenopathy, and kidney disorders. The disease progresses when the parasite passes through the blood–brain barrier producing neurologic signs such as confusion, incoordination, and daytime slumber and nighttime insomnia (hence the sleeping sickness name). Damage caused in the neurological phase can be irreversible and without treatment the disease is 100% fatal.

The stages of African trypanosomiasis are:

- *Primary lesion (trypanosomal chancre)*. Trypanosomes replicate at the inoculation site and after days or up to 2 weeks a tender, inflammatory nodule develops with a centrally located pustule. This nodule heals after 2 to 3 weeks and is more commonly seen in the Rhodesian form of disease and is more common in non-Africans.

- *Hemolymphatic stage*. The next phase of illness develops over several weeks in which the parasites are disseminated via the lymphatics and blood. In the Gambian form the parasite load is low and intermittent; in the Rhodesian form the parasite load is high and persistent. Bouts of high fever followed by afrebile periods, headaches, lymphadenopathy (especially with the Gambian form), rigors, endocrine problems (especially with the Rhodesian form), and transient edematous swellings occur when trypanosomes disseminate in blood, lymph nodes, and bone marrow. A characteristic, erythematous, pruritic rash occurs 6 to 8 weeks after infection. Winterbottom's sign (enlarged lymph nodes in the posterior cervical area) is characteristic of Gambian sleeping sickness.

- *Meningoencephalitic stage*. In the Gambian form, CNS involvement occurs months to several years after the onset of acute disease; in the Rhodesian form, CNS invasion occurs within a few months. Clinical signs of CNS involvement causes persistent headache, inability to concentrate, memory loss, personality changes, tremor, ataxia, and terminal coma. If untreated, death as a result of myocarditis, malnutrition, or secondary infection usually occurs within 9 months of disease onset in the Rhodesian form and during the second or third year in the Gambian form.

Diagnosis in Animals

Definitive diagnosis of Chagas' disease and tsetse-transmitted trypanosomiasis is by examining Giemsa-stained thick and thin blood smears. A wet preparation of the buffy coat area of a microhematocrit tube after centrifugation may show motile organisms. Serologic tests are available but are used more for herd screening than individual diagnosis. Necropsy findings vary and are nonspecific and may include extensive petechiation of the serosal membranes, swollen lymph nodes, and anemia. Histological sections of affected tissues may demonstrate the amastigote form of the parasite in animals with Chagas' disease.

Diagnosis in Humans

Chagas' disease is diagnosed in people by examining Giemsa-stained thick and thin blood smears for detection of the parasite. Parasite numbers are high during the acute phase making identification easier; few parasites are present in blood during latent infection or the chronic disease phase. When parasite numbers are low definitive diagnosis may require culture of blood or lymph node aspirate. Other diagnostic tests used include xenodiagnosis (examination of rectal contents of laboratory-raised reduviid bugs after they take a blood meal from a suspect person or inoculation of human blood into laboratory animals) or PCR tests. Serologic tests may yield false-positive results in people with leishmaniasis and are of limited use.

African trypanosomiasis is also diagnosed by examining Giemsa-stained thick and thin blood smears. The trypomastigote stage is the only stage found in infected humans. Areas to sample include chancre fluid, lymph node aspirates, blood, bone marrow, or CSF (in the late stages of disease). A wet preparation should also be examined for the motile trypanosomes. Concentration techniques using centrifugation followed by examination of the buffy coat will yield better results in blood samples. Finding trypanosomes in wet mounts or in Giemsa-stained thin or thick smears of peripheral blood is more useful in the Rhodesian type, whereas fluid aspirated from an enlarged lymph node is more useful in the Gambian type. Blood or body fluids can also be inoculated into animals, cultured, and used for serologic tests (immunofluorescent assay, ELISA, or card agglutination).

Treatment in Animals

Chagas' disease in animals is treated symptomatically to control signs of heart failure. Treatment is usually limited in animals.

Tsetse-transmitted trypanosomiasis in animals is treated with a variety of drugs including diminazen aceturate (cattle and dogs), quinapyramine sulfate (cattle, horses, camels, pigs, and dogs), suramin (horses, camels, and dogs), and prothidium (cattle).

Treatment in Humans

Chagas' disease in humans is treated with nifurtimox and benznidazole. Eflornithine (difluoromethylornithine) was developed in the 1970s and approved by the Food and Drug Administration (FDA) in 1990, but production was stopped in 1999. In 2001, it was manufactured again and donated for use in poorer countries. Treatment in the acute stage of disease reduces parasite load, shortens clinical illness, and reduces mortality, but it does not eradicate the infection. Treatment in the chronic stage is usually ineffective.

African trypanosomiasis is treated with either suramin (*Try. brucei rhodesiense* and *Try. brucei gambiense*) or pentamidine (*Try. brucei gambiense*). Suramin and pentamidine are effective against the blood forms of both subspecies but will not cure infection in the CNS. Advanced cases have been treated with melarsoprol or eflornithine. Melarsoprol is the drug of choice for CNS involvement.

Management and Control in Animals

The incidence of Chagas' disease can be reduced by controlling the reduviid bug population by use of insecticides especially in homes where they can burrow into crevices. Controlling exposure to reservoir hosts such as rodents and wild carnivores can also decrease the incidence of this disease in animals.

Control of tsetse-transmitted trypanosomiasis in animals includes eradication of tsetse flies (by spraying and dipping of animals, spraying insecticides on fly-breeding areas, using screens, and clearing of brush) and the use of prophylactic drugs in endemic areas. Some cattle breeds show innate resistance to trypanosomiasis and can be raised in endemic areas hoping to reduce the level of trypanosomiasis in those areas.

Management and Control in Humans

The incidence of Chagas' disease can be reduced by controlling the reduviid bug population. Reduviid bugs hide during the day in crevices and gaps in the walls and roofs of poorly constructed homes and even when insect colonies are eradicated in the house, they can reestablish themselves from the surrounding environment. Using a mosquito net wrapped under the mattress can protect people from adult reduviid bugs, but not from one of the five nymphal instars that could crawl up from the floor. Insect repellents can help prevent individual exposure. Transfusion-induced Chagas' disease can be prevented by screening blood for *Try. cruzi* or by adding gentian violet to blood.

African trypanosomiasis can be prevented by controlling the tsetse fly population. Regular active surveillance (detection and treatment of infected people and animals) can also help control this disease. Wearing substantial wrist- and ankle-length clothing (tsetse flies bite through thin clothes) and liberal use of insect repellents can also help control this disease. Prophylaxis with pentamidine offers some protection against the Gambian form of African trypanosomiasis; however, it can cause renal failure and diabetes making its use questionable.

Summary

Trypanosomiasis is caused by hemoflagellate protozoa of the *Trypanosoma* genus. Chagas' disease is caused by *Try. cruzi*, a large flagellate (17 to 20 µm × 2 µm) with an undulating membrane. Four forms can be identified during the *Try. cruzi* life cycle: trypomastigote, epimastigote, amastigote, and promastigote. Chagas' disease causes economic loss and human disease mainly in Mexico, Central America, and South America (30% of deaths are from Brazil). There are a few cases of Chagas' disease in the United States (Texas and California). *Try. cruzi* is mainly transmitted by the feces of the reduviid bug (80% of human infections). Reservoir hosts include humans, dogs, cats, opossums, mice, rats, Guinea pigs, and rabbits. *Try. cruzi* causes the development of organ lesions by inducing an inflammatory response, cellular lesions, and fibrosis frequently seen in the heart, esophagus, and colon. Domestic animals may become infected with *Try. cruzi* without showing clinical signs except in puppies and kittens. The only clinical sign in dogs may be acute death as a result of myocarditis. Chagas' disease in humans occurs in acute and chronic stages. The acute stage starts shortly after infection in which a localized skin nodule called a chagoma appears at the inoculation site. If the inoculation site is the conjunctiva, unilateral periorbital edema and conjunctivitis may occur producing a lesion called Romaña's sign. The acute stage is usually asymptomatic, but may manifest itself with fever, anorexia, lymphadenopathy, mild hepatosplenomegaly, and myocarditis. The chronic stage may not develop for years or decades after the initial infection and affects the nervous system, gastrointestinal system, and cardiovascular system. Clinical signs include dementia, heart damage, megacolon, megaesophagus, and weight loss. Definitive diagnosis of Chagas' disease in animals is by examining Giemsa-stained thick and thin blood smears. A wet preparation of the buffy coat area of a microhematocrit tube after centrifugation may show motile organisms. Chagas' disease is diagnosed in people by examining Giemsa-stained thick and thin blood smears for detection of the parasite. Other diagnostic tests used include xenodiagnosis or PCR tests. Chagas' disease in animals is treated symptomatically to control signs of heart failure. Chagas' disease in humans is treated with nifurtimox and benznidazole. The incidence of Chagas' disease can be reduced by controlling the reduviid bug population using insecticides.

African trypanosomiasis is caused by *Try. brucei*, a large flagellate (15 to 20 μm × 3 μm) with an undulating membrane. Two forms can be identified during the *Try. brucei* life cycle: trypomastigote and epimastigote. The two variants of African trypanosomiasis are the Gambian (West African) strain, caused by the subspecies *Try. brucei gambiense*, and the Rhodesian (East African) strain, caused by the subspecies *Try. brucei rhodesiense*. These two subspecies of *Trypanosoma* are isolated based on the different ecological niches of the tsetse fly. African trypanosomiasis has geographic isolation based on the subspecies involved. The Gambian form is found in central and western Africa (transmitted by *Gl. palpalis* and *Gl. tachinoides*) causing a chronic condition that can extend in a passive phase for months or years before symptoms emerge. The Rhodesian form has a much more limited range being found in southern and eastern Africa (transmitted by *Gl. morsitans*, *Gl. pallidipes*, and *Gl. swynnertoni*) and is the acute form of the disease. The main reservoir for *Try. brucei gambiense* is humans, but it can also be found in pigs and other animals. The main reservoirs of *Try. brucei rhodesiense* are wild game animals and cattle. Trypanosomes live in blood, lymph nodes, the spleen, and cerebrospinal fluid. The pathophysiology of trypanosomiasis is directed at lymphatics and areas surrounding blood vessels. Animals infected with trypanosomes mount a vigorous immune response and these immune complexes cause inflammation that produce the signs and lesions of the disease. Some animals, such as horses, mules, donkeys, some ruminants, and dogs suffer acute disease and death. After a typical 1- to 4-week incubation period, the primary clinical signs in animals are intermittent fever, anemia, and weight loss. Within a few days the animals become emaciated, uncoordinated, and paralyzed. In human African trypanosomiasis, the first sign of disease is a primary, localized inflammatory lesion (chancre) at the inoculation site. Symptoms begin with fever, headaches, and joint pain. If untreated, the disease slowly spreads to produce anemia, endocrine problems, cardiovascular problems, edema, lymphadenopathy, and kidney disorders. As the disease progresses neurologic signs such as confusion, incoordination, and daytime slumber and nighttime insomnia develop. Damage caused in the neurologic phase can be irreversible and without treatment the disease is 100% fatal. Definitive diagnosis of tsetse-transmitted trypanosomiasis in animals is by examining Giemsa-stained thick and thin blood smears. A wet preparation of the buffy coat area of a microhematocrit tube after centrifugation may show motile organisms. Serologic tests are available but are used more for herd screening than individual diagnosis. African trypanosomiasis is diagnosed by examining Giemsa-stained thick and thin blood smears. A wet preparation should also be examined for the motile trypanosomes. Blood or fluids can also be inoculated into animals, cultured, and used for serologic tests (immunofluorescent assay, ELISA, or card agglutination). Tsetse-transmitted trypanosomiasis in animals is treated with a variety of drugs including diminazen aceturate (cattle and dogs), quinapyramine sulfate (cattle, horses, camels, pigs, and dogs), suramin (horses, camels, and dogs), and prothidium (cattle). African trypanosomiasis in people is treated with either suramin (*Try. brucei rhodesiense* and *Try. brucei gambiense*) or pentamidine (*Try. brucei gambiense*). Control of trypanosomiasis in animals includes eradication of tsetse flies and the use of prophylactic drugs in endemic areas. Some cattle breeds show innate resistance to trypanosomiasis and can be raised in areas hoping to reduce the level of trypanosomiasis in their area. African trypanosomiasis can be prevented in people by controlling the tsetse fly population. Wearing substantial wrist- and ankle-length clothing and liberal use of insect repellents can also help control this disease. (For less common zoonotic ameba, see Table 6-4.)

Table 6-4   Less Common Zoonotic Ameba

Ameba	Disease	Type of Ameba	Predominant Signs in People	Clinical Signs in Animals	Transmission	Animal Source	Geographic Distribution	Pathology	Diagnosis	Treatment
Balantidium coli	Balantidiosis	Ciliate: single-celled protozoa that move by means of cilia. Cilia beat in a rhythmic pattern and the trophozoite moves in a spiral path. *Bal. coli* has a direct life cycle with both the trophozoite and cyst stages containing a large macronucleus and smaller micronucleus. The trophozoite is oval and is covered with cilia. The cyst state is the infective stage.	Acute disease produces bouts of colitis (abdominal pain, headache, tenesmus, vomiting and anorexia) Chronic disease produces intermittent diarrhea, alternating with constipation and blood and/or mucus found in the stool	■ Pigs: usually harmless, but it can cause severe clinical disease with some pigs experiencing moderate to severe diarrhea. ■ Also seen in rats, sheep, cattle, horses, primates, and Guinea pigs which are typically asymptomatic	Mainly found in humans, but also in pigs (pigs are an important reservoir in some parts of the world). Humans acquire infection by ingestion of fecally contaminated food and water. Sows can shed many cysts thereby contaminating farrowing pens. Flies may transmit the organism mechanically.	Feces of pigs	■ Rare infection in the U.S. ■ Worldwide especially in areas with a temperate climate and in people who have contact with pigs ■ Endemic in China, Japan, the South Pacific, the Philippines, and some South American countries	Cysts are resistant to digestion in the stomach and small intestine. *Bal. coli* trophozoites are liberated in the large intestine where they burrow into the epithelium if the mucosa is already damaged. In the large intestine they cause erosions of the intestinal mucosa producing varying degrees of damage. *Bal. coli* rarely penetrates the intestine or enters the blood.	Identification of trophozoites from fresh fecal samples (cysts are rarely found in human feces)	Tetracycline, carbarsone, and diiodo-hydroxyquin
Encephalitozoon spp.	Microsporidiosis	Microsporidia: obligate, intracellular, spore-forming protozoa. Infective stage is an oval spore. Spores that are trichrome stained show diagonal or horizontal strips that represent polar tubule structures; some have a terminal vacuole.	May be asymptomatic; however, keratoconjunctivitis, intestinal infections (self-limiting or producing watery and bloodless diarrhea), and fever have been seen in people. Infection may be fatal in immuno-compromised people.	■ *Encephalitozoon cuniculi* has been identified in rabbits, laboratory mice, monkeys, dogs, birds, rats, Guinea pigs, and other mammals ■ *Encephalitozoon hellem* has been found in psittacines ■ *Encephalitozoon intestinalis* has been found in ducks, pigs, dogs, cattle, and goats ■ *Enterocytozoon bieneusi* has been found in cats, cattle, llamas, and pigs	Unknown, but probably due to contaminated water or inhalation of aerosols	Spores are common in the environment	Rarely diagnosed, but primarily seen in the tropics and in immunocompromised people	The spore penetrates the intestinal wall and the spore is injected into the cell by its polar tubule. The organism multiplies asexually in the of the host cell cytoplasm. The spores mature and the spores are released to infect other host cells or to be seen in feces and urine.	Detection of spores in feces, aspirates (bile, bronchial, or duodenal), and urine sediments	Albendazole Fumagillin topically for keratoconjunctivitis

Organism	Disease	Morphology	Clinical signs	Transmission	Source	Distribution	Life cycle	Diagnosis	Treatment
Sarcocystis bovihominis (also called *Sarcocystis hominis*); *Sarcocystis suihominis*	Sarcosporidiosis	Coccidia: thin-walled oocyst that contains 2 mature sporocysts (each containing 4 sporozoites)	Seen in 2 forms: ■ intestinal infection with oocysts and sporocysts shed in feces ■ extraintestinal form with cysts in striated muscles causing myositis and vasculitis ■ Asymptomatic disease of cattle and pigs ■ Under experimental conditions can cause abortions, stillbirths, and death in pregnant cattle	Humans: ingestion of raw or undercooked cyst-containing meat from cattle and pigs Carnivores can spread sporocysts after becoming infected	Cyst-containing raw or undercooked meat	Worldwide with prevalence in Central Europe	After eating cyst-contaminated meat, sexual reproduction takes place in the intestinal epithelium. Mature oocysts are shed in feces and are ingested by an intermediate host. The organism penetrates the intestine and gains access to vascular endothelial cells where it is spread to cardiac and skeletal muscles. The muscle cysts are eaten by a definitive host.	Direct detection in feces using zinc sulfate floatation or in biopsy sections.	No known treatment. Treatment is symptomatic.

CESTODE ZOONOSES

Diphyllobothriasis

Overview

Diphyllobothriasis is an intestinal parasitic disease caused by ingestion of raw or undercooked fish infected with the cestode *Diphyllobothrium* spp., commonly known as the broad fish tapeworm. *Diphyllobothrium* does not have hooks or suckers on its scolex like other tapeworms and gets its named from the Greek terms *di* meaning two, *phyllon* meaning leaf, and *bothrion* meaning pit or groove, which are all used to describe the two leaf-shaped grooves (bothria) found on the scolex of the adult tapeworm. There are at least 13 distinct species of *Diphyllobothrium* that parasitize humans, with *Di. latum* being the most common. *Diphyllobothrium* was first identified by the Swiss physician Felix Plater, who also first described the disease at the beginning of the 17th century even though the eggs of *Diphyllobothrium* worms have been found since Neolithic time (around 4,000 B.C.). Descriptions of larval infestations of *Di. latum* were found on Egyptian papyruses and reports of mass mortality of aquatic animals were reported in 488 B.C. and 396 B.C. In the first century, Virgil reports of outbreaks of diphyllobothriasis in fish. In France and Germany in the Middle Ages *Diphyllobothrium* eggs were found in the bathrooms of wealthy people's homes because they were the class of people who could afford to eat fish and beef. In 1240 along the English coast a massive fish die-off was described as a result of battles among the fishes leaving multitudes of dead fishes washed up on the beach. Other reports of diphyllobothriasis outbreaks in Europe occurred in 1680, 1722, and 1760. *Diphyllobothrium* spp. parasitized native South Americans long before Columbus discovered the New World and continue to infect approximately 9 million people worldwide.

Causative Agent

Cestodes have two orders that infect humans: Cyclophyllidea (have a scolex with four suckers) and Pseudophyllidea (have a scolex with two opposing sucking grooves). Unlike all other cestodes that infect humans, *Diphyllobothrium* belong to the order Pseudophyllida, which have genital pores on the ventral side of proglottids (other cestodes belong to the Cyclophyllida order, which have genital pores on the dorsal side of proglottids). The species that infect humans include *Di. latum*, *Di. dendriticum*, *Di. pacificum*, and *Di. nihonkaiense*. Adult *Di. latum* can reach a length of 10 meters and shed up to a million eggs daily, whereas *Di. dendriticum* can reach a length of 1 meter. Their scolex is finger shaped with dorsal and ventral bothria (grooves or sucking organs). Their proglottids are wider than they are long and contain numerous testes and vitelline glands (the glands in Platyhelminthes that produce yoke material and the egg shell) throughout each proglottid except for a narrow zone in the center. *Diphyllobothrium* eggs are ovoid, measure about 60 µm by 40 µm, have an operculum (lidlike part of a parasite egg through which the larva escapes) on one end, and a small knob on the other. The embryo inside the egg must reach water for development that occurs over several weeks depending on the water temperature (Figure 6-35).

Epizootiology and Public Health Significance

Di. latum occurs worldwide (endemic in Scandinavia, the Baltic states, and western Russia); *Di. dendriticum* occurs in Alaska and North and South America;

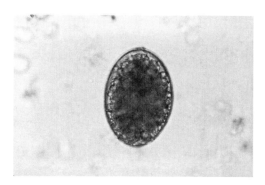

Figure 6-35 Unembryonated *Diphyllobothrium latum* egg. (Courtesy of CDC)

Di. pacificum is found on the Pacific coast of South America; *Di. nihonkaiense* is found in Japan.

In the United States, *Di. latum* infections have been reported in fish in the Great Lakes in the past, but recent reports of infections have not occurred. Six *Diphyllobothrium* species are found in Alaskan lakes and rivers. Infection with *Diphyllobothrium* species is widespread in American fish-eating birds and mammals. Pike, perch, and salmon are the most commonly infected fish. *Di. latum* infection is found in people in Europe, Africa, and the Far East as a result of the use of human excrement for fertilizer and poor sanitation.

Transmission

Humans and other mammals are infected with *Diphyllobothrium* after ingesting raw or undercooked freshwater fish (second intermediate host) containing the larva forms of the parasite. Humans do not typically eat undercooked minnows and small freshwater fish and these types of fish do not represent an important source of direct human infection; however, these small second intermediate hosts can be eaten by larger predator fish species such as trout, perch, and walleyed pike. In this case, the sparganum (larvae) can migrate to the muscles of the larger predator fish and mammals, which acquire the disease by eating these second intermediate host fish raw or undercooked. After ingestion of the infected fish, the plerocercoids develop into immature adults and then into mature adult tapeworms, which will reside in the small intestine. Adult tapeworms may infect humans, canines, felines, bears, pinnipeds (seals, sea lions, and walruses), and mustelids (weasels and wolverines).

> Both dogs and cats are susceptible to *Diphyllobothrium* infection after eating infected prey, but the parasite is not passed directly between dogs and cats.

Pathogenesis

Pseudophyllidean cestodes have indirect life cycles with two intermediate hosts in the life cycle of *Diphyllobothrium*. The adult tapeworm is in the small intestine and after self-fertilization (they are hermaphroditic) eggs are produced, and unembryonated eggs are passed in feces by the definitive host (humans, canines, felines, and other mammals). When the unembryonated egg contacts fresh water, the egg matures and disintegrates. A coracidium (ciliated embryo) is released into the water and is the infective form that can infect a copepod crustacean in the genus *Cyclops* (first intermediate host). The coracidium penetrates the intestinal wall and develops into a procercoid in the body cavity of the *Cyclops*. The infected *Cyclops* is then ingested by a small freshwater fish (second intermediate host) where the procercoid is released and penetrates the intestinal wall of the fish developing to a plerocercoid larva (sparganum) in the fish's muscles or viscera. Ingestion of small freshwater fish by larger predator fish occurs and the sparganum migrates to the

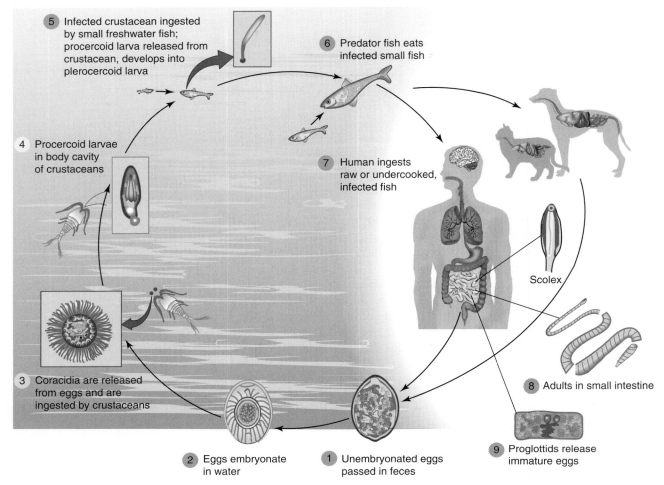

5 Infected crustacean ingested by small freshwater fish; procercoid larva released from crustacean, develops into plerocercoid larva

6 Predator fish eats infected small fish

4 Procercoid larvae in body cavity of crustaceans

7 Human ingests raw or undercooked, infected fish

Scolex

3 Coracidia are released from eggs and are ingested by crustaceans

8 Adults in small intestine

2 Eggs embryonate in water

1 Unembryonated eggs passed in feces

9 Proglottids release immature eggs

Figure 6-36   Life cycle of *Diphyllobothria*. Immature eggs are passed in feces❶. The eggs mature when exposed to fresh water ❷ producing oncospheres, which develop into a coracidia❸. After ingestion by a freshwater crustacean (the first intermediate host) the coracidia develop into procercoid larvae❹. Following ingestion of the copepod by a second intermediate host (minnows and other small freshwater fish), the procercoid larvae are released from the crustacean and migrate into the fish muscle where they develop into the infective plerocercoid larvae (sparganum)❺. The second intermediate hosts are eaten by larger predator species (trout, perch, walleyed pike)❻ and the sparganum can migrate to the musculature of the larger predator fish. Humans and animals acquire the disease by eating these larger fish raw or undercooked❼. After ingestion of the infected fish, the plerocercoid develop into immature adults and then into mature adult tapeworms, which will reside in the small intestine❽. Immature eggs are discharged from the proglottids❾ and are passed in the feces❶.

muscle of the predator fish. Humans or mammals ingest raw or undercooked fish. After ingestion of the infected fish, the scolex emerges and attaches to the intestinal wall. The sparganum develops into mature adults in the small intestine where proglottids release up to 1,000,000 eggs per day into feces. Eggs appear in human feces 5 to 6 weeks after infection. If humans ingest the *Cyclops* (first intermediate host) the procercoid larvae develop into a sparganum, which migrates into the subcutaneous tissue. In this case, humans serve as intermediate hosts and this is called sparganosis (Figure 6-36).

Cases of diphyllobothriasis may be asymptomatic or produce vague gastrointestinal signs as a result of the presence of the worm in the host's intestinal tract. *Diphyllobothrium* worms can also cause megaloblastic anemia, mainly in people of Finnish descent, as a result of the parasite's ability to absorb vitamin B_{12}, making the host deficient in this vitamin.

Clinical Signs in Animals

Dogs and cats with diphyllobothriasis show clinical signs associated with gastrointestinal disease such as diarrhea, weight loss, and vomiting. Dogs and cats may begin shedding of *Diphyllobothrium* eggs as early as 10 days postinfection.

Di. pacificum is commonly found in sea lions and heavy infections can cause intestinal obstruction.

Clinical Signs in Humans

In people, most cases of diphyllobothriasis are asymptomatic. Clinical signs may include abdominal pain, diarrhea, vomiting, and weight loss. Vitamin B_{12} deficiency causing anemia may occur, but is rare (about 2%). Heavy infections may result in intestinal obstruction. Migration of proglottids can cause cholecystitis.

> Diphyllobothriasis can last for decades in people if not treated.

Diagnosis in Animals

Diagnosis of diphyllobothriasis is by identification of operculated eggs on fecal flotation (Figure 6-35) or by identifying the long chains of proglottids segments of the adult worm (found in vomit or feces) (Figure 6-37). Adult tapeworms also have a ventral genital pore.

Figure 6-37 Gravid proglottids of *Diphyllobothrium latum.* (Courtesy of CDC/Dr. Mae Melvin)

Diagnosis in Humans

Diphyllobothriasis in people is diagnosed by identification of eggs or proglottids in feces.

Treatment in Animals

There are no approved anthelmintics for treatment of *Diphyllobothrium* infections in dogs and cats; however, praziquantel at higher doses and longer durations has been successful in treating this tapeworm.

> Some parasite eggs will not float in some less-dense floation solutions; therefore, a fecal sedimentation technique is the preferred fecal method to ensure visualization of all eggs in a sample.

Treatment in Humans

Nicolsamide and praziquantel are drugs of choice for this parasite.

Management and Control in Animals

Diphyllobothriasis can be controlled by preventing predation and scavenging activity of dogs and cats by keeping them indoors or confined to areas where they do not have access to intermediate hosts. Dogs and cats infected with *Diphyllobothrium* are not an immediate zoonotic risk because the coracidia that hatch from their eggs are only infectious to *Cyclops* (first intermediate host).

Management and Control in Humans

The best way to avoid human infection is not to eat raw or undercooked fish. Proper disposal of sewage (human feces can contain eggs) can reduce infection of fish and in turn reduce infection in humans.

Summary

Diphyllobothriasis is an intestinal parasitic disease caused by ingestion of raw or undercooked fish infected with the cestode *Diphyllobothrium* spp., commonly known as the broad fish tapeworm. Diphyllobothriasis is caused by several species of *Diphyllobothrium* including *Di. latum*, *Di. dendriticum*, *Di. pacificum*, and *Di. nihonkaiense*. Their scolex is finger shaped with dorsal and ventral bothria and their proglottids are wider than they are long. *Diphyllobothrium* eggs are ovoid and have an operculum on one end. *Di. latum* occurs worldwide (endemic in Scandinavia, the Baltic states, and western Russia); *Di. dendriticum* occurs in Alaska and North and South America; *Di. pacificum* is found on the Pacific coast of South America; *Di. nihonkaiense* is found in Japan. *Diphyllobothrium* has an indirect life cycle with two intermediate hosts. Humans and other mammals are infected with *Diphyllobothrium* after ingesting raw or undercooked freshwater fish (second intermediate host) containing the larva forms of the parasite. Dogs and cats show clinical signs associated with gastrointestinal disease such as diarrhea, weight loss, and vomiting. In people, most cases of diphyllobothriasis are asymptomatic. Clinical signs may include abdominal pain, diarrhea, vomiting, and weight loss. Vitamin B_{12} deficiency causing anemia may occur, but is rare (about 2%). Heavy infections may result in intestinal obstruction. Diagnosis of diphyllobothriasis is by identification of operculated eggs on fecal flotation or by identifying the long chains of proglottids segments of the adult worm. There are no approved anthelmintics for treatment of *Diphyllobothrium* infections in dogs and cats; however, praziquantel has been used to treat this tapeworm. Nicolsamide and praziquantel are drugs of choice for this parasite in humans. Diphyllobothriasis can be controlled by preventing predation and scavenging activity of dogs and cats. The best way to avoid human infection is not to eat raw or undercooked fish. Proper disposal of sewage (human feces can contain eggs) can reduce infection of fish and in turn reduce infection in humans.

> Infection with *Diphyllobothrium* may increase with the consumption of sushi.

Echinococcosis

Overview

Echinococcosis, also known as hydatid disease and hydatidosis, is a parasitic disease in which humans and animals are infected with the larvae of *Echinococcus* cestodes. *Echinococcus* is derived from the Greek *echinos* meaning hedgehog or sea urchin and *kokkus* meaning berry, which describes the numerous, spiny hooklets seen in the berrylike scolex of the larva form of the parasite. Large hydatid cysts, named from the Greek *hydatoeis* meaning watery, are fluid-filled cysts that can occur in the liver, lungs, or other organs as a consequence of infection by these cestodes. Hydatid cysts in the liver were known since ancient times with references made to these cysts in slaughtered animals for rituals in Babylonia and in animals slaughtered for food (by Hippocrates in the fourth century B.C., Arataeus in the 1st century A.D., and Galen in the second century A.D.). Francisco Redi in the 17th century first described the parasitic nature of these cysts and in 1766, German clinician Pierre Simon Pallas identified these cysts as the larva stages of tapeworms. In 1853, Carl von Siebold demonstrated that *Echinococcus* cysts from sheep produced adult tapeworms when fed to dogs and in 1863, Bernhard Naunyn found adult tapeworms in dogs fed with hydatid cysts from a human.

> Canines are the most common definitive hosts for *Echinococcus* and typically do not show clinical signs of infection.

Causative Agent

Species of the cestode genus *Echinococcus* cause echinococcosis with the two most important species being *Ec. granulosus* and *Ec. multilocularis*. All *Echinococcus*

Table 6-5 *Echinococcus* **Species in Animals**

Species	Definitive Hosts	Intermediate Hosts
Echinococcus granulosus	Canines (dogs, wolves, coyotes, foxes, jackals, hyenas, dingoes)	Mainly herbivores (sheep, goats, cattle, pigs, horses, buffalo, and a variety of wild herbivores)
Echinococcus multilocularis	Primarily foxes (red and arctic); others include wolves, coyotes, dogs, and cats	Small mammals such as voles, lemmings, shrews, and mice; humans and domestic animals rarely serve as intermediate hosts
Echinococcus vogeli	Bush dog in South America	South American rodents
Echinococcus oligarthrus	Wild felines (jaguars, lynxes, and pumas); domestic cats can also be definitive hosts	Wild rodents

species have an indirect life cycle, cycling between a definitive (causing intestinal infection) and an intermediate host (causing tissue invasion). *Echinococcus* species in animals are summarized in Table 6-5.

Ec. granulosus causes cystic echinococcosis (also called unilocular echinococcosis and cystic hydatid disease), a disease in which some animals (typically dogs and other canines) are definitive hosts. Other strains of *Ec. granulosus* can be found in sheep, cattle, pigs, horses, and reindeer. All strains of *Ec. granulosus* can infect humans other than the horse strain. Adult worms that are 3 to 6 mm long live in the small intestine of the definitive host. They typically contain a scolex, short neck, and three proglottids. The rostellum has a double crown of approximately 30 hooks. The gravid uterus is an irregular longitudinal sac that releases eggs capable of infecting an intermediate host. Eggs of *Echinococcus* spp. resemble those of the cestode *Taenia*.

Infection with the larva form of *Ec. multilocularis* causes alveolar echinococcosis (also known as alveolar hydatid disease, multilocular echinococcosis, and multivesicular hydatidosis), which clinically produces signs similar to those of a slow-growing malignant tumor. Metacestodes (juvenile forms) develop a thin outer wall, as opposed to *Ec. granulosus* metacestodes that develop a thick outer wall. The thin outer wall grows and slowly infiltrates into surrounding host tissue. Initially the lesion is located in the liver and then may metastasize to any other organ. *Ec. multilocularis* in Europe is predominantly sylvatic with red foxes as definitive hosts and rodents as intermediate hosts. In other countries, dogs and cats may be definitive hosts; however, all definitive hosts acquire the infection from the sylvatic cycle by ingestion of rodents infected with metacestodes of *Ec. multilocularis*. *Ec. multilocularis* adults are 1.0 to 4.5 mm long and have fewer testes than *Ec. granulosus* adults. *Ec. multilocularis* eggs are found in carnivore feces and resemble those of *Taenia*. A membrane protects *Echinococcus* eggs, making them extremely tolerant of environmental conditions.

Epizootiology and Public Health Significance

Echinococcal infections in humans occur worldwide. *Ec. granulosus* is most commonly found in temperate, sheep-raising areas such as the western United States (Utah, California, Arizona, and New Mexico). *Ec. granulosus* has also been found in Canada, Europe, Japan, Russia, India, Turkey, Iraq, China, and northern Africa. *Ec. multilocularis* has been found in the north central region of North America from eastern Montana to central Ohio, as well as Alaska and Canada. *Ec. vogeli* and *Ec. oligarthrus* have only been found in Central and South America and rarely cause human disease.

Most cases of echinococcosis are the cystic form (caused by *Ec. granulosus*) and are common in rural areas and areas with poor sanitation. Cystic echinococcosis is also prevalent in regions when canine intestines are part of the human diet. Hunters, fur trappers, wildlife veterinarians, and wildlife biologists have an increased risk for contracting cystic echinococcosis. Cystic echinococcosis is a potentially fatal disease; however, cysts are typically well tolerated and may be an incidental finding at necropsy. *Ec. granulosus* is common in dogs and livestock. In some regions more than 30% of dogs and 95% of sheep may be infected.

In the United States foxes infected with *Ec. multilocularis* are present in most of the northern and central states; however, only two cases of human infection in these areas have been described since the beginning of the 20th century. In Alaska from 1947 to 1990, 30 of 53 human cases of alveolar echinococcosis were found in Eskimos. Alveolar echinococcosis occurs only in the northern hemisphere, with an incidence of 1 to 20 cases per 100,000 people per year in these regions (overall worldwide incidence is 0.02 to 1.4 cases per 100,000 people). In endemic areas of China, prevalence averages 5% but may reach 10% in some villages. Untreated alveolar echinococcosis is typically fatal (70% to 100%). *Ec. multilocularis* is less common than *Ec. granulosus* in domestic animals, but appears to be spreading into new regions. Prevalence of *Ec. multilocularis* infection in foxes is 15% to 70% in endemic areas with percentages of infected foxes and distribution of those foxes increasing. Less than 1% of dogs and cats are infected worldwide, but approximately 12% of dogs are infected in China and Alaska.

Transmission

Dogs are infected with *Echinococcus* when they feed on offal of butchered animals, whereas herbivores are infected when they ingest pasture plants contaminated with dog feces.

Canines become infected with *Ec. granulosus* when they eat the viscera of infected intermediate hosts (mainly sheep). Once infected, they will pass the eggs in their stool that are directly infectious to other animals and humans. Wildlife reservoirs are important maintenance hosts (animals that can maintain an infection persistently in the population in the absence of the original host or external sources of reinfection).

Foxes, coyotes, dogs, and cats are definitive hosts and get infected when they ingest *Ec. multilocularis* larvae in infected intermediate rodent hosts (voles, shrews, and mice). Once definitive hosts become infected, they will pass the eggs in their stool that are directly infectious to other animals and humans.

Echinococcus is transmitted to humans in one of two ways: by ingesting food or drinking water that is contaminated with stool from an infected animal or by petting or having other contact with infected cats and dogs or their contaminated equipment. *Echinococcus* eggs have a sticky coat and will adhere to an animal's fur and other objects. Flies may be a mechanical vector as well.

Pathogenesis

Echinococcus granulosus

Different *Ec. granulosus* strains may be found in specific host-prey cycles; the dog-sheep cycle is most likely to cause human infection.

Canines (definitive host) become infected with *Ec. granulosus* when they ingest metacestodes (cysts) in the tissues of intermediate hosts such as sheep, goats, swine, cattle, and horses. After ingestion, protoscolices emerge from the cyst, evaginate, attach to the intestinal mucosa, and develop into adult tapeworms in the

Ec. granulosus

Ec. multilocularis

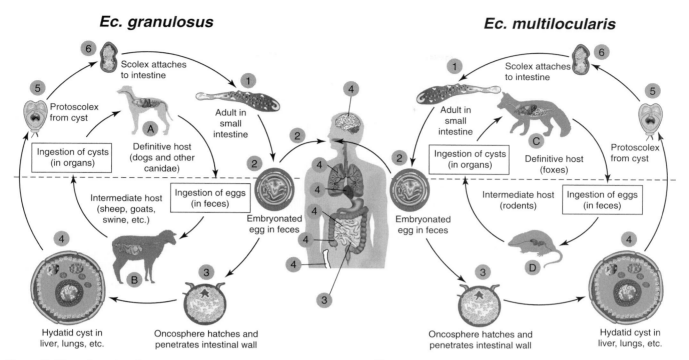

Figure 6-38   Life cycle of *Echinococcus*. Adult *Echinococcus granulosus*❶ resides in the small intestine of the definitive hosts (dogs). Gravid proglottids release eggs❷ that are passed in the feces (gravid proglottids may also be released). After ingestion by an intermediate host (sheep, goat, swine, cattle, horses), the egg hatches in the small intestine and releases an oncosphere❸ that penetrates the intestinal wall and migrates through the circulatory system into various organs (liver and lungs). In these organs, the oncosphere develops into a cyst❹ that enlarges gradually, producing protoscolices and daughter cysts that fill the cyst. The definitive host becomes infected by ingesting the cyst-containing organs of the infected intermediate host. After ingestion, the protoscolices❺ evaginate, attach to the intestinal mucosa❻, and develop into adult stages❶.

The life cycle of *Echinococcus multilocularis* differs in that the definitive hosts are mainly foxes; the intermediate host are small rodents; and larval growth (in the liver) continues indefinitely in the proliferative stage, resulting in invasion of the surrounding tissues.

host's small intestine. Gravid proglottids or eggs are shed in feces. Eggs are immediately infective and have a sticky coat that allows them to adhere to animal fur and objects. Proglottids move in rhythmic contractions that help disperse eggs over a large area of pasture. Intermediate hosts (sheep, goats, pigs, cattle, or horses) are infected when they ingest eggs. The egg hatches in the small intestine and releases an oncosphere. Oncospheres penetrate the intestinal wall and are carried in blood or lymph to various internal organs. Parasites develop into cysts in many different organs, but most often are found in the liver and then lungs. The fluid-filled cysts grow slowly and are surrounded by a fibrous wall from the host and two walls from the parasite. The innermost membrane of the parasitic walls contains one to several invaginated *Echinococcus* heads that can develop into adult tapeworms if ingested by definitive hosts. Other *Echinococcus* heads can float in the hydatid fluid and if the cysts rupture these heads (called hydatid sand) can develop into new cysts (Figure 6-38).

Echinococcus multilocularis
The life cycle of *Ec. multilocularis* is similar to that of *Ec. granulosus* with a variety of canines (mainly foxes) serving as definitive hosts.

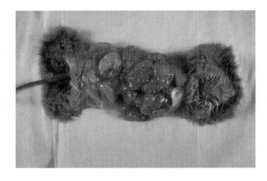

Figure 6-39 Multilocular cysts in a rat infected with *Echinococcus multilocularis.* (Courtesy of CDC/Dr. I. Kagan)

In intermediate hosts (small rodents) *Ec. multilocularis* metacestodes (juveniles) are found in the liver where the germinal membrane proliferates externally from multilocular cysts that embed in fibrous connective tissue. These cysts resemble tumors because they are not contained in a capsule and tend to be invasive. Cysts can metastasize to the CNS and lungs (Figure 6-39).

Clinical Signs in Animals

Ec. granulosus is most commonly found in canines (dogs, wolves, coyotes, dingoes, and jackals) that consume the viscera of infected sheep (Figure 6-40). Infection

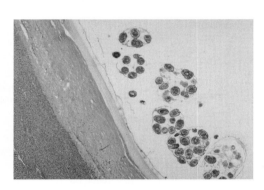

with *Ec. granulosus* produces cysts in the liver, lungs, kidney, and spleen and this condition is known as cystic hydatid disease. *Ec. multilocularis* is found in foxes, coyotes, dogs and cats. Infection with *Ec. multilocularis* results in the formation of tumors in the liver, lungs, brain, and other organs and this condition is called alveolar hydatid disease. *Ec. vogeli* and *Ec. oligarthrus* are found mainly in wildlife in Central and South America and rarely cause disease.

Figure 6-40 Histopathology of *Echinococcus granulosus* hydatid cyst in a sheep. (Courtesy of CDC/Dr. Peter Schantz)

Echinococcus spp. do not typically produce disease in their definitive hosts; if extremely large numbers of worms are present, animals may develop enteritis. Intermediate hosts infected with *Ec. granulosus* are typically asymptomatic when the cysts are growing slowly. Clinical signs may develop once the cysts are large enough to put pressure on tissues and organs. As a result of the different locations of the cysts, clinical signs will vary and reflect the affected organ. Intermediate hosts infected with *Ec. multilocularis* have liver masses that may metastasize to the lungs and brain. Infected rodents die within weeks of initial infection. Ascites, abdominal pain, hepatomegaly, dyspnea, diarrhea, vomiting, and weight loss may be seen in dogs.

Clinical Signs in Humans

The incubation period for echinococcosis in humans varies from months to years and may be as long as 20 to 30 years. Clinical signs vary with the size and location of the cysts. Cystic echinococcosis caused by *Ec. granulosus* produces

The incubation period for *Ec. granulosus* is 30 to 80 days and for *Ec. multilocularis* is 30 to 35 days.

very slowly growing masses that may become extremely large and contain liters of fluid. Clinical signs include anorexia, weight loss, and lethargy. If the cyst ruptures it can cause fever, pruritus, or anaphylaxis. Ruptured cysts may disseminate. Liver cysts produce clinical signs of abdominal pain, vomiting, or jaundice. Respiratory cysts produce chest pain and cough (Figure 6-41). Cysts may also be

Figure 6-41 Hydatid cysts from a human lung. (Courtesy of CDC/Dr. Kagan)

incidental findings on necropsy because they did not cause clinical disease during a person's lifetime.

Alveolar echinococcosis caused by *Ec. multilocularis* has a primary lesion in the liver that produces chronic disease that is typically asymptomatic in the early stages. In the later stages of disease ascites, jaundice, and liver failure may occur. These cysts invade tissues by growing outward resulting in a disease that is progressive and malignant.

Diagnosis in Animals

Diagnosis of echinococcosis in definitive hosts is by identification of adult worms or their proglottids in the definitive host. *Ec. granulosus* adults typically have three to four segments and are 3 to 6 mm in length. The scolex has four suckers and a double row of 25 to 50 hooks. *Ec. multilocularis* adults typically have two to six segments and are 1 to 4.5 mm in length. The scolex has four suckers and 26 to 36 hooks. Routine fecal examinations do not allow for distinction of *Echinococcus* eggs from other tapeworms eggs making them unreliable tests. Commercial ELISA tests and PCR tests are also available.

Diagnosis of echinococcosis in intermediate hosts is done at necropsy or surgery where metacestodes are identified by histology. PCR can identify *Ec. multilocularis* lesions. Ultrasonography is also helpful in diagnosis of echinococcosis. ELISA tests are available but demonstrate significant cross reactivity and are rarely used.

Diagnosis in Humans

Echinococcosis in people is diagnosed by imaging techniques such as ultrasonography, magnetic resonance imaging (MRI), or computed tomography (CT) scanning supported by positive serologic tests (ELISA, indirect immunofluorescence, immunoelectrophoresis, and immunoblotting). PCR tests are also available. An intradermal skin test (Casoni test) was historically used to diagnose echinococcosis in people, but is no longer used because of its low sensitivity, low accuracy, and potential for severe local allergic reaction (Figure 6-42).

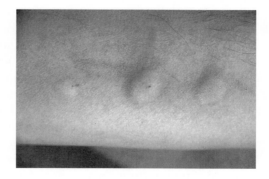

Figure 6-42 Man's arm showing a positive skin test (Casoni test) for hydatid disease (echinococcosis).
(Courtesy of CDC/Dr. I. Kagan)

Treatment in Animals

In definitive hosts, echinococcosis is treated with praziquantel and epsiprantel. In intermediate hosts, surgery is the best treatment. Long-term albendazole treatment following surgical removal of the mass is also effective.

Treatment in Humans

Most cysts are surgically removed. If surgery is not possible long-term albendazole or mebendazole may be needed to prevent the parasite from regrowing in the cysts. Liver transplants may be needed in cases of alveolar echinococcosis.

Management and Control in Animals

Echinococcosis can be controlled in animals by the following measures:

- Do not feed sheep viscera to dogs and avoid their access to sheep carcasses.
- Test and treat all pets for cestodes.

- Dogs should not be allowed to eat offal.
- Treatment of foxes with praziquantel in bait in heavily infected areas.
- Control and eradication of animals in endemic regions.
- Use of vaccination for *Ec. granulosus* in sheep.

Management and Control in Humans

Echinococcosis can be prevented in humans by the following measures:

- Always wash hands with soap and warm water after handling animals.
- Do not eat wild fruits or vegetables picked directly from the ground prior to washing or cooking them.
- Put fences around vegetable gardens to keep animals away.
- Thoroughly cook meat prior to eating it.
- Do not touch wild animals, dead or alive, without wearing gloves.
- Do not keep wild animals as pets or encourage them to come close to your house.
- Do not allow cats and dogs to eat rodents.

Summary

Echinococcosis is an intestinal parasitic disease caused mainly by two species of the cestode genus *Echinococcus* (*Ec. granulosus* and *Ec. multilocularis*). All *Echinococcus* species have an indirect life cycle, cycling between a definitive (causing intestinal infection) and an intermediate host (causing tissue invasion). *Ec. granulosus* causes cystic echinococcosis, a disease in which dogs and other canines are typically the definitive hosts. Adult worms that are 3 to 6 mm long live in the small intestine of the definitive host. Infection with the larva form of *Ec. multilocularis* causes alveolar echinococcosis, which clinically produces signs similar to those of a slow-growing malignant tumor. Initially the lesion is located in the liver and then may metastasize to any other organ. *Ec. multilocularis* adults are 1.0 to 4.5 mm long. Echinococcal infections in humans occur worldwide. *Ec. granulosus* is most commonly found in temperate, sheep-raising areas such as the western United States. *Ec. granulosus* has also been found in Canada, Europe, Japan, Russia, India, Turkey, Iraq, China, and northern Africa. Canines become infected with *Ec. granulosus* when they eat the viscera of infected intermediate hosts (mainly sheep). Once infected, they will pass the eggs in their stool. Wildlife reservoirs are important maintenance hosts who become infected when they ingest *Ec. multilocularis* larvae in infected intermediate rodent hosts (voles, shrews, and mice). *Echinococcus* is transmitted to humans in one of two ways: by directly ingesting food or drinking water that is contaminated with stool from an infected animal or by petting or having other contact with infected cats and dogs or their contaminated equipment. *Echinococcus* spp. do not typically produce disease in their definitive or intermediate hosts. The incubation period for echinococcosis in people varies from months to years and may be as long as 20 to 30 years. Clinical signs vary with the size and location of the cysts (liver cysts cause abdominal pain, vomiting, or jaundice, whereas respiratory cysts produce chest pain and cough). Cysts may also be incidental findings on necropsy because they did not cause clinical disease during a person's lifetime. Diagnosis of echinococcosis in definitive hosts is by identification of adult worms or their proglottids in the definitive host. Routine fecal examinations do not allow for distinction of *Echinococcus* eggs from other tapeworms eggs making them unreliable tests. Commercial ELISA tests and PCR tests are also available. Diagnosis of echinococcosis in intermediate hosts is done at necropsy or surgery. Echinococcosis in people is diagnosed by imaging techniques such as ultrasonography, MRI, or CT scanning supported by positive serologic tests (ELISA, indirect immunofluorescence, immunoelectrophoresis, and immunoblotting). PCR tests are also available. In definitive hosts, echinococcosis is treated with praziquantel and epsiprantel. In intermediate hosts, surgery is the best treatment. In people, most cysts are surgically removed and if surgery is not possible long-term albendazole or mebendazole is used. Echinococcosis can be controlled in animals by preventing access to infected animals and treating all pets. Echinococcosis can be prevented in humans by proper hygiene, proper food sanitation, and avoiding contact with wild animals.

Taenia Infections

Overview

Tapeworm infections caused by *Taenia* tapeworms have been known since the beginning of recorded history. The name *Taenia* was coined by the Greek writer Secundas (better known as Pliny the Elder who lived from A.D. 23 to 79) and is derived from the Latin term *taenia* meaning ribbon or tape. Hippocrates and Aristotle were aware of worms and described cucumber and melon seeds observed in human feces, most likely references to gravid proglottids of *Taenia saginata*. It is believed that about 2 million years ago, African hominids (our early ancestors) were exposed to tapeworms when they ate infected antelope and/or bovine meat. *Taenia* tapeworms cause human sickness and death worldwide and cause production losses in domestic livestock, including cattle and swine.

　　Taenia infections are caused mainly by two species of tapeworms, *Ta. saginata* and *Ta. solium*. The species name *saginata* is derived from the Latin sania meaning fattened animal (cattle) and the species name *solium* is Latin for a throne, which describes the ring of hooklets around the scolex of the worm. *Ta. saginata*, also known as the beef tapeworm, is a cestode parasite of cattle and humans, but only reproduces in humans. *Ta. saginata* occurs where cattle is raised, human feces are improperly disposed of, meat inspection programs are poor, and where meat is eaten without proper cooking. *Ta. solium*, also known as the pork tapeworm, is a cestode parasite of pigs and humans, but only reproduces in humans. Adult tapeworms live in the intestines of their definitive hosts. A newly recognized tapeworm found in Asia, *Ta. asiatica*, also causes human disease.

Causative Agent

Taenia spp. are long, segmented, parasitic tapeworms that have an indirect life cycle (cycling between a definitive and an intermediate host). Of the over thirty recognized species of *Taenia*, only a few species are zoonotic with humans serving as the definitive host, intermediate host, or both. When adult tapeworms live in the intestines of the definitive hosts the infection is called taeniasis. Humans are the definitive hosts for *Ta. solium*, *Ta. saginata*, and *Ta. asiatica* (which may be a subspecies of *Ta. saginata*, *Ta. saginata asiatica*). Animals are the definitive hosts for a variety of other *Taenia* species as described in Table 6-6.

　　Infection with the larvae of *Taenia* cestodes is either called **cysticercosis** or **coenurosis**. Infection with the larvae of *Ta. solium*, *Ta. saginata*, *Ta. crassiceps*, *Ta. ovis*, *Ta. taeniaeformis*, or *Ta. hydatigena* is called cysticercosis because the larvae are called cysticerci in these species. The larvae and adult tapeworms were once thought to be different species; therefore, the larval stages may be called a different name such as *Cysticercus cellulosae* (the larval stage of *Ta. solium*) or *Cy. bovis* (the larval stage of *Ta. saginata*). *Ta. solium* is the only *Taenia* species in which humans are definitive and intermediate hosts and is often the species found in humans. Cysticercosis caused by *Ta. saginata* is rare.

> Tapeworms require one intermediate host (always an herbivore) and a definitive host (always a carnivore) in which to reproduce.

Table 6-6   *Taenia* Spp. in Which Animals are Definitive Hosts

Taenia Species	Definitive Hosts	Intermediate Hosts
Taenia crassiceps	Foxes, wolves, coyotes, dogs	Rodents
Taenia ovis	Dogs, cats	Sheep, goats
Taenia taeniaeformis	Cats, dogs, lynx, wolves	Rats, mice, other rodents
Taenia hydatigena	Dogs, wolves, coyotes, lynx, cats (rare)	Domestic and wild cloven-hooved animals; hares, rodents (rare)
Taenia multiceps	Dogs, wolves, coyotes, foxes	Sheep, goats, other ruminants
Taenia pisiformis	Dogs, cats, foxes, wolves, coyotes, lynx	Rabbits, hares
Taenia serialis	Coyotes, wolves, dogs, foxes	Rabbits, hares, squirrels
Taenia brauni	Dogs, foxes	Gerbils
Taenia krabbei	Wolves, coyotes, dogs, bobcats	Moose, deer, reindeer

> Taeniasis is an adult tapeworm infection in the intestines, whereas cysticercosis (or coenurosis) is infection with *Taenia* larvae that eventually encyst in tissue. Larvae are more likely to cause disease than adult tapeworms.

Infection with the larvae of *Ta. multiceps*, *Ta. serialis*, and *Ta. brauni* is called coenurosis because the larvae are called coenurus in these species. The larval stage may have a different name then the adult such as *Coenurus cerebralis* (the larval stage of *Ta. multiceps*) or *Co. serialis* (the larval stage of *Ta. serialis*). Humans and animals can be intermediate hosts for these three species of *Taenia* tapeworm.

Ta. saginata adults are between 4 to 12 meters long (Figure 6-6B) and the cysticerci are approximately 4 to 6 mm in length. *Ta. saginata* has 15 to 30 lateral uterine branches in the gravid proglottids that are longer than they are wide and a scolex (head) with four suckers without a **rostellum** (projecting structure with hooks). Eggs are approximately 35 to 45 μm in diameter, have a thick embryophore that has numerous tiny pores that look like radial striations, and three pairs of hooklets in the embryo (Figure 6-43).

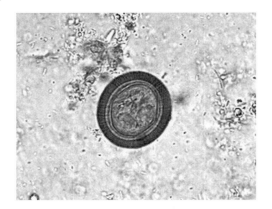

Figure 6-43   *Taenia* egg.
(Courtesy of CDC/Dr. Mae Melvin)

Ta. solium adults are between 1.5 to 8 meters long and the cysticerci are approximately 4 to 6 mm long when found in the muscles or subcutaneous tissues. The mature proglottids of *Ta. solium* have trilobed ovaries and with only approximately half the number of testes as *Ta. saginata*. The gravid proglottids have a uterus with between 8 to 12 lateral branches that are longer than they are wide (but not as long as *Ta. saginata*) and a scolex with four suckers and a rostellum containing a double crown of approximately 30 hooks that allows the worm to attach to the intestinal mucosa (Figure 6-44). Eggs are indistinguishable from those of *Ta. saginata*.

    Ta. asiatica is morphologically similar to *Ta. saginata*; it differs only in its intermediate host (pigs, not cattle).

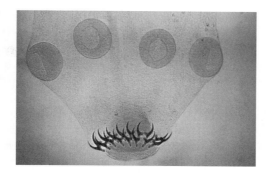

Figure 6-44 Scolex of *Taenia solium.* (Courtesy of CDC)

Epizootiology and Public Health Significance

Ta. saginata is found worldwide and is found in highest numbers in areas where beef is a major food source and sanitation is inadequate. Forty million people are believed to be infected worldwide with about 50,000 people dying annually of cysticercosis. There is a 1% prevalence of *Ta. saginata* in the United States because most cattle are free of the parasite. Between 1997 and 1999 approximately 3% of slaughtered cattle in the United States were condemned as a result of cysticerci. *Ta. saginata* infection is endemic in some areas such as East Africa and South America. Approximately 1,000 cases of cysticercosis are diagnosed annually in the United States with infection seen primarily in immigrants. Cysticercosis can also be contracted following close contact with recently immigrated, infected individuals.

    Ta. solium is found worldwide especially in regions where humans have close contact with pigs and eat undercooked pork (most common in Latin America, Southeast Asia, and Africa). Disease in the United States is generally seen in immigrants from Latin America, but people may get infected from contaminated human feces making the potential for infection ubiquitous. *Ta. solium* infection in pigs in the United States is rare due to laws restricting feeding garbage to pigs.

    Ta. asiatica is found in Southeast Asia and China with up to 20% of the human population infected. *Ta. multiceps* has been reported from the Americas and parts of Europe and Africa; *Ta. serialis* infections occur in the United States and Canada; and *Ta. brauni* has been reported in Africa.

Transmission

In people taeniasis is acquired by the consumption of raw or undercooked beef (*Ta. saginata*) or pork (*Ta. solium* and *Ta. asiatica*) containing cysticerci. In people cysticercosis and coenurosis is acquired by consumption of eggs on fruits or vegetables, through ingestion of contaminated water, or directly from soil. People who have *Ta. solium* in their intestinal tract can infect themselves with the eggs shed in their own feces.

    In animals taeniasis is acquired by eating tissues from intermediate hosts including ruminants, rabbits, and rodents. In animals cysticercosis and coenurosis is acquired by eating eggs or gravid proglottids that were shed in the feces of definitive hosts. Grazing animals acquire the eggs on pastures, in vegetation, or in contaminated water. Eggs can also be carried on fomites.

> One *Ta. saginata* tapeworm can survive in a human for 30 to 40 years. Under ideal conditions (adequate humidity with low to moderate temperatures), *Ta. saginata* eggs are viable for over a year. In a typical cattle barn, *Ta. saginata* eggs can survive approximately 18 months.

Pathogenesis

Definitive hosts for *Taenia* spp. are typically carnivores. A definitive host (person) becomes infected when they ingest raw or undercooked tissues of intermediate hosts (beef or pork) that contain cysticerci (larvae). The cysticercus becomes

> *Ta. solium* may survive up to 25 years in its host and there is typically only a single worm present in an infected person.

activated, attaches to the wall of the small intestine by the scolex, and becomes a mature tapeworm (a process that takes 10 to 12 weeks for *Ta. saginata* and 5 to 12 weeks for *Ta. solium*). Adult tapeworms live in the human small intestine. Humans pass eggs or gravid proglottids containing eggs in feces and these mature eggs contaminate pastures and barnyards, where cattle and pigs ingest them. Blowflies can also carry eggs from infected feces to meat, which is consumed by pigs. After ingestion, the eggs reach the intestine of infected animals (intermediate hosts), the oncospheres (larvae) are released, penetrate the intestinal wall, and enter the circulation. The oncospheres are filtered from the circulation and encyst in muscular tissue. Larvae become infectious within 2 to 3 months.

Humans can also be intermediate hosts for *Ta. solium* when they ingest eggs through fecal contamination or by autoinfection by reverse peristalsis of eggs or proglottids in the intestines. Cysticerci may develop in any organ and do not stimulate an inflammatory reaction while alive or dead (calcified), but cause inflammation when they are dying.

Cysticercosis occurs when eggs of *Ta. solium*, *Ta. saginata*, *Ta. crassiceps*, *Ta. ovis*, *Ta. taeniaeformis*, or *Ta. hydatigena* are ingested. Eggs hatch in the intestines after they have been exposed to the gastric secretions of the stomach followed by intestinal secretions. After hatching, the larvae penetrate the intestinal wall and are carried in the blood to tissues. In tissues larvae (metacestodes) develop into either cysticerci or coenuri. *Ta. saginata* and *Ta. ovis* larvae are found in muscle tissue, *Ta. asiatica* and *Ta. taeniaeformis* larvae are found in the liver, *Ta. hydatigena* larvae are found in the abdominal cavity, and *Ta. crassiceps* larvae are found in the subcutaneous tissue and peritoneal or pleural cavities. Coenurosis occurs when humans ingest *Ta. multiceps*, *Ta. serialis*, or *Ta. brauni* eggs usually from contaminated fruits or vegetables. Approximately 100 cases of coenurosis have been reported mainly in Africa. Adult tapeworms develop in dogs or other canines that ingest coenurus larvae in the tissues of various intermediate hosts (Figure 6-45).

Clinical Signs in Animals

In animals, the clinical signs associated with *Taenia* infections vary with the form of the disease.

> Animals can acquire infective eggs through direct contamination of pastures with human or animal feces, the use of waste water for fertilizer, improperly purified waste water, and floods.

- Taeniasis is usually asymptomatic with clinical signs limited to unthriftiness, lethargy, anorexia, and mild diarrhea or colic.
- Cysticercosis produces clinical signs as a result of inflammation caused by dying larvae and the mechanical effects of the parasite depending on the number and location of the larvae and the *Taenia* species involved (living cysticerci do not cause inflammation). There are three distinct types of cysticerci: cellulose cysticerci, which are fluid-filled vesicles containing a single inverted larva; intermediate cysticerci, which are fluid-filled vesicles containing a scolex; and racemose cysticerci, which consist of a grape-like mass with several connected fluid-filled vesicles and a dead scolex. Cellulose cysticerci are typically 0.5 to 1.5 cm in diameter, whereas intermediate and racemose cysticerci are larger and more dangerous. In most tissues most cysts are surrounded by a capsule of fibrous tissue.
  - *Ta. saginata*. The definitive hosts are typically humans. Intermediate hosts are cattle, buffalo, water buffalo, llamas, and wild ruminants such as giraffes. Animals are typically asymptomatic, but cattle may present with fever, weakness, anorexia, and muscle stiffness. In cattle *Ta. saginata* is usually found in muscles. Myocarditis as a result of *Ta. saginata* may cause death in cattle.

Taeniasis

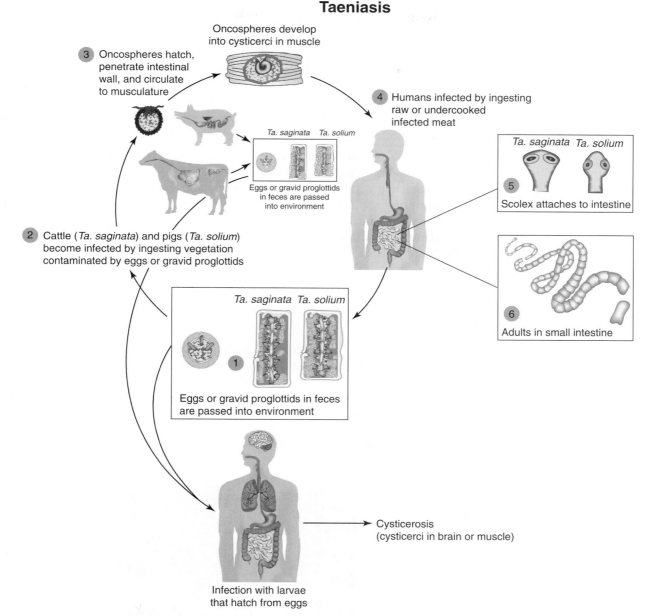

Figure 6-45 Life cycle of *Taenia*.

Eggs or gravid proglottids are passed with feces❶. Cattle (*Taenia saginata*) and pigs (*Taenia solium*) become infected by ingesting vegetation contaminated with eggs or gravid proglottids❷. In the animal's intestine, the oncospheres (larvae) hatch❸, invade the intestinal wall, and migrate to the striated muscles, where they develop into cysticerci. Humans become infected by ingesting raw or undercooked infected meat❹. In the human intestine, the cysticercus develops into an adult tapeworm. The adult tapeworms attach to the small intestine by their scolex❺ and reside in the small intestine❻. Adult tapeworms produce proglottids. Proglottids mature, become gravid, detach from the tapeworm, and migrate to the anus or are passed in the stool. The eggs contained in the gravid proglottids are released after the proglottids are passed with the feces.

- *Ta. solium*. The definitive hosts are typically humans. Intermediate hosts are usually pigs (occasionally sheep, dogs, cats, deer, camels, marine mammals, bears, and nonhuman primates are intermediate hosts). In pigs, *Ta. solium* are typically found in cardiac or skeletal muscle, liver, heart, and brain. *Ta. solium* can cause hypersensitivity of the snout, paralysis of the tongue, seizures, and fever in pigs; and encephalomyelitis in dogs.

- *Ta. asiatica.* Humans are the definitive hosts, and intermediate hosts include domestic and wild pigs (occasionally cattle, goats, or monkeys). *Ta. asiatica* cysticerci are typically found in the liver. Abdominal distension, lethargy, and weight loss are rare signs of *Ta. asiatica* seen in swine.
- *Ta. crassiceps, Ta. hydatingena, Ta. ovis,* and *Ta. taeniaeformis* rarely cause clinical disease in animals. The cysticerci of *Ta. crassiceps* (subcutaneous tissue), *Ta. hydatingena* (abdominal cavity), *Ta. ovis* (muscle), and *Ta. taeniaeformis* (liver) are found in a variety of tissues. Their definitive and intermediate hosts are summarized in Table 6-6.

- Coenurosis produces clinical signs as a result of inflammation caused by dying larvae and are similar to those of cysticercosis. *Ta. multiceps* may cause neurologic signs in ruminants. *Ta. serialis* and *Ta. brauni* rarely cause disease in animals. *Ta. multiceps* coenuri are typically 2 to 6 cm in diameter and are typically found in the brain, eye, or subcutaneous tissue. *Ta. serialis* coenuri are typically found in the subcutaneous tissues, muscles, and retroperitoneally. *Ta. brauni* coenuri are typically found in the subcutaneous tissues and eye.

Clinical Signs in Humans

In humans, the clinical signs associated with *Taenia* infections vary with the form of disease.

- Taeniasis is usually asymptomatic with occasional reports of mild abdominal pain, diarrhea, constipation, nausea, and weight loss. Infants and children typically have more clinical signs than adults. *Ta. saginata* proglottids may occasionally travel to other organs such as the appendix, uterus, bile ducts, and nasopharyngeal passages causing inflammatory disease. *Ta. solium* proglottids are less active than those of *Ta. saginata* and are less likely to be found in locations outside the intestine (Figure 6-46).

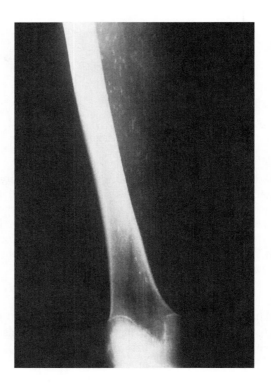

Figure 6-46 Cysticercosis of the muscle as a result of the presence of the cestode *Taenia saginata*, or beef tapeworm. (Courtesy of CDC)

- Cysticercosis symptoms vary with the location affected and number of larvae. Symptoms are typically associated with inflammation during larva death. Neurocysticercosis is the most serious form of cysticercosis with symptoms resulting from blockage of cerebrospinal fluid by a floating cysticercus. Headaches and seizures are the most common symptoms of neurocysticercosis. Cysts occurring in other body systems can cause blurred vision, blindness, muscle pain, fatigue, and subcutaneous nodules.
- Coenurosis symptoms are similar to those of cysticercosis and vary with the location and size of the coenurus. Coenuri are most frequently found in the CNS, subcutaneous tissue, and eye.

Diagnosis in Animals

Taeniasis can be diagnosed by detecting proglottids or eggs in the feces. Fecal flotation can demonstrate *Taenia* eggs; however, all *Taenia* eggs are similar and cannot be differentiated from each other. *Taenia* proglottids can be differentiated from each other based on their morphology by counting uterine branches after injection with India ink. Cysticercosis is difficult to diagnose in live animals; MRI can be used in small animals, or postmortem examination in large animals. Cysticerci can sometimes be palpated in the tongues of cattle or swine. Serology is not available in animals.

Diagnosis in Humans

Taeniasis can be diagnosed by examining three consecutive stool samples (direct and concentrated stool preparations) looking for eggs or proglottids in feces. Species identification based on egg examination is difficult because the eggs of *Ta. solium* and *Ta. saginata* are identical. Examining the gravid proglottids helps differentiate the *Taenia* species by counting the main uterine branches after injection with India ink. Examining the scolex helps differentiate the *Taenia* species because the scolex of *Ta. solium* has 4 suckers and a rostellum, whereas the scolex of *Ta. saginata* lack a rostellum. ELISA and PCR tests can also differentiate the species of *Taenia*.

Cysticercosis can be diagnosed using imaging studies such as CT scan and MRI to identify cysticerci in the brain. Calcified cysts can be seen in muscle and brain using X-ray. ELISA, complement fixation, and hemagglutination tests are also available. Coenurosis is uncommon and diagnosis is similar to that for cysticercosis.

Treatment in Animals

Taeniasis can be treated with praziquantel, espiprantel, mebendazole, febantel, and fenbendazole; valuable animals with coenuri or cysticerci may require surgical treatment.

Treatment in Humans

Most people with intestinal *Taenia* infection are asymptomatic or mildly symptomatic and do not require treatment. If anthelmintic therapy is needed, albendazole, niclosamide, buclosamide, mebendazole, or praziquantel can be used. These drugs provoke an anti-inflammatory response in the CNS and people with brain lesions should also be started on high-dose glucocorticosteroids. Ocular, cardiac, brain, and spinal lesions may require surgical treatment.

Management and Control in Animals

Taenia infections in livestock can be reduced by treating taeniasis in the definitive host. Dogs should not be allowed to eat the carcasses of animals with coenurosis and should be dewormed regularly. Animals should not be exposed to human feces. Hunting rodents and feeding of raw or undercooked carcasses should be avoided in all animals. No vaccines are currently available for *Taenia* prevention.

Management and Control in Humans

Taeniasis can be avoided by thoroughly cooking meat and preventing cross contamination of food with undercooked beef or pork. Freezing can also help decrease the number of taeniasis cases. Good hygiene, peeling fruits and vegetables, and drinking properly purified and filtered water reduces the risk of cysticercosis and coenurosis. Treating people infected with *Ta. solium* also prevents spreading of cysticercosis to themselves and others. No human vaccines are available for *Taenia* prevention.

Summary

Taenia infections are caused by cestodes in the genus *Taenia*, which are long, segmented, parasitic tapeworms that have an indirect life cycle. When adult tapeworms live in the intestines of the definitive hosts the infection is called taeniasis. Infection with the eggs of *Taenia* cestodes is either called cysticercosis (infection with the eggs of *Ta. solium, Ta. saginata, Ta. crassiceps, Ta. ovis, Ta. taeniaeformis,* or *Ta. hydatigena*) or coenurosis (infection with the eggs of *Ta. multiceps, Ta. serialis,* and *Ta. brauni*). *Ta. saginata* adults are between 4 to 12 meters long and the cysticerci are approximately 4 to 6 mm in length. Eggs are approximately 35 to 45 µm in diameter, have a thick embryophore that has numerous tiny pores that look like radial striations, and three pairs of hooklets in the embryo. *Ta. solium* adults are between 1.5 to 8 meters long and the cysticerci are approximately 4 to 6 mm long when found in the muscles or subcutaneous tissues. The different species of *Taenia* eggs are indistinguishable from each other. *Ta. saginata* is found worldwide and is found in highest numbers in areas where beef is a major food source and sanitation is subpar. *Ta. solium* is found worldwide especially in regions where humans have close contact with pigs and eat undercooked pork such as Latin America, Southeast Asia, and Africa. *Ta. asiatica* is found in Southeast Asia and China with up to 20% of the human population infected. The definitive hosts for *Taenia* spp. are typically carnivores (mainly humans) and are infected when they consume raw or undercooked beef or pork containing cysticerci. In animals taeniasis is acquired by eating tissues from intermediate hosts including ruminants, rabbits, and rodents. Adult tapeworms live in the human small intestine. Humans pass gravid proglottids containing eggs in feces and these mature eggs contaminate pastures and barnyards, where cattle and pigs ingest them. After ingestion, the eggs reach the intestine of infected animals (intermediate hosts), the oncospheres (larvae) are released, penetrate the intestinal wall, and enter the circulation. The oncospheres are filtered from the circulation and encyst in muscular tissue. In animals, the clinical signs associated with *Taenia* infections vary with the form of the disease. Taeniasis is usually asymptomatic with clinical signs limited to unthriftiness, lethargy, anorexia, and mild diarrhea or colic. Cysticercosis produces clinical signs as a result of inflammation caused by dying larvae and the mechanical effects of the parasite depending on the number and location of the larvae and the *Taenia* species involved. Coenurosis produces clinical signs as a result of inflammation caused by dying larvae and are similar to those of cysticercosis. In humans, the clinical signs associated with *Taenia* infections vary with the form of disease. Taeniasis is usually asymptomatic with occasional reports of mild abdominal pain, diarrhea, constipation, nausea, and weight loss. Cysticercosis symptoms vary with the location affected and number of larvae. Neurocysticercosis is the most serious form of cysticercosis with symptoms resulting from blockage of cerebrospinal fluid by a floating cysticercus (headaches and seizures). Coenurosis symptoms are similar to those of cysticercosis and most frequently found in the CNS, subcutaneous tissue, and eye. Taeniasis can be diagnosed by detecting proglottids or eggs in the feces (all *Taenia* eggs are similar and cannot be differentiated from each other). *Taenia* proglottids can be differentiated from each other based on their morphology after injection with India ink. Serology is not available in animals. Taeniasis in people can be diagnosed similarly to animals except that serologic tests such as ELISA, complement fixation, and hemagglutination are available. PCR tests are also available. Taeniasis can be treated with praziquantel, espiprantel, mebendazole, febantel, and fenbendazole in animals and humans; surgical treatment may be needed to remove some lesions. *Taenia* infections in livestock can be reduced by treating taeniasis in the definitive host. Dogs should not be allowed to eat the carcasses of animals with coenurosis and should be dewormed regularly. Animals should not be exposed to human feces. Hunting rodents and feeding of raw or undercooked carcasses should be avoided in all animals. Taeniasis in people can be avoided by thoroughly cooking meat and preventing cross contamination of food with undercooked beef or pork. Good hygiene, peeling fruits and vegetables, and drinking properly purified and filtered water reduces the risk of cysticercosis and coenurosis. No vaccines are available. See Table 6-7 for Less Common Zoonotic Cestodes.

NEMATODE ZOONOSES

Cutaneous Larva Migrans

Overview

Cutaneous larva migrans (CLM), also known as larva migrans cutanea and creeping eruption, is an acute skin syndrome caused by migrating larvae of parasitic nematodes. CLM manifests as an erythematous (red), serpiginous (twisting), pruritic skin eruption caused by accidental skin penetration and migration of nematode larvae. A variety of nematodes cause CLM, including hookworms (named for their shepherd's hook appearance as adults) and *Strongyloides* worms (*strongylo* is Greek for round). CLM is the most commonly acquired tropical skin disease whose earliest description can be found more than 100 years ago. Hookworm infection occurs predominantly among the world's poorest people (Mohandas Gandhi had hookworm infection in the later part of his life). Infection caused by *Ancylostoma duodenale*, the nonzoonotic hookworm that causes diseases known as ground itch worldwide and sandworms in the southern United States, was described as early as the 1840s during the building of the St. Gotthard tunnel between Switzerland and Italy. The post civil war era in the United States (1870s–1880s) led to poor conditions in the south where people were thought to be lazy until it was discovered that hookworm infection (caused by *Necator americanus*) caused anemia in these southerners. In the southern United States hookworm infection was believed to be a contributing factor in the slowing of this region's economic development during the early part of the 20th century. In the late 1800s and early 1900s many homes in rural American states did not have indoor plumbing and were often plagued with diseases such as hookworm that were directly linked to improper sanitary facilities. Before 1900, few American physicians were aware of hookworm disease

Table 6-7　Less Common Zoonotic Cestodes

Cestode	Disease	Predominant Signs in People	Clinical Signs in Animals	Transmission	Geographic Distribution	Diagnosis	Treatment and Control
Dipylidium caninum	Dipylidiosis	Infection is typically asymptomatic; heavy infections may cause abdominal pain, bloody diarrhea, and weight loss	Dogs and cats show unthriftiness, lethargy, increased appetite, and mild diarrhea	■ Ingestion of cysticercoids (infective stage) when the hand is licked by dogs with this stage in their mouth or ingestion of infected fleas	■ Worldwide (the most common cestode of dogs in Europe)	Identification of proglottids or eggs that are 40 μm in size in feces	Niclosamide and praziquantel Avoid hand licking by dogs or cats; routine fecal examinations for dogs and cats and appropriate treatment if positive
Hymenolepis nana	Hymenolepiasis	Large numbers of cestodes can cause atrophy of the intestinal villi and inflammation causing abdominal pain and diarrhea	■ Rodents are asymptomatic	■ Ingestion of eggs (rarely ingestion of infected intermediate hosts such as beetles) ■ Person-to-person autoinfection also possible (no intermediate host needed)	■ Worldwide	Fecal examination demonstrating colorless eggs that are 40 to 50 μm	Praziquantel Rodent and insect control; good personal hygiene
Spirometra spp.	Sparganosis	Slow growing, tender, painful nodule formation that typically occurs near the eye, producing edema	■ Dogs, cats, and other carnivores are final hosts and are asymptomatic	■ Ingestion of larvae in the second intermediate hosts (fish, frogs, reptiles)	■ Worldwide, especially China, Japan, and Southeast Asia	CT and MRI can identify lesions in the brain	Effective drugs not available Proper cooking of fish, frogs, and snails; avoid potentially contaminated water
Bertiella studeri and *Bertiella mucromata*	Intestinal cestode infections	Clinical signs rare; heavy infestations cause abdominal pain, diarrhea, and nausea	■ Nonhuman primates are asymptomatic	■ Ingestion of infected intermediate hosts (*Oribatidae* mites)	Africa, Asia, and South America	Identification of proglottids	Praziquantel Proper food hygiene such as thorough cooking of intermediate hosts and cleaning vegetables
Raillietina demerariensis and *Raillietina asiatica*	Intestinal cestode infections	Clinical signs rare; heavy infestations cause abdominal pain, diarrhea, and nausea	■ Rats and primates are asymptomatic	■ Ingestion of infected intermediate hosts (beetles and cockroaches)	Worldwide	Identification of proglottids	Praziquantel Proper food hygiene such as thorough cooking of intermediate hosts and cleaning vegetables

Inermicapsifer madagascariensis	Intestinal cestode infections	Clinical signs rare; heavy infestations cause abdominal pain, diarrhea, and nausea	■ Small rodents are asymptomatic	■ Ingestion of infected intermediate hosts (*Oribatidae* mites)	Africa, Madagascar, and the Caribbean	Identification of proglottids	Praziquantel Proper food hygiene such as thorough cooking of intermediate hosts and cleaning vegetables
Mesocestoides variabilis and *Mesocestoides lineatus*	Intestinal cestode infections	Clinical signs rare; heavy infestations cause abdominal pain, diarrhea, and nausea	■ Carnivores are natural hosts and are asymptomatic ■ First intermediate hosts are oribatid mites ■ Second intermediate hosts are rodents, birds, amphibians, and reptiles	■ Ingestion of undercooked infected intermediate hosts	Africa, Asia, and South and North America	Identification of proglottids	Praziquantel Proper food hygiene such as thorough cooking of intermediate hosts and cleaning vegetables
Diplogonoporus grandis	Intestinal cestode infections	Clinical signs rare; heavy infestations cause abdominal pain, diarrhea, and nausea	■ Marine mammals (especially whales) are final hosts ■ Larvae are found in saltwater fish	■ Ingestion of infected raw or undercooked saltwater fish	Japan	Identification of proglottids and eggs in fecal samples	Praziquantel Proper food hygiene such as thorough cooking of intermediate hosts and cleaning vegetables

until Dr. Charles Wardell Stiles learned about hookworms while assisting with animal autopsies in Europe in the late 19th century. When Stiles returned to the United States, he worked for the Bureau of Animal Industry of the Department of Agriculture in Washington, D.C., and taught at Johns Hopkins School of Medicine, spreading information about the parasite and health problems associated with this nematode. In 1910, Dr. Stiles convinced John D. Rockefeller to donate $1 million to found the Rockefeller Sanitation Commission for the Eradication of Hookworm Disease and worked state by state to establish ways to combat hookworm disease through health education (building of enclosed outhouses), patient treatment (thymol, enemas, and tetrachloroethylene), and community cooperation. As a result of these efforts, hookworm prevalence is lower in the southern United States. Worldwide the disease is still a major health problem.

Stro. stercoralis was first discovered in 1876 in French soldiers who had been in Vietnam (then known as Cochin China). It was considered to be a harmless parasite for years; however, in the last 40 years it has been known to cause severe inflammatory and ulcerative bowel disease in people. It is now widespread in tropical and temperate countries (especially in institutions with poor hygiene). It is one of the few worms that is more prevalent in adults than in children.

Causative Agent

CLM is caused by a variety of juvenile nematodes. The main nematode genus that causes CLM is *Ancylostoma*. *Ancylostoma* nematodes are stout with a curved anterior end giving the worm a hook-like appearance. There are two plates in their buccal capsule each with two large teeth that are fused at the base. A pair of small teeth is also found in this capsule (Figure 6-47). Adult males are about 8 to 11 mm in length and adult females are 10 to 13 mm in length. Each species has a unique bursa. Its normal life span is 1 year. Eggs are 50 × 30 µm and have a thin, smooth, colorless shell with two- to eight-cell stage of cleavage (Figure 6-48). Zoonotic species include:

> Nonzoonotic parasites such as *Ancylostoma duodenale* (*ankylo* is Greek for bent and *stoma* is Greek for mouth) and *Necator americanus* (American killer) can also cause CLM in people.

> CLM in people is caused by invasive juvenile nematodes that normally mature in animals other than humans.

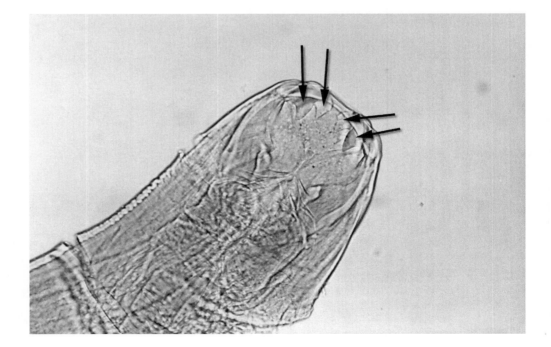

Figure 6-47 *Ancylostoma duodenale's* mouth parts are used to grasp firmly to the intestinal wall, allowing it to remain fastened in place while it ingests the host's blood obtaining its nutrients in this fashion.
(Courtesy of CDC)

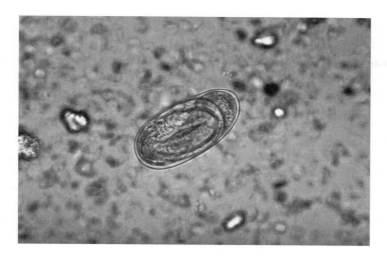

Figure 6-48   Embryonated egg of *Ancylostoma* and *Necator* hookworms are indistinguishable from each other.
(Courtesy of CDC)

- *An. braziliense* (the canine and feline hookworm) is the most important zoo-notic species causing CLM and is common in dogs and cats of the Gulf Coast and tropical areas in the Americas. It is also found in wild felines and various carnivores. *An. braziliense* is a parasite of dogs and cats whose third stage larvae can penetrate the superficial layers of human skin yet are unable to pass through the skin. Beaches, yards, and playgrounds frequented by infected dogs and cats may become heavily infested with larvae. *An. braziliense* is smaller than *An. caninum* and has only 2 ventral teeth on each side of the buccal cavity. The males are 5.0 to 7.5 mm long; females are 6.5 to 10.6 mm long. Eggs are 75 to 95 µm in length.

- *An. caninum* (the canine hookworm) is the most common hookworm of domesticated dogs, especially in the Northern Hemisphere. *An. caninum* is also found in foxes, coyotes, wolves, bears, wild carnivores, and cats. It is also an important cause of eosinophilic enteritis in northeastern Australia and the United States. *An. caninum* is stout-bodied, grayish or reddish in color, and has 3 pairs of ventral teeth in the buccal cavity (Figure 6-1B). Males are 11 to 13 mm long; females are 14.0 to 20.5 mm long. Eggs are 56 to 75 µm long.

- *An. ceylonicum* is a parasite found in dogs and cats in Sri Lanka, Southeast Asia, and the East Indies.

- *An. tubaeforme* (the feline hookworm) is a parasite found in cats and is generally distributed worldwide and throughout the United States. *An. tubaeforme* is intermediate in size between *An. caninum* and *An. braziliense* and has three teeth on each side of the buccal cavity. Males are 9.5 to 11.0 mm long; females are 12 to 15 mm long. Eggs are 55 to 76 µm long.

Another genus of nematode that causes CLM is *Strongyloides*. Species of *Strongyloides* are among the smallest nematodes and can maintain parasitic life cycles with only the female of the species as a result of its **parthenogenetic** (without fertilization of the male) nature. Parthenogentic females are 2 to 2.5 mm in length; nonparasitic males are up to 0.9 mm in length. Both sexes have a small buccal capsule. Eggs are 50 to 58 µm in length and are thin-shelled containing partially embryonated eggs. Zoonotic species include:

- *Stro. stercoralis* is the most common species worldwide and infects primates, dogs, cats, and other mammals. *Stro. stercoralis* is a small, slender nematode that is almost transparent and virtually impossible to see grossly at necropsy.

- *Stro. fuelleborni* occurs in Africa and is the hookworm of primates.
- *Stro. papillosus* is the hookworm of sheep, goats, and cattle.
- *Stro. westeri* is the hookworm of horses (known as the threadworm) and can cause bleeding and respiratory problems. In untreated foals it may cause diarrhea, weakness, weight loss and poor growth.

Other genera of nematodes that cause CLM include:

- *Uncinaria stenocephala* (the northern canine hookworm) is found in Europe. It can also be found in Canada and the northern parts of the United States (primarily in foxes and wolves). Its life cycle is similar to *An. caninum*. Percutaneous infection is rare with *U. stenocephala* in which 95% of the larvae fail to develop. *U. stenocephala* is a small hookworm, which does not have teeth, but two cutting plates in its mouth cavity. Males are 5 to 9 mm long; females are 7 to 13 mm long. Eggs are 63 to 76 µm long.
- *Bunostomum phlebotomum* is the cattle hookworm. The adult male is about 9 mm and the female is up to 18 mm in length. Infection in cattle is by ingestion or skin penetration of the distal limbs. It is a stout worm (about 20 mm) with a large mouth and a body that may be hook-shaped. It may cause diarrhea, anemia, and sore feet in cattle.
- *Gnathostoma* spp. (cat, dog, and pig roundworms), *Capillaria* spp. (whipworms found in rodents, cats, dogs, and poultry), and *Stro. myopotami* (found in the small intestine of mammals) are rare causes of CLM. These are summarized in the less common nematode zoonoses chart.

Epizootiology and Public Health Significance

Strongyloides larvae move more quickly than other nematodes in human skin and produce a disease sometimes referred to as larva currens (racing larva).

Ancylostoma and *Strongyloides* spp. occur worldwide with the greatest number of cases occurring in Asia and sub-Saharan Africa. *Ancylostoma* parasites live in sandy or loamy soil and cannot exist in clay or muck. *Ancylostoma* needs warm moist conditions where rainfall averages are more than 40 inches per year and the average temperature is greater than 50°F. *An. braziliense* is found mainly in tropical and subtropical areas; *An. ceylanicum* is found in Asia and the Middle East; *An. caninum* is found in the Northern Hemisphere and in northeastern Australia; *An. tubaeforme* is found in the United States. *Stro. fuelleborni* occurs more commonly in Africa. *Uncinaria* parasites are found mainly in Europe. *Bunostomum* parasites are found worldwide.

The prevalence of hookworm CLM in the United States is unknown; but approximately 7% of travelers to clinics specializing in travel-related disease present with CLM. In the United States, most cases occur in eastern and southern coastal areas from New Jersey to Texas with the highest incidence in Florida. Worldwide CLM is indigenous to the Caribbean, Central and South America, Africa, and Southeast Asia.

The prevalence of *Stro. stercoralis* is likely underestimated because infection is often asymptomatic. *Strongyloides* is endemic in the Appalachian United States, especially in eastern Tennessee, Kentucky, and West Virginia. Worldwide prevalence is estimated as 2% to 20% in endemic areas. At-risk people include those who have recently traveled to or immigrated from endemic areas and veterans of World War II and Vietnam. Currently 100 to 200 million people are estimated to be infected worldwide in 70 countries. The mortality rate for patients requiring hospitalization with *Strongyloides* infection is approximately 17%. In disseminated strongyloidiasis, the mortality rate can be as high as 70% to 90%.

The prevalence of CLM caused by the other genera of nematodes is less frequent than those listed above and varies by region.

Transmission

Hookworm (*Ancylostoma*, *Uncineria*, and *Bunostomum*) larvae of animals live in infected soil and penetrate human skin on contact (such as walking barefoot). In animals, transmission routes are percutaneous, prenatal, and transmammary. Young animals can be infected by ingestion of infective larvae from the environment or via the colostrum or milk of infected bitches (*An. caninum*). Infective larvae penetrate animal skin and migrate through the blood vessels to the lungs, enter the trachea, are coughed up, swallowed, and pass to the intestine where they complete their development. Colostral and lactogenic infection are also possible. In pregnant females, migrating larvae may pass to the placenta and enter the fetus. Transmission of *U. stenocephala* is similar to that of *An. caninum* except that dogs probably do not become infected by the prenatal or colostral routes.

Infection with *Strongyloides* nematodes occurs by the percutaneous route with filariform third-stage larvae. In animals, **autoinfections** and transmammary infections also occur. Infection can also occur when a horse ingests larvae when eating grass. Mares can pass the nematode in their milk.

> CLM usually presents as one of three scenarios: barefoot children in areas frequented by hookworm-infested pets; travelers on exotic vacation who walk barefoot on their travels; and laborers (plumbers, masons, and electricians) working in hookworm-infected crawl spaces and dirt.

Pathogenesis

The life cycle of *Ancylostoma* starts when infected animals defecate thin-walled, unembryonated eggs with the feces. In warm, moist, sandy soil the eggs hatch to rhabditiform larvae in 1 to 2 days and molt twice to become infective filariform (third-stage) larva after 5 to 10 days. Filariform larvae may remain viable in the environment for up to 3 weeks. The infective filariform larvae live in the top one-half inch of the soil, with their ends projecting upward from the surface, waiting for people or animals to pass by. Larval skin penetration requires skin contact with contaminated soil for 5 to 10 minutes. Using proteases, the larvae penetrate through follicles, fissures, or intact skin of the new host (another animal or human). After penetrating the stratum corneum layer of the epidermis, the larvae shed their natural cuticle. Larvae typically begin migration in four days. In animal hosts, the larvae are able to penetrate into the dermis and are transported via the lymphatic and venous systems to the lungs. They break through into the alveoli and migrate to the trachea, where they are coughed up and swallowed. In the intestine they mature sexually, and the cycle begins again with secretion of their eggs. A hookworm produces an anticoagulant in its saliva so the animal host's blood does not clot at the site the hookworm attaches. If the worm moves from that site to reattach itself at another, the first site may continue to bleed, sometimes seriously. Humans are accidental hosts, and the larvae lack the collagenase enzymes required to penetrate the epidermal basement membrane to invade the dermis. Therefore, in humans, the disease remains limited to the skin (Figure 6-49). Only *An. duodenalis* (nonzootic) spreads beyond the skin.

> *Strongyloides* is the only helminth to secrete larvae (and not eggs) in feces.

The *Strongyloides* life cycle is more complex than most nematodes because it alternates between free-living and parasitic cycles. In the free-living cycle rhabditiform larvae are passed in the stool and can either molt twice and become infective filariform larvae (direct development) or molt four times and become free-living adult males and females that mate and produce eggs from which rhabditiform larvae hatch. The filariform larvae penetrate the skin to initiate the parasitic cycle (ingestion, transmammary, and autoinfection are other transmission routes of *Strongyloides* but will not be discussed here). In the parasitic cycle filariform larvae in contaminated soil penetrate skin (the CLM manifestation) and are transported to the lungs where they penetrate the alveolar spaces. From the alveolar spaces they are carried through the bronchi to the pharynx, are coughed up, and are swallowed. Upon reaching the small intestine, they molt twice and become adult

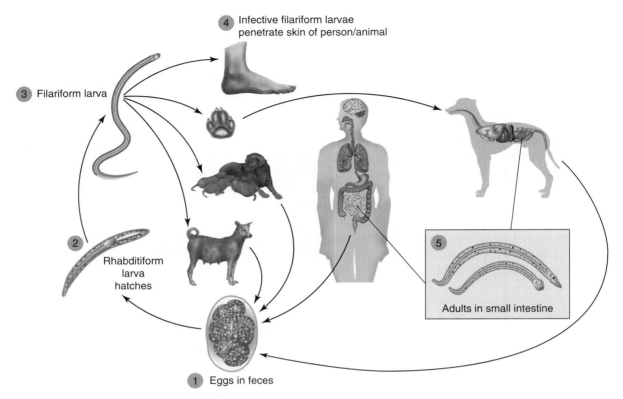

Figure 6-49 Life cycle of *Ancylostoma*.

Unembryonated eggs are passed in the stool❶, and under favorable conditions larvae hatch. The released rhabditiform larvae mature in the feces and/or the soil❷, and after two molts they become filariform (third-stage) infective larvae❸. Infective filariform larvae are transmitted to animals percutaneously, transmammary, or prenatally. Young animals can also be infected by ingestion of infective larvae. On contact with the human host, the larvae penetrate the skin❹ The nonzootic *An. duodenalis* can penetrate the skin and are carried through the veins to the heart and then to the lungs. They penetrate into the pulmonary alveoli, ascend the bronchi to the pharynx, and are swallowed❹. The larvae reach the small intestine, where they reside and mature into adults. Adult worms live in the lumen of the small intestine, where they attach to the intestinal wall❺.

female worms. The females live in the epithelium of the small intestine and by parthenogenesis produce eggs (which yield rhabditiform larvae). The rhabditiform larvae can either be passed in the stool or cause autoinfection (rhabditiform larvae become infective filariform larvae, which can penetrate either the intestinal mucosa (internal autoinfection) or the skin of the perianal area (external autoinfection)) (Figure 6-50).

The life cycle of *Uncinaria* begins when the host ingests an infective third stage larva. The larva matures in the small intestine where adults produce eggs that are passed out with the feces. The eggs hatch in the soil and the larvae molt twice to reach the infective third stage. Infective larvae penetrate skin and cause disease as described for *Ancylostoma*.

The life cycle of *Bunostomum* begins when larvae penetrate the principal host and are transported via the blood to the lungs before proceeding to their preferred sites where they mature.

Clinical Signs in Animals

Parasites that cause CLM in people have the ability to mature into adult worms in animals. These adult parasites tend to cause gastrointestinal disease, whereas juvenile parasites may cause cutaneous and systemic disease.

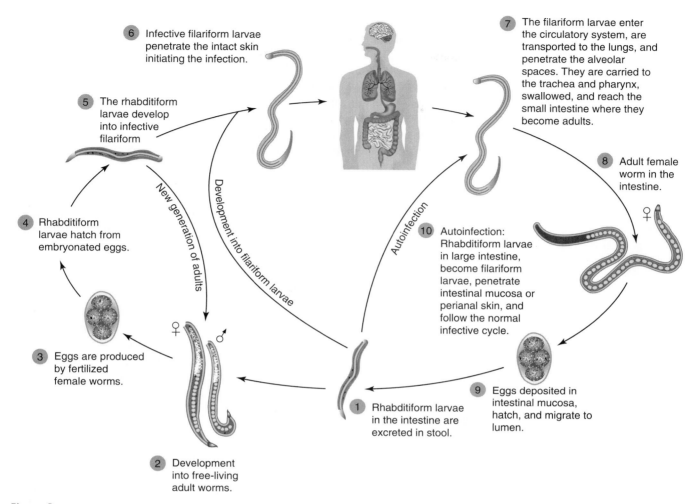

Figure 6-50 Life cycle of *Strongyloides*.

Free-living cycle: The rhabditiform larvae passed in the stool❶ can either molt twice and become infective filariform larvae (direct development)❻ or molt four times and become free-living adult males and females❷ that mate and produce eggs.❸ Rhabditiform larvae hatch from the eggs❹. Rhabditiform larvae can either develop❺ into a new generation of free-living adults❷, or into infective filariform larvae ❻. The filariform larvae penetrate the host skin to initiate the parasitic cycle❻.

Parasitic cycle: Filariform larvae in contaminated soil penetrate the skin❻, and are transported to the lungs where they penetrate the alveolar spaces. Then they are carried through the bronchi to the pharynx, are swallowed, and reach the small intestine ❼. In the small intestine they molt twice and become adult female worms❽. The females live in the epithelium of the small intestine and by parthenogenesis produce eggs❾, which yield rhabditiform larvae. The rhabditiform larvae can either be passed in the stool ❶ or can cause autoinfection❿. In autoinfection, the rhabditiform larvae become infective filariform larvae, which can penetrate either the intestinal mucosa (internal autoinfection) or the skin of the perianal area (external autoinfection).

Hookworms

Ancylostoma nematodes in dogs can cause an acute normocytic, normochromic anemia followed by hypochromic, microcytic anemia in young puppies and may be fatal. Debilitated and malnourished animals may continue to be unthrifty and suffer chronic anemia. Mature, well-nourished dogs may harbor a few worms asymptomatically; they are of primary concern as the direct or indirect source of infection for pups. Diarrhea with dark, tarry feces may accompany severe infections. Anemia, anorexia, emaciation, and weakness develop in chronic disease. Anemia rarely develops with *An. braziliense* or *U. stenocephala*. Dermatitis as a result of larval invasion of the skin may be seen with any hookworm but has been seen most frequently in the interdigital spaces with *U. stenocephala*; however, both genera can cause gastrointestinal disease including diarrhea and protein-losing enteropathies. Pneumonia may develop from overwhelming infections in puppies. In cats, *An. tubaeforme* causes weight loss and a regenerative anemia. Death can occur with heavy infestations. *An. braziliense* is less pathogenic in cats producing very little hemorrhage at the sites where adults attach. In cattle, *Bunostomum* adult worms can cause anemia, rapid weight loss, and alternating bouts of diarrhea and constipation. Hypoproteinemia may also be seen. Death can occur in calves.

Larval hookworms may cause clinical signs such as dermatitis where larvae penetrate the skin or from larval migration. The lesions may produce erythema, pruritus, and papules and are typically found on the feet and interdigital spaces. In most cases, these lesions resolve approximately 5 days after they appear. Some infections can be severe and result in self-inflicted trauma. Larval penetration of the distal limbs of cattle can cause them to stamp and there may be local skin lesions, edema, and scabs.

Strongyloides

Heavy infection with *Strongyloides* may produce a blood-streaked, mucoid diarrhea typically in young animals (horses, cattle, dogs, and cats) during hot humid weather. Emaciation (even with good appetite) and reduced growth rate may be early signs of disease. In advanced stages, animals may develop shallow, rapid breathing and fever. Autoinfection may be induced by the use of corticosteroids.

Clinical Signs in Humans

Hookworms

CLM in people presents with a stinging sensation upon initial penetration of the larvae in the skin that develops into an erythematous papule or a nonspecific dermatitis hours after penetration. Larval penetration occurs most commonly in the feet (39%), followed by the buttocks (18%) and the abdomen (16%). Larvae migrate to produce a 2- to 4-mm wide erythematous, elevated, vesicular

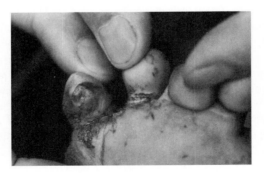

serpiginous track. Migration of the larvae through the skin varies with the nematode and ranges from a week to several months after initial penetration. The rate of larval migration ranges from 2 mm to 2 cm per day. An allergic response to the larvae or their byproducts may cause the pruritic, erythematous track. The actual location of the larvae is usually 1 to 2 cm beyond the tract (Figure 6-51).

Figure 6-51 A hookworm infection involving the toes of the foot.
(Courtesy of CDC)

Strongyloides

Most *Strongyloides* infections are asymptomatic and can be present for decades undiagnosed. If symptoms occur they typically involve the gastrointestinal, pulmonary, and integumentary systems. Specific signs of disease by body system include:

- Skin penetration by larvae typically occurs on the feet and can produce cutaneous eruptions that are pruritic and vesicular. Larva currens (racing larvae) is the rash produced by *Strongyloides* infection that can creep 5 to 15 cm per hour. This rash is most likely an allergic response to the migrating larvae and looks like a pruritic wheal or linear urticaria. This skin manifestation may last hours to days.

- Gastrointestinal symptoms occur after larvae enter crypts of intestinal mucosa where they mature and invade tissue. Symptoms are vague and include anorexia, weight loss, nausea, chronic diarrhea, constipation, and bloating. Malabsorption may occur in chronic infections. As eggs hatch in the intestine, autoinfection is possible.
- Pulmonary symptoms occur when larvae migrate to and damage lung tissue. Initial symptoms are wheezing and a mild cough. Disseminated disease is associated with wheezing, dyspnea, cough, tachypnea, and hemoptysis.

Diagnosis in Animals

Ancylostoma, *Uncinaria*, and *Bunostomum* infections are diagnosed by identification of thin-shelled, oval eggs on flotation of fresh feces from infected animals.

Strongyloides larvae are identified from fresh fecal material using the Baermann technique to separate larvae from fecal material. Adult female worms can also be identified by scraping the mucosa of the small intestine (the presence of eggs in the uterus easily differentiates them from larvae of other nematodes). An ELISA test is also available.

> Unlike in animals, the larvae are unable to penetrate the epidermal basement membrane of human skin resulting in the roaming of larvae haphazardly in the epidermis because they are unable to complete their life cycle.

Diagnosis in Humans

Diagnosis of hookworm CLM in people is based on the classic clinical appearance of the eruption. A minority of people demonstrates peripheral eosinophilia on a CBC and increased IgE levels on serum immunoglobulin tests. A skin biopsy sample, taken just ahead of the leading edge of a tract, may show a larva using periodic acid-Schiff stain.

Strongyloides larvae are seen in stool approximately 1 month after skin penetration. Baermann technique (larvae migrate from stool samples to warmed water and are detected in the centrifuged fluid), agar plate culture, direct fecal smear, or using Harada-Mori filter paper are recommended techniques for identification of *Strongyloides*. Larvae can be found in extraintestinal sites such as sputum, bronchoalveolar lavage, urine, semen, ascites, gastric biopsy, skin biopsy, and cerebrospinal fluid. ELISA tests are also available.

Treatment in Animals

In dogs, *An. caninum* and *U. stenocephala* infections can be treated with a variety of antinematodal drugs such as dichlorvos, fenbendazole, mebendazole, piperazine, pyrantel, pyrantel/febantel, and praziquantel/pyrantel/febantel. Milbemycin is also licensed for treatment of *An. caninum* infections. If anemia is severe, blood transfusions or supplemental iron and a high-protein diet may be needed until the hemoglobin levels are normal. Heartworm prevention with milbemycin, milbemycin/lufenuron, and diethylcarbamazine/oxibendazole controls *An. caninum*, whereas pyrantel/ivermectin controls *An. caninum*, *An. braziliense*, and *U. stenocephala*. In cats, drugs approved for treatment of *An. tubaeforme* include dichlorvos, mebendazole, milbemycin, piperazine, pyrantel, pyrantel/praziquantel, and selamectin. Dichlorvos, mebendazole, and piperazine are also approved for treatment of *U. stenocephala*. Heartworm prevention with ivermectin, milbemycin, and selamectin controls *An. tubaeforme*, whereas ivermectin also controls *An. braziliense*. *Bun. phlebotomum* in cattle can be treated with febantel or ivermectin-based products.

Stronglyoides infections in dogs can be treated with ivermectin or thiabendazole. In cats, fenbendazole is used.

Treatment in Humans

Thiabendazole is currently considered the agent of choice for treating hookworm CLM in people. Topical application is used for early, localized lesions and oral treatment is used for widespread lesions or unsuccessful topical treatment. Other effective alternative treatments include albendazole, mebendazole, and ivermectin. Antibiotics are indicated if secondary bacterial infections are present. Untreated lesions resolve after the larvae die typically within weeks to months.

Strongyloides infections should be treated even in the absence of symptoms with ivermectin or thiabendazole. Disseminated strongyloidiasis requires treatment for at least 7 days or until the parasite can no longer be identified in clinical samples.

Management and Control in Animals

Bitches should be free of hookworms prior to breeding and should be kept out of potentially contaminated areas during pregnancy. Whelping should occur in sanitary quarters. Concrete runways that can be washed are best for housing dogs. Prompt isolation of healthy dogs from dogs that appear unhealthy can prevent infection. Thorough washing of surfaces with steam or concentrated salt or lyme solutions, followed by rinsing with hot water, effectively destroys *Strongyloides*. In areas where both ascarids (nematodes in genera such as *Ascaris*, *Toxocara*, *Toxascaris*, and *Baylisascaris*) and hookworms are common, puppies and their mothers should be treated with an age-appropriate anthelmintic at 2, 4, 6, and 8 weeks of age (some recommend extending this to 12 weeks and then treating monthly until the pet is 6 months old). Because prenatal infection does not occur in kittens, preventive treatment should begin at 3 weeks of age, and be repeated at 5, 7, and 9 weeks. Nursing dogs and queens should be treated concurrently with their offspring because they often develop patent infections along with their young.

Management and Control in Humans

Human can prevent CLM by avoiding skin to soil contact (do not walk barefoot), defecating in proper facilities, avoiding the use of human and animal excrement or raw sewage as manure/fertilizer, deworming pets and livestock, and seeking medical care if mosquito-like bites on the sole of the foot turn into lines.

Summary

Cutaneous larva migrans (CLM) is caused by a variety of juvenile nematodes mainly in the genus *Ancylostoma*. Zoonotic species include *An. braziliens*, *An. caninum*, *An. ceylonicum*, and *An. tubaeforme*. Another genus of nematode that causes CLM is *Strongyloides*. Species of *Strongyloides* include *Stro. stercoralis*, *Stro. fuelleborni*, *Stro. papillosus*, and *Stro. westeri*. Other genera of nematodes that cause CLM include *Uncinaria*, *Bunostomum*, *Gnathostoma*, and *Capillaria*. *Ancylostoma* and *Stronglyoides* spp. occur worldwide with the greatest number of cases occurring in Asia and sub-Saharan Africa. *Ancylostoma* parasites live in sandy or loamy soil and cannot exist in clay or muck. *Ancylostoma* needs warm moist conditions where rainfall averages are more than 40 inches per year and the average temperature is greater than 50°F. *Strongyloides* parasites are found worldwide, although some species are found in higher proportions in certain regions. *Uncineria* parasites are found mainly in Europe. *Bunostomum* parasites are found worldwide.

Ancylostoma and *Uncineria* larvae live in infected soil and penetrate human skin on contact (such as walking barefoot). In animals, transmission may result from skin penetration, ingestion of infective larvae from the environment or via the colostrum or milk of infected dams (*An. caninum*) and transplacentally. Transmission of *U. stenocephala* is similar to that of *An. caninum* except that dogs probably do not become infected by the prenatal or colostral routes. Infection with *Strongyloides* nematodes occurs by the percutaneous route with filariform third-stage larvae. Autoinfections and transmammary infections also occur. Infection can also occur when a horse ingests larvae when eating grass.

Ancylostoma nematodes in dogs can cause an acute normocytic, normochromic anemia followed by hypochromic, microcytic anemia in young puppies that may be fatal. In cats, *An. tubaeforme* causes weight loss and a regenerative anemia. Heavy infection with *Strongyloides* may produce a blood-streaked, mucoid diarrhea typically in young animals during hot humid weather. CLM in people presents with a stinging sensation upon initial penetration of the larvae that develops into an erythematous papule or a nonspecific dermatitis hours after penetration. Larval penetration occurs most commonly in the feet, followed by the buttocks, and the abdomen. Most *Strongyloides* infections are asymptomatic and can be present for decades undiagnosed. If symptoms occur they typically involve the gastrointestinal, pulmonary, and integumentary systems. *Ancylostoma*, *Uncinaria*, and *Bunostomum* infections are diagnosed by identification of thin-shelled, oval eggs on flotation of fresh feces from infected animals. *Strongyloides* larvae are identified from fresh fecal material using the Baermann technique to separate larvae from fecal material. Diagnosis of hookworm CLM in people is based on the classic clinical appearance of the eruption. Baermann technique, agar plate culture, direct fecal smear, or using Harada-Mori filter paper are recommended techniques for identification of *Strongyloides*.

In dogs, *An. caninum* and *U. stenocephala* infections can be treated with a variety of antinematodal drugs. In cats, many antinematodal drugs are approved for treatment of *An. tubaeforme*. *Bunostomum* infections in cattle are treated with febantel and ivermectin products. *Stronglyoides* infections in dogs can be treated with ivermectin or thiabendazole. In cats, fenbendazole is used. Thiabendazole is currently considered the agent of choice for treating hookworm CLM in people. Topical application is used for early, localized lesions and oral treatment is used for widespread lesions or unsuccessful topical treatment. *Strongyloides* infections should be treated even in the absence of symptoms with ivermectin or thiabendazole. Many control measures are recommended for these nematodes including testing bitches for hookworms prior to breeding, allowing whelping only in sanitary quarters using concrete runways for housing dogs, and isolation of healthy dogs from dogs that appear unhealthy can prevent infection. Humans can prevent CLM by avoiding skin to soil contact, defecating in proper facilities, avoiding the use of human or animal excrement or raw sewage as manure/fertilizer, deworming pets and livestock, and seeking medical care if mosquito-like bites on the sole of the foot turn into lines.

Visceral Larva Migrans

Overview

Visceral larva migrans (VLM), also known as larva migrans visceralis, is a syndrome caused by invasion of internal organs of the paratenic host (transport host where larvae do not undergo any development) by second-stage nematode larvae. When nematode larvae gain entry into an improper host, they do not complete the normal migration but instead have arrested development and begin an extended, random wandering through various body organs. VLM is caused by a variety of nematodes, but common species include *Toxocara canis*, *To. cati*, *Baylisascaris procyonis*, and *Ascaris suum*. *To. canis* is by far the most common species causing VLM and for many years it was believed that dog and cat nematodes could not infect humans. This was proven false in the early 1950s when Wilder discovered nematode larvae in eye tissue while performing histological studies at the Armed Forces Institute of Pathology. Wilder named the disease nematode endophthalmitis and concluded that it was an unrecognized cause of childhood blindness. Two years later Beaver demonstrated the larval form of *To. canis* as the causative agent of systemic disease causing cough, fever, and chronic eosinophilia in three children. As a result of the internal organ involvement the disease was termed visceral larva migrans to differentiate it from cutaneous larva migrans; however, an ocular form is also recognized.

Bay. procyonis is a common intestinal parasite of raccoons in North America with other species of this genus found in bears, skunks, and badgers. *Baylisascaris* was first described as *As. procyonis* by Stefanski and Zarnowski in 1951, but was later reclassified by Sprent in 1968 (named after H. A. Baylis of the British Museum). In many regions of the United States large populations of raccoons infected with *Bay. procyonis* live in close proximity to humans. Although documented cases of human baylisascariasis remain relatively uncommon, environmental contamination by infected raccoons suggests that the risk of exposure and human infection is substantial.

Ascaris worms, named from the Greek *asketos* meaning fidgety describing their movement, are among the largest nematodes and have been known since the times of Ancient Greece and Rome, Mesopotamia, and China. Descriptions of this parasite were recorded on papyrus in 1550 B.C. by the ancient Egyptians. When swine were domesticated and began living in close association with humans, the adaptation of the swine parasite into people was presumed to occur. The life cycle of *Ascaris* was not known until 1916 and is cosmopolitan in distribution especially prevalent in poor children in tropical countries with overcrowded slums and poor sanitation.

Causative Agent

VLM is caused by a variety of juvenile nematodes. The main nematode genus that causes VLM is *Toxocara*. Zoonotic species include:

Toxocara species with no reported zoonotic potential include *To. tanuki*, *To. alienate*, and *To. mackerrasae*. Two newer species with unresolved zoonotic status are *To. malayasiensis* (domestic cats) and *To. lyncus* (caracals).

- *To. canis*. This species is the most important species causing VLM with dogs and other canines serving as definitive hosts. Mature worms are found in the intestines of the definitive host that sheds large numbers of unembryonated eggs into feces. Eggs become embryonated in the environment in about 9 to 15 days under optimal humidity and temperature (25°C to 30°C). *To. canis* adult males are about 4 to 6 cm in length and females are up to 15 cm in length. The tail of the male decreases in diameter and has five papillae on each side. Three lips and prominent cervical alae (cuticle expansion) are present in the adult worm. Eggs are 90 × 75 μm long and are brownish, almost spherical with thick, finely pitted shells, and are unembryonated when laid.

- *To. cati* is widely distributed among domestic cats and other felines. The cervical alae of *To. cati* are shorter and broader than those of *To. canis* and the eggs are slightly different in their appearance. *To. cati* is more commonly seen in cases of human ocular larva migrans.

- *To. vitulorum* is a parasite found in cattle, is believed to be a low level zoonosis, and mainly affects children in the tropics. Young calves are infected by their mother's milk.

- *To. pteropodis* is a parasite of fruit bats (flying foxes) in Australia and is passed by the transmammary route.

Another genus of nematode that causes VLM is *Bay. procyonis*, also known as the raccoon roundworm. *Bay. procyonis* is the most common and widespread cause of clinical larva migrans in animals. Adult male worms are about 12 cm in length and females are 23 cm in length. Eggs are smaller than *Toxocara* eggs and are about 62 to 70 µm long. Paratenic hosts (animals acting as substitute intermediate hosts of a parasite) include rodents, birds, and rabbits with human infections being rare. Other species of *Baylisascaris* are found in skunks (*Bay. columnaris*), badgers (*Bay. melis* and *Bay. columnaris*), bears (*Bay. transfuga*), fishers (*Bay. devosi*), and martins (*Bay. devosi*); no human infections have been reported from these species.

Another genus of nematodes that cause VLM is *Ascaris*. *As. lumbricoides* is a human parasite and *As. suum* is an intestinal nematode of pigs. Adults have three prominent lips and do not have alae. Adult males are between 15 to 30 cm in length and females are between 20 and 49 cm in length. Fertilized eggs are oval to round, are 45 to 75 µm long, and have a thick, bumpy outer shell.

> *As. suum* is morphologically identical to *As. lumbricoides* and are considered by some to be the same species.

Epizootiology and Public Health Significance

To. canis and *To. cati* occur worldwide in soil. *To. vitulorum* is found mainly in the tropics (also is present in the United States), whereas *To. pteropodis* is found in Australia. The prevalence of *Toxocara* VLM in the United States is unknown because it is not a reportable disease; but it is believed that approximately 10,000 human cases occur annually. Most cases are seen in children ages 1 to 7 years. Antibody titers to *Toxocara* in U.S. children are 4.6% to 7.3% (compared to 83% in the Caribbean). Titers are highest in the southeastern United States and Puerto Rico. Antibodies to *To. canis* worldwide are between 2% and 14%. *Toxocara* in dogs in the United States has been reported between 2% and 79% (10% to 85% in cats).

Baylisascaris parasites have been reported only from the United States and Europe. In the United States infected raccoons are more commonly found in the mid-Atlantic, northeastern and Midwestern states, and in parts of California. The prevalence of *Baylisascaris* is rare with only 25 cases reported in the United States in 2003. Five of the 25 cases were fatal and many survivors have permanent neurologic damage.

As. suum parasites are found worldwide, but particularly affect people living in warm, moist environments. *Ascaris* infections were estimated to be 643 million in 1947 and by 1979 between 800 million and 1 billion people were infected (it ranked third among human infections). *Ascaris* infections are common throughout Asia and are prevalent in Africa. Endemic regions exist in Canada and the Gulf Coast and southern Appalachian states of the United States (30% of the people may be infected).

Transmission

To. canis infection in humans and dogs occurs by ingestion of embryonated eggs. Sources of embryonated eggs include sandboxes and playgrounds contaminated with dog feces. Dogs can also become infected by eating tissues containing hypobiotic (dormant) larvae found in prey (these larvae mature in the dog's intestines but do not migrate). Hypobiotic larvae also serve as a reservoir of infection in pregnant dogs as they become reactivated during the last third of pregnancy. Larvae enter the uterus or mammary glands where they can infect the fetus or puppy. *To. cati* infection in humans and cats occurs by ingestion of embryonated eggs. Cats can become infected after ingesting hypobiotic larvae. *To. cati* is not transmitted to cat fetuses through the placenta; however, kittens can be infected through the milk or colostrum. Humans can be infected with *To. vitulorum* by ingestion of embryonated eggs. *To. vitulorum* in cattle occurs by ingestion of embryonated eggs; calves become infected mainly through milk (larvae are in high numbers in milk during the first week after calving).

Baylisascaris infections in humans and raccoons occur from ingestion of embryonated eggs in soil, water, or on fomites. Raccoons can acquire the disease by ingestion of larvae in tissue of intermediate hosts (rodents).

Ascaris infections in humans and pigs occur from ingestion of embryonated eggs usually eaten with contaminated food and water.

Pathogenesis

Toxocara

> Adult *To. canis* nematodes survive approximately 4 months in the intestines and most of the parasites have been expelled in the feces within 6 months of infection.

The life cycle of *Toxocara* starts when the embryonated eggs are ingested and larvae hatch in the intestine (it takes unembryonated eggs 14 to 35 days to become embryonated eggs). Larvae invade the intestinal wall and are spread by the blood to the liver, heart, and lung. In young dogs the larvae penetrate the alveolar wall; migrate to the bronchioles, bronchi, and trachea; reach the pharynx; are swallowed into the esophagus, and reach the small intestine. When the larvae reach the intestines the second time, they develop into adults, mate, and produce eggs that are passed in the feces. Occasionally immature larvae may also be found in feces. In older puppies and adult dogs, many larvae do not complete migration to the lungs and the larvae travel to the muscles, liver, kidneys, and other viscera where they become dormant. In humans, larval migration ends on route to the lung or ends in the lung or the larvae enter the arterial system and are disseminated throughout the body. Larvae leave the blood vessels where they generally become encapsulated by connective tissue. Encapsulated larvae may survive for years and may become reactivated. Larvae can be transmitted between people by cannibalism.

Any age of dog can also develop patent infections if they ingest tissues (rodent or other mammal) containing hypobiotic larvae. These larvae mature in the dog's intestines without further migration. Hypobiotic larvae may also serve as a reservoir of infection in pregnant dogs. Hypobiotic larvae become reactivated during the last third of pregnancy, enter the uterus or mammary glands, and infect the fetus or puppy. Fetuses acquire the parasites in utero where they then enter the liver, migrate through the lungs, and develop into adults (in about 3 weeks). Larvae acquired through milk do not migrate through tissues instead they complete their development in the intestines. Bitches can acquire infections during lactation from movement of hypobiotic larvae to the intestines or from ingestion of larvae from contaminated puppy feces. These infections in bitches spontaneously resolve in 4 to 10 weeks after parturition (Figure 6-52).

The life cycle of *To. cati* is similar to *To. canis* except that *To. cati* is not transmitted in utero.

Baylisascaris

Raccoons become infected with *Bay. procyonis* by ingesting embryonated eggs or larvae in tissue of intermediate hosts. Larvae hatch in the intestine, develop in the intestinal wall, and complete their maturation in the intestinal lumen. Extraintestinal migration does not appear to occur in raccoons. Unembryonated eggs are excreted in raccoon feces, develop for at least 2 to 4 weeks in the environment, and become infective. Humans become infected when they ingest embryonated eggs. Migrated larvae may be found in the liver, heart, lungs, CNS, and eyes.

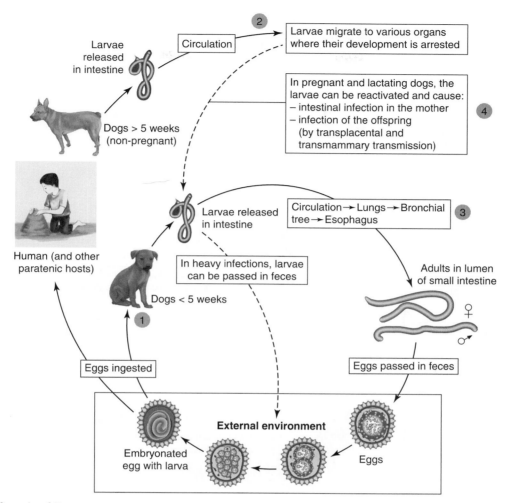

Figure 6-52 Life cycle of *Toxocara*.

Toxocara canis accomplishes its life cycle in dogs, with humans acquiring the infection as accidental hosts. Following ingestion by dogs❶, the infective eggs hatch and larvae penetrate the intestinal wall and migrate into various tissues, where they encyst if the dog is older than 5 weeks❷. In younger dogs, the larvae migrate through the lungs, bronchi, and esophagus❸; adult worms develop and oviposit in the small intestine. In the older dogs, the encysted stages are reactivated during pregnancy, and infect puppies by the transplacental and transmammary routes❹. Humans are accidental hosts who become infected by ingesting embryonated eggs in contaminated soil. After ingestion, the eggs hatch and larvae penetrate the intestinal wall and are carried by the circulation to a wide variety of tissues (liver, heart, lungs, brain, muscle, eyes). The larvae do not undergo any further development in these sites; however, they can cause severe local reactions.

Bay. procyonis is different from other nematodes that cause larva migrans because of its aggressive tissue migration and invasion of the CNS, its continued growth of larvae to a large size within the CNS, and its resistance to dying once in the paratenic host (Figure 6-53).

Ascaris suum

Ascaris infection is acquired through ingestion of embryonated eggs that hatch in the cecum and proximal colon, penetrate the intestinal wall, and enter lymphatics or venules. The larvae pass through the right side of the heart, enter the pulmonary circulation, penetrate the alveolar wall, move to the bronchioles to the bronchi to the trachea, reach the pharynx, are swallowed, and reach the small intestine where

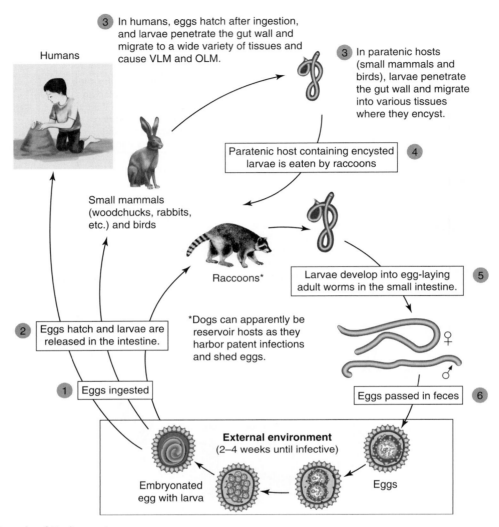

Figure 6-53 depicts the life cycle of *Baylisascaris*, with the following numbered steps:

3 In humans, eggs hatch after ingestion, and larvae penetrate the gut wall and migrate to a wide variety of tissues and cause VLM and OLM.

Humans

3 In paratenic hosts (small mammals and birds), larvae penetrate the gut wall and migrate into various tissues where they encyst.

Paratenic host containing encysted larvae is eaten by raccoons **4**

Small mammals (woodchucks, rabbits, etc.) and birds

Larvae develop into egg-laying adult worms in the small intestine. **5**

Raccoons*

*Dogs can apparently be reservoir hosts as they harbor patent infections and shed eggs.

2 Eggs hatch and larvae are released in the intestine.

♀

♂

1 Eggs ingested

Eggs passed in feces **6**

External environment (2–4 weeks until infective)

Embryonated egg with larva

Eggs

Figure 6-53   Life cycle of *Baylisascaris*.

Baylisascaris procyonis completes its life cycle in raccoons with humans acquiring the infection as accidental hosts. Following ingestion by many different hosts (over 50 species of birds and mammals have been identified as intermediate hosts)❶ eggs hatch❷ and larvae penetrate the intestinal wall and migrate into various tissues, where they encyst❸. The life cycle is completed when raccoons eat these hosts❹. The larvae develop into adult worms in the small intestine❺ and unembryonated eggs are eliminated in raccoon feces❻. People become accidentally infected when they ingest embryonated eggs from the environment (takes 2 to 3 weeks for eggs to embryonate). After ingestion, the eggs hatch and larvae penetrate the intestinal wall and migrate to a wide variety of tissues (liver, heart, lungs, brain, eyes), and cause visceral (VLM) and ocular (OLM) larva migrans syndromes. In contrast to *Toxocara* larvae, *Baylisascaris* larvae continue to grow during their time in the human host.

they mature. In humans, *As. suum* larvae can enter the liver and lungs producing an eosinophilic pneumonia and liver lesions (Figure 6-54).

Clinical Signs in Animals

Toxocara

Toxocara nematodes in young puppies can cause poor growth, an enlarged abdomen, diarrhea, constipation, vomiting, flatulence, and coughing. Chronic enteritis can produce intestinal wall thickening or intussusception. In severe cases, gall bladder obstruction, bile or pancreatic duct obstruction, or intestinal rupture may

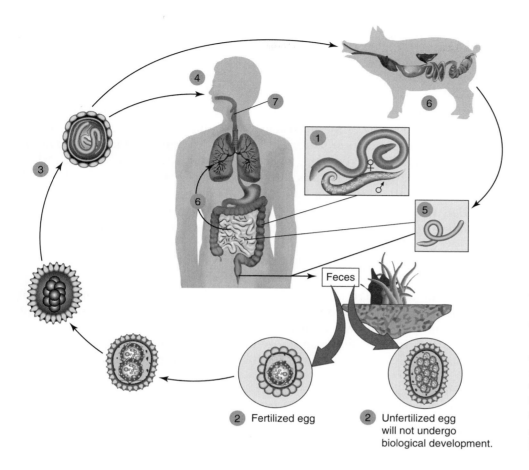

Figure 6-54 Life cycle of *Ascaris*. Adult worms❶ live in the lumen of the small intestine and females produce eggs which are passed with the feces❷. Unfertilized eggs may be ingested but are not infective. Fertile eggs embryonate and become infective❸. After infective eggs are swallowed❹, the larvae hatch❺, invade the intestinal mucosa, and are carried via the portal, then systemic circulation to the lungs❻. The larvae mature in the lungs, penetrate the alveolar walls, ascend bronchi to the throat, and are swallowed❼. When they reach the small intestine, they develop into adult worms❶. In humans, *Ascaris sius* larvae can also enter the liver. In pigs, *Ascaris suis* can also enter the liver and bile ducts.

(within figure)
❷ Fertilized egg

❷ Unfertilized egg will not undergo biological development.

Feces

occur. As the larvae pass through the liver and lungs inflammation and respiratory distress may be observed. Elevated liver enzymes may be seen during larval migration and ocular signs such as retinal disease have been described. Clinical signs are rare in adult dogs.

In cats, *To. cati* infection does not occur in utero; therefore, kittens are older when larvae are migrating. Many infections in kittens are asymptomatic, but heavy infections can cause abdominal distension, rough coat, diarrhea, and dehydration. Adult cats tend to be asymptomatic. In calves, *To. vitulorum* causes anorexia, abdominal pain, diarrhea, constipation, dehydration, weight loss, and poor gain.

Baylisascaris

As the primary hosts, raccoons appear to tolerate *Bay. procyonis* in their small intestine, and extensive migration of the parasite beyond the intestinal lumen does not seem to take place. Heavy infections in young raccoons can cause intestinal obstruction. In dogs, infection does not cause overt clinical signs and is typically discovered on routine fecal examinations.

Ascaris

Young pigs infected with *As. suum* may show clinical signs of reduced growth rate and in heavy infections may have intestinal obstruction. Adult worms can migrate into the bile ducts producing icterus especially in hogs in transit to slaughter who have long periods of time between feedings. Larval migration to the liver can cause hemorrhage and fibrosis; in the lungs pulmonary edema may occur.

Clinical Signs in Humans

Toxocara

Larva migrans as a result of *Toxocara*, also known as toxocariasis, in people presents in three forms:

- *Visceral*. Most cases of VLM are asymptomatic and are discovered via routine blood work demonstrating persistent eosinophilia. Signs in children include lethargy, fever, hepatomegaly, and cranial abdominal pain. Less common signs include nausea, vomiting, wheezing, coughing, dyspnea, pruritic rashes, lymphadenopathy, muscle pain, and neurologic signs. Symptoms may persist for months with deaths being rare.

- *Ocular*. Cases of ocular larva migrans can cause a variety of clinical disease including retinal granulomas, retinal detachment, uveitis, keratitis, vitreous abscesses, and endophthalmitis. Infection is typically unilateral and a single larva is responsible for the symptoms. Vision loss may be progressive or sudden and may be permanent.

- *Covert*. In the covert form, antibodies to *Toxocara* cause symptoms that cannot be related to the other syndromes. The most common clinical sign is abdominal pain followed by hepatomegaly, coughing, sleep alteration, headaches, rash, pruritus, and respiratory distress. The covert form is not always associated with eosinophilia and this form can last for months or years.

Baylisascaris

Two syndromes have been described for people with *Baylisascaris* infections:

- *Visceral*. *Bay. procyonis* visceral larva migrans (also called neural larva migrans or cerebrospinal nematodiasis) presents as acute fulminant eosinophilic meningoencephalitis producing symptoms that include incoordination, ataxia, torticollis, nystagmus, and mentation changes. These symptoms may lead to stupor and coma. Once invasion of the CNS has occurred, the prognosis is grave with or without treatment.

- *Ocular*. Ocular larva migrans is typically unilateral and occurs when larvae migrate to the eye. Clinical signs include photophia, retinitis, and blindness.

Ascaris

Clinical signs in humans caused by *Ascaris* infections include pneumonitis (as a result of hemorrhage at the site causing blood to pool in the alveoli), abdominal pain, rashes, ocular pain, and restlessness. Heavy infections can cause intestinal blockage that may become fatal.

Diagnosis in Animals

A single *To. canis* female worm can produce 200,000 eggs per day.

All nematodes causing VLM can be diagnosed via fecal floatation and identification of eggs. *Toxocara* eggs are between 90 and 75 μm in length, contain a single dense cell mass and a thick, brown outer shell. The shell has distinctive fine pits in its proteinaceous coat. Larvae may also be seen in feces or vomitus. Eggs may be shed intermittently in dogs. *Baylisascaris* eggs are similar to *Toxocara* eggs but are smaller (62 to 70 μm in length). *Ascaris* eggs are oval to round, 45 to 75 μm in length and have a thick, bumpy outer shell (Figure 6-55).

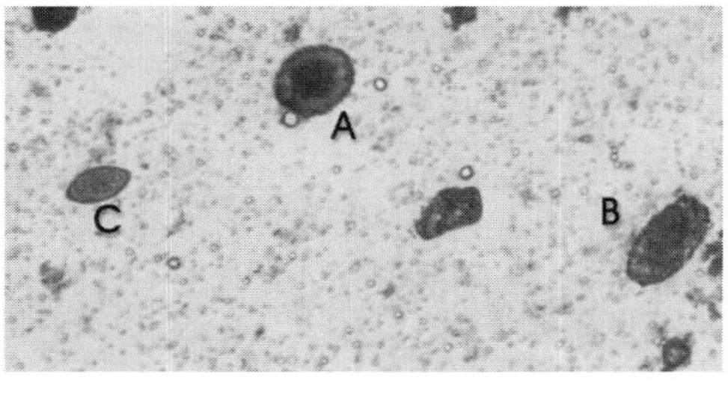

Figure 6-55 Image of both fertilized (A) and unfertilized (B) *Ascaris* eggs, and a *Trichuris* egg (C). Mag. 125X. (Courtesy of CDC)

An ELISA test has been used to detect nonpatent *Toxocara* infections in dogs. *Baylisascaris* is difficult to diagnose in live animals, but the larvae may be found in tissue biopsy. *Baylisascaris* larvae are 50 to 70 μm in diameter and have a prominent single lateral alae. *Ascaris* larvae may be found in sputum.

Diagnosis in Humans

People with *Toxocara* infections are typically diagnosed by clinical signs, ophthalmic examination, and routine blood tests (eosinophilia). Histopathology is not routinely used for diagnosis. Humans infected with VLM do not produce or excrete eggs; therefore, fecal examination is not a useful diagnostic tool. Serologic tests such as ELISA and immunoblot assays are used in humans. PCR tests have been developed but are not currently used in the United States.

Baylisascariasis is diagnosed by complete blood count (CBC) and CSF examination and biopsy. Serologic tests are not commercially available but are available in research laboratories (ELISA, indirect immunofluorescence, and immunoelectrotransfer).

Ascariasis is diagnosed by fecal floatation and identification of eggs. Larvae may be seen in sputum, but are difficult to accurately identify by most technicians.

Treatment in Animals

In dogs and cats, *Toxocara* infections can be treated with a variety of antinematodal drugs such as fenbendazole, mebendazole, and piperazine. Treatment of puppies/kittens and nursing bitches/queens should be done at the same time. Because most antinematodal drugs kill the adult worm, these drugs should be repeated at 2- to 4-week intervals. The CDC recommends puppies receive anthelmintic drugs at 2, 4, 6, and 8 weeks of age. The Companion Animal Parasite Council recommends year-round treatment with broad-spectrum heartworm anthelmintics that also have activity against parasites with zoonotic potential.

Dogs or raccoons infected with *Baylisascaris* can be treated with a variety of antinematodal drugs including piperazine, pyrantel, ivermectin, moxidectin, albendazole, and fenbendazole. *Ascaris* infections in swine can be treated with piperazine, dichlorvos, fenbendazole, and pyrantel.

Treatment in Humans

Anthelmintic drugs such as thiabendazole, mebendazole, and albendazole can be used to treat severe cases of VLM caused by *Toxocara*. Anti-inflammatory drugs such as corticosteroids may be needed to counteract the severe hypersensitivity reactions caused by dying larvae. Treatment of ocular disease may require surgery.

Baylisascaris infections can be treated with thiabendazole, fenbendazole, levamisole, and ivermectin, but the clinical outcome is poor. Corticosteroids can be used to counteract hypersensitivity reactions. Laser photocoagulation has been used for ocular larva migrans. *Ascaris* infections are treated with mebendazole, ivermectin, or pyrantel pamoate.

Management and Control in Animals

Puppies and kittens should be dewormed to eliminate shedding of *Toxocara* eggs. Puppies should begin deworming programs at 2 weeks of age, repeated at 2-week intervals. In kittens, prenatal infection does not occur and egg excretion begins later than in puppies. Deworming for kittens can begin at 3 weeks of age and repeated at 5, 7, and 9 weeks. Removal of feces and cleaning of kennels is essential in preventing *Toxocara* infections. Sand boxes should be covered when not in use.

Baylisascaris infections in dogs can be reduced by avoiding contact with raccoons and their feces (eggs may remain infective for years under ideal conditions). Raccoons should be discouraged from visiting backyards and farms, including preventing access to food, garbage, attics, and basements. Sand boxes should be covered when not in use. Removal of brush may discourage raccoons from making a den on someone's property.

Ascaris infections in swine can be reduced by strategic deworming and waste removal. The antibiotic hygromycin has been effective as a feed additive in reducing *Ascaris* levels.

Management and Control in Humans

Humans can prevent VLM caused by *Toxocara* by having pets strategically dewormed, removing feces from the environment before they become embryonated, enforcing leash laws and collection of feces by pet owners, proper hygiene (washing hands and raw foods), controlling pica in children, and public education. Families may want to consider not getting a pet until children are past the toddler stage.

VLM as a result of *Baylisascaris* infections can be reduced by avoiding contact with raccoons and their feces (not feeding raccoons, removing brush near homes, keeping food and garbage in raccoon-proof containers). Raccoons tend to use latrines where they regularly defecate (usually at the base of trees, near fallen logs, large rocks, woodpiles, decks, and rooftops). Raccoons should not be kept as pets (even very young raccoons are often infected with *Baylisascaris*).

VLM as a result of *Ascaris* infections can be reduced by limiting contact with pigs, strategic deworming of pigs, and proper hygiene.

Summary

Visceral larva migrans (VLM) is a syndrome caused by invasion of internal organs of the paratenic host by second-stage nematode larva. VLM is caused by a variety of nematodes, but common species include *To. canis*, *To. cati*, *Bay. procyonis*, and *As. suum*. *To. canis* is the most important species causing VLM with dogs and other canines serving as definitive hosts. Mature worms are found in the intestines of the definitive host that shed large numbers of unembryonated eggs into feces. *To. cati* is widely distributed among domestic cats and other felines. *Bay. procyonis* is the most common and widespread cause of clinical larva migrans in animals. *As. lumbricoides* is a human parasite and *A. suum* is an intestinal nematode of pigs. *To. canis* and *To. cati* occur worldwide in soil. *Baylisascaris* parasites have been reported only from the United States and Europe. *As. suum* parasites are found worldwide, but particularly affect people living in warm, moist environments. *To. canis* infection in humans and dogs occurs by ingestion of embryonated eggs. Dogs can also become infected by eating tissues containing dormant larvae found in prey. In pregnant dogs reactivation of dormant larvae may occur during the last third of pregnancy and larvae may enter the uterus or mammary glands where they can infect the fetus or puppy. *To. cati* infection in humans and cats occurs by ingestion of embryonated eggs. Cats can become infected after ingesting dormant larvae (*To. cati* is not transmitted to cat fetuses through the placenta). Kittens can be infected through the milk or colostrum. *Baylisascaris* infections in humans and raccoons occur from ingestion of embryonated eggs in soil, water, or on fomites. Raccoons may also acquire the disease by ingestion of larvae in tissue of intermediate hosts. *Ascaris* infections in humans and pigs occur from ingestion of embryonated eggs usually eaten with contaminated food and water. *Toxocara* nematodes in young puppies can cause poor growth, an enlarged abdomen, diarrhea, constipation, vomiting, flatulence, and coughing. Chronic enteritis can produce intestinal wall thickening or intussusception. Many infections in kittens are asymptomatic, but heavy infections can cause abdominal distension, rough coat, diarrhea, and dehydration. Adult cats tend to be asymptomatic. Larva migrans as a result of *Toxocara* in people presents in three forms: visceral, ocular, and covert. As the primary hosts, raccoons appear to tolerate *Bay. procyonis* in their small intestine, and extensive migration of the parasite beyond the intestinal lumen does not seem to take place. Two syndromes have been described for people with *Baylisascaris* infections: visceral (mainly neurologic) and ocular. Young pigs infected with *Ascaris suum* may show clinical signs of reduced growth rate and in heavy infections may have intestinal obstruction. Clinical signs in humans caused by *Ascaris* infections include pneumonitis, abdominal pain, rashes, ocular pain, and restlessness.

All nematodes causing VLM can be diagnosed via fecal floatation and identification of eggs. An ELISA test has been used to detect nonpatent *Toxocara* infections in dogs. Identification of larvae may be helpful in diagnosis. People with *Toxocara* infections are typically diagnosed by clinical signs, ophthalmic examination, and routine blood tests (eosinophilia). Serologic tests such as ELISA and immunoblot assays are used in humans. PCR tests have been developed but are not currently used in the United States. Baylisascariasis is diagnosed by CBC and CSF examination and biopsy. Ascariasis is diagnosed by fecal floatation and identification of eggs. In dogs and cats, *Toxocara*, *Baylisascaris*, and *Ascaris* infections can be treated with a variety of antinematodal drugs. Anthelmintic drugs can be used to treat severe cases of human VLM. Anti-inflammatory drugs such as corticosteroids may be needed to counteract the severe hypersensitivity reactions caused by dying larvae. To prevent VLM, puppies and kittens should be dewormed to eliminate shedding of *Toxocara* eggs. Removal of feces and cleaning of kennels is essential in preventing *Toxocara* infections. Sand boxes should be covered when not in use. *Baylisascaris* infections in dogs can be reduced by avoiding contact with raccoons and their feces. *Ascaris* infections in swine can be reduced by strategic deworming and waste removal. Humans can prevent VLM caused by *Toxocara* by having pets strategically dewormed, removing feces from the environment before they become embryonated, enforcing leash laws and collection of feces by pet owners, proper hygiene, controlling pica in children, and public education. VLM as a result of *Baylisascaris* infections can be reduced by avoiding contact with raccoons and their feces. VLM as a result of *Ascaris* infections can be reduced in people by limiting contact with pigs, strategic deworming of pigs, and proper hygiene.

Dracunculiasis

Overview

Dracunculiasis, also known as Guinea worm infection, is a disease characterized by inflammation and cutaneous ulcers in the distal limbs as a result of the presence of *Dracunculus medinensis* (the Guinea worm) in the subcutaneous tissue. *Dracunculus* comes from the Greek *drakontion* meaning little dragon, which describes the serpent-like appearance of the worm in cutaneous tissue. For centuries, *D. medinensis* was associated with the cities of Medina (Saudi Arabia) and Guinea (West Africa), whose populations had an unusually high incidence of the disease and where the species and common names of the nematode are derived. The Guinea worm has been known since ancient times when the disease and its treatment were described in the Ebers papyrus around 1550 B.C. Calcified Guinea worms were found in Egyptian mummies and the description of the plague of the fiery serpents during the Hebrews' exodus from Egypt referred to Guinea worm infection. Early Greek and Roman physicians associated Guinea worm infection with particular water sources. During the Middle Ages a Persian physician identified the worm and during the 11th century Avicenna described dracunculiasis, its treatment, and complications. Drawings of the worm were made in 1598, winding the worm out on a stick as a cure was described in 1674 by Velschius, descriptions of the adult worm were made by Linnaeus in 1758, and by 1871 its detailed life cycle was identified by Fedchenko. European physicians were not aware of the Guinea worm until the 19th century when the British army medical officers started to see signs of the worm in soldiers serving in India. Slaves transported from the Gulf of Guinea to the New World from the 17th century onward were often infected with Guinea worms; however, dracunculiasis never became a problem in the Americas with only a few areas in northern South America establishing the disease. Dracunculiasis has been identified in the early 1980s by the WHO as a disease on its eradication list and by 1996 eradication has been 97% completed. In 1947 there were an estimated 48 million people infected with *D. medinensis*; in 1998 there were an estimated 70,000 people infected with the Guinea worm. If global elimination of Guinea worm infection is achieved, dracunculiasis will be the third disease eradicated in world history (smallpox was the first; SARS was the second).

Causative Agent

> *D. medinensis* is also known as the Medina worm, Guinea worm, serpent worm, dragon worm, pharaoh worm, and Avicenna worm.

Dracunculiasis is caused by *D. medinensis*, one of the largest known nematodes, and is a parasite of humans, dogs, horses, cattle, wolves, leopards, monkeys, and baboons. Males may reach a length of 4 cm and females may reach a length of 120 cm. The adult worm has a small, triangular mouth without lips and has both dorsal and ventral papillae. In females the gravid uterus has an anterior and posterior branch containing hundreds of thousands of embryos. Males are typically 12 to 40 mm long with unequal spicules. Larvae are 500 to 700 µm long. *D. medinensis* is **ovoviviparous** with larvae developing within eggs that remain within the female until they hatch (larvae are provided with a sheltered environment). Eggs are not used in the identification of this organism.

    D. insignis is the Guinea worm of North American wildlife in which the male worm is approximately 20 mm long and the female is 300 mm long. *D. insignis* has been reported in the raccoon, mink, striped skunk, fox, muskrat, fisher, short-tailed weasel, opossum, badger, Bonaparte weasel, and dog in the United States and Canada. Raccoons are the most favorable definitive host for *D. insignis* in North America. This species was first described by Leidy in a short account in 1858 and confirmed as a separate species by Chitwood in 1950; however, this species is not accepted by all authorities.

    D. lutrae is the species of Guinea worm that has only been found in river otters in Ontario, New York, and Michigan. Other species of *Dracunculus* are summarized in Table 6-8.

Epizootiology and Public Health Significance

Dracunculiasis was widespread in Africa and Asia prior to eradication programs begun by the WHO in 1986. Dracunculiasis is currently reported in 13 countries in Central, East, and West Africa. In 1986, there were an estimated 3.5 million cases of dracunculiasis in 20 countries, with 120 million persons at risk for contracting the disease. By the end of 2004, Asia was free from dracunculiasis. The remaining countries where dracunculiasis is endemic are all in Africa; however, there has been a reported 50% reduction in the number of cases from 2003 to 2004 (from 32,193 to 16,026). In 2005, Ghana and Sudan have reported 95% of the world's cases; however, Ghana reduced its cases by 53% during the first half of 2005.

Transmission

> In general, Guinea worms in mammals are regarded as *D. medinensis* in the Old World and South America and as *D. insignis* in North America.

Infection with *D. medinensis* in people occurs by ingestion of the intermediate host (the crustacean *Cyclops* spp.) in water. Dogs and other animals can become infected with *D. insignis* by ingestion of contaminated water or a paratenic host (frogs or fish). Transmission of *D. insignis* in raccoons is confined to only a few weeks of the

Table 6-8   Different Species of *Dracunculus*

Species	Host	Geographical Distribution
Dracunculus alii	Snakes	India
Dracunculus coluberensis	Snakes	India
Dracunculus dahomensis	Snakes	West and Central Africa
Dracunculus doi	Snakes	India
Dracunculus houdemeri	Snakes	Vietnam
Dracunculus ophidensis	Snakes	Italy, U.S.
Dracunculus oesophageus	Snakes	Italy, Madagascar
Dracunculus globocephalus	Turtles	U.S.
Dracunculus fuellebornius	Opossums	Brazil
Dracunculus insignis	Dogs, wild carnivores	Canada, U.S.
Dracunculus lutrae	Otters	Canada
Dracunculus medinensis	Many mammals	Africa, Americas, Asia

Source: Cairncross, S., Muller, R., and Zagaria, N. Dracunculiasis (Guinea Worm Disease) and the Eradication Initiative, American Society for Microbiology, Clin. Microbiol Rev. 2002, April; 15(2).

year. Adult worms are usually patent in late spring or early summer corresponding with changes in the raccoon's food habits. Mink infections are not as seasonal as raccoon infections as a result of the mink's year-round feeding on aquatic life.

D. *lutrae* is transmitted to river otters through ingestion of contaminated water or a paratenic host, but there is not the seasonal infection rate because of the river otter's year-round food habits.

Pathogenesis

Inside the definitive host (human, dog, raccoon), the ingested *Cyclops* is destroyed by stomach acids. Once the larvae are freed they penetrate the duodenal lining, enter the lymphatic system, and migrate to subcutaneous tissues. Larvae molt at about 20 days postinfection and molt a final time at approximately 43 days. Once in subcutaneous tissue the worms mature slowly, reaching full maturity in one year. After the adults mate (by the third month after infection), the male dies, becomes encysted, and degenerates. Gravid females migrate to the skin between the eighth and tenth months (embryos in the uterus are fully mature) causing an allergic reaction as a result of the release of metabolic wastes into the skin. Between 10 and 14 months postinfection the female worm produces a pruritic blister in the skin (90% of lesions are found in the feet and legs). When these blisters are exposed to water (such as a river or lake) the female's uterus ruptures and releases large numbers of larvae into the water. The larvae can live for 6 days in clean water and 2 to 3 weeks in muddy water. Larvae are ingested by the crustacean *Cyclops*. Once

Dracunculus infections in domestic animals in endemic areas are possibly of human origin.

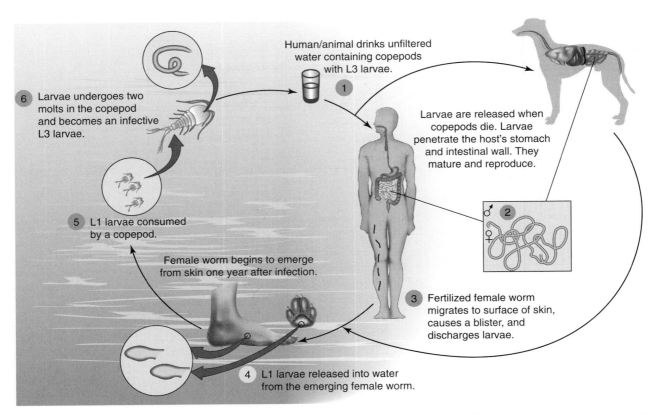

Figure 6-56   Life cycle of *Dracunculus*. Humans and animals become infected by drinking unfiltered water containing copepods that are infected with larvae of *Dracunculus*❶. Following ingestion, the copepods die and release the larvae, which penetrate the host stomach and intestinal wall and enter the abdominal cavity and retroperitoneal space❷. After maturation into adults and copulation, the male worms die and the females migrate in the subcutaneous tissues toward the skin surface❸ where the female worm induces a blister on the skin that eventually ruptures. When this lesion comes into contact with water, the female worm emerges and releases larvae❹. The larvae are ingested by a copepod❺ and after two molts have developed into infective larvae❻ Ingestion of the copepods ends the cycle❶.

ingested, the larvae mature into their infective stage in approximately 14 days and can then reinfect humans (Figure 6-56).

Clinical Signs in Animals

D. medinensis infections although rare have been reported in dogs in endemic regions. The main clinical sign in dogs is the development of cutaneous ulcers. *D. insignis*, a species found in North America, can infect the subcutaneous connective tissues of the legs of raccoons, mink, opossums, muskrats, and other animals, including dogs. They produce 3 to 5 cm cutaneous, nonhealing ulcers through which the female worm's anterior end is protruded upon contact with water. Infections in animals are rare but are occasionally found in animals that spend time around small lakes and shallow, stagnant water. Lesions are typically found on the limbs. Dehydration, vomiting, diarrhea, and dyspnea are also common symptoms of infection.

Dracunculus is the only parasite known to be solely transmitted by water consumption.

Clinical Signs in Humans

D. medinensis infections are asymptomatic in people until the female worm reaches the skin approximately 10 to 12 months after acquiring the worm. Once the worm is near the skin surface people develop a blister in the epidermis and may develop a fever. Prior to blister formation, allergic-type symptoms, such as wheezing and pruritus, are often present. As the blister grows it becomes erythematous at the edges, edema occurs, and inflammation causes pruritus and a burning sensation (Figure 6-57). Typically the blister ruptures in a few days (relieving the pain and swelling) and the female worm releases larvae-containing fluid.

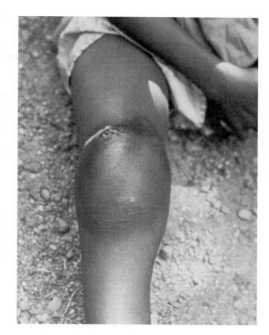

Figure 6-57 A severely swollen right knee because of an infection caused by the subcutaneous emergence of an adult female Guinea worm. (Courtesy of CDC)

After the blister ruptures an ulcer forms around the blister site as the adult worm continues to emerge. Secondary bacterial infections can occur as bacteria are drawn under the skin by a retreating worm. Worms that fail to reach the skin can cause complications in deeper tissues when they begin to degenerate and release antigens. Complications include aseptic abscesses and arthritis. Other worms that do not reach the skin may calcify or be resorbed. Calcified worms can cause problems as a result of the pressure they can put on tissues.

D. insignis does not parasitize humans; although some people believe that *D. medinensis* infections in the United States may actually be caused by *D. insignis*.

Diagnosis in Animals

Dracunculiasis is typically diagnosed by palpation of the worm in a skin nodule or by examination of larvae from an ulcerated skin nodule.

Diagnosis in Humans

Clinical diagnosis of dracunculiasis in people is made when larvae can be found in ulcers and identified by microscopic examination. When the adult female worm emerges from the ruptured nodule, it can be identified as a probable Guinea worm infection.

Treatment in Animals

Treatment in animals is by careful extraction of the parasite. Repeated wetting of the wound helps release most of the larvae. Administration of anthelmintics such as niridazole or benzimidazole may be useful in facilitating the worm's removal.

Treatment in Humans

Treatment in people is by winding the anterior end of the worm around a stick (Figure 6-58). One to 2 weeks is necessary to complete removal. Rupturing of the worm during extraction may cause severe inflammation. Metronidazole or niridazole can be used in treating

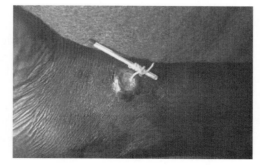

Figure 6-58 One method used to extract a Guinea worm from the leg vein of a human patient. (Courtesy of CDC)

Dracunculus infections although these drugs do not kill the worm but rather facilitate its removal. Analgesics and wound care are also advised.

Management and Control in Animals

Although dracunculiasis is rare in animals, preventing animals from using or coming in contact with contaminated water may help prevent this disease. Controlling wildlife infections is more difficult and has not been attempted in a large scale way.

Management and Control in Humans

The WHO declared in early 1980s that one of its goals was the eradication of dracunculiasis by 1995. Although the disease is not eradicated, it had a rapid and sustained reduction in all countries other than Sudan. Strategies for controlling this disease include providing safe drinking water (boiled and filtered), preventing infected persons from entering the water of ponds and stepwells, health education (including instruction on the use of cloth water filters), early treatment and bandaging of lesions, and vector control using temephos that kills *Cyclops* crustaceans.

Summary

Dracunculiasis, also known as Guinea worm infection, is caused by *D. medinensis*, one of the largest known nematodes, and is a parasite of humans, dogs, horses, cattle, wolves, leopards, monkeys, and baboons. Males may reach a length of 4 cm and females may reach a length of 120 cm. *D. medinensis* is ovoviviparous with larvae developing within eggs that remain within the female until they hatch (larvae are provided with a sheltered environment). Dracunculiasis was widespread in Africa and Asia prior to eradication programs begun by the WHO in 1986. Dracunculiasis is currently reported in 13 countries in Central, East, and West Africa. Infection with *D. medinensis* occurs by ingestion of the intermediate host (the crustacean *Cyclops* spp.) in water. Dogs and other animals can become infected with *D. insignis* (which is not believed to be zoonotic) through ingestion of contaminated water or a paratenic host (frogs or fish). Inside the definitive host (human, dog), the ingested *Cyclops* is destroyed by stomach acids, larvae are freed and penetrate the duodenal lining, enter the lymphatic system, and migrate to subcutaneous tissues. Between 10 and 14 months postinfection the female worms produces a pruritic blister in the skin and these blisters are exposed to water (such as a river or lake) the female worm's uterus ruptures and releases large numbers of larvae into the water. Larvae are ingested by the crustacean *Cyclops* and mature into their infective stage in approximately 14 days and can then reinfect humans. *D. medinensis* infections although rare have been reported in dogs in endemic regions producing cutaneous ulcers. *D. medinensis* infections are asymptomatic in people until the female worm reaches the skin approximately 10 to 12 months after acquiring the worm. Once the worm is near the skin surface people develop a blister in the epidermis and may develop a fever. Typically the blister ruptures in a few days and the female worm releases larvae-containing fluid. Dracunculiasis in animals and people is typically diagnosed by palpation of the worm in a skin nodule or by examination of larvae from an ulcerated skin nodule. Larvae can be found in ulcers and identified by microscopic examination. Treatment in animals and people is by careful extraction of the parasite. Administration of anthelmintics (niridazole or benzimidazole) may be useful in facilitating the worm's removal. Preventing animals from using or coming in contact with contaminated water may help prevent this disease. Strategies for controlling this disease in people include providing safe drinking water, preventing infected persons from entering the water of ponds and stepwells, health education, early treatment and bandaging of lesions, and vector control using temephos that kills *Cyclops* crustaceans.

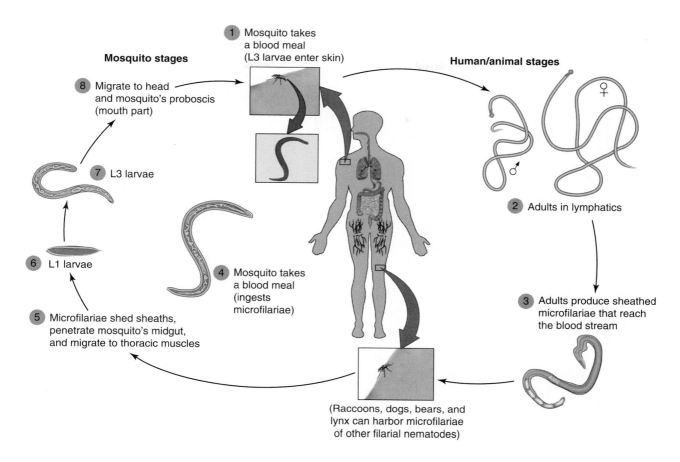

Figure 6-59 Life cycle of a filarial nematode. Infective larvae are transmitted by infected biting arthropods during a blood meal❶. Larvae migrate to the appropriate site of the host's body❷, where they develop into microfilariae-producing adults❸. The adults dwell in various tissues where they can live for several years. Most female worms produce microfilariae which circulate in the blood. The microfilariae infect biting arthropods (mosquitoes, blackflies, midges, and deerflies)❹. Inside the arthropod, the microfilariae develop into infective filariform (third-stage) larvae❺, ❻, ❼, ❽. During a subsequent blood meal by the insect, the larvae infect the vertebrate host and migrate to the appropriate site of the host's body, where they develop into adults.

Filarial Nematode Zoonoses

Overview

Filarial nematodes are a group of parasitic worms found in the blood and tissue of its host. Filarial nematodes rely on an arthropod vector (intermediate host) to complete its life cycle and to be transmitted to another animal (Figure 6-59). Larval filarial nematodes are known as microfilariae and are found in the blood or in subcutaneous tissue (Figure 6-60). There are

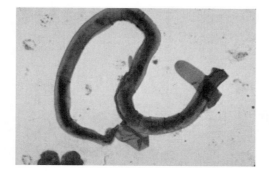

Figure 6-60 Microfilaria of *Brugia malayi* in a thick blood smear, hematoxylin stain.
(Courtesy of CDC/Dr. Mae Melvin)

Table 6-9 Filarial Zoonotic Nematodes

Nematode	Disease	Predominant Signs in People	Clinical Signs in Animals	Transmission	Geographic Distribution	Diagnosis	Treatment and Control
Brugia malayi, Brugia timori	*Brugia* filarias; lymphatic filariasis; Malayan filariasis	Often asymptomatic; clinical signs include lymphangitis mainly in the legs and groin area; fever, headache, and backache may also be seen	■ Dogs and cats are main reservoirs of the nocturnally periodic strain of *Brugia malayi* (intermediate hosts are *Aedes* and *Mansonia* mosquitoes) ■ Dogs, cats, and wild felines are reservoirs of subperiodic strain of *Brugia malayi* (transmitted by *Mansonia* mosquitoes)	Infected mosquitoes transmit infective, third-stage larvae Adult worms live in lymph vessels and lymph nodes; females release first-stage larvae (microfilariae) that circulate in blood, are taken up by mosquitoes, and develop to third-stage larvae. After transmission larvae migrate to an area and develop into adult worms within 3 months	■ South, Southeast, and East Asia (India, Burma, Thailand)	Identification of microfilariae measuring 220 μm in length from Giemsa-stained blood smears; filtration techniques may be used to concentrate larvae; serologic tests are available but in some endemic areas positive titers can be seen in over 50% of people	Diethylcarbamazine and ivermectin Mosquito repellents and nets should be used to prevent mosquito bites
Dirofilaria immitis	Dirofilariasis	Typically asymptomatic; occasionally localized vasculitis and pulmonary infarcts are found producing chest pain, cough, and hemoptysis. Granulomas around the dead worm may be seen as a nodule in the lung on radiography and needs to be differentiated from cancer	■ Dogs show weight loss, decreased exercise tolerance, and cough; in advanced cases dyspnea, fever, ascites, cyanosis, and periodic collapse may occur	Third-stage larvae are transmitted by infected mosquitoes and blackflies. Microfilariae are not found in humans.	■ Worldwide especially in warm climates (southern United States, Central and South America, Mediterranean)	Radiographs Serologic assays	Surgical excision of parasite Animals are treated with adulticide (melarsomine) followed by microfilaricide (ivermectin, milbemycin, or levamisole) Insect repellents to protect from biting insects Preventative drugs such as ivermectin, milbemycin, selamectin, or diethylcarbamazine is given to serologically negative dogs
Dirofilaria repens, Dirofilaria tenuis, Dirofilaria ursi, and Dirofilaria striata	Dirofilariasis; connective tissue dirofilariasis	Typically asymptomatic; occasionally settle in subcutaneous tissue producing painful, pruritic nodules	■ *Dirofilaria repens* is in dogs who may show weight loss, cough, or may be asymptomatic ■ *Dirofilaria tenuis* is in raccoons in the southern United States and is typically asymptomatic ■ *Dirofilaria ursi* is found in bears and lynxes in the United States and is asymptomatic ■ *Dirofilaria striata* is found in felines in the United States and is asymptomatic	Third-stage larvae are transmitted by infected mosquitoes and blackflies. Microfilariae are not found in humans.	■ Worldwide especially in warm climates (southern United States, Central and South America, Mediterranean)	Radiographs Excision and identification of parasites from nodules Serologic assays	Surgical excision of parasite Insect repellents to protect from biting insects

many human or primate only filarial nematodes such as *Onchocera volvulus* (river blindness), *Wuchereria bancrofti* (elephantiasis), and *Loa loa* (loiasis). Zoonotic filariasis occurs when humans are infected with larva that normally infect animals and cannot complete their life cycle in humans. Zoonotic filarial nematodes are summarized in Table 6-9.

Trichinosis

Overview

Trichinosis, also known as trichinellosis, trichinelliasis, and trichina infection, is caused by the intestinal nematode *Trichinella spiralis*. *Trich. spiralis* is named from the Latin *tricho* meaning hair and *spira* meaning coil used to describe the appearance of the nematode in muscle tissue. Trichinosis is common in carnivorous mammals primarily on North American and Eurasian continents. Humans typically acquire trichinosis by eating pork infected with the encysted larvae and may be the basis for the tradition of avoiding pork in Mosaic and Islamic religions. The association between *Trichinella* infections and pigs has been recognized for a long time; however, the encysted larvae in the muscle were not seen until 1821. The discovery of the worm in humans in 1835 was made by James Paget, a first-year medical student at St. Bartholomew's Hospital in London and later knighted as a distinguished physician. Paget described *Trichinella* infection in muscle tissue of the diaphragm in a paper presented at the Abernethien Society February 6, 1835. Richard Owen, who is attributed with the discovery of trichinosis, actually presented Paget's paper to the Zoological Society 18 days after Paget's presentation. Adult *Trichinella* worms were discovered by Rudolf Virchow in 1859. In 1860, Friedrich Zenker demonstrated the clinical significance of eating infected raw pork and the development of *Trichinella* infection. By the late 1860s, trichinosis was well recognized as a disease spread through infected pigs, leading to a cultural aversion to certain pork products such as German and Dutch sausage. Since 1895 the causative agent of trichinosis was known as *Trich. spiralis*, but since then there have been several species, subspecies, or strains of *Trichinella*; however, no morphological differences exist among some of these different *Trichinella* strains. Trichinosis is easily prevented by thorough cooking of meat and is uncommon in countries such as the United States and France where there are laws prohibiting the feeding of raw meat scraps to hogs. In the United States most cases of trichinosis are attributed to ingestion of undercooked wild game. In 1990, one of the last large outbreaks of trichinosis in the United States occurred in Southeast Asian immigrants who ate raw pork sausage at a wedding reception.

Causative Agent

Trichinella spp. are the smallest parasitic nematodes of humans. Adult males are between 1.4 and 1.6 mm in length with a more slender anterior portion. Males do not have a copulatory spicule. Adult females are twice the size of males with a more slender anterior portion. Females have a vulva located approximately one-third of the body length from the anterior end. Larvae are approximately 0.1 mm in length.

> *Trichinella* nematodes are unique in that the same animal serves as both definitive and intermediate host (larvae and adults are located in the same animal, but in different organs).

Trichinella spp. are considered to be a complex of species, with a variety of genotypes identified by DNA analysis. There are few distinct morphologic differences between species and species identification is based on characteristics such as

Table 6-10   Different Species and Genotypes of *Trichinella*

Species	Distribution	Major Host Reservoir	Infectivity to Humans	Resistance to Freezing
Trichinella spiralis (T1)	Worldwide; most common species affecting humans	Swine, wild boar, bear, horse, fox	High High infectivity for pigs and rodents	Low to none
Trichinella nativa (T2)	Arctic	Bear, horse	High Low infectivity to rats and pigs	High
Trichinella britovi (T3)	Temperate; southern Europe	Wild boar, horse	Moderate Moderate infectivity for pigs	Low
Trichinella pseudospiralis (T4)	Worldwide; lacks cyst in muscle	Birds, omnivorous mammals	Moderate Moderate infectivity for pigs	None
Trichinella murreli (T5)	Temperate, near arctic; United States	Bear	Low Low infectivity for pigs	Low
T6	Northern temperate; arctic and subarctic regions	Bear	Low Low infectivity for pigs	High
Trichinella nelsoni (T7)	Tropical; southern hemisphere; Africa south of the Sahara	Warthog; wild carnivores	High Low infectivity for rats and pigs	None
T8	South Africa	Lion	Low	None
T9	Japan	Bear	Low	None
Trichinella papuae (T10)	Papua New Guinea; not encapsulated larvae	Pigs, wild boars	? Moderate infectivity for pigs	?
Trichinella zimbabwensis (T11)	East Africa	Crocodiles	? (zoonotic potential unknown) Moderate infectivity for animals	?

Adapted from Murray, C. and Lowry, K. Trichinosis, www.emedicine.com, December 16, 2005 and Krauss, H. Weber, A. et al., Zoonoses infectious diseases transmissible from animals to humans, 3rd ed., American Society of Microbiology, 2003.

reproductive isolation, infectivity to certain hosts, and resistance to freezing. The species of *Trichinella* include:

- *Trich. spiralis*; the primary species associated with domesticated animals.
- *Trich. britovi*; the species seen frequently in wild boars, horses, and free-ranging swine. *Trich. britovi* has also been reported in bear (Japan) where it has been given a separate classification by some (T9).
- *Trich. nelsoni*; the species seen in various large carnivores of tropical Africa.
- *Trich. nativa*; the species documented in cougars, walruses, whales, and bears causing more prolonged diarrhea and fewer muscle symptoms.
- *Trich. pseudospiralis*; the species documented in birds that does not form a capsule in the muscle therefore causing less muscle inflammation and pain.
- Less common species are summarized in Table 6-10.

Epizootiology and Public Health Significance

Trichinosis occurs worldwide (Australia does not have any documented cases that have originated within its borders), but is seen more frequently in temperate climates than in the tropics. Trichinosis has relatively high rates of infection in the United States and Europe as a result of the ethnic customs of eating raw or rare pork dishes or wild animal meats.

In the United States trichinosis is largely limited to sporadic cases or small clusters related to consumption of home-processed meats from noncommercial farm-raised pigs and wild game. Trichinosis has been a reportable disease in the United States since 1966. The CDC surveillance system has data as far back as 1947 demonstrating a significant decrease in cases from a peak of nearly 500 in 1948 to the current rate of fewer than 50 per year. The fatality rate of trichinosis in the United States is 0.003. The USDA conducts periodic surveillance of farm-raised pigs to monitor diseases such as trichinosis.

Over the past 20 years, there has been an increase in the number of trichinosis outbreaks in developing countries. International trade of meats and rising affluence in countries without well-established monitoring systems has contributed to the increasing incidence of trichinosis. Trichinosis is rare in countries with laws prohibiting the feeding of raw garbage/animal byproducts to commercially raised pigs and in countries with well-managed slaughterhouse surveillance systems. Home raising of pigs, with feeding of raw garbage instead of grain, is still a common practice in the developing world. China has some of the highest recorded case numbers worldwide.

Although trichinosis is generally considered a disease of omnivorous or carnivorous animals, herbivores have become infected most likely from prepared feed that contained remnants of infected animals. In France, imported horse meat has become the most common source of trichinosis (more than a dozen outbreaks infected more than 3,000 people since 1976). Mutton and goat meat have become a recognized source of infection in countries where pig consumption is restricted for religious or economic reasons.

Tasting of raw homemade pork sausage for proper seasoning is a common source of Trichinella *infection.*

Transmission

The life cycle of *Trichinella* is spent entirely in the mammalian body host. In nature, the parasite is maintained in an encysted (encapsulated) larval form in the muscles of animal reservoirs and is transmitted when other animals prey on them. Human trichinosis infections are established by consumption of insufficiently cooked infected meat, typically pork or bear (other animal species have also been implicated). Humans are dead-end hosts (unless the practice of cannibalism occurs). Most mammals are susceptible to trichinosis through consumption of meat containing the encysted larvae. Rats are important in maintaining *Trichinella* cycles because farm pigs will eat rats, which are commonly found near garbage sites.

Trichinella is the world's largest intracellular parasite.

Pathogenesis

Trichinella infection occurs by ingestion of encysted larvae in muscle. The cyst wall is digested by gastric acid and pepsin in the stomach, allowing freed larvae to penetrate the duodenal and jejunal mucosa. In approximately 4 days, the larvae develop into sexually mature adults that begin mating. After mating, females penetrate deeper into the mucosa and discharge living larvae (up to 1,500) over a 4- to 16-week period. After mating, noninfective adult males are expelled in the stool. Eventually the females die and pass out of the host. Larvae migrate into the lymphatic system, are carried by the portal circulation to the peripheral

Encysted (encapsulated) Trichinella *larvae survive and remain infectious for years (maybe greater than 30 years) even if the capsule is calcified.*

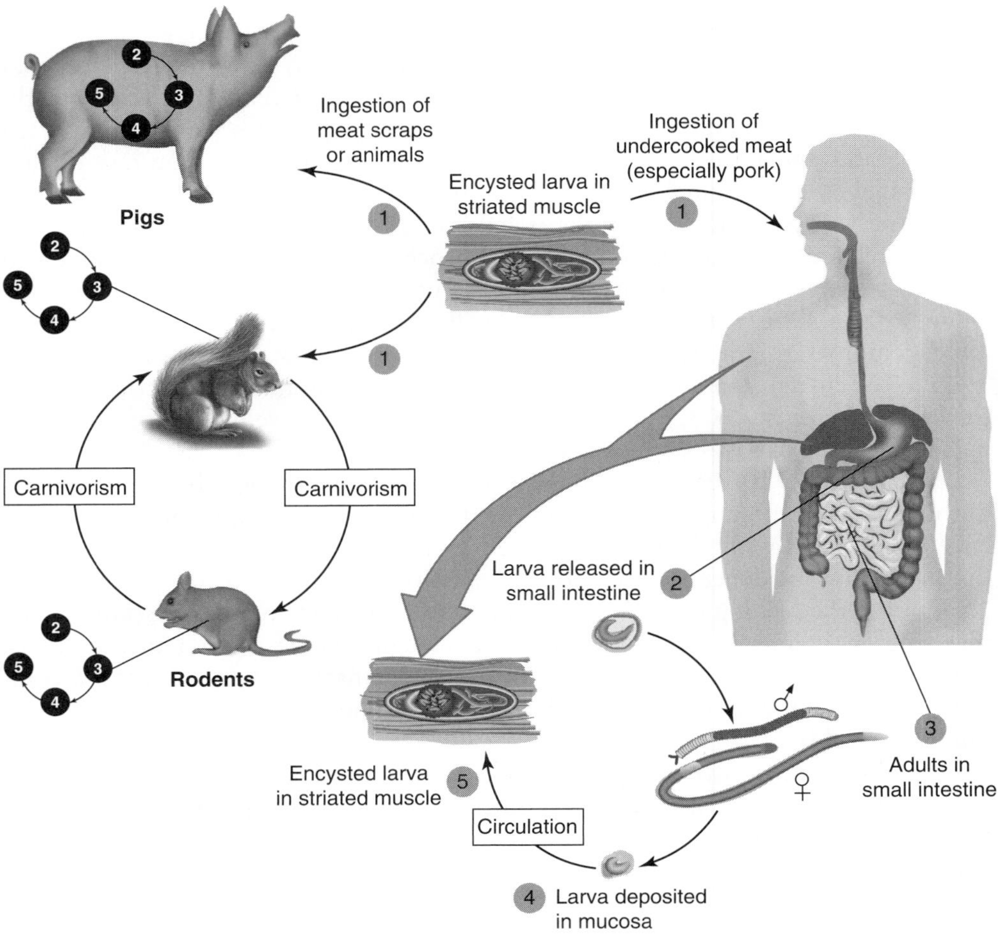

Figure 6-61 Life cycle of *Trichinella.*

Trichinosis is acquired by ingesting meat containing encysted larvae❶ of *Trichinella.* After exposure to gastric acid and pepsin, the larvae are released❷ from the cysts and invade the small intestinal mucosa where they develop into adult worms❸. After approximately 1 week, the female worms release larvae❹ that migrate to the striated muscles where they encyst❺. Ingestion of the encysted larvae perpetuates the cycle. Rats and rodents are primarily responsible for maintaining the endemicity of this infection. Carnivorous/omnivorous animals, such as pigs or bears, feed on infected rodents or meat from other animals.

Trichinosis requires two hosts in its life cycle: female worms produce larvae that encyst in muscle of the first host and the new, second host becomes infected when muscle is eaten.

In trichinosis, there is typically one larva per muscle cell.

---

circulation, reach striated muscle, and penetrate individual muscle cells. The diaphragm, tongue, masseter, and intercostal muscles are among the most common muscles involved in pigs. Larvae grow rapidly and begin to coil within the cell. Capsule formation begins approximately 15 days after infection and is finished by 4 to 8 weeks postinfection. Once capsule formation is complete the larvae are infective. The muscle cell will degenerate as the larva grows, at which time calcification begins. If immature larvae pass through the intestine and are eliminated in the feces, they are infective to other animals (Figure 6-61).

Clinical Signs in Animals

Most *Trichinella* infections in domestic and wild animals are asymptomatic.

Clinical Signs in Humans

In humans, the usual incubation period of trichinosis is 8 to 15 days. Heavy *Trichinella* infections may produce serious illness with 3 successive clinical phases:

- *Intestinal.* The first signs of trichinosis occur between 12 and 48 hours after ingestion of infected meat with initial symptoms as a result of the invasion of the intestinal wall by the juvenile larvae. In this phase vague symptoms such as nausea, vomiting, and diarrhea are seen. Dyspnea and red blotchy rashes may erupt on the skin. At the end of this phase facial edema and fever is seen 5 to 7 days after the first appearance of symptoms.

- *Muscle invasion.* This phase is caused by the migration of numerous larvae into muscle. Clinical signs include periorbital edema (classic sign), intense muscle and joint pain, and shortness of breath. Neurologic complications such as deafness and seizures may be seen. A pronounced eosinophilia is seen during this phase.

- *Convalescent.* Clinical signs may resolve after a few days, but may persist for 5 to 6 weeks. In severe infection, muscle pain can lead to difficulty swallowing, speaking, breathing, and chewing. The most frequent cause of death is myocarditis as a result of invasion of cardiac muscle and heart failure. If acute signs go into remission most people become asymptomatic even without treatment and with the existence of larvae in the musculature. Late sequelae include rheumatoid arthritis and continued muscle pain.

Diagnosis in Animals

Trichinosis in animals may be diagnosed by microscopic examination of a muscle biopsy sample (usually tongue). A negative muscle biopsy does not necessarily rule out trichinosis. Serologic testing using ELISA technique is a reliable test to detect anti-*Trichinella* antibodies. Seroconversion in animals may occur with as few as 0.01 larvae per gram of meat, but may not occur for weeks after infection.

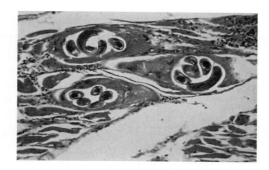

Figure 6-62 *Trichinella* cysts within human muscle tissue. (Courtesy of CDC)

Diagnosis in Humans

Trichinosis in people may be suspected in patients with eosinophilia (CBC), myoglobinuria (UA), and elevated creatine kinase values (serum chemistry). Calcified densities in muscles may be seen on radiographs, but are not present in acute infections. CT brain scans may show focal defects in the cerebral cortex if neurologic involvement is present. Parasite-specific indirect IgG ELISA titers are approximately 100% accurate in diagnosing trichinosis. Anti–larvae antibodies are also available and are approximately 30% accurate. Western blot analysis is used to confirm positive ELISA results. Muscle biopsy is the definitive diagnostic test and is done by crushing a portion of muscle tissue between two slides and viewing directly (Figure 6-62).

Treatment in Animals

Treatment in animals is generally impractical. Symptomatic and supportive care to alleviate pain until the infection is resolved may be attempted.

Treatment in Humans

In human cases of trichinosis anthelmintic therapy is considered effective only during the intestinal phase of infection. Thiabendazole and mebendazole are commonly used in people. Corticosteroid treatment can reduce the immunologic response to the larvae.

Management and Control in Animals

Control of trichinosis in pigs is accomplished with good management (controlling rodents, cooking garbage for 30 minutes at 212°F if feeding to it to pigs, and preventing cannibalism (tail biting) and access to wildlife carcasses).

Management and Control in Humans

Trichinosis in people is preventable by thoroughly cooking meat. Most developed countries have surveillance programs to monitor meats entering the commercial market; however, these controls have been documented to fail. Inspection of meat at the time of slaughter, either by microscopic examination of muscle or by digestion methods, is effective in preventing human infection. In North America, commercial products that are labeled ready to eat must be processed by adequate heating, freezing, or curing to kill *Trichinella* before marketing. Pork should be cooked to an internal temperature of 137°F (58°C) or greater. Freezing pork is also effective at 5°F (–15°C) for 20 days, –9.4°F (–23°C) for 10 days, or –22°F (–30°C) for 6 days. Freezing cannot be relied on to kill *Trichinella* organisms in meat other than pork. Pickling and smoking meat does not kill the larvae.

Summary

Trichinosis is caused by *Trichinella* spp., a parasitic nematode with adult males measuring between 1.4 and 1.6 mm in length and adult females measuring twice the size of males. *Trichinella* spp. are considered to be a complex of species, with a variety of genotypes identified by DNA analysis. The species of *Trichinella* include *Trich. spiralis*, *Trich. britovi*, *Trich. nelsoni*, *Trichi. nativa*, *Trich. pseudospiralis*, and a few less common species. Trichinosis occurs worldwide and is seen more frequently in temperate climates than in the tropics. The life cycle of *Trichinella* is spent entirely in the mammalian body host. In nature, the parasite is maintained in an encysted (encapsulated) larval form in the muscles of animal reservoirs and is transmitted when other animals prey on them. *Trichinella* infection occurs by ingestion of encysted larvae in muscle. Most *Trichinella* infections in domestic and wild animals are asymptomatic. In humans, heavy *Trichinella* infections may produce serious illness with three successive clinical phases, which include intestinal (nausea, vomiting, diarrhea, dyspnea, red blotchy rashes, facial edema, and fever), muscle invasion (periorbital edema (classic sign), intense muscle and joint pain, and shortness of breath), and convalescent (resolution of clinical signs, death, or chronic infection). Trichinosis in animals may be diagnosed by microscopic examination of a muscle biopsy sample (usually tongue). Serologic testing using ELISA technique is a reliable test to detect anti- *Trichinella* antibodies. Trichinosis in people is diagnosed with parasite-specific indirect IgG ELISA titers. Anti–larvae antibodies are also available and are approximately 30% accurate. Western blot analysis is used as to confirm positive ELISA results. Muscle biopsy is the definitive diagnostic test and is done by crushing a portion of muscle tissue between two slides and viewing directly.

Treatment of trichinosis in animals is generally impractical. In human cases of trichinosis anthelmintic therapy using thiabendazole or mebendazole is considered effective only during the intestinal phase of infection. Control of trichinosis in pigs is accomplished with good management. Trichinosis in people is preventable by thoroughly cooking meat and meat inspection.

Trichuriasis

Overview

Trichuriasis, also known as whipworm infection, is caused by a variety of *Trichuris* nematodes commonly called whipworms. *Trichuris* nematodes are named from the Greek *trichos* meaning hair and *oura* meaning tail because the anterior two-thirds of the worm is thin and hair-like, whereas the posterior one-third is fat like the handle of a whip. Evidence of whipworm infection can be traced to ancient civilizations. *Trichuris* eggs have been found in the fossilized feces of stone-age humans from 10,000 years ago. Mummified bodies dated around 2400 B.C. from Nubia (Northeast Africa) had *Trichuris* eggs in visceral contents kept in Canopic jars during the mummification process. As Egyptians and Nubians maintained trade during this time period whipworm infections were probably a common infection in Egyptians as well. Eggs of *Tric. trichiura*, the primate parasite, have been found in glacier mummies more than 5,000 years old. *Trichuris* eggs were found in the Neolithic wetland villages in Western Europe. Otzi the Iceman, found on September 19, 1991 in the Alps, believed to have lived around 3200 B.C. and had been infected with whipworms. Today, *Tric. trichiura* is one of the most common human intestinal parasites in parts of the world where sanitary conditions are poor.

Causative Agent

Most human cases of trichuriasis are caused by *Tric. trichiura*, the nematode of humans and nonhuman primates. *Tric. trichiura* is about 30 to 50 mm in length, with males being shorter than females. Its mouth is tiny, lacks lips, and contains a small spear. Adult worms have a long esophagus, an anus near the tip of the tail, and lack an excretory system. Both sexes have a single gonad; males have a single spicule surrounded by a sheath and females have a uterus that contains many unembryonated, lemon-shaped eggs with an operculum at each end.

> The cat whipworm, *Tric. campanula*, is not found in the United States.

Zoonotic species of *Trichuris* include *Tric. vulpis* (the dog whipworm, which is commonly found in the United States yet is difficult to detect at times because low numbers of eggs are shed and are shed in waves) and *Tric. suis* (the pig whipworm). There are approximately 60 to 70 other species of *Trichuris* in a variety of animals, but their zoonotic potential has not been proven. For example, *Tric. ovis* is an important pathogen of sheep and cattle and *Tric. muris* is found in rats and mice.

Epizootiology and Public Health Significance

Trichuriasis is found worldwide, but is rarely found in arid, extremely hot, or extremely cold regions of the world. For trichuriasis to become a serious health problem in a region, two requirements are needed: proper environmental conditions (a warm climate, high rainfall and humidity, dense shade (allows for egg survival and development), and moisture-retaining soil) and poor sanitation.

In the United States, whipworm infection is rare with the highest incidence occurring in the rural Southeast, where 2.2 million people are infected. Internationally there are more than 500 million people infected with *Tric. trichiura* nematodes. The prevalence of zoonotic trichuriasis is unknown, but is typically associated with poor socioeconomic standards of living. In the United States, approximately 10% to 20% of dogs are positive for *Tric. vulpis* (40% in stray dogs). *Tric. suis* occurs in 2% to 5% of adult swine and 15% to 40% of nursing piglets.

Transmission

Trichuris nematodes have a direct life cycle and mature in a single host that becomes infected when it ingests embryonated eggs from contaminated soil. *Trichuris* eggs are sticky and may be carried to the mouth by hands, other body parts, fomites, or foods. Using manure as a fertilizer can be a source of whipworm infection. House flies can serve as mechanical vectors of whipworms.

Pathogenesis

Trichuris eggs are unembryonated and noninfectious when they are excreted. Development to an egg containing first-stage larvae (the infectious form) takes

> Both larval and adult whipworm are found only in the intestines and do not undergo tissue migration.

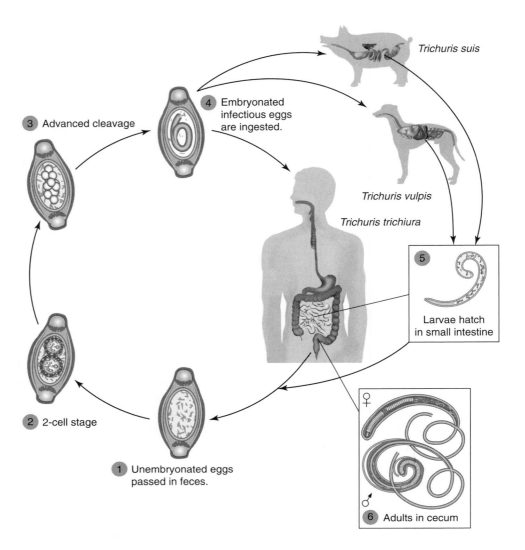

Figure 6-63 Life cycle of *Trichuris*. Unembryonated eggs are passed with the stool❶. In the soil, the eggs develop into various stages❷, ❸ and then embryonate❹. After ingestion (soil-contaminated hands or food), the eggs hatch in the small intestine, and release larvae❺ that mature and establish themselves as adults in the colon❻. The adult worms live in the cecum and ascending colon. Female worms in the cecum shed eggs, which are passed in stool.

10 to 14 days. The host ingests the embryonated eggs from soil (mainly in children who eat dirt) or soil-contaminated objects (typically food or water) and the eggs hatch in the small intestine. In the small intestine the larvae penetrate the cells in the base of the crypts, begin to develop, and tunnel in the intestinal epithelium back toward the luminal surface. This development in the small intestine lasts for up to 14 days with final maturation occurring in the large intestine. Adults are found in the cecum and colon where the thinner, anterior end of the worm burrows into the large intestine and the thicker, posterior end hangs into the lumen and mates with nearby worms. Eggs are produced and shed with the feces. *Tric. vulpis* females produce eggs in approximately 70 to 90 days in dogs; *Tric. suis* females produce eggs in approximately 41 to 45 days in pigs; *Tric. trichiura* females produce eggs in 30 to 90 days in primates. Adult females lay eggs for up to 5 years (Figure 6-63).

Clinical Signs in Animals

Tric. vulpis is found in dogs and wild canines. Most cases of trichuriasis are asymptomatic with occasional signs being poor condition or performance. Heavy parasite loads can cause weight loss, anemia, and diarrhea that may be mucoid or bloody. Chronic infections causing intestinal irritation may lead to intussusceptions.

Tric. suis is found in pigs and wild boars with a shorter incubation period than the other species (10 to 12 days). Most cases of trichuriasis are asymptomatic in adult swine. In pigs up to 3 months of age, severe diarrhea containing mucus and blood, anorexia, and death may be seen. Pigs with *Tric. suis* may be more prone to other intestinal infection especially *Ca. jejuni*.

Tric. trichiura is found in humans, nonhuman primates, and rarely swine. Most cases in animals are asymptomatic. Heavy infections can lead to diarrhea.

Clinical Signs in Humans

People with trichuriasis are typically asymptomatic. Heavy infections tend to be seen in small children or people who eat a lot of dirt. Symptoms in people with heavy infections include chronic bloody diarrhea, abdominal pain and distension, nausea, vomiting, flatulence, headache, weight loss, and anemia. Children with severe trichuriasis may develop dystenery, anemia, growth retardation, finger clubbing, and rectal prolapse. Extremely rare cases of visceral larva migrans caused by *Tric. vulpis* have been reported in humans.

Diagnosis in Animals

Trichuriasis is diagnosed by identifying *Trichuris* eggs by fecal examination (Figure 6-64). Eggs can be seen using fecal flotation, but centrifugation is the preferred method. Eggs are oval, yellowish-brown, thick-shelled, and contain two polar plugs. Eggs vary in size with *Tric. vulpis* eggs approximately 72 to 90 μm × 32 to 40 μm (twice the size of the other two species); *Tric. suis* eggs approximately 50 μm × 25 μm; *Tric. trichiura* eggs approximately 50 μm × 23 μm (*Tric. suis* and *Tric. trichiura* eggs are very similar in their appearance). *Trichuris* eggs can be shed intermittently making repeated fecal testing or proctoscopy helpful in identifying positive cases.

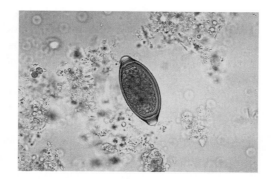

Figure 6-64 *Trichuris vulpis* egg.
(Courtesy of CDC/Dr. Mae Melvin)

Diagnosis in Humans

Trichuriasis in people is diagnosed by fecal identification of eggs as described for animals. Fecal examination can detect eggs and charcot-leyden (C-L) crystals (crystals are formed from the breakdown of eosinophils and may be seen in the stool or sputum of animals/people with parasitic disease).

Treatment in Animals

Trichuriasis can be treated with a variety of anthelmintics such as fenbendazole, febantel, mebendazole, and dichlorvos. Milbemycin alone or in combination with heartworm preventative is effective treatment in dogs. Hygromycin as a feed additive is used to control *Tric. suis* in pigs.

Treatment in Humans

Trichuriasis in people can be treated with mebendazole, alendazole, and oxantel, but may be difficult to treat as a result of the worm's location in the cecum and colon.

Management and Control in Animals

Trichuriasis can be prevented in animals by conducting regular fecal examinations in domestic pets, properly treating positive animals, and utilizing proper hygiene in positive animals. Sanitation and eliminating moist areas in the environment can reduce the survival of *Trichuris* worms. Pigs should be housed on slatted concrete floors to aide in sanitation of their housing. Regular pen cleaning and disinfection between use is recommended. Outdoor lots and pastures should be well-drained so that moisture cannot increase in the environment. Land rotation is also recommended. Dogs should be housed on cement or gravel runs in contrast to dirt runs. Runs should be cleaned daily and regularly disinfected. Lawns should be kept short and not over watered if dogs are allowed to defecate on them.

Management and Control in Humans

Trichuris infections in humans can be avoided by proper disposal of human feces, educating people about the problems associated with eating dirt or crops fertilized with night soil, preventing animals from having access to children's playground areas, practicing good hygiene, washing raw foods before eating them, boiling or filtering drinking water, and postponing acquiring new pets until children are past toddler age.

Summary

Trichuriasis, also known as whipworm infection, is caused by species of the nematode *Trichuris*. Most human cases of trichuriasis are caused by *Tric. trichiura*. Zoonotic species of *Trichuris* include *Tric. vulpis* (the dog whipworm) and *Tric. suis* (the pig whipworm). Trichuriasis is found worldwide, but is rarely found in arid, extremely hot, or extremely cold regions of the world. *Trichuris* infection occurs when embryonated eggs are ingested. *Trichuris* eggs are sticky and may be carried to the mouth by hands, other body parts, transport hosts, fomites, or foods. *Trichuris* eggs are unembryonated and noninfectious when they are excreted. Development to an egg containing the infectious first-stage larvae takes 10 to 14 days. *Tric. vulpis*, found in dogs and wild canines, is typically asymptomatic with occasional signs being poor condition or performance. Heavy parasite loads can cause weight loss, anemia, and diarrhea that may be mucoid or bloody. Chronic infections causing intestinal irritation may lead to intussusception. *Tric. suis* is found in pigs and wild boars causing severe diarrhea in young pigs, whereas adults may be asymptomatic. *Tric. trichiura* is found in humans, nonhuman primates, and rarely swine and is usually asymptomatic. Heavy infections can lead to diarrhea. People with trichuriasis are typically asymptomatic. Symptoms in people with heavy infections include chronic bloody diarrhea, abdominal pain and distension, nausea, vomiting, flatulence, headache, weight loss, and anemia. Children with severe trichuriasis may develop dystenery, anemia, growth retardation, finger clubbing, and rectal prolapse. Trichuriasis is diagnosed by identifying *Trichuris* eggs by fecal examination. Trichuriasis can be treated with a variety of anthelmintics. Trichuriasis in people can be treated with mebendazole, alendazole, and oxantel. Trichuriasis can be prevented in animals by conducting regular fecal examinations in domestic pets, properly treating positive animals, and utilizing proper hygiene in positive animals. Infection in people can be avoided by proper disposal of human feces, educating people about the problems associated with eating dirt or crops fertilized with night soil, preventing animals from having access to children's playground areas, practicing good hygiene, washing raw foods before eating them, boiling or filtering drinking water, and postponing acquiring new pets until children are past toddler age. (See Table 6-11 for less common zoonotic hematode.)

Table 6-11 Less Common Zoonotic Nematodes

Nematode	Disease	Predominant Signs in People	Clinical Signs in Animals	Transmission	Geographic Distribution	Diagnosis	Treatment and Control
Angiostrongylus cantonensis	Angiostrongyliasis; Cerebral angiostrongyliasis; Eosinophilic meningoencephalitis; Eosinophilic meningitis	Severe headache, fever, vomiting, neck stiffness, seizures, and occasionally paresis. Accumulation of eosinophils in cerebrospinal fluid causing increased pressure. Symptoms persist for days to months; may be lethal.	■ Cardiopulmonary disease in rats. Mature worms live in pulmonary arteries of infected rats. Eggs hatch, larvae penetrate the alveoli, are coughed up, and are passed in feces. Larvae invade snails, develop into infective larvae, and intermediate or paratenic hosts ingested by final hosts.	■ Ingestion of raw or undercooked paratenic hosts (crabs, crayfish, or freshwater fish) ■ Ingestion of intermediate host (snails) ■ Ingestion of free infectious larvae	■ Australia, China, India, Southeast Asia, and Oceana (also reported in southern United States)	Eosinophilia in CSF; serologic tests available	Mebendazole and corticosteroids Avoid ingestion of raw or undercooked snails, crabs, or crayfish; proper cleaning of vegetables; avoid ingestion of raw or undercooked internal organs of animals.
Anisakis simplex and *Pseudoterranoa decipiens*	Anisakiasis; herring worm disease	Abdominal pain, nausea, and vomiting; may lead to gastric ulceration and intestinal perforation; pruritus and urticartia; may be self self-limiting	■ Final hosts (whales, seals, and dolphins) are asymptomatic. ■ Intermediate hosts (herring, mackerel, cod, and flounder) are asymptomatic.	■ Ingestion of undercooked saltwater fish and squid ■ Killed larvae can cause anaphylactic reactions	■ Worldwide, especially Spain, Japan, France, the Netherlands, and North America	Endoscopic exam demonstrating larvae; ELISA and immunoblotting techniques	Endoscopic extraction of larvae; mebendazole In endemic areas avoid ingestion of saltwater fish; cook fish to 60°C, proper hygiene when handling fish.
Capillaria hepatica; renamed *Calodium hepatica*	Hepatic capillariasis	Light infections are asymptomatic; heavy infections cause acute liver disease (death possible in children).	■ Rodents (can be as high as 80%) develop liver infection.	■ Ingestion of embryonated eggs from environment	■ North and South America, Africa, Asia, and central Europe	Liver biopsy show eggs 30 to 50 μm	Mebendazole Avoid food contamination by rats; proper hygiene when handling food and utensils that may be exposed to rodents; disposal of animal livers.
Capillaria philippinensis; renamed *Paracapillaria philippinensis*	Intestinal capillariasis	Abdominal pain, anorexia, diarrhea; in severe cases vomiting, weight loss, diarrhea with electrolyte and protein imbalances	■ Fish are asymptomatic intermediate hosts. ■ Fish-eating birds are asymptomatic definitive hosts.	■ Ingestion of raw or undercooked freshwater or brackish water fish	■ Southeast Asia (endemic in Philippines) and Western pacific areas	Fecal examination shows bipolar, thick-shelled eggs; larvae and adults may also be found in feces	Mebendazole Avoid consumption of undercooked fish or crustaceans from endemic areas.

(Continued)

Table 6-11　(Continued)

Nematode	Disease	Predominant Signs in People	Clinical Signs in Animals	Transmission	Geographic Distribution	Diagnosis	Treatment and Control
Capillaria aerophila; renamed *Eucoleus aerophila*	Pulmonary capillariasis	Bronchitis, coughing, bloody sputum, dyspnea	■ Young dogs and cats may develop chronic, nonresponsive cough.	■ Ingestion of embryonated eggs; after ingested larvae hatch, invades intestinal wall, migrates through lymph and blood to lung and finally reside in trachea or bronchi	■ Worldwide	Fecal examination or airway cytology specimens show double-operculated eggs	Thiabendazole or fenbendazole Proper food hygiene by washing soil-contaminated food; avoid areas with animal feces
Ancylostoma caninum; dog hookworm	Eosinophilic enteritis	Ulceration and strictures of ileum, abdominal pain, vomiting, and diarrhea	■ Young dogs show patient infection (diarrhea, weight loss).	■ Third-stage larvae (infective form) from dog feces are transmitted to people via skin penetration	■ All tropical and subtropical regions, but higher rates in Australia	ELISA, immunoblots, fecal examination	Mebendazole Avoid walking barefoot in areas containing dog feces; treat infected dogs.
Gnathostoma spinigerum	Gnathostomiasis	Abdominal pain, pruritus, development of painful, red, migrating swellings of skin (subcutaneous gnathostomiasis); larvae may break through skin, inflammation of internal organs (visceral gnathostomiasis); CNS invasion can be fatal	■ Dogs and cats are asymptomatic final hosts. ■ Freshwater fish, frogs, and snails are asymptomatic second intermediate hosts. ■ Chickens are asymptomatic paratenic hosts.	■ Ingestion of third stage larvae of second intermediate host (freshwater fish, frogs, and snails) or paratenic hosts (chickens); ingestion of liberated stages or stages in first intermediate hosts (water fleas); third-stage larvae can be acquired percutaneously	■ Southeast Asia, Central America, and Australia	Excision and examination of larvae; serologic tests available	Surgery; albendazole Avoid raw or undercooked fish, frog, or poultry meat; boil water.
Oesophagostomum bifurcum, Oesophagostomum aculeatum, and *Oesophagostomum stephanostomum*	Oesophagostomiasis	Infections may be inapparent; nodules in the intestinal wall may lead to large bowel obstructions. Multinodular oesophagostomiasis produces hundreds of small nodules in the large intestine. Uninodular oesophagostomiasis, also called Dapaong tumour, produces painful granulomatous masses in the abdominal wall or within the abdominal cavity.	■ Primates have caseous granulomatous lesions in the colon.	■ Ingestion of third stage larvae adhering to vegetation; first stage larvae develop in the egg, hatch, and develop into third-stage larvae (the remnant of the second stage is sheath)	■ West Africa (Ghana and Togo)	Identification of 40 to 90 μm oval, thin-shelled eggs; ELISA; laparoscopy can show nodules in intestinal wall	Mebendazole, albendazole, or surgery Personal hygiene following handling of primates or potentially contaminated material

Thelazia callipaeda and *Thelazia californiensis*	Thelaziasis	Inflammation of conjunctiva; most frequent in small children	■ Dogs and wild canines are main hosts with clinical signs of increased tear production, light sensitivity, and occasionally conjunctivitis. ■ Cats, cattle, badgers, rabbits, foxes, and monkeys may also be affected.	■ Transmission of *Thelazia californiensis* is by the face fly, *Fannia canicularis*. ■ Transmission of *Thelazia callipaea* is by *Musca* and *Fannia* flies. The adult female worm lays her eggs in the tears, eggs develop into larvae that are ingested by the fly. Larvae develop in the fly, move to the mouth of the fly, and when the fly feeds near the eye, the larvae move from the fly's mouth to the eye of the new host.	*Thelazia callipaeda* is in Southeast Asia *Thelazia californiensis* is in western United States	Visual identification of worm in conjunctival sac	Removal of worm and irrigation Fly control
Trichostrongylus spp.	Trichostrongylidiasis	People are usually asymptomatic or have mild gastrointestinal signs; severe infections cause diarrhea and abdominal pain	■ Ruminants carry the worm in their abomasums and small intestine. ■ *Trichostrongylus axei* is the small stomach worm of cattle, sheep, and goats. It causes young calves, lambs, and kids to develop watery diarrhea.	■ Ingestion of infective third-stage larvae that has contaminated plant material. Eggs are excreted by infected ruminants (most kept on pasture are infected). Third-stage larvae develop within one to several weeks depending on temperature.	Worldwide; human infection is high in the Middle East (25%).	Identification of thin-shelled eggs 50 × 90 μm; third-stage larvae can also be identified	Mebendazole and albendazole are effective; pyrantel pamoate is used in the United States Proper personal hygiene; rinsing and proper cooking of vegetables

TREMATODE ZOONOSES

Fascioliasis

Overview

Fascioliasis, mainly caused by the liver fluke *Fasciola hepatica*, has been known as a parasite of sheep and cattle for hundreds of years. The first record of *F. hepatica* was in 1379 by Jean de Brie who described large, flat worms as the cause of the disease liver rot in sheep. The fluke was first illustrated in 1668 by Francisco Redi, but it was not until 1737 when the microscopic cercariae were described by Jan Swammerdam. In 1758 Linnaeus gave the worm its scientific name (*fasciola* is Greek for band, which describes the characteristic shoulders of this fluke), but he considered the organism a leech. The first human infection with *F. hepatica* was described by Pallas in 1760. The role of mollusks in the life cycle of *F. hepatica* was not described until the mid-1800s when both A. P. Thomas and Rudolph Leuckart independently traced the development of the fluke to the same species of snail. In 1892, Brazilian scientist Adolph Lutz determined that ruminants (the definitive hosts) become infected by *Fasciola* flukes by ingesting juvenile flukes encysted on vegetation. Lutz was working with *F. gigantica*, a larger fluke of the genus *Fasciola*, which has the same biology as *F. hepatica*.

Fascioliasis in humans is characterized by fever, eosinophilia, and abdominal pain, although many people are asymptomatic. Humans are incidental hosts for *F. hepatica*, commonly known as the sheep liver fluke or common liver fluke, and *F. gigantica*, a rare and geographically isolated cause of fascioliasis. Fascioliasis in ruminants produces a variety of clinical signs ranging from devastating disease in sheep to asymptomatic disease in cattle.

Causative Agent

> *F. hepatica* is the most common disease-causing liver fluke in temperate areas of the world and the most important trematode of domestic ruminants.

F. hepatica is one of the largest flukes (30 mm × 13 mm) and is leaf shaped with a pointed posterior end and a wide anterior end. Its oral sucker is small and located at the end of a cone-shaped projection at the anterior end. The body has a marked widening called an oral core that gives the fluke the appearance of shoulders. The testes are large and greatly branched, arranged in tandem behind the ovary. The smaller ovary is on the right side and the uterus is short. Vitelline follicles are extensive and fill most of the lateral body. *F. hepatica* cercariae have a simple, club-shaped tail about twice their body length. Operculated eggs are 90 × 150 µm. The ova of *F. hepatica* are indistinguishable from that of *Fasciolopsis buski*.

F. gigantica is a longer and more slender fluke than *F. hepatica*, but otherwise is very similar. Its operculated egg is 90 × 190 µm.

Epizootiology and Public Health Significance

> There is not a second intermediate host in the life cycle of *Fasciola* flukes.

F. hepatica is found worldwide and is especially common in moist regions with high rainfall (favors the development of amphibious snails that serve as intermediate hosts). In the United States, *F. hepatica* is endemic along the Gulf Coast, the West Coast, the Rocky Mountain region, and other areas. It can also be found in eastern Canada, British Columbia, and South America, the British Isles, western and eastern Europe, Australia, and New Zealand.

F. gigantica is found in Africa, Asia, and Hawaii and is relatively common in herbivorous animals especially cattle. Its morphology, biology, and pathology are nearly identical to those of *F. hepatica* except different snail hosts are needed for each.

Fascioliasis causes considerable morbidity worldwide in temperate regions, except Oceania, with disease prevalence high in specific regions of Bolivia, Ecuador, Egypt, and Peru. Human fascioliasis is rare in the United States; however, an estimated 2 million cases exist worldwide. The incidence of human fascioliasis has increased since 1980.

Transmission

Infection occurs by ingestion of plants with metacercariae attached to them. The most common plants that have metacercariae attached are watercress, water lettuce, mint, parsley, khat, and other vegetables grown in water. Drinking surface water containing floating metacercariae can also lead to infection. Infection can also occur from eating raw liver containing immature liver flukes.

Pathogenesis

After oral ingestion of contaminated plants or water, the larvae excyst in the duodenum, migrate through the intestinal wall, and enter the liver from the peritoneal cavity. The young flukes penetrate the liver capsule and move in the parenchyma for several weeks, growing and destroying tissue. After 6 to 8 weeks of migration through the liver, the flukes penetrate the bile ducts (their final destination). In ruminants, sexual maturity is reached in approximately 10 weeks postinfection and in humans sexual maturity is reached in 3 to 4 months postinfection. Mature flukes consume hepatocytes and may reside for years in the hepatic and common bile ducts and occasionally in the gall bladder (causing the chronic adult biliary stage of infection). Adult fluke worms produce eggs about 4 months postinfection. Adult fluke eggs in the bile ducts are released into the common bile duct into the duodenum. Eggs are shed with feces into the environment. To develop further the egg must reach tepid (22°C to 26°C) surface water, where miracidia (larvae) develop and hatch within approximately 1 to 2 weeks. These miracidia then search for and invade many species of freshwater snails as intermediate hosts. After several developmental stages (sporozoites, redia, and cercariae) free-swimming cercariae leave the intermediate hosts, swim actively through water, and adhere to plants, where they encyst to form metacercariae (the infectious stage). Free-swimming cercariae may remain suspended in the water and encyst over a few hours (Figure 6-65).

> Encysted cercariae (metacercariae) may remain viable for many months unless they become desiccated.

Clinical Signs in Animals

As immature *F. hepatica* flukes wander through the liver they destroy liver tissue and cause hemorrhage. In sheep and cattle, fascioliasis can be seen in three forms (determined by the number of metacercariae ingested over a period of time):

- *Chronic.* This form is often fatal in sheep and rarely fatal in cattle. Chronic fascioliasis is seen in all seasons; signs include anemia, unthriftiness, submandibular edema, and reduced milk secretion, but even heavily infected cattle may show no clinical signs. Heavy chronic infection is fatal in sheep. Chronic liver damage in sheep is cumulative over several years; fibrosis of liver tissues, cirrhosis, and calcification of bile ducts may be seen in cattle. Flukes may also be found in other sites such as the lungs.

- *Acute.* This form occurs primarily in sheep and is often fatal. In acute fascioliasis, there is extensive liver damage and the liver becomes enlarged and friable. In sheep, acute fascioliasis occurs seasonally and clinical signs in sheep include a distended, painful abdomen, anemia, and sudden death. Death can occur within 6 weeks of infection.

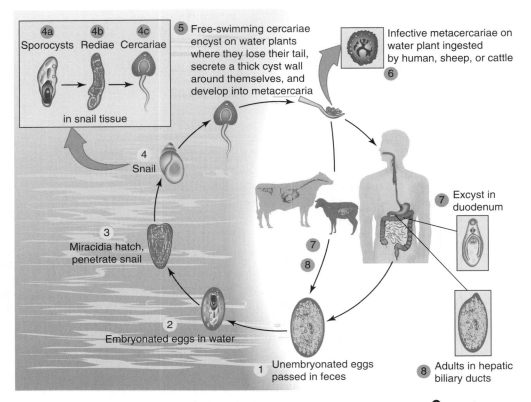

Figure 6-65   Life cycle of *Fasciola*. Immature eggs are released from the biliary ducts into the stool❶. Eggs become embryonated in water❷, eggs release miracidia❸, which invade a suitable snail intermediate host❹. In the snail the parasites undergo several developmental stages (sporocysts❹ᵃ, rediae❹ᵇ, and cercariae❹ᶜ). The cercariae are released from the snail❺ and encyst as metacercariae on aquatic vegetation or other surfaces. Mammals acquire the infection by eating vegetation containing metacercariae. Humans can become infected by ingesting metacercariae-containing freshwater plants❻. After ingestion, the metacercariae excyst in the duodenum❼ and migrate through the intestinal wall, the peritoneal cavity, and the liver into the biliary ducts, where they develop into adults❽.

- *Subacute*. This form occurs in cattle and sheep. In subacute disease, survival is longer (typically 7 to 10 weeks) than the acute form. Clinical signs are associated with hepatic damage; death may occur as a result of hemorrhage and anemia.

Clinical Signs in Humans

In humans, fascioliasis may be asymptomatic (approximately 50%) or cause clinical disease with such clinical signs as intermittent fever, lethargy, weight loss, hepatomegaly (abdominal enlargement), abdominal pain, hives, cough, and gastrointestinal signs. In children, signs that resemble pancreatitis may be seen.

Diagnosis in Animals

Fascioliasis is diagnosed by the identification of oval, operculated, golden eggs (130 × 75 µm) via fecal sedimentation techniques (simple cup sedimentation using

> *Fasciola hepatica* infection in cattle causes liver condemnation at slaugher (termed liver rot).

tap water, formaldehyde-ether concentration) (Figure 6-66). *F. hepatica* eggs cannot be observed in feces during acute fascioliasis and in subacute or chronic disease eggs are shed intermittently making repeated fecal examination necessary. ELISA tests are also available that can detect antibody levels about 2 to 3 weeks after infection. Necropsy samples show extensive liver damage and visualization of adult flukes in the bile ducts (immature stages may be squeezed or teased from the cut surface).

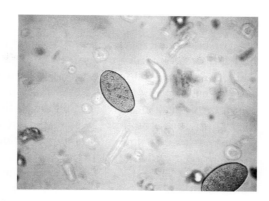

Figure 6-66 Eggs of *Fasciolopsis buski* and *Fasciola hepatica* are almost identical. Below are eggs of *Fasciolopsis buski.*
(Courtesy of CDC/Dr. Mae Melvin)

Diagnosis in Humans

Fascioliasis is diagnosed in humans using fecal stool samples (the small number of eggs shed intermittently in stool requires multiple specimens) and serologic methods (complement fixation, indirect hemagglutination, and ELISA).

Treatment in Animals

In ruminants several drugs are available to treat *F. hepatica* including clorsulon (cattle and sheep only) and albendazole in the United States and triclabendazole, netobimin, closantel, rafoxanide, and oxyclozanide in other countries (resistance to triclabendazole has developed in Australia and Europe). Most of these drugs have long withdrawal periods and need to be given at correct times to result in the optimal removal of flukes (each drug has varying efficacy against different ages of fluke). Treatments are determined by environmental factors (Gulf Coast states treat cattle before the fall rainy season and again in the late spring; northwestern states treat cattle at the end of the pasture season and in late January or February; European countries treat sheep in September or October, January or February, and again in April or May).

Treatment in Humans

People are treated with triclabendazole or bithionol.

Management and Control in Animals

Under ideal circumstances removing flukes in affected animals, reducing the snail population, and preventing livestock access to snail-infested pasture would be ways to control *Fasciola hepatica* in animals. Routine use of flukicides is practiced; however, molluscicide used to remove snails can be toxic to grazing animals and the ability to limit livestock access to snail-infested pasture may be cost prohibitive.

Management and Control in Humans

Avoiding ingestion of plants such as watercress found on moist ground in animal grazing areas and in areas with temporary flooding that may contain metacercariae will lower the cases of human fascioliasis. Avoiding consumption of unboiled or unfiltered surface water will also decrease the number of human fascioliasis cases.

Summary

Fascioliasis is caused by *F. hepatica*, one of the largest flukes that is found world-wide, especially in moist regions with high rainfall. In the United States, *F. hepatica* is endemic along the Gulf Coast, the West Coast, the Rocky Mountain region, and other areas. Infection occurs by ingestion of plants with metacercariae attached to them, drinking surface water containing floating metacercariae, or from eating raw liver containing immature liver flukes. In sheep and cattle, fascioliasis can be seen in three forms: chronic (anemia, unthriftiness, and potentially fatal in sheep), acute (distended abdomen and anemia in cattle and sheep, and sudden death in sheep), and subacute (hepatic disease, hemorrhage, and anemia in sheep and cattle). In humans, fascioliasis may be asymptomatic or cause clinical disease with such clinical signs as intermittent fever, lethargy, weight loss, hepatomegaly, abdominal pain, hives, cough, and gastrointestinal signs. Fascioliasis is diagnosed in animals by the identification of oval, operculated, golden eggs via fecal sedimentation techniques. ELISA tests are also available. Necropsy samples show extensive liver damage and visualization of adult flukes in the bile ducts. Fascioliasis is diagnosed in humans using fecal stool samples (the small number of eggs shed intermittently in stool requires multiple specimens) and serologic methods (complement fixation, indirect hemagglutination, and ELISA). In ruminants several drugs are available to treat *F. hepatica* infection including clorsulon (cattle and sheep only) and albendazole in the United States. People are treated with triclabendazole or bithionol. Fluke control includes removal of flukes in affected animals, reduction of the snail population, and prevention of livestock access to snail-infested pasture. Avoiding ingestion of plants found on moist ground in animal grazing areas and consumption of unboiled or unfiltered surface water will decrease the number of human fascioliasis cases.

Fasciolopsis

Overview

Fasciolopsis, caused by the trematode *Fasciolopsis buski* (commonly known as the giant intestinal fluke), is a clinically variable disease of the intestine. Fasciolopsis comes from the Latin *fasciola* meaning band (used to name *Fasciola* spp.) and the Greek *opsis* meaning appearance. The eggs of *Fa. buski* and *F. hepatica* are very similar in their appearance and their life cycles parallel each other. Busk first described *Fa. buski* in London after finding the flukes in the duodenum of a sailor in 1843. Barlow determined its life cycle in humans in 1925 and in 1947 Stoll estimated that worldwide there were 10 million human infections. Today fasciolopsis is found mainly in Asia and on the Indian subcontinent where humans raise pigs and consume freshwater plants.

Causative Agent

Fa. buski lives in the small intestine of its definitive host and is elongated, oval, and approximately 50 mm x 20 mm in size (it is one of the largest parasitic trematodes). The adult fluke has unbranched ceca, dendritic testes in its posterior half, branched ovaries that lie in the midline (anterior to the testes), extensive vitelline follicles, and a short uterus. The eggs are almost identical to the eggs of *F. hepatica*, being operculated and about 80 to 130 μm in size (Figure 6-66).

> There are more than 50 different species of intestinal trematodes; however, only a few species cause disease.

Epizootiology and Public Health Significance

Fa. buski is found in southeastern and eastern Asia. Endemic areas include China, Taiwan, India, Bangladesh, Indonesia, Thailand, and Vietnam where people and pigs eat fecal-contaminated water plants and their fruits.

> *Fa. buski* flukes can produce about 25,000 eggs daily.

Infection with intestinal flukes in the United States is typically seen in immigrants from endemic areas. Intestinal flukes are endemic in the Far East and Southeast Asia. Death from infection is rare and is usually observed only in people with a heavy worm burden.

Transmission

Infection with *Fa. buski* occurs after oral ingestion of metacercariae adhered to raw water plants and their fruits such as water nuts and walnuts or ingestion of contaminated surface water. Water vegetation, particularly *Trapa natans* (water caltrop) or the water chestnut, thrives in ponds fertilized by night soil (human feces). Only the inner part of water caltrop is eaten, but human infection occurs if the outer covering of the infected plant is peeled using the person's teeth.

> The definitive host of *Fa. buski* is the pig.

Pathogenesis

Immature eggs are shed with feces into fresh water where miracidia hatch from eggs after 3 to 7 weeks. Miracidia invade the soft tissues of snails (intermediate hosts), asexually form sporocysts, then rediae, and finally cercariae. The cercariae are released from the snails and encyst (become metacercariae) on various water plants such as water caltrop, water chestnut, lotus (on the roots), and water bamboo. People and pigs become infected when they ingest water plants with metacercariae. After ingestion of metacercariae, *Fa. buski* attaches to the duodenal and jejunal mucosa (in severe infections, they may attach to the ileum or colon) and become adult flukes in approximately 3 months. At the point of attachment, intestinal flukes cause inflammation, ulceration, and mucus secretion. Severe infections may also cause intestinal obstruction or malabsorption. Some of the fluke's metabolites are toxic and when absorbed by the host can be fatal (Figure 6-67).

> Disease conditions resulting from *Fa. buski* infection are immuno-pathologic, obstructive, and traumatic.

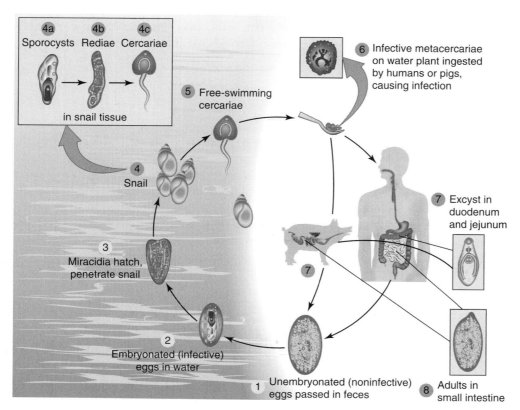

Figure 6-67 Life cycle of *Fasciolopsis*. Immature eggs are released into the intestine and stool❶. Eggs become embryonated in water❷, eggs release miracidia❸, which invade a suitable snail intermediate host❹. In the snail the parasites undergo several developmental stages (sporocysts❹ᵃ, rediae❹ᵇ, and cercariae❹ᶜ). The cercariae are released from the snail❺ and encyst as metacercariae on aquatic plants❻. The mammalian hosts become infected by ingesting metacercariae on the aquatic plants. After ingestion, the metacercariae excyst in the duodenum and jejunum❼ and attach to the intestinal wall.

Clinical Signs in Animals

Pigs are the definitive hosts of *Fa. buski* and may be asymptomatic or may present with gastrointestinal disease such as weight loss, lethargy, abdominal pain, and diarrhea. Dogs may also be infected with a similar clinical picture.

Clinical Signs in Humans

Many people infected with *Fa. buski* are asymptomatic. People with moderate infection may have occasional loose stools, weight loss, lethargy, and generalized abdominal pain. People with severe infection have diarrhea alternating with constipation, hunger pangs, facial edema, anemia, and ascites. Cachexia may follow these signs (possibly as a result of the toxic metabolic products) and death may occur.

Diagnosis in Animals

Diagnosis is made by fecal examination of operculated eggs.

Diagnosis in Humans

Diagnosis is made by fecal examination of operculated eggs filled with yolk material; however, *Fa. buski* eggs can easily be mistaken for *F. hepatica* eggs.

Treatment in Animals

Fasciolopsis is treated with praziquantel, diminazene aceturate, and imidocarb dipropionate. These drugs are not available in all endemic countries or their use may be restricted. Supportive treatment may include the use of anti-inflammatory drugs, antioxidants, and corticosteroids.

Treatment in Humans

Fasciolopsis in people is treated with praziquantel or triclabendazole (not available for humans in United States). Supportive care may include the use of anti-inflammatory drugs, antioxidants, corticosteroids, or blood transfusions.

Management and Control in Animals

Animal disease can be prevented by adequate waste removal of human and pig feces. Snail control may also be attempted.

Management and Control in Humans

Human disease can be prevented by immersing vegetables in boiling water for a few seconds to kill the metacercariae. Hands may become contaminated while handling the contaminated plants and should be thoroughly washed in clean water. People should avoid cracking water nuts and chestnuts with their teeth. Adequate waste removal of human and pig feces as well as not using human or pig feces for fertilizer would also control disease. Snail control may be attempted.

Summary

Fasciolopsis is caused by *Fa. buski*, a fluke that lives in the small intestine of its definitive host. *Fa. buski* is elongated, oval, and approximately 50 mm × 20 mm in size. *Fa. buski* is found in southeastern and eastern Asia. Infection with *Fa. buski* occurs after oral ingestion of metacercariae adhered to raw water plants and their fruits such as water nuts and walnuts or ingestion of contaminated surface water. Pigs are the definitive hosts of *Fa. buski* and may be asymptomatic or may present with gastrointestinal disease such as weight loss, lethargy, abdominal pain, and diarrhea. Dogs may also be infected with a similar clinical picture. Many people infected with *Fa. buski* are asymptomatic. People with moderate infection may have occasional loose stools, weight loss, lethargy, and generalized abdominal pain. People with severe infection have diarrhea alternating with constipation, hunger pangs, facial edema, anemia, and ascites. Diagnosis is made by fecal examination of operculated eggs. Fasciolopsis in animals is treated with praziquantel, diminazene aceturate, and imidocarb dipropionate. Fasciolopsis in people is treated with praziquantel or triclabendazole (not available in United States). Human and animal disease can be prevented by adequate waste removal of human and pig feces as well as not using human or pig feces for fertilizer. Snail control may be attempted. Human disease can be prevented by immersing vegetables in boiling water for a few seconds to kill the metacercariae. Hands may become contaminated while handling the contaminated plants and should be thoroughly washed in clean water. People should avoid cracking water nuts and chestnuts with their teeth.

Schistosomiasis

Overview

Schistosomiasis is the second most prevalent tropical disease in the world (first is malaria) with 200 million people infected (120 million are symptomatic) and 600 million people in 74 countries at risk of contracting the disease. Schistosomiasis, mainly caused by three species of *Schistosoma* blood trematodes, is also known as bilharzia in honor of Theodore Bilharz who first identified the etiological agent in Egypt in 1851. *Schistosoma* is derived from the Greek *skhizein* meaning to split or cleft and *soma* meaning body describing the male fluke, which has a deep cleft extending the length of his body where the female is held during copulation.

Schistosomiasis has been known since antiquity dating from about 1200 B.C. in Egyptian mummies to the curse Joshua placed on Jericho in the Bible. The first Europeans to contract schistosomiasis were surgeons in Napoleon's army in Egypt in 1799. These surgeons observed many cases of hematuria but the cause was not identified until Bilharz identified and named the parasite *Distomum haematobium* in 1851. In 1858, Weinland changed the name to *Schistosoma* and 3 months later it was named *Bilharzia* by Cobbold. Although many people still identify the parasite as *Bilharzia*, its official name is *Schistosoma*.

The three main species that cause disease are *Sc. haematobium*, *Sc. mansoni*, and *Sc. japonicum*. *Sc. haematobium* was the single species for some time; however, some infected people had eggs with terminal spines while other people had eggs with lateral spines. In 1905, Sir Patrick Manson proposed that intestinal and urinary bladder schistosomiasis were distinct diseases caused by distinct flukes. Another scientist named Sambon agreed with the two species concept and named the fluke that produced laterally spined eggs *Sc. mansoni*. In 1915 Leiper was working in Egypt and discovered that the laterally spined eggs came from a different snail than the eggs with terminal spines solidifying the argument for two different species of fluke. Meanwhile Japanese scientists were discovering another species on their own and in 1904 Katsurada recognized adult flukes in dogs and cats that resembled those found in humans and named the flukes *Sc. japonicum*. Additional species that are rarer causes of human disease are *Sc. intercalatum*, *Sc. mekongi*, *Sc. mattheei*, and *Sc. bovis*. Schistosomes are also called blood flukes because they live in the vascular system of humans and other vertebrates.

Schistosome cercarial dermatitis is another disease manifestation caused by species of *Schistosoma*. This disease has been known for a long time and has been referred to by a variety of names including swimmer's itch, rice paddy itch, clam diggers itch, sawah (Malaysia), kubure or kobanyo (Japan), hoi con (Thailand), and duckworms (New Jersey). Schistosome cercarial dermatitis has been around as long as mankind and has been recorded as early as the 19th century, particularly in people involved in the logging industry. In 1855 a description of a disease by LaValette producing erythemous maculopapular eruptions is presumed to have been cercarial dermatitis. The disease was first identified in the United States (Douglas Lake, MI) in 1928 by Cort who determined that the cause of swimmer's itch was the larva stage of trematodes whose adult form lives in the blood vessels of birds and mammals (prior to this discovery it was believed that the disease was only caused by human schistosomes). Originally the disease was called schistosome dermatitis by Cort in 1928; however, in 1930, Vogel used the term cercarial dermatitis. Since then cases of cercarial dermatitis have been increasing and occur commonly in people who swim in the Great Lakes and other recreational waters.

Causative Agent

> The pathophysiology of schistosomiasis is a result of the immune response against the schistosome eggs.

Schistosoma spp. are flukes that have considerable sexual dimorphism with males being shorter and stouter than females. Males have a ventral, longitudinal groove called a gynecophoral canal where the female normally resides. The mouth has a strong oral sucker with the suckers of females being smaller and not so muscular. Paired intestinal ceca meet and fuse at the worm's midpoint and continue as a single gut to the posterior end. Variation in the number of testes and length of uterus varies among the species as described in Table 6-12.

The eggs of schistosomes are nonoperculate, possess a spine, and contain a miracidium. The microscopic appearance of the egg allows diagnostic differentiation of the species as described in Table 6-12.

Adult worms live in veins that drain abdominal organs of the host: *Sc. haematobium* lives mainly in the veins of the urinary bladder, *Sc. mansoni* lives in the veins draining the large intestine, and *Sc. japonicum* lives in the veins of the small intestine. *Sc. intercalatum* and *Sc. malayensis* are found in Asia and are rarer causes of schistosomal disease.

Cercarial dermatitis is caused by the cercariae of certain species of schistosomes whose normal hosts are birds and mammals other than humans. These cercariae seem to have a chemotrophic reaction to secretions from the skin and

Table 6-12   Morphology of the Three Most Common Schistosomes in People

Schistosoma Species	Papillae	Size	Number of Testes	Ovary Position	Uterus	Egg
Sc. haematobium	Small tubercles	Male: 10 to 15 mm × 0.8 to 1.0 mm Female: 20 mm × 0.25 mm	4 to 5	Midbody	Averages 50 eggs	Elliptical with sharp terminal spine
Sc. mansoni	Large papillae with spines	Male: 10 to 15 mm × 0.8 to 1.0 mm Female: 20 mm × 0.25 mm	6 to 9	Anterior half	Short; few eggs at a time	Elliptical with sharp lateral spine
Sc. japonicum	Smooth	Male: 12 to 20 mm × 0.5 mm Female: 26 mm × 0.3 mm	7	Posterior to midbody	Long; may contain up to 300 eggs	Oval to spherical; rudimentary lateral spine

are not as host-specific as other types of schistosomes. Species of schistosomes that infect animals are described in the clinical signs in animals section. One species of schistosome often implicated in cases of cercarial dermatitis is *Austrobilharzia variglandis*, whose normal hosts are ducks.

Epizootiology and Public Health Significance

Humans are the main host for the three main species of schistosomes; however, some animal reservoirs exist for each species. *Sc. haematobium* is prevalent in North Africa and is also present throughout Africa and in the Middle East. Urinary schistosomiasis caused by *Sc. haematobium* affects 54 countries in Africa and the eastern Mediterranean. Animal reservoir hosts of *Sc. haematobium* are baboons, monkeys, pigs, and rodents. *Sc. mansoni* is found in Egypt, the Middle East, northern and western South America, the Caribbean islands, and East, Central, and West Africa. Animal reservoir hosts of *Sc. mansoni* are monkeys and rodents. *Sc. japonicum* is found in Southeast and East Asia. *Sc. japonicum* is endemic in China, the Philippines, Indonesia, and Thailand. Animal reservoir hosts of *Sc. japonicum* include domestic ruminants, dogs, and rodents.

> Female *Schistosoma* worms live in the gynecophoral canal (groove) of males whose muscles help the paired worms work their way up into smaller veins where the female deposits eggs.

The number of people with schistosomiasis in the United States is estimated at more than 400,000 infected persons. The appropriate snail intermediate host is not endemic to the United States and most U.S. cases are found in immigrants or in someone who acquired the infection outside the United States. Intestinal schistosomiasis caused by *Sc. mansoni* occurs in 52 countries, including Caribbean countries, eastern Mediterranean countries, South American countries, and most countries in Africa.

Schistosomal cercarial dermatitis occurs worldwide with cases reported from every continent except Antarctica. In the United States, cases are most commonly reported from the Great Lakes region.

Transmission

The infectious forms of schistosomes are the cercariae swimming in water. Invasion of cercariae into the final host is percutaneous.

Pathogenesis

The life cycle of schistosomes involves a sexual stage in the human and an asexual stage in the fresh water snail host (intermediate host). Adult worms mate and lay eggs, which are excreted with feces or urine (depending on the species). When the eggs reach the fresh water, the miracidia (ciliated larva) hatch from the egg and the miracidia penetrate the snail. In the snail the fork-tailed cercariae asexually develop following two sporocyst stages (within 3 to 5 weeks) and leave the snail. The cercariae swim close to the water surface until they meet a suitable host (human or animal), where they adhere to the skin, shed their tail (become schistosomulae), and penetrate the skin within minutes with the help of secreted enzymes. Once inside the host, cercariae travel to the heart, the lungs, and through the systemic circulation to reach the portal veins where they develop into adult worms. They then travel to their final destination of intestinal veins (*Sc. mansoni* and *Sc. japonicum*) or urinary bladder veins (*Sc. haematobium*). The time between cercariae penetration and the first egg production is 4 to 6 weeks. The spined eggs reach the capillaries where miracidia develop in the eggs. The eggs get stuck in capillaries where they induce granulomatous reactions. If these reactions occur close to the mucosal surface, eggs break through into the lumen of the organ. Otherwise miracidia die within 3 to 4 weeks and the granulomas get organized by

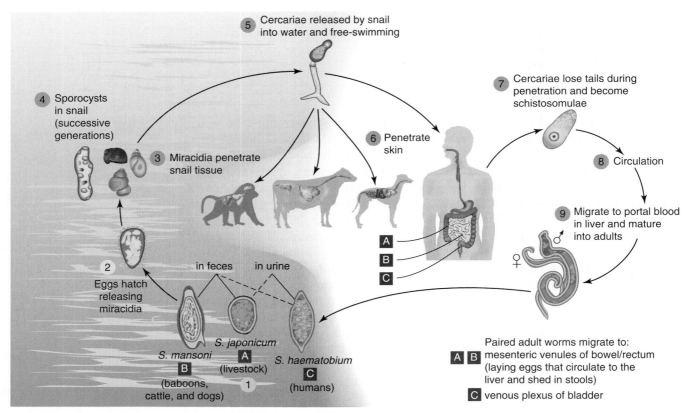

Figure 6-68   Life cycle of *Schistosoma*. Eggs are eliminated with feces or urine❶. The eggs hatch and release miracidia❷, which swim and penetrate specific snail intermediate hosts❸. The stages in the snail include two generations of sporocysts❹ and the production of cercariae❺. Upon release from the snail, the infective cercariae swim, penetrate the skin of the human host❻, and shed their forked tail, becoming schistosomulae❼. The schistosomulae migrate through several tissues and stages before taking up residence in the veins❽,❾. The females deposit eggs in the small venules of the portal and perivesical systems. The eggs are moved progressively toward the lumen of the intestine (*Sc. mansoni* and *Sc. japonicum*) and of the bladder and ureters (*Sc. haematobium*), and are eliminated with feces or urine, respectively❶. Human/animal contact with water is thus necessary for infection by schistosomes.

connective tissue. Humans excrete approximately 50% of the eggs, whereas the rest are trapped in various parts of the body (Figure 6-68).

The life cycle of avian schistosomes varies from other schistosomes in that the cycle is typically maintained in birds and humans are accidental, dead-end hosts. Adult worms (*Trichobilharzia*, *Gigantobilharzia*, *Austrobilharzia*, *Ornithobilharzia*, *Microbilharzia*, and *Heterobilharzia*) are found in the gastrointestinal tract of the bird host. Adult worms release eggs that are shed in the bird's feces and upon immersion in water, a miracidium emerges. Miracidia infect snails and develop into sporocysts, which in turn undergo asexual reproduction producing cercariae. Cercariae swim to the surface of the water to infect another bird. After infecting a bird, the cercaria develops into a schistosomulum and migrates to the gastrointestinal tract where it matures and mates producing eggs starting a new cycle (Figure 6-69).

Clinical Signs in Animals

Most species of pathogenic schistosomes of animals are found in the hepatic portal system. The principal clinical signs seen in animals are associated with passage

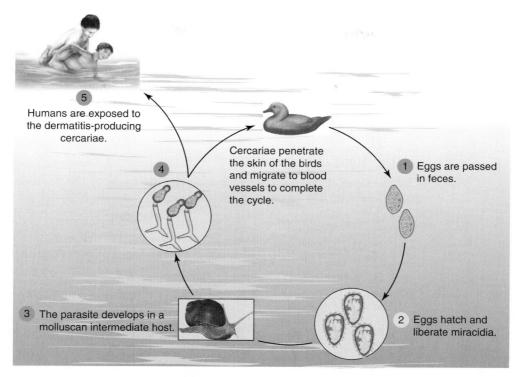

Figure 6-69   Life cycle of avian schistosomes. Adult worms are found in the blood vessels of migratory water birds and produce eggs that are swallowed and passed in the feces❶. On exposure to water, the eggs hatch and liberate a ciliated miracidium that infects a suitable snail (intermediate host)❷. The parasite develops in the intermediate host❸, to produce free-swimming cercariae that are released and penetrate the skin of the birds to complete the cycle❹. Humans are accidental, dead-end hosts; cercariae may penetrate the skin but do not develop further❺. A number of species of dermatitis-producing cercariae have been described from both freshwater and saltwater environments.

of the spined eggs through the tissues to the lumen of the gastrointestinal tract. Species of *Schistosoma* in animals include:

- *Sc. bovis* is the predominant species infecting ruminants in northern and eastern Africa, southern Europe, and the Middle East.
- *Sc. mattheei* is the predominant species infecting ruminants in southern and central Africa.
- *Sc. nasale* found in the veins of the nasal mucosa of ruminants and horses in India, Sri Lanka, and Burma, where it may cause upper respiratory disease.
- *Sc. curassoni*, *Sc. margrebowiei*, and *Sc. leiperi* are other species infecting ruminants in Africa.
- *Sc. spindale*, *Sc. nasale*, *Sc. indicum*, *Sc. incognitum*, and *Sc. japonicum* are widespread in livestock in Asia.
- *Sc. mansoni* can infect rodents, baboon, cattle, and dogs and is widespread in Africa, the eastern Mediterranean, the Caribbean, and South America.
- *Sc. mekongi* can infect dogs and monkeys and is found in Southeast Asia.
- *Sc. intercalatum* is found in cattle, sheep, antelope, and goats in Central Asia.
- Species of avian schistosomes include *Trichobilharzia*, *Gigantobilharzia*, *Austrobilharzia*, *Ornithobilharzia*, *Microbilharzia*, and *Heterobilharzia* and are found worldwide.

Clinical signs associated with the intestinal and hepatic forms of schistosomiasis in ruminants include bloody diarrhea, anemia, and emaciation, which develop after the onset of egg excretion. Severely affected animals may die within a few months of infection, whereas those less heavily infected develop chronic disease that may present as slowed growth. Nasal schistosomiasis is a chronic disease of cattle, horses, and occasionally buffalo producing large amounts of mucopurulent discharge and dyspnea with milder cases frequently asymptomatic.

Clinical Signs in Humans

Clinical signs in people correlate with the life cycle of the parasite. Clinical signs associated with the parasitic form include:

- *Cercariae* (Figure 6-70). The syndrome caused by cercarial penetration of skin is called schistosome cercarial dermatitis (also called swimmer's itch or paddy field dermatitis). Skin penetration of cercariae produces an allergic dermatitis at the site of entry (Figure 6-71). Within minutes to days after swimming in cercariae contaminated water, tingling, burning, or itching of the skin occurs. Vesicles appear within 12 hours after the first clinical signs. Scratching the areas may result in secondary bacterial infections. Itching may last up to a week or more, but is self-limiting. Upon second exposure, a pruritic papular rash occurs. Schistosome cercarial dermatitis is caused by avian schistosomes (*Trichobilharzia, Gigantobilharzia, Austrobilharzia, Ornithobilharzia, Microbilharzia,* and *Heterobilharzia* in North America) found in ducks, geese, gulls, and swans, and by schistosomes of certain aquatic mammals (such as muskrats and beavers). This larval form searches for a suitable host (bird, muskrat), but will burrow into human skin if a suitable host is not found. These larvae cannot develop inside a human and typically die rapidly. Cercarial dermatitis typically occurs on the exposed skin outside of close-fitting garments. Other organisms that can cause cercarial dermatitis are *Sc. spindale* and *Sc. bovis*.

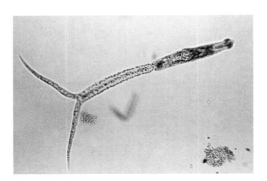

Figure 6-70 A schistosomal cercaria, which is the larva stage of a parasite that causes swimmer's itch.

(Courtesy of CDC/Minnesota Dept. of Health, RN Barr Library, Librarians Melissa Rethlefsen and Marie Jones, Prof. William A. Riley)

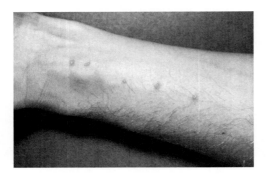

Figure 6-71 Schistosome dermatitis, or "swimmer's itch" occurs when skin is penetrated by an infective cercaria.

(Courtesy of CDC)

- *Schistosomula*. These are tailless cercariae that are transported through blood or lymphatic vessels to the right side of the heart and lungs producing symptoms such as cough and fever.

- *Adult worm*. Adult worms do not multiply inside the body, however, the adult male and female worms mate in the veins producing eggs 4 to 6 weeks after cercarial penetration. Adult worms are rarely pathogenic.

- *Eggs*. Eggs cause Katayama fever (named after an endemic region of Japan) and schistosomiasis. Katayama fever is a condition caused by the high worm and egg antigen stimulus that results from immune complex formation and leads to a serum sickness–like illness, Schistosomiasis, a condition caused by immunologic reactions to *Schistosoma* eggs trapped in tissues stimulating a granulomatous reaction resulting in inflammation, collagen deposition and fibrosis, and organ damage.

Chronic disease and its manifestations depend on the species of schistosome causing infection, the duration and severity of infestation, and the immune response to the eggs.

Sc. mansoni and *Sc. japonicum* cause intestinal tract (fatigue, abdominal pain, diarrhea, and dysentery) and liver disease (abdominal pain, flatulence, anemia, weakness, edema, and melena); *Sc. haematobium* only rarely causes intestinal or liver disease but characteristically causes urinary tract disease (dysuria, urinary frequency, and hematuria) (Figure 6-72). Spread to other organs may also be seen including the respiratory system (cough, wheezing, and low-grade fever), cardiovascular system (palpitations, dyspnea, and hemoptysis), and CNS (seizures, headache, and paresthesia).

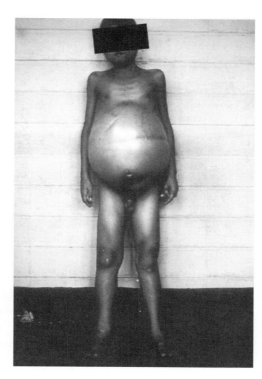

Figure 6-72 Child infected with *Schistosoma japonicum*, showing signs of ascites. (Courtesy of CDC)

Diagnosis in Animals

Schistosomiasis in animals is diagnosed via identification of eggs in feces, rectal scrapings, or nasal mucus of infected animals. Other species of *Schistosoma* that may be found in animals include *Sc. bovis* (eggs are 200 × 60 μm and spindle shaped), *Sc. mattheei* (eggs are 177 × 55 μm and spindle shaped), *Sc. spindale* (eggs are 380 × 70 μm and are elongated and flattened on one side), and *Sc. nasale* (eggs are 455 × 65 μm and are boomerang shaped). At necropsy adult flukes can be found in blood vessels.

> The female adult worm lives for approximately 3 to 8 years and lays eggs throughout her life span.

Diagnosis in Humans

Diagnosis of schistosomiasis is by fecal or urine detection of eggs (Figure 6-73). The urine is most likely to yield positive results for *Sc. haematobium* from 10 A.M until 2 P.M. Quantification of egg excretion determines the severity of infection and is calculated by collecting 24-hour urine or fecal samples, homogenizing the sample, and counting the eggs in a measured sample (less than 100 eggs per gram is a light infection, 100 to 400 eggs per gram is a moderate infection, and greater than 400 eggs per gram is a heavy infection). The egg viability test is used to assess the effectiveness of treatment and involves mixing feces or urine with room temperature distilled water and observing for hatching miracidia.

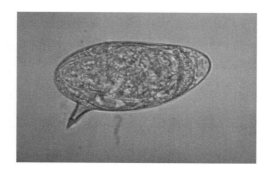

Figure 6-73 *Schistosoma mansoni* egg. (Courtesy of CDC)

Active infection produces viable eggs, whereas treated or past infection results in nonviable eggs and an absence of miracidia. Serologic tests can be used in nonendemic areas using antibody detection methods specific to the worm species (mansoni adult worm microsomal antigen [MAMA], haematobium adult worm microsomal antigen [HAMA], and japonicum adult worm microsomal antigen [JAMA]) have been used in the falcon assay screening test (FAST), enzyme-linked immunoassay (ELISA), and immunoblot assays.

Treatment in Animals

Praziquantel is highly effective treatment in animals and requires two treatments 3 to 5 weeks apart.

Treatment in Humans

Praziquantel is the treatment of choice for all species of schistosomiasis causing systemic disease.

Schistosome cercarial dermatitis can be treated symptomatically with corticosteroid cream or calamine lotion, applying cool compresses to the affected areas, and bathing in epson salts, baking soda, or oatmeal.

Management and Control in Animals

Control measures are mainly used in China, where infected livestock are important reservoirs of human infection. Transmission can be reduced by controlling snails using molluscicides such as niclosamide, by modifying snail habitats, or by fencing off contaminated bodies of water and providing clean drinking water.

Infected snails produce cercariae their entire life.

Management and Control in Humans

People should avoid contact with surface water in endemic areas or areas with snails, towel dry or shower immediately after leaving the water, avoid attracting birds to areas where people swim, and encourage health officials to post signs on shorelines where swimmer's itch is a current problem.

Summary

Schistosomiasis is the second most prevalent tropical disease in the world (first is malaria) with 200 million people infected (120 million are symptomatic) and 600 million people in 74 countries are at risk of contracting the disease. Schistosomiasis is mainly is caused by three species of *Schistosoma* blood trematodes: *Sc. haematobium*, *Sc. mansoni*, and *Sc. japonicum*. Adult worms live in veins that drain abdominal organs of the host: *Sc. haematobium* lives mainly in the veins of the urinary bladder, *Sc. mansoni* lives in the veins draining the large intestine, and *Sc. japonicum* lives in the veins of the small intestine. Humans are the main host for the three main species of schistosomes with *Sc. haematobium* prevalent in North Africa (its reservoir hosts are baboons and other monkeys, pigs, and rodents; some believe it is only found in humans); *Sc. mansoni* prevalent in Egypt, the Middle East, northern and western South America, the Caribbean islands, and East, Central, and West Africa (its reservoir hosts are monkeys and rodents); and *Sc. japonicum* prevalent in Southeast and East Asia (its reservoir hosts include domestic ruminants, dogs, horses, cats, swine, deer, and rodents). The infectious forms of schistosomes are the cercariae swimming in water. Invasion of cercariae into the final host is percutaneous. The life cycle of schistosomes involves a sexual stage in the human and an asexual stage in the fresh water snail host. The cercariae swim close to the water surface until they meet a suitable host (human or animal), where they adhere to the skin, shed their tail, and penetrate the skin. Once inside the host, cercariae travel through veins to a variety of organs where they develop into adult worms. Species of *Schistosoma* in animals include *Sc. bovis*, *Sc. mattheei*, *Sc. nasale*, *Sc. curassoni*, *Sc. margrebowiei*, *Sc. leiperi*, *Sc. spindale*, *Sc. nasale*, *Sc. indicum*, *Sc. incognitum*, and *Sc. japonicum*. Cercarial dermatitis (swimmer's itch) is caused by several genera of avian schistosomes including *Trichobilharzia*, *Gigantobilharzia*, *Austrobilharzia*, *Ornithobilharzia*, *Microbilharzia*, and *Heterobilharzia*. Clinical signs associated with the intestinal and hepatic forms of schistosomiasis in ruminants include bloody diarrhea, anemia, and emaciation. Clinical signs in people correlate with the life cycle of the parasite and are associated with the parasitic form: cercariae (cercarial dermatitis), schistosomula (producing symptoms such as cough and fever), adult worm (rarely pathogenic), and eggs (causing Katayama fever and schistosomiasis). Schistosomiasis in animals is diagnosed via identification of eggs in feces, rectal scrapings, or nasal mucus of infected animals. Diagnosis of schistosomiasis in people is by fecal or urine detection of eggs, quantification of egg excretion, and egg viability tests. Serologic tests such as the falcon assay screening test (FAST), enzyme-linked immunoassay (ELISA), and immunoblot assays are also available. Praziquantel is the treatment of choice for all species of schistosomiasis. Control measures include controlling snails using molluscicides such as niclosamide, by modifying snail habitats, or by fencing off contaminated bodies of water and providing clean drinking water. People should avoid contact with surface water in endemic areas or areas with snails, towel dry or shower immediately after leaving the water, avoid attracting birds to areas where people swim, and encourage health officials to post signs on shorelines where swimmer's itch is a current problem. (For less common zoonotic trematodes, see Table 6-13.)

ARTHROPOD DISEASE

Overview

Arthropods have been known since ancient history with fossils present in the geological record as far back as the Paleozoic age. The phylum Arthropoda contains over 80% of all animal species making it the largest group of living organisms. Arthropod is derived from the Latin *arthro* meaning joint and *podos* meaning foot, which describes one of this groups key evolutionary features that allow these organisms to move quickly from one place to another. Arthropods are characterized by jointed chitinous exoskeletons (protects the organism and provides an area for muscle attachment), segmented bodies (compartmentalization and specialization of function), and jointed appendages (rapid movement). Arthropods have a true coelom (filled with fluid that supplies nutrients), a small brain, an extensive nervous system, and separate sexes. Any of numerous invertebrate animals of the phylum Arthropoda is divided into three classes: insects (95% of all arthropods), arachnids, and crustaceans. Insects have three body regions (head, thorax, and abdomen), three pairs of legs (6 total), and highly specialized mouthparts. Examples of insects are mosquitoes, flies, true bugs, lice, and fleas. Arachnids have two body regions (cephalothorax and abdomen), four pairs of legs (8 total), and mouthparts that are used for capturing and tearing. Examples of arachnids are ticks, mites, spiders, and scorpions. Crustaceans are typically aquatic arthropods that have a pair of appendages associated with each segment (such as crabs and crayfish). Crustaceans such as *Cyclops* spp. are intermediate hosts of parasites such as *Di. latum* and *Paragonimus westermani*. These diseases were covered in previous chapters and will not be covered here.

Table 6-13   Less Common Zoonotic Trematodes

Trematode	Disease	Type of Trematode	Predominant Signs in People	Clinical Signs in Animals	Transmission	Animal Source	Geographic Distribution	Pathology	Diagnosis	Treatment
Clonorchis senensis (named for its branched testicles; *klon* is Greek for branch, *orchis* is Greek for testicle, and *sinos* for oriental); also known as the Chinese liver fluke	Clonorchiasis First discovered in the bile ducts of a Chinese carpenter in Calcutta in 1875	Liver fluke Transparent, lancet-shaped fluke that is 1 to 5 mm by 8 to 25 mm. It has paired testes in the posterior part of its body. Eggs contain a well-developed miracidium.	Minor infection is common and may be asymptomatic. Inflammation of the bile ducts with the severity depending on the number of flukes and the duration of their persistence. Symptoms include fatigue, anorexia, jaundice, and gastrointestinal problems. Persistent infections produce liver cirrhosis, hepatomegaly, edema, and ascites.	■ Dogs and cats are reservoir hosts ■ Clinical signs in dogs and cats include biliary disease and pancreatitis because of pancreatic duct obstruction; carcinomas of the bile and pancreatic ducts have been seen in chronic cases.	People are infected by ingesting raw or undercooked freshwater fish (mainly carp). Feces of infected vertebrates containing eggs can contaminate surface water.	Snails are the first intermediate host. Freshwater fish such as carp are the second intermediate host. Dogs, cats, and wild fish-eating mammals are reservoir hosts.	■ Endemic and widely distributed in Japan, Korea, China, Taiwan, and Vietnam ■ Prevalence rates in humans in these areas is up to 50%	Infection occurs when an animal or person eats raw or partially cooked freshwater fish or dried, salted, or pickled fish infected with the metacercariae (the developmental form in encapsulated cercariae). In the duodenum, the cyst is digested and an immature larva is released. The larva enters the biliary duct, where it develops and matures into an adult worm. The adult worm feeds on the mucosal secretions and lays fully embryonated operculated eggs (which are excreted in the feces). Upon reaching freshwater and upon ingestion by a suitable species of snail (first intermediate host), the eggs hatch to produce a miracidium. Inside the snail, the miracidia multiply asexually through a single generation of sporocysts and two generations of rediae to fork-tailed cercariae. The cercariae escape from the snail to the water and penetrate under scales of freshwater fish (second intermediate host). In the fish, the cercariae lose their tails and encyst in the scale or muscle of the fish to the metacercariae (the form, which is infectious to humans and animals). When ingested, the infected fish cause infection in humans and animals.	Identification of eggs in feces and duodenal secretions. Concentration techniques (such as formol-ether sedimentation) may be needed due to the small size of the eggs. Serologic tests are available in people.	Praziquantel or albendazole. Control of the disease can be achieved by cooking fish or the use of gamma irradiation of fish to kill the metacercariae. Fecal contamination of water should also be avoided.

Heterophyes heterophyes	Heterophyiasis Also known as metagonimiasis or intestinal dwarf fluke infection	Intestinal fluke Small flukes that are less than 1 mm in size	People are typically asymptomatic; only rarely do people show gastrointestinal symptoms (diarrhea and abdominal pain) and even rarer do they show extraintestinal signs (infarcts to the heart and CNS)	■ Main hosts are piscivorous birds and mammals such as dogs and cats that are usually asymptomatic	Ingestions of raw or undercooked fish (second intermediate hosts) that contain metacercariae	Spores are common in the environment	■ Worldwide, but found predominantly in tropical climates of Southeast and East Asia, Northern Africa, and the Middle East	Eggs are shed with feces and are taken up by freshwater snails. Cercariae emerge from the snails, invade freshwater fish, and encyst to metacercariae. After ingestion by the final host they are released in the intestinal tract, enter the mucosa, and develop into adult flukes in 1 to 2 weeks.	Microscopic examination of operculated, brownish eggs of 15 to 30 μm in stool	Praziquantel or niclosamide Disease can be prevented by avoiding consumption of raw or undercooked fish.
Opisthorchis felineus; Opisthorchis viverrini	Opisthorchiasis	Liver fluke Eggs are small (15 to 30 μm), operculated, and yellow-brown in color.	Liver disease similar to that seen with clonorchiasis: cholangitis and in later stages liver cirrhosis and pancreatitis	■ Mild to severe fibrosis of the bile ducts and gallbladder	Ingestion of raw or undercooked freshwater fish containing metacercariae	Snails are the first intermediate host. Fish are the second intermediate host. Dogs, cats, and wild fish-eating mammals are reservoir hosts.	■ Worldwide with prevalence in Central Europe	Eggs are released with the miracidia (larvae) infected feces of infected hosts. Miracidia must reach water for further development; in water they are ingested by snails (first intermediate host). In the snails several asexual replications take place producing cercariae. Cercariae leave the snails and enter fish (second intermediate host). In the fish the parasites encyst to become metacercariae. Metacercariae are ingested by suitable hosts where young flukes are released into the duodenum. From the duodenum they migrate to the bile ducts, where they mature to adults.	Direct detection in feces using zinc sulfate floatation. Serologic tests are available.	Praziquantel Disease can be controlled by eating properly cooked fish or fish that have been irradiated.

(Continued)

Table 6-13 (Continued)

Trematode	Disease	Type of Trematode	Predominant Signs in People	Clinical Signs in Animals	Transmission	Animal Source	Geographic Distribution	Pathology	Diagnosis	Treatment
Paragonimus westermani and eight other known species	Paragonimiasis, pulmonary distomatosis	Lung fluke Eggs are 60 to 90 μm and are operculated	Migrating parasites cause inflammation of the peritoneum, pleura, and lungs producing coughing, fever, and chest pain	■ Chronic, deep, intermittent cough, and lethargy in dogs, cats, and other animals	Ingestion of undercooked meat of freshwater crabs and other crustaceans	Snails are the first intermediate host. Crustaceans are the second intermediate host. Dogs, cats, and wild fish-eating mammals are reservoir hosts.	■ Central, Southeast, and East Asia; China; Japan; Far East ■ *Paragonimus kellicotti* occurs mainly in North America	Adult parasites live in pairs in cysts in the connective tissue of lung. Eggs are coughed up, swallowed, and excreted in feces. Eggs reach water for further development to miracidia. Miracidia hatch and enter snails (first intermediate host). After asexual replication, cercariae develop and leave the snails to enter crustaceans (second intermediate host). In crustaceans the parasite encysts in muscles and internal organs. The final hosts (humans and other mammals) ingest the contaminated meat, young flukes are liberated in the gastrointestinal tract, penetrate the intestinal mucosa, and migrate to the peritoneal cavity. In the lung the parasites become encapsulated by connective tissue and begin egg laying.	Direct detection in sputum and occasionally feces using zinc sulfate floatation. Serologic tests are available.	Praziquantel or triclabendazole Surgery may be needed in some cases. Disease can be controlled by eating properly cooked crustaceans.

Arthropods are involved in nearly every kind of parasitic relationship. Arthropods may affect human and animal health directly (by bites, stings, or infestation of tissues) or indirectly as vectors, definitive hosts, and intermediate hosts (disease transmission). The direct effects of arthropod bites are typically irritation, pruritus, and secondary infections. The most significant diseases produced by arthropods are through vector-borne transmission. Arthropods can serve as means of biological transmission (blood-feeding arthropods such as mosquitoes and ticks acquire the pathogen, it multiplies within the arthropod vector, and is transmitted when the arthropod takes a blood meal) or mechanical transmission (arthropods physically carry pathogens from one host or place to another).

Causative Agents

Adult Diptera

Arthropods in the order Diptera are commonly known as true flies and include insects such as mosquitoes, blackflies or gnats, midges, sand flies, tsetse flies, and house flies. The name Diptera was coined by Aristotle in the 4th century B.C. and is derived from the Greek words *di* meaning two and *ptera* meaning wings in reference to the fact that true flies have only a single pair of wings (two total). The Diptera are divided into three suborders: Nematocera (small, delicate insects with long, multisegmented antennae), Brachycera (compact, robust flies with short, stylate (pointed) antennae), and Cyclorrhapha (compact, robust flies with short, aristate (flagellum-like) antennae). Flies can be found worldwide except Antarctica. Many species are particularly important as vectors of disease. The earliest fossil flies are from the Mesozoic geological period (225 million years ago) and over time they have become one of the largest most diverse groups of organisms.

Flies have a life cycle that involves a major change from a soft-bodied, wingless larval stage to a hardened, winged adult. The larvae typically have a variety of common names such as wriggler and maggot. Fly larvae of different species have different feeding habits with some species having very precise requirements. Many fly larvae consume decaying organic matter, whereas others are parasitic on other insects and organisms. Adult flies are typically free-living, fly during the day or night depending on the species, and consume liquid food (nectar) or decomposing organic matter.

The major morphological feature, which distinguishes flies from other insects is their reduced hind wings (halteres). The halteres are small, club-like structures that help the fly balance during flight. Adult flies have only one pair of functional wings (the forewings). The mesothorax of Diptera species has become enlarged to contain the enormous flight muscles as a result of their reliance on the forewings for flight. The mouthparts of flies may be modified for stabbing and piercing other insects. Table 6-14 summarizes some diseases caused by adult Diptera arthropods.

Larval Diptera or Myiasis

Myiasis, also known as fly-strike or fly-blown, is the infestation of live vertebrate animals with dipteran larvae (first-stage that develop into third-stage) that feed on the host's dead or living tissue. Flies causing myiasis can be obligate parasites that can develop only on live hosts or facultative parasites that can develop on either live hosts or organic debris. The adult flies are not parasitic, but when they lay their eggs in open wounds the eggs hatch into their larval stage (maggots or grubs), the larvae feed on live and/or necrotic tissue, causing myiasis to develop

Table 6-14 **Zoonotic Diseases Caused by Diptera Adult Arthropods**

Diptera Suborder	Scientific Name	Features	Disease Transmitted	Bite Reaction	Prevention
Nematocera	■ *Culicidae* mosquitoes (*Anopheles, Culex, Aedes*) (Figure 6-76)	■ Mosquitoes need stagnant or slowly moving water for breeding ■ Eggs may be deposited on damp soil or vegetation, tree holes, containers, pools, swamps, or directly in water.	■ Equine encephalitis, dengue fever, yellow fever, and a variety of viral diseases	■ Bites cause local signs such as swelling, pruritus, pain, and erythema.	■ Insect repellent ■ Long pants and sleeves ■ Bed netting ■ Remove water source for breeding ■ Insecticides for buildings
	■ *Phlebotomidae* sand flies (*Phlebotomus*)(Figure 6-74) ■ *Simuliidae* blackflies (*Simulium, Odagmia*) (Figure 6-75)	■ Sand flies are hairy insects with long legs and are inactive during the day. ■ Blackflies need running water for breeding sites (breeding occurs in temperate zones in later spring and early summer). ■ Blackflies fly long distances and attack during the day in the open, but will not enter dwellings.	■ Leishmaniasis ■ Onchocerciasis		
	■ Ceratopogonidae midges (*Culicoides*) (Figure 6-77)	■ Midges are small and lay eggs in damp places (edge of ponds and swamps).	■ Loiasis		
Brachycera	■ *Tabanidae* (horseflies) arthropods (*Tabanus, Haematopota, Chrysops*)	■ Blood-sucking insects ■ Females bite and feed on warm-blooded animals ■ Most live near waters ■ Strong fliers ■ Breed in moist earth or leaf mold	■ Tularemia ■ Trypanosomiasis	■ Bites cause relatively large, painful bite wounds and secondary hemorrhages	■ As above ■ Avoid infested areas during the daytime
Cyclorrhapha	■ *Glossinidae* arthropods (*Glossina*)	■ *Glossina* are found only in Africa south of the Sahara (other Glossinidae are found worldwide). ■ Many species of *Glossina* are bloodsucking, but only a few attack humans.	■ Trypanosomiasis	■ Bites are not immediately recognized but in time become tender, hard swellings; some may produce anaphylactic reaction	■ As above ■ Avoid infested areas during the daytime

Note: Mosquitoes and flies find their hosts by chemical signals such as CO_2 or valerianic acid in sweat.

Figure 6-74 Female *Phlebotomus* spp. sandfly, a vector of the parasite responsible for Leishmaniasis.

(Courtesy of CDC/World Health Organization)

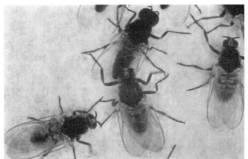

Figure 6-75 *Simulium* spp. of flies, or "black flies," a vector of the disease onchocerciasis also known as river blindness.

(Courtesy of CDC/World Health Organization)

Figure 6-76 *A female* Aedes aegypti *mosquito.*

(Courtesy of CDC/James Gathany)

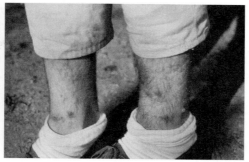

Figure 6-77 *Erythematous papules after having been bitten by midges.*

(Courtesy of CDC/Margaret Parsons)

(Figure 6-78). Accidental infestations with fly larvae can occur if eggs or larvae are inhaled or swallowed inadvertently with food and are referred to as pseudomyiasis. Myiases are often classified according to the area of the animal that the larvae infest or the appearance of the lesion they cause. When larvae are deposited into open wounds the myiasis is known as traumatic, when the appearance is boil-like the myiasis is termed furuncular,

Figure 6-78 *Third instar screwworm larvae demonstrating their dark tracheal tubes.*

(Courtesy of Foreign Animal Diseases, "The Grey Book," USAHA)

when the larvae migrate beneath the skin and its path can be traced the myiasis is called creeping, and when the larvae are bloodsucking the myiasis is called sanguinivorous.

Most myiasis-causing flies belong to one of three major families: Oestridae (botflies), Sarcophagidae (fleshflies), or Calliphoridae (blowflies) (Figure 6-79). The

Figure 6-79 *Horse botfly larvae are parasitic maggots that attach themselves to the stomach lining of horses.*

(Courtesy of Leland S. Shapiro)

family Oestridae has approximately 150 species of flies. All botflies (*Gastrophilus* spp.) cause myiasis and are obligate parasites typically with a high degree of host specificity. The family Sarcophagidae has only two genera of medical importance: *Sarcophaga* and *Wohlfahrtia*. Members of Sarcophagidae are larviparous (deposit first-stage larvae instead of laying eggs). The family Calliphoridae can be divided into two groups: the nonmetallic flies such as the Congo floor maggot and tumbu fly and the metallic flies such as the blowflies (bluebottles and greenbottles). The zoonoses caused by Diptera myiasis are summarized in Table 6-15.

Fleas

Fleas belong to the order Siphonaptera, one of the major groups of blood-sucking insects with more than 2,500 species and subspecies of fleas. The name Siphonaptera is derived from the Greek words *siphon* meaning a tube or pipe and *aptera* meaning wingless. Adult fleas are obligatory hematophages (blood feeders) that parasitize warm-blooded animals (94% parasitize mammals and 6% parasitize birds). Fleas are wingless insects that have a flattened body. Flea bodies have three major parts: head (triangular in shape with club-shaped antennae and downward pointed mouthparts), thorax (contains three pairs of legs and has three distinct segments called the prothorax, mesothorax, and metathorax), and abdomen (contains 10 segments in adults with the caudal segment containing a dorsal sensilium that detects vibrations and temperature changes for host detection). The flea body may be light yellow, yellowish black, brown black, or jet-black in color.

Fleas have four phases of development: the egg, the free-living larva, the pupa, and the adult. Eggs are deposited in debris such as nests, burrows, cracks, or crevices (sometimes they are deposited on the host where the eggs typically fall to the ground) (Figure 6-80). Eggs hatch in about 5 days depending on species, temperature, and humidity. A legless, wormlike larva with a small blackish eyeless head containing

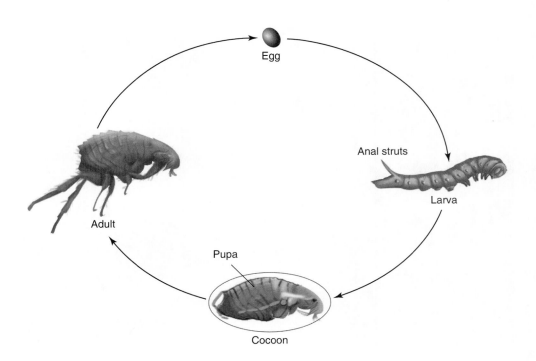

Figure 6-80 Life cycle of a flea.

Table 6-15   Zoonotic Diseases Caused by Diptera Larvae

Type of Myiasis	Scientific Name	Animal Host	Features	Treatment and Prevention
Dermal, cutaneous, or subdermal myiasis	■ *Cordylobia anthropophaga* (Mango fly) ■ *Cordylobia rodhaini* (New World screwworm) ■ *Cochliomyia hominivorax* (New World screwworm) ■ *Chrysomya bezziana* (Old World screwworm) ■ *Dermatobia hominis* (human botfly)	■ Dogs (Africa) ■ Rodents (Africa) ■ Cattle (Central and South America) ■ Cattle (Africa, Asia, and Middle East) ■ Cattle mainly; other mammals (Central and South America)	■ Larvae burrow several centimeters deep causing subdermal lesions on various parts of the body. Larvae develop from first to third stage in 5 to 8 days. ■ Larvae cause painful lesions in skin typically on head, back, abdomen, arm, thigh, axilla, or orbit of eye. ■ Larvae feed for 4 to 12 weeks on host.	■ Remove maggots ■ Treat any secondary bacterial infection ■ Personal hygiene ■ Male flies may be irradiated so that they are sterile and cannot fertilize females
Wound myiasis	■ *Lucilia sericata* and *Lucilia cuprina* ■ *Sarcophaga* spp. (fleshflies) ■ *Calliphora* spp. (blowflies)	■ Sheep (worldwide) ■ Facultative myiasis organism (worldwide) ■ Facultative myiasis organism (worldwide)	■ Adults deposit eggs (*Sarcophaga* spp. are larviparous and deposit first stage larvae) into wounds and under bandages where larvae stage for approximately 5 days during which time they feed on necrotic tissue and secrete bacteriostatic chemicals. ■ May be used therapeutically.	■ Remove maggots ■ Treat any secondary bacterial infection ■ Personal hygiene
Ophthalmomyiasis (ocular myiasis)	■ *Oestrus ovis* (botflies) ■ *Rhinoestrus purpurea* ■ *Wohlfahrita magnifica* (Old World fleshflies) ■ *Wohlfahrita nuba*	■ Sheep (worldwide) ■ Equine (Europe, Africa, and Asia) ■ Rats and rabbits (Mediterranean) ■ Camels (Africa to Pakistan)	■ Adult flies deposit eggs or larvae on eyelids, in the conjunctival sac, or on adjacent skin. ■ Clinical signs appear quickly producing pain and irritation. ■ Clinical signs range from conjunctivitis to anterior uveitis. ■ In natural hosts maggots invade the nasopharynx.	■ Remove maggots (may need to be surgically removed) ■ Treat any secondary bacterial infection ■ Personal hygiene
Creeping myiasis	■ *Gasterophilus* spp. (horse botflies) ■ *Hypoderma* spp. (cattle botflies)	■ Equine (worldwide) ■ Cattle (worldwide)	■ Hatched larvae penetrate the subcutaneous tissue and migrate. ■ In definitive hosts larvae migrate extensively, develop in subcutaneous boils (cattle botflies) or in the gastrointestinal tract (equine botflies) after being licked off fur by host. ■ In humans, larvae penetrate skin, cause swelling, and migrate in the epidermis where they persist for long periods of time.	■ Remove maggots ■ Treat any secondary bacterial infection ■ Personal hygiene

(Continued)

Table 6-15   (Continued)

Type of Myiasis	Scientific Name	Animal Host	Features	Treatment and Prevention
Sanguinivorous myiasis	▪ Auchmeromia luteola (Congo floor maggot)	▪ Pigs (Africa)	▪ Larvae cause skin lesions and feed on blood (mainly at night) ▪ Larvae repeatedly attack requiring 5 to 20 blood meals for development during a 10-week period	▪ Remove maggots ▪ Treat any secondary bacterial infection ▪ Personal hygiene
Auricular myiasis	▪ Wohlfahrtia spp. ▪ Cochliomyia hominivorax ▪ Musca domestica	▪ Rodents (worldwide) ▪ Cattle (Central and South America) ▪ Facultative myiasis organism (worldwide)	▪ Larvae burrow several centimeters deep causing subdermal disfiguring lesions and discharge in the auditory canal and nasal cavity (may extend to bone, cartilage, and brain)	▪ Remove maggots ▪ Treat any secondary bacterial infection ▪ Personal hygiene
Urogenital and rectal myiasis	▪ Calliphora spp. (bluebottle flies) ▪ Fannia spp. ▪ Musca domestica	▪ Sheep; many animal species; facultative myiasis organism (worldwide) ▪ Facultative myiasis organism (worldwide) ▪ Facultative myiasis organism (worldwide)	▪ Larvae invade the vagina, ureter, or rectum	▪ Remove maggots ▪ Treat any secondary bacterial infection ▪ Personal hygiene
Gastrointestinal myiasis, enteric myiasis, or pseudomyiasis	▪ Calliphora spp. ▪ Sarcophaga spp. ▪ Fannia spp. ▪ Musca domestica	▪ Many animals (worldwide) ▪ Facultative myiasis organism (worldwide) ▪ Facultative myiasis organism (worldwide) ▪ Facultative myiasis organism (worldwide)	▪ Eggs or larvae are ingested with contaminated food ▪ Larvae survive transport through the gastrointestinal tract causing irritation, intestinal pain, vomiting, and diarrhea ▪ Larvae may cause ulceration ▪ Larvae may be found in vomit or feces	▪ Remove maggots ▪ Treat any secondary bacterial infection ▪ Personal hygiene

Note: Facultative myiasis organisms can feed on organic debris.

small antennae emerges from the egg. Larvae are very active, avoid light, cannot tolerate humidity extremes, and seek protection in cracks, crevices, burrows, and nest debris. Larvae feed on organic debris with the larval stage lasting approximately 2 to 3 weeks. There are typically three larval instars (stages). At the end of the larval period the larva empties the alimentary canal and spins a white silken cocoon from silk produced by its salivary glands. After 2 to 3 days in the cocoon the larva pupates. Adults emerge from the pupa after approximately 5 to 14 days depending on temperature. The adult flea requires a stimulus (vibrations or chemicals) to cause it to escape from the cocoon. Adult fleas avoid light and are therefore found among animal hair or feathers or on people's clothing (Figure 6-81).

Figure 6-81 A larva (left) and adult female (right) *Xenopsylla cheopis* flea.
(Courtesy of CDC/World Health Organization)

Fleas are attracted by vibrations, sound, or chemical signals. Most species of fleas have preferred hosts, but they are not entirely host-specific. Typically people are not attacked by fleas if there is a more suitable host available. Fleas move by jumping (fleas can jump up to 150 times their body length). Ideally fleas feed daily for 2 to 15 minutes, but may survive for months between feedings. Fleas are capillary feeders that produce bites frequently occurring in groups or on a line. The bites are pruritic within a few minutes. Hypersensitivity reactions may occur and severe itching may produce large papules. Secondary infections may also occur as a result of intense scratching. Flea bites can be treated with anti-itch creams or systemic antipruritic drugs such as antihistamines or corticosteroids. Antibiotics may be used to treat secondary infections. Fleas and flea bites can be prevented by avoiding dark places where fleas prefer to live, by controlling fleas on animals, and environmental hygiene. Zoonoses caused by fleas are summarized in Table 6-16.

> Both sexes of fleas take blood meals and are equally important a vectors of disease.

Heteroptera

The suborder Heteroptera is a diverse insect group with approximately 30,000 species. The name Heteroptera is derived from the Greek *hetero* meaning

Table 6-16   Zoonotic Diseases Caused by Fleas

Scientific Name	Host	Disease Transmitted
■ *Ctenocephalides canis*	■ Dogs, humans, foxes, mammals	■ *Diphylidium* larvae
■ *Ctenocephalides felis*	■ Cats, dogs, humans	■ Cestode larvae
■ *Pulex irritans*	■ Humans, pigs, dogs, cats	■ *Yersinia pestis*, cestode larvae
■ *Ceratophyllus gallinae*	■ Poultry, humans, cats	■ Mechanical transmission of pathogens
■ *Echidnophaga gallinacean*	■ Chickens, dogs, humans	■ Mechanical transmission of pathogens
■ *Spilopsyllus cuniculi*	■ Rabbits, humans	■ *Francisella tularensis*
■ *Nosopsyllus fasciatus*	■ Rats, rodents, humans	■ *Yersinia pestis*, *Rickettsia typhi*, bacteria
■ *Archaeopsylla erinacei*	■ Hedgehogs, humans, mammals	■ Mechanical transmission of pathogens
■ *Xenopsylla cheopis*	■ Rats, mice, humans, domestic animals	■ Cestode larvae, *Yersinia pestis*, rickettsiae
■ *Tunga penetrans*	■ Humans, large mammals	■ Bacteria

Note: Fleas are temporary ectoparasites of humans and animals and transmitters of disease.

Figure 6-82 The common bedbug, *Cimex lectularius*, uses its mouth parts to pierce skin and obtain its blood meal.
(Courtesy of CDC/World Health Organization)

Figure 6-83 *Triatoma infestans*, also known as the kissing bug and assassin bug, is a vector for Chagas' disease.
(Courtesy of CDC/World Health Organization)

different and *ptera* meaning wings referring to the difference in texture of the front wings (leathery) from those at the apex (membranous). At rest, these wings cross over one another to lie flat along the insect's back. Heteroptera insects also have elongate, piercing mouthparts, which arise from either the ventral (hypognathous) or anterior (prognathous) part of the head. Antennae are slender with 4 to 5 segments. The mandibles and maxillae interlock with one another and are long and thread-like to form a flexible feeding tube (proboscis). The immature forms look similar to the adults and always are wingless. Heteroptera insects live in a broad range of habitats including land, water, and semi-aquatic environments. Bed bugs, members of the ectoparasites family Cimicidae, live as ectoparasites on birds and mammals (including humans) (Figure 6-82). *Triatoma* bugs, members of the family Reduviidae, colonize habitats built by people or nests of animals (Figure 6-83).

Blood-sucking Heteroptera are attracted by warm temperature and CO_2. Blood is required by all stages for molting and egg deposition. Eggs are laid in bunches in cracks and crevices and hatch after about 8 to 15 days. Nymphs emerge from the eggs depending on environmental temperature. Nymphs are similar in appearance to the adults and feed on blood for approximately 10 minutes. There are five nymph stages and the nymphal period lasts 2 to 7 weeks. Adults can survive for up to 4 years and more than 1 year without a bloodmeal (bedbugs) or 4 to 6 months without a bloodmeal (Triatomine bugs). The zoonoses caused by Heteroptera (bugs) are summarized in Table 6-17.

Both sexes of bedbug and kissing bug take blood meals and are equally important as pests.

Table 6-17   Zoonotic Diseases Caused by *Heteroptera* (Bugs)

Common Name	Scientific Name	Host	Disease	Treatment & Prevention
Bedbugs	■ *Cimex lectularius* (worldwide) ■ *Cimex hemipterus* (tropics) ■ *Leptocimex boueti* (West Africa)	■ Mammals, bats, birds, humans	■ Bites on unprotected body parts (face, neck, forearm, leg) develop into severely pruritic, hemorrhagic urticaria within hours ■ Lesions may occur in linear pattern	■ Anti-inflammatory creams to reduce pruritus ■ Insecticidal sprays or foggers with pyrethroids can be used in buildings
Kissing bug, assassin bug, cone-nose bug	■ *Triatoma* spp. (tropics) ■ *Rhodnius* spp. (tropics) ■ *Panstrongylus* spp. (tropics) ■ *Dipetalogaster* spp. (tropics)	■ Marsupials, rodents, carnivores, bats, birds, humans	■ Bites on unprotected body parts (face, neck, forearm, leg) are usually painless and go recognized ■ Triatomine bugs are intermediate hosts and transmit *Trypanosoma cruzi* (Chagas' disease)	■ Anti-inflammatory creams to reduce pruritus ■ Insecticidal sprays or foggers with pyrethroids can be used in buildings

Ticks

Ticks are covered in chapter 4.

Mites

Acariasis is disease caused by a variety of mites. Mites are in the Acari subclass in the Arachnidea class of arthropods that contain approximately 30,000 species worldwide (Figure 6-84). Acari is derived from the Greek word *akari* meaning mite. Mites are not insects and are small with six legs as larvae and eight legs as nymphs and adults.

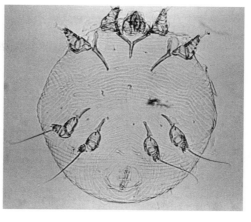

Figure 6-84 A ventral view of a *Sarcoptes scabei* mite. (Courtesy of CDC/World Health Organization)

Mites are wingless, lack antennae, and usually have flat or round bodies. There are numerous free-living, ectoparasitic and endoparasitic species.

Female mites deposit eggs either on the host or in the environment. Eggs hatch into larvae and then typically pass through two nymphal stages before becoming adults. All mites causing acariasis are transmitted by direct contact. Indirect transmission via fomites is important in some, but not all, species of mites (Figure 6-85).

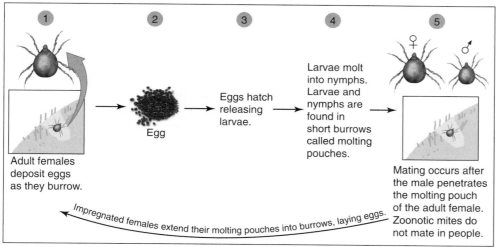

Figure 6-85 Life cycle of *Sarcoptes* mite. *Sarcoptes scabei* undergoes four stages in its life cycle; egg, larva, nymph, and adult. Females burrow through the skin❶ and deposit eggs at 2- to 3-day intervals❷. After the eggs hatch❸, the larvae (three pairs of legs) migrate to the skin surface and burrow into the intact stratum corneum to construct short burrows called molting pouches. After larvae molt, the resulting nymphs have four pairs of legs❹. This form molts into slightly larger nymphs before molting into adults. Mating occurs after the male penetrates the molting pouch of the adult female❺. Impregnated females extend their molting pouches into the characteristic serpentine burrows, laying eggs in the process. The impregnated females burrow into the skin and spend the remaining 2 months of their lives in tunnels under the surface of the skin. Males are rarely seen.

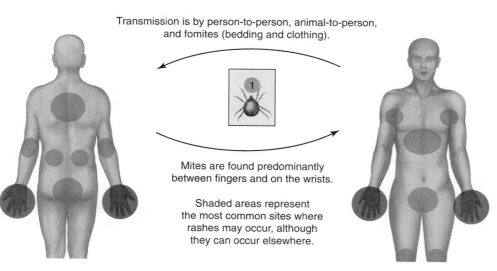

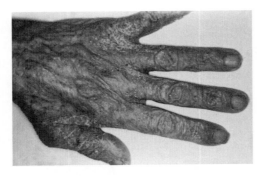

Figure 6-86 The scabies rash manifests itself as pimple-like eruptions.
(Courtesy of CDC)

All acariases are highly contagious for susceptible species. Close contact may be needed to spread mites if their numbers are low. Scabies is the classic mite disease whose name was derived from the Latin word for itch (*scabies*) and the verb *scabere* meaning to scratch (Figure 6-86). Zoonotic mites do not reproduce on people; therefore, the clinical signs they produce are self-limiting. Disease in people tends to be temporary (pruritus and discomfort) and are not fatal. The zoonoses caused by mites are summarized in Table 6-18.

Lice

Demodex mites cause significant disease in puppies (spread from bitch to nursing pups during the first days of life), but are not zoonotic.

Figure 6-87 (A) The sucking louse has a head that is more slender than its thorax, with mouthparts designed to suck its host's tissue, body fluids, and blood. (B) The biting louse has no eyes, and its body is broad and flat. The head of a biting louse is wider than its thorax and as wide as its abdomen. The mouthparts of a biting louse are designed to bite and chew. They feed on debris on the surface of the skin rather than suck like the other group of lice. (Courtesy of Shapiro)

Lice are highly host specific and lice found on animals do not cause pediculosis in people.

Louse infestation is prevalent throughout the animal kingdom. Lice are divided into two orders: Mallophaga (chewing lice, which are common in birds and domestic animals with humans being accidental hosts) and Anoplura (sucking lice, which only parasitize mammals with the three types of human lice belonging to this group). Examples of Mallophaga lice include *Menopon gallinae* (fowl louse), *Menacanthus stramineus* (chicken and turkey louse), *Bovicola* spp. (cattle, horses, sheep, and goat species), *Trichodectes canis* (dog louse), and *Felicola subrostratus* (cat louse). Mallophaga lice are significant pests of animals and birds and may rarely cause short-term irritation to humans (animal lice may live for only a few hours on humans). The three types of human lice are in the Anoplura order producing disease called pediculosis because of the genera of lice involved (*Pediculus humanus capitis* (head louse), *Pediculus humanus corporis* (body louse), and *Pthirus pubis* (crab louse). These lice are not zoonotic, but may rarely be transmitted to animals (particularly swine). Lice can transmit disease such as trench fever (*Ba. quintana*), relapsing fever (*Bo. recurrentis*), and epidemic typhus (*Rickettsia prowazekii*) (Figure 6-87).

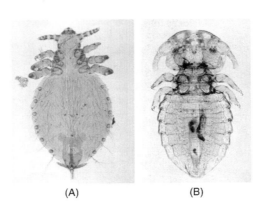

(A)   (B)

Lice are permanent ectoparasites (spend their entire life on the host) and are highly host specific. Lice eggs are called nits and are glued to the host hair (Figure 6-1A). The egg stage lasts approximately 5 to 10 days (longer if the eggs are on clothing or are in a cool environment). The louse hatches from the egg as a nymph and passes through three nymphal instars (stages) before becoming an adult. They nymph stage lasts from 7 to 12 days and is also temperature dependant. The entire louse life cycle is approximately 2 to 3 weeks.

Summary

The phylum Arthropoda contains over 80% of all animal species making it the largest group of living organisms. Arthropods are characterized by jointed chitinous exoskeletons, segmented bodies, and jointed appendages. Arthropods have a true coelom, a small brain, an extensive nervous system, and separate

Table 6-18 Zoonotic Diseases Caused by Mites

Family	Scientific Name	Animal Host	Animal Disease	Human Disease	Treatment and Prevention
Sarcoptidae These are burrowing mites that live in tunnels in the skin. They complete their life cycle on the host (do not survive for long periods in the environment).	■ *Sarcoptes scabiei* Sarcoptic mange is caused by several subtypes of *Sarcoptes* that may have host preference. Subtypes include *Sarcoptes scabiei* var. *hominis* (people), *Sarcoptes scabiei* var.*canis* (dogs), *Sarcoptes scabiei* var.*suis* (pigs), *Sarcoptes scabiei* var.*equi* (horses), *Sarcoptes scabiei* var.*bovis* (cattle), and *Sarcoptes scabiei* var.*ovis* (sheep). Most are found worldwide with some species rarely found in some countries.	■ Dogs, horses, cattle, pigs, humans	■ Dogs: lesions are first found on ventral chest and abdomen. Other affected areas include ears, periorbital area, elbows, and legs. Lesions are pruritic with thick yellowish crusts. ■ Pigs: lesions appear first on the head (periorbital area, snout, and ears) then spread to hind legs. Affected skin is pruritic, red, and inflamed with scabs, erosions, ulcers, or cracks. Skin eventually becomes wrinkles with clinical signs resolving in 12 to 18 weeks. Some pigs may become unthrifty. ■ Cattle: lesions are first found on the head and neck or above the scrotum or udder. Lesions are pruritic and may appear as papules, crusts, and alopecia. Skin may thicken and develop large folds. ■ Sheep: lesions are similar to those in cattle and are typically found on nonwooly skin areas. ■ Horses: lesions first appear on head, neck, and shoulders. Initial lesions are pruritic, papules, and vesicles which later become crusts. Alopecia, crusting, and skin thickening may be seen as the disease progresses.	■ Severe pruritus, papular rash (particularly on shoulders, spaces between fingers and toes, and skin folds), burrows on skin are pathognomonic. Secondary bacterial infections may be seen. ■ Zoonotic scabies is almost always self-limiting with mites disappearing in a few days and clinical signs resolving in 1 to 3 weeks. ■ Host specific species may cause temporary dermatitis in other hosts, but do not reproduce except on their preferred host.	■ Animals: acaricides include lyme sulfur, amitrax, pyrethrins, coumaphos, malathion, rotenone, or carbaryl. Ivermectin and doramectin are effective for treating mites. ■ People: topical lotions containing permethrin or lindane may be used, but most human cases are self-limiting. Ivermectin is also effective. ■ Sarcoptic mange in cattle is a reportable disease in the United States. ■ Isolation of affected animals and proper disinfection is important in preventing disease spread. ■ Zoonotic disease can be prevented by treating affected animals and wearing protective clothing and gloves when handling animals.
Sarcoptidae These mites are burrowing mites with a life cycle similar to *Sarcoptes scabiei*.	■ *Notoedres cati* (worldwide)	■ Felines	■ Notoedric mange is intensely pruritic in most felines. Lesions start on pinna, face, eyelids, dorsum of neck, and paws. Initial papular rash may become red and have areas of partial or complete alopecia. Crusts, thickened skin, and enlarged lymph nodes may also be seen.	■ Presents with signs of scabies but is self-limiting in several weeks.	■ Same as for *Sarcoptes*
Cheyletiellidae These are nonburrowing mites that remain on the surface of skin and feed on scales or tissue fluids/blood. They can survive for short periods in the environment.	■ *Cheyletiella yasguri* (causes clinical disease in puppies, but not adults) ■ *Cheyletiella blakei* ■ *Cheyletiella parasitovorax* (all are found worldwide)	■ Dogs, cats, rabbits, wild mammals	■ Dogs and cats: lesions appear on dorsum as dry, scaly dermatitis with dandruff. Pruritus may be mild to moderate. Occasionally there may be red, excoriated lesions with alopecia. Cats may develop miliary dermatitis. Adults may also be subclinical carriers. ■ Rabbits: clinical signs vary from loose body fur; thick reddish-brown crusts in ears; scaly, red, oily patches on dorsum and head; dandruff; to absence of clinical signs.	■ Humans are aberrant hosts that can get a self-limiting, mildly pruritic dermatitis on abdomen, chest, arms, and legs.	■ Same as for *Sarcoptes*

(Continued)

Table 6-18   (Continued)

Family	Scientific Name	Animal Host	Animal Disease	Human Disease	Treatment and Prevention
Macronyssidae These are nonburrowing mites that can survive for longer periods of time in the environment (hence they can be transmitted by fomites).	■ *Ornithonyssus bacoti* (worldwide) ■ *Ornithonyssus bursa* (tropical areas) *and Ornithonyssus sylviarum* (Australia, New Zealand, and temperate zones of northern hemisphere)	■ Rodents and small marsupials ■ Birds	■ Laboratory and pet rodents: anemia, debility, weakness, pruritus, and death. *Ornithonyssus bacoti* is a vector for murine typhus, Q fever, and plague. ■ Birds: blackened feathers, cracks and scabs around the cloaca, decreased productivity	■ Painful or pruritic lesions may cause irritation and localized dermatitis. Most of these mites cause papules, but may also produce vesicles, urticaria, and hemorrhagic necrosis. ■ *Ornithonyssus bursa* can carry Western equine encephalitis virus.	■ Same as for *Sarcoptes*
Dermanyssidae These are nonburrowing mites. Zoonotic species can spend most of their life in the environment (drop off the host after feeding).	■ *Dermanyssus gallinae* (also known as the chicken mite and poultry mite; found worldwide) ■ *Liponyssoides sanguineus* (Asia, Europe, United States, and northern Africa)	■ Birds ■ Small rodents	■ Poultry: anemia (feeds on blood at night) and lowered productivity (decreased weight gain or egg production) ■ Pet birds: restlessness, anemia, excessive preening, pruritus, and death ■ Rodents: may develop papules in area of bite. Mice and gerbils tend to be asymptomatic.	■ Painful or pruritic lesions may cause irritation and local dermatitis. Most of these mites cause papules, but may also produce vesicles, urticaria, and hemorrhagic necrosis. ■ *Liponyssoides sanguineus* can transmit *Rickettsia kauri* (human rickettsialpox)	■ Same as for *Sarcoptes*
Trombiculidae Also known as chiggers, these mites are only parasitic in their larval stage. These mites deposit their eggs on the ground or on low bushes. Hatched larvae feed on tissue juices then drop to the ground to develop into nymphs.	■ *Eutrombicula* spp. ■ *Neotrombicula* spp. ■ *Schoengastia* spp. ■ *Euschoengastia* spp. ■ *Acomatacarus* spp. ■ *Siseca* spp. ■ *Blankaartia* spp. (all are found worldwide)	■ Mammals and birds	■ Dogs and cats: lesions are variable with bites usually on area that has contact with the ground. Bites result in pruritic papules, alopecia, crusts, and scabs. ■ Horses: lesions are in areas in contact with the ground. Bites lesions are similar to dogs and cats, but wheals and larvae in the lesions may also be seen. ■ Birds: large numbers of mites may produce anorexia, lethargy, and death in birds.	■ Dermatitis and allergic reactions are possible. The first lesion is typically a small red papule with intense pruritus; a wheal develops that may bleed. ■ Some bites remain painful for weeks. ■ Some species of chiggers can transmits scrub typhus (*Orientia tsutsugamushi*).	■ Same as for *Sarcoptes*

Notes: Diagnosis is skin scraping or collection of mite for examination.

Psoroptic mange (*Psoroptes ovis* in the family Psoroptidae) only affects domestic animals such as sheep and cattle and is not zoonotic. Psoroptic mange is a reportable disease in the United States.

sexes. The phylum Arthropoda contains three classes: insects, arachnids, and crustaceans. Arthropods may affect human and animal health directly (by bites, stings, or infestation of tissues) or indirectly as vectors, definitive hosts, and intermediate hosts (disease transmission). Arthropods in the order Diptera are commonly known as true flies and include insects such as mosquitoes, blackflies or gnats, midges, sand flies, tsetse flies and house flies. Diptera adults cause zoonotic disease by serving as vectors of a variety of diseases. Myiasis is the infestation of live vertebrate animals with dipteran larvae (first-stage that develop into third-stage) that feed on the host's dead or living tissue. The adult flies are not parasitic, but when they lay their eggs in open wounds the eggs hatch into their larval stage (maggots or grubs), the larvae feed on live and/or necrotic tissue, causing myiasis to develop. Most myiasis-causing flies belong to one of three major families: Oestridae (botflies), Sarcophagidae (fleshflies), or Calliphoridae (blowflies). Fleas belong to the order Siphonaptera, one of the major groups of blood-sucking insects with more than 2,500 species and subspecies of fleas. Adult fleas are obligatory hematophages that parasitize warm-blooded animals (94% parasitize mammals and 6% parasitize birds). Most species of fleas have preferred hosts, but they are not entirely host-specific. Typically people are not attacked by fleas if there is a more suitable host available. The suborder Heteroptera is a diverse insect group that includes bed bugs (that live as ectoparasites on birds and mammals) and *Triatoma* bugs (that colonize habitats built by people or nests of animals). Acariasis is disease caused by a variety of mites. Mites are in the Acari subclass in the Arachnidea class of arthropods that contain approximately 30,000 species worldwide. All acariases are highly contagious for susceptible species. Close contact may be needed to spread mites if their numbers are low. Scabies is the classic mite disease. Zoonotic mites do not reproduce on people; therefore, the clinical signs they produce are self-limiting. Louse infestation is prevalent in animals. Lice are divided into two orders: Mallophaga (chewing lice, which are common in birds and domestic animals with humans being accidental hosts) and Anoplura (sucking lice, which only parasitize mammals with the three types of human lice belonging to this group). Lice do not produce zoonotic disease. Lice rarely cause skin irritation in people (lice can only live a few hours on a person's skin and lice may rarely be transmitted to animals). Lice are permanent ectoparasites (spend their entire life on the host) and are highly host specific.

Review Questions

Multiple Choice

1. Parasites are organisms that
   a. provide nourishment and protection to other living organisms.
   b. are found on the exterior of the host.
   c. live independently until they are exposed to an optimal host.
   d. live in association with, and at the expense of, other organisms.

2. Protozoa are types of parasites that
   a. are unicellular, eukaryotic organisms.
   b. cannot make their own food.
   c. are found typically in moist environments.
   d. all of the above.

3. Helminths are
   a. unicellular worms that cannot make their own food.
   b. divided into groups called Platyhelminthes (flatworms) and Nematoda (roundworms).
   c. parasites typically found in moist environments outside of an animal's body.
   d. all of the above.

4. Protozoa are classified based upon their
   a. movement.
   b. digestive system.
   c. segmented body.
   d. type of nutritional status.

5. Trematodes are commonly referred to as _____ and cestodes are commonly referred to as _____.
   a. flukes, roundworms
   b. roundworms, tapeworms
   c. flukes, tapeworms
   d. tapeworms, roundworms

6. What two key evolutionary adaptations do arthropods possess?
   a. jointed appendages and an exoskeleton
   b. jointed appendages and wings
   c. wings and an exoskeleton
   d. antennae and wings

7. Instar is a term used in arthropods to describe a/an
   a. egg.
   b. nymph.
   c. larval stage.
   d. adult insect.

8. What type of parasitic life cycle involves a definitive host and one or more intermediate hosts?
   a. direct
   b. indirect
   c. vertical
   d. horizontal

9. What best defines a reservoir host?
   a. A reservoir host is the host in which asexual reproduction of the parasite occurs.
   b. All hosts in the parasitic life cycle are reservoir hosts.
   c. A reservoir host is an organism that harbors the same stage of the parasite that is found in humans.
   d. A reservoir host is an organism that accidentally becomes infected with a parasite that normally infects another species.

10. A stool sample is collected from a patient suspected of having an *En. histolytic* infection. The stool sample is liquid. What stage of the life cycle of *Entamoeba* is most likely to be observed in the specimen?
    a. cyst
    b. trophozoite
    c. sporozoite
    d. tryptomastigote

11. A protozoan cyst measuring 15 μm is observed on examination of a concentrated fecal specimen. It has four nuclei, with even peripheral chromatin, a central small endosome, and rounded chromatin bars and ingested erythrocytes in the cytoplasm. The organism is most likely

a. *Endolimax nana.*
b. *Entamoeba coli.*
c. *Iodamoeba butschlii.*
d. *Entamoeba histolytica.*

12. This organism causes severe, chronic, and potentially fatal diarrhea in immunocompromised people. It also produces mild to moderate diarrhea in calves. The disease in humans usually results from drinking contaminated water. This organism is identified in fecal samples by using the acid-fast staining method. The organism is
    a. *Pneumocystis carinii.*
    b. *Cryptosporidium parvum.*
    c. *Giardia duodenalis.*
    d. *Toxoplasma gondii.*

13. People involved in what occupation are most likely to become infected with *Balantidium coli*?
    a. sewage workers
    b. beef ranchers
    c. nurses
    d. pork ranchers

14. The tsetse fly (*Glossina*) serves as the vector for
    a. leishmaniasis.
    b. African sleeping sickness.
    c. Chagas' disease.
    d. filariasis.

15. The definitive host for *Toxoplasma gondii* is the
    a. dog.
    b. cat.
    c. pig.
    d. flea.

16. What parasite has the greatest risk to unborn human fetuses?
    a. *En. coli*
    b. *G. lamblia*
    c. *Cry. parvum*
    d. *T. gondii*

17. An infection with this sporozoan is characterized by the gradual onset of headache, chills, sweating, and fatigue following a tick bite. The disease is usually self-limiting and is caused by
    a. *Babesia microti.*
    b. *Borellia burgdorferi.*
    c. Colorado tick fever parasite.
    d. *Trypanosoma brucei.*

18. Heavy infections with _____ (nematode) can cause rectal prolapse (especially in children).
    a. *Taenia saginata*
    b. *Trichuris trichiura*
    c. *Angiostrongylus cantonensis*
    d. *Paracapillaria philippinensis*

19. The infective stage of *Strongyloides stercoralis* is the
    a. embryonated egg.
    b. rhabditiform larva.
    c. filariform larva.
    d. adult worm.

20. The most common agent of cysticercosis is
    a. *Diphyllobothrium latum.*
    b. *Ascaris lumbricoides.*
    c. *Trichuris trichiura.*
    d. *Taenia solium.*

21. Human infection with *Di. latum* results from consuming infected
    a. freshwater fish.
    b. saltwater fish.
    c. pork.
    d. snails.

22. Cystic hydatid disease is caused by
    a. *Echinococcus granulosus.*
    b. *Taenia solium.*
    c. *Taenia saginata.*
    d. *Schistosoma mansoni.*

23. A 7-year-old girl is small for her age and anemic. She has experienced alternating bouts of diarrhea and dysentery for at least 6 months. A fecal examination reveals football-shaped eggs with polar plugs at both ends. How was the child most likely infected?
    a. By eating raw fish
    b. By eating contaminated snails
    c. Oral-fecal route
    d. By playing in contaminated soil

24. You are an epidemiologist working for the WHO in charge of investigating cystic hydatid disease in Spain. Your most effective advice to the local authorities to help prevent the spread of this disease to humans would include which of the following? (1) To increase the chlorine content of local drinking water; (2) To thoroughly educate the public on personal hygiene measures; (3) To stop the feeding of animal organs to other domestic animals; (4) To trap and sample wild carnivores for disease, then to recommend extensive hunting of carnivores if they carry the disease.
    a. 1, 2
    b. 3, 4
    c. 1, 4
    d. 2, 3

25. Which parasites are commonly associated with domestic dogs? (1) *Toxocara canis*; (2) *Dirofilaria* spp.; (3) *Ancylostoma caninum*; (4) *Echinococcus granulosus.*
    a. 1, 2, 3
    b. 2, 3, 4
    c. 1, 3, 4
    d. 1, 2, 3, 4

26. The parasites that cause cutaneous larva migrans
    a. cause diarrhea in animals and skin lesions in people.
    b. include *Ancylostoma* and *Strongyloides.*
    c. can be transmitted transplacentally and through colostrum/milk in animals.
    d. all of the above.

27. Visceral larva migrans
    a. tend to cause organ and ocular disease.
    b. is as a result of organisms such as *Toxocara, Baylisascaris,* and *Ascaris.*

c. can cause gastrointestinal disease in animals or animals may be asymptomatic.

d. all of the above.

28. Infection with Diptera larvae is called
    a. mange.
    b. myiasis.
    c. taeniasis.
    d. infestation.

29. What is very host specific?
    a. fleas
    b. lice
    c. mosquitoes
    d. flies

30. Arthropods cause disease by what method?
    a. biological transmission
    b. mechanical transmission
    c. irritation from bites and stings
    d. all of the above

Matching

31. _____ *Echinococcus granulosus*

32. _____ *Dracunculus medinensis*

33. _____ *Taenia solium*

34. _____ *Ancyclostoma caninum*

35. _____ *Entamoeba histolytica*

36. _____ *Babesia microti*

37. _____ *Cryptosporidium parvum*

38. _____ *Giardia duodenalis*

39. _____ *Leishmania* spp.

40. _____ *Toxoplasma gondii*

41. _____ *Trypanosoma cruzi*

42. _____ *Trypanosoma brucei*

43. _____ *Diphyllobothrium latum*

44. _____ *Echinococcus multilocularis*

45. _____ *Trichinella spiralis*

46. _____ *Trichuris trichiura*

A. hookworm causing CLM

B. protozoan that parasitizes erythrocytes causing anemia

C. hemoflagellate transmitted by sandflies

D. cystic hydatid cyst

E. alveolar hydatid cyst

F. cestode acquired from eating fish

G. Guinea worm larvae released from skin blisters

H. fluke that may cause liver disease in sheep and cattle and abdominal pain and fever in humans

I. burrowing mites that tunnel in skin

J. avian schistosomes that may cause cercarial dermatitis

K. cysticercus

L. protozoan causing amebic colitis

M. common protozoan in calves that is typically transmitted to people via contaminated water

N. flagellated protozoan that may cause chronic diarrhea in animals and humans

O. mite that causes walking dandruff in animals and a mild pruritic dermatitis in people

P. coccidian protozoan that can cause congenital defects in humans if the mother contracts the disease while pregnant

47. _____ *Fasciola hepatica*

48. _____ *Trichobilharzia* and *Microbilharzia*

49. _____ *Sarcoptes scabiei*

50. _____ *Cheyletiella* spp.

Q. flagellate that may cause myocarditis in puppies and children

R. flagellate that may cause neurological disease in animals and sleeping sickness in humans

S. nematode acquired by ingestion of encysted larvae in muscle tissue (especially pork)

T. whipworm that may cause diarrhea and anemia in dogs and rectal prolapse in children

Case Studies

51. A 4-year-old boy was taken to his physician for evaluation of diarrhea. One month prior to the examination, he had been having chronic abdominal pain and gas to frank diarrhea for the preceding 1 to 2 weeks. On the morning of the examination, the boy's diarrhea was yellow, foul-smelling, and frothy. Stool was collected for culture and ova and parasite examination. After 48 hours, the stool culture was reported as negative for enteric pathogens and the ova and parasite examination was positive for pear-shaped trophozoites that were 12 to 15 μm long with two bilateral nuclei.
    a. What parasite was found in the stool specimen?
    b. How is this parasite transmitted?
    c. How can infection with this organism be prevented?

52. A 10-year-old boy was admitted to the hospital with fever of unknown orgin and nasal discharge of 1-week duration. On numerous occasions he had been observed eating dirt in the yard. The family has no house pets; however, the family raised pigs on their land for their own use. On admission the boy was pale with a temperature of 101°F, and had a slightly enlarged spleen. His hemoglobin was 7.2 g/dl, WBC count was 32,000/μl, with 64% eosinophils. Radiographs of the chest showed abnormal shadows in the right cranial and caudal lobes. A tentative diagnosis was made of iron deficiency anemia and possible visceral larva migrans. He was treated and sent home. Three weeks later the boy was readmitted in a semicomatose state with a 10-day history of shaking spells and ataxia. His WBC count was 9,100/μl with 4% eosinophils. The boy developed convulsions that were difficult to control and he died 3 weeks later (7 weeks following his first admission). On autopsy granulomatous lesions were found in the liver, heart, and brain. The liver and heart lesions were considered to be chronic, but the brain lesions appeared to be more recent in origin.
    a. What do you think is the most likely causative agent?
    b. Explain why this boy had such severe clinical symptoms.

53. A 20-year-old man developed a rash on his forearm that progressed to vesicles and crusts. The skin lesions were pruritic and he was developing breaks in the skin as a result of his itching of the lesions. This man had recently purchased two hunting dogs that also have pruritis. A skin scraping was taken and an oval mite that had eight legs was found.
    a. What parasite do you think this person has?
    b. How did this person acquire the infection?
    c. How is this disease treated in both animals and people?
    d. What should be done with the animals until they clear the infection?

References

Ackers, J. P. 2002. The diagnostic implication of the separation of *Entamoeba histolytica* and *Entamoeba dispar*. *Journal of Biosciences* 27(6 Suppl 3):573–8.

Alcamo, I. E. 1998. *Schaum's Outlines Microbiology*. New York: McGraw-Hill.

American Veterinary Medical Association. 2003. What you should know about toxoplasmosis. http://www.avma.org/communications/brochures/toxoplasmosis/toxoplasmosis_brochure.asp (accessed October 15, 2003).

Arnold, L. K. 2005. Trichinosis. http://www.emedicine.com/emerg/topic612.htm (accessed December 16, 2005).

Bauman, R. 2004. *Microbiology*. San Francisco, CA: Pearson Benjamin Cummings.

Becker, H. 2005. Out of Africa: the origins of tapeworms. http://www.ars.usda.gov/is/AR/archive/may01/worms0501.htm (accessed March 10, 2006).

Biddle, W. 2002. *A Field Guide to Germs*. New York: Anchor Books.

Bimi, L., A. Freeman, M. Eberhard, E. Ruiz-Tiben, and N. J. Pieniazek. 2005. Differentiating *Dracunculus medinensis* from *D. insignis*, by the sequence analysis of the 18S rRNA gene. *Annals of Tropical Medicine & Parasitology* 99(5):511–7.

Black, J. 2005. *Microbiology Principles and Explorations*, 6th ed. Hoboken, NJ: John Wiley & Sons.

Bray, R. 1987. Note on the history of cutaneous leishmaniasis in the Mediterranean and Middle East area. *Parassitologia* 29(2–3):175–9.

Burton, R., and P. Engelkirk. 2004. *Microbiology for the Health Sciences*, 7th ed. Baltimore, MD: Lippincott, Williams & Wilkins.

Bush, A., J. Fernandez, G. Esch, and J. R. Seed. 2001. *Parasitism: The Diversity and Ecology of Animal Parasites*. New York: Cambridge University Press.

Cambridge University Schistosome Research Group. 1998. Tapeworms of man of the genus *Taenia*. http://www.path.cam.ac.uk/~schisto/Tapes/Taenia.html (accessed March 11, 2006).

Carlier, Y., and A. Luguetti. 2003. Chagas' Disease (American Trypanosomiasis). http://www.emedicine.com/med/topic327.htm (accessed February 27, 2003).

Carpenter, R. E., and A. Richard. 2006. *Strongyloides stercoralis*. http://www.emedicine.com/emerg/topic843.htm (accessed March 27, 2006).

Center for Food Security and Public Health. Iowa State University. 2005. Baylisascariasis. http://www.cfsph.iastate.edu/Factsheets/pdfs/baylisascariasis.pdf (accessed May 1, 2005).

Center for Food Security and Public Health. Iowa State University. 2005. Echinococcosis. http://www.cfsph.iastate.edu/Factsheets/pdfs/echinococcosis.pdf (accessed May 1, 2005).

Center for Food Security and Public Health. Iowa State University. 2005. Larva migrans. http://www.cfsph.iastate.edu/Factsheets/pdfs/larva_migrans.pdf (accessed May 1, 2005).

Center for Food Security and Public Health. Iowa State University 2005. Taenia infections. http://www.cfsph.iastate.edu/Factsheets/pdfs/taenia.pdf (accessed May 1, 2005).

Center for Food Security and Public Health. Iowa State Univeristy. 2005. Toxocariasis, http://www.cfsph.iastate.edu/Factsheets/pdfs/toxocariasis.pdf (accessed May 1, 2005).

Center for Food Security and Public Health. Iowa State University. 2005. Trichuriasis. http://www.cfsph.iastate.edu/Factsheets/pdfs/trichuriasis.pdf (accessed May 6, 2005).

Centers for Disease Control and Prevention. 1999. Cutaneous larva migrans. http://www.cdc.gov/ncidod/dpd/parasites/hookworm/factsht_hookworm.htm (accessed April 17, 2006).

Centers for Disease Control and Prevention. 2004. Amebiasis. http://www.cdc.gov/ncidod/dpd/parasites/amebiasis/factsht_amebiasis.htm (accessed February 27, 2006).

Centers for Disease Control and Prevention. 2004. Giardiasis. http://www.cdc.gov/ncidod/dpd/parasites/giardiasis/factsht_giardia.htm (accessed March 6, 2006).

Centers for Disease Control and Prevention. 2004. Schistosomiasis. http://www.cdc.gov/ncidod/dpd/parasites/schistosomiasis/factsht_schistosomiasis.htm (accessed March 14, 2006).

Centers for Disease Control and Prevention. 2004. Toxoplasmosis. http://www.cdc.gov/ncidod/dpd/parasites/toxoplasmosis/factsht_toxoplasmosis.htm (accessed March 15, 2006).

Centers for Disease Control and Prevention. 2004–2005. Progress toward global eradication of Dracunculiasis. *Morbidity and Mortality Weekly Report* 54(42):881–3.

Centers for Disease Control and Prevention. 2005. Cryptosporidiosis. http://www.cdc.gov/ncidod/dpd/parasites/cryptosporidiosis/factsht_cryptosporidiosis.htm (accessed March 14, 2006).

Centers for Disease Control and Prevention. 2006. Babesiosis. http://www.dpd.cdc.gov/DPDx/HTML/Babesiosis.htm (accessed March 14, 2006).

Centers for Disease Control and Prevention. 2006. East African trypanosomiasis. http://www.cdc.gov/ncidod/dpd/parasites/trypanosomiasis/factsht_ea_trypanosomiasis.htm (accessed March 14, 2006).

Centers for Disease Control and Prevention. 2006. Leishmaniasis. http://www.cdc.gov/healthypets/diseases/leishmania.htm (accessed October 14, 2006).

Centers for Disease Control and Prevention. 2006. Taenia infections. http://www.dpd.cdc.gov/DPDx/HTML/Taeniasis.htm (accessed March 6, 2006).

Centers for Disease Control and Prevention. 2006. *Toxocara* infection (toxocariasis) and Animals. http://www.cdc.gov/healthypets/diseases/toxocariasis.htm (accessed April 24, 2006).

Centers for Disease Control and Prevention. 2006. West African trypanosomiasis. http://www.cdc.gov/ncidod/dpd/parasites/trypanosomiasis/factsht_wa_trypanosomiasis.htm (accessed March 14, 2006).

Chacon-Cruz, E., and D. Mitchell. 2006. Intestinal protozoal diseases. http://www.emedicine.com/ped/topic1914.htm (accessed October 4, 2006).

Clinton White, A. and D. Eisen. 2006. Cryptosporidiosis. http://www.emedicine.com/med/topic484.htm (accessed July 1, 2006).

Companion Animal Parasite Council (CAPC). 2005. 2005 Companion Animal Parasite Council (CAPC) Guidelines. http://www.capcvet.org/?p=Guidelines_Introduction&h=0&s=0 (accessed April 7, 2006).

Cowan, M., and K. Talaro. 2006. *Microbiology: A Systems Approach*. New York: McGraw-Hill.

Cox, F. E. G. 2002. History of human parasitology. *Clinical Microbiology Reviews* 15(4):595–612.

Cunha, B., and B. Barnett. 2006. Babesiosis. http://www.emedicine.com/med/topic195.htm (accessed August 14, 2006).

Elston, D. 2007. Lice. http://www.emedicine.com/derm/topic229.htm (accessed February 28, 2007).

Farrell, C. 2002. Hookworms and *Stronglyoides stercoralis*. http://www.cumc.columbia.edu/dept/ps/2007/para/old/transcript_02_pd02.pdf (accessed October 17, 2002).

Forbes, B., D. Sahm, and A. Weissfeld. 2002. *Bailey & Scott's Diagnostic Microbiology*, 11th ed. St. Louis, MO: Mosby.

Gavin, P., K. Kazacos, and S. Shulman. 2005. Baylisascariasis. *Clinical Microbiology Reviews* 18(4):703–18.

Ghosh, S., and R. Wientzen. 2006. Dracunculiasis. http://www.emedicine.com/ped/topic616.htm (accessed June 20, 2006).

Go, C. H. U., and B. Cunha. 2003. Intestinal flukes. http://www.emedicine.com/med/topic1177.htm (accessed March 19, 2003).

Haubrich, W. S. 1997. *Medical Meanings: A Glossary of Word Origins*. Philadelphia, PA: American College of Physicians.

Hawn, R. 2006. Zoonosis: how lifestyle alters risks to people and pets, techniques in practice focus: zoonosis. *Veterinary Technician Techniques in Practice* 27(8 (A)):4–6, 8.

Heelan, J., and F. Ingersoll. 2002. *Essentials of Human Parasitology*. Albany, NY: Delmar Publishing.

Henderson, S., and R. Magana. 2006. Babesiosis. http://www.emedicine.com/emerg/topic49.htm (accessed March 8, 2006).

Hökelek, M., and A. Safdar. 2006. Toxoplasmosis. http://www.emedicine.com/med/topic2294.htm (accessed May 16, 2006).

Homer, M., and D. Persing. 2005. Human Babesiosis. In *Tick-Borne Diseases of Humans*, edited by J. Goodman, D. Dennis, and D. Sonenshine, pp. 343–60. Washington, DC: American Society of Microbiology Press.

Horga, M. and T. Naparst. 2006. Amebiasis. http://www.emedicine.com/ped/topic80.htm (accessed March 31, 2006).

Juzych, L., and M. Douglas. 2006. Cutaneous larva migrans. http://www.emedicine.com/derm/topic91.htm (accessed April 10, 2006).

Kenner, J., and P. Weina. 2005. Leishmaniasis. http://www.emedicine.com/derm/topic219.htm (accessed April 14, 2005).

Kimura, K., M. Stoopen, and R. Moncada. 2000. Ascariasis: Tropical Medicine Central Resource. http://tmcr.usuhs.mil/tmcr/chapter1/intro.htm (accessed April 21, 2006).

Kogulan, P., and D. Lucey. 2002. Schistosomasis. http://www.emedicine.com/med/topic2071.htm. (accessed October 10, 2005).

Krauss, H., A. Weber, M. Appel, B. Enders, H. D. Isenbergy, H. G. Schiefer, W. Slenczka, A. von Graevenitz, and H. Zahner. 2003. Parasitic zoonosis. In *Zoonoses: Infectious Diseases Transmissible from Animals to Humans*, 3rd ed. Washington, DC: American Society of Microbiology Press.

Lafferty, K. D., J. W. Porter, and S. E. Ford. 2004. Are diseases increasing in the ocean? *Annual Review of Ecology Evolution Systematics* 5:31–54.

Laufer, M. 2004. Toxocariasis. http://www.emedicine.com/ped/topic2270.htm (accessed February 12, 2004).

Linklater, D., and G. Holmes. 2005. Diphyllobothriasis. http://www.emedicine.com/med/topic571.htm (accessed June 20, 2005).

Merck Veterinary Manual. 2006. Amebiasis: introduction. http://www.merckvetmanual.com/mvm/index.jsp?cfile=htm/bc/21100.htm&word=Amebiasis (accessed February 27, 2006).

Merck Veterinary Manual. 2006. Babesiosis: overview. http://www.merckvetmanual.com/mvm/index.jsp?cfile=htm/bc/10402.htm&word=Babesiosis (accessed March 14, 2006).

Merck Veterinary Manual. 2006. Cryptosporidiosis. http://www.merckvetmanual.com/mvm/index.jsp?cfile=htm/bc/200802.htm&word=Cryptosporidiosis (accessed March 13, 2006).

Merck Veterinary Manual. 2006. *Fasciola hepatica*. http://www.merckvetmanual.com/mvm/index.jsp?cfile=htm/bc/22702.htm (accessed March 14, 2006).

Merck Veterinary Manual. 2006. Giardiasis: introduction. http://www. merckvetmanual.com/mvm/index.jsp?cfile=htm/bc/21300.htm&word =Giardiasis (accessed March 6, 2006).

Merck Veterinary Manual. 2006. Hookworms. http://www.merckvetmanual.com/ mvm/index.jsp?cfile=htm/bc/23507.htm&word=ancylostoma (accessed April 17, 2006).

Merck Veterinary Manual. 2006. Roundworms (Ascariasis). http://www. merckvetmanual.com/mvm/index.jsp?cfile=htm/bc/23505.htm&word =Ascariasis (accessed April 21, 2006).

Merck Veterinary Manual. 2006. Schistosomiasis. http://www.merckvetmanual. com/mvm/index.jsp?cfile=htm/bc/10408.htm&word=Schistosoma (accessed March 14, 2006).

Merck Veterinary Manual. 2006. Tapeworms (Cestodes). http://www. merckvetmanual.com/mvm/index.jsp?cfile=htm/bc/23512.htm&word= Taenia (accessed March 6, 2006).

Merck Veterinary Manual. 2006. Toxoplasmosis. http://www.merckvetmanual. com/mvm/index.jsp?cfile=htm/bc/52200.htm&word=toxoplasmosis (accessed March 15, 2006).

Merck Veterinary Manual. 2006. Trypanosomiases. http://www.merckvetmanual. com/mvm/index.jsp?cfile=htm/bc/10412.htm&word=Trypanosomiasis (accessed March 18, 2006).

Merck Veterinary Manual. 2006. Visceral leishmaniasis: introduction. http:// www.merckvetmanual.com/mvm/index.jsp?cfile=htm/bc/57400.htm&word =Leishmaniasis (accessed March 15, 2006).

Meyer, J. 2006. Entomology 425: Hemiptera. http://www.cals.ncsu.edu/course/ ent425/compendium/hetero~1.html (accessed June 1, 2006).

Nuwer, D. 2002. The importance of wearing shoes: hookworm disease in Mississippi. http://mshistory.k12.ms.us/features/feature31/hookworm.html (accessed September 5, 2002).

Petersen, E., and J. P. Dubey. 2001. Biology of toxoplasmosis. In *Toxoplasmosis: A Comprehensive Clinical Guide*, edited by Joynson, D. and T. Wreghitt, pp. 1–42. New York: Cambridge University Press.

Pitelli, R. 2006. Visceral larva migrans. http://www.emedicine.com/ped/topic2407. htm (accessed April 10, 2006).

Prescott, L. M., D. A. Klein, and J. P. Harley. 2002. *Microbiology*, 5th ed. New York: McGraw-Hill.

Puiu, D., S. Enomoto, G. Buck, M. Abrahamsen, and J. Kissinger. 2002. CryptoDB: the *Cryptosporidium* genome resource. *Journal of Eukaryotic Microbiology* 49:433–40.

Rai, A., and M. Weisse. 2002. *Diphyllobothrium latum* infections. http://www. emedicine.com/ped/topic597.htm (accessed November 1, 2002).

Roberts, L., and J. Janovy, Jr. 2005. *G. D. Schmidt and L. S. Roberts' Foundations of Parasitology*, 7th ed. New York: McGraw-Hill.

Sarver, S. 2005. Protista. http://msc.bhsu.edu/~ssarver/protista.htm (accessed April 15, 2005).

Schantz, P., and J. Stehr-Green. 1995. Toxocaral larva migrans: Zoonosis Updates. *Journal of the American Veterinary Medical Association.* http:// www.avma.org/reference/zoonosis/zntoxcar.asp (accessed February 27, 2004).

Sciammarella, J. 2002. Toxoplasmosis. http://www.emedicine.com/emerg/topic601. htm (accessed July 8, 2002).

Service, M. 2004. *Medical Entomology for Students*, 3rd ed. New York: Cambridge University Press.

Shimeld, L. A. 1999. *Essentials of Diagnostic Microbiology*. Albany, NY: Delmar Publishers.

Stark, C., and G. Wortmann. 2003. Leishmaniasis. http://www.emedicine.com/med/topic1275.htm. (accessed January 6, 2006).

Tolan, R. 2005. Fascioliasis. http://www.emedicine.com/ped/topic760.htm (accessed August 17, 2005).

Upton, S. 2005. Taxonomic chronology of *Cryptosporidium:* some historical milestones (good or bad). http://www.k-state.edu/parasitology/taxonomy (accessed August 18, 2005).

U.S. Food and Drug Administration. 1991. *Diphyllobothrium* spp: Bad bug book. http://www.cfsan.fda.gov/~mow/chap26.html (accessed April 7, 2006).

U.S. Food and Drug Administration. 1992. *Cryptosporidium parvum*: Bad bug book. http://www.cfsan.fda.gov/~mow/chap24.html (accessed March 13, 2006).

VanDenEden, E. 2004. Leishmaniasis: illustrated lecture notes on tropical medicine. http://www.itg.be/itg/DistanceLearning/LectureNotesVandenEndenE/index.htm (accessed March 15, 2006).

Vohra, R. and R. Vohra. 2006. *Taenia* infection. http://www.emedicine.com (accessed May 26, 2006).

Vuitton, D., and S. Bresson-Hadni. 2006. Echinococcosis. http://www.emedicine.com/med/topic326.htm (accessed March 5, 2006).

Wang, J., and K. Wang. 2006. Cutaneous larva migrans. http://www.emedicine.com/ped/topic1278.htm (accessed February 28, 2006).

Weiss, E. 2005. *Trichuris trichiura*. http://www.emedicine.com/emerg/topic842.htm (accessed June 8, 2005).

White, A., and D. Eisen. 2006. Cryptosporidiosis. http://www.emedicine.com/med/topic484.htm (accessed July 22, 2006).

Wiegmann, B., and D. Yeates. 1996. Diptera. http://www.tolweb.org/Diptera (accessed May 31, 2004).

Wikipedia.org. 2006. Hookworm. http://en.wikipedia.org/wiki/Hookworm (accessed April 17, 2006).

Wikipedia.org. 2006. *Strongyloides*. http://en.wikipedia.org/wiki/Strongyloides (accessed April 17, 2006).

Wikipedia.org. 2006. *Taenia saginata*. http://en.wikipedia.org/wiki/Taenia_saginata (accessed March 11, 2006).

Wikipedia.org. 2006. *Taenia solium*. http://en.wikipedia.org/wiki/Taenia_solium (accessed March 11, 2006).

Zoological Institute, St. Petersburg. 2003. Fleas (*Siphonaptera*: St. Petersburg fleas). http://www.zin.ru/Animalia/Siphonaptera/index.htm (accessed June 1, 2006).

Objectives

After completing this chapter, the learner should be able to

- Describe why a virus is considered noncellular based upon its structure
- Describe how viruses replicate
- Briefly describe the history of specific viruses
- Describe the causative agent of specific viral zoonoses
- Identify the geographic distribution of specific viral zoonoses
- Describe the transmission, clinical signs, and diagnostic procedures of specific viral zoonoses
- Describe methods of controlling viral zoonoses
- Describe protective measures professionals can take to prevent transmission of viral zoonoses

Key Terms

accidental host
amplifying host
arbovirus
bacteriophage
capsid
capsomeres

dead-end host
envelope
negative sense
plaque assay
positive sense
reservoir host

robovirus
variant
virion
virus

OVERVIEW

Many epidemics caused by viral diseases occurred before anyone understood the nature of viruses. *Virus* is Latin for poison and was originally used to describe diseases of unknown origin. Viruses were originally thought to be nonreproducing toxins.

Viruses cannot reproduce independently and are therefore considered nonliving by some. Viruses are noncellular infectious agents consisting of a single type of nucleic acid (either ribonucleic acid [RNA] or deoxyribonucleic acid [DNA]) surrounded by a protein coat. Viruses lack independent metabolism and must reproduce within living host cells. Viruses are generally specific for a given host, although all organisms (including other microbes) are susceptible to viral attack. Viruses are typically classified according to their host range, size, structure, and life cycle. Host range is determined by the presence of receptors on the cell's surface. Each virus type attaches to a particular type of receptor, so each type of virus can only attack those cells that have a receptor for it. The size of viruses varies and can be as small as 25 nanometer (nm) or as large as 300 nm (from one-tenth to one-third the size of a bacterial cell). The structure of a virus (one infectious virus

Table 7-1   Properties of Viruses Versus Other Living Microorganisms

Characteristic	Viruses	Other Living Microorganisms
Size	Between 25 — 300 nm (usually less than 200 nm)	Usually greater than 200 nm
Nuclei acid	DNA or RNA (not both)	DNA and RNA
Outer covering	Protein coat; some viruses have an envelope	Complex cell membrane, cell wall, or both
Reproduction	Requires host for reproduction and metabolic needs	Usually self-replicating
Cultivation	Must be grown on cell containing media	Usually grown on nutrient media that is cell-free

particle is a **virion**) is a nucleic acid core surrounded by protein. Some viruses may have additional structural proteins or a surrounding membrane that makes them different from other viruses. All viruses have the same basic life cycle; however, some viruses may enter a host cell intact, whereas other viruses disassemble and only the nucleic acid enters the host cell. Table 7-1 compares the properties of viruses to living microorganisms.

Most viruses consist of nothing more than nucleic acid surrounded by a protein shell. The outer protein is referred to as the **capsid**, which is made up of repeating protein subunits called **capsomeres**. The architectural arrangement of the capsid determines the virus' general form, classified as being either polyhedral (equilateral triangles that fit together to form a dome), helical (capsomeres spiral to form a rod-shaped structure), or complex (a combination of structures with a helical portion called a tail attached to a polyhedral portion called a head that is typically seen with bacteriophages) (Figure 7-1).

Other viruses, particularly those with animal hosts, may have a membranous **envelope** that surrounds the capsid. Viruses that contain an envelope are termed enveloped, whereas those that lack an envelope are called naked. The envelope is connected to the capsid by a layer of matrix proteins, which serve to strengthen the envelope. The envelope is composed of glycoproteins (a protein-carbohydrate complex) embedded within lipid from the host's cell membrane (enveloped viruses acquire their envelopes during the release stage of replication when the virus moves to the host cell membrane and force their way through the membrane in a process called budding). Glycoproteins may also project from the envelope's surface to form spikes of various lengths and shapes. These spikes serve an important role in attachment and infection. Because the origins of each component determine the virus' resistance and survival, it is important to distinguish between components that are coded by the virus' genes (genes are segments of RNA or DNA) versus those that are coded by the host's genes. Only the proteins of the viral envelope are coded by the virus. The carbohydrate and lipids are coded by the host cell. Therefore, the type of lipid and carbohydrate within a given viral envelope is dependent on the particular host.

Bacteriophages (or simply phages) are bacterial viruses that usually lyse bacterial cells.

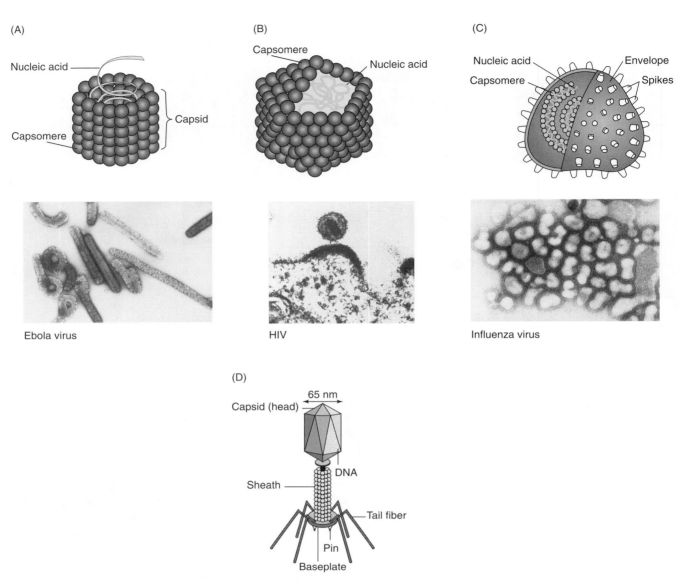

(A)

Nucleic acid

Capsid

Capsomere

Ebola virus

(B)

Capsomere

Nucleic acid

HIV

(C)

Nucleic acid

Capsomere

Envelope

Spikes

Influenza virus

(D)

65 nm

Capsid (head)

DNA

Sheath

Tail fiber

Pin

Baseplate

Figure 7-1 Some examples of viral morphology. (A) a helical virus may look like long or coiled threads; (B) a polyhedral virus (icosahedral); (C) an enveloped helical virus; (D) bacteriophage.
(Photos for A & C courtesy Centers for Disease Control and Prevention; photo for B courtesy National Institute of Allergy and Infectious Diseases.)

TYPES OF VIRUSES

Virus orders end with –virales, viral families end with –viridae, and genera end with –virus.

Animal viruses (viruses that require animal hosts) are classified based on:

- *Size.* Viruses range in size from about 25 to 300 nm. The smallest viruses are a little larger than ribosomes, whereas the largest viruses are about the same size as small bacteria.

- *The nature of the nucleic acid* (Table 7-2). Viruses may contain DNA or RNA. RNA viruses may be single-stranded or double-stranded (most are single-stranded). Most eukaryotic cells do not have the enzymes to copy viral RNA molecules; therefore, the RNA viruses must either carry the enzymes or have

Table 7-2   Selected RNA and DNA Viruses of Animals

Family	Properties	Virus (Species or Genus)	Diseases
DNA Viruses			
Herpesviridae	Enveloped dsDNA	Herpes simplex types 1 and 2 (HSV-1, HSV-2) Varicella zoster virus	Cold sores, genital herpes Chickenpox/shingles
Poxviridae	Enveloped dsDNA	Herpes B virus Smallpox virus Contagious ecthyma virus	Herpes B infection Smallpox (variola) Contagious ecthyma
RNA Viruses			
Retroviridae	Enveloped, single-stranded, positive sense RNA	Human immunodeficiency viruses (HIV-1 and HIV-2) Human T-cell leukemia viruses (HTLV-1 and HTLV-2)	AIDS T-cell leukemia
Togaviridae	Enveloped, single-stranded, positive sense RNA	Eastern, Western, and Venezulan encephalitis viruses	Equine encephalitis
Flaviviridae	Enveloped, single-stranded, positive sense RNA	Yellow fever virus SLE virus Dengue fever virus WNV	Yellow fever St. Louis encephalitis Dengue fever West Nile fever
Picornaviridae	Nonenveloped, single-stranded, positive sense RNA	FMD virus SVD virus Poliomyelitis virus	Foot-and-mouth disease Swine vesicular disease Polimyelitis
Coronaviridae	Enveloped, single-stranded, positive sense RNA	Coronavirus	SARS
Orthomyxoviridae	Enveloped, single-stranded negative sense RNA	Influenza virus	Influenza
Rhabdoviridae	Enveloped, single-stranded, negative sense RNA	Rabies virus Vesicular stomatitis virus	Rabies Vesicular stomatitis

(Continued)

Table 7-2   **(Continued)**

Family	Properties	Virus (Species or Genus)	Diseases
RNA Viruses			
Paramyxoviridae	Enveloped, single-stranded, negative sense RNA	Newcastle disease virus Nipah virus	Newcastle disease Nipah virus encephalitis
Bunyaviridae	Enveloped, single-stranded, negative sense RNA	*Hantavirus* La Crosse virus	Hantavirus-associated respiratory distress syndrome La Crosse encephalitis
Reoviridae	Nonenveloped, dsRNA	*Coltivirus*	Colorado tick fever
Arenaviridae	Enveloped, single-stranded, negative sense RNA	Lymphocytic choriomeningitis Lassa virus	Lymphocytic choriomeningitis Lassa fever
Filoviridae	Enveloped, single-stranded, negative sense RNA	Filovirus	Marburg hemorrhagic fever Ebola hemorrhagic fever

the genes for those enzymes as part of their genome. There are two types of single-stranded RNA viruses: **positive sense** and **negative sense**. Some are positive in that they have a "sense" strand of RNA (coded information about how to build proteins) as their genetic material. During an infection the RNA acts like mRNA and can be translated by the host's ribosomes. Other RNA viruses are negative in that they have an "antisense" strand (the paired opposite of the coded information). In negative sense viruses the RNA acts as a template during transcription to make a complementary positive sense mRNA after a host cell has been entered.

- *The type of viral replication.* Viral replication varies depending on the type of nucleic acid, type of participating enzymes, and location of replication within the host cell.
- *Capsid structure.* Viruses have a capsid structure that may be polyhedral, helical, or complex.
- *Presence or absence of viral envelope.* Viruses may be enveloped or naked (see Table 7-2).
- *Routes of transmission* (Table 7-3).

Table 7-3   Types of Viral Transmission

Type of Transmission	Examples
Fecal-oral	Poliovirus, rotavirus
Respiratory	Influenza virus, measles virus
Zoonotic	Rabies virus, cowpox virus
Sexual contact	Human immunodeficiency virus, human papilloma virus

HOW DO VIRUSES REPLICATE?

Viruses are obligate intracellular parasites. In general, DNA viruses mature in the nucleus of the host cell, whereas RNA viruses replicate in the cytoplasm (Table 7-4). The replication cycle of a virus usually results in host cell death and lysis; therefore, this type of replication is termed lytic replication. In order to understand how viruses cause disease, one must first know the basics of viral replication.

Replication of Animal Viruses

There are six steps of replication, the details of which are varied and complex for a given type of virus. The type of disease and its timing are influenced by each stage of the replication cycle.

1. Attachment (adsorption)—the receptors of the virion attach to the host cell receptors.
2. Entry—the virion may penetrate through the cell wall and/or membrane or may enter the cell by receptor-mediated endocytosis.

Table 7-4   Patterns of Animal Virus Replication

Nucliec Acid in Virion	Replication of Nuclei Acid	Transcription/Translation	Examples
dsDNA	Replication of DNA begins after some viral protein has been made.	dsDNA is transcribed by the same mechanisms as chromosomal DNA	Herpesviruses, poxviruses
ssDNA	In host cell nucleus, ssDNA is concerted to double-strand replicative form, which is copied and generates ssDNA.	Host-cell RNA polymerase transcribes double stranded replicative-form DNA.	Parvoviruses
Positive sense RNA	Virus-encoded RNA-dependent RNA polymerase makes negative sense RNA, which serves as a template for more positive sense.	Translation of positive sense RNA begins immediately, forming capsid protein and RNS-dependent RNA polymerase.	Picornaviruses, togaviruses
Negative sense RNA	RNA-dependent RNA polymerase makes positive sense RNA, then more negative sense.	Transcription is catalyzed by RNA-dependent RNA polymerase in virion.	Orthomyxoviruses, rhabdoviruses
dsRNA	RNA is replicated by virus-encoded RNA polymerase.	dsRNA is converted to mRNA by RNA-dependent RNA polymerase.	Reoviruses

3. Uncoating—the virion releases its nucleic acid from the protein coat allowing viral replication to begin.

4. Replication (eclipse phase)—the virion uses the host's cellular mechanisms for producing more viral nucleic acid and viral proteins.

5. Assembly (maturation)—viral nucleic acids and proteins are assembled.

6. Release—virions are released from the host cell; this may occur when the viruses lyse the host cell resulting in cell and tissue damage and further activation of the immune system or by budding, which releases viral particles.

Viral replication typically destroys the host cells. As the host cell dies, thousands of new viruses are produced to infect and destroy other cells.

Replication of Bacteriophages

A bacteriophage (commonly called phage) is a virus that infects bacteria. Phages are obligate intracellular microbes and must enter the bacterial cell to replicate. There are three categories of phages based on their shape: icosahedron (almost spherical in shape), filamentous (long tubes in a helical structure), and complex (icosahedron heads attached to helical tails). Phages can also be categorized by the type of nucleic acid they possess: single-stranded DNA, double-stranded DNA, single-stranded RNA, and double-stranded RNA. Phages can also be categorized based on the events that occur following injection of their nucleic acid into the bacterial cell: virulent phages (those that cause the lytic cycle to occur) and temperate phages (those that do not immediately cause the lytic cycle to occur but rather integrate their DNA into the bacterial cell chromosome).

Virulent Phages and the Lytic Cycle

Virulent phages initiate the lytic cycle, which always ends with destruction of the bacterial cell (Figure 7-2). This whole process typically takes less than one hour to complete.

1. Attachment (absorption)—the phage attaches to the surface of a bacterial cell that possesses the appropriate receptor (most phages are species and strain-specific).

2. Entry—the phage injects its DNA into the bacterial cell. From this point on, the phage DNA dictates what occurs in the bacterial cell.

3. Synthesis—phage genes are expressed resulting in the production of viral pieces. During this step the host cell's enzymes, nucleotides, amino acids, and ribosomes are used to make viral DNA and viral proteins.

4. Assembly—the phage pieces are assembled to produce complete viral particles (virions). Viral DNA is packaged into capsids.

5. Release—the host cell lyses and all of the new virions escape from the cell. The phage gene codes for the production of an enzyme that causes lysis of the cell wall.

Temperate Phages and Lysogeny

Temperate phages do not initially cause bacterial cell lysis as virulent phages do. After temperate phages inject their DNA into the bacterial cell, the phage DNA becomes part of the bacterial chromosome and does not cause the lytic cycle to occur. In this process, known as lysogeny, the only thing that remains of the phage is its DNA (known as a prophage) (Figure 7-3). Each time the bacterial cell containing the prophage (known as a lysogenic cell) undergoes binary fission, the

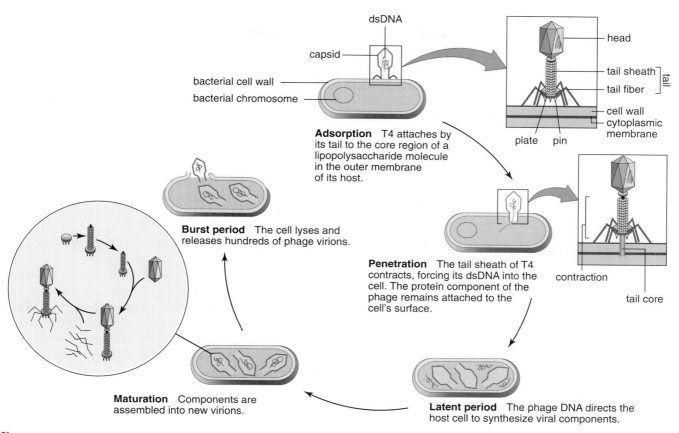

Figure 7-2   Life cycle of a virulent phage. (T4)
(Art by Carlyn Iverson)

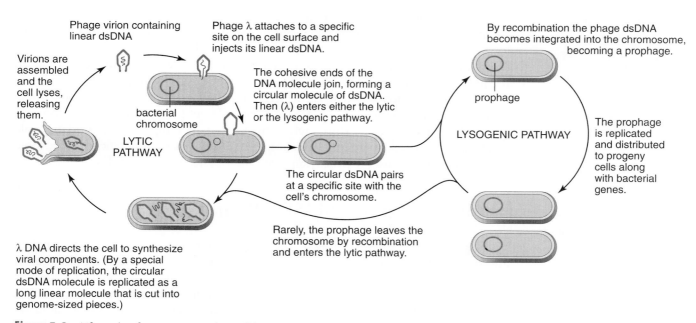

Figure 7-3   Life cycle of a temperate phage (λ).
(Art by Carlyn Iverson)

phage DNA is replicated and passed on to each daughter cell along with the bacterial DNA. These daughter cells are therefore lysogenic cells. Sometimes when the prophage is integrated into the bacterial chromosome, the bacterial cell produces new gene products that are coded for by the prophage genes. The bacterial cell will display new properties (known as lysogenic conversion) and is now able to produce one or more gene products that it was previously unable to produce. Examples of lysogenic conversion occurs in certain strains of *Streptococcus pyogenes* carrying a prophage capable of producing erythrogenic toxin (the toxin that causes scarlet fever) and certain strains of *Vibrio cholerae* that carry a prophage that can produce cholera toxin.

HOW DO VIRUSES CAUSE DISEASE?

Some viruses cause acute infections in which the virions usually remain localized and the signs of disease are short-lived. The infected host cells die upon release of new virions, resulting in cell and tissue damage. Typically the host's defense mechanisms will eliminate the virus over a period of several days or months. Acute infections often stimulate the host organism's immunity to protect against future infection. Examples of acute infections include measles, mumps, and influenza (the flu).

Other viruses cause persistent infections in which virions persist within the host organism without any disease signs. There are several different categories of persistent infections, which are influenced by the process of replication. Persistent infections may be the result of an acute infection. For example, measles has relatively short infectious and disease phases, yet late complications result in a fatal brain disorder that occurs within 10 years of having the disease. The new condition is the result of a defective viral replication with brain cells.

Other persistent infections are termed latent because they exhibit a long noninfectious stage between the original disease and the subsequent disease. The best example of a latent infection is chickenpox and shingles, both of which are caused by the varicella-zoster virus. Shingles or zoster appears after the virus becomes reactivated by unknown circumstances. Herpes virus (both herpes simplex type 1 and 2) is another example of a latent virus.

Persistent infections may also result in continuous chronic infectious stages following a relatively brief disease stage. Examples of chronic infections include hepatitis B and C. Initial symptoms may include nausea, fever, and jaundice; however, the patient typically recovers, but remains infectious. Over time the disease slowly manifests as hepatitis, cirrhosis of the liver, or cancer. In contrast to a chronic infection a slow infection has no disease signs initially yet the infected person becomes more infectious over time. Ultimately a disease becomes apparent. A good example of a disease caused by a slow infection is acquired immunodeficiency syndrome (AIDS), which is caused by the human immunodeficiency virus (HIV).

HOW ARE VIRAL DISEASES DIAGNOSED?

Viruses cannot be grown on agar like bacteria and fungi. Viruses are cultivated and their numbers increased in cultures of living cells called tissue or cell cultures. Viruses can also be cultivated in living animals and fertilized chicken eggs.

In a laboratory, viruses cannot be seen using standard light microscopy. Clinically, indirect techniques (such as hemagglutination) and tissue destructions (such as **plaque assays**) are used to detect viruses. Bacteriophages (viruses that infect bacteria) can cause zones of destruction (clear zones) called plaques when cultivated in host cells. Plaque assays involve plating dilutions of virus particles on an area of host cells (Figure 7-4). Results are expressed in plaque-forming units (PFU). In a plaque assay, each plaque is assumed to have come from the reproduction of a single virus. Therefore, the number of plaques at a given dilution will give the number of infectious virions, or plaque forming units.

Cell cultures make it possible to count animal viruses quickly and inexpensively. A single layer of animal cells is grown in a Petri dish and the cells are infected with a sample of animal

Figure 7-4 Plaques formed by a bacteriophage.

virus. A virion that kills the cultured cells causes a region of the cell layer to die, forming a plaque. These plaques are visible when stained with a vital stain (one that stains only living cells) and are counted to determine viral numbers.

Viruses can also cause cells to clump (agglutinate); therefore, agglutination techniques can serve as another detection method. Immunoassays (antigen/antibody testing such as enzyme-linked immunosorbent assay (ELISA), hemagglutination, etc.) are commonly used to detect viral infections clinically. Antigen/antibody tests may give qualitative (+ or −) results or quantitative (numeric value) results. In addition to antigen/antibody testing, gene amplification techniques like polymerase chain reaction (PCR) are becoming popular methods for the identification of viruses. Both ELISA techniques and PCR techniques are replacing older methods of detecting viral antigens because of the ease of performing/interpreting these types of tests and their high levels of sensitivity.

ARBOVIRUSES

Overview

Vertebrates are hosts to more than 500 viruses transmitted mainly by blood-sucking arthropods. **Arbovirus** is a named derived from *ar*thropod-*bo*rne virus and these viruses exist in cycles between vertebrate hosts and primary vectors. The major arbovirus families are the Togaviridae (the genus *Alphavirus*), Flaviviridae (the genus *Flavivirus*), Bunyaviridae (all genera except *Hantavirus*), and Reoviridae (the genera *Coltivirus* and *Orbivirus*).

Figure 7-5 Arboviruses are transmitted by mosquitoes. This image is of a female *Aedes aegypti* mosquito, the vector of Dengue fever.

(Courtesy of CDC/James Gathany)

Mosquitoes are the most important arthropod vector with ticks, sand flies, and gnats also playing a role in viral transmission (Figure 7-5). Humans are typically accidental, dead-end hosts for arboviruses and not important as intermediate hosts in the maintenance of the viral cycle.

The geographic location of arboviruses is based on several factors including the number and immune status of the vertebrates that represent the reservoir host, longevity of the vector, availability of food and breeding sites, climatic conditions in which the vector population must reproduce, the ability of the vector to over-winter (survive in winter) in a given area, feeding conditions especially for ticks that have a 2-year cycle in temperate climates, and the ability of multiple viruses to use the same reservoir host and vector. Infections usually show peak incidence when the arthropod is actively feeding and reproducing, usually from late spring through early fall. Warm-blooded vertebrates also maintain the virus during inclement periods of the year such as the cold and dry seasons. Humans can serve as **dead-end hosts, accidental hosts,** or **reservoir hosts.** Birds are more crucial to the survival of arboviruses than humans. Examples of arboviral diseases include Colorado tick fever, the equine encephalitides, St. Louis encephalitis, yellow fever, dengue fever, and West Nile fever. Arboviral infections that are zoonotic will be covered under their respective viral family chapters.

Controlling Arboviral Diseases

Mosquito-borne disease can be controlled by reducing the numbers of mosquitoes by the following methods (Figure 7-6):

- Removing their habitat.
- Eliminating standing water in rain gutters, old tires, buckets, toys, or any other container where mosquitoes can breed.

> Another viral name derived from its transmission is **robovirus** (*rodent-bo*rne virus). Hantaviruses and arenaviruses are also known as roboviruses.

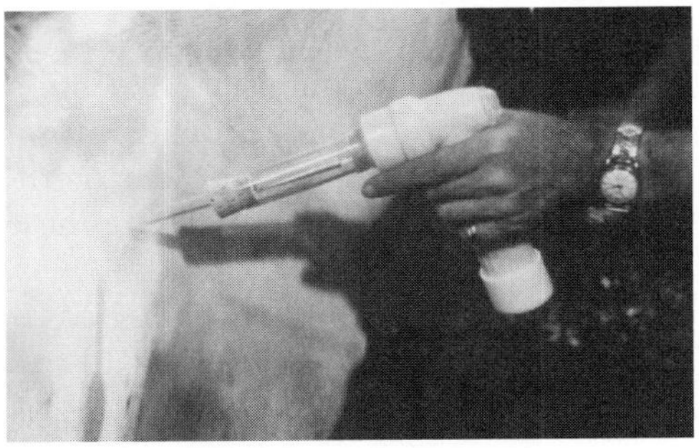

Figure 7-6 Mosquito collection and monitoring is important in tracking and controlling mosquito-borne diseases. Mosquitoes are being collected on the side of a horse, using a mechanical hand-held aspirator.
(Courtesy of CDC)

- Emptying and changing the water in bird baths, fountains, wading pools, rain barrels, and potted plant trays weekly to destroy potential mosquito habitats.
- Draining pools of water or filling them with dirt.
- Keeping swimming pool water treated and circulating.
- Using Environmental Protection Agency (EPA)-registered mosquito repellents when necessary and following label directions and precautions closely.
- Using head nets, long sleeves, and long pants when in areas with high mosquito populations.
- Staying inside during the evening when mosquitoes are active if there is a mosquito-borne disease warning in effect.
- Making sure window and door screens are intact to prevent mosquitoes from entering buildings.
- Replacing outdoor lights with yellow lights, which tend to attract fewer mosquitoes than ordinary lights.
- Spraying of mosquito-infested areas to prevent disease and nuisance caused by large mosquito numbers.

State or local mosquito specialists conduct surveillance of mosquito-borne diseases by evaluating larval and adult mosquito populations. These agencies may use the following prevention methods:

- Controlling water levels in lakes, marshes, ditches, or other mosquito breeding sites;
- Eliminating small breeding sites;

- Stocking bodies of water with fish species that feed on larvae;
- Using chemical and biological measures to kill immature mosquitoes during larval stages.

ARENAVIRUSES

Overview

Arenaviruses, in the order Mononegavirales and the Arenaviridae family, tend to cause hemorrhagic disease. The Arenaviridae family includes Lassa virus (Africa) and several rare South American hemorrhagic fevers such as Machupo, Junin, Guanarito, and Sabia. In 1933 lymphocytic choriomeningitis virus (LCMV) was the first arenavirus isolated during an investigation of an epidemic of St. Louis encephalitis (LCMV is not the cause of St. Louis encephalitis [a flavivirus], but LCMV was found to be a cause of meningitis). In 1958 Junin virus was shown to cause Argentine hemorrhagic fever, a disease found in the pampas in Argentina. In 1963, Machupo virus was isolated in the remote savannas of Bolivia. The next member of this virus family associated with an outbreak of human illness was Lassa virus in Nigeria in 1969. Lassa virus is the most clinically significant of the Arenaviridae and has been found in all countries of West Africa posing a significant public health problem in endemic areas that results in serious morbidity and mortality. Guanarito (Venezuela) and Sabia (Brazil) were more recently added to this family. Arenaviruses and the diseases they cause are summarizing in Table 7-5.

Arenaviruses are divided into two groups based on location. Old World arenaviruses are found in the Eastern Hemisphere and New World arenaviruses are found in the Western Hemisphere. LCMV is the only arenavirus to exist in both areas; however, it is classified as an Old World virus.

Arenaviruses tend to cause persistent subclinical infections in their natural hosts (mainly rodents) and severe disseminated disease in humans (accidental hosts). Transmission of the virus to humans occurs in areas where humans come in contact with virus-containing rodent urine. In humans, the disease is acute. There are five pathogens of humans: four cause severe hemorrhagic fever (Lassa virus in West Africa, Junin virus in the Argentine pampas, Machupo virus in Bolivia, and Guanarito virus in Venezuela) and the fifth, lymphocytic choriomeningitis (LCM) virus, causes milder infections.

The rodent hosts of arenaviruses are chronically infected with the viruses. Some Old World arenaviruses appear to be passed from dams to offspring during pregnancy; remaining in the rodent population generation after generation. Some New World arenaviruses are transmitted among adult rodents usually from fighting and inflicting bites. The viruses are shed into the environment in the urine or droppings of their infected hosts.

Human infection with arenaviruses is incidental to the natural cycle of the viruses and occurs when an individual comes in contact with the excretions or materials contaminated with the excretions of an infected rodent, such as ingestion of contaminated food or by direct contact of abraded/broken skin with rodent excrement. Infection can also occur by inhalation of tiny particles soiled with rodent urine or saliva. Some arenaviruses, such as Lassa and Machupo viruses, are associated with secondary person-to-person (direct contact with virus-contaminated blood or other excretions) and nosocomial transmission (including indirect transmission via contaminated medical equipment). The epidemiology of

Table 7-5   **Categories of Arenaviruses**

Virus	Rodent Host	Location	Source of Human Contact
Old World Arenaviruses			
Lymphocytic choriomeningitis virus	*Mus musculus* and *Mus domesticus* (house mouse) and *Mesocricetus auratus* (Syrian hamster)	Europe, Asia, and the Americas	Primarily within households
Lassa virus	*Mastomys natalensis* (multimammate mouse)	West Africa	Primarily within households
Mopeia virus	*Mastomys natalensis*	Southern Africa	Unclear
Mobala virus	*Praomys* spp. (soft-furred rat)	Central African Republic	Unclear
Ippy virus	*Arvicanthus* spp. (Nile grass rat)	Central African Republic	Unclear
Classic New World Arenaviruses			
Junin virus	*Calomys masculinus* (corn mouse), *Akodon azarae* (grass field mouse), *Bolomys obscurus* (dark field mouse)	Argentina	Occupational hazard from working in fields
Machupo virus	*Calomys callosus* (vesper mouse)	Bolivia	Primarily within households
Guanarito virus	*Sigmodon alstoni* (cane mouse)	Venezuela	Primarily within households
Sabia virus	Unknown	Brazil	Associated with human cases
Selected Less Common New World Arenaviruses			
Tacaribe virus	*Artibeus* spp. (fruit-eating bat)	Trinidad	Unclear
Amapari virus	*Oryzomys goeldii* (rice rat), *Neacomys guianae* (bristly mouse)	Brazil	Unclear
Tamiami virus	*Sigmodon hispidus* (hispid cotton rat)	Florida	Unclear
Whitewater Arroyo	*Neotoma albigula* (white-throated wood rat)	California, New Mexico	Unclear

arenaviral diseases depends on patterns of infection in reservoir hosts and on the factors which bring man into contact with rodents or their saliva and urine.

Causative Agent

> The Arenaviridae family of viruses is generally associated with rodent-transmitted disease in humans.

The Arenaviridae family (*arenosus* is Latin for sand) was named as a result of the sandy appearance of ribosome structures within the virion that are acquired from the host during the budding process (their function is unknown). New viral particles are created by budding from the surface of their hosts' cells. Arenaviruses are enveloped, single-stranded, negative sense RNA viruses. Their genome has only two segments. The virus particles are spherical and have an average diameter of 110 to 130 nm.

See Table 7-6 for less common zoonotic arenaviruses.

Table 7-6   Less Common Zoonotic Arenaviruses

Arenavirus	Disease	Predominant Signs in People	Animal Source and Clinical Signs	Transmission	Geographic Distribution	Diagnosis	Control
Lassa virus	Lassa fever First described in 1969 when three American missionary sisters became infected in Lassa, Nigeria	High fever, lethargy, muscle pain, headaches, ulcerations on mucous membranes, enlarged cervical lymph nodes, coughing, and vomiting; approximately one-third of people develop hemorrhagic signs such as hemoptisis, melena, hematuria, and hematomas (once hemorrhagic signs develop disease is typically fatal).	▪ Mice (*Mastomys* genus) are asymptomatic	▪ Rodent-excrement contaminated food and dust ▪ Ingestion of infected rodents ▪ Person-to-person via direct contact or sexual contact ▪ Mechanical transmission via contaminated needles	African countries south of the Sahara (West Africa, Nigeria, Liberia, and Sierra Leone)	Virus isolation (BSL-4 precautions), RT-PCR, ELISA, indirect immunofluorescence test	Rodent control; barrier protection when caring for infected individuals; isolation of infected people. Early administration of ribavirin has helped in the treatment of some arenaviral diseases such as Lassa virus.
Junin virus	Argentinian hemorrhagic fever	Fever, muscle pain, hemorrhagic disorders, shock, and neurologic disease	▪ Rodents are asymptomatic	Direct contact with *Calomys laucha* and *Calomys musculinus* mice	Argentina	Virus isolation (BSL-4 precautions), RT-PCR, ELISA	Rodent control, administration of serum from recovered people; isolation of infected people, vaccine is available.
Machupo virus	Bolivian Hemorrhagic Fever	Fever, muscle pain, hemorrhagic disorders, shock, and neurologic disease	▪ Rodents are asymptomatic	Direct contact with *Calomys callosus* mice	Bolivia (providence of Beni)	Virus isolation (BSL-4 precautions), RT-PCR, ELISA	Rodent control, administration of serum from recovered people; isolation of infected people, vaccine for Argentinian hemorrhagic fever offers cross protection.
Guanarito virus	Venezuelan hemorrhagic fever	Fever, muscle pain, hemorrhagic disorders, shock, and neurologic disease	▪ Rodents are asymptomatic	Direct contact with *Sigmodon alstoni* cotton rats	Venezuela	Virus isolation (BSL-4 precautions), RT-PCR, ELISA	Rodent control, administration of serum from recovered people; isolation of infected people.
Lymphocytic chorio-meningitis virus (LCM) There are three strains: ▪ WE ▪ Armstrong ▪ Traub	Lymphocytic choriomeningitis	LCM begins with influenza-type signs such as high fever, lethargy, headaches, photophobia, and bronchitis. The disease may not progress from this point or following a brief recovery meningitis may appear which people developing neck stiffness, confusion, and nausea. LCM may be a cause of congenital lesions of the CNS and impaired vision.	▪ Rodents such as mice, hamsters, and rats are asymptomatic. ▪ The natural reservoir is the house mouse (*Mus musculus*).	Infected rodents can spread LCMV through bite wounds, smear infection, or by aerosol. Food contaminated with mouse excretions is an important source of human infection. Arthropod transmission has been demonstrated experimentally.	Worldwide, but is best documented in Europe and North America	RT-PCR, indirect immunofluorescence, neutralization, and immunoenzyme tests are available.	Lumbar spinal tap can give relief for LCM patients Rodent control, testing of laboratory mice and hamsters, careful handling of rodents and their excrements, educating women of childbearing years about the potential risk in handling rodents.

BUNYAVIRUSES

Overview

Bunyaviruses, in the Bunyaviridae family, are named after Bunyamwera, Uganda, where the first virus in this group was found in mosquitoes. Bunyaviruses consist of arthropod-borne viruses (arboviruses) and rodent-borne viruses (roboviruses). The family is divided into four genera that affect animals: *Nairovirus*, *Bunyavirus*, and *Phlebovirus* (all transmitted in nature by arthropods) and *Hantavirus* (the only genus that is transmitted by rodents). The specific vectors and life cycles of bunyaviruses differ from one another, but typically involve mammalian and avian reservoir hosts with arthropod vectors.

Most bunyaviruses cause hemorrhagic fevers and encephalitis; hantaviruses cause severe respiratory distress syndromes. Diseases caused by bunyaviruses tend to be endemic to a particular region with each region having its own specific virus, reservoir, and vector. Native populations in endemic regions have therefore been exposed to the virus throughout their lives. Outbreaks occur when nonnative people enter the area, typically through war. Hantavirus killed nearly 10% of the United Nations soldiers it infected in Korea in 1951; sandfly fever viruses caused epidemics in Allied troops in World War II and the soldiers of the Napoleonic wars.

Bunyaviruses cause diseases that include Rift Valley fever (a virus found in Africa that is transmitted to humans and livestock by mosquitoes) and Crimean-Congo hemorrhagic fever (a bunyavirus found in Africa, Asia, and Europe that is carried by ticks and causes a highly pathogenic hemorrhagic fever).

Causative Agent

The Bunyaviridae family of viruses are spherical, single-stranded, enveloped, negative sense RNA viruses whose genome has three segments (designated small, medium, and large, or S, M, and L). Bunyaviruses are typically 90 to 100 nm in length. These viruses replicate in the cytoplasm and assembly is by budding through the smooth membrane of the Golgi apparatus.

La Crosse Encephalitis

Overview

La Crosse encephalitis, also known as California encephalitis, is an acute mosquito-borne viral infection causing encephalitis or encephalomeningitis. The first viruses of the California serogroup were isolated in a person in Kern County, in the Central Valley of California in 1943 (caused by a virus called California virus). Two years later, three human cases of encephalitis were linked to this new virus with all three cases occurring in residents of Kern County, California. Since the original cases from California, no further cases of human disease caused by the California encephalitis virus have been reported. Other closely related viruses were isolated, but were not associated with human disease until 1960. Another virus in this serogroup, La Crosse virus, was first isolated in 1963 from the brain of a 4-year-old boy who died of encephalitis in La Crosse County, Wisconsin. The viruses are now classified in the California encephalitis complex (with 14 serotypes) with La Crosse virus being one of the serotypes. The most serious and only fatal disease in this group is La Crosse encephalitis (LAC encephalitis) caused by the La Crosse virus. Historically, most cases of LAC encephalitis occur in the upper Midwestern states (Minnesota, Wisconsin, Iowa, Illinois, Indiana, and Ohio); however, more cases are being reported from states in the mid-Atlantic (West Virginia, Virginia, and North Carolina) and southeastern (Alabama and Mississippi) regions of the United States. Most cases of LAC encephalitis occur in children under 16 years of age.

> La Crosse virus is the most common cause of pediatric arboviral encephalitis in the United States.

Causative Agent

La Crosse virus (LAC virus) is part of the California serogroup in the family Bunyaviridae in the genus *Bunyavirus*. LAC virus is a negative sense, single-stranded RNA viruses with an envelope (Figure 7-7). Other human pathogens such as Jamestown Canyon virus and Tahyna virus also belong to the California serogroup. Four proteins are associated with LAC virus: two of the proteins are glycoproteins (G1 and G2), one is a component of the viral nucleocapsid (N), and the other is associated with the viral nucleocapsid (L).

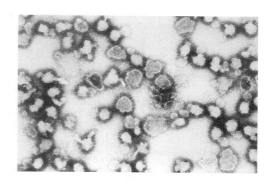

Figure 7-7 Electron micrograph of La Crosse virus. (Courtesy of CDC)

Epizootiology and Public Health Significance

LAC encephalitis is most commonly seen in the hardwood deciduous forest areas of the upper Midwestern United States (Minnesota, Wisconsin, Iowa, Illinois, Indiana, and Ohio) and in the Appalachian region of the United States (West Virginia, North Carolina, Tennessee, and Virginia). The number of inapparent human infections is high with 40% of people in Wisconsin having antibodies to LAC virus. Epidemics are frequent in Wisconsin, Minnesota, Iowa, Ohio, Illinois, and Indiana.

In the United States about 75 cases of LAC encephalitis are reported annually to the CDC. From 1996 to 1998, approximately three times as many human cases of arboviral encephalitis were caused by California serogroup viruses then were reported for all other viruses combined. Midwestern states carry the highest incidence of LAC encephalitis in the United States with most cases occurring in the late summer to early fall. Increased incidence occurs with outdoor activities, especially in woodland areas. LAC encephalitis has been reported in 28 states. Most patients with clinical signs recover completely; however, 20% of patients develop behavioral problems or recurrent seizures. Mortality rates are low (<1%).

Transmission

California viruses are transmitted by numerous mosquitoes with several small mammals serving as reservoirs (Figure 7-8). The daytime-biting treehole mosquito, *Aedes triseriatus*, and vertebrate amplifier hosts such as chipmunks and squirrels are responsible for transmitting and maintaining LAC virus. Mosquitoes acquire the virus after a blood meal from hosts who are in the viremic stage. LAC virus is maintained over the winter by transovarial transmission in mosquito eggs. If the female mosquito is infected, she may lay eggs that carry the virus, and the adults originating from those virus-infected eggs may be able to transmit the virus to small mammals and humans.

> *Ae. triseriatus* is a daytime biter and only travels a few hundred yards from its breeding location. Breeding locations include tires, tree holes, and other types of containers.

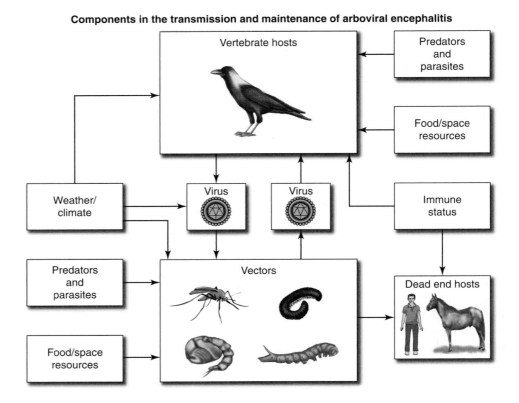

Components in the transmission and maintenance of arboviral encephalitis

Figure 7-8 Transmission and maintenance of arboviral encephalitis, such as La Crosse encephalitis.

Pathogenesis

After inoculation via a mosquito bite, the virus undergoes focal replication at the original skin site. A primary viremia occurs and the virus travels through the reticuloendothelial system (mainly the liver, spleen, and lymph nodes). As the virus continues to replicate a secondary viremia occurs allowing the virus to infect the CNS. The virus invades the CNS through either cerebral capillary endothelial cells or the choroid plexus. CNS infection depends on the efficiency of viral replication and the degree of viremia. The host's immune system may try to contain the infection by production of antibodies against the G1 part of the virus. These antibodies neutralize the virus, block fusion, and inhibit hemagglutination in an attempt to clear the virus, allow for recovery, and prevent reinfection.

Clinical Signs in Animals

Many wildlife small mammals are serologically positive for LAC virus without showing clinical signs of disease. The incubation period in animals is short (24 to 48 hours) and clinical signs of disease have only been seen in experimental infections.

Clinical Signs in Humans

The incubation period for LAC encephalitis is approximately 2 to 7 days with the disease typically occurring during the summertime. LAC encephalitis results in subclinical infection more commonly than clinical infection. Clinical infection with LAC virus initially presents as a nonspecific illness with clinical signs of fever, headache, nausea, vomiting, and lethargy. A more severe form of the disease occurs in children younger than the age of 16 and is characterized by seizures, coma, paralysis, and a variety of neurologic sequelae after recovery (poor memory, poor balance, and speech problems). Over 50% of children who have had a severe case of La Crosse encephalitis and are older than 5 years of age have signs of attention-deficit/hyperactivity disorder and nearly one-third had borderline intelligence or mental retardation. During the acute phase, frequent seizures can occur and recur in about 20% to 50% of people. Death from LAC encephalitis occurs in less than 1% of clinical cases. Recovery can be slow, especially in children.

Diagnosis in Animals

Diagnosis of LAC encephalitis in animals is not done clinically, but may be done to monitor the virus in wildlife. Virus isolation during the viremic stages in an animal or a rise in antibody titer by complement fixation methods are available in animals. Reverse transcriptase polymerase chain reaction (RT-PCR) is used for testing mosquitoes.

Diagnosis in Humans

LAC encephalitis can be diagnosed in people by hemagglutination inhibition (greater than 320) or paired antibody serum tests (a fourfold increase in antibody titer between the acute and the convalescent samples). Virus isolation from living patients has not been accomplished, but the virus has been isolated from brain tissue postmortem. A group-specific RT-PCR test is available for Simbu-California-Bunyamwera serogroup testing.

Treatment in Animals

Animals are not treated for LAC encephalitis.

Treatment in Humans

Treatment of LAC encephalitis in people is symptomatic to manage seizures and increased intracranial pressure. Ribavirin may be used in severe cases and is believed to block viral replication of bunyaviruses.

Management and Control in Animals

Mosquito control may help decrease the level of viremia in small mammals, but a control program has not been established.

Management and Control in Humans

There is not a currently approved vaccine for the prevention of LAC virus infection. LAC encephalitis can also be reduced by mosquito control as previously described.

Summary

La Crosse encephalitis (LAC encephalitis) is an acute mosquito-borne viral infection causing encephalitis or encephalomeningitis. LAC encephalitis is caused by La Crosse virus (LAC virus), a virus that is part of the California serogroup in the family Bunyaviridae in the genus *Bunyavirus*. LAC encephalitis is most commonly seen in the hardwood deciduous forest areas of the upper Midwestern United States and in the Appalachian region of the United States. Epidemics are frequent in Wisconsin, Minnesota, Iowa, Ohio, Illinois, and Indiana.

California viruses are transmitted by numerous mosquitoes with several small mammals serving as reservoirs. The daytime-biting treehole mosquito, *Ae. triseriatus*, and vertebrate amplifier hosts such as chipmunks and squirrels are responsible for transmitting and maintaining LAC virus. LAC virus is maintained over the winter by transovarial transmission in mosquito eggs. Many wildlife small mammals are serologically positive for LAC virus without showing clinical signs of disease. In people, LAC encephalitis typically occurs during summertime and produces subclinical infection more commonly than clinical infection. Clinical infection with LAC virus initially presents as a nonspecific illness with clinical signs of fever, headache, nausea, vomiting, and lethargy (more severe disease occurs in children younger than the age of 16). Death from LAC encephalitis occurs in less than 1% of clinical cases. Diagnosis of LAC encephalitis in animals is not done clinically, but may be done to monitor the virus in wildlife. RT-PCR is used for testing mosquitoes. LAC encephalitis can be diagnosed in people by hemagglutination inhibition or paired antibody serum tests. Animals are not treated for LAC encephalitis. Treatment of LAC encephalitis in people is symptomatic to manage seizures and increased intracranial pressure. Ribavirin may be used in severe cases. Mosquito control may help reduce LAC virus in animals and people. There is not a currently approved vaccine for the prevention of LAC virus infection.

Hantavirus Infection

Overview

Hantaviruses are a group of antigenically distinct viruses that are carried by rodents and are transmitted to humans when they inhale vapors from contaminated rodent urine, saliva, or feces. There are at least 25 strains of hantavirus with some strains being pathogenic and other strains being nonpathogenic. A disease that appeared to have been caused by hantavirus was first described in the 1930s in the Amur basin (Russia) and in Manchuria. During the Korean War in the early 1950s about 3,000 U.S. and U.N. forces were infected with a mysterious viral illness that is now believed to have been caused by a hantavirus. In 1976 the first hantavirus isolated from Korean field mice found near the Han River near Seoul, South Korea was the Hantaan virus. Hantaan virus and its related strains, Seoul virus, Dobrava virus, and Puulmala virus, cause Korean hemorrhagic fever (KHF) in the Far East; hemorrhagic fever with renal syndrome (HFRS), nephrosonephritis, or Tula fever in Russia and China; and nephropathia epidemica (NE) in Scandinavia. These diseases produce a condition in which the capillaries of the circulatory system begin to leak blood with outbreaks of hemorrhagic fevers occurring in Russia (1913), Scandinavia (1932–1935), and Finland (1945). A second disease, hantavirus pulmonary syndrome (HPS), was identified in the United States in 1993 and is caused by many strains of the virus (most cases in the United States are caused by the Sin Nombre virus). It is known to be carried by deer mice, white-footed mice, and cotton rats. HPS is much more deadly than HFRS, causing flu-like symptoms that can lead to pneumonia and death. The Sin Nombre strain was named for a Spanish massacre of Native Americans that occurred in the Muerto Canyon where the virus was discovered. The Sin Nombre strain was the cause of a 1993 outbreak in the Four Corners area of New Mexico in the Southwestern United States that killed 32 of 53 infected people (Four Corners is the intersection formed by the borders of Utah, New Mexico, Arizona, and Colorado). That outbreak occurred in part from weather changes caused by El Niño, which led to increases in the rodent populations. Other hantavirus strains have caused disease in South America in 1996 and 1997 and in Africa in 2006 (that may be named Sangassou virus, for the region in which it was found).

The *Hantavirus* genus is unique in the Bunyaviridae because there are no arthropod vectors in its transmission cycle.

Causative Agent

HFRS and HPS are both caused by viruses in the genus *Hantavirus* in the family Bunyaviridae (the virus was added to this family in 1987) (Figure 7-9). Hantaviruses are enveloped viruses with three single-stranded RNA segments: S (small), M (medium), and L (large). The S RNA encodes the nucleocapsid (N) protein, the M RNA encodes a polyprotein that is split into the envelope glycoproteins G1 and G, and the L RNA encodes the L protein, which helps with viral replication. Hantaviruses replicate in the host cell cytoplasm. The hantaviruses that cause HFRS are known as Old World Hantaviruses, whereas the hantaviruses that cause HPS are known as New World Hantaviruses. Hantaviruses (and arenaviruses) are known as roboviruses because they are transmitted by rodents.

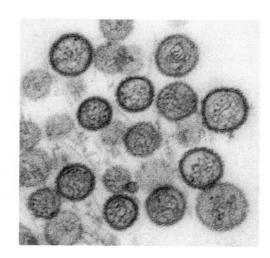

Figure 7-9 Electron micrograph of Sin Nombre virus.
(Courtesy of CDC/Cynthia Goldsmith, Luanne Elliott)

Epizootiology and Public Health Significance

Hantaviruses are found worldwide in rodents. HFRS viruses occur in Europe and Asia. The Seoul virus (SEOV) can be found worldwide (it has been associated with cases in the United States). The strains seen in various regions include Hantaan (HTNV), which occurs in China, Russia, and Korea; Puumala (PUUV), which occurs in Europe, Russia, and Scandinavia; and Dobrava (DOBV), which occurs in the Balkans.

HPS viruses occur only in North, Central, and South America. The strains that are seen in the United States are Sin Nombre (SNV), New York (NYV), Black Creek (BCC), and Bayou (BAY). In South America the strains seen are Andes (Argentina and Chile), Bermejo (Bolivia and Argentina), Oran (Argentina), Lechiguanas (Argentina), Laguna Negra (Paraguay and Bolivia), Choclo (Panama), Araraquara (Brazil), and Casteol dos Sonhos (Brazil).

The United States typically sends over $7 million to fight Hantavirus infections annually (funds through the CDC). Since hantavirus was first identified in 1993, a total of 379 laboratory-confirmed cases have been reported nationally. HPS has been reported in 31 states, the majority of cases in the southwestern United States. About three-fourths of cases have been in rural areas. The fatality rate in people with HPS is approximately 50%.

Transmission

Transmission of hantaviruses can occur when dried materials contaminated by rodent excreta are disturbed and inhaled, directly introduced into broken skin or conjunctivae, or when ingested in contaminated food or water. Rodent bites have also transmitted the virus. Person-to-person transmission has only been documented once. Table 7-7 summarizes hantaviruses and their vectors. HFRS-causing hantaviruses are associated with murine and arvicoline rodents. In Europe and Asia the Hantaan viruses are found in murine field mice (*Apodemus agrarius*), Dobrava viruses are found in a different species of murine field mice (*Ap. flavicollis*), Seoul viruses are found in the Norway rat (*Rattus norvegicus*), and Puumala virus is found in the bank vole (*Clethrionomys glareolus*), an arvicoline rodent.

Table 7-7   **Hantaviruses and Their Vectors**

Disease Form	Hantavirus Strain	Vector
HFRS Vectors	HTNV	Striped field mouse (*Apodemus agrarius*)
	DOBV	Yellow-necked field mouse (*Apodemus flavicollis*)
	PUUV	Bank vole (*Clethrionomys glareolus*)
	SEOV	Commensal rats (*Rattus norvegicus* and *Rattus rattus*)
HPS Vectors	SNV	Deer mice (*Peromyscus maniculatus*)
	NYV	White-footed mouse (*Peromyscus leucopus*)
	Black Creek Canal	Cotton rat (*Sigmodon hispidus*)
	BAY	Rice rat (*Oryzomys palustris*)

Wet weather allows more food plants to grow in the spring and summer, which in turn leads to larger rodent populations. Denser rodent populations spread the disease by facilitating mouse-to-mouse transmission and by increasing the competition for food (sends more infected mice inside peoples' homes).

All HPS-causing hantaviruses are carried by the New World rats and mice (family Muridae, subfamily Sigmodontinae). Sigmodontinae contains at least 430 species of mice and rats, which are widespread in North and South America. These wild rodents are not encountered by humans as frequently as house mice and the black and Norway rats (all of which are in the murid subfamily Murinae). In the United States hantaviruses associated with Sigmodontinae rodents include New York virus found in the white-footed mouse, *Peromyscus leucopus*; Black Creek Canal virus found in the cotton rat, *Sigmodon hispidus*; Bayou virus found in the rice rat, *Oryzomys palustris*; and Sin Nombre virus found in the deer mouse (*Peromyscus maniculatus*). Sin Nombre virus is the primary causative agent of HPS in the United States. Different species of Sigmodontinae rodents are associated with HPS in South America. Sigmodontinae rodents are restricted to the Americas; therefore, HPS is restricted to the Americas.

Pathogenesis

Infection with hantaviruses occurs by inhalation of aerosolized virus from feces, urine, or saliva of infected rodents. Hantaviruses replicate in endothelial cells (kidneys for HFRS, pulmonary vasculature for HPS) initiating an immune response to viral antigens that have penetrated the endothelium causing vascular damage. The onset of clinical signs correlates with the development of specific antibodies to the virus. This vascular damage increases capillary permeability that can lead to nephron damage or pulmonary edema.

Clinical Signs in Animals

Rodent species infected with hantavirus are asymptomatic (Figure 7-10). Once infected with hantavirus rodents remain infected and can excrete virus particles for life.

Figure 7-10 The cotton rat is a hantavirus carrier that becomes a threat when it enters human habitation in rural and suburban areas. Hantavirus pulmonary syndrome (HPS) is a deadly disease transmitted by infected rodents through urine, droppings, or saliva. Humans can contract the disease when they breathe in aerosolized virus.

(Courtesy of CDC/James Gathany)

Clinical Signs in Humans

The incubation period for hantavirus infections is 14 to 17 days with symptoms appearing after 14 to 30 days. People with HFRS usually present with an acute onset of disease with clinical signs of kidney disease including headache, fever, chills, and backache. People may also have facial flushing or a petechial rash. Fever typically lasts 3 to 8 days and is followed by a proteinuric period. During this stage of HFRS, hypotension, nausea, vomiting, and death may result from shock. This stage is followed by an oliguric stage as kidney function improves. If the disease is fatal, death typically occurs during the hypotensive or oliguic phases. In severe cases, kidney failure, pulmonary edema, or disseminated intravascular coagulation (DIC) may occur. Hantaan virus and Dobrava virus typically produce severe disease, whereas Seoul virus produces moderate disease and Puumala virus produces mild disease. Convalescence may take weeks or months.

HPS shows clinical signs of pulmonary disease with the initial phase with clinical signs of fever, lethargy, nausea, vomiting, and diarrhea. Respiratory disease (cough, tachypnea, pulmonary edema (Figure 7-11), and hypoxia) and

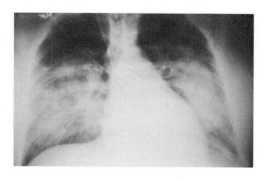

Figure 7-11 Bilateral pulmonary effusion as a result of hantavirus pulmonary syndrome (HPS).

(Courtesy of CDC/D. Loren Ketai, MD)

hypotension occurs abruptly. Cardiac signs such as bradycardia, ventricular tachycardia, or fibrillation may be seen. Convalescence may take weeks or months. Mild infections are rare.

Diagnosis in Animals

Diagnosis of hantavirus in animals is not done clinically. Identification of infection is done to assess infection rates in a given area. RT-PCR tests are available for several strains of hantavirus.

Diagnosis in Humans

Isolation of hantavirus in tissue culture is difficult and a biosafety issue (hantaviruses are BSL 4 organisms). Specific diagnosis of hantavirus infection may be achieved by serologic techniques (ELISA IgM and IgG titers), immunoblotting (Western blotting), and immunohistochemistry (IHC) studies. PCR testing is also available. A rapid immunoblot strip assay (RIBA) has been developed to detect antibodies to Sin Nombre virus.

Treatment in Animals

Animals are not treated for hantavirus infection.

Treatment in Humans

HPS occurs in the United States primarily in the fall when small rodents come indoors to protect themselves from the cold weather.

Antiviral medication such as ribavirin can shorten the course of HFRS-hantavirus infection if initiated early enough (it has not been affective with cases of HPS-hantavirus infection). Symptomatic treatment for kidney disease (HFRS) and pulmonary disease (HPS) is warranted.

Management and Control in Animals

Transmission of hantavirus between rodents can be attempted by controlling rodent populations; however, the wide distribution of Sigmodontinae rodents in North America and their importance in the function of natural ecosystems makes this option not feasible.

Management and Control in Humans

Controlling human disease transmitted by rodents is best accomplished through environmental modification and hygiene practices that deter rodents from colonizing home and work environments. The safe cleanup of rodent waste and nesting materials is also necessary to control rodent-borne disease. Some ways to control the interaction of rodents and humans include:

- Inspect both the interior and exterior of the home at least twice per year for any openings where rodents could enter the home and for conditions that could support rodent activity.
- Keep food (human and pet) and water covered and stored in rodent-proof containers.
- Wash dishes, pans, and cooking utensils immediately after use.
- Remove leftover food and clean up all spilled food from cooking and eating areas.
- Dispose of trash and garbage on a frequent and regular basis, and pick up or eliminate clutter.

■ Keep items such as boxes, clothes, and blankets off the floor to prevent rodents from nesting in them.

■ Repair water leaks and prevent condensation from forming on cold water pipes by insulating them.

■ Keep exterior doors and windows closed unless protected by tight-fitting screens.

■ Use traps to lower the rodent population in homes. When disposing of traps or trapped animals wear rubber, latex, vinyl, or nitrile gloves. Spray the dead rodent with a disinfectant or chlorine solution. Dispose of the rodent in the double bag.

■ Place woodpiles and stacks of lumber, bricks, stones, or other materials over 100 feet from the house.

■ Mow grass closely, and cut or remove brush and dense shrubbery to a distance of at least 100 feet from the home.

■ Use raised cement foundations in new construction of sheds, barns, and out-buildings.

Summary

Hantaviruses are a group of antigenically distinct viruses that are carried by rodents. Hantaan virus and its related strains (Seoul virus, Dobrava virus, and Puulmala virus) cause Korean hemorrhagic fever in the Far East; hemorrhagic fever with renal syndrome (HFRS) in Russia and China; and nephropathia epidemica in Scandinavia. A second disease, HPS, was identified in the United States in 1993 and most cases in the United States are caused by the Sin Nombre virus. It is known to be carried by deer mice, white-footed mice, and cotton rats. (See Table 7-8 for less common zoonotic Bunyaviruses.)

Hantaviruses are found worldwide in rodents. HPS viruses occur only in North, Central, and South America. HFRS viruses occur in Europe and Asia. Transmission of hantavirus can occur when dried materials contaminating rodent excreta are disturbed and inhaled, directly introduced into broken skin or conjunctivae, or when ingested in contaminated food or water. Rodent bites have also transmitted the virus. Rodent species infected with hantavirus are asymptomatic. Once infected with hantavirus rodents remain infected and can excrete virus particles for life. People with HFRS usually present with an acute onset of disease with clinical signs of kidney disease (headache, fever, chills, and backache). During the acute stage of HFRS, hypotension, nausea, vomiting, and death may result from shock. This stage is followed by an oliguric stage as kidney function improves. People with HPS show clinical signs of pulmonary disease in the initial phase with clinical signs of fever, lethargy, nausea, vomiting, and diarrhea. Respiratory disease and hypotension occurs abruptly.

Diagnosis of hantavirus in animals is not done clinically. Specific diagnosis of hantavirus infection in people may be achieved by serologic techniques (enzyme-linked immunoabsorbant assay [ELISA] IgM and IgG titers), immunoblotting (Western blotting), and immunohistochemistry (IHC) studies. PCR tests are also available. A RIBA has been developed to detect antibodies to Sin Nombre virus. Animals are not treated for hantavirus infection. Antiviral medication can shorten the course of HFRS-hantavirus infection if initiated early enough; however, it has not been affective with cases of HPS-hantavirus infection. Rodent control and the safe cleanup of rodent waste and nesting materials may help reduce rodent-borne disease.

Table 7-8　Less Common Zoonotic Bunyaviruses

Bunyavirus	Disease	Predominant Signs in People	Animal Source and Clinical Signs	Transmission	Geographic Distribution	Diagnosis	Control
Rift Valley Fever virus	Rift Valley Fever Rift Valley fever was first identified in the 1930s when it infected people of Kenya's great Rift Valley.	Flu-like symptoms of fever, weakness, muscle pain, headache, nausea, and photophobia. Occasionally a hemorrhagic condition develops 2 to 4 days after fever producing jaundice, hematemesis, melena, petechiae, and death.	▪ Sheep and cattle are the primary hosts and amplifiers of RVF virus. ▪ Adult sheep have fever, mucopurulent nasal discharge, and occasionally vomiting. Lambs have fever, anorexia, weakness, and death (fatality rate 90%). ▪ Adult cattle have fever, weakness, anorexia, excessive salivation, and occasionally diarrhea. Calves have fever and lethargy (fatality rate 10% to 70%). ▪ Goats and dogs are highly susceptible to disease producing abortion in pregnant animals; horses and pigs are resistant.	Infection may be acquired by *Aedes* mosquitoes or exposure to aerosols.	Africa	Virus isolation, immunofluorescence, complement fixation, immunodiffusion, ELISA, virus neutralization, and hemagglutination inhibition	Mosquito control; notification of state and federal veterinarians in suspected cases; carcasses should be buried or burned; an inactivated vaccine is available for cattle and an attenuated live vaccine is used for sheep and goats.
Oropouche virus	Oropouche Fever; also called *febre de Mojiu* (after the village of Mojui in northern Brazil). Virus was first isolated in 1955 in a forest worker in Trinidad.	Onset is sudden with fever, rigors, headaches, muscle pain, joint pain, and photophobia; rarer signs include bronchitis, nausea, and vomiting. Disease usually lasts 2 to 7 days	▪ Monkeys and sloths are asymptomatic.	▪ Urban cycle (human to human) is via *Culicoides paraensis*. ▪ Sylvan cycle (monkeys and sloths to humans) is via *Aedes albopictus* and *Culex quinquefasciatus* mosquitoes.	Brazil, Colombia, Panama, Peru, Tobago, and Trinidad	Virus isolation, RT-PCR, ELISA	Mosquito control

Crimean-Congo hemorrhagic fever virus; genus *Nairovirus* with at least 3 subtypes	Crimean-Congo hemorrhagic fever (CCHF) First described in 1944 on the Crimean peninsula of the Ukraine	Abrupt fever, shivering, lethargy, backache, anorexia, vomiting, and nausea. Recurring fever and skin rash on face and neck is common. Bleeding on mucosal membranes, hematemesis, melena, and urogenital bleeding may be seen in 75% of people and is fatal in about 30% to 50% of these.	■ Domestic ruminants (cattle more so than sheep and goats) are asymptomatic. ■ Birds (ostriches and chickens), hedgehogs, horses, and rodents are asymptomatic.	Infection may be acquired by ticks predominantly of the genus *Hyalommao*. CCHF can also be contracted by handling animals at slaughter, while performing castrations, branding animals, and during birthing procedures. Nosocomial infections are common.	The former USSR, Eastern Europe, the Middle East, China, parts of Africa, and Australia.	Virus isolation (BSL-4 level procedures), RT-PCR, ELISA	Vaccine is available in Russia; Wearing gloves when handling animals; Tick prevention; Proper lab handling of specimens
Sandfly fever virus, genus *Phlebovirus*	Sandfly fever, also called phlebotomus fever and pappataci fever. First observed in 1886 in solders in Herzegovina and was isolated from American soldiers in WWII (more than 10,000 soldiers infected).	Abrupt, high fever, headaches, photophobia, muscle and joint pain, vomiting, anorexia, and facial flushing; occasionally meningoencephalitis may occur.	■ Sheep, cattle, squirrels, and forest mice are asymptomatic. ■ Gerbils, ground squirrels, and hedgehogs may be amplification hosts.	Infection is acquired via a sandfly species (*Phlebotomus papatasii, Phlebotomus perfilieri, and Phlebotomus perniciosus*)	Mediterranean countries (Italy, Portugal, Spain, and Cyprus), Northern Africa, South America, and West and Central Asia	Virus isolation; PCR, ELISA, hemagglutination inhibition, and virus neutralization	Sandfly control (especially from May to October) and disruption of their breeding areas.

CORONAVIRUSES

Overview

Coronaviruses, in the Coronaviridae family, are named for the crownlike surface projections seen when the virus is examined by electron microscopy (*corona* is Latin for crown). Coronaviruses were first isolated from chickens in 1937 and by the 1950s approximately 50% of human colds in which the etiologic agent was not identified were attributed to coronaviruses. In 1965, Tyrrell and Bynoe cultivated the first human coronavirus (HCoV) in vitro. There are now approximately 15 species in the Coronaviridae family, which infect humans, cattle, pigs, rodents, cats, dogs and birds. Coronaviridae are divided into two genera: *Coronavirus* and *Torovirus*. Coronaviruses are further divided into three groups based on antigenic features as a result of the presence or absence of hemagglutinin-esterase protein and gene arrangement, which are summarized in Table 7-9. Coronaviruses primarily infect the upper respiratory and gastrointestinal tract. Coronaviruses are

Table 7-9   Coronavirus Groups and Examples of Virus

Coronavirus Group	Virus	Species
I	Transmissible gastroenteritis virus	Pigs
	Porcine respiratory coronavirus	Pigs
	Porcine epidemic diarrhea virus	Pigs
	Canine coronavirus	Dogs
	Feline coronavirus	Cats
	Feline infectious peritonitis virus	Cats
	Rabbit coronavirus	Rabbits
	Human coronavirus 229E (cold virus)	Humans
II	Murine hepatitis virus	Mice
	Sialodacryoadenitis virus	Rats
	Porcine hemagglutinating encephalomyelitis virus	Pigs
	Bovine coronavirus	Cattle
	Human coronavirus HCoV-OC43 (cold virus)	Humans
III	Infectious bronchitis virus	Chickens
	Turkey coronavirus	Turkeys

common in domesticated animals causing respiratory, enteric, and neurologic diseases primarily in pigs, dogs, cats, and poultry. Coronaviruses fall into three antigenic groups: I (causing human coronavirus HcoV-229E, transmissible gastroenteritis (TGE) and porcine respiratory virus in swine, canine coronavirus, feline coronavirus (causing feline infectious peritonitis), and rabbit coronavirus); II (human coronavirus HCoV-OC43, murine hepatitis virus in mice, sialodacryoadenitis virus in rats, porcine hemagglutinating encephalomyelitis virus in pigs and bovine coronavirus in cattle); and III (infectious bronchitis virus in chickens and turkey coronavirus in turkeys).

For many years, scientists knew about the existence of two human coronaviruses (HCoV-229E and HCoV-OC43), which are responsible for 30% of human colds. The discovery of the virus (SARS-CoV) that causes severe acute respiratory syndrome (SARS) by Carol Urbani in February 2003 added another human coronavirus to the list. By the end of 2004, the discovery of a fourth human coronavirus (NL63, NL or the New Haven coronavirus) was made and by early 2005, a research team at the University of Hong Kong reported finding a fifth human coronavirus (HKU1) in two pneumonia patients. Table 7-9 summarizes the different coronavirus groups.

Causative Agent

Coronaviruses are enveloped, single-stranded, positive sense RNA viruses with a helical capsule (Figure 7-12). They are the largest RNA viruses and infect a wide range of hosts (rodents, cats, pigs, cattle, birds, and people) and cause a variety of diseases (respiratory infections, gastroenteritis, and hepatitis). A characteristic feature of coronaviruses is a high rate of genetic mutation. Proteins that contribute to the overall structure of all coronaviruses are the spike (S) protein, a major antigen that helps with receptor binding; envelope (E) protein, a small, envelope-associated protein; membrane (M) protein, a protein responsible for budding and envelope formation; and nucleocapsid (N) protein, a protein in which the genome is associated. Coronaviruses are transmitted by aerosols of respiratory secretions, by the fecal-oral route, and by mechanical transmission. Clinically, most coronavirus infections cause a mild, self-limited disease, but there may be rare neurologic complications. SARS is unique among coronaviruses in that it produces a form of viral pneumonia in which infection encompasses the lower respiratory tract.

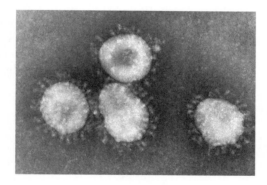

Figure 7-12 Coronaviruses are a group of viruses that have a crown-like (corona) appearance when viewed microscopically. (Courtesy of CDC/Dr. Fred Murphy)

SARS

Overview

Severe acute respiratory syndrome or SARS is a type of viral pneumonia in people that produces a fever, dry cough, dyspnea (difficulty breathing), headache, and hypoxemia (low blood oxygen concentration). In November 2002, people in the Guangdong province of China began becoming sick with a new respiratory disease (305 people became ill) and at least five people died. The initial case was diagnosed on November 16, 2002, in the city of Foshan; however, the Chinese government did not report the disease until February 2003 when they described

SARS has been linked to civet cats making it the single zoonotic coronavirus.

the disease as atypical pneumonia. The disease was spread in February 2003 by an American businessman who came down with pneumonia-like symptoms while on an airline flight from China to Singapore. The plane had to stop at Hanoi, Vietnam, for the victim to receive medical help where he died in a hospital. Several of the doctors and nurses who treated him soon came down with the same respiratory disease and several of them died. The rapid progression of symptoms and the infection of hospital staff alarmed global health authorities who feared the emergence of a new pneumonia epidemic. Initial reports claimed that the new disease, named severe acute respiratory syndrome, was caused by a paramyxovirus; however, the true cause of SARS was eventually linked to a coronavirus. SARS was first described in February 2003 by Carol Urbani as the cause of clinical cases of atypical pneumonia that she was treating at a hospital in Hanoi, Vietnam. Retrospectively, she found out that cases of SARS were occurring in the Guangdong province of China since 2002. The suppression of the news of the SARS outbreak by the Chinese government allowed the disease to spread rapidly. In March 2003 the WHO issued a global SARS alert and the CDC issued a health alert. By May, China was threatening to kill or jail anyone who broke their SARS quarantine orders. By the time the SARS epidemic was under control (the last outbreak occurred in June 2003), it had spread to 29 countries, had killed 774 people, and had infected 8,098 people.

SARS is the first serious and highly contagious disease to emerge in the 21st century, making it this century's first pandemic. Until the 2002 outbreak, the SARS virus had only been found in certain wild animals that are part of the traditional diet in Southern China and are purchased live in crowded, relatively unsanitary live markets. In an attempt to control the disease and its spread, the Chinese government killed over 10,000 civet cats (weasel-like cats sold in wild-game markets as a delicacy in many restaurants) (Figure 7-13). The rapid spread of SARS and its many unknown features in the early days of the outbreak made it difficult for health care workers to know the most effective methods of infection control to prevent its transmission resulting in a wide variety of infection control responses.

Figure 7-13 Civet cat.
(Courtesy of Tassilo Rau)

The SARS outbreak in humans was striking because of the high morbidity and mortality associated with it.

Causative Agent

SARS is caused by a coronavirus that is a enveloped, single-stranded, positive sense RNA virus that has a halo of spikes extending from its envelope. Coronaviruses are in the family Coronaviridae and are best known as the second most common cause of the common cold. Prior to 2003 coronaviruses were not believed to be zoonotic. Coronaviruses have a high frequency of mutation and a high frequency for recombination, which can result in the rapid development of new strains within an individual. The coronoavirus that emerged in China in November 2002 was caused by a novel coronavirus currently known as SARS-CoV (SARS-associated coronavirus) and was detected in lungs, nasopharyngeal aspirates, and feces of infected patients. SARS-CoV is most like the group II coronaviruses. Coronaviruses from two wild animal species (civet cats and raccoon dogs) have been characterized genetically as members of the SARS-CoV group and are believed to be the source of the SARS outbreak.

Epizootiology and Public Health Significance

SARS has been reported worldwide. The 2002–2003 SARS outbreak predominantly affected mainland China, Hong Kong, Singapore, and Taiwan. Because of international travel, SARS was distributed to other cities and countries all over the world such as Toronto, Vancouver, Manila, Singapore, Hanoi, Taiwan, Inner Mongolia, and Hong Kong. Canada experienced a significant outbreak in the area around Toronto, Ontario and the United States had 8 people contract the disease while working in laboratories (as of September 2004). Worldwide numbers of SARS cases from the original outbreak (November 2002 through July 2003) included 8,098 cases and 774 deaths. Most cases were from Mainland China (5,327 cases, 349 deaths), Hong Kong (1,755 cases, 299 deaths), Taiwan (346 cases, 37 deaths), and Canada (primarily around Toronto, Ontario with 251 cases, 43 deaths).

The mortality rate of SARS is higher than that of influenza or other common respiratory tract infections with an overall mortality rate of approximately 10% (below 1% for people age 24 or younger, 6% for those 25 to 44, 15% in those 45 to 64 and more than 50% for those older than 65).

Transmission

The SARS virus spreads from person to person in saliva droplets either by sneezing or coughing. People acquire the virus through the respiratory or gastrointestinal tract. SARS mainly affects adult humans and is rarely seen in children younger than 15 years of age (except in infants younger than 12 months). Thirty percent to 50% of people infected with the SARS virus have been health care workers.

Wild animals sold as food in the local market in Guangdong, China were sampled and it was found that the SARS coronavirus could be isolated from civet cats. This finding suggested that the SARS virus crossed the xenographic (species) barrier from civet cats. Civet cats are closely related to mongooses, are indigenous in Asia and Africa, and were sold in a Guangdong marketplace as a delicacy. Close contact with the animals themselves, or with their saliva or feces, could have transmitted a mutated form of the virus to humans. A 2005 study discovered that 40% of horseshoe bats tested near Hong Kong were infected with a close genetic relative of SARS, raising the possibility that SARS originated in bats and spread to humans either directly, or through civet cats.

> In May 2003, the World Health Organization (WHO) reported that only 16 people infected with SARS-CoV had contracted the virus on airplanes.

Pathogenesis

Human and animal coronaviruses are transmitted by the respiratory or enteric routes and initially infect epithelial cells found in these tissues. Coronaviruses of many animal species (pigs, cats, dogs, mice, and cattle) infect gastrointestinal tissue producing diarrhea and gastrointestinal distress that may be fatal. When coronaviruses cause gastrointestinal disease they tend to cause edema, petechial hemorrhages, and villous atropy of the epithelial cells. As virus particles accumulate in the cell, the cell ruptures, releases the viruses, and causes cellular death. In adult animals, coronavirus infections that cause respiratory disease can be severe if there are high exposure doses, respiratory coinfections, stress related to shipping or commingling with animals, and immunosuppression (disease conditions or treatment with corticosteroids). When coronaviruses cause respiratory disease they tend to cause inflammation, cellular edema, fibrinous exudate, and infiltration by neutrophils, lymphocytes, and connective tissue cells.

In humans, SARS-CoV typically infects respiratory tissue and causes interstitial pneumonia with fever. Infiltrates and exudates are found in the lung tissue of people infected with SARS-CoV. SARS-CoV can also infect gastrointestinal tissue causing diarrhea in people.

> Coronaviruses demonstrate high rates of mutation and RNA-RNA recombinations to produce viruses that are able to adapt to changing conditions and to acquire and regain virulence.

Clinical Signs in Animals

Civet cats can be infected with SARS-CoV and do not show any clinical signs.

Clinical Signs in Humans

The mean incubation time for SARS is 7 days (ranges from 4 to 12 days). Clinical signs of SARS range from mild to moderate to severe respiratory disease. All people with SARS have fever and lethargy. The other clinical signs may be different depending on the severity of the disease and may include dry cough, dyspnea (difficulty breathing), headache, and hypoxemia (low blood oxygen concentration). Nausea, vomiting, anorexia, and diarrhea may be seen in about 30% of cases. The typical clinical course of SARS involves an improvement in symptoms during the first week of infection, followed by a worsening during the second week. Death may result from progressive respiratory failure as a result of alveolar damage.

Diagnosis in Animals

SARS is not typically diagnosed in animals other than to determine if an animal may be the source of human infection. Viral isolation and RT-PCR tests are available to identify the SARS virus.

Diagnosis in Humans

People can be diagnosed with SARS by ELISA tests to detect antibodies (appear 21 days after the onset of symptoms), an immunofluorescence assay that can detect antibodies 10 days after the onset of the disease (labor and time intensive), and PCR to detect genetic material of the SARS virus in blood, sputum, tissue samples, and stool. Viral isolation of SARS-CoV can be done in cell culture.

Treatment in Animals

Animals are not treated for SARS-CoV. Treatment for other coronavirus infections is symptomatic.

Treatment in Humans

People diagnosed with SARS are isolated (preferably in negative pressure rooms) and may be treated with antiviral drugs such as ribavirin (usually in conjunction with steroids). Interferon has been used in conjunction with other drugs and may have an anti-SARS effect. Supportive care with antipyretics, supplemental oxygen, and ventilatory support may be needed.

Management and Control in Animals

Coronavirus infections in animals can be decreased by limiting animal-to-animal contact. The link between civet cats and human SARS brings the selling of live animals in crowded markets into question. Other coronavirus in animals can be controlled via vaccination.

Management and Control in Humans

As a result of the high infectivity rate of SARS, confirmed SARS patients should limit interactions outside the home until 10 days after their fever has resolved. All members of a household with a SARS patient should carefully follow proper hand hygiene such as frequent hand washing and the use of alcohol-based hand sanitizers and wearing disposable gloves when handling the body fluids of a SARS patient. SARS-infected people should cover their mouth and nose with a facial tissue when coughing or sneezing and wear surgical masks during close contact with uninfected persons in order to prevent the spread of infectious respiratory droplets. Eating utensils, towels, and bedding should not be shared between SARS patients and others. Environmental surfaces soiled by body fluids of SARS patients should be cleaned with a household disinfectant. Clinical trials were planned in China for testing of a SARS-CoV vaccine in 2004 and an experimental vaccine in the United States in December 2004 began testing at the NIH Clinical Center in Bethesda, Maryland.

Summary

SARS is caused by a coronavirus that prior to 2003 were not believed to be zoonotic. Coronaviruses cause important respiratory and enteric diseases of humans and many animal species. SARS has been reported worldwide. SARS originated in China, but spread to other cities and countries around the world such as Toronto, Vancouver, Manila, Singapore, Hanoi, Taiwan, Inner Mongolia, and Hong Kong during the 2002–2003 outbreak. The SARS virus spreads from person to person in saliva droplets either by sneezing or coughing. People acquire the virus through the respiratory or gastrointestinal tract. SARS mainly affects adult humans and is rarely seen in children younger than 15 years of age (except in infants younger than 12 months). Thirty percent to 50% of people infected with the SARS virus have been health care workers. Wild animals sold as food were found to have the SARS-associated coronavirus suggesting that the SARS virus crossed the xenographic (species) barrier from animals (civet cats in particular). Civet cats can be infected with SARS-CoV and not show any clinical signs. Clinical signs of SARS in people range from mild to moderate to severe respiratory disease. All people with SARS have fever and lethargy. Nausea, vomiting, anorexia, and diarrhea may be seen in about 30% of cases. SARS is not typically diagnosed in animals other than to determine if an animal may be the source of human infection. People can be diagnosed with SARS by ELISA tests, an immunofluorescence assay, PCR methods, and viral isolation.

Animals are not treated for SARS-CoV. People diagnosed with SARS are isolated and may be treated with antiviral drugs such as ribavirin. Interferon and supportive care may also be used. Coronavirus infections in animals can be decreased by limiting animal-to-animal contact and via vaccination. SARS in people can be controlled by limiting interactions outside the home in confirmed cases, having households with a SARS patient follow proper hand hygiene/respiratory precautions, and disinfecting anything that had contact with the infected person. Clinical trials are being conducted for the approval of a SARS vaccine for humans.

> In May 2005, SARS was declared eradicated by the WHO becoming the second disease in mankind to receive this label (the other was smallpox).

FILOVIRUSES

Overview

Viruses in the Filoviridae family are in the order Mononegavirales and have a nonsegmented, single-stranded, negative sense RNA genome. The order includes four families: Bornaviridae, Rhabdoviridae, Filoviridae, and Paramyxoviridae. Filoviruses were named as a result of their thread-like structure (*filum* is Latin for thread) and are characterized by elongated, branched, curved, or spherical virions.

The family Filoviridae has two genera: *Marburgvirus* (with one species) and *Ebolavirus* (with four species). These viruses cause viral hemorrhagic fevers, which are characterized by extensive bleeding from many body orifices. These diseases have very high mortality rates (between 30% and 90% depending on the virus) and are classified as Biosafety level 4 organisms (meaning they are among the most lethal and destructive viruses). All human filoviral hemorrhagic fever outbreaks have been traced to an African origin.

Marburg virus, named after the German town where it first was reported in 1967, is a virulent pathogen of the Filoviridae family whose origins can be traced to central Africa. The natural host for Marburg virus is unknown. Marburg virus has been endemic since 1998 in the Democratic Republic of the Congo.

Ebola virus was first described in 1976 after febrile, rapidly fatal hemorrhagic illness outbreaks were reported along the Ebola River in Zaire (now the Democratic Republic of the Congo) and Sudan. Sporadic outbreaks have continued since that time (usually in isolated areas of central Africa) and have produced high mortality rates. The natural reservoir of Ebola virus is unknown. Ebola has four distinct subtypes: Ebola-Zaire, Ebola-Sudan, Ebola-Ivory Coast, and Ebola-Reston (a form that causes illness only in nonhuman primates and is not considered zoonotic).

> Filoviruses target primates.

Causative Agent

Filoviruses are enveloped, nonsegmented, single-stranded, negative sense RNA viruses (Figure 7-14). Filoviruseses are filamentous, circular, or spheroidal and

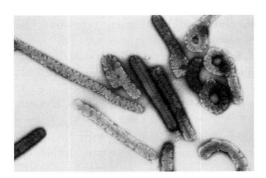

may be curved, coiled, or straight. They are enveloped with a single glycoprotein projection on the envelope surface. The nucleocapsid contains the RNA and four viral structural proteins including the virus-encoded polymerase. Filoviruses can be transmitted by contact with blood, semen, body secretions, and fomites. (See Table 7-10 for less common zoonotic Filoviruses.)

Figure 7-14 The filamentous Marburg virus has the appearance of a shepherd's crook.
(Courtesy of CDC)

FLAVIVIRUSES

Overview

The Flaviviridae family is a large family of viruses including yellow fever virus, from which this family gets its name (*flavus* in Latin means yellow). Flaviviruses are divided into three genera: *Flavivirus* (containing zoonotic viruses such as yellow fever virus, dengue fever virus, and Japanese encephalitis (JE) virus), *Pestivirus* (containing three serotypes of bovine viral diarrhea, but no known human pathogens), and *Hepacivirus* (consisting of hepatitis C virus and related viruses that are transmitted between humans). Members of the *Flavivirus* genus are transmitted by arthropods (mosquitoes and ticks) and are capable of reproducing in the vector by transovarial and transstadial means. Flaviviruses have worldwide distribution and there appears to be little overlap between endemic areas as a result of the ability of specific antibodies to neutralize other flaviviruses.

Table 7-10   **Less Common Zoonotic Filoviruses**

Filovirus	Disease	Predominant Signs in People	Animal Source and Clinical Signs	Transmission	Geographic Distribution	Diagnosis	Control
Marburg virus	Marburg virus hemorrhagic fever First discovered in 1967 in Belgrade, Yukoslavia; and Frankfurt and Marburg, Germany	Sudden onset of high fever, lethargy, headaches, hyperesthesia, vomiting, abdominal pain, diarrhea, and rash. Hemorrhagic disorders occur in less than half of the cases. Encephalitis, apathy, and depression may occur towards the end of the acute phase. Morality rate is 30%.	■ Vervet monkeys (*Cercopithecus aethiops*) are asymptomatic	■ Contact with monkey blood or organs ■ Aerosol transmission possible ■ Nosocomial via accidental inoculations	East Africa	Virus identification via electron microscopy or virus isolation (BSL-4 precautions), ELISA and PCR testing available	Barrier techniques when handling monkeys or infected people; cremation of cadavers instead of customary touching of dead people in some African cultures; proper laboratory handling techniques
Ebola virus (four different types: Maridi, Zaire, Ivory Coast, and Reston)	Ebola virus hemorrhagic fever Maridi virus was isolated in 1976 from an epidemic in Sudan. Zaire virus was isolated in Yambuku (northern Zaire) in 1976. Ivory Coast was isolated in 1994. Reston virus was isolated in monkeys in Reston, Virginia in 1989 and 1996 and in Italy in 1991 (cynomolgus monkeys originated from the Philippines).	Sudden onset of fever, leghargy, headaches, muscle pain, abdominal pain, and diarrhea. Hemorrhagic signs appear in 75% of people with typically fatal outcomes. Reston virus is nonpathogenic for humans.	■ Chimpanzees and gorillas have signs similar to humans	■ Person-to-person transmission by intimate contact is the main route of infection (direct contact with infected blood, secretions, organs, or semen) ■ Contaminated syringes and needles ■ Droplets and small particle aerosols from infected monkeys (low rate of transmission) ■ Ingestion of infected monkeys	Africa (Equatorial Africa, East Africa, Democratic Republic of the Congo, Gabon, Sudan, Uganda) Reston virus is found in Java cynomolgus monkeys (*Macaca fascicularis*) in southern Asia	Virus isolation (BSL-4 precautions) ELISA and RT-PCR testing available	Cremation of people during epidemics; safe laboratory handling techniques; proper disposal of nonhuman primates found acutely dead

Causative Agent

Flaviviruses are enveloped, polyhedral, single-stranded, positive sense RNA viruses that are approximately 40 to 60 nm in length (Figure 7-15). These viruses contain

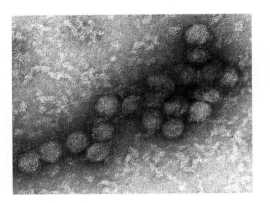

three structural proteins and a host-derived lipid bilayer. Both *Pestivirus* and *Hepacivirus* genera contain internal ribosomal entry sites that provide a translation initiation site for host ribosomes; this is in contrast to the *Flavivirus* genus that uses ribosomal scanning to begin protein synthesis.

Flaviviruses of clinical significance fall into three antigenic complexes that are determined by vector and pathogenicity. Flaviviruses have only one glycoprotein that is genus specific that can result in an immune response against a particular virus and will provide protecting immunity against related flaviviruses.

Figure 7-15   Electron micrograph of the West Nile virus, a flavivirus. (Courtesy of CDC)

Dengue Fever

Overview

Dengue fever (DF), a disease that can progress to dengue hemorrhagic fever (DHF) and dengue shock syndrome (DSS), is the most important and most frequently seen mosquito-borne (*Aedes aegypti*) viral infection in people worldwide. Other names for DF include breakbone fever, dandy fever, seven-day fever, duengero, and *Ki denga pepo*. Dengue fever-like illnesses were described in Chinese medical writings dating back to A.D. 265 with numerous outbreaks recorded throughout history. Dengue is a Spanish word derived from the Swahili *Ki denga pepo* (which was described in English literature during an 1827–1828 outbreak) a term used to describe a severe cramp that people develop with this disease. Dengue in Spanish also means affectation, which is used to describe the mincing walk that untreated people afflicted with this disease develop to ease the pain in their joints. Dengue virus has been recognized since the late 18th century as causing epidemics in tropical and subtropical parts of the world. DF probably originated in Africa where the *Ae. aegypti* mosquitoes are prevalent and moved to the Americas via the slave trade. The first description of DF in English was in 1780 when Benjamin Rush (also a signer of the Declaration of Independence) called it breakbone fever when the disease appeared in Philadelphia. The name breakbone described the pain associated with the disease. People with DF develop fever and severe headache, back pain, and limb pain. During the Spanish-American War, there were large numbers of United States casualties as a result of DF resulting in the development of a Dengue Commission in 1900. In 1906, Ashburn and Craig discovered the causative agent of DF as ultramicroscopic and nonfilterable agent (virus) and confirmed that the disease was spread by mosquitoes. In 1922 an epidemic of DF occurred in the United States infecting more than 1 million people. In 1943, Japanese scientists Hotta and Kimura isolated the Den-1 virus and a few months later Sabin and Schlesinger isolated Den-2 while working in Hawaii and New Guinea. A global pandemic of DF began in Southeast Asia after World War II and continues today. In 1995 the last U.S. outbreak of DF occurred in South Texas.

Dengue hemorrhagic fever (DHF) first emerged in 1949 in Thailand (known as Thai hemorrhagic fever) and the Philippines (known as Philippine hemorrhagic fever) and presented as an acute, infectious disease characterized by fever, muscle pain, and rash. The first epidemic of DHF was described in Manila in 1953. DHF caused an epidemic in Bangkok between 1958 and 1963 infecting more than 10,000 people and killing close to 700. Other outbreaks of DHF occurred in Havana (1981), New Delhi (1982), and Venezuela (1990). Currently DHF only occurs in the United States when brought into this country by travelers to endemic areas (the Caribbean, Central and South America, Africa, and Southeast Asia). In endemic areas billions of people are at risk of contracting DF.

Causative Agent

DF is caused by DEN virus, a virus in the family Flaviviridae, genus *Flavivirus.* DEN virus is a single-stranded, positive sense RNA virus with three structural proteins and seven nonstructural proteins. The structural proteins are capsid (C), membrane (M), and envelope (E). The E protein helps with binding of the virus to a receptor and allows the virus to enter a host cell. There are four closely related serotypes of DEN virus that are designated DEN-1, DEN-2, DEN-3, and DEN-4. DEN-1 and DEN-4 are the most closely related viruses antigenically (and can provide cross protection in people infected with either serotype).

Ae. albopictus (the Asian tiger mosquito) can also transmit dengue virus and was introduced into the United States in a tire shipment from Japan in 1985.

Epizootiology and Public Health Significance

DF is found on all continents except Europe and Antarctica. The virus is found more commonly in tropical and subtropical climates such as those in Asia, Africa, South America, and the Caribbean. DF is predominantly found in urban areas with epidemics occurring seasonally (during and shortly after the rainy season). DF is present in the United States; however, most cases are found imported from travelers.

Between 1990 and 1992, there were reports of 10 imported cases of DF in the United States and in 1998, 90 confirmed or probable cases of DF were imported into the United States with one fatality. Although DF is rare in the United States it is on the rise (only 1 case was reported during the period from 1987 to 1989). The current estimate of DF cases in the United States is 100 cases annually. Outside the United States DEN virus causes about 100 million cases of acute febrile disease annually, including more than 500,000 reported cases of DHF/DSS. Annually, approximately 24,000 deaths are globally attributed to DF. Recovery from dengue infection usually is complete. Currently, DF is endemic in 112 countries. The world's largest known epidemic of DHF/DSS occurred in Cuba in 1981, when as many as 11,000 cases were reported in a single day. The fatality rate for DSS varies by country from 12% to 44%. The mortality rate of DF is less than 1%. Untreated DHF/DSS is associated with a 50% mortality rate.

Transmission

DEN viruses are transmitted to humans through the bites of infective female *Aedes* mosquitoes. *Ae. aegypti* mosquitoes (Figure 7-5) are the vectors of the urban cycle of DF, whereas *Ae. albopictus, Ae. scutellaria,* and *Ae. africanus* mosquitoes are the vectors of the rural cycle of DF. *Ae. aegypti* mosquitoes are most active during the day and are typically found near human dwellings. Elevated temperatures shorten the incubation period for DEN virus in mosquitoes, which increases the rate of mosquito-to-human transmission. *Ae. aegypti* cannot survive in temperatures below 48°F. *Ae. albopictus* mosquitoes were introduced into the United States in 1982 (Houston, Texas) and are currently found in at least 18 states.

Ae. albopictus can carry a variety of viruses including dengue, yellow fever, and the equine encephalitis viruses. More recently a jungle cycle (sylvan cycle) has been identified in which the virus is transmitted by rain forest *Aedes* mosquito species (*Ae. niveus*) between monkeys. Transovarial transmission occurs in mosquitoes. Mosquitoes typically acquire the virus while feeding on the blood of an infected person. After virus incubation for 8 to 10 days, an infected mosquito is capable of transmitting the virus for the rest of its life.

Humans are the main **amplifying host** (host that contributes to the buildup and transmission of a disease-causing agent by transmitting the agent among vertebrate hosts) of DEN virus, although in some parts of the world monkeys serve as a source of virus for uninfected mosquitoes.

Pathogenesis

Once inside the host DEN virus replicates in regional lymph nodes and is spread via the lymphatics or blood to various parts of the body. In classical DF there is little if any plasma leakage and patients recover after 7 to 10 days. In DHF and DSS DEN virus is able to replicate in macrophages producing the virus-antibody complexes that stimulate viral replication. Antibodies attach to the virus's outer envelope, which signals macrophages to engulf the virus. After it is engulfed the virus replicates inside the macrophage and it is then transported throughout the body by the macrophage. The antibodies present do not neutralize the virus, but instead form an antibody-antigen complex that binds to macrophages. These complexes activate cytokines, which increase capillary permeability, cause perivascular edema, and result in circulatory collapse.

Clinical Signs in Animals

Clinical disease from DEN virus only occurs in humans. Nonhuman primates such as chimpanzees, macaques, and gibbons develop high levels of viremia that lasts long enough for mosquitoes to become infected with virus. Other monkey species develop only a low-level viremia that is short-lived and do not serve as a significant source of infection.

Clinical Signs in Humans

There are two forms of DF: classic DF and hemorrhagic dengue fever (DHF)/dengue shock syndrome (DSS).

- Classic DF has a 4- to 7-day incubation period with mild cases lasting 2 to 3 days. Clinical signs of classic DF are sudden onset of fever, weakness, severe headaches, retro-orbital pain, muscle and joint pain, nausea, and vomiting. A petechial rash appears 3 to 4 days after the onset of fever and typically appears initially on the trunk before spreading peripherally. Symptoms typically last for 7 days (hence the name seven-day fever). Classic DF is usually nonfatal. Classic DF primarily occurs in nonimmune, nonindigenous adults and children.
- DHF/DSS is indistinguishable from classic DF in the initial stages, but progresses to produce clinical signs such as bleeding (especially gastrointestinal) and shock. DHF is defined as classic DF with thrombocytopenia (low platelet count) and hemoconcentration (as a result of fluid losses). The WHO identifies four clinical manifestations of DF complications:
  - Grade I—thrombocytopenia and hemoconcentration
  - Grade II—spontaneous bleeding
  - Grade III—hypotension
  - Grade IV—shock

Grades I and II are milder complications and are called DHF, whereas grades III and IV are more severe and are called DSS. DHF/DSS usually occur during a second dengue infection in people with pre-existing active (had previous infection) or passive (maternal antibodies) immunity to a dengue virus serotype. DHF/DSS begins abruptly after 2-4 days of mild signs of classic DF followed by rapid deterioration with restlessness, abdominal pain, and hypotension. Increased vascular permeability, bleeding, and possible disseminated intravascular coagulation (DIC) may occur as a result of circulating dengue antigen-antibody complexes, activation of complement, and release of immune mediators.

Diagnosis in Animals
Diagnosis of dengue fever in animals is not done clinically.

Diagnosis in Humans
Tentative diagnosis of DF is based on clinical signs and a history of travel to endemic areas. Confirmation of DF is by serology, PCR, and viral isolation. Serologic tests such as IgM capture ELISA, complement fixation, neutralization test, hemagglutination inhibition, and IgG ELISA are used.

Treatment in Animals
Animals are not treated for DF.

Treatment in Humans
Classic DF is a self-limited illness and only supportive care is required (acetaminophen may be used to treat fever and pain; aspirin and nonsteroidal anti-inflammatory drugs and corticosteroids should be avoided as a result of their platelet effects). People who develop DHF/DSS may develop dehydration, tachycardia, altered mental status, and decreased urine output, which may require intravenous fluid administration. Whole blood transfusions or platelets may be needed in patients with severe bleeding.

Management and Control in Animals
Mosquito control may help decrease the level of viremia in monkeys.

Management and Control in Humans
There is not a currently approved vaccine for the prevention of DEN virus infection; however, tetravalent, attenuated, live vaccines have been developed and are undergoing clinical trials. The problem with developing a vaccine for DF is that immunity to a single DEN virus serotype puts a person at risk for developing DHF/DSS upon exposure to another DEN virus serotype; therefore, a vaccine must provide high levels of immunity to all four dengue strains to be clinically useful. DF can also be reduced by mosquito control such as using insect repellent containing DEET, wearing protecting clothing impregnated with permethrin insecticide, using screens in dwellings (mosquito netting is of limited benefit because *Aedes* are day-biting mosquitoes), eliminating the breeding ground of the mosquitoes, and supporting community-based vector control programs.

Summary
DF is the most important and most frequently seen mosquito-borne viral infection in people worldwide. DF is caused by DEN virus, a virus in the family Flaviviridae, genus *Flavivirus*. There are four closely related serotypes of DEN virus that are

designated DEN-1, DEN-2, DEN-3, and DEN-4. DF is found on all continents except Europe and Antarctica. DF is present in the United States; however, most cases are imported from travelers. DEN viruses are transmitted to humans through the bites of infective female *Aedes* mosquitoes. Clinical disease from DEN virus only occurs in humans. Nonhuman primates such as chimpanzees, macaques, and gibbons develop high levels of viremia that lasts long enough for mosquitoes to become infected with virus. Other monkey species develop only a low-level viremia that is short-lived and do not serve as a significant source of infection. There are two forms of DF in humans: classic DF and dengue hemorrhagic fever (DHF)/dengue shock syndrome (DSS). Classic DF has a sudden onset of fever and produces weakness, severe headaches, retro-orbital pain, muscle and joint pain, nausea, and vomiting. DHF/DSS is indistinguishable from classic DF in the initial stages, but progresses to produce clinical signs such as bleeding and shock. DHF is defined as classic dengue with thrombocytopenia and hemoconcentration. DHF/DSS usually occur during a second dengue infection in people with preexisting active or passive immunity to a dengue virus serotype. DHF/DSS begins abruptly after 2 to 4 days of mild signs of classic DF followed by rapid deterioration with restlessness, abdominal pain, and hypotension. Increased vascular permeability, bleeding, and possible DIC may occur as a result of circulating dengue antigen-antibody complexes, activation of complement, and release of immune mediators. Diagnosis of DF in animals is not done clinically. DF is diagnosed in people by serology, PCR, and viral isolation. Serologic tests such as IgM capture ELISA, complement fixation, neutralization test, hemagglutination inhibition, and IgG ELISA are used. Animals are not treated for DF. Classic DF in people is a self-limited illness and only supportive care is required. People who develop DHF/DSS may require intravenous fluid administration, whole blood transfusions, or platelets. Mosquito control may help decrease the level of viremia in monkeys and prevent disease in people. There is not a currently approved vaccine for the prevention of DEN virus infection in humans.

St. Louis Encephalitis

Overview

SLE is the most common mosquito-transmitted human pathogen in the United States.

St. Louis encephalitis (SLE), also known as type C lethargic encephalitis, is an arboviral infection that is seen in the Americas and is transmitted from wild birds and domestic fowl to people via mosquitoes. SLE virus was first identified as the cause of human disease in North America in Paris, Illinois, in 1932; however, it was not identified until after a large urban epidemic in which more than 1,000 people in St. Louis, Missouri, were infected during the summer of 1933. Since then, numerous outbreaks of SLE have occurred throughout North America. SLE virus is found throughout the lower 48 states; however, periodic SLE epidemics have only occurred in the Midwest and southeast United States. Between 1964 and 1997 there have been approximately 4,440 confirmed cases of SLE in the United States with an average of 193 cases per year (many cases are not reported and the annual rate varies considerably). Worldwide approximately 10,000 human cases of SLE have been documented with about 1,000 fatalities. In 1975 there were 3,000 cases of SLE and the last big outbreak of SLE occurred in Florida in 1990. SLE produces clinical signs in less than 1% of infections with signs ranging from headache and fever to meningoencephalitis.

Causative Agent

SLE virus belongs to the Flaviviridae family in the *Flavivirus* genus. Flaviviruses are enveloped, single-strand, positive sense RNA viruses. There are more than

50 antigenically related *Flaviviruses* with SLE virus falling into the group formerly known as group B arbovirus along with West Nile virus and Japanese encephalitis virus. The lipid-containing envelope has one surface glycoprotein that helps with attachment, fusion, and penetration of the virus. SLE virus also has an internal matrix protein. The nucleocapsid contains capsid protein and RNA. Virion maturation occurs at intracytoplasmic membranes.

Epizootiology and Public Health Significance

SLE epidemics have occurred throughout North America, including Mexico and Canada, with the exception of some New England states. SLE occurs seasonally as a result of its transmission by mosquitoes. SLE virus transmission to birds and to humans during outbreaks occurs most frequently between July and November. The disease is more prevalent in the Ohio-Mississippi Valley. The annual rate of SLE cases varies greatly with outbreaks occurring more frequently in years with hot summers.

The United States spends approximately $150 million on mosquito control annually. SLE virus is widely distributed in the United States and infections occur as periodic focal outbreaks of encephalitis in the midwestern, western, and southwestern United States that typically occur between July and November. During the last 50 years, 10,000 cases have been reported. The annual incidence of SLE is 0.003 to 0.752 cases per 100,000 people. Outbreaks of SLE have occurred in Canada and Mexico with sporadic cases occurring in South America and the Caribbean. The mortality rate ranges from 2% to 20% and is higher in older people. Twenty percent of people with clinical disease develop sequelae such as irritability, memory loss, various types of movement disorders, and motor deficits.

Transmission

SLE virus is transmitted by several *Culex* spp. of mosquito (Figure 7-16). During the summer, SLE virus is maintained in a mosquito-bird-mosquito cycle, with periodic amplification by birds and *Culex* mosquitoes. The species of mosquito that transmits SLE virus varies regionally with the principal vector in Florida being *Culex nigripalpus*, in the Midwest the principle vectors are *Cu. pipiens pipiens* and *Cu. pipiens quinquefasciatus*, and in the western United States the principle vectors are *Cu. tarsalis* and members of the *Cu. pipiens* complex.

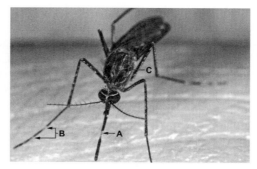

Figure 7-16 A *Culex tarsalis* mosquito is the vector of Western Equine Encephalitis (WEE), St. Louis Encephalitis (SLE), and California Encephalitis, and is currently the main vector of West Nile virus in the western United States. The dark-scaled proboscis is labeled A, the light-colored bands wrapped around its distal appendages are labeled B, and its scutum is labeled C. (Courtesy of CDC)

The transmission of SLE virus involves an amplification phase in which the number of wild birds infected with SLE virus rapidly increases during seasons with favorable rainfall patterns, mosquito activity, and bird nesting activity. A bird bitten by a mosquito with SLE virus can later produce enough virus in its blood to infect other mosquitoes that feed upon it. There is a narrow time frame for which mosquitoes can acquire the virus and spread it. Birds typically produce large quantities of virus in their blood 1 to 2 days after infection; however, SLE virus numbers rapidly decrease 1 to 3 days later as the bird recovers from the infection. Bird populations can amplify SLE virus allowing its chance transmission to humans and other incidental hosts. SLE virus can produce severe disease in humans; however, they are poor viral hosts and produce little SLE virus in the blood making them dead end hosts. Birds and people will clear the viral infection but mosquitoes remain infected for life.

Pathogenesis

SLE virus replicates in the mosquito host's midgut and proceeds to the salivary glands. As the mosquito bites the bird or animal, it transmits viral particles to the vertebrate host. SLE virus replicates at the site of inoculation and enters the bloodstream producing viremia. In subclinical cases of SLE, the virus is cleared by the reticuloendothelial system (the liver, spleen, and lymph nodes) before invasion of the CNS can take place. In clinical cases viral replication continues producing a secondary viremia. The reticuloendothelial system will become saturated with virus and excess virus will invade the CNS. Entry into the CNS depends on the efficiency of viral replication and the degree of viremia. The virus enters the CNS either through the blood–brain barrier or across fenestrated endothelium in areas of the CNS such as the choroid plexus. Focal necrosis of neurons may occur, which leads to the development of glial nodules. As these lesions heal spongiform changes take place. Perivascular infiltrates made up of activated T cells and macrophages may also occur.

Clinical Signs in Animals

Domestic and wild animals can be infected with SLE virus but do not produce clinical disease. Infected birds and mosquitoes do not show clinical signs of SLE. SLE virus can be found in a variety of wild birds. An urban cycle exists between urban-dwelling bird species (House Sparrow, Rock Dove, Blue Jay, American Robin, Northern Cardinal, Mourning Dove and Northern Mockingbird). Birds act as amplifiers of infection, each infected bird potentially infecting many mosquitoes.

Clinical Signs in Humans

Humans are the only animal to have clinical disease from SLE virus.

The incubation period for SLE is typically 5 to 15 days. Clinical signs occur in only 10% of people infected with SLE virus. Most cases are mild, last for a few days, and show signs of fever, photophobia, sore throat, muscle pain, and headache. Severe disease produces meningitis or encephalitis that can occur a few days after infection and lasts for up to 2 weeks. CNS signs such as reduced consciousness, seizures (especially in children), neck stiffness, and nystagmus can occur with severe disease. Persistent fever with CNS signs is a grave symptom. Convalescence may take a long time with headaches, insomnia, fatigue, and irritability possible during the recovery period.

Diagnosis in Animals

SLE is not clinically diagnosed in animals. ELISA tests and virus isolation can be done on birds and virus isolation can be done on mosquitoes for surveillance purposes.

Diagnosis in Humans

Humans are diagnosed with SLE by clinical signs and serologic methods. SLE virus can be detected via antibody titers using hemagglutination inhibition (higher than 1:320), complement fixation (higher than 1:128), immunofluorescence (higher than 1:256), or plaque reduction neutralization test (higher than 1:160). Virus isolation can be done from tissue, blood, or CSF. An IgM antibody-capture ELISA test is also available. A group-specific RT-PCR test is available for detection of all flavirviruses, but does not distinguish one flavivirus from another.

Treatment in Animals

Animals have subclinical disease and are not treated for SLE.

Treatment in Humans

Treatment for SLE in humans is symptomatic including management of seizures or any neurologic symptoms.

Management and Control in Animals

Animals have subclinical disease making controlling the disease in them important as a result of their role in causing human disease. Reducing the transmission of SLE virus includes mosquito control methods previously described.

Management and Control in Humans

There is no vaccine for SLE. Control of SLE in people is focused on mosquito control.

Summary

St. Louis encephalitis (SLE) is an arboviral infection that is seen in the Americas and is transmitted from wild birds and domestic fowl to people via mosquitoes. SLE virus belongs to the Flaviviridae family in the *Flavivirus* genus. SLE epidemics have occurred throughout North America, including Mexico and Canada, with the exception of some New England states. SLE occurs seasonally due its transmission by mosquitoes. SLE virus is transmitted by several *Culex* spp. of mosquito. During the summer, SLE virus is maintained in a mosquito-bird-mosquito cycle, with periodic amplification by birds and *Culex* mosquitoes. The species of mosquito that transmits SLE virus varies regionally. Domestic and wild animals, birds, and mosquitoes can be infected with SLE virus but do not produce clinical disease. Human cases of SLE in humans are mild, last for a few days, and show signs of fever, photophobia, sore throat, muscle pain, and headache. Severe disease produces meningitis or encephalitis that can occur a few days after infection and last for up to 2 weeks. ELISA tests and virus isolation can be done on birds and virus isolation can be done on mosquitoes for surveillance purposes. Humans are diagnosed with SLE by clinical signs and serologic methods (hemagglutination inhibition, complement fixation, immunofluorescence, or plaque reduction neutralization test). Virus isolation can be done from tissue, blood, or CSF. An IgM antibody-capture ELISA test is also available. Animals have subclinical disease and are not treated for SLE. Treatment for SLE in humans is symptomatic including management of seizures or any neurologic symptoms. There is no vaccine for SLE. Control of SLE is focused on mosquito control.

West Nile Fever

Overview

West Nile fever (WNF), also known as West Nile encephalitis, is an arboviral infection first identified in 1937 and is named after the Ugandan province where it was discovered. West Nile virus (WNV) is a mosquito-borne virus that infects humans, horses, birds, and occasionally other animals producing encephalitis.

West Nile virus is the most widespread flavivirus.

Alexander the Great died when he was 32 years old in Babylon in May 323 B.C. following a 2-week febrile illness. Plutarch described a flock of ravens exhibiting unusual behavior and dying at Alexander's feet prior to his own death making some people believe Alexander the Great died of WNF. Until the early 1990s, the virus was largely confined to Africa, Europe, and Asia. Since the mid-1990s, WNF outbreaks of increasing frequency and severity have occurred in Algeria in 1994, Romania in 1996–1997, the Czech Republic in 1997, the Democratic Republic of the Congo in 1998, Russia in 1999, the United States in 1999–2003, and Israel in 2000. These outbreaks were accompanied with a large number of bird deaths. WNV was first detected in the United States in 1999 in flamingos and pheasants from the Bronx Zoo and crows from the New York City area. The unexpected appearance of WNV in New York City was first thought to be an act of bioterrorism; however, the virus and its vectors have been shown to survive intercontinental airline flights. WNF as an equine disease was first noted in Egypt and France in the early 1960s. Epizootics of WNF in horses occurred in Morocco in 1996, Italy in 1998, the United States in 1999–2001, and France in 2000; and in birds in Israel in 1997–2001 and in the United States in 1999–2002. In the United States since 1999, human, bird, animal, or mosquito WNV infection has been reported from all states except Hawaii, Alaska, and Oregon.

Causative Agent

WNV belongs to the Flaviviridae family in the *Flavivirus* genus. Flaviviruses are enveloped, single-strand, positive sense RNA viruses. WNV falls into the same category of viruses as St. Louis encephalitis virus and Japanese encephalitis virus. The viral envelope protein is composed of envelope E and membrane M proteins embedded in a lipid bilayer. The E protein helps with virus assembly, cell receptor recognition, agglutination of red blood cells, and induction of B and T cell responses. There are different strains of WNV, which are typically named for the location and year that they caused disease. For example, NY99 is the strain of WNV that was found in birds in New York City in 1999.

Epizootiology and Public Health Significance

WNV has been found in Africa, Europe, the Middle East, west and central Asia, and North America. WNV seropositivity of children in Egypt is approximately 50% with WNV being the most common cause of viral meningitis or encephalitis in that country. Most cases in children from Asia and Africa have a benign course and only rarely result in fatality. Cluster outbreaks in Romania, Russia, the Congo, and the United States are characterized by a high rate of encephalitis and death. In 2002 there were 4,156 laboratory-confirmed human cases of infection worldwide. The actual number of WNV infections is probably much higher than reported because WNF mimics many other self-limiting febrile illnesses.

Since 1999, WNV found a distribution cycle in the northeastern United States and has spread throughout the United States. In 2005, WNV caused 2,949 cases of disease in the United States, including 116 deaths. Wild birds with WNV have been found in many states and in Canada. The United States spends approximately $150 million on mosquito control annually. In the United States, most fatal cases have occurred in elderly patients.

Transmission

WNV is spread by infected female mosquitoes as they take a blood meal from a susceptible vertebrate. Seventeen species of wild birds transmit WNV to humans via the *Culex* (Figure 7-16), *Aedes* (Figure 7-5), and *Anopheles* mosquitoes. Birds are the primary reservoir host for WNV and serve as amplifying hosts with the degree of amplification depending on bird species and environmental conditions. Viremic birds spread the virus to mosquitoes as the arthropods take blood meals. These mosquitoes may spread WNV to humans. Bird migration helps spread this virus. Animal-to-animal transmission has not occurred naturally.

Pathogenesis

WNV replicates in the mosquito host's midgut, proceeds to the salivary glands, and is transmitted to the vertebrate host when the mosquito takes a blood meal. WNV replicates at the site of inocula-tion and enters the bloodstream pro-ducing viremia. WNV enters the CNS either through the blood-brain barrier and infects the brain tissue producing encephalitis (Figure 7-17). WNE may also affect the leptomeninges with a clinical presentation of aseptic men-ingitis (viral meningitis). WNV infec-tions may present with features of both encephalitis and aseptic meningi-tis (meningoencephalitis).

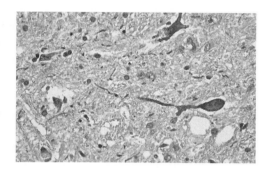

Figure 7-17   Brain tissue from a West Nile encephalitis patient, showing antigen-positive neurons and neuronal processes (dark areas).
(Courtesy of CDC/W. J. Shieh and S. Zaki)

Clinical Signs in Animals

Most animal species naturally infected with WNV are asymptomatic. Antibodies to WNV have been found in asymptomatic dogs, horses, donkeys, and mules. Latent infections are possible in animals. Encephalitis may be seen in cattle, horses, and other mammals. The incu-bation period for WNV in horses is 3 to 15 days. Horses infected with WNV may show clinical signs of encepha-litis such as fever (less than 25% of infected horses), lethargy, weakness, ataxia, head pressing, partial paraly-sis, or death (Figure 7-18). During outbreaks, 20% to 43% of infected horses appear to develop acute neuro-logic signs. The mortality rate in horses showing clinical signs is approximately 33%. Diseased birds may show clin-ical signs including abnormal head and neck posture, weight loss, ataxia, tremors, circling, disorientation, and impaired vision. Most birds with clin-ical signs will die. Chickens and turkeys are typically asymptomatic.

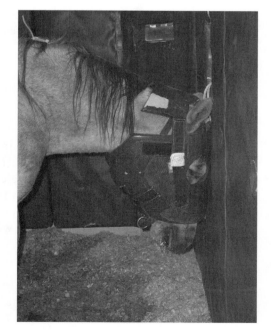

Figure 7-18   Horse with West Nile virus demonstrating head pressing.
(Courtesy of Laura Lien, B.S., CVT Moraine Park Technical College)

Clinical Signs in Humans

The severity of clinical signs in people varies with the patient's age. Children and young adults typically develop mild disease, whereas adults show more severe symptoms. Less than 1% of people infected with WNV develop serious clinical signs. The incubation period is 3 to 6 days. Clinical signs typically last less than 1 week and begin abruptly with a biphasic fever, lethargy, headache, muscle and joint pain, lymphadenopathy, and a rash that appears on the torso. In severe cases signs include neck stiffness, stupor, disorientation, coma, tremors, seizures, paralysis, and rarely death. Neurologic signs are seen in 25% of cases in the elderly with a fatality rate of 10%.

Diagnosis in Animals

WNF is diagnosed in animals by virus isolation or serologic testing for the presence of antibody. Blood or serum samples for virus isolation can be collected late in the incubation period or up to 5 days after clinical signs begin. Serologic tests available for WNV include complement fixation, virus neutralization, hemagglutination inhibition, ELISA, and IgM capture ELISA (currently the most accurate). Cross-reactivity with other flaviviruses may occur with these testing methods.

Diagnosis in Humans

WNF in people is diagnosed via virus isolation from blood on the first day of illness throughout the course of the febrile phase. WNV is not usually cultured from CSF. Serologic tests such as enzyme immunoassay (EIA) with plaque reduction neutralization test are available for antibody determination (fourfold or greater rise between acute and convalescent titer is diagnostic of WNF). A WNV-specific IgM antibody appears in blood about 1 week into the infection and remains detectable for at least two months (and occasionally for 1 year). PCR is also available.

Treatment in Animals

There is no treatment for WNF. Animals with neurologic signs may be given anticonvulsant medication to control seizures or are anesthetized to avoid injury.

Treatment in Humans

Treatment for WNF in people is symptomatic.

Management and Control in Animals

Reducing the transmission of WNV includes mosquito control and surveillance methods as previously described. Horses should be vaccinated to protect against infection with WNV. There are currently two fully approved West Nile virus vaccines available for horses each requiring an initial series of at least two vaccinations (at 3 to 4 months and repeated in 1 month), followed by booster injections (frequency depending upon whether the horse is in an endemic area or not).

Management and Control in Humans

In November 2003, vaccine development began for the first human clinical trial of a WNV vaccine (under a fast-track project). The vaccine is a chimeric virus vaccine, containing genes from two different viruses (yellow fever and West Nile). The West Nile vaccine performed well in hamsters, mice, monkeys, and horses, and has entered human clinical trials.

Mosquito control and surveillance are also important in people as previously described.

Summary

West Nile Fever (WNF) is an arboviral infection transmitted by mosquitoes and infects humans, horses, birds, and occasionally other animals producing encephalitis. West Nile virus (WNV) belongs to the Flaviviridae family in the *Flavivirus* genus. (See Table 7-11 for less common zoonotic Flaviviruses.) WNV is spread by infected mosquitoes as they take a blood meal from a susceptible vertebrate. Most animal species naturally infected with WNV are asymptomatic. Antibodies to WNV have been found in asymptomatic dogs, horses, donkeys, and mules. Encephalitis may be seen in cattle, horses, and other mammals. Horses infected with WNV may show clinical signs of encephalitis such as fever, lethargy, weakness, ataxia, partial paralysis, or death. Diseased birds may show clinical signs including abnormal head and neck posture, weight loss, ataxia, tremors, circling, disorientation, and impaired vision. In people, the severity of WNF varies with the patient's age. Children and young adults typically develop mild disease, whereas adults show more severe symptoms. Clinical signs typically last less than 1 week and begin abruptly with a biphasic fever, lethargy, headache, muscle and joint pain, lymphadenopathy, and a rash that appears on the torso. In severe cases signs include neck stiffness, stupor, disorientation, coma, tremors, seizures, paralysis, and rarely death. WNF is diagnosed in animals by virus isolation or serologic testing for the presence of antibody. Serologic tests available for WNV include complement fixation, virus neutralization, hemagglutination inhibition, ELISA, and IgM capture ELISA. WNF in people is diagnosed via virus isolation. Serologic tests such as enzyme immunoassay (EIA) with plaque reduction neutralization test are available. PCR is also available. There is no treatment for WNF; however, animals with neurologic signs may be given anticonvulsant medication to control seizures or are anesthetized to avoid injury. Treatment for WNV in people is symptomatic. Reducing the transmission of WNV includes mosquito control and surveillance methods. A vaccination is available for horses and a human vaccine for WNV is in human clinical trials.

HERPESVIRUSES

Overview

Herpesviruses, in the Herpesviridae family, are large, enveloped, double-stranded DNA viruses that are widely distributed in nature with most animals infected with one or more of the 100 species discovered. The term herpes comes from Greek work *herpein* meaning to creep slowly and has been used for thousands of years to describe the slow spreading diseases caused by viruses in this family. Herpesviruses were named for the characteristic creeping rashes that were produced by these viruses. Herpes simplex virus (HSV) type 1 was first described by Herodotus, a Roman doctor who around A.D. 100 documented herpetic eruptions, which appear around the mouth during simple fevers (hence the name fever blisters). HSV type 2 causes genital blisters and was first publicly reported in the early 1700s by French physician Jean Astruc. These two herpes viruses share about half of their genetic codes and can infect similar body sides under certain circumstances. Hippocrates described the cutaneous spreading of herpes simplex lesions and Shakespeare is believed to have described herpes lesions in Romeo and Juliet. In 1893, Vidal first recognized that herpes virus was transmitted from one person to another. In 1919, Lowenstein confirmed the infectious nature of herpes simplex virus and in the 1930s, the concept of latency was described. By the 1940s and 1950s, research abounded on the many diseases caused by HSV. Another herpesvirus, varicella-zoster virus

Table 7-11   Less Common Zoonotic Flaviviruses

Flavivirus	Disease	Predominant Signs in People	Clinical Signs in Animals	Transmission	Animal Source	Geographic Distribution	Diagnosis	Control
Tick-borne flaviviruses								
Tick-borne encephalitis virus (TBE complex)	■ Central European Tick encephalitis (CEE); biphasic milk fever, biundulant milk fever, spring-summer meningoencephalitis. First described in Austria in 1931 by Schneider. ■ Eastern Subtype tick encephalitis; Far East encephalitis; Russian spring-summer meningoencephalitis (RSSE). First described in 1937 in Russia.	■ Biphasic course with nonspecific influenza-like symptoms, followed by an asymptomatic period, followed by a second disease stage including clinical signs of meningitis, approximately half of infected people have permanent neurologic damage. ■ Fatality rate is higher for RSSE than CEE.	■ Dogs develop clinical signs of meningo-encephalitis. ■ Other animals are reservoir hosts	■ CEE is acquired by *Ixodes ricinus* ■ RSSE is acquired by *Ixodes persulcatus, Dermacentor marginatus, and Dermacentor silvarum* ■ CEE can also be transmitted by consumption of contaminated fresh milk and nonpasteurized milk products.	■ Hedgehogs, shrews, and moles for both CEE and RSSE. Less frequently seen in waterfowl, bats, cattle, goats, sheep, and dogs.	■ Western and Central Europe, Scandinavia, countries that made up the former Soviet Union, and Asia	Cell culture, PCR, and ELISA tests	Preexposure vaccination is available (postexposure vaccination and treatment with hyperimmunoglobin is contraindicated). Tick control in endemic areas
Louping III virus (part of TBE complex)	Louping ill Louping is derived from the Scottish dialect term that indicates the hyperactivity and jumping gait of infected sheep.	■ Biphasic course that begins with influenza-like signs lasting 2-11 days; followed by asymptomatic period; followed by a second fever that initiates the CNS signs; rarely fatal.	■ Sheep: biphasic fever, depression, ataxia, muscular incoordination, tremors, posterior paralysis, coma, and death in 20% to 50% of sheep (also called ovine encephalomyelitis, infectious encephalomyelitis of sheep, trembling-ill).	■ Primary vector is the sheep tick *Ixodes ricinus*. ■ A majority of human cases are the result of laboratory infection	■ Sheep ■ Low levels of virus are found in cattle, horses, pigs, dogs, deer, shrews, woodmice, voles, and hares (but are not considered to be infection sources for humans).	■ Endemic in rough upland areas in Scotland, Northern England, Wales and Ireland	Rising antibody titers (ELISA); RT-PCR	Vaccine available for sheep; tick control for sheep and humans

Agent	Description	Clinical signs	Transmission and reservoir	Host	Geographic distribution	Diagnosis	Prevention and control
Powassan virus	Powassan virus encephalitis First isolated in 1957 from a human case of fatal encephalitis.	■ Human disease is rare; begins as a nonspecific influenza-type infection followed by meningitis and encephalitis. Typically seen in 20 + year old males involved in outdoor activities ■ Dogs can develop CNS disease after infection. ■ Asymptomatic in cattle, grouse, red deer, hares, rabbits, bats, and hedgehogs.	■ Infection is acquired via a variety of tick species (*Ixodes marxi, I. cookie, I. spinnipalpus,* and *I. anderson*) ■ Squirrels and groundhogs serve as reservoir hosts. ■ Other possible reservoir hosts include rabbits, dogs, skunks, and foxes. ■ Goats shed the virus in milk.	■ Squirrels and groundhogs (woodchucks)	■ Canada, Russia, northeastern United States, and Wisconsin	Virus isolation from biopsy samples; RT-PCR, ELISA	Tick control; vaccines have not been developed
Kyasanur forest virus (TBE complex)	Kyasanur Forest disease Named after the Kyasanur Forest in Mysore, South India where the disease was discovered in 1957 in langur and macaque monkeys.	■ Biphasic disease in which the first phase manifests with fever, headache, muscle pain, gastrointestinal signs, and mucous membrane hemorrhages (lasting about 1 week); clinical signs then subside with encephalitis appearing 1 to 3 weeks later. ■ The disease is typically seen in spring.	■ Infection is acquired via several species of *Haemaphysalis* ticks (especially *Haemaphysalis spingera*) ■ Monkeys may develop encephalitis and die. ■ Cattle, rhesus monkey, langur monkeys, small rodents, and birds are asymptomatic.	■ Monkeys ■ Many reservoir hosts (cattle, rodents, birds) that do not show clinical signs.	■ India	Virus isolation from cell culture, PCR, ELISA, and indirect immunofluorescence	Tick control; vaccination is used in endemic area (both human vaccines and bovine vaccines are available).
Omsk Hemorrhagic virus (TBE complex)	Omsk Hemorrhagic fever First described in 1944 near Novosibirsk and Kurganski, Russia	■ Sudden onset of headache, lethargy, vomiting, and fever (lasting 2 to 15 days); hyperemia of mucous membranes and face is common. ■ Found commonly in muskrat hunters	■ Infection is acquired via *Dermacentor reticulates* and *Dermacentor apronophorus* ticks or by direct contact with infected muskrat blood, urine, or feces ■ Muskrats have fatal disease from this virus. ■ Small water rats are asymptomatic.	■ Muskrats, small water rats	■ Forest regions of western Siberia	Mild cases are not diagnosed; virus isolation, PCR, ELISA	Limit contact with muskrats; tick control; cross protection from CEE or RSSE vaccine.

(Continued)

Table 7-11 (Continued)

Flavivirus	Disease	Predominant Signs in People	Clinical Signs in Animals	Transmission	Animal Source	Geographic Distribution	Diagnosis	Control
Mosquito-borne flaviviruses								
Japanese encephalitis virus	Japanese encephalitis; also called Japanese B encephalitis and JE Encephalitis epidemics were described in Japan from the 1870s onward; Japanese encephalitis virus was first isolated in the 1930s. JE has spread over the past 50 years across Southeast Asia, India, southern China, and the Pacific reaching Australia in 1998.	■ Most infections are asymptomatic (only 1 in 300 infected people develop clinical signs); clinical signs include fever, hadache, lethargy, respiratory distress, and gastrointestinal problems. Hemorrhage and melena are common. Neurologic signs include seizures, ataxia, paralysis, paresis, and reduced consciousness. Permanent neurologic signs are seen in up to 40% of affected people; many symptomatic cases are fatal.	■ Sows may abort and piglets may die.	■ Infection is acquired via *Culex* spp. *and Aedes* spp. mosquitoes (overwinter survival is possible in temperate climates); transplacental infection can occur and cause fetal death or abortion.	■ Swine and frogs are main reservoirs.	■ Widespread through Asia, southern and eastern India, Bali, and Australia	RT-PCR, ELISA	Vaccine available in China and Japan; altering pig farming techniques, insecticide use, and decreasing fertilizer use for rice crops.
Murray Valley encephalitis virus	Murray Valley encephalitis; also called Australian encephalitis First reported in 1917, MVE is found across the Australian mainland and is passed from water birds to humans via mosquitoes.	■ Varies from mild to severe with permanent brain damage or sometimes death. Only one person in 500 infected becomes noticeably ill, although if they do the common symptoms are fever, anorexia and headache sometimes along with vomiting, nausea, diarrhea and dizziness. ■ In severe cases, brain function may be impaired after a few days and lethargy, irritability, drowsiness and confusion may set in. Convulsions and fits come next leading to the possibility of coma and death.	■ Waterfowl such as herons and pelicans are asymptomatic. ■ Horses and wild boars may carry the virus, but do not have clinical signs of disease.	■ Infection is acquired via *Culex annulirostris* mosquitoes.	■ Waterfowl are main reservoirs; mammals such as wild boars and horses may harbor the virus.	■ Australia and New Guinea	ELISA, post-mortem virus isolation	JE vaccine has some protective effect for MVE; mosquito control

Virus	Description	Signs in humans	Signs in animals	Transmission	Distribution	Diagnosis	Prevention/Control
Wesselsbron Fever virus	Wesselsbron Fever Named derived from Wesselsbron, the South African town where causative agent first isolated.	Humans experience fever, headache, muscular pains, joint pain, and mild rash that typically resolves favorably. In severe cases encephalitis and visual disturbances may be seen.	Sheep may abort or fetal mummification; infected lambs have high mortality rate	■ Infection is acquired via *Aedes* spp. mosquitoes ■ Farmers and veterinarians have contracted the disease via obstetrics and necropsies of infected sheep.	■ Virus has been isolated from sheep, cattle, ducks, and coyotes	■ Central and Southern Africa, Madagascar, and Thailand	Virus isolation, RT-PCR	Mosquito control
Yellow Fever virus	Yellow Fever; urban Yellow Fever is when the disease is spread from one person to another via mosquitoes; sylvanic or jungle Yellow Fever is the form spread from infected monkeys to people via mosquitoes. The shift from jungle yellow fever to urban yellow fever is thought to be the result of humans entering the sylvan setting and becoming part of the yellow-fever cycle. Yellow fever was first recognized in an outbreak occurring in the New World in 1648. The term yellow fever was first used during an epidemic in Barbados in 1750 and refers to the jaundiced appearance of people who contract the disease. The viral cause of yellow fever was not discovered until after 1928. The last epidemic of yellow fever in North America occurred in New Orleans in 1905 (more than 3000 cases with 452 deaths); last urban epidemic in the Americas was in Trinidad in 1954.	Clinical signs range from inapparent infection to severe disease. Initial signs are sudden fever, bradycardia, headache, muscle pain, epistaxis, nausea, and conjunctivitis. Second phase is reappearing fever, bradycardia, jaundice, hematemesis, urogenital bleeding, and kidney failure. Fatality rate is 10% to 50%.	Monkeys are asymptomatic	■ Infection is acquired via *Aedes aegypti* and *Aedes simpsoni* mosquitoes in Africa (typically urban cycle) and *Haemagogus* spp. in South American (typically jungle cycle).	■ Monkeys are reservoir hosts; most cases in urban cycle are from humans	■ Middle Africa (between 15°N and 10°S), South America (20°N to 25°S). Eighty percent of South American cases are from Bolivia and Peru.	Virus isolation, immunofluorescence, PCR, ELISA	Vaccine required for visitors and immigrants to endemic areas and for Yellow Fever-free countries for citizens from endemic areas or those who have visited endemic areas.

(the causative agent of chickenpox), has historically been difficult to distinguish from small pox; however, in the late 18th century Heberden clinically differentiated between these two diseases. In 1888, von Bokay reported that chickenpox and herpes zoster were caused by very similar etiologic agents. Another herpesvirus Epstein-Barr virus has been discovered more recently when in 1964, Epstein and Barr isolated virus particles from a patient with lymphoma. The virus they isolated was named after them and became known as Epstein-Barr virus (EBV). In humans, EBV has been linked to cancer development such as Burkitt's lymphoma, nasopharyngeal carcinoma, and B cell lymphomas. Cytomegalovirus, another herpesvirus, was first found in patients with congenital disease. It is almost never symptomatic in immunocompetent people. All of the diseases associated with cytomegalovirus are characterized by enlarged cells, hence the name cytomegalovirus.

Causative Agent

Herpesviruses are host specific, persist for a lifetime, and may be reactivated from a latent form.

The herpesviruses are divided into three subfamilies: Alphaherpesvirinae (herpes simplex virus), Betaherpesvirinae (cytomegalovirus), and Gammaherpesvirinae (Epstein-Barr virus). Alphaherpesvirinae have a relatively short replication cycle that is typically less than 24 hours, have a wide host range, cause rapid destruction of cultured cells, and establish latent infections in neural cells. Most herpesviruses of veterinary importance are found in the subfamily *Alphaherpesvirinae*, genus *Varicellovirus*. Betaherpesvirinae have a relatively slow replication cycle that is typically greater than 24 hours, have a narrow host range, cause slow destruction of cultured cells, produce infected cells that are enlarged and may contain cytoplasmic and nuclear inclusions, and establish latent infections in lymphatic tissue and secretory gland cells. The Betaherpesvirinae subfamily contains the genera *Cytomegalovirus*, *Muromegalovirus*, and *Roseolovirus*, which are of little veterinary significance. Gammaherpesvirinae contains the genera *Lymphocryptovirus* found in marine and fresh water fish and *Rhadinovirus* found in marmosets and monkeys.

There are eight original herpesviruses pathogenic for humans known as herpesvirus hominis type 1 through type 8 (herpes simplex virus (HSV) type 1 causes cold sores, HSV-2 causes genital herpes, HSV-3 is known as varicella-zoster virus causing chickenpox and shingles, HSV-4 is known as Epstein-Barr virus causing mononucleosis, HSV-5 is known as cytomegalovirus causing congenital disease in newborns, HSV-6 and HSV-7 cause roseola, and HSV-8 causing Kaposi's sarcoma seen in AIDS patients). Examples of herpesviruses in animals include bovine herpesvirus 1 (causes infectious bovine rhinotracheitis), porcine herpesvirus 1 (causes pseudorabies), equine herpesvirus 1 (causes abortion), canine herpesvirus 1 (causes a severe hemorrhagic disease in puppies), feline herpesvirus 1 (causes feline viral rhinotracheitis and keratitis), and duck herpesvirus 1 (causes duck plague). There are more than thirty herpesvirus species in monkeys; however, only herpes B virus (*Herpesvirus simiae*) is zoonotic with the virus being endemic in macaque monkeys. HSV-1 and HSV-2 are highly pathogenic for monkeys producing reverse zoonotic disease.

Herpesviruses are large, enveloped DNA viruses with a polyhedral capsule (Figure 7-19). There are over 100 herpesviruses found in mammals, birds, and reptiles. A herpesvirus attaches to

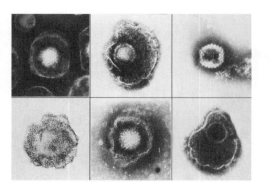

Figure 7-19   Various Herpesviridae family virus seen using an electron microscope. (Courtesy of CDC)

a host cell's receptor and enters the cell through fusion of its envelope with the plasma membrane. After the viral genome is replicated, the virus acquires its envelope from the nuclear membrane and leaves the cell via exocytosis. Herpesviruses have the ability to become latent (remain inactive inside infected cells), often times for years. Latent viruses are reactivated under certain conditions such as stress, immunosuppression, and age and may be inserted into a host's chromosomes, leading to genetic changes and the potential development of cancer. The zoonotic herpesvirus of concern is herpes B virus. Herpes B virus, also known as *Herpesvirus simiae*, causes simian B disease in macaque monkeys producing mild cold-sore type lesions at mucocutaneous junctions and the tongue. Lesions can be found in the mouth and genital areas. In monkeys the disease may present as only conjunctivitis or it can be asymptomatic in these animals. Human infection with herpes B virus produces encephalitis, vesicular rash, and fever with a mortality rate of 70%.

Herpes B Virus Infection

Overview

Herpes B virus infection, also referred to as monkey herpes infection, *Herpesvirus simiae* infection, and simian B disease, is a viral infection acquired from monkeys that causes ascending encephalitis and myelitis in greater than 90% of the people who contract it (70% of cases are fatal). Herpes B virus infects a broad range of mammalian and avian species, including New World monkeys, Old World monkeys, and humans. The natural hosts of herpes B virus are macaque monkeys (Old World monkeys) such as *Macaca mulatta* (rhesus monkeys), *Macaca fascicularis* (cynomolgus monkeys), *Macaca fuscata* (Japanese macaque), and *Macaca arctoides* (stump-tail macaque) all of which naturally exist in various regions of Asia. Both wild and captive macaque monkeys typically have high rates of infection (some populations may have infection rates of 100%), but most individual monkeys show few or any clinical signs. In 1934 Sabin and Wright described the first human case of herpes B infection in a 29-year-old laboratory worker with the initials W. B. from which the B in herpes B is derived. The laboratory worker developed fatal meningoencephalitis and myelitis 18 days after being bitten on the hand from a seemingly healthy rhesus monkey. There have only been 40 cases of herpes B virus infection in humans reported worldwide with most being fatal.

> The zoonotic herpes virus, herpes B virus, belongs to the Alphaherpesvirinae subfamily along with HSV-1, HAV-2, and varicella-zoster virus.

> Herpesvirus causes subclinical or latent infection in reservoir hosts and typically fatal in nonreservoir hosts. For example, *Herpesvirus suis* is latent in swine and fatal in cattle.

Causative Agent

Herpesviruses are large (160 to 180 nm), enveloped DNA viruses that can be divided into three subfamilies (Alphaherpesvirinae, Betaherpesvirinae, and Gammaherpesvirinae). Herpes B virus belongs to the Alphaherpesvirinae subfamily, genus *Simplexvirus*, and is closely related to herpes simplex I and herpes simplex II. There are many different herpesviruses and for each herpesvirus there exists a host for which the virus is almost always fatal and reservoir hosts in which the virus exists in a subclinical or latent infection. For example, Herpes simplex I is latent in humans and fatal in marmoset monkeys; *Herpes suis* is latent in swine and fatal in cattle; and *Herpesvirus simiae* (Herpes B virus) is latent in rhesus monkeys and fatal in humans.

> There are greater than 35 different herpesviruses of nonhuman primates most of which are not zoonotic.

Epizootiology and Public Health Significance

Herpes B virus is found naturally in macaque monkeys in Asia. Young monkeys are typically not infected, whereas older adult monkeys are infected. Approximately

70% of monkeys in captivity and in the wild are infected with herpes B virus. Animals housed together have significantly higher titers to herpes B virus than animals individually caged. Approximately 2% to 3% of infected monkeys shed virus in body secretions with stress, immunosuppression, and pregnancy enhancing viral shedding.

There have only been approximately 40 human cases of herpes B virus found in people who have contact with nonhuman primates such as laboratory workers or zoo workers. Of the roughly 40 well-documented cases of human infection, two-thirds of the cases occurred in the United States and the others were reported in Canada and Great Britain. The majority of the well-documented human cases of herpes B virus infection occurred in the 1950s and 1960s, when large numbers of rhesus macaques were used in the production and testing of poliomyelitis vaccines. In the late 1980s, several cases occurred in Pensacola, Florida, and Kalamazoo, Michigan, coinciding with an increased use of macaques in retroviral research. In countries where macaques are prevalent and commonly interact with humans (such as Japan) there are not reported cases of herpes B infection in humans. This may be a result of the limited availability of B virus diagnostic facilities (actual cases not identified) or the use of different animal-handling procedures.

Transmission

> Once a monkey is infected with herpes B virus, it should be considered infected for life.

Herpes B virus can be transmitted by direct contact with virus-containing secretions (bite wounds in contact with infected saliva, scratches, or needle/scalpel injuries in people), indirect transmission (monkey-to-monkey transmission when placed in a cage that previously housed an infected monkey or fomites from monkey to person), and aerosols.

Macaques transmit the herpes B virus to each other through oral, ocular, or genital contact of mucous membranes or nonintact skin. Herpes B virus can be shed in animals without clinical signs through bodily fluids such as semen, milk, saliva, and in aerosol form. Animals most frequently become infected at the onset of sexual activity; however, younger animals can become infected through contact with another virus-shedding animal. Although most macaques test positive for herpes B virus antibodies, only a small percentage (0% to 2%) shed the virus at any given time. For most macaques, herpes B virus is latent in the trigeminal ganglia and becomes reactivated only when the macaque experiences psychological stress, pharmacological stress (such as corticosteroid use), or immunosuppression.

Pathogenesis

> Asymptomatic shedding of herpes B virus may occur and is more likely during breeding season or times of stress.

In monkeys clinical disease is characterized by vesicles and ulcers that are typically seen on the dorsal surface of the tongue and on the mucocutaneous junction of the lip. The lesions are characterized by necrosis of epithelial cells and the presence of intranuclear inclusion bodies. Inclusion bodies may also be found in macrophages and endothelial cells. Visceral lesions may also develop and are characterized by focal necrosis in the liver. In the central nervous system (CNS), neuronal necrosis and perivascular cuffing with lymphocytes may be found. Intranuclear inclusion bodies occur in glial cells and neurons.

When herpes B virus is inoculated in humans, the virus disseminates to the CNS; however, the virus can also produce local infection in the skin at the inoculation site producing local and regional inflammatory changes. Histologically, lymph nodes draining the area where the virus was inoculated into the human host can be hemorrhagic and focally necrotic. Once the virus is in the spinal cord it can cause necrosis in the spinal cord before it ascends to the brain.

Clinical Signs in Animals

Many macaques infected with herpes B virus are asymptomatic. If clinical signs occur, they are similar to those caused by herpes simplex virus. Clinical manifestations include fluid-filled vesicles on the dorsal aspect of the tongue, lips, and in the mouth. Occasionally vesicles may appear on the skin. When the vesicles rupture, they produce ulcers and fibronecrotic scabs, which may lead to secondary bacterial and fungal infections. Scabs typically heal within 7 to 14 days. Conjunctivitis may also be seen in these animals. Systemic disease associated with herpes B virus in macaques is rare and if present may produce ulcerative lesions in the mouth, esophagus, and stomach as well as necrosis of the liver, spleen, and adrenal glands. Macaques may never develop encephalitis or myelitis as other monkey species may. In other species of monkeys herpes B virus can cause fatal disease.

Clinical Signs in Humans

In humans infected with herpes B virus, local signs will begin to appear within 48 hours after exposure. Local signs include erythema, vesicles, ulcers, and localized pain at the inoculation site (Figure 7-20). Regional lymphadenopathy follows and systemic signs appear 1 to 3 weeks later. Systemic signs include muscle weakness, paralysis of the infected limbs, conjunctivitis, and dysphagia (difficulty eating). As the disease progresses to encephalitis clinical signs such as fever, headaches, nausea, vomiting, neck stiffness, vision problems, ataxia, speech alteration, paralysis, and coma may be seen. Death is usually caused by respiratory paralysis and occurs in 70% of infected people.

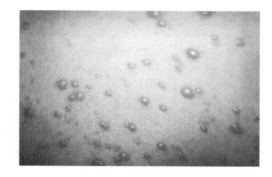

Figure 7-20 A pustulo vesicular rash commonly seen with a generalized herpes outbreak as a result of the Varicella-zoster virus (VZV) pathogen is similar to that seen with herpes B virus infection. (Courtesy of CDC/Joe Miller)

Diagnosis in Animals

Serodiagnosis of herpes B virus in animals is difficult because of cross-reactivity with herpes simplex I and herpes simplex II virus. An ELISA test with monoclonal antibodies has been used to determine the serologic status of monkeys for herpes B virus. Virus isolation using epithelial cell lines from monkeys, rabbits, and hamsters can also be performed. Serial serologies can be obtained from the primate because a rise suggests primary infection and a higher risk of viral shedding.

> Herpes B virus is a biosafety level 4 pathogen and cultures should be performed only at an authorized laboratory.

Diagnosis in Humans

Herpes B infection in people should be suspected in people showing the appropriate clinical signs in conjunction with contact with monkeys, their tissues, or cell cultures prepared from them. Human exposure to herpes B virus can be determined using serologic methods. If the exposed person has a frozen serum sample from the past six months it should be used to collect baseline data or if frozen serum is not available from the last 6 months a serum sample should be taken and frozen. A follow-up serum sample should be collected approximately 3 weeks after exposure or after the onset of illness to compare with the initial specimen for herpes B virus seroconversion. Rapid diagnosis can be made by IFA, electronmicroscopy, or immunofluorescence with antisera to human herpes simplex I. Viral isolation can be done using epithelial lines from humans, monkeys, rabbits, and hamsters. PCR tests are also available. Postmortem diagnosis in people can be done based on history and histologic CNS lesions in addition to immunofluorescence or PCR.

Treatment in Animals

In monkeys that have been bitten or scratched it is important to delay wound cleansing and potentially forcing the virus deeper into the wound. A sample should be obtained for exposure-directed primate virus cultures from buccal mucosa (for cases of saliva exposure), conjunctiva, and the urogenital area (for cases of urine exposure).

Treatment for herpes B infection in monkeys is usually not attempted. Establishing a herpes B virus-free colony is desired and is done through serologic testing of monkeys, performing cesarean sections on dams, and immediately separating dam from offspring to avoid infection in the newborn.

Treatment in Humans

Before treatment is attempted in humans suspected of herpes B virus exposure, a wound swab should be taken for virus isolation. A blood sample should be drawn for baseline data and a mucous membrane scraping from the monkey should be taken as wound decontamination is being performed.

Humans who are bitten or scratched by a monkey should practice wound decontamination protocols including cleansing the exposed area within minutes of obtaining the wound because herpes B virus can enter host cells within 5 minutes. The wound should be scrubbed and/or irrigated for at least 15 minutes. Sterile saline or rapidly flowing water is used for eye decontamination and soap solution, povidone-iodine, or chlorhexidine can be used at skin sites. Dakin solution (0.25% hypochlorite) has been used to clean high-risk deep lacerations or needle sticks. Antiviral therapy including acyclovir or acyclovir alternatives (such as valacyclovir) may be helpful.

Management and Control in Animals

All macaque monkeys should be considered infectious unless proven otherwise. Macaques should not be housed near seronegative monkeys or near monkeys of other species. New monkeys added to a colony should be tested prior to introduction into a facility. Establishing herpes B virus-free macaques should be attempted. Cesarean section and immediate separation of dam from offspring should be done to avoid infection in the newborn.

Management and Control in Humans

Tetanus and rabies prophylaxis should be considered in any cases involving animal bite wounds.

People handling monkeys should treat them as though they were infected with herpes B virus and use proper handling protocols (wearing protective clothing, gloves, and face mask). Any bite wounds or scratches should be cleaned appropriately and within 5 minutes of the exposure as described in the treatment section.

Summary

Herpes B infection is a viral infection acquired from monkeys that causes ascending encephalitis and myelitis in greater than 90% of the people who contract it. Herpes B virus belongs to the Alphaherpesvirinae subfamily, genus *Simplexvirus*, and is closely related to herpes simplex I and herpes simplex II. Herpes B virus infects a broad range of mammalian and avian species, including New World monkeys, Old World monkeys, and humans. The natural hosts of herpes B virus are macaque monkeys, which naturally exist in various regions of Asia. Young monkeys are typically not infected, whereas older adult monkeys are infected. Approximately 70% of monkeys in captivity and in the wild are infected with herpes B virus. Herpes B virus can be transmitted by direct contact with virus-containing secretions, indirect transmission, and aerosols.

Macaques transmit the herpes B virus to each other through oral, ocular, or genital contact of mucous membranes or nonintact skin. Many macaques infected with herpes B virus are asymptomatic. If clinical signs occur they include the development of fluid-filled vesicles on the dorsal aspect of the tongue, lips, and in the mouth. Systemic disease associated with herpes B virus in macaques is rare and if present may produce ulcerative lesions in the mouth, esophagus, and stomach as well as necrosis of the liver, spleen, and adrenal glands. In humans infected with herpes B virus, local signs include erythema, vesicles, ulcers, and localized pain at the inoculation site with regional lymphadenopathy appearing 1 to 3 weeks later. Systemic signs include muscle weakness, paralysis of the infected limbs, conjunctivitis, and dysphagia that may progress to encephalitis. Death is usually caused by respiratory paralysis and occurs in 70% of infected people. Serodiagnosis in animals can be done using an ELISA with monoclonal antibodies to determine the serologic status of monkeys for herpes B virus. Virus isolation using epithelial cell lines from monkeys, rabbits, and hamsters can also be performed. Human exposure to herpes B virus can be determined using serologic methods including IFA, electronmicroscopy, or immunofluorescence with antisera to human herpes simplex I. Viral isolation and PCR tests are also available. Treatment for herpes B infection in monkeys is usually not attempted. Humans who are bitten or scratched by a monkey should practice wound decontamination protocols including cleansing the exposed area within minutes of obtaining the wound because herpes B virus can enter host cells within 5 minutes. Antiviral therapy including acyclovir or acyclovir alternatives may be helpful. Herpes B infection in macaques can be reduced by not housing them near seronegative monkeys or near monkeys of other species, testing them prior to introduction into a facility, and establishing a herpes B virus-free colony. People handling monkeys should treat them as though they were infected with herpes B virus and use proper handling protocols.

ORTHOMYXOVIRUSES

Overview

Orthomyxoviruses, named from the Greek *myxa* for mucus and *ortho* for straight, have a high affinity for mucus with influenza being the major disease in this group. Influenza is an acute and very contagious disease that spreads rapidly and was probably the epidemic catarrhs associated with seasonal periods described by Hippocrates around the year 400 B.C. In Florence, in the 14th century, this disease started to be called influenza di catarro then simply influenza (*influencia* in Italian means a fluid given off by stars that governed human behavior). English scientists named it influenza and in France it was called grippe. Between the 12th and the 18th century, there had been a number of influenza pandemics including the 1889–1891 pandemic that started in Siberia and subsequently spread to Russia and to the rest of Europe. The etiology of influenza remained unknown for a long time; however, considerable progress in understanding the disease occurred during the 1889–1890 epidemic. Pfeifer, a German scientist, isolated a bacterium in people's lungs that was similar to a bacterium found in pigs with swine influenza. The link between influenza in people and animals and the ability to develop bacterial infections secondary to these "filterable viruses" was proposed. The major influenza pandemic of the 20th century was the Spanish Flu (1918–1919) and caused over 20 million deaths. This pandemic started during World War I with the outbreak taking place on the European western front with other outbreaks occurring in Western Africa (Sierra Leone) and China. A H1N1 shift crossed over to humans causing the Spanish Flu pandemic. Other 20th century pandemics occurred in 1957 (Asiatic Flu), 1968 (Hong Kong Flu), and 1977 (Russian Flu).

Influenza virus was first isolated from poultry in 1878 (the cause of fowl plague) and as early as 1901 the causative agent of influenza in man was shown to be an ultra-filterable agent. In 1930 in the United States Shope discovered that a virus and a bacterium were involved in swine influenza. In 1933 Smith, Andrews, and Laidlaw isolated the virus in ferrets. In the 1940s the property of haemagglutination was observed, followed by the discovery that the virus could be grown in hen's eggs making influenza one of the best studied viruses. Viruses with similar properties continued to be added to the group, until it was split into two families (Paramyxoviridae and Orthomyxoviridae) in the 1970s. Orthomyxoviruses are the influenza viruses that affect various animal species most with their own specific influenza strains. Some animal strains may produce infections in humans. There are three general types of influenza viruses: types A, B, and C. Influenza type A and type B cause human epidemics and immunity against one does not provide protection against the others. The influenza type A virus is ubiquitous in nature and can infect a variety of animal species including man. The influenza type B virus is only carried by humans and produces infections similar to those caused by type A. Virus C plays a minor role in causing the disease.

Causative Agent

Orthomyxoviruses (influenza viruses are the only viruses in this group) are medium-sized, enveloped, single-stranded, negative sense RNA viruses that vary in shape from spherical to helical. Their genome is segmented into eight pieces. The external surface is covered with glycoprotein spikes that are the main antigenic structures of the virus. Three influenza virus antigens (the nucleoprotein, the hemagglutinin, and the neuraminidase) are used in classification of these viruses. The nucleoprotein antigen is stable and is used to differentiate the three influenza virus types (A, B, or C). The hemagglutinin and neuraminidase antigens are variable and are responsible for immunity to infection. The surface antigens hemagglutinin and neuroaminidase display 16 subtypes (H1-H16) and 9 subtypes (N1-N9) respectively. Hemagglutinin allows the virus to adhere to the membrane and it is important for the development of infection. Neuroaminidase contributes to the release of viral particles from infected cells and helps with virus adhesion to the host's cells. Influenza subtypes are labeled with both the hemagglutinin and neuroamidase number. There are only three known A subtypes of influenza viruses (H1N1, H1N2, and H3N2) currently circulating among humans. It is believed that some genetic components of human influenza A virus originally came from birds. The subtypes responsible for human pandemics are H1N1 (1947), H2N2 (1957), H3N2 (1968) and the reappearance of H1N1 (1977). Mammals and birds (both domestic and migratory) can be infected with influenza virus and certain pandemics have been associated with viruses of an animal origin (1918, swine strain; 1957 and 1968, avian strains).

Influenza

Overview

Influenza, also known as flu or grippe (French for attack, which describes the acute febrile disease), is a contagious viral respiratory infection that has produced epidemics dating back to the Plague of Athens in 430 B.C. Charlemagne's army may have contracted influenza during an epidemic in 876 B.C. The first recorded pandemic was in 1580 when influenza spread across Africa and Europe. By 1647, influenza had reached North America when an epidemic from the Caribbean moved north into New England. Influenza was formally named in 1733 by John Huxham, an English doctor who used an Old Italian folk term that linked the symptoms of influenza to the astrological influence of the stars. Prior to that, influenza was known by a variety of names such as la grippe, jolly rant, and the new acquaintance. There have been at least 17 global outbreaks of influenza between 1175 and 1920. Although influenza was known for some time, it was not believed to be a serious disease until the great pandemic of Spanish flu in 1918–1919. The influenza pandemic of 1918 was the 20th century's worst and most deadly pandemic in modern Western society. This pandemic may have originated in the United States (although many people believe it originated at the Western Front and was carried to the United States in soldiers arriving back from World War I) and is believed to have killed approximately 21 million people worldwide. The first United States fatalities were from army barracks in Denver and South Carolina. The flu spread quickly through surrounding communities, killing over 10,000 people per week in United States cities. An estimated 675,000 people died in the United States from this pandemic with a significant number of deaths occurring in young adults. Since 1919, other outbreaks of influenza have occurred: the Asian flu killed 70,000 Americans in 1957–1958 and the Hong Kong flu killed about 34,000 Americans in 1968–1969. The **variants** of influenza (a variant has something that differs in its characteristic from the classification to which it belongs) are named according to where they first strike such as Bangkok (Bangkok flu) or New Jersey (New Jersey flu).

Influenza virus was first isolated from chickens in 1901, but it was not recognized as a cause of influenza until 1955. There are three types of influenza virus. A is the most common and was isolated in ferrets by Smith, Andrews, and Laidlaw in 1933. Influenza B was discovered in 1936 and influenza C was isolated in 1947. In 1931 an influenza virus was isolated from pig snouts and was named swine flu making some people believe that pigs contracted the virus from humans during the 1918–1919 pandemic, while others believed it caused the 1918–1919 pandemic. The most recent outbreak of swine influenza virus in people started on February 5, 1976 when an army recruit at Fort Dix, New Jersey, became tired and weak. He died the next day and health officials determined that swine flu was the cause of death. Public health officials urged people in the United States to be vaccinated for the disease; however, the vaccine was blamed for 25 deaths (more people died from the vaccine than died from the swine flu itself) and resulted in an increase in the incidence of Guillain-Barré syndrome (a side effect of the vaccine).

Other animals can also contract influenza. Avian influenza virus is found worldwide causing varying degrees of illness in poultry. There are two types of avian influenza viruses called low pathogenic avian influenza (causes little or no clinical signs in infected birds) and highly pathogenic avian influenza (causes serious and often fatal disease in birds). Low pathogenic avian influenza may mutate into highly pathogenic avian influenza. Outbreaks of highly pathogenic avian influenza occurred in late 2003 to early 2004 in southeast Asia and again in

> Influenza viruses may display different degrees of variations: these can be classified as minor variations (drifts) or major variations (shifts).

June 2004 in southeast Asia that spread into Eurasia by late 2005, causing both human and bird deaths. Equine influenza virus was isolated from horses in 1956. There are two main types of virus called equine-1 (H7N7), which commonly affects horse heart muscle and equine-2 (H3N8), which is usually more severe. Horse flu is endemic throughout the world with nearly 100% infection rate in an unvaccinated horse population that has not been previously exposed to the virus. Equine influenza is not believed to be zoonotic. Canine influenza occurs in dogs and is caused by the equine influenza virus (H3N8). Canine influenza is highly contagious among dogs. In recent outbreaks, dogs were exposed to the virus at horse racing tracks where dog racing also occurs.

Causative Agent

Influenza is caused by influenza virus, a member of the Orthomyxoviridae family and there are three serotypes of influenza virus (A, B, and C) (Figure 7-21). Influenza virus is a helically-shaped, single-stranded, negative sense RNA virus with its type determined by its nuclear material. The outer surface of the virus particle consists of a lipid envelope from which two types of glycoprotein spikes are found (hemagglutinin and neuraminidase). Hemagglutinin (HA) is an envelope spike antigen that can attach to erythrocytes and cause agglutination. HA is responsible for the attachment of the virus to cell surface receptors (this process can be blocked by antibody). Neuraminidase (NA) is an envelope protein whose enzyme activity liquefies mucus, which contributes to viral spread. Internal proteins consist of nucleocapsid protein (NC), some matrix proteins (M1) and three polymerases (PA, PB1, and PB2). NC and M1 determine type specificity. The types of influenza viruses include:

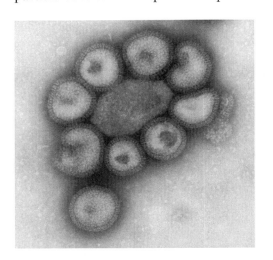

Figure 7-21 An electron micrograph of a number of influenza virus particles (courtesy of CDC/Dr. F.A. Murphy) and the influenza virus nomenclature.

- Type A, which has subtypes determined by the surface antigens hemagglutinin and neuraminidase. There are 15 hemagglutinin antigens (H1-H15) and 9 neuraminidase antigens (N1-N9). Three types of hemagglutinin in humans (H1, H2, and H3) have a role in viral attachment to cells and two types of neuraminidase (N1 and N2) have a role in viral penetration into cells. The nomenclature used to describe influenza A viruses is expressed in the order of virus type, followed by geographic site where it was first isolated, followed by strain number, followed by year of isolation, and ending with virus subtype (Figure 7-22). For example A/Moscow/21/99 (H3N2) is an influenza A virus, first isolated in Moscow, strain 21, isolated in 1999 and is the subtype containing hemagglutinin 3 and neuraminidase 2. Type A influenza causes moderate to severe illness in all age groups in humans and animals. Influenza A virus infects pigs, birds, horses, dogs, and seals. Influenza A viruses frequently change with new strains and subtypes causing epidemics and pandemics. New strains evolve through two types of antigenic changes. One process of producing antigenic variation occurs as the virus accumulates point mutations during viral replication (a process known as antigenic drift). Antigenic drifts are minor changes in genes coding for H and N glycoproteins. Antigenic shifts are major changes in the

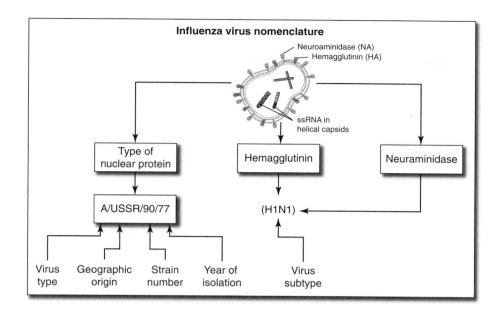

Figure 7-22 Glycoprotein spikes on influenza virus are neuraminidase (NA) and hemagglutinin (HA). Variations in NA and HA glycoproteins determine the strain of the virus.

genome as a result of reassortment of genome segments. In the reassortment process, entire segments of RNA that code for a single protein are exchanged between two viruses that infect the same host. If two different viral strains infect a cell at the same time, genetic reassortment can occur producing a third, new strain. Reassortment can result in the emergence of new strains and can occur between avian, swine, equine, and human influenza A viruses. This reassortment can result in a hybrid virus with proteins from each species.

- Type B produces milder epidemics and is found in humans, especially children. Influenza B virus has also been found in ferrets, seals, pigs, dogs, and horses. Disease caused by influenza B virus is not as severe as that caused by type A. Influenza B viruses do not have distinct serotypes, but do have strains. Influenza B viruses undergo antigenic drift but not antigenic shift. Influenza B virus has caused human epidemics, but has not produced any pandemics.

- Type C infects humans (and perhaps swine) and produces mild, sporadic respiratory disease. Dogs can be experimentally infected with influenza C virus and antibody levels to this virus have been found in pigs, dogs, and horses. Influenza C viruses do not have distinct serotypes, but do have antigenically-stable strains.

Epizootiology and Public Health Significance

Influenza occurs worldwide. Influenza A virus is the subtype of most concern and there are influenza A viruses in people, birds, swine, equine, canines, and felines. In the past 100 years, there have been four antigenic shifts that have led to major influenza pandemics (1889–1891, 1918–1919, 1957–1058, and 1968–1969). Pandemics start from a single focus and spread along travel routes with high attack rates affecting all age groups. Pandemics produce high morbidity and high mortality. In influenza epidemics, attack rates are lower than pandemics producing high morbidity and a rise in mortality.

Influenza epidemics in people occur in winter months (December through March) and vary in severity. Millions of people develop infection every year. In the United States significant influenza activity occurred in the winters of 1999–2000 and 2003–2004. During 2000–2003 influenza activity was low. Typically 20,000 U.S.

> Influenza viruses are highly unstable, genetically labile, and well-adapted to elude host defenses.

deaths occur annually from influenza. Women who contract influenza A or B during the third trimester of pregnancy can abort. The CDC receives weekly surveillance reports from states showing the extent of influenza activity. More than 100 WHO Collaborating Laboratories in the United States submit reports on the number of specimens and the number and type of influenza viruses isolated for each week from early October to mid-May. Influenza pandemics have produced substantial human disease and death throughout history. Human avian influenza outbreaks have occurred in Asia. Since 1959, human infections with avian influenza have occurred on only eleven occasions (six of these were documented in 2003). From 1959 to 2004 there have been 70 human cases of avian influenza caused by the H5N1 virus with 43 human fatalities. In 1997 there was an outbreak of avian influenza in humans in Hong Kong (18 infections with 6 of them being fatal); however, the absence of disease in high risk groups such as poultry workers and cullers indicated that the H5N1 virus did not cross easily from birds to humans. The outbreak of avian influenza in Hong Kong ended after 1.5 million poultry were slaughtered within 3 weeks. On January 5, 2004, there was a cluster of respiratory disease in children in a Hanoi hospitial. The Vietnamese government found H5N1 virus in dead chickens in their country on January 8, 2004. Other countries that reported the H5N1 virus in 2004 were Japan (in poultry on January 12) and Thailand (in humans and poultry on January 23). The outbreaks that occurred in June 2004, in Southeast Asia spread into Eurasia by late 2005.

The occurance of animal influenza varies with the species. Swine influenza is common in North America, South America, Europe, and Asia. Equine influenza is seen worldwide except in New Zealand, Australia, and Iceland. Canine influenza has only been seen in the United States and is not believed to be zoonotic. From January to May 2005, outbreaks of canine influenza occurred at 20 racetracks in 10 states (Florida, Texas, Arkansas, Arizona, West Virginia, Kansas, Iowa, Colorado, Rhode Island, and Massachusetts). Feline influenza has been reported in Germany and Asia. Avian influenza in birds is seen in highest proportion in Asia. Sometime prior to 1997, the H5N1 strain of avian influenza began circulating in poultry in parts of Asia causing mild disease in birds such as ruffled feathers and decreased egg production. After months of circulating in chickens, the virus mutated to a highly pathogenic form that could kill birds within 48 hours with a mortality rate close to 100%. The highly pathogenic form of avian influenza erupted in 1997, but then did not reappear until 2003. In December 2003, the highly pathogenic form was seen in Seoul, South Korea, for the first time. Another resurfacing of the virus occurred in January 2004 in Vietnam, Japan, and Thailand. Prior to 2004, outbreaks of highly pathogenic avian influenza were considered rare in poultry. In July 2004, another outbreak of the highly pathogenic avian influenza virus occurred in Asia. It is now believed that this virus has found an ecological niche in Asia and is considered endemic in poultry in Asia. In the United States from 1997 to 2005, there were 16 outbreaks of low pathogenic avian influenza A viruses (H5 and H7 subtype) and one outbreak of highly pathogenic avian influenza A (H5N2) in poultry. The H5N1 virus is now more deadly in poultry and experimentally in mice; is being seen in new species such as cats and tigers; is killing some migratory birds; and is seen in asymptomatic domestic ducks that are shedding large numbers of the organisms.

Transmission

In mammals, influenza virus is transmitted by aerosols and by direct or indirect contact with nasal discharge. Close, closed environments help the spread of influenza virus. Influenza virus can survive for several hours in dried mucus. In birds, avian influenza virus is shed in the feces, saliva, and nasal secretions making fecal-oral transmission the most common means of viral spread in birds (shared drinking water and fomites are common sources of disease transmission). Avian influenza virus has also been found in the yolk and albumen of eggs from infected hens (broken shells can transmit the virus). Avian influenza has been reported in domestic and zoo cats in Asia that are believed to have acquired the virus through ingestion of raw infected poultry. Migratory waterfowl are the natural reservoir for avian influenza.

Transmission of influenza virus between species typically occurs from direct contact with infected animals or fomites (some believe swine influenza can be transmitted to people by aerosol). Human infections with swine influenza virus have occurred sporadically with the most recent example occurring in 500 military recruits in Fort Dix, New Jersey, in 1976. Pigs can be infected with human influenza A and B virus. The H5N1 avian influenza virus likely undergoes cross-species transmission. Human-to-human transmission is not believed to have occurred because no H5N1 cases have been detected in health care workers despite close contact with patients with the virus.

> Avian influenza virus can survive for up to 2 weeks in feces and on cages.

Pathogenesis

Following inhalation of virus particles, the virus may be neutralized by secretory antibodies (clearing the infection) or the virus attaches to and penetrates respiratory epithelial cells in the trachea and bronchi. Viral replication begins causing cellular dysfunction resulting in destruction of the host cell. Host cell destruction releases viruses so that they can infect other cells. Viremia has been rarely documented in cases of influenza and the virus is shed in respiratory secretions for 5 to 10 days.

Clinical Signs in Animals

In animals clinical signs of influenza are caused by influenza type A viruses and are unique to each virus and species.

> Both avian influenza and velogenic viserotrophic Newcastle disease can produce rapid mortality of birds with minimal clinical signs.

- Avian influenza (H5N1) is found in a variety of domestic and wild birds including chickens, ducks, turkeys, quail, pheasants, and waterfowl. Avian influenza is highly contagious (spread by aerosol and by direct and indirect contact), spreads rapidly, and consists of numerous strains or subtypes. Avian influenza virus affects the respiratory, nervous, and gastrointestinal systems of birds and the incubation period is typically 3 to 14 days. Most avian influenza strains are low pathogenic avian influenza (LPAI) strains and produce subclinical or mild to moderate respiratory infections that produce coughing, sneezing, sinusitis, ruffled feathers, and decreased egg production in infected birds (Figure 7-23). Migratory waterfowl infected with avian influenza virus have subclinical, gastrointestinal infections that can excrete virus for extended periods of time serving as a frequent source of LPAI infection for domestic poultry. Highly pathogenic avian influenza (HPAI), also known as fowl plaque, is a severe, often fatal disease in domestic poultry (morbidity and mortality may reach 100%).

Figure 7-23 Swollen sinus in a turkey with avian influenza.
(Courtesy of University of Minnesota, College of Veterinary Medicine)

Only subtypes that contain H5 or H7 hemagglutinins have been highly pathogenic. Clinical signs in birds infected with HPAI include ruffled feathers, excessive thirst, anorexia, and watery diarrhea that progresses from bright green to white. Sudden death may be seen in 24 to 48 hours after the onset of clinical signs or may occur with few clinical signs. Mature chickens may have swollen combs and wattles and have cyanotic comb tips. Conjunctival swelling or hemorrhages can also be seen. In broiler chickens, clinical signs include facial and neck edema, torticollis, ataxia, and death. Similar clinical signs are seen in turkeys. Ducks and geese develop viremia producing minimal clinical signs.

- Swine influenza (H1N1, H1N2, and H3N2), also known as swine flu and hog flu, is found mainly in pigs, but can also infect humans. Swine influenza commonly appears in the colder months and is spread by aerosol and direct and indirect contact. In herds where the virus is endemic young pigs are continually infected and maintain the virus in the herd. Subtypes H1N1 (found in United States pigs in 1918) and H3N2 (circulating in pigs in Europe and Asia since the 1970s) are the most frequent causes of swine influenza. Morbidity in pigs with swine influenza is high, but mortality is approximately 2%. Clinical signs are mild without secondary bacterial infection and include fever, tachypnea, dyspnea, coughing, and anorexia (Figure 7-24). The disease runs its course in 2 to 6 days. Secondary bacterial infection, especially with the bacterium *Hemophilus suis*, produces more serious illness and occasionally death.

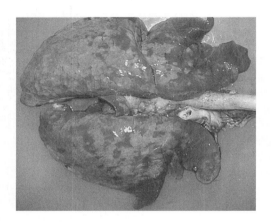

Figure 7-24 Lungs from a pig with swine influenza.
(Courtesy of Dr. Daniel Harrington, Prof. Emeritus, Dept. Comparative Pathology, School of Vet. Medicine, Purdue University)

- Equine influenza is a highly contagious disease of horses, mules, and donkeys worldwide except in Australia, New Zealand, and Iceland. The serotypes involved in producing disease are H7N7 (equine-1) and H3N8 (equine-2). H3N8 has recently jumped species and has infected dogs. Equine influenza virus is primarily spread by aerosol (direct and indirect contact is also possible) with clinical signs appearing 1-3 days after infection. Clinical signs include fever, anorexia, and coughing with some degree of ocular discharge and photophobia. Leg edema is also possible. Pneumonia is occasionally seen when the viral disease is complicated with secondary bacterial infection. Recovery in cases without bacterial involvement typically occurs in 7 to 10 days. Young horses are particularly susceptible to equine influenza.

- Canine influenza, also known as dog flu, is frequently caused by the equine subtype H3N8 and was first seen in greyhounds in Florida in 2004. Thirty percent of dogs infected during this outbreak died; however, the current mortality rate is 5% to 8%. Canine influenza is spread by aerosol and indirect contact with older dogs and puppies more severely affected. The incubation period is 2 to 5 days and 80% of dogs infected develop mild respiratory disease. Clinical signs include high fever, coughing, dyspnea, anorexia, and mucoid nasal discharge. Pneumonia may occur if the dog develops a secondary bacterial infection and some dogs have acutely died with evidence of hemorrhages in the respiratory tract. Recovery occurs in 1 to 3 weeks in dogs with mild disease. Dogs have been experimentally infected with influenza C virus that produced clinical signs of nasal discharge and conjunctivitis that lasted for 10 days.

- Feline influenza occurred in 2004 in a few zoo tigers and leopards in Thailand. These felines were infected with H5N1 avian influenza virus and develop a severe, fatal pneumonia. Domestic cats have also contracted influenza virus experimentally and naturally (in 2005 in Germany believed to have been acquired from a wild bird). Cross-species infection from bird to cat requires a large viral infective dose. Cats are susceptible to avian influenza; however, evidence suggests that they are unlikely to act as a reservoir for the virus.

- Influenza can occur in other species such as ferrets, mink, and marine mammals. Ferrets are susceptible to human influenza virus and display clinical signs such as fever, anorexia, lethargy, sneezing, coughing, and purulent nasal discharge. Infection in adult ferrets typically resolves in 5 to 14 days; however, the disease can be fatal in neonatal ferrets. Influenza in minks was seen on Swedish mink farms in 1984 and was caused by an avian influenza strain. Mink developed sneezing, coughing, anorexia, and nasal and ocular discharge. Some mink died. Influenza in marine mammals (seals and whales) produces respiratory disease and is caused by avian influenza virus.

Clinical Signs in Humans

The incubation for influenza in people is 1 to 4 days. Human influenza A or B viral infections are characterized by upper respiratory disease producing fever, chills, anorexia, headache, muscle pain, sneezing, sore throat, nonproductive cough, and weakness. Diarrhea, nausea, vomiting, and abdominal pain may be seen, especially in children, but are not hallmarks of influenza. Most people recover in 1 to 5 days, but symptoms may continue for 2 weeks or longer. More severe clinical signs may be seen in people with underlying respiratory or cardiac disease who may develop pneumonia, encephalopathy, myocarditis, or Reye's syndrome. Influenza C virus can cause a mild respiratory disease in children and young adults that may progress to bronchitis or pneumonia.

Human infections with avian influenza have been reported in healthy children and adults as well as people with chronic medical conditions. In people, avian influenza can cause conjunctivitis, typical influenza signs, or can be fatal. In 1997, 18 human infections with H5N1 avian influenza virus were reported in Hong Kong producing clinical signs of fever, sore throat, cough, and occasionally respiratory distress. Six of these people died. In 1999, H9N2 avian influenza was found in two children in Hong Kong producing mild disease. In 2003, two H5N1 avian influenza virus infections were reported in Hong Kong (one person died) and 347 H7N7 avian influenza virus infections were reported in the Netherlands (associated with an outbreak in poultry; produced 1 death). In 2004 two cases of H7N3 avian flu infections occurred in Canada producing conjunctivitis. In 2004 and 2005 a widespread outbreak of H5N1 avian influenza occurred in Asia with 218 infections and 124 deaths occurring by May 2006.

Human infections with swine influenza produces high fever, lethargy, and bronchitis, which may lead to secondary bacterial infections and pneumonia. A 1976 outbreak in Fort Dix, New Jersey, resulted in the death of 1 recruit from pneumonia and infection in 500 people that is believed to have been spread person-to-person. Some people feel that the 1918–1919 flu pandemic was from the H1N1 strain of swine influenza.

Human infections with equine influenza virus have been reported in people, but have not produced significant clinical disease. Virus could be isolated in these people for up to 10 days.

Diagnosis in Animals

Diagnosis in animals for each influenza virus is described as follows:

There is not a chronic carrier state for influenza.

- Avian influenza virus is identified by virus isolation in embryonated eggs, virus neutralization, hemagglutination inhibition tests, ELISA, RT-PCR, and agar gel diffusion. Allantoic fluid from infected chicken embryos will agglutinate erythrocytes. All positive samples are sent to and confirmed by the CDC. Clinical samples include whole birds, lung, trachea, air sac, kidney, spleen, and serum. Tracheal or cloacal swabs from live or dead birds are commonly used for virus isolation and identification.
- Swine influenza virus is identified by virus isolation in embryonated eggs, immunofluorescence, RT-PCR, or seroconversion by complement fixation, hemagglutination inhibition, or ELISA test methods. Clinical samples include nasal swabs, lung tissue, and serum.
- Equine influenza virus is identified by virus isolation in embryonated chicken eggs or an increase in acute and convalescent antibody titers using hemagglutination inhibition tests. A rapid human influenza A kit can be used to diagnose equine influenza. An ELISA test that can distinguish between natural and vaccine-induced antibodies is being developed. Clinical samples include nasal and ocular swabs and serum.
- Canine influenza virus is identified by virus isolation in embryonated chicken eggs or antibody detection via hemagglutination inhibition or virus neutralization tests. The rapid human influenza A kit can diagnose the equine influenza virus strain that infects dogs. Clinical samples include nasopharyngeal swabs (taken within 72 hours of clinical signs) and serum.
- Feline influenza virus is identified by methods used to identify avian influenza virus.

Diagnosis in Humans

In people, influenza A and B viruses are diagnosed by virus isolation from throat and nasopharyngeal swabs within 3 days of disease onset, serologic tests such as complement fixation, immunodiffusion, hemagglutination inhibition, and ELISA rapid detection kits, and PCR.

Treatment in Animals

The treatments for influenza in animals include:

- Avian influenza virus can be reduced in some avian species by using amantadine hydrochloride, although some resistance problems have occurred. Poultry flocks with HPAI are depopulated and not treated.
- Swine influenza virus infections are treated symptomatically.
- Equine influenza virus can be treated in valuable horses with antiviral drugs.
- Canine influenza virus infections are treated symptomatically including the use of antibiotics to prevent secondary bacterial infections.
- Feline influenza virus infections are treated symptomatically.
- Other species with influenza virus infections are treated symptomatically if at all.

Treatment in Humans

There are four antiviral drugs that can be used for treatment of influenza in the United States. Amatadine and rimantadine interfere with the function of the M2 protein blocking viral penetration and viral uncoating. These drugs are used against influenza A virus in the first 48 hours of disease. Zanamivir and oseltamivir are neuraminidase inhibitors and are effective against influenza A and B. Treatment typically shortens the duration of illness by one day. Drug resistance to amantadine and rimantadine has occurred. Symptomatic treatment for pain and fever reduction may also be used in people to lessen the signs of clinical disease.

Management and Control in Animals

Inactivated influenza vaccines are available for pigs, horses, and in some countries birds. The vaccine may not prevent disease but produce milder disease. In the United States avian influenza vaccine is used frequently in turkeys to prevent LPAI infections. An inactivated H5 vaccine is licensed in the United States for emergency use in outbreaks. Vaccines are costly and do not offer cross protection between the 15 serotypes of avian influenza. Influenza vaccines change periodically in response to current subtypes and strains in a given area.

Avian influenza in poultry can be reduced by practicing an all-in/all-out flock management style. Preventing contact between domestic poultry and wild birds and their water sources may also reduce disease potential. Biosecurity measures and proper disinfection also play roles in preventing disease among birds. Avian influenza viruses are relatively unstable in the environment and are inactivated by extremes in pH, heat, and dryness. After removal of organic matter, several classes of disinfectants are effective at destroying avian influenza virus including phenolics, quaternary ammonium compounds, oxidizing agents, and dilute acids. In the presence of organic matter, avian influenza virus can be inactivated by aldehydes. During outbreaks, quarantines and isolation of infected animals can prevent spread of disease. Poultry with HPAI are eradicated (quarantine, depopulation, cleaning, disinfection, and surveillance around affected flocks). Other animal species should not be housed with poultry because of the ability of avian influenza to cross species barriers.

Swine and equine influenza can be minimized by isolating new animals to a farm or facility. In horses, rest decreases viral shedding. Infected swine herds may need to be depopulated. Ferrets should avoid contact with people with respiratory signs consistent with influenza. Cats should not be fed poultry infected with avian influenza or uncooked poultry. Keeping cats indoors to prevent them from eating infected birds is also advised.

Management and Control in Humans

Inactivated influenza vaccines have been produced since the 1950s and the first live attenuated influenza vaccine was licensed in 2003. Annual vaccines are available for influenza A and B in inactivated, injectable and inactivated, intranasal forms (vaccines have 2 influenza A and 1 influenza B inactivated strains in them). These vaccines contain the viral strains thought to be the likely strain to produce epidemics during the following winter and are updated annually. Amantadine, rimantadine, and oseltamivir can be used prophylactically in high risk people (elderly and immunocompromised). Avoiding contact with symptomatic people, thorough hand washing, and good hygiene practices can limit exposure to influenza virus.

Avian influenza in people can be reduced by controlling epidemics in birds, requiring poultry workers to practice proper hygiene procedures (wearing boots, coveralls, gloves, facemasks, and headgear), and obtaining human influenza vaccines to prevent infection with human influenza virus which may allow the virus to co-mingle with avian influenza viruses. Avoiding contact with sick birds also limits a person's exposure to avian influenza.

Summary

Influenza is a contagious viral respiratory infection that has produced epidemics throughout history. Influenza is caused by influenza virus, a member of the Orthomyxoviridae family and there are three serotypes of influenza virus (A, B, and C). Influenza occurs worldwide. Influenza A virus is the subtype of most concern and there are influenza A viruses in people, birds, swine, equine, canines, and felines. Swine influenza is common in North America, South America, Europe, and Asia. Equine influenza is seen worldwide except in New Zealand, Australia, and Iceland. Canine influenza has only been seen in the United States. Feline influenza has been reported in Germany and Asia. Avian influenza is endemic in poultry in Asia. In mammals, influenza virus is transmitted by aerosols and by direct or indirect contact with nasal discharge. In birds, avian influenza virus is shed in the feces, saliva, and nasal secretions making fecal-oral transmission the most common means of viral spread in birds. Avian influenza virus has also been found in the yolk and albumen of eggs from infected hens. Transmission of influenza virus between species typically occurs from direct contact with infected animals or fomites. The H5N1 avian influenza virus likely undergoes cross-species transmission. Human infections with swine influenza virus have occurred sporadically. Pigs can be infected with human influenza A and B virus. In animals clinical signs of influenza are caused by influenza type A viruses and are unique to each virus and species. Avian influenza is found in a variety of domestic and wild birds and is highly contagious, spreads rapidly, and consists of numerous strains or subtypes. Most avian influenza strains are low pathogenic avian influenza (LPAI) strains and produce subclinical or mild to moderate respiratory infections that produce coughing, sneezing, sinusitis, and decreased egg production in infected birds. Highly pathogenic avian influenza (HPAI), also known as fowl plaque, is a severe, often fatal disease in domestic poultry. Clinical signs in birds infected with HPAI include ruffled feathers, excessive thirst, anorexia, and watery diarrhea that progresses from bright green to white. Sudden death may be seen in 24 to 48 hours after the onset of clinical signs or may occur with few clinical signs. Swine influenza, also known as swine flu and hog flu, is found mainly in pigs, but can also infect humans. Clinical signs are mild without secondary bacterial infection and include tachypnea, coughing, and anorexia. Equine influenza is a highly contagious disease of horses, mules, and donkeys. Clinical signs include fever, anorexia, and coughing with some degree of ocular discharge and photophobia. Canine influenza, also known as dog flu, is frequently caused by the equine subtype H3N8 and produces high fever, coughing, dyspnea, anorexia, and mucoid nasal discharge. Feline influenza is caused by the H5N1 avian influenza virus and cats develop a severe, fatal pneumonia. Human influenza A or B viral infections are characterized by upper respiratory disease producing fever, chills, anorexia, headache, muscle pain, sneezing, sore throat, nonproductive cough, and weakness. Influenza C virus can cause a mild respiratory disease in children and young adults that may progress to bronchitis or pneumonia. Human infections with avian influenza virus have been reported causing conjunctivitis, typical influenza signs, or fatalities. Human infections with swine influenza virus produces high fever, lethargy, and bronchitis, which may lead to secondary bacterial infections and pneumonia. Human infections with equine influenza virus have been reported in people, but have not produced significant clinical disease. Diagnosis in animals varies for each influenza virus and may include virus isolation in embryonated eggs, virus neutralization, hemagglutination inhibition tests, complement fixation, ELISA, RT-PCR, agar gel diffusion, and immunofluorescence. In people, influenza A and B are diagnosed by virus isolation from throat and nasopharyngeal swabs and serologic tests (complement fixation, immunodiffusion, hemagglutination inhibition, and ELISA rapid detection kits). PCR is also available. The treatments for influenza in animals include using amantadine (in birds), depopulation, symptomatic treatment, and antibiotics for secondary bacterial infections. There are four antiviral drugs (amatadine, rimantadine, zanamivir, and oseltamivir) that can be used for treatment of influenza in people in the United States. Inactivated influenza vaccines are available for pigs, horses, and in some countries birds. Avian influenza in poultry can be reduced by practicing an all-in/all-out flock management style. Preventing contact between domestic poultry and wild birds and their water sources may also reduce disease potential. Biosecurity measures and proper disinfection also play roles in preventing disease among birds. Annual vaccines are available for human influenza A and B in inactivated, injectable and inactivated, intranasal forms. Avoiding contact with symptomatic people, thorough hand washing, and good hygiene practices can limit exposure to influenza virus. Avian influenza in people can be reduced by controlling epidemics in birds, requiring poultry workers to practice proper hygiene procedures, and obtaining human influenza vaccines to prevent infection with human influenza virus which may allow the virus to comingle with avian influenza virus. Avoiding contact with sick birds also limits a person's exposure to avian influenza.

PARAMYXOVIRUSES

Overview

The order Mononegavirales contains viral species that have a RNA genome and include the four families Bornaviridae, Rhabdoviridae, Filoviridae, and Paramyxoviridae. The family Paramyxoviridae has two subfamilies: Paramyxovirinae and Pneumovirinae. The Paramyxovirinae include the genera *Parainfluenzavirus* (such as Sendai virus), *Rubulavirus* (such as mumps in humans), *Morbillivirus* (such as measles in humans), and *Henipavirus* (such as Hendra virus). *Parainfluenzavirus* and *Morbillivirus* consist of strictly human pathogens and related viruses that only infect animals (such as canine distemper). There are only a few members of Paramyxovirinae that are truly zoonotic; however, most members are capable of changing hosts and have the potential to become zoonotic. Paramyxoviruses are named from the Latin *para* for near and the Greek *mxyo* for mucus. Paramyxoviruses cause respiratory diseases that have been associated with the production of mucus. The subfamily Pneumovirinae (such as respiratory syncytial virus) does not contain zoonotic viruses.

Causative Agent

Paramyxoviruses are enveloped, nonsegmented, single-stranded, negative sense RNA viruses. Paramyxoviruses are roughly spherical (about 150 to 300 nm in diameter) but they can be much larger and more pleomorphic. The virus envelope is a lipid bilayer with spikes of haemagglutinin-neuraminidase (HN) and fusion (F) glycoproteins that function in cell attachment, cellular membrane fusion, pathogenicity, and immune system stimulation. The nucleocapsid associates with the matrix protein (M) at the base of a double-layered lipid envelope. The various genera of paramyxoviruses can be distinguished from each other by the viral proteins and by the biochemical properties for their viral attachment proteins. All paramyxoviruses are labile and quickly inactivated by heat, organic solvents, detergents, and ultraviolet light.

Newcastle Disease

Overview

Newcastle disease, also called atypical fowl plaque and pseudo fowl pest, is a viral disease that produces a wide range of clinical signs. Newcastle disease (ND) is a highly contagious viral infection of poultry, with Exotic Newcastle disease being the most severe form of the disease producing neurologic and gastrointestinal signs. ND is a highly contagious zoonotic bird disease affecting many domestic and wild avian species. Cases of Newcastle disease were described in chickens in the Western Isles of Scotland as early as 1896. ND was discovered in Java, Indonesia, in 1926, but its name is derived from Newcastle-upon-Tyne, England, where it was rediscovered in 1927. Other less severe infections caused by very similar viruses have been described including an outbreak of mild respiratory disease occurring in the United States reported in the 1930s (termed pneumoencephalitis). Since then numerous Newcastle disease viruses that produce a mild disease in chickens have been seen worldwide. The emergence of ND as a highly pathogenic disease in poultry since 1926 suggests that major changes occurred either in the virus or in its hosts. The commercialism of the poultry industry, the incursion of humans into the habitats of some host species, and viral mutation have all been suggested as causes for the increased pathogenicity of the disease.

Several panzootics of ND have occurred in poultry since 1926. The first progressed very slowly across the globe from the Far East taking over 20 years to become a true panzootic. The beginning of the second panzootic was recognized at the end of the 1960s and within four years had spread worldwide. The varying rates of spread of the two panzootics were in part as a result of the development of the world poultry industry and the commercialization of poultry production leading to greater contact between farms (delivery vehicles moving from one farm to another as well as the increase in worldwide transportation). The third panzootic ND outbreak occurred in poultry in California in 1971 most likely as a result of the importation of exotic birds. The outbreak threatened not only the California poultry industry but the entire United States poultry and egg supply. During this outbreak 1,341 infected flocks were identified and almost 12 million birds were destroyed costing taxpayers $56 million in 1971 dollars. This outbreak severely disrupted the operations of many poultry producers and increased the prices of poultry and poultry products to consumers. Another ND panzootic occurred in the 1980s in racing and show pigeons that probably began in the Middle East in the late 1970s. In Western Europe there was a marked increase in ND outbreaks during the early 1990s, peaking with 239 outbreaks in European Union countries in 1994. The last outbreak of ND in the United States was in 2003 in southern California.

Causative Agent

Newcastle disease viruses (NDV) are in the family Paramyxovirdiae, genus *Rubulavirus* and are classified in the serotype group avian paramyxovirus (APMV). It is an enveloped, single-stranded, negative sense RNA virus (Figure 7-25). Two specific virus proteins, hemagglutinin-neuraminidase and fusion protein, are the main proteins found in the outer coat of NDV particles. There are nine serotype groups based on RNA fingerprinting. The serotypes are designated APMV-1 to APMV-9.

NDV strains can be categorized into several strains including:

- velogenic (highly virulent) strains that spread rapidly and cause up to 90% mortality. There are viscerotropic velogenic (VVND) and neurotropic velogenic types (NVND). The viscerotropic form produces hemorrhagic intestinal

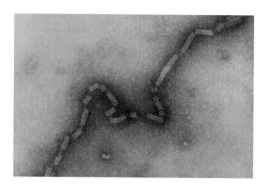

Figure 7-25 Electron micrograph of paramyxoviruses with their herringbone-shaped RNA cores.
(Courtesy of CDC/Dr. Fred Murphy)

lesions and the neurotropic form produces respiratory and neurologic signs. VVND is also known as Exotic Newcastle disease (END) and is currently eradicated from the United States.

- mesogenic (intermediate virulence) strains that cause coughing, reduce egg quality and production, and result in up to 10% mortality.
- lentogenic (nonvirulent) strains that produce mild signs with negligible mortality. Lentogenic strains are also known as respiratory strains.
- asymptomatic enteric strains that are typically subclinical.

Epizootiology and Public Health Significance

ND occurs worldwide in domestic fowl and several types of wild birds. END, the most severe form, is endemic in Asia, the Middle East, Africa, and Central and South America. Modern methods of slaughter of commercial poultry, marketing of poultry meat, and veterinary inspection have reduced the movement of live commercial poultry as well as the risk of spreading diseases such as ND. The last major outbreak of Newcastle disease in the United States occurred in 2003 in California and cost taxpayers approximately $200 million dollars to contain. ND has major economic impact to people, but does not pose a major public health risk to people.

Transmission

NDV is transmitted to people and birds by direct contact with bird droppings (feces) and respiratory discharges of infected birds. Conjunctival and mucosal infections occur after direct contact with infected birds. Fertile eggs laid by infected hens can carry virus although they rarely hatch. Contaminated food, water, equipment, and clothing are also sources of transmission as a result of the high concentration of virus in the birds' bodily discharges. The disease is often spread by vaccinating and debeaking workers, manure haulers, rendering-truck drivers, feed-delivery personnel, poultry buyers, egg service people, and poultry farm owners and employees. NDV can survive for long periods of time in the environment (especially in feces and feathers). Virus tends to be shed during the incubation period and for a short time after recovery. Intermittent viral shedding may occur in some psittacine species for more than one year. Virus particles can be found in all parts of an infected bird. The disease spreads rapidly among birds kept in confinement such as commercially-raised chickens. Veterinarians and poultry farms may also get infected from the live virus Newcastle disease vaccine, which is nonpathogenic for poultry but pathogenic for humans. Migratory wild birds may introduce infection, but most ND isolates from wild birds have been of low virulence. Migratory birds may play a significant role in the spread of ND once it is present in an area. In the past birds have become infected with NDV from the live vaccine.

Newcastle disease is spread primarily through direct contact between healthy birds and the bodily discharges of infected birds.

Pathogenesis

NDV initiates infection by fusion of the viral membrane with host cell plasma membranes. Most NDVs replicate in host cells producing new viral proteins and copies of the viruses' genetic material in the host cell's cytoplasm. Once the virus gains access to the host, viral nucleic acid are detected in multiple tissues but most prominently in macrophages associated with lymphoid tissue. Tissue affected and clinical signs vary greatly depending on the viral strain. Clinical signs are most severe with VVND strains producing acute systemic illness with extensive necrosis of lymphoid areas (mainly macrophages) in the spleen and intestine. NVND isolates produce CNS disease with only minimal amounts of viral nucleic acid detected in neural tissue. Mesogenic and lentogenic strains do not produce overt disease; however, viral nucleic acid is found in myocardium and air sac epithelium producing compromise of the air sacs and heart function, which may predispose these birds to secondary infection and/or decreased meat and egg production. Asymptomatic enteric strains do not produce disease or compromised function in infected birds.

Clinical Signs in Animals

NDVs have been reported to infect animals other than birds, but are mainly considered to cause avian disease. NDV infections have been established in at least 240 species of birds in 27 of the 50 orders of birds. NDV isolates have been frequently isolated from migratory waterfowl and other aquatic birds with these isolates having low virulence for chickens. The most significant outbreaks of NDV in wild birds occurred in double-crested cormorants (*Phalacrocorax auritus*) in North America during the 1990s. Virulent NDV isolates have been identified in captive caged birds who probably acquired the infections in the countries of origin. Severe ND in pet birds occurred from infected cage birds in six states in United States in 1991. Racing and show pigeons have also been infected with NDV and in the late 1970s a panzootic originating from the Middle East occurred in these birds.

The clinical signs in infected birds depend on the host species, age of host, co-infection with other organisms, environmental stress, and immune status. Viscerotropic velogenic viruses cause acute lethal infections typically producing hemorrhagic lesions in the intestines of dead birds. Edema around the eyes and hemorrhagic lesions of the head may also be seen. Neurotropic velogenic viruses cause disease characterized by respiratory (outstretched neck, gasping, and coughing) and neurologic disease (muscle tremors, drooping wings, dragging legs, and circling) with high mortality. Mesogenic viruses cause clinical signs consisting of respiratory and neurologic signs, with low mortality. Lentogenic viruses cause mild infections of the respiratory tract. Asymptomatic enteric viruses cause avirulent infections in which viral replication occurs primarily in the gastrointestinal tract. Clinical signs associated with various strains may present differently in different avian species; for example, psittacines and pigeons may show neurologic signs when infected with the viscerotropic strain, whereas finches and canaries may show no signs of disease.

In general, ND may produce clinical signs such as diarrhea, head and wattle edema (Figure 7-26), neurologic signs such as paralysis and torticollis, and respiratory signs. Decreases in egg production may precede more overt signs of disease and deaths in egg-laying birds. Deaths occur within 24 to 43 hours after the initial drop in egg production and may continue for 7 to 10 days. Birds that survive an outbreak past 14 days typically live but may have permanent neurologic damage (paralysis) and reproductive damage (decreased egg production and production of thin-shelled eggs).

> Smuggled pet birds, especially Amazon parrots from Latin America, pose a great risk of introducing ND into U.S. poultry flocks. Amazon parrots can be carriers of the disease, not showing clinical signs, yet shedding the virus for at least 400 days.

Figure 7-26 Marked hemorrhage of the comb and head, with cyanosis of the margin of the comb is seen in this bird with Newcastle disease.

(Courtesy of California Animal Health & Food Safety Laboratory System, San Bernardino Branch)

Clinical Signs in Humans

Humans that are exposed to infected birds (typically in poultry processing plants) can develop mild conjunctivitis that is usually unilateral. Influenza-like symptoms such as mild fever and headache may also be seen. The incubation period is 1 to 2 days and recovery occurs without complication.

Diagnosis in Animals

Prior to collecting or sending any samples from animals suspected of having foreign animal disease, proper authorities should be contacted. Samples must be sent under secure conditions to authorized laboratories.

There are not any gross or microscopic lesions considered pathognomonic for any form of ND. Carcasses of birds dying as a result of virulent ND usually have a dehydrated appearance. Velogenic NDV typically causes hemorrhagic lesions of the intestinal tract predominantly in the duodenum, jejunum, and ileum. Birds showing neurologic signs prior to death typically do not have gross lesions in the CNS. Birds that have respiratory signs may have hemorrhagic lesions and congestion of the lung and air sacculitis may be evident. Egg peritonitis is often seen in laying hens infected with velogenic NDV.

Laboratory tests for identification of ND include virus isolation in 9- to 11-day-old embryonated chicken eggs and hemagglutination and hemagglutination inhibition tests on chorioallantoic fluid of dead embryos. Swabs can be taken for virus isolation from the trachea and cloaca of live birds, or tissue samples from dead birds (lung, trachea, spleen, cloaca, and brain). Feces can be collected for cell culture. Culture tubes are kept on ice if they can reach the laboratory within 24 hours or quick-frozen for transport longer than 24 hours. Serologic tests using clotted blood or serum are available and include hemagglutination-inhibition and ELISA tests; however, previous natural exposure or vaccination may affect serologic results. Rapid detection of the virus in chickens using RT-PCR is important in commercial poultry flocks because of the high economic impact of ND.

Diagnosis in Humans

The clinical signs of unilateral (less frequently bilateral) conjunctivitis with a history of bird contact usually indicate ND in people. Virus isolation in cell culture or in embryonated chicken eggs can be done on swabs from pharyngeal mucosa, tears, ocular discharge, or throat washes.

High concentrations of exotic ND (END) virus are in birds' discharges and can be spread easily on shoes and clothing, by vaccination and beak trimming crews, manure haulers, and poultry farm employees. The virus can also survive for several weeks in a warm, humid environment on birds' feathers, manure, and other materials.

No known infections with ND have occurred in humans from handling or consuming poultry products.

Treatment in Animals

There is no treatment for ND. Recommendations for control and eradication of the disease include strict quarantine, slaughter and disposal of all infected and exposed birds, and disinfection of the property. New birds should not be introduced into the facility for 30 days.

Treatment in Humans

There is no treatment for ND in humans other than symptomatic therapy.

Management and Control in Animals

NDV has the ability to efficiently replicate in human cancer cells and is being investigated for its human anticancer effects.

The only way to eradicate ND from commercial poultry is by destroying all infected animals and imposing strict quarantine and surveillance programs. Biosecurity practices are important to protect commercial, backyard, and hobby flocks and should include permitting only essential workers and vehicles on the premises, providing clean clothing and disinfection facilities for employees, cleaning and disinfecting vehicles entering and leaving the premises, avoiding visiting other poultry operations, maintaining an all-in/all-out flock management strategy, controlling the movement of all poultry and poultry products from farm to farm, cleaning and disinfecting poultry houses between each lot of birds, not keeping pet birds on the farm, limiting access to vaccination crews, catching crews, and other service personnel who may have been in contact with other poultry operations within 24 hours, protecting flocks from wild birds that may try to nest in poultry houses or feed with domesticated birds, controlling movements associated with the disposal and handling of bird carcasses, litter, and manure, and taking diseased birds to a diagnostic laboratory for examination. Some countries require poultry feed to be heat treated to reduce the possibility of contracting NDV through ingestion. NDV is readily destroyed by heat, chemicals such as hypochlorites and gluteraldehyde, and direct sunlight for 30 minutes. The virus can survive in cool weather in manure and contaminated poultry may shed virus for many weeks. A minimum core temperature of 80°C for one minute destroys the virus in meat products.

Since 1973, birds entering the United States have been quarantined for 30 days for ND. Pet bird owners should request certification from suppliers that birds are legally imported or are of U.S. stock, are healthy prior to shipment, and will be transported in new or thoroughly disinfected containers. Newly purchased birds should be isolated for at least 30 days. The USDA–APHIS requires that all imported birds (poultry, pet birds, birds exhibited at zoos, and ratites) be tested and quarantined for diseases before entering the country.

Vaccination with live and/or oil emulsion vaccines greatly reduces the loss of poultry in commercial flocks. Live B_1 and La Sota vaccine strains are administrated in drinking water, as a coarse spray, intranasally or intraocularly. Healthy chickens may be vaccinated as early as day 1 to 4 of life, but delaying vaccination until the second or third week increases the vaccine's efficiency. Poultry that are vaccinated may shed the virus for up to 15 days, which may serve as an infection source for humans during this time.

Poultry or pet bird owners and veterinarians who suspect a bird may have ND should immediately contact state or federal animal health authorities.

Management and Control in Humans

In 1977, both velogenic strains of NDV were not known to exist in the United States and were classified as exotic.

Humans can prevent contracting NDV by practicing proper hygiene, using protective gear when working with poultry (goggles and nose and mouth protection), and avoiding sick and diseased birds.

Summary

Newcastle disease (ND) is primarily a disease of birds caused by Newcastle disease viruses (NDV) in the viral family Paramyxovirdiae. NDV strains can be categorized into several strains including velogenic, mesogenic, lentogenic, and asymptomatic enteric strains. ND occurs worldwide in domestic fowl and several species of wild birds. END, the most severe form, is endemic in Asia, the Middle East, Africa, and Central and South America. END is transmitted to people and birds by direct contact with bird droppings (feces) and respiratory discharges of infected birds. Contaminated food, water, equipment, and clothing are also sources of transmission as a result of the high concentration of virus in the birds' bodily discharges. Viscerotropic velogenic viruses cause acute lethal infections typically producing hemorrhagic lesions in the intestines of dead birds. Neurotropic velogenic viruses cause disease characterized by respiratory (outstretched neck, gasping, and coughing) and neurologic disease (muscle tremors, drooping wings, dragging legs, and circling) with high mortality. Mesogenic viruses cause clinical signs consisting of respiratory and neurologic signs, with low mortality. Lentogenic viruses cause mild infections of the respiratory tract. Asymptomatic enteric viruses cause avirulent infections in which viral replication occurs primarily in the gastrointestinal tract. In general, ND may produce clinical signs such as diarrhea, head and wattle edema, neurologic signs such as paralysis and torticollis, and respiratory signs. Decreases in egg production may precede more overt signs of disease and deaths in egg-laying birds. Deaths occur within 24 to 43 hours after the initial drop in egg production and may continue for 7 to 10 days. Humans may develop mild conjunctivitis that is usually unilateral and influenza-like symptoms such as mild fever and headache. Prior to collecting or sending any samples from animals suspected of having foreign animal disease, proper authorities should be contacted and samples must be sent under secure conditions to authorized laboratories. Laboratory tests for identification of ND include virus isolation in 9- to 11-day-old embryonated chicken eggs and hemagglutination and hemagglutination inhibition tests on chorioallantoic fluid of dead embryos. Serologic tests such as hemagglutination-inhibition and ELISA tests are also available. Rapid detection of the virus in chickens using RT-PCR is important in commercial poultry flocks because of the high economic impact of ND. In people, virus isolation in cell culture or in embryonated chicken eggs is used to diagnose ND. There is no treatment for ND. Recommendations for control and eradication of the disease include strict quarantine, slaughter and disposal of all infected and exposed birds, and disinfection of the property. Vaccination with live and/or oil emulsion vaccines greatly reduces the loss of poultry in commercial flocks. Humans can prevent contracting NDV by practicing proper hygiene, using protective gear when working with poultry, and avoiding sick and diseased birds.

Nipah Virus Encephalitis

Overview

Nipah virus encephalitis is caused by Nipah virus, a previously unknown virus of the family Paramyxoviridae. Nipah virus was first isolated from a farm worker during the 1998–1999 epidemic in the village of Sungai-Nipah in a northern peninsula of Malaysia. The first human cases of disease attributed to Nipah virus occurred in late September 1998 and were first attributed to the Japanese encephalitis virus. In all, more than 276 cases of encephalitis, including 106 deaths, had been reported during this epidemic with nearly a 40% fatality rate. In March 1999, Malaysian researchers identified the virus as a previously unknown paramyxovirus, which was later confirmed by the CDC. The virus was first called Hendra-like virus because it is similar to the Hendra virus first identified in horses in Australia in 1994. The virus is now named after the village near Kuala Lumpur from where it was first isolated. Nipah virus is found mainly in pigs and in humans who have direct contact with pigs and their body fluids. Epizootics caused by Nipah virus can cause serious economic losses to swine farmers. The outbreak of Nipah virus in Malaysia was controlled following the culling of over 1 million pigs.

Causative Agent

Nipah virus encephalitis is caused by Nipah virus in the family Paramyxovirdiae, genus *Henipavirus*. Henipaviruses are pleomorphic, are large (ranging in size from 40 to 600 nm), and have a lipid membrane overlying a shell of viral matrix protein (containing spikes of F (fusion) protein and G (attachment) protein) (Figure 7-27).

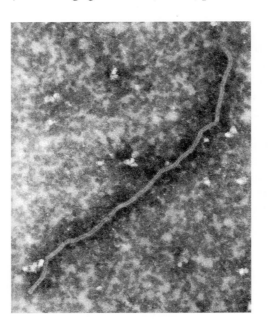

The G protein functions in the attachment of the virus to the surface of a host cell. The F protein functions in the fusion of the viral membrane with the host cell membrane, releasing the virion contents into the cell. *Henipavirus* RNA is tightly bound to N (nucleocapsid) protein and is associated with the L (large) and P (phosphoprotein) proteins, which provide RNA polymerase activity during replication. Nipah virus is a large, enveloped, single-stranded, negative sense RNA virus that is closely related to Hendra virus. Antibodies to Hendra virus cross-react with Nipah virus. The virulence of the Nipah virus is believed to be a result of its interferon-antagonist activity in some of its proteins (V, W, and C).

Figure 7-27   Electron micrograph of Nipah virus. (Courtesy of CDC)

Epizootiology and Public Health Significance

Nipah virus has only been reported in Malaysia and in pigs exported from that country. Several human cases occurred in Singapore in slaughter house workers; however, the source of infections is unknown. The reservoir host is the fruit bat (*Pteropus*), which is found in Malaysia, Australia, Indonesia, the Philippines, and some Pacific Islands. It is currently believed that Nipah virus may be endemic in Southeast Asia.

Prior to the 1999 outbreak and mass culling of pigs, Nipah virus is present is about 6% of all pig farms in Malaysia. In the Malaysian outbreak, morbidity in 4-week to 6-month-old pigs was nearly 100%, but mortality was between 1% and 5%.

Transmission

Fruit bats appear to be the reservoir host of Nipah virus. Pigs probably become infected with Nipah virus from infected fruit bat urine as these bats hang in nearby trees at night. Nipah virus can be transmitted directly from bats to humans or to pigs and then to humans; however, most human infections are from close direct contact with pigs and their body fluids. Nipah antibodies have been detected in pigs and other domestic and wild animals; however, their role in transmitting infection to other animals has not yet been determined. The risk of transmission of Nipah virus from sick animals to humans is believed to be low. Viral transmission through a break in the skin is theoretically possible; therefore, it has been categorized as a Biosafety level 4 organism. Human-to-human transmission of Nipah virus has not been reported.

Pathogenesis

Nipah virus infects host cells via a cell receptor called Ephrin-B2, which is found on brain cells and endothelial cells of blood vessels. Ephrin-B2 is found in humans, horses, pigs, and bats, which may explain why the infection can jump so easily between these species. Nipah virus has been found in the lungs, upper and lower respiratory system, CNS, and kidneys. Pathologic findings include pneumonic changes such as lung lobe consolidation, hyperemia in the kidneys, and petechial hemorrhages in the brain. Widespread cerebral vascular problems lead to local thromboses and localized demyelination. Vasculitis is found in small vessels of the brain and other organs.

Clinical Signs in Animals

In pigs, asymptomatic Nipah virus infections appear common. Clinical infections in piglets produce open mouth breathing, leg weakness, and twitching. Infection in 4-week to 6-month-old pigs produce signs such as tachypnea, dyspnea, fever, a loud and nonproductive cough; and neurologic changes including lethargy, trembling, lameness, ataxia, or aggressive behavior. In boars and sows neurologic disease such as head pressing, nystagmus, tetanus-like spasms, seizures, and muscle paralysis is more common. The disease has a high morbidity and low mortality rate in pigs.

Other animals such as horses, goats, dogs, and cats may also contract Nipah virus. Clinical signs in dogs are similar to those observed in pigs. Nipah virus is highly pathogenic in pregnant mares. Infected reservoir hosts (fruit bats) are asymptomatic (Figure 7-28).

Figure 7-28 Fruit bats are reservoir hosts of Nipah virus. (Courtesy of Jan Sevelk)

Clinical Signs in Humans

The incubation period of Nipah virus encephalitis is between 4 and 18 days. Subclinical, mild, and severe forms of the disease exist in humans. Subclinical cases are asymptomatic, mild cases produce mild influenza-like symptoms (high fever and muscle pains), and severe cases produce highly febrile encephalitis (drowsiness, disorientation, seizures, and coma). Fifty to 70% of clinically apparent cases result in fatality.

Diagnosis in Animals

Prior to collecting or sending samples from animals suspected of having foreign animal disease, the proper authorities should be contacted. Samples should only be sent under secure conditions to authorized laboratories. Samples can be collected from the lungs, upper and lower respiratory tract, CNS, and kidneys. Nipah virus infections can be diagnosed by serum neutralization, PCR tests, and virus isolation (African green monkey kidney (Vero), baby hamster kidneys, and porcine spleen cells). An ELISA test is available in Malaysia.

Diagnosis in Humans

Nipah virus infections in people can be diagnosed by Vero cell culture (using CSF or blood), immunofluorescence, and RT-PCR.

Treatment in Animals

There is no treatment for Nipah virus in animals.

Treatment in Humans

There is no treatment for Nipah virus in people and therapy relies on intensive supportive care. The antiviral drug ribavirin may reduce the duration of fever and severity of disease.

Management and Control in Animals

Nipah virus infections must be reported to state and federal authorities (federal and state veterinarians) upon diagnosis or disease suspicion. Strict quarantine followed by mass culling of seropositive animals and exposed pigs has been used to control epizootics of Nipah virus. Control of fruit bats in areas inhabited by pigs and disinfection of property/equipment are important measures in preventing disease in pigs. Nipah virus is readily inactivated by detergents and routine cleaning and disinfections with sodium hypochlorite, povidone iodine, phenols, and gluteraldehyde. Spills can be treated with 10,000 ppm chlorine.

Management and Control in Humans

People can avoid Nipah virus infection by limiting contact with pigs and their bodily fluids and using proper hygiene when handling pigs (protective clothing, gloves, disposable gowns, and face shield). There is no vaccine for Nipah virus.

Summary

Nipah virus encephalitis is caused by Nipah virus in the family Paramyxovirdiae, genus *Henipavirus*. Nipah virus is a large, enveloped, single-stranded, negative sense RNA virus that is closely related to Hendra virus. Nipah virus has only been reported in Malaysia and in pigs exported from that country. It is currently believed that Nipah virus may be endemic in Southeast Asia. Fruit bats appear to be the reservoir host of Nipah virus. Pigs probably become infected with Nipah virus from infected fruit bat urine as these bats hang in nearby trees at night.

> Nipah virus is classified as a Biosafety level 4 organism.

Nipah virus can be transmitted directly from bats to humans or to pigs and then to humans; however, most human infections are from close direct contact with pigs and their body fluids. In pigs, asymptomatic Nipah virus infections appear common. Clinical infections in piglets produce open mouth breathing, leg weakness, and twitching. Infection in 4-week to 6-month-old pigs produce signs such as tachypnea, dyspnea, fever, a loud and nonproductive cough; and neurologic changes including lethargy, trembling, lameness, ataxia, or aggressive behavior. In boars and sows neurologic disease such as head pressing, nystagmus, tetanus-like spasms, seizures, and muscle paralysis is more common. The disease has a high morbidity and low mortality rate in pigs. Other animals such as horses, goats, dogs, and cats may also contract Nipah virus. Infected reservoir hosts (fruit bats) are asymptomatic. Subclinical, mild, and severe forms of the disease exist in humans. Subclinical cases are asymptomatic, mild cases produce mild influenza-like symptoms, and severe cases produce highly febrile encephalitis. Fifty to 70% of clinically apparent cases result in fatality. Nipah virus infections in animals can be diagnosed by serum neutralization, PCR tests, and virus isolation. An ELISA is available in Malaysia. Nipah virus infections in people can be diagnosed by Vero cell culture, immunofluorescence, and RT-PCR. There is no treatment for Nipah virus in animals or people. Recommendations for control and eradication of the disease include strict quarantine, slaughter and disposal of all infected and exposed pigs, and disinfection of the property. Nipah virus infections must be reported to state and federal authorities (federal and state veterinarians) upon diagnosis or disease suspicion. Control of fruit bats in areas inhabited by pigs and disinfection of property/equipment are important measures in preventing disease in pigs. People can avoid Nipah virus infection by limiting contact with pigs and their bodily fluids and using proper hygiene when handling pigs. There is no vaccine for Nipah virus. For less common zoonotic Paramyxoviruses, see Table 7-12.

PICORNAVIRUSES

Overview

Picornaviruses are among the smallest (*pico* is Greek for small and rna for its RNA core), most diverse (more than 200 serotypes), and oldest known viruses. Poliomyelitis virus, a member of the Picornaviridae family, was one of the first recorded infections with evidence found in an Egyptian tomb on a stone tablet of a drawing depicting a man with deformities typical of paralytic poliomyelitis. No civilization or culture has escaped polio infection with people such as Sir Walter Scott, Franklin Delano Roosevelt, and Itzhak Perlman afflicted with this disease. Poliomyelitis was first recognized to be caused by a virus by Landsteiner and Popper in 1909, but the virus was not isolated until the 1930s. The polio epidemics of the 1930s and 1940s aided in the understanding of picornaviruses in general.

Viruses in the Picornaviridae family cause an extraordinarily wide range of diseases, have narrow host ranges and are typically transmitted horizontally (are not passed on from parent to offspring). Syndromes associated with these agents include asymptomatic infection, aseptic meningitis syndrome, colds, febrile illnesses with rashes, conjunctivitis, and hepatitis. There are nine genera of picornaviruses: *Enterovirus* (poliovirus, swine vesicular disease, coxsackievirus (which causes hand-foot-and-mouth disease in humans), and enterovirus A), *Rhinovirus* (human and bovine rhinoviruses),

Table 7-12   Less Common Zoonotic Paramyxoviruses

Paramyxovirus	Disease	Predominant Signs in People	Animal Source and Clinical Signs	Transmission	Geographic Distribution	Diagnosis	Control
Hendra virus	Hendra virus disease First discovered in 1994 in horses in Brisbane, Australia (Hendra is a suburb of Brisbane)	Acute dyspnea, hemorrhagic disorders, and neurologic tissue necrosis	▪ Horses develop hemorrhagic fever and respiratory distress with high fatality rates ▪ Highly pathogenic in pregnant mares	▪ Close contact with infected horses ▪ Horses contract disease indirectly from fruit bats (flying foxes of the genus *Pteropus*). Flying foxes stay in trees overnight and infectious excrement is shed onto pastures where horses can be exposed.	Australia	Virus isolation in cell culture, RT-PCR, ELISA	Use barrier protection when handing suspicious horses (gloves, goggles, masks with respirators, etc.); avoid contact with flying foxes and their excrement; do not plant flowering or fruit trees near farm buildings; isolation of affected animals.
Menangle virus	Menangle First discovered in a swine operation in Menangle near Sydney in New South Wales, Australia in 1997.	Influenza-like illness with a macular rash; no fatalities reported	▪ Pigs (typically 10 to 16 weeks old) harbor active infection. ▪ Infected sows may abort, have mummified or autolysed piglets, or they may be born with skeletal or nervous system deformities.	▪ Close association with infected pigs (especially during necropsy or birthing procedures) ▪ Pigs may become infected by flying foxes (genus *Pteropus*).	Australia (only outbreak was in New South Wales)	Virus isolation, virus neutralization, ELISA	Use barrier protection when handling suspicious pigs (gloves, goggles, masks with respirators, etc.); avoid contact with flying foxes and their excrement; do not plant flowering or fruit trees near farm buildings; isolation of affected animals.
Tioman virus	Tioman virus disease First isolated on Tioman island, Malaysia in 2000 while trying to discover the natural host of Nipah virus	It is not proven that Tioman virus can cause illness in humans or animals; its relationship to other disease-causing paramyxoviruses suggests a possibility that it may cause disease when it crosses the species barrier.	▪ Bats are asymptomatic.	▪ Contact with the urine of fruit bats (*Pteropus hypomelanus*).	Malaysia	Western blot	Avoid contact with flying foxes.

Source: Kraus, Center for Food Security and Public Health, Menangle, Iowa State University, 1/24/06.

Hepatovirus (hepatitis A virus), *Aphthovirus* (foot-and-mouth disease virus in cattle), *Cardiovirus* (encephalomyocarditis virus in mice), *Erbovirus* (equine rhinitis B virus), *Kobuvirus* (Aichi kobuvirus), *Teschovirus* (porcine teschovirus), and *Parechovirus* (human parechovirus, which was formerly known as *Echovirus* 22 and 23). Of these viruses, only swine vesicular disease virus, foot-and-mouth disease virus, and encephalomyocarditis virus are zoonotic. In 1898, Friedrick Loeffler and Paul Frosch isolated foot-and-mouth disease virus in cattle, one of the first viruses isolated. Foot-and-mouth disease is a highly contagious, acute infection of cloven-hooved animals producing vesicular lesions and erosions in the mouth, nares, muzzle, feet, teats, udder, and rumen. Swine vesicular disease is a transient disease mainly found in cattle, pigs, and horses that produces vesicular lesions in the mouth and on the feet. Encephalomyocarditis virus causes neurologic and cardiac disease in mice and rats and can be transmitted to swine. Although there have been outbreaks of poliomyelitis virus in wild monkeys, monkeys are not a source of human poliomyelitis and is not considered zoonotic.

Causative Agent

All viruses of the Picornaviridae family are small (22 to 30 nm), single-stranded, positive sense RNA viruses that lack a lipid envelope (nonenveloped) and are icosahedral in shape. Areas of viral replication vary with the genus and include the epithelium of the nasopharynx and regional lymphoid tissue, conjunctiva, and intestines (*Enterovirus*), the nasopharyngeal epithelium and regional lymph nodes (*Rhinovirus*), and the intestinal epithelium progressing to viremia with blood transport of the virus to the liver (*Hepatovirus*). The areas of viral replication affect the route of transmission of picornaviruses and include fecal-oral (*Enterovirus*) and respiratory droplets (*Rhinovirus*).

Foot-and-Mouth Disease

Overview

Foot-and-mouth disease (FMD, *Aphtae epizooticae* in Latin, *Fiebre Aftosa* in Spanish, hoof-and-mouth disease) is a highly contagious and sometimes fatal viral disease mainly of cattle and pigs but can affect other cloven-hooved animals such as deer, goats, and sheep, as well as elephants, rats, and hedgehogs. There has been a high incidence of FMD in animals; however, human disease is rare. When FMD was endemic in Central Europe many vesicular diseases in humans with mouth and hand lesions were called FMD that may or may not have been FMD. The first human infection with FMD was reported in 1695 by Valentini in Germany. In 1897 Friedrich Loeffler determined that the cause of FMD was viral by passing the blood of an infected animal through a fine porcelain-glass filter and found that the filtered fluid could still cause the disease in healthy animals. In 1834, Hertwig infected three veterinarians with FMD. These veterinarians drank 250 milliliters of milk from infected cows on four consecutive days and then developed clinical disease. The first U.S. outbreak of FMD occurred in 1870 in New England and originated from imported livestock, as did outbreaks in 1880 and 1884 (all of these outbreaks were mild and contained). In October 1914 the largest U.S. outbreak of FMD spread rapidly throughout the Chicago stockyards and other markets affecting more than 3,500 livestock herds in 22 States and the District of Columbia before the outbreak was contained in September 1915. There have been nine U.S.

> As a result of their extremely small size, approximately 500,000,000 picornaviruses could fit side by side on the head of a pin.

outbreaks of FMD in 1870, 1880, 1884, 1902, 1905, 1908, 1914, 1924, and 1929 before being eradicated from the United States. The last United States outbreak of FMD occurred in California in 1929. FMD first appeared in the United Kingdom in 1839. After World War II, FMD was widely distributed throughout the world. In 1967 and 1968 FMD outbreaks in the U.K. led to the slaughter of more than 430,000 animals and by 1989 following a successful vaccination program Europe was deemed FMD-free. Once Europe was declared FMD-free vaccination of animals ceased. In February 2001 an epizootic in the United Kingdom that originated in Little Warley in Essex spread across Britain in 2 weeks and resulted in the destruction of more than 1 million animals. This epizootic began on a farm where restaurant scraps from a Chinese restaurant were fed to pigs (which was illegal).

Causative Agent

Horses are not susceptible to FMD.

FMD is caused by a virus in the *Aphthovirus* genus of the viral family Picornaviridae. The members of the Picornaviridae family are small, nonenveloped, icosahedral, positive sense viruses that contain single-stranded RNA (Figure 7-29). There are seven different FMD serotypes, O, A, C, SAT-1 (SAT stands for South African Territories), SAT-2, SAT-3 and Asia-1, that show some regionality. Serotypes A, O, and C are referred to as European types because they were first isolated in France and Germany; however, they occur in other regions such as South America. Type O is the most common serotype. There is no serologic cross-reaction between these serotypes. There are numerous variants and subtypes within each serotype.

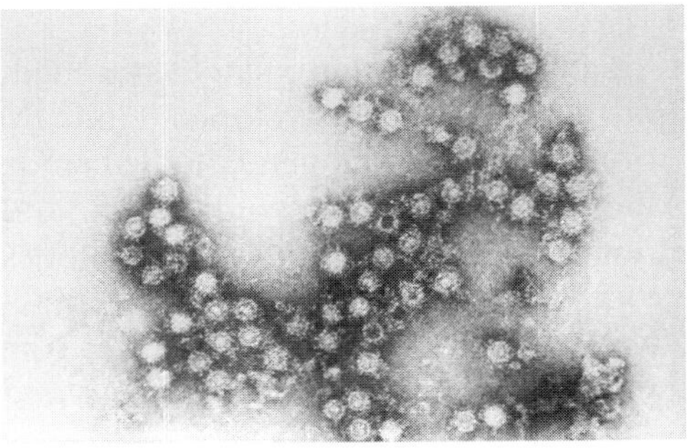

Figure 7-29 Electron micrograph of picornaviruses. (Courtesy of CDC)

Epizootiology and Public Health Significance

FMD occurs worldwide with the exception of Australia, New Zealand, and North America. FMD occurs most commonly in Asia, Africa, the Middle East, and South America. SAT-1 to SAT-3 are predominantly found in Africa; Asia-1 is only found in Asia, and types A, O, and C are found in Europe and South America. Many countries in Europe have been FMD-free for almost 30 years; the last major outbreaks in Europe occurred in Denmark in 1988 and in England and France in 2001.

Foot-and-mouth disease is not to be confused with hand, foot, and mouth disease, which affects people producing a rash on the hands and feet and in the mouth. FMD is often confused with Coxsackie A virus (hand, foot, and mouth disease), herpes simplex, and vesicular stomatitis.

FMD is endemic in animals in parts of Asia, Africa, the Middle East and South America. In 2001 FMD spread through farms in the United Kingdom leading to quarantines, the mass slaughter of animals, and the accumulation of agricultural damages amounting to an estimated $4.4 billion. The last large-scale outbreak of FMD in the United States occurred in 1929 and neither Canada nor Mexico has had an outbreak since the 1950s. However, a single FMD case would instantly shut down the export of all animal products. Morbidity can be 100% in a susceptible population. The mortality rate in adult cattle is 2% to 3%; in calves the mortality rate can be up to 40%. FMD is rare in humans and its economic impact is seen with trade embargoes rather than human disease.

Transmission

FMD is acquired by inhalation and ingestion. It is spread by contact (directly by animals or indirectly by fomites) or passively by vectors, aerosol, and contaminated water. Rodents, birds, animals, and humans can spread the virus as a result of the persistence of the virus in animals (6 months or more) and the high stability of the virus. Humans are typically infected by direct handling of infected animals or animal contact during slaughter. FMD virus can survive up to 24 hours in the human respiratory tract and is the reason people are kept away from infectious farms and their animals. Laboratory infections are possible and human-to-human transmission is rare.

Sheep and goats are maintenance hosts showing very mild signs. Pigs exhale FMD virus in higher concentrations than other species (thousands of times more concentrated) and are considered amplifying hosts. Cattle can be indicators of disease because they are typically the first species to show clinical signs of FMD.

Pathogenesis

When a picornavirus, like all RNA viruses, come in contact with a host cell, it binds to a receptor site and enters the host cell through invagination of the cell membrane. Once the virus is inside the host cell, its protein coat dissolves and new viral RNA is synthesized in large quantities and assembled to form new viruses. After assembly, the host cell lyses and releases the new viruses. The specific lesions of FMD are microscopic in the early stages and are found in the epithelium. Lesions start with degeneration of cells in the middle of the epithelial layers with epithelial cells becoming round and detached from one another. Edema forms between the cells and separates them. Neutrophils infiltrate the epithelium, serum accumulates in the area, and leukocytes (WBCs) produce fluid filled vesicles. The small vesicles coalesce to form bullae causing large areas of epithelium to detach and to be shed (or rubbed off). Epithelial loss is most common on the dorsal surface of the rostral two-thirds of the tongue in cattle. Epithelial loss results in a raw, red, painful area that causes anorexia in affected animals. Later in the disease the vesicles can contain necrotic epithelial cells, leukocytes, erythrocytes, and sometimes bacteria. Lesions can also be seen in the myocardium (young calves and lambs) and skeletal muscle.

Clinical Signs in Animals

After contact with infected animals, clinical signs will appear in newly infected animals in 3 to 5 days. FMD is characterized by high fever and blisters inside the mouth that can lead to excessive secretion of stringy or foamy saliva. Blisters are typically 2 mm to 10 cm in diameter and can also appear on the nares, muzzle, feet, or teats. Blisters on the feet may rupture and cause lameness. Adult animals may suffer weight loss that can last for several months, pregnant animals may abort as a result of the high fever, and testicles of mature males can swell. Most animals eventually recover from FMD in approximately 2 weeks with mortality rates being low. Complications can occur and include myocarditis and death, especially in newborn animals. Some infected animals become carriers of FMD and can transmit it to other animals. Clinical signs in specific species include:

- *Cattle.* A significant drop in milk production can occur in cows. In cattle, oral vesicular lesions are common on the tongue, dental pad, gums, soft palate, nares, and muzzle (Figure 7-30). Oral lesions result in excessive drooling and serous nasal discharge. Hoof lesions are seen in the coronary band and interdigital space and may make the animal reluctant to move.
- *Pigs.* Hoof lesions are severe with vesicles on the coronary band, heel, and interdigital space. Vesicles may also appear on the snout. Oral lesions and drooling are not as common in pigs as they are in cattle.
- *Sheep and goats.* FMD is mild in sheep and goats producing low grade fever, oral lesions, and lameness.

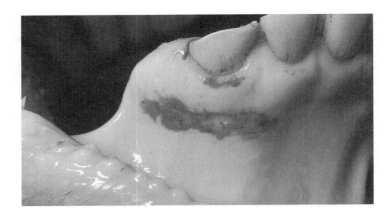

Figure 7-30 An elongated erosion representing a ruptured vesicle is seen ventral to the incisors in the mouth of this bovine with foot-and-mouth disease.

(Courtesy of Plum Island Animal Disease Center (PIADC), Greenport, NY)

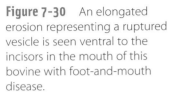

FMD virus does not cross the placenta.

Since 1921, approximately 40 human cases of FMD have been diagnosed.

Clinical Signs in Humans

FMD in humans is biphasic. After an incubation period of 2 to 8 days, symptoms including malaise, fever, nausea, vomiting, red painful ulcerative lesions of the oral mucous membranes, and sometimes vesicular lesions of the skin are seen. Erosions remain after the vesicles dry up. Most skin lesions heal in 5 to 10 days. The virus always enters through defects in the skin or oral mucous membranes producing a primary vesicle at the infection site. Viral dissemination occurs after the initial infection.

Diagnosis in Animals

FMD is suspected based on clinical signs of salivation and lameness associated with vesicles. Before collecting or sending samples from animals with vesicular disease, the proper authorities should be contacted. All samples should be collected and handled with appropriate precautions. Acceptable clinical specimens include vesicular fluid, affected mucous membranes, pharyngeal and esophageal fluid, blood, and serum. FMD is confirmed clinically by detecting antibodies in serum using ELISA and complement fixation serologic testing. Virus isolation to detect antigen or RT-PCR tests need to be done to confirm the first case of FMD within a facility. Virus isolation is done by inoculation of bovine thyroid cells, inoculation of pig, calf, and lamb kidney cells, or inoculation of mice. Mouse inoculation is done on suckling mice inoculated intraperitoneally with liquid from vesicles and if the material is positive the mice die in a few days. There is not an acceptable method for distinguishing between vaccinated and naturally-infected animals.

Diagnosis in Humans

FMD is diagnosed in people using virus-neutralizing antibodies or ELISA tests.

Treatment in Animals

There is no treatment for FMD in animals. State and federal veterinarians need to be contacted immediately if any vesicular disease is observed in animals. Adult animals may need to be culled if the hoof wall becomes separated. Recommendations for control and eradication of the disease include strict quarantine and disinfection of the property.

Treatment in Humans

There is no treatment for FMD in people and therapy is symptomatic. Antibiotics may be needed for secondary infections.

Management and Control in Animals

One way to control FMD in animals is by vaccination. FMD vaccines given either prophylactically or to control an outbreak must closely match the serotype and subtype the animal will be exposed to. With 7 serotypes and more than 60 sub-types there is no universal FMD vaccine. The United States, Canada, and Mexico maintain the North American FMD Vaccine Bank, which contains vaccine strains for the most prevalent serotypes worldwide. The North American FMD Vaccine Bank is located at the USDA Foreign Animal Disease Diagnostic Laboratory (FADDL) at Plum Island Animal Disease Center. The Chief Veterinary Officer of each country would make the decision as to the use of vaccination to control an outbreak. A formalin-inactivated vaccine against serotypes A, C, and O was used in Europe prior to 1990; however, it was discontinued because the seroconversion was not sufficient to provide protection against contracting FMD. Ethylenimine-inactivated virus vaccines grown in BHK (baby hamster kidney) cells are currently used in at-risk countries. There are currently three disease states regarding FMD: FMD present with or without vaccination, FMD free with vaccination, and FMD free without vaccination. Countries designated as FMD free without vaccination have the greatest access to export markets, making this status the most desirable.

The USDA bans the import of animals and animal products from known FMD-infected areas. The USDA also monitors all ports of entry to the United States, requiring passengers who have traveled in affected areas to disinfect their shoes. They also search luggage and cargo for the presence of products that could spread FMD.

When FMD outbreaks occur, the movement of animals and animal products is halted, infected animals and contact animals are slaughtered and their carcasses disposed of, vehicles and personnel leaving the infected area are disinfected, and ring vaccination of nearby animals may be done to create a buffer zone. Vaccination is only allowed in some countries during outbreaks and only with a special permit.

Management and Control in Humans

People can avoid FMD virus infection by proper individual hygiene (wearing gloves and protective clothing). FMD is not a serious public health concern as a result of the low number of cases seen to date.

On June 18, 1981, the U.S. government announced the creation of vaccine targeted against FMD, which was the world's first genetically engineered vaccine.

Summary

Foot-and-mouth disease (FMD) is a highly contagious and sometimes fatal viral disease mainly of cattle and pigs but can affect other cloven-hooved animals such as deer, goats, and sheep. There has been a high incidence of FMD in animals; however, human disease is rare. FMD is caused by a virus in the *Aphthovirus* genus of the viral family Picornaviridae. There are seven different FMD serotypes, O, A, C, SAT-1, SAT-2, SAT-3 and Asia-1. There is no serologic cross-reaction between these serotypes and there are numerous variants and subtypes within each serotype. FMD occurs worldwide with the exception of Australia, New Zealand, and North America. FMD is acquired by inhalation and ingestion. It is spread by contact or passively by vectors, aerosol, and contaminated water. Humans are typically infected by direct handling of infected animals or animal contact during slaughter. After contact with infected animals, clinical signs will appear in newly infected animals in 3 to 5 days. FMD is characterized by high fever and blisters inside the mouth that can lead to excessive secretion of stringy or foamy saliva. Blisters are typically 2 mm to 10 cm in diameter and can also appear on the nares, muzzle, feet, or teats. Most animals eventually recover from FMD in approximately 2 weeks with mortality rates being low. Complications can occur and include myocarditis and death, especially in newborn animals. Clinical signs in specific species include a significant drop in milk production in cows, oral lesions resulting in excessive drooling and serous nasal discharge in cattle, vesicles on the snout in pigs, oral lesions and drooling in pigs (less than in cattle), and mild signs such as low grade fever, oral lesions, and lameness in sheep and goats. FMD in humans is biphasic producing symptoms such as malaise, fever, nausea, vomiting, red painful ulcerative lesions of the oral mucous membranes, and sometimes vesicular lesions of the skin. Before collecting or sending samples from animals with vesicular disease, the proper authorities should be contacted. FMD in animals is confirmed using ELISA and complement fixation serologic testing (antibody detection), virus isolation to detect antigen, or RT-PCR tests to detect nucleic acids. FMD is diagnosed in people using virus-neutralizing antibodies or ELISA tests. There is no treatment for FMD in animals. State and federal veterinarians need to be contacted immediately if any vesicular disease is observed in animals. Recommendations for control and eradication of the disease include strict quarantine and disinfection of the property. There is no treatment for FMD in people and therapy is symptomatic. One way to control FMD in animals is by vaccination. The USDA bans the import of animals and animal products from known FMD infected areas. People can avoid FMD virus infection by proper individual hygiene.

Swine Vesicular Disease

Overview

Swine vesicular disease (SVD), also known as porcine enterovirus infection, is a contagious viral disease of pigs that produces clinical signs similar to foot-and-mouth disease. SVD produces vesicles on the coronary band, heels of the feet and occasionally on the lips, tongue, snout, and teats. SVD was first reported in feeder pigs in Lombardy, Italy, in 1966. The origin of this viral disease is unknown, but it has been suggested that it is derived from a human enterovirus. Since 1966 SVD has been seen in Hong Kong, Japan, and Western European countries. The first outbreak of SVD in Great Britain was in 1972 resulting in 532 cases involving a total of 322,081 pigs over a 10-year period. SVD was eradicated from Great Britain in 1982. SVD has persisted in Italy especially in an enzootic form in the southern part of the country, where there were 171 outbreaks in 2002. The rest of Europe is currently free of SVD except for Portugal who had 2 cases early in 2004. Since 2000 SVD has only been seen in Portugal and Italy. In the United States there was an outbreak of SVD in 1982 to 1983 in which 600 premises in 14 states were affected. In 1995 another SVD outbreak occurred involving 365 premises in 5 states.

Causative Agent

Swine vesicular disease is caused by a virus in the *Enterovirus* genus of the viral family Picornaviridae. The members of the Picornaviridae family are small, non-enveloped, positive sense, icosahedral viruses that contain single-stranded RNA. SVD virus is antigenically similar to the human enterovirus Coxsackie B-5 (which is classified as the human enterovirus B). Despite consisting of a single serotype, swine vesicular disease virus (SVDV) shows a high genetic and antigenic variability. The

SVD presents clinically identical to foot-and-mouth disease; however, SVD only occurs in pigs.

virion capsid consists of a nonenveloped protein shell, which is made up of one copy of each of the four structural proteins VP1, VP2, VP3 and VP4. SVDV isolates are classified in four groups based on nucleotide sequencing with groups 1 and 2 being more closely related to each other than groups 3 and 4. Some SVDV isolates are nonpathogenic, whereas virulent isolates produce typical lesions in infected pigs.

Epizootiology and Public Health Significance

SVD was seen in Italy, England, Scotland, Wales, Malta, Austria, Belgium, France, the Netherlands, Germany, Poland, Switzerland, Greece, and Spain, but has been eradication from all European countries. SVD is still seen in countries in the Far East.

> SVDV is unrelated to other known porcine enteroviruses.

The monetary losses from weight loss in pigs and piglet mortality from SVD do not produce a significant economic impact. The cost of SVD surveillance and control is greater than loss to farmers; however, surveillance and control is needed to avoid any confusion between SVD and foot-and-mouth disease. SVD rarely causes human disease and does not pose a significant threat on human health. Vesicular diseases in animals are summarized in Table 7-13.

Transmission

SVD virus is spread via ingestion of contaminated meat scraps and contact with infected animals or their infected feces. Virus is spread from the pig's nose, mouth, and feces for up to 48 hours before clinical signs are seen. Viral shedding from the feces can occur for up to 3 months following infection. The virus can survive for long periods in the environment and for up to 2 years in lymphoid tissue contained in dried, salted, or smoked meat.

Pathogenesis

Once SVDV gets into the body it has a predilection for epithelial tissue producing changes in the stratified epithelium of the skin of the coronary band, the metatarsals and metacarpals, snout, tongue, and tonsil. Coagulation necrosis results in vesicle formation and sloughing followed by hyperplasia of epithelial tissue. Necrosis of intradermal sweat glands may result in leukocyte accumulation.

Clinical Signs in Animals

The incubation period for SVD is 2 to 7 days following exposure to infected pigs and 2 to 3 days after ingestion of contaminated feed. Pigs with SVD have clinical

Table 7-13   Vesicular Diseases in Animals

Disease	Type of Virus	Major Species Affected	Transmission
Foot-and-mouth disease	Picornavirus	Swine, cattle, sheep, goats, African buffaloes	Aerosol, contact, fomites
Swine vesicular disease	Picornavirus	Swine	Contact, fomites
Vesicular exanthema of swine	Calicivirus	Swine, marine mammals	Contact, fomites
Vesicular stomatitis	Rhabdovirus	Horses, cattle, swine, wildlife, llamas, alpacas	Insects, direct contact

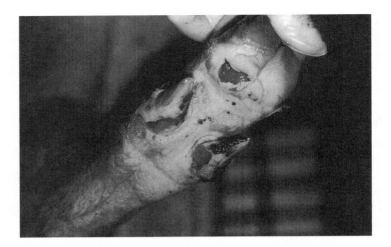

Figure 7-31   A claw and both dewclaws have ulcers at the coronary bands in this pig with swine vesicular disease.
(Courtesy of Iowa State University, College of Veterinary Medicine, Ames, IA)

signs similar to foot-and-mouth disease including fever, salivation, and lameness. Vesicles and erosions may be seen on the snout, mammary glands, coronary band, and interdigital areas of the feet (Figure 7-31). Vesicles are rarely found in the oral cavity. The infection in pigs may be subclinical, mild, or severe. Severe signs may be seen in pigs housed on damp concrete and in young pigs. Neurologic signs may be seen with the severe form of SVD producing clinical signs such as shivering, ataxia, and chorea (rhythmic jerking) of the limbs. Abortion in pregnant animals is rare. Recovery typically occurs in 2-3 weeks.

Clinical Signs in Humans

Human infections of SVD rarely produce clinical signs. Clinically inapparent infection can be seen in people who handle pigs. Seroconversion and mild clinical disease has been seen in laboratory workers. One human case of meningitis has been seen in a laboratory worker.

Diagnosis in Animals

SVD is suspected based on clinical signs of salivation and lameness associated with vesicles. Before collecting or sending samples from animals with vesicular disease, the proper authorities should be contacted. All samples should be collected and handled with appropriate precautions. SVD is confirmed using ELISA, direct complement fixation, virus neutralization, and virus isolation in pig-derived cell culture. RT-PCR is also available.

Diagnosis in Humans

SVD is suspected in people with close contact with pigs and is confirmed by virus isolation or antibody detection. Serologic testing for SVD in people is complicated by its close relationship to coxsackievirus B-5.

Treatment in Animals

There is no treatment for SVD in animals.

Treatment in Humans

There is no treatment for SVD in people and therapy is symptomatic.

Management and Control in Animals

When SVD outbreaks occur, infected farms or areas are quarantined. State and federal veterinarians need to be contacted immediately if any vesicular disease is observed in animals. Animals in contact with infected animals should be culled and properly disposed of. Recommendations for control and eradication of the disease include strict quarantine and disinfection of the property. Sodium hydroxide (1% with detergent), oxidizing agents, and iodophors work well against SVD virus.

Management and Control in Humans

People can avoid contracting SVD virus by using proper individual hygiene (wearing gloves and protective clothing) when handling infected pigs. SVD is not a serious public health concern because of the low number of cases seen to date.

Summary

Swine vesicular disease (SVD) is a contagious viral disease of pigs that produces clinical signs similar to foot-and-mouth disease. SVD produces vesicles on the coronary band, heels of the feet and occasionally on the lips, tongue, snout and teats. SVD is caused by a virus in the *Enterovirus* genus of the viral family Picornaviridae. SVD virus is antigenically similar to the human enterovirus Coxsackie B-5 (which is classified as the human enterovirus B). SVD was seen in many European countries, but has only been seen in Italy and Portugal since 2000. SVD is still seen in countries in the Far East. SVD virus is spread via ingestion of contaminated meat scraps and contact with infected animals or their infected feces. Pigs with SVD have clinical signs similar to foot-and-mouth disease including fever, salivation, and lameness. Vesicles and erosions may be seen on the snout, mammary glands, coronary band, and interdigital areas of the feet. The infection in pigs may be subclinical, mild, or severe. Human infections of SVD rarely produce clinical signs. Clinically inapparent infection can be seen in people who handle pigs. Seroconversion and mild clinical disease has been seen in laboratory workers. SVD is suspected in animals based on clinical signs of salivation and lameness associated with vesicles. SVD is confirmed using ELISA, direct complement fixation, virus neutralization, and virus isolation in pig-derived cell culture. RT-PCR is also available. SVD is suspected in people with close contact with pigs and is confirmed by virus isolation or antibody detection. There is no treatment for SVD in animals. State and federal veterinarians need to be contacted immediately if any vesicular disease is observed in animals. There is no treatment for SVD in people and therapy is symptomatic. When SVD outbreaks occur, infected farms or areas are quarantined. Infected pigs and pigs in contact with infected pigs should be slaughtered and properly disposed of. The premises should be disinfected. People can avoid contracting SVD virus by using proper individual hygiene when handling infected pigs. For information on less common zoonotic Picornaviruses, see Table 7-14.

POXVIRUSES

Overview

Viruses in the family Poxviridae produce disease associated with the production of pox on the skin. Pox are small eruptive skin pustules that may scar upon healing. Poxvirus infections have been present since antiquity with the first evidence

Table 7-14   Less Common Zoonotic Picornaviruses

Picornavirus	Disease	Predominant Signs in People	Animal Source and Clinical Signs	Transmission	Geographic Distribution	Diagnosis	Control
Encephalo-myocarditis virus	Encephalomyo-carditis There are 4 strains; originally isolated in 1940 from a monkey EMC virus has the ability to cause interspecies infections and had led to numerous outbreaks in zoos in Australia and the United States.	Fever, severe headaches, vomiting, stiff neck, and hyperactive reflexes; rare in humans and not fatal	■ Pigs (less than 2 months of age) may die ■ Nonhuman primates may die suddenly ■ Rodents are likely reservoir hosts ■ Virus has been isolated from horses, cattle, pheasants, and mosquitoes.	■ Virus is shed in feces and transmitted in contaminated food, water, and infected cadavers.	Virus exists worldwide, but natural disease only seen in Europe.	Virus isolation, virus neutralization, hemaggluti-nation, PCR	Proper hygiene and using caution when handling diseased animals

of smallpox (the best known poxvirus) found in Egyptian mummies of the 18th Dynasty (1580–1350 B.C.). Smallpox endemics occurred in India in the first millennium B.C. and spread to Asia and, ultimately, to Europe in the 8th century. The introduction of smallpox to the New World in the 15th and 16th centuries decimated the Native American populations and had been used by the British as a biological weapon in the French-Indian wars. Smallpox was a worldwide problem into the 20th century (causing up to a half million deaths in Europe annually). In the 20th century an intense vaccination program eradicated smallpox. The last outbreak of smallpox occurred in 1977 in Somalia, and the WHO certified eradication in 1980. Two known stocks of variola virus exist: one at the CDC in Atlanta, Georgia, and the other in the former USSR.

Smallpox virus was used in vaccines when it was discovered that cutaneous exposure to dried smallpox lesions caused a milder smallpox infection in these people (this process was called variolization and unfortunately led to disease and death in addition to immunologic protection). In the 19th century, Jenner discovered that inoculation of people with cowpox virus led to smallpox immunity. This observation about immunity established the practice of vaccination. There are many other types of poxviruses that infect animals and humans. These other poxvirus infections are important to understand because their lesions must be differentiated from those of smallpox.

Causative Agent

Poxviruses are enveloped, linear, double-stranded DNA viruses in the order Caudovirales and family Poxviridae. Poxviruses are the largest and most complex of all viruses and are divided into two subfamilies: the Chordopoxvirinae (with eight genera) and the Entomopoxvirinae (with three genera). The Chordopoxvirinae are poxviruses of the vertebrates and the Entomopoxvirinae are the poxviruses of insects. The eight Chordopoxvirinae genera include *Orthopoxvirus*, *Parapoxvirus*, *Yatapoxvirus*, *Molluscipoxvirus*, *Avipoxvirus*, *Capripoxvirus*, *Leporipoxvirus*, and *Suipoxvirus*.

Poxviruses are the largest animal viruses and can be visualized via light microscopy. Poxvirus is oval or brick-shaped, between 200 to 400 nm in length, and composed of an external coat containing lipid and tubular or globular protein structures enclosing one or two lateral bodies and a core, which contains the genome (Figure 7-32). Poxviruses contain more than 30 structural proteins and several viral enzymes. Hemagglutinin is produced only by orthopoxviruses. Instead of a capsid, poxviruses have a nucleosome, which contains DNA and is surrounded by its own membrane. Poxviruses have a large number of NS (nonstructural) proteins, which are important for replication and pathogenesis. These viruses multiply in specialized parts of the host cell cytoplasm called the viroplasm, where they can cause skin lesions.

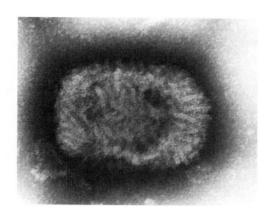

Figure 7-32 Electron micrograph of a poxvirus. (Courtesy of CDC)

Contagious Ecthyma

Overview

Contagious ecthyma (CE) is a highly contagious, acute viral disease of sheep, goats, and wild ungulates that is known by a variety of other names such as orf, soremouth, scabby mouth, contagious pustular dermatitis, infectious labial dermatitis, and contagious pustular stomatitis. The term ecthyma comes from the Greek word *ekthyma* meaning pustule, which describes the deeper lesions associated with this disease. Orf is the name given to the human form of the disease, but will occasionally be used to describe the disease in animals. Infection by the contagious ecthyma virus was first described in sheep by Steeb in 1787 and in goats in 1879. The first case in humans was reported by Newson and Cross in 1934. CE is seen in countries that raise sheep and may be seen in any age sheep, but lambs are more susceptible. There is an antigenically distinct virus that causes CE in goats with more severe clinical signs than those found in sheep. CE can also be found in other ungulates such as camels, reindeer, and oxen.

> Contagious ecthyma should not be confused with ecthyma, an ulcerative pyoderma usually caused by a *Streptococcus* bacterial infection at the site of minor trauma.

Causative Agent

CE is caused by the contagious ecthyma virus (also known as orf virus and *Parapoxvirus ovis*) in the Poxviridae family and *Parapoxvirus* genus (which also includes the milker's nodule virus (pseudocowpox) and bovine papular stomatitis virus). Parapoxviruses are large, enveloped, double-stranded DNA viruses with a regular surface structure that has spiral criss-crossing tubules (often described as a ball of wool structure), which differs from the other poxviruses. Parapoxviruses cause skin diseases and consists of five species and three tentative species including bovine papular stomatitis virus (BPSV), contagious ecthyma virus or *Parapoxvirus ovis* (PPVO), parapoxvirus of red deer in New Zealand (PVNZ), pseudocowpox virus (PCPV), and squirrel parapoxvirus (SPPV), as well as the tentative species auzdyk disease virus, chamois contagious ecthyma virus, and sealpox virus. PPVO is ovoid in shape and is one of the smallest viruses in the Poxviridae family.

Epizootiology and Public Health Significance

CE is found worldwide in countries that raise sheep. In the United States CE occurs most often in the Western states. A higher frequency of CE is found in Europe and New Zealand compared with North America. CE is an extremely infectious disease in animals with up to 90% of a flock showing clinical signs. Most infected animals will have mild loss of condition caused by anorexia from the painful condition of their mouths. Young lambs and kids may not eat and may die from malnutrition as a result of painful oral lesions or to abandonment by their mothers who have painful teat and udder lesions. Morbidity rates in unvaccinated flocks may be approximately 80%. CE is more severe in goats than sheep.

CE is more commonly seen in people who have close contact with sheep and goats (herd workers, shearers, veterinarians, butchers, and slaughter house workers). Most infected people develop a single lesion that will resolve in 3 to 6 weeks. No human deaths have been reported from CE.

Transmission

PPVO is found in skin lesions and scabs of infected animals and is transmitted by direct or indirect contact through cuts and abrasions. People acquire the infection from contact with infected animals/carcasses or fomites. Sheep can be carriers of PPVO and the virus can remain viable on wool and hides for approximately 1 month after lesions have healed. PPVO is very stable in the environment and has been recovered from dried lesions after 12 years. Nursing lambs can transmit the virus to their dam. There is no transmission of the virus to cattle. Human-to-human transmission does not occur.

Pathogenesis

CE is primarily a disease involving the skin with lesions appearing as papules, vesicles, and then pustules. Occasionally the virus is found in the lymph nodes. The epidermis may show marked hyperplasia. Necrosis of the epidermis with ulceration occurs in the center of the lesion that eventually sloughs. The viral infection causes intranuclear and intracytoplasmic inclusion bodies in keratinocytes and the superficial layers of the epidermis undergo degenerative changes. A dense inflammatory infiltrate (plasma cells, macrophages, histiocytes, and lymphocytes) is also observed. Microscopically the lesions are sharply delineated. In lambs, secondary infections with *Staphylococcus* bacteria are common.

Clinical Signs in Animals

CE occurs in sheep, goats, alpacas, camels, reindeer, musk oxen, bighorn sheep, deer, pronghorn antelope, and wapitis. Rare cases of dogs being infected after eating infected carcasses have been reported. Within 2 to 3 days of contacting the virus, the first clinical signs of CE are papules, pustules, and vesicles on the lips, nares, ears, eyelids, and sometimes feet of infected animals (Figure 7-33). Lesions may rarely be found in the esophagus, stomach, intestines, or respiratory tract. Teat and udder lesions can be found in dams that have acquired the virus from their nursing lambs. In time the pustules break and raised brown scabs form over these areas. If a number of pustules rupture close to each other, large scabs may result. These scabs are friable and bleed easily. CE lesions are painful and if found in the mouth may lead to anorexia and starvation, especially in lambs who refuse to nurse. Lambs may also be unable to nurse from ewes with udder or teat lesions because the dam will not let them nurse as a result of the pain. Mortality in lambs may be high as a result of malnutrition.

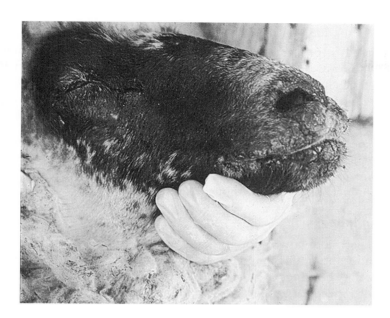

Figure 7-33 Contagious ecthyma in a sheep.
(Courtesy of Ron Fabrizius, DVM, Diplomate ACT)

Secondary bacterial infections may occur in the areas of the lesions. If there is no further infection of the lesions, CE lesions will heal in 1 to 4 weeks. Animals that contact the disease typically develop strong immunity and do not become reinfected for at least 1 year.

Clinical Signs in Humans

In people, CE typically appears as a single skin lesion (commonly as a small papule on the dorsum of the index finger) or a few lesions. The lesion is small, firm, and red that in time develops from a papule into a hemorrhagic pustule. Pustules may contain a central crust. In later stages, the lesion develops into a nodule containing fluid or may be covered by a thin crust. In time this lesion becomes covered by a thick crust. Low-grade fever and lymphadenopathy may occur with CE, which usually subsides within 3 to 4 days. In immunocompromised people, large lesions may develop. CE goes through six clinical stages in people with each stage lasting about 1 week.

- Stage 1 (maculopapular)—A red elevated lesion
- Stage 2 (targetoid)—A bulla with a red center, a white middle ring, and a red periphery
- Stage 3 (acute)—A weeping nodule
- Stage 4 (regenerative)—A firm nodule covered by a thin crust through which black dots are seen
- Stage 5 (papillomatous)—Small papillomas appearing over the surface
- Stage 6 (regressive)—A thick crust covering the resolving elevation

Diagnosis in Animals

Animals are typically diagnosed with CE based on clinical signs. Definitive diagnosis is confirmed by electron microscopy of the scabs (collected early in the disease). Other tests available include virus isolation, PCR, and serologic methods such as serum neutralization, agar gel immunodiffusion, complement fixation, agglutination, and ELISA tests.

Diagnosis in Humans

CE is diagnosed in people by electron microscopy with negative staining of the crust, a small biopsy, or fluid from the lesion. Viral isolation, PCR, RT-PCR, and serologic methods are also available.

Treatment in Animals

There is no specific treatment for CE. Supportive treatment such as antibiotics to prevent secondary bacterial infection, fly repellents to prevent maggots, and tube feeding for nutritional support may be used. If infected animals are kept clean, CE is self-limiting in 1 to 4 weeks without treatment.

Treatment in Humans

There is no specific treatment for CE. Supportive treatment with moist dressings, antiseptics, and antibiotics may be warranted. Large lesions in immunocompromised people may be removed by surgery, curettage, or electrocautery. Cryotherapy may be used to speed the recovery of CE in people.

Management and Control in Animals

The best preventative measure in animals is vaccination. A live virus vaccine should be used on farms where infections have occurred in the past. Recently vaccinated animals pose an increased risk of disease transmission to humans and other animals; therefore, recently vaccinated animals should be isolated from people and unvaccinated animals. Outbreaks have occurred in vaccinated animals. Isolation of infected animals may prevent disease spread on a farm; however, once the virus has entered a flock or herd it is difficult to eradicate.

Management and Control in Humans

People with skin cuts or abrasions should avoid infected animals and their wool/hides. Gloves should be worn if vaccinating animals for CE and contact with these animals should be temporarily avoided because the live vaccine may cause viral shedding.

Summary

Contagious ecthyma (CE) is a highly contagious, acute viral disease of sheep, goats, and wild ungulates. CE is caused by the contagious ecthyma virus in the Poxviridae family and *Parapoxvirus* genus. CE is found worldwide in countries that raise sheep. In the United States CE occurs most often in the Western states. CE virus is found in skin lesions and scabs of infected animals and is transmitted by direct or indirect contact through cuts and abrasions. The first clinical signs of CE are papules, pustules, and vesicles on the lips, nares, ears, eyelids, and sometimes feet of infected animals. Teat and udder lesions can be found in dams that have acquired the virus from their nursing lambs. In time the pustules break and raised brown scabs form over these areas. CE lesions are painful. CE lesions will heal in 1 to 4 weeks if not secondarily infected. In people, CE typically appears as a single skin lesion or a few lesions. The lesion is small, firm, and red that in time develops from a papule into a hemorrhagic pustule. In later stages, the lesion develops into a nodule containing fluid or may be covered by a thin crust. In time this lesion becomes covered by a thick crust. Low-grade fever and lymphadenopathy may occur with CE. Animals are typically diagnosed with CE based on clinical signs. Definitive diagnosis in animals and people is confirmed by electron microscopy of the scabs. Other tests available include virus isolation, PCR, and serologic methods such as serum neutralization, agar gel immunodiffusion, complement fixation, agglutination, and ELISA tests. There is no specific treatment for CE.

Supportive treatment such as antibiotics to prevent secondary bacterial infection, fly repellents to prevent maggots, and tube feeding for nutritional support may be used in animals. Supportive treatment with moist dressings, antiseptics, and antibiotics may be warranted in people. The best preventative measure in animals is vaccination. Recently vaccinated animals pose an increased risk of disease transmission to humans and other animals; therefore, recently vaccinated animals should be isolated from people and unvaccinated animals. Isolation of infected animals may prevent disease spread on a farm; however, once the virus has entered a flock or herd it is difficult to eradicate. People with skin cuts or abrasions should avoid infected animals and their wool/hides. Gloves should be worn if vaccinating animals for CE.

Poxvirus Infection

Overview

Poxviruses produce eruptive skin pustules called pox, a term derived from the French *poque* meaning pouch or little pocket in the skin. Pox lesions leave small, depressed scars upon healing, which are called pockmarks. The best known poxvirus is variola virus, the agent that causes smallpox. *Variola* is Latin for smallpox and was used to describe a variety of mottled rashes as early as the 6th century. The first recorded cases of smallpox were in ancient Egypt on the mummy of Ramses V. The Roman Empire lost more than one-third of its subjects during a 15-year epidemic that began in A.D. 165. Crusaders brought smallpox back with them from the Holy Land. Variola became endemic in India in the first millennium B.C. and spread to Asia and, ultimately, to Europe in the 8th century. Hernando Cortez carried smallpox to the New World in 1520 where it killed many people in the Aztec and Inca empires. Smallpox was differentiated from other diseases (such as syphilis, which was known as the large pox) in the 10th century by Persian physician Abu Bakr Muhammed Ibn Zalariya. In America, smallpox allowed the Europeans to eradicate indigenous people because they had developed immunity to it and the natives had not. Pocahontas died of smallpox in 1617 after visiting London and epidemics in Europe in 1600 to 1700 killed royalty such as Queen Mary II of England (1694), Emperor Joseph I of Austria (1711), and King Louis XV of France (1774). In 1714 a Greek physician named Emanuel Timoni described how he prevented smallpox during an epidemic by embedding a knife in an infected person's rash and then scratching a healthy person with the blade. By 1717 this method of immunization was used in an outbreak in Boston in which 6,000 people contracted smallpox (900 died), but of the 287 people who were vaccinated only 6 died. Smallpox was responsible for at least 130,000 American deaths during the Revolutionary War because vaccination had been declared illegal. In 1796, Edward Jenner, an English doctor, noticed that milkmaids who contracted cowpox while milking infected cows were immune to smallpox. Jenner injected fluid from a cowpox pustule into 8-year-old James Phipps who did not contract smallpox. Vaccination did not catch on for a while (Abraham Lincoln contracted smallpox several hours after delivering the Gettysburg Address); however, by the 1950s vaccination of schoolchildren for smallpox eradicated the disease from the United States. In 1975 the last case of smallpox caused by wild variola major was in a 3-year-old Bangladeshi girl (last case in Asia) and the last case of smallpox caused by wild variola minor was in a 23-year-old Somalian in December 1977. In December 1979 the world was declared free of smallpox (certified by WHO in 1980) except for that found in research labs.

Smallpox was the most destructive disease in history and the first to be eradicated from nature in 1979.

There are a variety of poxviruses that infect animals and these must be ruled out as a differential diagnosis when the smallpox virus was found in nature. There is also concern that these animal poxviruses can cause an epidemic in people who are no longer protected from the smallpox vaccine or naturally occurring smallpox immunity.

Causative Agent

The poxviruses that cause diseases similar to smallpox are in the *Orthopoxvirus* genus and Poxviridae family. Orthopoxviruses are large (can be seen by light microscopy) and have a rectangular shape. Orthopoxviruses are large, enveloped, double-stranded DNA viruses that have an external surface ridged in parallel rows. These virus particles are extremely complex containing many proteins (more than 100). The outer surface is composed of lipid and protein, which surrounds the biconcave core. The core is composed of a tightly compressed nucleoprotein. The extracellular virus has two envelopes, whereas the intracellular virus has only one envelope. The orthopoxviruses cause disease in animals and people such as smallpox (variola), monkeypox, vaccinia, and cowpox viruses.

There are other poxviruses that cause disease in animals, but are not believed to be zoonotic. These poxviruses include the following genera and examples: *Avipoxvirus* (fowlpox); *Capripoxvirus* (sheep pox, goatpox, and bovine lumpy skin disease); *Suipoxvirus* (swinepox); and *Parapoxvirus* (pseudocowpox).

Epizootiology and Public Health Significance

Animal poxvirus infections occur worldwide and cause sporadic disease in people (rarely produce epidemics). Smallpox was eradicated from the world as of 1979 after a $330 million 10-year campaign against smallpox was completed. Smallpox eradication was first proposed by the Soviet Union in 1958. Smallpox has a mortality rate of approximately 30% in populations without immunity and is a concern of governments in regards to bioterrorism activity. In 1999, the WHO recommended that all existing laboratory stock of smallpox virus be destroyed by the end of 2002, but the bioterrorism acts of 2001 stopped this destruction. The last smallpox viruses are stored frozen in liquid nitrogen at the CDC in Atlanta and the Center for Research for Virology and Biotechnology in Koltsovo, Russia.

Chickenpox is a herpesvirus, not a poxvirus.

Figure 7-34 In June 2003 the first outbreak of monkeypox was reported among several people in the United States. Most of these people became ill after having contact with pet prairie dogs that were sick with the virus.

(Courtesy of CDC, Susy Mercado)

Monkeypox infection poses some risk to people in Africa where the disease is seen mainly in children who have ingested infected monkey meat. The cross protection to monkeypox offered by smallpox vaccine is decreasing as a result of the cessation of smallpox vaccination in the 1970s. Case fatality rates for people from monkeypox in Africa range from 1% to 10% (mainly in children). An outbreak of monkeypox occurred in 2003 in the United States from exposure to infected prairie dogs (Figure 7-34).

The other zoonotic poxviruses rarely occur and do not pose as great a threat to human health as smallpox (nonzoonotic) and monkeypox.

Transmission

Transmission of most animal poxvirus to humans occurs by direct or indirect contact of nonintact skin or wounds with lesions of animal pox. Monkeypox can be transmitted by bite, aerosols, ingestion of infected meat, or direct contact with lesions. Rarely, human-to-human transmission of animal poxviruses has occurred. Smallpox is transmitted person-to-person by aerosol.

Pathogenesis

Pathogenesis varies with the poxvirus as described:

- Smallpox infection begins by inhalation of virus and has an incubation period of 10 to14 days. The virus replicates locally, spreads to the regional lymph nodes, and an asymptomatic viremia is observed by day 3 or 4 of the infection. The virus then spreads to the bone marrow and spleen. A secondary viremic period begins on approximately day 8 and is associated with symptoms of fever and toxicity. The virus in leukocytes localizes in the blood vessels of the dermis and the characteristic rash of smallpox follows. Papules appear on the cheek and pharyngeal mucosa and on the face and extremities before progression to the trunk. Over several days, these papules form vesicles then slowly form pustules. Approximately 8 days after initial infection, the pustules rupture and scabs form. In time the smallpox lesions heal; however, they lead to significant scarring.

- Other poxviruses are introduced by cutaneous inoculation and replicate at the site of inoculation, leading to the formation of local red papules. These papules form vesicles then pustules, rupture, scar, and heal over 10 to 14 days. Poxvirus may also spread to regional lymph nodes, which become tender and produce fever. Other poxviruses generally follow the same pattern of evolution, with primarily localized disease.

- Unlike the other animal poxviruses, monkeypox infection leads to a clinical syndrome similar to smallpox. Infection is usually caused by invasion through broken skin and in most cases remains localized. Human monkeypox is acquired by contact or by airborne transmission to the respiratory mucosa. Initial viremia during the incubation period spreads infection to internal organs; a second viremia then spreads the virus to the skin.

Clinical Signs in Animals

There are a variety of poxviruses that infect a variety of animals (Figure 7-35). Zoonotic poxviruses are summarized in Table 7-15. Nonzoonotic poxviruses in animals include variola virus (only infects humans), volepox, ectromelia virus (rodents), raccoonpox, skunkpox, and taterapox (gerbils) and are not covered.

Figure 7-35 Monkeypox lesions are generally seen on the face, digits, and perineum in nonhuman primates.

(Courtesy of Armed Forces Institute of Pathology)

Clinical Signs in Humans

Smallpox in people is not zoonotic and presents in two clinical forms, variola major (25% to 30% fatality rate with signs of shock, toxemia, and intravascular coagulation) and a similar but milder disease known as variola minor (less than 1% fatality rate). Clinical signs include fever followed by the slow development of a papular rash typically on the face and extremities that then spreads to the trunk. The rash evolves rapidly into vesicles, followed by pustules, scabs, and healing.

Table 7-15   Zoonotic Animal *Orthopoxviruses* and the Diseases They Cause

Virus	Species Affected	Clinical Signs in Animals	Clinical Signs in People
Monkeypox virus Originated in Zaire in 1970–1971 and was transmitted from monkeys to unvaccinated humans.	New and Old World monkeys; rodents (rats, mice, squirrels, and prairie dogs); rabbits	Nonhuman primates have self-limiting rash and fever (mortality may be seen in young monkeys); rodents and rabbits have fever, conjunctivits, nasal discharge, cough, lymphadenopathy, anorexia, and lethargy.	Resembles smallpox and may be fatal
Vaccinia virus Vaccinia virus has been used to immunize people against smallpox and was the original virus obtained from cattle; assumed to be a hybrid between coxpox and smallpox.	Unknown		Used for vaccination in humans; originally produced nodular lesions but now appears to be hybrid strain
Buffalopox virus First characterized in 1971 in India	Waterbuffalo and cattle	Buffalo and cattle develop skin eruptions	Nodular lesions typically on hand or face
Camelpox virus Seen in Africa and Asia Most closely related to smallpox	Camels	Camels (especially young) develop pustules around nares, lips, and hairless areas. Young camels have 25% mortality rate	Produces nodules on the hands of camel drivers
Cowpox virus	Cattle (also cats, anteaters, and rodents)	Cattle develop nodules of the udder and teats; decrease in milk production may occur due to the soreness of affected teats or secondary bacterial infection.	Produces single lesion on hand or face of person in close contact with affected cattle
Feline cowpox This virus is identical to cowpox virus and is believed to cause about 50% of cowpox infections in people.	Domestic and zoo cats (lions, cheetahs, pumas, etc.)	Felines develop chronic disease with multiple skin lesions that present as papules, followed by vesicles, then pustules, and finally scabs. Lesions are randomly distributed over the body.	Produces single lesion on hand or face of person in close contact with affected felines
Horsepox Also known as grease-heal and contagious pustular stomatitis	Horses	Horses develop papules, vesicles, and pustules on the skin of lips, nares and oral mucous membranes. Other clinical signs include nasal discharge, fever, and salivation. "Leg" form has pustules on leg and lameness	Rare in humans
Elephantpox virus	Circus or zoo elephants	Elephants develop fever, lameness, conjunctivitis, and pustules.	Produces nodular lesions

Zoonotic poxvirus infections include the following:

- Monkeypox infection can produce a disease similar to variola minor that clinically cannot be distinguished from smallpox (Figure 7-36). Cases of monkeypox infection generally occur in villages in tropical regions of western and central Africa, but an outbreak occurred in 2003 in the midwestern United States from exposure to infected prairie dogs. During the 2003 outbreak, 71 people were infected with one person becoming severely ill after developing encephalitis.
- Cowpox infection causes a localized pustular skin lesion on hands, arms, and face. A febrile lymphangitis, conjunctivitis, and meningoencephalitis are rarely seen.

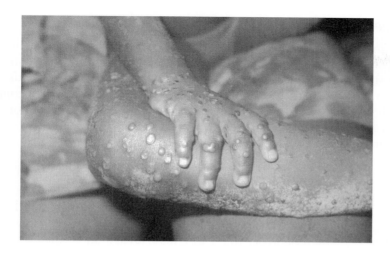

Figure 7-36 Monkeypox lesions in a child.
(Courtesy of CDC)

- Buffalopox, camelpox, and elephantpox produce a localized nodular lesion that is self-limiting.

Diagnosis in Animals

Poxvirus infections in animals may be suspected in animals with pustular lesions. Definitive diagnosis is done via histopathology, virus isolation, PCR, and ELISA tests.

Diagnosis in Humans

Most poxvirus infections can be recognized clinically with definitive diagnosis consisting of electron microscopy, virus isolation, PCR, and serologic tests such as hemagglutination inhibition, complement fixation, and ELISA tests.

Treatment in Animals

Animals are typically not treated for poxvirus infections. Antiretroviral drugs have been used in experimental animals with monkeypox virus. During the 2003 outbreak of monkeypox in the midwestern United States, the CDC recommended that all suspect animals be euthanized to prevent disease spread.

Treatment in Humans

Treatment of poxvirus infections in people is supportive. Cidofovir, an antiretroviral drug, has been used in animal studies as a possible treatment of monkeypox in humans.

Management and Control in Animals

The best preventative measure in animals is vaccination. The standard smallpox vaccine protects monkeys against monkeypox. An attenuated smallpox vaccine protects camels against camelpox. There is a vaccine for elephantpox that is a live attenuated vaccine. Vaccination for cowpox, feline cowpox, buffalopox, and horsepox is not practiced. Quarantine of infected animals or isolation of animals new to a facility can also decrease the outbreak of poxvirus among animals.

Eczema is a risk factor for vaccinia infection.

Management and Control in Humans

All *Orthopoxviruses* are immunologically related; therefore, vaccination for one orthopoxvirus protects against the rest. Smallpox vaccination may offer protection against animal poxvirus infection. The CDC currently recommends smallpox vaccination only to people exposed to monkeypox. People who have contact with animals should protect skin cuts or abrasion from having direct contact with infected or suspect animals. Aerosol protection is recommended for people handling nonhuman primates.

Summary

Poxviruses produce eruptive skin pustules called pox that leaves small, depressed scars upon healing. The best known poxvirus is variola virus, the agent that causes smallpox. The poxviruses that cause diseases similar to smallpox are in the *Orthopoxvirus* genus and Poxviridae family. Animal poxvirus infections occur worldwide and cause sporadic disease in people. Transmission of most animal pox virus to humans occurs by direct or indirect contact of injured skin or wounds with lesions of animal pox. Monkeypox can be transmitted by bite, aerosols, ingestion of infected meat, or direct contact with lesions. There are a variety of poxviruses that infect a variety of animals that typically produce pustular lesions. Zoonotic poxvirus infections include monkeypox, cowpox, buffalopox, feline cowpox, horsepox, camelpox, and elephantpox. Poxvirus infections in animals may be suspected in animals with pustular lesions. Definitive diagnosis is done via histopathology, virus isolation, PCR, and ELISA tests. Most poxvirus infections in people can be recognized clinically with definitive diagnosis consisting of electron microscopy, virus isolation, PCR, and serologic tests such as hemagglutination inhibition, complement fixation, and ELISA tests. Animals are typically not treated for poxvirus infections. Treatment of poxvirus infections in people is supportive. The best preventive measure in animals is vaccination (if available for that animal species). Quarantine of infected animals or isolation of animals new to a facility can also decrease the outbreak of poxvirus among animals. Smallpox vaccination in people may offer protection against animal poxvirus infection. People who have contact with animals should protect skin cuts or abrasion from having direct contact with infected or suspect animals. Aerosol protection is recommended for people handling nonhuman primates.

REOVIRUSES

Overview

The Reoviridae, nonenveloped double-stranded RNA viruses, are a family of viruses that affect the gastrointestinal system and the respiratory system. The Reoviridae family began in 1959 when Albert Sabin reclassified an echovirus group (echo is an acronym for enteric cytopathic human orphan), a group of viruses not known to cause any human disease. The new viral family was called reo, an acronym for respiratory enteric orphan (orphan viruses are viruses not associated with any known disease). Although Reoviridae have since been identified with specific diseases, the original name is still currently used. The family Reoviridae is divided into nine genera with only four (*Orthoreovirus*, *Coltivirus*, *Rotavirus*, and *Orbivirus*) infecting humans and animals. Four other genera infect only plants and insects, and the last one only infects fish.

The genus *Coltivirus*, named for *Colorado tick* fever (the major disease in this group), contains two known zoonotic viruses, Colorado tick fever virus and Eyach virus (two other viruses in this group that may be zoonotic are Sunday canyon virus and Banna virus). Colorado tick fever is an arbovirus found in the Rocky Mountains of the United States and Canada. Its vector is *Dermacentor andersoni* and the disease in humans is characterized by sudden fever, muscle pain, leukopenia, and retroorbital eye pain. Eyach virus is found in Europe and is a tick-borne disease causing encephalitis and polyneuritis.

The genus *Orbivirus* was named for the ring-shaped capsomers making up the inner capsid (*orbi* is Latin for ring) and contains Kemerovo virus (an arbovirus transmitted by ticks causing febrile illness in Siberia), Orungo virus (a mosquito-borne virus that causes febrile illness in Africa), Lebombo virus (a rare, possibly mosquito-borne virus that causes febrile illness in Nigeria), and Changuinola virus (a fly transmitted virus that causes febrile illness in Panama).

The genus *Orthoreovirus* was named for the straight genome (*ortho* is Latin for straight) and does not cause significant disease in humans, but can infect a variety of mammals. Bluetongue is an example of a disease caused by an *Orthoeovirus* that is transmitted by the insect vector *Culicoides varripennis* and can infect cattle sheep, goats, and wild ruminants.

The genus *Rotavirus* was named for the wheel-like appearance (*rota* is Latin for wheel) and was first isolated from infant diarrhea by Stanley, Dorman, and Ponsford in 1951; however, it was not until 1973 when it was identified as the cause of gastroenteritis. Rotaviruses are found worldwide and occur in infants and young children causing fever, vomiting, diarrhea, and occasionally dehydration. Rotaviruses are not believed to be zoonotic.

Causative Agent

Viruses in the Reoviridae family have a double-stranded RNA, which is unlike any other RNA virus (Figure 7-37). This genome is linear and segmented with the number of segments varying with the genera of the particular virus (*Orthoreovirus* and *Orbivirus* have 10 segments, *Rotavirus* has 11 segments, and *Coltivirus* has 12 segments). Replication occurs in the cytoplasm within a nearly intact virion particle (in most other viral families, the virion disassembles and uncoats completely before

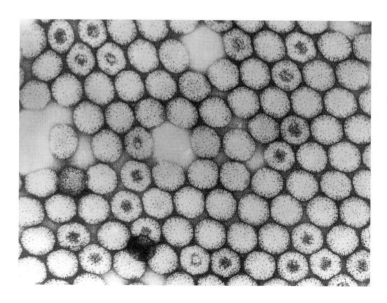

Figure 7-37 Electron micrograph of reoviruses.
(Courtesy of CDC/Dr. Erskine Palmer)

it replicates). Reoviridae viruses are nonenveloped, spherical in appearance, and the capsid shape is icosahedral with two concentric capsid shells. They form distinctive inclusions that stain with eosin. Reoviruses are heat stable, stable through a wide pH range, and are stable in aerosols, particularly when the relative humidity is high. Reoviruses can be inactivated by 95% ethanol, phenol, and chlorine.

Orthoreoviruses are ubiquitous in nature and cause many inapparent infections based on the fact that most people have serum antibodies to orthoreoviruses by early adulthood. They have an inner protein shell surrounded by an outer protein shell. *Reoviruses* replicate in a large number of tissue culture systems of both primate and other animal origin and easily cause infection in laboratory animals making them good viruses for laboratory studies.

There are more than 100 serotypes of *Orbiviruses*, most infecting insects, and many transmitted by insects to vertebrates. Animal diseases caused by *Orbiviruses* include bluetongue virus of sheep and African horse sickness virus. Antibodies are found in many vertebrates, including humans.

Coltiviruses resemble the *Orbiviruses* in size and include such viruses as Colorado tick fever virus, Salmon River virus (Idaho), Eyach virus (in Europe), isolate S6-14-03 (California), Banna virus, Beijing virus, and Gansu virus (China). Colorado tick fever is the most common arboviral disease of humans in the United States (reported from at least 11 Western states in the United States and from Alberta and British Columbia in Canada).

Rotavirus causes a large portion of gastroenteritis cases in children aged 6 months to 2 years. *Rotavirus* particles have a double-stranded RNA genome enclosed in a double-shelled capsid. *Rotavirus* has been found on all continents since its discovery in 1973. The transmission of rotavirus infections is fecal-oral with little evidence of airborne transmission.

The *Reoviridae* are summarized in Table 7-16.

> Viruses in the Reoviridae family are also called diplornaviruses (*diplo* is Greek for double) because of their double-stranded RNA genomes.

Colorado Tick Fever

Overview

Colorado tick fever, also known as Colorado fever, is a nonfatal, tick-borne viral infection seen in the Rocky Mountains. Reports of mountain fever were first seen in the mid-19th century in settlers in the Rocky Mountains. Mountain fever probably included many different infectious diseases with a mild form of fever endemic to the area most likely being Colorado tick fever (CTF). William Clark of Lewis and Clark fame may have contracted CTF in July 1805 while traveling through the western United States. In 1930, Becker associated the mild form of mountain fever with exposure to the Rocky Mountain wood tick (*Dermacentor andersoni*) and named the disease Colorado tick fever. CTF was more thoroughly described by Topping in 1940 and the viral cause of CTF was discovered in both ticks and people by Florio in 1945. CTF is typically a relatively mild disease producing fever, headache, joint pain, and muscle pain in the spring and early summer months in endemic areas. There are many mammals that have been found to be naturally infected with CTF virus including hibernating animals that may help the virus survive winter and initiate a new cycle in spring.

Causative Agent

Colorado tick fever is caused by a virus in the family Reovirdiae in the genus *Coltivirus*. Reoviruses are nonenveloped, double-stranded RNA viruses with icosahedral symmetry and three capsid layers. Most *Coltivirus* proteins have yet to be characterized. Historically, *Coltiviruses* have been divided into subgroup A (North

Table 7-16   Various Types of Reoviruses and Their Properties

Virus Name	Genus	Geographic Distribution	Vector	Disease Signs	Transmission Route
Rotavirus	*Rotavirus*	Worldwide	None known	Gastroenteritis (infant)	Fecal-oral; possibly respiratory
Orthoreoviruses	*Orthoreovirus*	Worldwide	None known, possible zoonotic	Possible respiratory illness (mild)	Fecal-oral and respiratory
Colorado Tick Fever virus	*Coltivirus*	Rocky Mountains, North America	Wood tick–*Dermacentor andersoni*	Febrile illness–encephalitis, hemorrhagic fever	Arboviral; arthropod vector
Sunday canyon virus	*Coltivirus*	North America	Tick	Febrile illness	Arboviral; arthropod vector
Eyach	*Coltivirus*	Europe	Tick	Possibly encephalitis	Arboviral; arthropod vector
Banna virus	*Coltivirus*	China	Not known	Febrile illness, encephalitis	Vector
Orungo	*Orbivirus*	Africa	Mosquitoes	Febrile illness	Arboviral; arthropod vector
Lebombo	*Orbivirus*	Africa	Mosquitoes	Febrile illness	Arboviral; arthropod vector
Changuinola	*Orbivirus*	Panama	Phlebotomine fly	Febrile illness	Arboviral; arthropod vector
Kemerovo	*Orbivirus*	Russia (Siberia)	Ticks	Febrile illness, encephalitis	Arboviral; arthropod vector

Modified from Loerke, C. www.stanford.edu

American and European species) and subgroup B (Asian species). Currently, subgroup A is known as the *Coltivirus* genus and subgroup B is separated into the *Seadornavirus* genus. The genus *Coltivirus* contains only CTF virus, California hare coltivirus, Eyach virus (found in Central Europe), and Salmon River virus.

Epizootiology and Public Health Significance

CTF is seen where its principal tick vector *Dermacentor andersoni* is found (Rocky Mountains including the western provinces of Canada). CTF is found in California, Colorado, Idaho, Montana, Nevada, Oregon, Utah, Washington, Wyoming, British Columbia, and Alberta. There is a seasonal occurrence of CTF with infections occurring mainly in April, May, and June, when adult ticks are abundant.

CTF causes four human cases of disease per 100,000 people with 50% of infected people developing clinical signs. The infection rate is more prevalent in Colorado where 15% of people who spend time outdoors are seropositive for the virus. Most western states include CTF on their reportable disease list; however, CTF is not a nationally reportable disease in the United States. Surveillance data is limited, but is it believed that there are approximately 200 case reports annually. Since the disease is usually mild and self-limiting control measures are not emphasized (however, tick control for other diseases is emphasized).

CTF virus can live in red blood cells for the life of the cell (120 days).

Transmission

CTF virus is transmitted mainly by the Ixodes (hard-shelled) tick *Dermacentor andersoni*. *De. andersoni* is found at elevations between 1,500 to 3,000 meters (greater than 4,000 feet). CTF virus is maintained in nature in a cycle involving ticks and a vertebrate reservoir in rodents and small mammals. The virus over-winters in nymph stage of ticks, which will feed on and infect small mammals in spring. Important reservoirs are ground squirrels, western chipmunks, wood rats, and deer mice. As the animals become viremic, they in turn infect larval ticks. The larvae metamorphose during the summer and they overwinter as infected nymphs. Ticks do not transmit virus to humans until reaching the adult stage, which may be 1 or 2 years after infection.

Human-to-human transmission has occurred by blood transfusion.

Pathogenesis

CTF virus is introduced into an animal or human by the bite of a blood sucking tick. For approximately 2 weeks free virus can be isolated from blood. Virus then travels to the regional lymph nodes and eventually gets to the bone marrow where it replicates in bone marrow cells. In the bone marrow the virus stops the maturation of the neutrophils, eosinophils, and basophils and sometimes thrombocytes. Erythrocytes are infected with CTF virus as erythroblasts and are later detected in large numbers in erythrocytes in the peripheral blood. The virus is found only briefly in serum. Antibodies can be detected about 2 weeks after the onset of clinical signs and virus can be isolated from peripheral blood cells for up to 6 weeks. In less than 15% of infected children younger than 10 years of age, CTF virus invades the CNS and causes encephalitis.

Clinical Signs in Animals

CTF virus is not known to cause illness in mammalian species (they serve as amplifying hosts), although disease resembling human disease has been produced experimentally in laboratory mice, hamsters, Guinea pigs, and rhesus monkeys. Infection does not appear to affect survival of the *Dermacentor andersonii* tick.

Clinical Signs in Humans

The incubation period for CTF in people is 1 to 14 days. CTF in people appears abruptly, with initial clinical signs of high fever, chills, joint and muscle pains, severe headache, ocular pain, nausea, and occasional vomiting persisting for a few days. A transitory rash is seen in 5% to 12% of patients. In about 50% of people there is biphasic illness, in which acute symptoms last for several days, followed by fever remission, followed by a relapse lasting 2 to 3 days. This biphasic fever is also known as saddleback fever. A small proportion of people have more severe illness, including anorexia, continuing fatigue, and convalescence for several more weeks. CTF is typically more severe in children, who may have hemorrhagic manifestations such as DIC and gastrointestinal bleeding. CNS involvement (headache, neck stiffness, and photophobia) has been seen in severely affected children.

Diagnosis in Animals

Diagnosis of CTF in animals is not done clinically, but surveillance of disease can be done by virus isolation and neutralizing antibody tests.

Diagnosis in Humans

CTF can be diagnosed in people by virus isolation (from blood 2 to 4 weeks after infection) or serologic tests such as IgM capture ELISA, immunofluorescent assay, plaque reduction neutralization, and complement fixation tests. Antibodies to CTF virus are frequently seen in people who frequent endemic areas; therefore, a single elevated IgG titer does not indicate acute infection. RT-PCR can be used for diagnosis before antibodies appear.

Treatment in Animals

Animals are not treated for CTF.

Treatment in Humans

There is no specific treatment for CTF. Antipyretics and drugs to help control secondary infection may be used in cases of CTF.

Management and Control in Animals

Tick control may help decrease the level of viremia in small mammals, but a control program has not been established specifically for CTF.

Management and Control in Humans

Inactivated and modified live virus vaccine are available for CTF, but have not been widely used as a result of the mildness of disease produced by the CTF virus. Tick control will help decrease the spread of virus from ticks to animals and people. Recovered patients should not donate blood for a minimum of 6 months after recovery.

Summary

Colorado tick fever (CTF) is a nonfatal, tick-borne viral infection seen in the Rocky Mountains. CTF is caused by Colorado tick fever virus in the family Reovirdiae in the genus *Coltivirus*. CTF is seen where its principal tick vector *De. andersoni* is found (Rocky Mountains including the western provinces of Canada). There is a seasonal occurrence of CTF with infections occurring mainly in April, May, and June, when adult ticks are abundant. CTF virus is maintained in nature in a cycle involving ticks and a vertebrate reservoir in rodents and small mammals. Human-to-human transmission has occurred by blood transfusion. CTF virus is not known to cause illness in mammalian species (they serve as amplifying hosts). CTF in people appears abruptly, with initial clinical signs of high fever, chills, joint and muscle pains, severe headache, ocular pain, nausea, and occasional vomiting persisting for a few days. In about 50% of people there is biphasic illness, in which acute symptoms last for several days, followed by fever remission, followed by a relapse lasting 2 to 3 days. CTF is typically more severe in children, who may have hemorrhagic manifestations such as DIC, gastrointestinal bleeding, and CNS involvement. Diagnosis of CTF in animals is not done clinically, but surveillance of disease can be done by virus isolation and neutralizing antibody tests. CTF can be diagnosed in people by virus isolation (from blood, 2 to 4 weeks after infection) or serologic tests such as IgM capture ELISA, immunofluorescent assay, plaque reduction neutralization, and complement fixation tests. RT-PCR can be used for diagnosis before antibodies appear. Animals are not treated for CTF. There is no specific treatment for CTF in people. Tick control may help decrease the level of viremia in small mammals and help decrease the spread of virus from ticks to animals and people. Inactivated and modified live virus vaccine are available for CTF in people, but have not been widely used as a result of the mildness of disease produced by the CTF virus. Recovered patients should not donate blood for a minimum of 6 months after recovery.

RHABDOVIRUSES

Overview

The order Mononegavirales contains viral species that have a nonsegmented, single-stranded negative sense RNA genome. The order includes four families: Bornaviridae, Rhabdoviridae, Filoviridae, and Paramyxoviridae. The family Rhabdoviridae includes two genera of plant viruses and three genera of animal viruses. The three animal virus genera are *Lyssavirus*, *Vesiculovirus*, and *Ephemerovirus*; only *Lyssavirus* and *Vesiculovirus* are zoonotic. Tables 7-17 and 7-18 list the various zoonotic viruses in the Rhabdoviridae family.

The two main zoonotic rhabdoviral diseases are rabies and vesicular stomatitis. In general, rhabdoviruses have a worldwide distribution with some virus found in specific areas such as Africa, Europe, and Australia.

Causative Agent

Rhabdoviruses are medium-sized, enveloped, single-stranded, negative sense RNA viruses. Although these viruses have an envelope, their capsid is helical, which makes the viruses appear bullet-shaped. Rhabdoviruses carry genes encoding for five proteins: nucleocapside protein (N), phosphoprotein (P), matrix protein (M), receptor binding protein (G), and polymerase (L). The N, P, and L are attached to the nucleocapsid within the virus particle. The M protein is located between the nucleocapside and the outer membrane. The G protein is the only glycoprotein,

Table 7-17   Lyssaviruses

Virus Type	Host	Geographic Location
Rabies virus (lyssavirus type 1)	Mammals (especially canine species)	Worldwide
Lagos bat virus (lyssavirus type 2)	Bats	Africa
Mokola virus (lyssavirus type 3)	Bats	Africa
Duvenhage virus (lyssavirus type 4)	Bats	South Africa
European bat virus type 1 (lyssavirus type 5)	Bats	Europe
European bat virus type 2 (lyssavirus type 6)	Bats	Europe
Australian bat virus (lyssavirus type 7)	Flying foxes	Australia

Table adapted from Krauss.

Table 7-18 Vesiculoviruses

Virus Type	Geographic Location
Vesicular Stomatitis Virus type 1 (Indiana)	U.S.
Vesicular Stomatitis Virus type 2 (New Jersey)	U.S.
Vesicular Stomatitis Virus type 3 (Cocal)	U.S.
Vesicular Stomatitis Virus type 4 (Alagoas)	U.S.
Chandipura virus	India
Isfahan virus	Iran
Piry virus	Brazil

which combines the functions of receptor binding and fusion. The nucleocapsid consists of an RNA and N protein complex together with an NS (M1) protein and is surrounded by a lipid envelope containing M (M2) protein. The nucleocapsid contains transcriptase activity and is infectious. The shape of the virus that infects vertebrates is a bullet (*rhabdo* is Greek for rod) with one blunt and one pointed end.

Rabies

Overview

Rabies is an ancient viral infection of the CNS and is considered the oldest communicable disease of humans. The word rabies originates from the term *rabhas*, which means to do violence. The first recorded reference to rabies was in the 23rd century B.C. when Babylonian law described the financial penalty for the owner of a rabid animal that bites a human. The writers Homer (approximately 700 B.C.), Democritus (420 B.C.), and Aristotle (approximately 300 B.C.) both used descriptions of rabies in their works.

The first outbreak of dog rabies extensively recorded occurred in Italy in 1708. Rabies did not appear in the Western Hemisphere until the mid-18th century, when dogs caught the virus in Virginia in 1753. During the 19th century, rabies ravaged Europe and fear of this condition escalated to hysteria. Calming the terror people felt about this disease soon became the job of Louis Pasteur. Rabies virus was first isolated by Pasteur in 1885 (although other scientists were testing it before this time). Pasteur and his group of scientists began injecting the "disease" into rabbits to determine the amount of time it took for the virus to affect the spinal cords of these rabbits. Pasteur then began injecting dogs with weakened versions of the virus and observed that these injections could prevent the disease before and after exposure. In time, Pasteur produced the first rabies vaccine by inactivating the virus using ultraviolet light (the first vaccine was given to 9-year-old Alsatian boy named Joseph Meister, who later survived a rabid bite). Rabies has been reduced in many countries through pet control and careful quarantines, but it is still enzootic in various wild animals in the Americas and in domestic and wild animals throughout Africa and Asia.

Rhabdo means rod-shaped; viruses in the rhabdovirus family have a distinct bullet shape.

All mammals can get rabies; however, small rodents (squirrels, rats, mice, hamsters, Guinea pigs, gerbils, and chipmunks) and lagomorphs (rabbits and hares) are rarely infected with rabies. Birds, fish, reptiles, and insects do not get rabies.

Causative Agent

Rabies is a preventable neurologic disease caused by a rhabdovirus. The Rhabdoviridae family is in the order Mononegavirales and is a nonsegmented, single-stranded, negative sense RNA virus. Rabies virus is neurotropic (seeks out nervous cells) and is typically found in one species in a given area. There are different variants of this virus and each variant is responsible for rabies transmitted between members of the same species in a given area.

Epizootiology and Public Health Significance

Rabies has a worldwide distribution in both wild and domestic animals with the exceptions of Australia, the United Kingdom, Ireland, New Zealand, Japan, Antarctica, Scandinavia, and Hawaii; however, the reservoir varies geographically. In the United States and Canada, rabies in wildlife populations typically includes skunks, raccoons, and bats. Distinct variants are responsible for rabies in different species; there is a distinct variant for dogs and coyotes, a different variant for skunks, and a different variant for foxes. In general, each variant found in bats belongs to a predominant bat species. In the United States, most cases of rabies in humans in the past decade have been caused by bat rabies variants. Cats, cattle, and dogs are also frequently infected with rabies in the United States. Wildlife that are common sources of rabies virus are raccoons, skunks, foxes, and coyotes. The number of postexposure prophylaxis treatments given in the United States is not tabulated, but is estimated to be around 40,000 per year. The estimated annual cost of rabies vaccination in the United States is over $300,000 (the majority of the cost is for dog vaccinations). Worldwide, 40,000 to 50,000 human fatalities from rabies (especially from wild animal rabies) occur annually, mainly in rabies-endemic areas such as Asia and Africa.

Transmission

Rabies virus may be present in saliva and transmitted by an infected animal several days before the onset of clinical signs (dogs can shed virus 5 to 7 days before clinical signs and cats can shed virus 3 days before clinical signs). Rabies is almost always transmitted by introduction of virus-laden saliva into tissues. The virus-laden saliva usually is introduced into tissue by a rabid animal bite; however, the virus can also be introduced into fresh wounds (a scratch or abrasion that is exposed to infective saliva) or through intact mucous membranes (eyes, nose, and by ingestion of the virus through the mouth). Aerosol transmission in a laboratory setting and in caves where bats roost has also been shown to transmit the virus, but a high concentration of suspended viral particles is needed for these types of transmission. Viral spread via blood is extremely rare. Rabies-contaminated transplants are also a source of rabies infection. The only documented cases of rabies caused by human-to-human transmission are through corneal transplants and solid organ transplants.

Pathogenesis

Rabies is a fatal disease that usually causes death within 10 days after the onset of clinical signs. The incubation period from exposure to clinical signs is extremely variable. The incubation period for rabies virus is unusually long and the virus can remain at the inoculation site for a long time. In dogs most cases of rabies occur approximately 20 to 80 days after viral exposure, but the incubation period may be longer or shorter. Rabies virus replicates in muscle cells near the inoculation site and then travels to the spinal cord via the peripheral nerves. The animal does not appear abnormal at this time. From the spinal cord the virus travels to the gray matter of the brain (mainly in neurons of the limbic system, midbrain, and hypothalamus) and finally to the salivary glands via the peripheral nerves. After the virus has multiplied in the brain, most animals begin to show signs of rabies. Within 3 to 5 days, the virus has caused enough brain damage that the animal begins to show neurologic signs of rabies.

Clinical Signs in Animals

Rabid animals exhibit CNS signs that vary slightly among species. All species infected with rabies virus exhibit behavioral changes and paralysis. The first sign to appear in all animals is a change in behavior, such as anorexia, irritability, and nervousness. These changes may be difficult to distinguish from early infectious disease, injury, an oral foreign body causing excessive salivation, or gastrointestinal disorders. Change in body temperature is usually not significant. Animals infected with rabies virus usually stop eating and drinking. Some animals seek solitude. An irritated urogenitial system may result in frequent urination and increased sexual desire.

Rabies is often described as either being furious or dumb. Animals display either the furious or dumb phase of rabies 1 to 3 days after infection. Furious rabies, often described as the classical "mad-dog" rabies or vicious rabies, refers to animals that have a pronounced aggressive phase. During this phase the animal becomes irrational and with slight provocation becomes aggressive. The animal is alert, anxious, has mydriasis (dilated pupils), and will attack with auditory stimulation. Specific species examples include:

- Rabid dogs may show behavioral changes and may chew cages and other restraint devices, and attempt to bite (Figure 7-38). Some dogs may eat dirt, sticks, and other objects. Some dogs are dysphagic (have difficulty eating), which will cause profuse salivation. Seizures, paralysis (including facial and pharyngeal), and death may follow the hypersalivation phase.

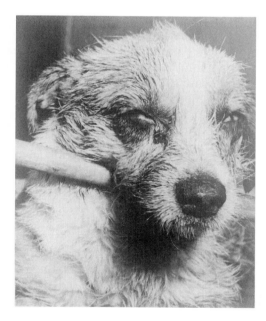

Figure 7-38 Dog's face during late-stage "dumb" or paralytic rabies.
(Courtesy of CDC/Barbara Andrews)

- Rabid cats may show behavioral changes and will show an excitatory phase where they may attack suddenly by biting and scratching. The pronounced excitatory phase is usually followed by progressive paralysis.

- Rabid equines (such as horses and mules) often present with signs of spinal cord disease such as ataxia (incoordination). Some rabid equines will bite with the slightest provocation. Equines frequently show signs of distress and agitation, and may roll (may be interpreted by some as a sign of colic). Increased vocalization is seen in some rabid equines.

- Rabid cattle typically begin with hind limb ataxia and paresis (weakness). Cattle may butt any moving object, attacking and pursuing humans and other animals. In dairy cattle, lactation stops abruptly. Yawning and tail paralysis may be seen in some rabid cattle. Facial expressions change from a placid expression to one of alertness. Eye and ear movements follow sound. Abnormal bellowing may continue intermittently until just prior to death.

■ Rabid pigs will show behavior changes, which typically include biting and becoming aggressive.

■ Rabid goats and sheep will show increases in bleating, hind-leg weakness, difficulty in walking, aggression, excessive sexual activity, and paddling.

■ Rabid wildlife act abnormally. Rabid foxes and coyotes will invade yards and may attack pets and people. Rabid raccoons and skunks typically do not fear people and are ataxic, aggressive, and active during the day (even though they are nocturnal animals in nature). Rabid bats can be seen flying during the day, resting on the ground, attacking people and animals, or fighting.

Dumb rabies, often described as the paralytic form of rabies, manifests itself as paralysis of the throat and chewing muscles. In dogs dropping of the mandible, profuse salivation, and inability to swallow may be seen. Owners should not attempt to examine the mouth of the animal or administer medication to these animals as they are exposing themselves to the rabies virus. Animals displaying the dumb form of rabies are not vicious and do not attempt to bite. Paralysis will progress rapidly to all parts of the animal's body with coma and death occurring in a few hours.

> Furious rabies is most commonly seen in carnivores, equine, and wild animals; dumb rabies is most often seen in ruminants. Each form occurs about 50% of the time in swine.

Clinical Signs in Humans

People who contract rabies usually have a history of an animal bite, although some people are unaware of their bite wounds if the wound is inflicted by bats. The site of the wound initially feels painful followed by a period of burning and increased sensitivity to temperature changes. The first symptoms of rabies may be nonspecific flu-like signs such as fever or headache. In time drinking becomes extremely painful as a result of laryngeal spasm and the person refuses to drink (hydrophobia). Behavior changes such as restlessness or extreme excitability may occur. Muscle spasm and laryngospasm are followed by convulsions at which time large amounts of thick saliva are present.

Diagnosis in Animals

The pathology of rabies infection involves the development of encephalitis and myelitis. Lymphocytes, segmented neutrophils, and plasma cells infiltrate perivascular areas throughout the CNS. Rabies virus frequently produces round or oval cytoplasmic inclusion bodies in neurons. These inclusions are known as Negri bodies, named after Dr. Adelchi Negri who identified them in 1903. Negri bodies are the sites of active viral replication and range in size from 0.25 to 0.27 µm (Figure 7-39). Negri bodies are found in neurons of the brain (especially the cerebellum), salivary glands, tongue, and other organs.

It is difficult to clinically diagnose rabies because in its early stages rabies can be confused with other diseases. When rabies is suspected in an animal, laboratory confirmation of fresh brain tissue is warranted. The laboratory test of choice for rabies identification in animals is a fluorescent antibody (FA) test that has been utilized for approximately forty years. A benefit of the rabies FA test is that it provides a highly specific diagnosis within a few hours. Fresh brain tissue including the hippocampus, cerebellum, and medulla oblongata must be obtained for testing. The brain tissue must be preserved by refrigeration with wet ice or cold packs; not frozen. The FA test for rabies identification is based on the fact that rabies-infected animals have rabies virus proteins (antigens) in their tissues. Fluorescent labeled antibody to this viral protein is incubated with suspect brain tissue. If rabies viral protein is present, the fluorescent labeled antibody will bind to it. Any unbound

> The outcome of rabies exposure depends on the variant of the virus, the amount of virus inoculum, the route of exposure, the location of exposure, and host factors.

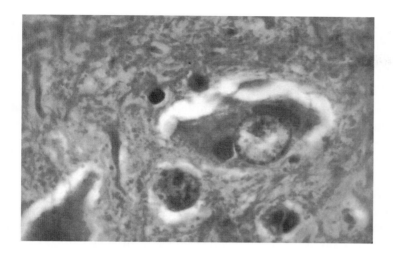

Figure 7-39 Microscopic appearance of Negri bodies.
(Courtesy of CDC/Dr. Daniel P. Perl)

labeled antibody is washed away. The sample is then examined with a fluorescent microscope and fluorescent labeled antibody bound to the antigen will be seen as a fluorescent green area if present in the tissue.

Diagnosis in Humans

Brain tissue and meninges that are infected with rabies virus have a variety of changes seen on histopathologic analysis. These changes include Negri bodies, Babes nodules of glial cells, lymphocyte foci, perivascular cuffing of lymphocytes or segmented neutrophils (Figure 7-40), and mononuclear infiltration. Stains such as Mann's, Giemsa, or Sellers stains can differentiate rabies inclusions from other intracellular inclusions.

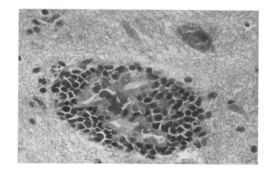

Figure 7-40 Perivascular cuffing seen histologically in a rabies positive animal.
(Courtesy of CDC/Dr. Daniel P. Perl)

Immunohistochemistry methods for rabies detection are used to detect rabies in formalin-fixed brain samples. These methods are more sensitive than histologic stain methods. These methods use specific antibodies and enzyme-labeling to detect rabies virus inclusions.

Electron microscopy can also be used to examine the ultrastructure of rabies virus and its inclusions. Rabies viruses are seen as bullet-shaped particles using electron microscopy. Amplification methods can be used to identify rabies virus in samples suspected of containing small amounts of virus. Rabies virus replicates in cell culture using mouse neuroblastoma cells or baby hamster kidney cells (BHK cells). Cell culture techniques increase the numbers of viral antigens to help identify positive cases that have low viral load. Another method for increasing the viral concentration is by using reverse transcriptase-polymerase chain reaction (RT-PCR). In this technique, rabies virus RNA from saliva or skin biopsy samples is enzymatically amplified as DNA copies. Rabies RNA is first copied into a DNA molecule using reverse transcriptase. The rabies DNA copy is then amplified using PCR techniques.

Rapid and accurate laboratory diagnosis of rabies in humans is essential for timely administration of postexposure rabies prophylaxis. Antemortem (before death) testing for rabies in humans requires several tests that are performed on a variety of samples including salvia, serum, spinal fluid, and skin biopsies of hair

Fixed sample preparations can still be infectious and need to be handled appropriately.

Histologic examination of brain tissue from clinically rabid animals shows Negri bodies in approximately 50% of the samples. In contrast, FA tests show binding of fluorescent-labeled antibody in nearly 100% of the samples.

follicles at the nape of the neck. Saliva is tested by viral cell culture or RT-PCR. Serum and spinal fluid are tested for antibodies to rabies virus, whereas skin biopsy samples are examined for rabies antigen.

Treatment in Animals

Treatment of rabid animals is not attempted as a result of the zoonotic and usually fatal nature of the disease. Rabid animals are euthanized and their brains are tested.

Treatment in Humans

Treatment of human rabies depends upon the point of disease recognition and symptoms present in the person. Any animal bite that may have come from a potentially rabid animal needs to be seen by medical personnel that should perform vigorous washing and first aid of the wound. State or local health departments, veterinarians, and animal control officers need to be made aware of a possible rabies exposure. Post-exposure rabies prophylaxis (PEP) is indicated for people who have been potentially exposed to a rabid animal (bite wound, mucous membrane contamination, transplantation of contaminated tissue, etc.) or a person with rabies (exposed to the person 14 days preceding the appearance of clinical signs until the person's recovery or death). Post-exposure prophylaxis is begun as soon as possible after exposure and when given promptly has been successful. Currently, postexposure prophylaxis is a regimen of one dose of immune globulin (antibody) that provides immediate but temporary protection, a dose of rabies vaccine, which starts the body producing its own antibodies, and four additional doses of rabies vaccine over a 28-day period. Additional doses of rabies vaccine are given on days 3, 7, 14, and 28 after the first vaccination. Human rabies immune globulin is also infiltrated around the wound. Current vaccines are relatively painless and given intramuscularly in the arm. Adverse reactions to the rabies vaccine and immune globulin are not common and may include pain, redness, swelling, or itching at the injection site. Low-grade fever may be seen following injection with rabies immune globulin.

There are only six documented cases of human survival from clinical rabies and five of these included a history of either pre- or postexposure vaccination. The first case of rabies survival without pre- or postexposure vaccination was documented in Wisconsin in 2004 in which a 15-year-old girl contracted rabies from a bat bite and developed symptoms of rabies. Blood, spinal fluid, saliva, and skin samples were confirmed rabies positive by the CDC and she was put into a drug-induced coma for 1 week. During that one week's time she produced massive amounts of antibodies against rabies virus and no longer showed signs of infection. After 10 weeks of hospitalization she was discharged from the hospital and is regaining neurologic function (she graduated from high school in June 2007).

Management and Control in Animals

Rabies in the United States has changed over the past 100 years and now more than 90% of all animal cases reported to the CDC occur in wildlife (before 1960 the majority of cases were in domestic animals). In the United States, human fatalities from rabies have declined from over 100 annually to approximately 1 to 2 annually (usually as a result of people unaware of their exposure). Raccoons are the most frequently reported rabid wildlife species (about 40%), followed by skunks (about 30%), bats (about 17%), foxes (about 6%), and other wild animals. Cases of rabies in raccoons are especially high along the eastern coast of the United States. The Compendium of Animal Rabies Control, updated annually by the National Association of State Public Health Veterinarians (NASPHV), monitors

> Any animal showing neurologic signs should be considered a rabies suspect and veterinary staff should protect themselves against exposure (by wearing gloves and protective clothing).

current USDA-licensed rabies vaccines and their recommended protocols. Both modified live virus and killed vaccines are available for use in dogs worldwide. Currently available vaccines in the United States are effective for either 1 or 3 years and several types are available for use in dogs, cats, ferrets, horses, cattle, and sheep. The increased incidence of rabies in cats has made cat vaccination extremely important. As of June 2007 a newly approved vaccine for cats labeled for use every four years has been approved. Oral vaccines distributed in baits have been effective in Europe and Canada for control of fox rabies.

Any animal bitten or scratched by a wild, carnivorous mammal or bat is considered to have been exposed to rabies. The NASPHV recommends that any unvaccinated dog or cat exposed to rabies be immediately destroyed and tested. If the owner does not consent to this, the animal is strictly quarantined for 6 months and vaccinated for rabies one month before release. If the exposed animal is currently vaccinated, it should be revaccinated and observed for 45 days. Vaccinated or unvaccinated animals that have bitten someone, yet are considered to have not been exposed to rabies and are healthy, are quarantined for 10 days. If the animal develops any signs of rabies during this time period, it is humanely euthanized and its brain submitted for rabies testing. If the animal is a stray or unwanted, it may be euthanized and its brain submitted for rabies testing.

Guidelines for control of rabies in dogs prepared by WHO include the following:

- Notification of suspected cases and destruction of dogs with clinical signs as well as dogs bitten by a rabies-suspect animal,
- Reduction of contact with susceptible dogs by introduction of leash laws and quarantine,
- Immunization of dogs,
- Stray dog control and destruction of unvaccinated dogs, and
- Dog registration.

In May 2007, an updated rabies vaccination certificate was released by the National Association of State Public Health Veterinarians which includes a space for microchip number and more space for animal and vaccine information (the form is known as Form 51).

Management and Control in Humans

Pre-exposure vaccination for rabies is recommended for high-risk people such as veterinarians, animal handlers, wildlife workers, and laboratory workers. International travelers likely to be exposed to enzootic dog rabies should consider pre-exposure prophylaxis. Pre-exposure prophylaxis consists of three doses of rabies vaccine given intramuscularly on days 0, 7, and either day 21 or 28. There are currently three types of rabies vaccines made from killed rabies virus: human diploid cell vaccine (HDCV), rabies vaccine adsorbed (RVA), and purified chick embryo cell culture (PCEC). Human rabies vaccines are effective when given properly. High-risk individuals who have received the pre-exposure prophylactic vaccine series should routinely have antibody titers drawn to make sure their antibody numbers are at a protective level. If antibody levels are low, a booster vaccine is given.

Summary

Rabies is a preventable viral disease of mammals. Rabies virus, a rhabdovirus, is an RNA virus that causes acute encephalitis and myelitis in all warm-blooded animals including people. The outcome is almost always fatal. All species of mammals

Some time is needed following the initial inoculation of rabies vaccine to when the body mounts a sufficient response to achieve protection. Studies have shown that this time period is 28 days following the initial vaccination. Clients need to be aware that their vaccinated pets are not protected during this time and risks of contracting rabies need to be avoided.

are susceptible to rabies infection; however, only a few species are important as reservoirs for the disease (such as raccoons, skunks, foxes, coyotes). In addition to terrestrial reservoirs, several species of insect-eating bats are reservoirs for rabies.

Transmission of rabies virus is usually through virus-laden saliva following an animal bite. Other routes of transmission include contaminated scratches, contaminated mucous membranes, aerosol transmission, and transplants (Figure 7-41).

Rabid animals exhibit CNS signs that vary slightly among species. Rabies virus infects the CNS, causing encephalitis and death, usually within 10 days of developing clinical signs. Rabies is often described as either being furious or dumb. Animals display either the furious or dumb phase of rabies 1 to 3 days after infection. People who contract rabies usually have a history of an animal bite, although some people are unaware of their bite wounds if the wound is inflicted by bats. The site of the wound initially feels painful followed by a period of burning and increased sensitivity to temperature changes. The first symptoms of rabies may be nonspecific flu-like signs such as fever or headache. Behavior changes such as restlessness or extreme excitability may occur. Muscle spasm and laryngospasm are followed by convulsions at which time large amounts of thick

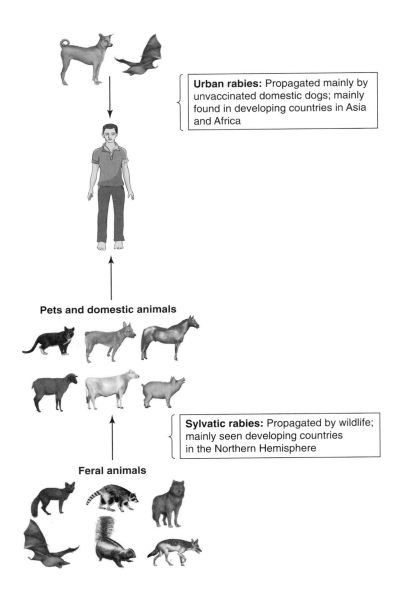

Urban rabies: Propagated mainly by unvaccinated domestic dogs; mainly found in developing countries in Asia and Africa

Pets and domestic animals

Sylvatic rabies: Propagated by wildlife; mainly seen developing countries in the Northern Hemisphere

Feral animals

Figure 7-41 Transmission of rabies.

saliva are present. The laboratory test of choice for rabies identification in animals is a fluorescent antibody (FA) test on unfrozen brain tissue. Immunohistochemistry methods for rabies detection are used to detect rabies in formalin-fixed human brain samples. Electron microscopy can also be used to examine the ultrastructure of rabies virus and its inclusions. Animals are not treated for rabies. Post-exposure rabies prophylaxis is begun in people as soon as possible after exposure and when given promptly has been successful. Currently, postexposure prophylaxis is a regimen of one dose of immune globulin (antibody) that provides immediate but temporary protection, a dose of rabies vaccine, which starts the body producing its own antibodies, and four additional doses of rabies vaccine over a 28-day period. Additional doses of rabies vaccine are given on days 3, 7, 14, and 28 after the first vaccination. Human rabies immune globulin is also infiltrated around the wound. Vaccines are available for animals and humans. Animal control and vaccination laws are used to manage rabies.

Vesicular Stomatitis

Overview

Vesicular stomatitis (VS), also called sore mouth of cattle, is a highly contagious, sporadic disease of livestock characterized by vesicular (blisters and ulcers) lesions on the tongue, mouth, teats, or coronary bands. In addition to vesicular lesions, classic signs in horses, cattle, and swine include excessive salivation and lameness. VS was first reported in 1897 by Theiler who described an outbreak in horses and mules in Transvall, South Africa, in 1884. Affected animals had a fever, were anorexic, and had excessive salivation. Vesicles appear on the gums, tongue, and lips; ruptured-producing reddish ulcerations; and healed within 6 to 7 days. Although the disease has not been a problem in Africa since 1900 (unconfirmed cases may have occurred in 1934, 1938, and 1943), it sporadically appears in North and South America. The first officially reported case of VS in the United States was described in the veterinary literature in 1916 in the Denver stockyards; however, cavalry horses in the Civil War were described as having a disease similar to VS as early as 1862. J. R. Mohler in 1904 described a disease occurring in summer and fall, which affected the mouths (causing drooling and inability to eat) and feet (producing painful fissures near the coronary band) of cattle in some eastern and central western states. In 1906 Heiny described cases of VS in western Colorado and in 1907 horses in the Chicago stockyards were also described as having a disease similar to VS. In 1926 a major outbreak of VS occurred in New Jersey in mid-October and lasted until mid-November affecting 752 cattle on 33 farms in a 300 square mile area and 12 horses. In 1937 VS appeared in Wisconsin, Minnesota, the Dakotas, and Manitoba in both cattle and horses. VS was reported for the first time in South America in 1939 in horses and cattle of the La Plata region of Argentina. Two years later a major outbreak occurred in Venezuela involving 715 cattle, 195 horses, and 48 pigs (the first account of swine infection). In 1942 another VS strain appeared in Colorado (Indiana type) affected cattle and horses and was the last time this strain was seen in the United States. The Indiana serotype has been responsible for outbreaks of VS in 1942, 1956, 1964, 1965, and 1997 to 1998. The other VS serotype, the New Jersey serotype, has occurred in the United States in 1944, 1949, 1957, 1959, 1963, 1982 to 1983, 1995, and 2005 (470 cases during this outbreak).

Causative Agent

Vesicular stomatitis virus (VSV) is a member of the Rhabdoviridae family in the genus *Vesiculovirus*. It is a large (70 to 175 nm) bullet-shaped, helical, single-

The major concerns with vesicular stomatitis are its similar appearance to foot-and-mouth disease, its highly contagious nature, and the trade restrictions associated with it.

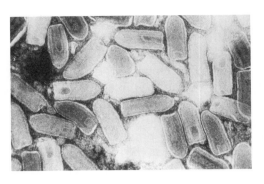

Figure 7-42 Electron micrograph of the *Vesiculovirus* that causes vesicular stomatitis. (Courtesy of CDC)

stranded, negative sense RNA virus (Figure 7-42). Its envelope has 10 nm spikes and is comprised of two virus-specific proteins (an internally situated nucleocapsid membrane protein (M) and an externally located, type-specific glycoprotein (G)). There are more than twenty serotypes in this viral genus, but only four are important as human pathogens (2 strains of VSV, Chandipura virus, and Piry virus). The two strains of VSV are the Indiana (VSV-I, which has subtypes 1 (Alagoas), 2 (cocal or Argentina), and 3 the New Jersey strains (VSV-NJ). The Indiana-1 and New Jersey strains of VSV are endemic in the Americas (Central and South America and Mexico) and have been responsible for the U.S. outbreaks. Three strains are found in South America: Indiana-2, Indiana-3, and Piry.

Epizootiology and Public Health Significance

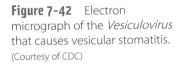

The most important serotypes of VSV are the Indiana and New Jersey variants.

VS Indiana virus was the first arbovirus identified in the United States.

VS is found in the warmer regions of North, Central, and South America. Outbreaks have occasionally occurred in the more temperate regions of the Western hemisphere. In Central and South America, VS can occur anytime but is especially common at the end of the rainy season. In the southwestern United States, VS is more common during the warmer months. The Indiana and New Jersey serotypes occur in North and South America and are responsible for most outbreaks. VS occurs occasionally in the southern United States with small outbreaks in horses occurring in 1998 and 2004. Other subtypes of VSV have been identified in Brazil (Alagoas or subtype 1) and South America (cocal, Argentina, or subtype 2). Outbreaks have been reported in France, but those cases were seen in horses shipped from the United States to Europe.

VS in people is rare and typically a self-limiting disease. In the United States, VS can affect the economy through reduced milk production, increased animal culling, increased morbidity, and veterinary and labor costs. Morbidity in animals is approximately 90%; however, mortality is rare.

Transmission

The transmission of VS remains unclear, but is believed to be transmitted by insect vectors because of its typical occurrence during the warm months along rivers and in valleys. The sand fly and blackfly are the most important vectors of VSV. VSV-New Jersey has been isolated from several possible vectors including biting insects (*Culicoides* [biting midges], *Simuliidae* [black flies] (Figure 6-75), *Aedes* [mosquitoes], *Lutzomyia* [sand flies]) and nonbiting insects (*Chloropidae* [eye gnats], *Anthomyiidae*, *Musca* [house flies]). Vectors for VSV-Indiana include *Aedes* mosquitoes (Figure 6-76), *Simuliidae* (black flies), and *Lutzomyia* (sand flies). Transovarial transmission occurs in these arthropods.

Once in a herd or group of animals, VSV is transmitted between animals by direct contact (oral wounds and abrasions), fomites contaminated with saliva (milking machines, trailers, feed, bedding, cleaning equipment, etc.), indirect transmission from the hands of contaminated workers, or from the fluid of ruptured vesicles. Humans become infected by contact with vesicular fluid or saliva from infected animals. In laboratories aerosol transmission is possible. Arthropod transmission is also possible in people.

Pathogenesis

VS has an incubation period of 2 to 8 days in animals with clinical signs typically appearing in 3 to 5 days. Once the virus is inside a host cell VSV will shut down a host cell's system for protein synthesis. VSV then takes over the cell and makes its own proteins so it can replicate and spread. Histologic changes include intercellular edema of the middle layers of the epidermis, necrosis of epithelial cells in the middle layers, and inflammatory cellular infiltration of monocytes into necrotic epithelial areas. Edema in the skin can escape through vertical cracks in some layers of the epidermis. Large vesicles lose their overlying epithelial layers giving the lesions their eroded appearance. Healing of these lesions occur by regrowth of the epithelium from the basement membrane.

Clinical Signs in Animals

VS can occur in a variety of animals including horses, cattle, pigs, donkeys, mules, South American camelids, and some experimentally infected animals such as deer, raccoons, bobcats, and monkeys. Sheep and goats appear to be resistant to VS. The first clinical sign in most animals is excessive salivation, followed by development of vesicles. Just prior to or at the time vesicles appear, the affected animal may develop a fever. The characteristic vesicular lesions are raised blisters that may be seen on the lips, nares, hooves, teats, or in the mouth. Vesicles may be small or large. In horses, vesicles are typically on the dorsal surface of the tongue, the gums, lips, nares, and corners of the mouth. In cattle, vesicles are typically found on the hard palate, lips, gums, nares, and muzzle (Figure 7-43). In horses and cattle, hooves may have secondary lesions. In pigs, vesicles typically appear first on the feet and the pig may be lame. The muzzle is also commonly affected in pigs.

The vesicles associated with VS will eventually swell, break open, become painful, and produce erosions. At this point, animals may become anorexic, refuse to drink as a result of vesicular pain, and develop lameness. Weight loss and a drop in milk production may be seen in dairy animals. Animal recover in approximately 2 weeks unless secondary bacterial infections develop. Heart and rumen lesions may be seen on necropsy.

Clinical Signs in Humans

The incidence of VS in people is low and is most frequently contracted in the laboratory rather than from animal sources. Most people with VS are

> Vesicular lesions with VS are typically found in only one area of the body.

> VS is not as contagious as foot-and-mouth disease.

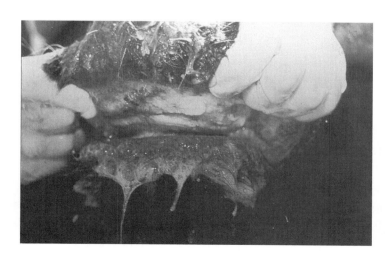

Figure 7-43 There is extensive ulceration of the dental pad and severe salivation in this bovine with vesicular stomatitis.
(Courtesy of Iowa State University, College of Veterinary Medicine, Ames, IA)

asymptomatic or may have influenza-like symptoms. The incubation period is 1 to 6 days. If people develop clinical signs they may develop a high fever, headaches, muscle and joint pain, ocular pain, and nausea. The formation of vesicles in people is rare and if found occur on the oral mucosa, lips, and nose. The disease is self-limiting in people. Recovery in humans may be long.

Diagnosis in Animals

VS cannot be distinguished visually from other vesicular diseases such as foot-and-mouth disease and swine vesicular disease; therefore, laboratory diagnosis is needed. Federal and state veterinarians need to be contacted of any suspected vesicular disease. Proper authorities should be contacted prior to sample collection and submission. Samples must be sent under secure conditions and sent to authorized laboratories. Acceptable clinical specimens should be collected early in the disease and include vesicular fluid, saliva, and affected mucous membranes. Rapid diagnosis of VS is needed to exclude the diagnosis of foot-and-mouth disease. Virus detection from vesicle material is diagnostic through propagation of the virus in cell culture. A presumptive diagnosis can be achieved quickly by the electron microscopic demonstration of rhabdovirus in distilled H_2O lysates of lesion material. Viral antigen is detected by serologic tests such as ELISA, complement fixation, and virus neutralization; antibody levels are detected by paired serum serologic tests using ELISA, complement fixation, and virus neutralization. Virus isolation has been used (tissue culture, embryonated chicken eggs, or unweaned mice) but has been unsuccessful.

> Precautions such as wearing gloves should be taken when examining animals for evidence of VS.

Diagnosis in Humans

Human disease typically occurs during animal outbreaks. Human VS infections can be diagnosed with virus isolation from throat swabs or from blood using RT-PCR techniques or serologic tests.

Treatment in Animals

> Paired serum test taken 1 to 2 weeks apart are used in the United States to determine if an outbreak is occurring nationwide; after an outbreak is verified, a single positive complement fixation test is adequate for VS diagnosis.

There is no anti-viral treatment for VS. Therapy is aimed at elevating pain and making sure the animal eats during the period in which vesicles are present. Oral ulcers can be swabbed with a 1% to 2% solution of Lugol's iodine and foot ulcers can be treated with copper sulfate to help prevent secondary infections. Feeding soft feeds may reduce mouth discomfort, administering anti-inflammatory drugs to minimize swelling and pain, and treating secondary bacterial infection of ulcerated areas are all measures that can help animals feel more comfortable during VS outbreaks.

Treatment in Humans

There is no anti-viral treatment for VS in humans. Therapy is symptomatic to relieve pain and to prevent secondary infections.

Management and Control in Animals

VS can spread rapidly within a herd. The following precautions can be taken to limit the spread of VSV.

- Clean feed bunks and water sources daily.
- Use foot protection, different boots, or disinfectant footbaths when moving between clean and infected areas. Phenolic disinfectants work best.
- Use individual rather than communal feeders and equipment.

- Practice arthropod control.
- Limit grazing during peak insect feeding times.
- Avoid the use of rough, coarse feed during outbreaks to prevent pain. Any leftover feed should be properly disposed of daily (burning).
- Know whether new animals added to a herd had clinical signs of VS within the past 3 months.
- Isolate newly purchased animals for at least 21 days. Animals can become reinfected with VSV after only a few weeks.
- Spray carcasses around the mouth, teats, and feet with disinfectant and treat them with insecticide if they cannot be disposed of immediately.
- Avoid livestock contact with other animals, such as dogs, cats, rodents, birds, and insects that can serve as vectors of VSV.
- Keep service personnel, farriers, and other visitors entering the premises to a minimum. Showers and clothes changes are recommended for these people. If possible, prevent feed, delivery, supplies, and other trucks from directly entering the unit.
- Limit use of farm vehicles transporting cattle to slaughter or driving to places where other cattle-hauling trucks and producers congregate.
- Vaccinate cattle for vesicular stomatitis (available intermittently during outbreak years).

During an outbreak, state or federal regulations restrict movement of animals and quarantines are placed on facilities with positive animals. Animals with clinical signs are isolated from the rest of the herd. Animals cannot be moved from an infected property for at least 21 days after all lesions have healed unless they are going to slaughter.

Management and Control in Humans

People can avoid infection with VSV by practicing good hygiene including the use of gloves when handling animals and animal samples. Insect control is important for both humans and animals. Both formalin-inactivated and attenuated live virus vaccines are available for human use, but are rarely administered.

Summary

Vesicular stomatitis (VS) is a highly contagious, sporadic, viral disease of livestock characterized by vesicular lesion on the tongue, mouth, teats, or coronary bands. Vesicular stomatitis virus (VSV) is a member of the Rhabdoviridae family in the genus *Vesiculovirus* and is a large, bullet-shaped, helical, single-stranded, negative sense RNA virus. There are more than twenty serotypes in this viral genus, but only two strains of VSV are important in North America: the Indiana (VSV-I, which has subtypes 1 (Alagoas), 2 (cocal or Argentina), and 3) and the New Jersey strains (VSV-NJ). VS is found in the warmer regions of North, Central, and South America. In the southwestern United States, VS is more common during the warmer months. The transmission of VSV remains unclear, but is believed to be transmitted by insect vectors because of its typical occurrence during the warm months along rivers and in valleys. Once in a herd or group of animals, VSV is transmitted between animals by direct contact, fomites contaminated with saliva, indirect transmission from the hands of contaminated workers, or from the fluid of ruptured vesicles. Humans become infected by contact with vesicular fluid or saliva from infected animals. In laboratories aerosol transmission is possible. Arthropod transmission is also possible in people. VS can occur in a variety of animals with the first clinical sign in most animals is excessive salivation, followed by development of vesicles. The characteristic vesicular lesions are raised blisters that may be seen on the lips, nares, hooves, teats, or in the mouth.

The incidence of VS in people is low and is most frequently contracted in the laboratory than from animal sources. Most people with VS are asymptomatic or may have influenza-like symptoms. VS cannot be distinguished visually from other vesicular diseases such as foot-and-mouth disease and swine vesicular disease; therefore, laboratory diagnosis is needed to confirm cases of VS. Virus detection from vesicle material is diagnostic through propagation of the virus in cell culture. Viral antigen is detected by serologic tests such as ELISA, complement fixation, and virus neutralization; antibody levels are detected by paired serum serologic tests using ELISA, complement fixation, and virus neutralization. Human VS infections can be diagnosed with virus isolation from throat swabs or from blood using RT-PCR techniques or serologic tests. There is no anti-viral treatment for VS. The spread of VSV can be limited by proper disinfection, proper animal management including isolation of animals and strict rules when adding animals to a herd, arthropod control, avoiding the use of rough, coarse feed during outbreaks to prevent pain, avoiding livestock contact with other animals, and keeping people and vehicles entering the premises to a minimum. People can avoid infection with VSV by practicing good hygiene including the use of gloves when handling animals and animal samples. Both formalin-inactivated and attenuated live virus vaccines are available for human use, but are rarely administered.

TOGAVIRUSES

Overview

Togaviruses, derived from the Greek *toga* meaning gown or cloak, are named in reference to the loose cloak appearance of the envelope surrounding the virus (the envelope has been shown to be quite tightly bound with the use of electron microscopy). Togaviruses used to be classified into four genera; however, in 1984 it was divided into three families (Flaviviridae, Pestiviridae, and Togaviridae). The Togaviridae family currently consists of two genera: *Alphavirus* (containing 27 genera of which 11 are recognized to be pathogenic for man) and *Rubivirus* (containing one genus *Rubella* the causative agent of Rubella or German Measles that is transmitted by direct human contact, inhalation of viral infected aerosol, or congenitally). Alphaviruses will be the only Togaviruses covered in this section. All alphaviruses of clinical significance infect invertebrate hosts (mosquitoes) and vertebrate hosts (many bird and mammal species) and are geographically distributed mainly in the new world. Viruses normally replicate in animal reservoirs and only occasionally spread to humans by insect vectors. Vector transmission from person to person has not been documented as a result of the low viral concentration in the human host making transmission ineffective. The alphavirus' natural cycle usually involves birds and mammals and rarely humans.

All clinically relevant alphaviruses are transmitted by mosquitoes with bird migration playing a critical role in the import of virus into northern hemisphere countries. Alphaviruses predominate in the southern hemisphere; however, different variants of Eastern and Western equine encephalitis are found in South and North America suggesting that the virus may be able to survive during winter in cooler climates. The survival of alphavirus in certain regions depends on the ability of the vector and the vertebrate to develop viremic infections with low pathogenicity.

Amplification hosts include birds, rodents, and monkeys. The diseases caused by alphaviruses include those caused by the agents of American encephalitides (EEE, WEE, and VEE) that produce a noncharacteristic febrile disease and those caused by a group that produce mild to severe arthropathies and rashes (Chikungunya virus, Sindbis virus, and Barmah Forest virus). Subclinical infections are possible with all alphavirus infections.

Alphavirus is the zoonotic genus in this group with the major diseases being Eastern, Western, and Venezuelan equine encephalitis. All of these encephalitides occur following the bite of an infected mosquito resulting in an asymptomatic viremia unless the virus invades neural tissue resulting in encephalitis. Venezuelan equine encephalitis (VEE), in contrast to Eastern equine (EEE) and Western equine encephalitis (WEE), has more systemic manifestations with less neural involvement. WEE was first isolated in the United States in 1930 and is typically seen in the western plain states. WEE is spread by the *Culex tarsalis*, a mosquito that breeds in ditches. This virus can multiply at cool temperatures and may be seen in northern areas. EEE was first isolated in the United States in 1933 and is transmitted by *Culiseta melanura*, a mosquito that lives in fresh water swamps and rarely bites animals. EEE is rare, but fatal in up to 90% of cases. VEE virus was isolated in 1938 and is found in rodents in the forests and marshes of the more tropical parts of America. It is transmitted by *Cu. melanoconion*. VEE does not cause significant disease in humans and the last epidemic of VEE in horses was in 1971 in southern Texas.

Causative Agent

There are 27 species of *Alphavirus* that are spread mainly by mosquitoes and ticks. Alphaviruses are small, enveloped, polyhedral, single-stranded, positive sense RNA viruses that multiply in the cytoplasm of host cells. The RNA is enclosed in an icosahedral capsid and a surrounding envelope containing two viral glycoproteins: E1 (the protein associated with hemagglutination) and E2 (the receptor binding protein). Members of this genus contain a nucleoprotein capsid surrounded by a lipid bilayer envelope that is derived from the host cell membrane. Alphaviruses enter host cells by pinocytosis and they replicate in the cytoplasm. Translation of RNA produces a large protein that is cleaved to yield an RNA polymerase and the structural proteins. Once assembled, the virions bud from the host cell from which they acquire their envelope. Within their hosts they replicate in a wide variety of cells including neurons and glial cells, skeletal and smooth muscle cells, cells of lymphoid origin, and synovial cells.

> Alphaviruses of the family Togaviridae are viruses that utilize insect vectors.

Equine Encephalitides

Overview

Encephalitis is an acute inflammatory process primarily involving the brain and meninges. Bacteria, fungi, and autoimmune disorders can cause encephalitis, but most cases are viral in origin. The equine encephalitides are a group of diseases cause by four viruses in the *Alphavirus* genus (Eastern Equine Encephalitis (EEE), Western Equine Encephalitis (WEE), Venezulan Equine Encephalitis (VEE), and Everglades). Clinical disease caused by these four viruses varies from an asymptomatic infection to clinical illness similar to mild influenza to advanced neurologic disease. The probability of contracting these viruses and developing encephalitis vary widely among the different viruses with the incidence highest for the EEE virus and lowest for the VEE virus. The mortality rates for each of the encephalitides

> Alphaviruses only infect humans incidently.

vary from 5% for WEE, 35% for VEE, to 50% for EEE. Everglades encephalitis virus causes very mild, influenza-like illness with no known fatalities.

WEE virus was first isolated from the brains of dead horses in the San Joaquin Valley of California in the 1930s. Karl Meyer and his associates investigated this outbreak and were able to isolate a virus from the brain of an infected horse. This initial outbreak affected over 6,000 horses with mortality exceeding 50%. In 1938, WEE virus was isolated from a child's brain that had died from encephalitis making a link between the disease in animals and people. In 1941, the virus was isolated from *Cu. tarsalis* mosquitoes in the Yakima Valley of Washington and in the San Joaquin Valley of California. A large epidemic/epizootic of WEE spread over the western portion of the United States and Canadian prairies in that same year, infecting an estimated 300,000 horses and mules (50,000 developed the disease with 15,000 of them dying) and causing infection in over 3,000 people. WEE is still a threat to horses, but cases have declined as a result of proper vaccination.

EEE was the cause of epizootics in North American horses since 1831. EEE was first recognized by Dr. Alfred Large at the New York City Veterinary College who described a cerebrospinal meningitis occurring in horses on Long Island over a 20-year period. EEE was commonly referred to as staggers or putrid fever prior to a series of outbreaks that occurred on the eastern coastal area in 1933 (Delaware, Maryland, New Jersey and Virginia). EEE virus was first isolated from a dead horse's brain during that epizootic. In 1934 and 1935 outbreaks of EEE occurred in Virginia and North Carolina. The disease was discovered to occur annually in horses along the eastern seaboard of the United States (occasionally making its way as far inland as the Midwest and as far north as southeastern Canada). The virus continues to affect horses in North America, but the incidence of disease has decreased significantly as a result of effective vaccination protocols.

In 1936, VEE was first recognized as a disease in horses in Venezuela following a major outbreak. From 1936 to 1968, outbreaks of VEE continued to occur in equines in several South American countries. In 1969, the disease spread north throughout Central America and spread to Mexico and Texas in 1971. Since then major outbreaks of VEE occurred in 1993 (southern Mexico) and in 1995 (Colombia). Prolonged and heavy rainy seasons that cause increases in mosquito populations are largely responsible for outbreaks of VEE. The VEE virus strain that is pathogenic for humans is amplified in horses with equine disease occurring prior to human disease. This is in contrast to the EEE and WEE viruses, where horses appear to be dead-end hosts.

The Everglades encephalitis virus (also known as VEE sylvatic subtype II virus) is an alphavirus in the VEE serocomplex and its geographic location is restricted to the state of Florida. Everglades encephalitis virus was first recognized in South Florida in the 1960s, when Seminole Indians living north of Everglades National Park were shown to have as high as a 58% seropositive rate. Everglades encephalitis virus circulates among rodents and mosquitoes and infects humans, causing a febrile disease occasionally accompanied by mild neurologic signs. In most people clinical signs include very mild, influenza-like illness that begins with a gradual onset of respiratory problems and gradual recovery. No new cases of Everglades encephalitis virus causing human disease have been reported since 1971. In horses it is considered nonpathogenic. Many people consider Everglades encephalitis virus to be an enzootic subtype of VEE.

EEE and WEE are maintained in nature between mosquitoes and birds; VEE is maintained in nature between mosquitoes and rodents.

Causative Agent

All equine encephalitis viruses (EEE, WEE, VEE, and Everglades) are members of the genus *Alphavirus* and the family Togaviridae. Alphaviruses are single-stranded, positive sense RNA viruses in which replication takes place in the cytoplasm of the infected host cell (Figure 7-44). EEE virus typically cycles among passerine birds and the bird-feeding mosquito *Culiseta melanura*. A less harmful variant of EEE virus exists in Central and parts of South America. EEE causes disease in humans, horses, and some species of birds. WEE virus typically cycles among birds and the mosquito *Culex tarsalis* (the same mosquito that transmits SLE virus). Five subtypes (WEE, Buggy Creek, Fort Morgan, Highland J (all in North America), and Aura in South America) are found in the WEE complex. WEE causes disease in humans, horses, and some species of birds. VEE virus typically cycles among rodents and the mosquito *Culex melanoconion* (birds

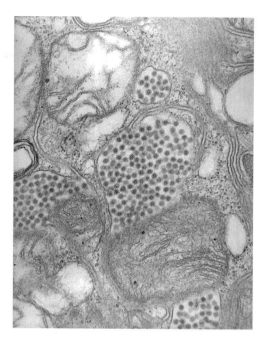

Figure 7-44 Electron micrograph of the Eastern equine encephalitis virus. (Courtesy of CDC/Fred Murphy, Sylvia Whitfield)

may also be involved in some cycles). There are at least 8 subtypes that are divided into epizootic and enzootic groups. Epizootic subtypes are responsible for most epidemics and are highly pathogenic for horses (can cause mild illness in people). Enzootic (sylvatic) subtypes (Everglades encephalitis virus) are found in limited geographic areas occurring naturally between rodents and mosquitoes. Enzootic types can cause human disease and are usually nonpathogenic in horses.

Epizootiology and Public Health Significance

All of the equine encephalitis viruses are found in North, Central, and South America. Geographic differences between the viruses include:

- EEE virus is found in eastern Canada, all states east of the Mississippi River, Arkansas, Minnesota, South Dakota, Texas, the Caribbean, and regions of Central and South America (particularly along the Gulf coast). EEE occurs in the United States in approximately 12 to 17 people annually with an infection rate of about 33% and morbidity rate of 90%. Most cases of EEE are seen in people older than 55 years and children younger than 15 years. Case fatality rates vary from 33% to 70% with permanent neurologic deficits occurring in survivors.

- WEE virus is found in western Canada, Mexico, parts of South America, and west of the Mississippi River. WEE occurs in the United States at a variable rate. In 1941 there was an epidemic in which 3,000 cases occurred. Most infections in adults are asymptomatic or mild; therefore, accurate case numbers are difficult to attain. Overall mortality rates are 3% to 4%.

- VEE virus is endemic in South and Central America and Trinidad. VEE in humans typically follows an epizootic in horses. Ninety percent to 100% of exposed people will contract the disease with almost 100% of infections being symptomatic. Less than 1% develops encephalitis and approximately 10% of these cases are fatal. Very young or very old people tend to develop severe infections.

- Everglades encephalitis virus (enzootic subtypes of VEE) is found in Florida. A few isolated cases have been seen in the Rocky Mountains and northern plains of the United States.

Transmission

The equine encephalitis viruses are transmitted by mosquitoes. The differences between modes of transmission revolve around the types of mosquito and the animals in which the virus cycles. EEE and WEE viruses cycle between birds and mosquitoes with humans and horses being incidental, dead-end hosts. EEE virus can be isolated from 27 species of mosquito in the United States with the most important vector being *Culiseta melanura* (a mosquito that feeds primarily on birds). WEE cycles between passerine birds and culicine mosquitoes with the most important vector being *Cu. tarsalis* (Figure 7-16). In birds, EEE and WEE can be spread by feather picking and cannibalism. VEE virus cycles between rodents and mosquitoes with the most important vectors coming from the *Culex* spp. (*Culex melanoconion*). Humans and horses are incidental hosts of VEE virus. Enzootic subtypes (Everglades) can also have birds involved in their cycle. Epizootic subtypes can be found in other mammals such as cattle, pigs, and dogs, but they do not become ill nor spread the virus. Horses are the main amplifiers of epizootic subtypes during epidemics. In some cases, humans have developed VEE from being exposed to laboratory rodent debris.

Pathogenesis

After being bitten by an infected mosquito vector, muscle tissue is the primary site of initial viral replication. From the muscle, the virus travels to the regional lymph nodes and multiplies in neutrophils and macrophages. If the immune system cannot rid the body of the virus, the viral particles can be found in several tissues including the liver and spleen, where they continue to replicate. As viral quantities increase, a fever typically develops. Approximately 3 to 5 days postinfection the virus reaches the brain causing a wide variety of lesions and pathologic processes such as cerebral edema, petechial hemorrhage, focal brain necrosis, and edema of the meninges. In contrast to EEE and WEE, VEE virus causes a systemic infection with viremia rather than a localized brain infection. Viremia as a result of VEE virus typically leads to respiratory and gastrointestinal signs.

Clinical Signs in Animals

Eastern Equine Encephalitis

EEE virus can infect horses, pigs, birds, bats, reptiles, amphibians, and rodents. The incubation period for EEE is about 1 to 3 days with signs of clinical illness lasting from 2 to 9 days. EEE in horses typically presents with fever and anorexia (prodromal (initial) stage). In severe cases, this initial stage is followed by neurologic signs such as impaired vision, aimless wandering, head pressing, circling, ataxia, paralysis, and seizures. Other signs that can be seen include periods of excitement, pruritus, and lateral recumbency with paddling. Death can occur with EEE viral infections (average mortality rate of about 90% in clinically affected horses) (Figure 7-45). Most EEE virus infections in birds are asymptomatic; however, disease can be seen in partridges, pheasants, psittacine birds, ratites, and whooping cranes. Disease in these birds can be fatal.

Western Equine Encephalitis

WEE virus can infect birds, horses, and a variety of mammals. The incubation period for WEE is 2 to 9 days and the clinical signs of disease will last between 2 and 9 days. WEE in horses is clinically similar to EEE. Both EEE and WEE can be asymptomatic, cause mild disease, or cause severe neurologic signs (Figure 7-46).

Figure 7-45 Death in a horse as a result of Eastern equine encephalitis (EEE).
(Courtesy of CDC)

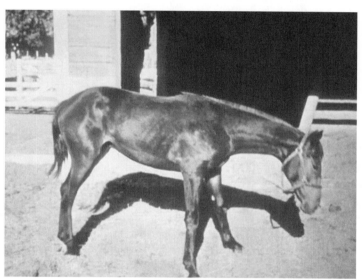

Figure 7-46 This horse is displaying the typical stance having been afflicted with Western equine encephalitis (WEE) virus.
(Courtesy of CDC/James Stewart)

WEE typically does not progress beyond the general, milder signs. Death as a result of WEE virus ranges from 5% to 20%. Most WEE infections in birds are asymptomatic.

Venezuelan Equine Encephalitis

VEE virus can infect rodents, horses, mules, burros, and donkeys. Birds, cattle, pigs, and dogs can be infected with the virus without producing clinical signs. The incubation period for VEE is 1 to 5 days and the clinical signs of disease will last between 2 and 9 days. In horses, the epizootic subtypes can cause either asymptomatic signs; a febrile, prodromal disease followed by neurologic disease (and occasionally diarrhea and colic with death occurring within hours after the onset of neurologic signs); or a generalized acute febrile disease without neurologic signs (Figure 7-47).

Everglades Equine Encephalitis

Enzootic VEE (Everglades) virus can infect rodents, equine, and a variety of laboratory animals. The enzootic subtypes of VEE are usually asymptomatic in horses.

Figure 7-47 Horse with neurologic signs as a result of Venezuelan equine encephalitis (VEE).
(Courtesy of CDC/Mr. J. Bagby)

The equine encephalidites cause illness only in horses and humans; although these viruses can infect a variety of other animals, they are often asymptomatic.

Clinical Signs in Humans

EEE and WEE present similarly in people with an incubation period of 4 to 15 days. EEE usually starts acutely with clinical signs of fever, muscle pain, headache, and occasionally nausea and vomiting. These initial signs may be followed by neurologic signs such as confusion, neck stiffness, depression, coma, seizures, and paralysis. Mortality rates for EEE in people are high. WEE can present like EEE, but is usually asymptomatic in adults (nonspecific signs of illness that present acutely such as fever, headache, vomiting, and lethargy with few fatalities). WEE can be severe in children resembling the more severe signs of EEE.

VEE in people typically presents with an acute, mild illness with clinical signs such as fever, lethargy, severe headache, and muscle pain (especially in the legs and lumbosacral region). These initial symptoms can last 24 to 72 hours and may be followed by a cough, nausea, vomiting, and diarrhea. The disease course is typically 1 to 2 weeks. In pregnant women, VEE virus can cause fetal damage including encephalitis, abortion, or congenital neurologic abnormalities. Encephalitis from VEE virus is about 4% in children and approximately 0.4% in adults. Encephalitis typically occurs with the second peak of the biphasic fever. Enzootic VEE (Everglades) does not typically cause clinical disease in people.

Diagnosis in Animals

In horses, EEE and WEE are diagnosed by serologic methods such as hemagglutination inhibition, ELISA, and complement fixation. Cross-reaction between EEE and WEE antibodies may occur with some tests. Clinical infections in birds are typically diagnosed by virus isolation. Virus isolation in horses (in brain, hepatic, or splenic tissue) is used for cases of EEE, but is not accurate for cases of WEE. EEE and WEE viruses can be isolated from animal culture such as mice or chicks.

VEE is diagnosed in horses via virus isolation or serologic methods. The virus can be isolated from blood collected during the febrile state and occasionally from the brains of animals with encephalitis. Serologic methods used for VEE virus identification include complement fixation, hemagglutination inhibition, ELISA, and immunofluorescent assays. Everglades encephalitis virus is identified by the same methods as VEE virus.

Postmortem gross lesions in horses are nonspecific; EEE cases may demonstrate brain and meningeal congestion; VEE cases may show no nervous system lesions while some show necrosis with hemorrhages. Microscopic brain tissue examination of all the equine encephalitides viruses can demonstrate inflammation of the gray matter and perivascular cuffing (Figure 7-48).

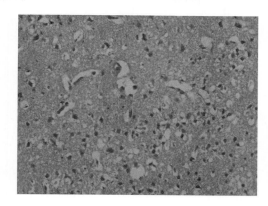

Figure 7-48 Brain tissue affect by Venezuelan encephalitis virus reveals neural necrosis and edema.
(Courtesy of CDC/Dr. F. A. Murphy)

Diagnosis in Humans

EEE, WEE, and VEE viruses can be diagnosed in people using virus isolation and serologic methods such as ELISA, complement fixation, immunofluorescent antibodies, and hemagglutination inhibition. PCR testing is also available.

Treatment in Animals

Treatment of EEE, WEE, and VEE is supportive in animals. Fever reduction is important with EEE and WEE.

Treatment in Humans

Treatment of EEE, WEE, and VEE is supportive. Physical therapy is important during the recovery phase of EEE. Fever reduction is important with EEE and WEE. VEE and Everglades is usually mild and does not require treatment.

Management and Control in Animals

Equine vaccines are available for EEE, WEE, and VEE (have been available for about 30 years). EEE vaccines are also available for birds. Mosquito control is also important to control the spread of disease.

Management and Control in Humans

There is no human vaccine available against these viruses. Prevention of these diseases in people depends on the surveillance of these viruses in mosquitoes, birds, rodents, and horses. Mosquito control is indicated in areas with a high mosquito population including the use of insect repellants and environmental spraying.

Summary

Encephalitis is an acute inflammatory process primarily involving the brain and meninges. The equine encephalitides are a group of diseases cause by four viruses in the *Alphavirus* genus, Togaviridae family (Eastern Equine Encephalitis (EEE), Western Equine Encephalitis (WEE), Venezulan Equine Encephalitis (VEE), and Everglades). (For less common zoonotic Alphaviruses, see Table 7-19.) Clinical disease caused by these four viruses varies from an asymptomatic infection to clinical illness similar to mild influenza to advanced neurologic disease. All of the equine encephalitis viruses are found in North, Central, and South America. The

Table 7–19 Less Common Zoonotic Alphaviruses

Alphavirus	Disease	Predominant Signs in People	Clinical Signs in Animals	Transmission	Animal Source	Geographic Distribution	Diagnosis	Control
Semliki Forest virus	Semliki forest fever First isolated from mosquitoes in the Semliki Forest, Uganda in 1944	Fever, headaches, joint and muscle pain. Occasional diarrhea, abdominal, and conjunctivitis.	■ Horses develop encephalitis	Infection is acquired via a mosquito bite; aerosol contamination possible	Horses, birds, wild rodents, and a variety of domestic animals	■ African countries and parts of Asia	Cell culture, PCR, and ELISA tests	Mosquito control; vaccines have not been developed.
Sindbis virus	Sinbis fever First isolated in 1955 from Culex mosquitoes in Sinbis, Eygpt	Disease begins with low-grade fever, headaches, and joint pain of hands and feet. A rash develops on the body. Acute disease last for about 10 days. High percentage of seropositive people in the Nile Valley and other parts of Africa.	■ Wild birds are asymptomatic	Infection is acquired via a variety of mosquito species (Anopheles, Mansonia, Aedes, and Culex).	Wild birds Latent infection seen in cloven-hoofed animals	■ Nile Valley and parts of Africa, Asia, Europe, and Australia	RT-PCR and ELISA tests available Rarely by virus isolation	Mosquito control; vaccines have not been developed.
O'Nyong-Nyong virus	O'Nyong-Nyong fever First time disease was a major epidemic in Uganda in 1959 O'Nyong-Nyong means very painful and weak in Acholi language	Acute onset of fever, chills, and epistaxis. Occasionally have back and joint pain, headache, and ocular pain. A pruritic rash on face and body occurs later in the disease. Disease usually resolved in 5 days. Morbidity during epidemics can be as high as 70%.	■ Proof of animal disease is unsubstantiated	Infection is acquired via a variety of mosquito species (Anopheles gambiae and Anopheles funestus).	Natural reservoir unknown	■ East Africa	Cell culture from blood during febrile stage, RT-PCR, neutralization test based on 80% plaque reduction	Mosquito control; vaccines have not been developed.
Mayaro virus	Mayaro fever Disease information based on three epidemics with fewer than 100 cases per epidemic; each outbreak was associated with close human contact with forests.	Fever, headaches, abdominal pain, joint pain, chills, and a rash.	■ New World monkeys are asymptomatic	Infection is acquired via a variety of mosquito species (Haemagogus spp., Culex spp., etc).	South American monkey species, marmosets are amplification hosts in tropical rain forests, wild birds.	■ Brazil, Trinidad, Bolivia, Suriname, Guyana, Columbia, Peru, Panama	Cell culture, RT-PCR, ELISA	Mosquito control; vaccines have not been developed.

Virus	Description	Signs in humans	Signs in animals	Transmission	Reservoir	Distribution	Diagnosis	Prevention/Control
Ross River virus	Epidemic polyarthritis, Ross River fever Named after the river in northern Queensland where it was first identified	Initially mild fever and polyarthritis (especially small joints of hands and feet), generalized rash, and enlarged lymph nodes. Encephalitis in rare cases. Cases resolve in 4 to 7 months.	▪ Animals and birds are asymptomatic	Infection is acquired via a variety of mosquito species (*Aedes vigilax* and *Culex annulirostris*)	Wild and domestic animals (cattle, pigs, sheep, horses, kangaroos, wallabies rodents, dogs); wild birds	Australia (widespread) and Oceania	Cell culture, RT-PCR, ELISA	Mosquito control; vaccines have not been developed.
Barmah Forest virus	Barmah Forest fever First isolated in 1974 in Barmah Forest area of Murray River in northern Victoria, Australia	Similar to Ross River fever except milder	▪ Animals and birds are asymptomatic	Infection is acquired via a variety of mosquito species	Wild and domestic animals (cattle, pigs, sheep, horses, kangaroos, wallabies rodents, dogs); wild birds	Western Australia (widespread)	Cell culture, RT-PCR, ELISA	Mosquito control; vaccines have not been developed.
Chikungunya virus	Chikungunya fever Chikungunya is Swahili for what bends (joints)	Sudden fever, joint pain, and rash. Fever lasts 3 to 10 days and is typically biphasic	▪ Monkeys develop Dengue-like signs (fever, headaches, joint and muscle pain, rash) ▪ Birds are asymptomatic	Infection is acquired via a variety of mosquito species (*Aedes aegypti* is main vector in Asia; *Aedes aegypti* and *Culex* spp. and other *Aedes* spp. are main vectors in Africa)	Non-human primates	Africa (all countries south of the Sahara), Southern and Southeastern Asia Found in areas near the jungle	Cell culture, RT-PCR	Mosquito control; vaccines have not been developed.

equine encephalitis viruses are transmitted by mosquitoes. EEE and WEE viruses cycle between birds and mosquitoes with humans and horses being incidental, dead-end hosts. VEE virus cycles between rodents and mosquitoes with humans and horses being incidental hosts of VEE virus. In animals, EEE virus can infect horses, pigs, birds, bats, reptiles, amphibians, and rodents. EEE in horses typically present with fever and anorexia that may progress to neurologic signs such as impaired vision, aimless wandering, head pressing, circling, ataxia, paralysis, and seizures. Death can occur in about 90% of EEE viral infections in horses. WEE virus can infect birds, horses, and a variety of mammals. WEE in horses is clinically similar to EEE with death being rarer in cases of WEE. VEE virus can infect rodents, horses, mules, burros, and donkeys. In horses, the epizootic subtypes of VEE can cause asymptomatic signs; a fever followed by neurologic disease; or a generalized acute febrile disease. Enzootic VEE (Everglades) virus can infect rodents, equine, and a variety of laboratory animals. The enzootic subtypes of VEE are usually asymptomatic in horses. In people, EEE and WEE have similar clinical signs. EEE usually starts acutely with clinical signs of fever, muscle pain, headache, and occasionally nausea and vomiting that may be followed by neurologic signs such as confusion, neck stiffness, depression, coma, seizures, and paralysis. WEE is usually asymptomatic in adults, but can resemble the more severe signs of EEE in children. VEE in people typically presents with an acute, mild illness with clinical signs such as fever, lethargy, severe headache, and muscle pain (especially in the legs and lumbosacral region). In horses, EEE and WEE are diagnosed by serologic methods such as hemagglutination inhibition, ELISA, and complement fixation. VEE is diagnosed in horses via virus isolation or serologic methods. Microscopic brain tissue examination of all the equine encephalitides viruses can demonstrate inflammation of the gray matter and perivascular cuffing. EEE, WEE, and VEE viruses can be diagnosed in people using virus isolation and serologic methods such as ELISA, complement fixation, immunofluorescent antibodies, and hemagglutination inhibition. PCR testing is also available. Treatment of EEE, WEE, and VEE is supportive in animals and people. Equine vaccines are available for EEE, WEE, and VEE. EEE vaccines are also available for birds. Prevention of these diseases in people depends on the surveillance of these viruses in mosquitoes, birds, rodents, and horses. Mosquito control is indicated in areas with a high mosquito population including the use of insect repellants and environmental spraying. There is no human vaccine for these viruses.

Review Questions

Multiple Choice

1. Viruses are
   a. prokaryotic cells.
   b. eukaryotic cells.
   c. animal cells.
   d. noncellular.

2. In general, DNA viruses mature in the _____, whereas RNA viruses replicate in the _____.
   a. nucleus, nucleoplasm
   b. cytoplasm, ribosomes
   c. nucleus, ribosomes
   d. nucleus, cytoplasm

3. Persistent viral infections that exhibit a long noninfectious stage between the original disease and the subsequent disease are called
   a. acute.
   b. chronic.
   c. virions.
   d. latent.

4. Common ways to identify viruses are
   a. culture on agar and staining methods.
   b. staining methods and biochemical tests.
   c. antigen/antibody testing and gene amplification.
   d. cell culture and staining methods.

5. Emerging viral infections occur because of
   a. a relaxation of public health standards.
   b. changes in weather patterns.
   c. airline travel and its effect on travel to disease endemic areas.
   d. all of the above.

6. Hantavirus pulmonary syndrome
   a. can result in death as a result of kidney failure much as in HFRS.
   b. is an emerging disease resulting from mutations of hantavirus the causative agent of HFRS.
   c. occurs in the desert southwest of the United States especially during years of heavy rainfall.
   d. has flying foxes as its reservoir hosts.

7. Negri bodies are associated with
   a. dengue fever.
   b. hantavirus pulmonary syndrome.
   c. rabies.
   d. influenza.

8. One characteristic of influenza viruses is the tendency to cause pandemics. New epidemics arise because of the organism's ability to undergo major structural changes and genetic reassortment. This phenomenon is called
   a. antigenic shift.
   b. continental drift.
   c. Reye's syndrome.
   d. evolution.

9. Dengue hemorrhagic fever is as a result of
   a. preformed antibodies to the infecting serotypes of dengue virus.
   b. acute respiratory distress/shock syndrome.
   c. development of autoantibodies.
   d. preformed antibodies to a different serotype of dengue virus.

10. What is the most common pediatric arboviral encephalitis in the United States?
    a. West Nile encephalitis
    b. St. Louis encephalitis
    c. La Crosse encephalitis
    d. Everglades encephalitis

11. What played a major role in the emergence of SARS?
    a. SARS is caused by a coronavirus, which are viruses with a high frequency of mutation and a high frequency of recombination so that new strains can form to adapt to changing conditions.

    b. Civet cats are asymptomatic when they first acquire the virus and spread the virus to people prior to the development of clinical signs (respiratory and neurologic signs are common).

    c. People can carry the SARS virus for years and not show clinical signs while they are transmitting the virus to other people.

    d. The disease is contracted most commonly on airplanes (95%) and with the increase in air travel the virus spread quickly through travelers.

12. What mosquito-borne flavivirus causes sows to abort and piglets to die; has spread over the past 50 years across Southeast Asia, India, southern China, and the Pacific reaching Australia in 1998; and is of great concern of spreading worldwide?

    a. Powassan virus

    b. Louping ill virus

    c. Tick-borne encephalitis virus

    d. Japanese encephalitis virus

13. What arbovirus is transmitted from wild birds and domestic fowl to people via mosquitoes?

    a. *Coltivirus*

    b. St. Louis encephalitis virus

    c. Hantavirus

    d. Fowlpox virus

14. What viral encephalitis was introduced into the United States in 1999 and can cause severe neurologic disease in horses?

    a. Eastern equine encephalitis

    b. Western equine encephalitis

    c. Everglades encephalitis

    d. West Nile fever

15. What is the appropriate treatment of monkeys following a nonhuman primate bite?

    a. Wound cleaning and debridement with chlorhexidine solution.

    b. Wound cleaning and flushing with saline.

    c. Delayed wound cleaning that could force viruses like Herpes B deeper into the wound.

    d. Wound cleaning with chlorhexidine solution, flushing the wound with saline, and suturing of the wound with nonabsorbable suture.

16. What is true regarding avian influenza?

    a. Most avian influenza strains are highly pathogenic strains that cause fowl plaque, a severe, often fatal disease in domestic poultry.

    b. Most avian influenza strains are low pathogenic strains that produce subclinical or mild respiratory infections in infected birds.

    c. Most avian influenza strains are the same strain as swine influenza and can mutate causing disease in birds, pigs, and humans.

    d. Most avian influenza strains are severely pathogenic in ducks and geese who develop viremia producing death in 24−48 hours.

17. For what disease are birds quarantined for upon entry into the United States?

    a. influenza

    b. fowl plaque

    c. chickenpox

    d. exotic Newcastle disease

18. What is the reservoir host of Nipah virus?
    a. pigs
    b. cattle
    c. fruit bats
    d. birds

19. Federal authorities need to be contacted whenever a vesicular disease is seen in animal because of the fear of
    a. swine vesicular disease.
    b. foot-and-mouth disease.
    c. equine herpes virus.
    d. feline herpes virus.

20. What type of sample is submitted for rabies virus determination in animals?
    a. serum
    b. plasma
    c. whole blood
    d. brain

21. What form of equine encephalitis is most deadly in horses?
    a. Eastern
    b. Western
    c. Venezuelan
    d. Everglades

22. What form of equine encephalitis is most deadly in people?
    a. Eastern
    b. Western
    c. Venezuelan
    d. Everglades

23. The vector for Colorado tick fever is
    a. *Aedes* mosquitoes.
    b. *Phlebotomus* sandflies.
    c. *Ixodes scapularis* ticks.
    d. *Dermacentor andersonii* ticks.

24. What viral disease causes 80% morbidity in unvaccinated flocks of sheep, malnutrition in lambs because ewes with sore teats will not let them nurse, and typically presents with papules/pustules/vesicles on the lips, nares, ears, and eyelids?
    a. foot-and-mouth disease
    b. vesicular stomatitis
    c. herpes virus
    d. contagious ecthyma

25. What viral diseases are transmitted by mosquitoes and have monkeys as reservoir hosts?
    a. Dengue fever and yellow fever
    b. Rift Valley fever and herpes B virus
    c. Ebola virus and Marburg virus
    d. Lassa fever and hantavirus

26. What virus was isolated in monkeys in Reston, Virginia, in 1989?
    a. Marburg virus
    b. Ebola virus
    c. Hantavirus
    d. Bunyavirus

27. An example of a robovirus is
    a. St. Louis encephalitis virus.
    b. Hantavirus.
    c. Herpes B virus.
    d. Colorado tick fever virus.
28. What virus was seen for the first time in the United States in 2003 in prairie dogs?
    a. Rift Valley fever virus
    b. monkeypox virus
    c. Epstein-Barr virus
    d. robovirus
29. Chickens that have clinical signs of diarrhea, head and wattle edema, neurologic signs, and decreased egg production should be tested for
    a. rabies.
    b. vesicular disease.
    c. coronavirus.
    d. Newcastle disease.
30. The first pandemic of the 21st century was
    a. influenza.
    b. SARS.
    c. hantavirus.
    d. dengue shock syndrome.

Matching

31. _____ Meurto Canyon virus

32. _____ Ebola virus

33. _____ SARS

34. _____ Newcastle disease virus

35. _____ Nipah virus

36. _____ lymphocytic choriomeningitis virus

37. _____ foot-and-mouth disease virus

38. _____ swine vesicular disease

39. _____ dengue fever

40. _____ contagious ecthyma

41. _____ La Crosse encephalitis virus

A. Bunyavirus that typically causes severe encephalitis and neurologic sequelae in children younger than 16 years of age

B. soremouth in sheep

C. Flavivirus that is spreading from Asia to other parts of the world and is found in swine (abortion may occur in infected animals)

D. virus that causes small, depressed scars upon healing

E. Hantavirus more commonly known as the Sin Nombre strain

F. Filoviridae virus that causes hemorrhagic fever in humans, chimpanzees, and gorillas

G. most widespread flavivirus

H. disease caused by a coronavirus that mutated from civet cats

I. reovirus that may cause a biphasic fever

J. Arenavirus whose natural host is the house mouse (*Mastomys musculus*)

K. bullet-shaped virus that produces neurologic signs

42. _____ Colorado tick fever virus

43. _____ herpes B virus

44. _____ St. Loius encephalitis virus

45. _____ poxvirus

46. _____ rabies virus

47. _____ Japanese encephalitis virus

48. _____ vesicular stomatitis

49. _____ influenza virus

50. _____ West Nile virus

L. disease that causes excessive salivation, followed by the development of oral blisters in animals

M. most frequently seen mosquito-borne viral infection worldwide

N. orthomyxovirus that has three serotypes: A (zoonotic), B, and C

O. picornavirus spreads rapidly through cattle herds having great economic impact on a country

P. paramyxovirus that causes unilateral conjunctivitis and influenza-like symptoms in people

Q. paramyxovirus that is asymptomatic in pigs and has a reservoir host of fruit bats

R. disease that presents clinically identical to foot-and-mouth disease except it is only in pigs

S. virus of concern for laboratory workers and zoo workers who work with nonhuman primates

T. Flavivirus that is mosquito-borne, uses birds as amplifying hosts, and is seen in North America

Case Studies

51. Three weeks after returning from a camping trip to Arizona, a 25-year-old female developed a fever and severe muscle aches. She also had a dry cough, headaches, nausea, and vomiting. Four days later the symptoms became progressively worse as fluid began to accumulate in her lungs and she went to the emergency room. She developed dyspnea and her blood pressure dropped. She was intubated and put on a ventilator. Microbiological tests on various bacterial pathogens were negative. The patient's symptoms worsened and she had cardiopulmonary arrest and died. An epidemiologist was called in to investigate this person's cause of death.
    a. Based on this person's travel history and clinical signs, what disease may she have?
    b. What type of sample should be run from this patient?
    c. What would the epidemiologist be looking for at and near the campsite?
    d. What other investigation should be done?

52. A 7-year-old boy had recently started taking horseback riding lessons during his summer vacation. During a particularly rainy period he did not take lessons for 2 weeks, but started taking lessons again the following month. At the end of the summer he started developing neck stiffness, fever, and mental confusion. He was taken to the hospital where a spinal tap was done on the boy (results were negative for bacterial meningitis). He got progressively worse and developed seizures over the next few days. He was put on anticonvulsant medication and antipyretics to reduce his fever. Over the next few days he started to improve with symptomatic care.
    a. What may have been a cause of this boy's neurologic problems?

b. What in the history leads you to this conclusion?

c. What should be asked of the owners of the horseback riding establishment?

d. What samples could be tested on this boy for confirmation of this disease?

e. What should this boy do next time he goes horseback riding?

53. On September 1, a 10-year-old boy from rural Mississippi developed a fever and headache. He was evaluated by a pediatrician three days later and had a temperature of 102.6°F and was noted to have sensations he described as an "itchy" scalp. The pediatrician diagnosed viral illness and the patient was advised to return if symptoms worsened. Two days later the boy's condition worsened and he was taken to the emergency room. All laboratory tests and chest radiography were within normal limits and he was sent home. The boy's clinical signs worsened throughout the day, and he returned to the emergency room that evening with symptoms of fever, insomnia, urinary urgency, paresthesia of the right side of the scalp and right arm, dysphagia, disorientation, and ataxia. He was admitted to the hospital for suspected encephalitis. Shortly after admission, the patient's neurologic status deteriorated rapidly with his speech becoming slurred and he began to hallucinate. He became increasingly agitated and combative and required sedation. In his agitated state, the patient bit a family member. The next morning the patient was transferred to a tertiary care facility. Within hours after transfer, he became lethargic and was intubated. During the next 10 days, the boy continued to worsen and experienced wide fluctuations in blood pressure and temperature. On the eleventh day of his symptoms he had onset of cerebral edema and subsequent brain herniation. His life support was withdrawn, and the patient died the next day.

a. What should the medical staff check for on this patient's body?

b. What samples should be tested in this patient?

c. Assuming the boy may have contracted rabies, what should be investigated?

d. Pending test results, what should happen to the family member who the boy had bitten?

References

Alcamo, I. E. 1998. *Schaum's Outlines: Microbiology*. New York: McGraw-Hill.

Alexander, D. J., J. G. Bell, and R. G. Alders. 2004. Newcastle disease virology and epidemiology. In *A technology review: Newcastle Disease with special emphasis on its effects on village chickens*, edited by Food and Agriculture Organization of the United Nations. http://www.fao.org/docrep/006/y5162e/y5162e02.htm (accessed May 1, 2006).

American College of Veterinary Pathologists. 2006. Foot and mouth disease factsheet http://www.acvp.org/news/factsheet/factfmd.php (accessed May 9, 2006).

AVMA. 2006. Foot and mouth disease backgrounder. http://www.avma.org/public_health/fmd_bgnd.asp (accessed January 25, 2006).

AVMA. 2007. Updated rabies vaccination certificate released, http://www.avma.org/onlnews/javma/may07/070515l.asp (accessed June 10, 2007).

Barraviera, S. 2005. Diseases caused by poxvirus: contagious ecthyma and milker's nodules—a review. *Journal of Venomous Animals and Toxins including Tropical Diseases* 11(2):102–8.

Bauer, K. 1997. Foot and mouth disease as zoonosis. *Archives of Virology Supplement* (13):95–7.

Bauman, R. 2004. *Microbiology*. San Francisco, CA: Pearson Benjamin Cummings.

Biddle, W. 2002. *A Field Guide to Germs*. New York: Anchor Books.

Black, J. 2005. *Microbiology: Principles and Explorations*, 6th ed., Hoboken, NJ: Wiley.

Blackwell, L. 2000. Togaviridae. http://www.stanford.edu/group/virus/toga/2000/a.html (accessed February 11, 2006).

Bridson, E. 2003. The great influenza pandemic of 1918. *Culture* 24(2):1–8.

Buller, R., and G. Palumbo. 1991. Poxvirus pathogenesis. *Microbiological Reviews* 55(1):80–122.

Burton, R., and P. Engelkirk. 2004. *Microbiology for the Health Sciences*, 7th ed. Baltimore, MD: Lippincott, Williams & Wilkins.

Canadian Cooperative Wildlife Health Center. 2000. Western equine encephalitis. http://wildlife1.usask.ca/wildlife_health_topics/arbovirus/arbowee.php (accessed March 20, 2006).

Canadian Cooperative Wildlife Health Center. 2000. Eastern equine encephalitis. http://wildlife1.usask.ca/wildlife_health_topics/arbovirus/arboeee.php (accessed March 20, 2006).

Carter, G., D. Wise, and E. Flores. 2005. Coronaviridae: a concise review of veterinary virology. http://www.ivis.org/advances/carter/Part2Chap24/chapter.asp?LA=1 (accessed May 25, 2006).

Carter, G., D. Wise, and E. Flores. 2005. Herpesviridae: a concise review of veterinary virology. http://www.ivis.org/advances/carter/Part2Chap11/chapter.asp?LA=1 (accessed May 25, 2006).

Carter, G., D. Wise, and E. Flores. 2005. Picornaviridae: a concise review of veterinary virology. http://www.ivis.org/advances/carter/Part2Chap22/chapter.asp?LA=1 (accessed May 25, 2006).

Carter, G., and D. Wise. 2005. Poxviridae: a concise review of veterinary virology, International Veterinary Information Service. http://www.ivis.org/advances/carter/Part2Chap10/chapter.asp?LA=1 (accessed May 25, 2006).

Carter, G., and D. Wise. 2005. Rhabdoviridae: a concise review of veterinary virology. http://www.ivis.org/advances/carter/Part2Chap19/chapter.asp?LA=1 (accessed May 25, 2006).

Centers for Disease Control and Prevention. 2005. Arenaviruses http://www.cdc.gov/ncidod/dvrd/spb/mnpages/dispages/arena.htm (accessed March 3, 2006).

Centers for Disease Control and Prevention. 2005. Arboviral encephalitis. http://www.cdc.gov/ncidod/dvbid/arbor/arbofact.htm (accessed March 3, 2006).

Centers for Disease Control and Prevention. 2006. Hantavirus. http://www.cdc.gov/nceh/ehs/Topics/Hantavirus.htm (accessed May 17, 2006).

Centers for Disease Control and Prevention. 2004. Hendra virus disease and Nipah virus encephalitis. http://www.cdc.gov/ncidod/dvrd/spb/mnpages/dispages/nipah.htm (accessed May 9, 2006).

Centers for Disease Control and Prevention. 1999. Outbreak of Hendra-like virus—Malaysia and Singapore, 1998–1999. *Morbidity and Mortality Weekly Review* 48(3):265–9.

Centers for Disease Control and Prevention. 2003. The rabies virus. http://www.cdc.gov/ncidod/dvrd/rabies/the_virus/virus.htm (accessed March 3, 2005).

Centers for Disease Control for Prevention. 2005. St. Louis encephalitis. http://www.cdc.gov/ncidod/dvbid/arbor/slefact.htm (accessed May 15, 2006).

Centers for Disease Control and Prevention. 1999. Update: outbreak of Nipah virus—Malaysia and Singapore, 1999. *Morbidity and Mortality Weekly Review* 48(16):335–7.

Center for Food Security and Public Health, Iowa State University, 2007. Avian influenza: In Depth. http://www.cfsph.iastate.edu/Feature/aiInDepth.htm (accessed June 11, 2007).

Center for Food Security and Public Health, Iowa State University. 2005. Contagious ecthyma. http://www.cfsph.iastate.edu/Factsheets/pdfs/contagious_ecthyma.pdf (accessed May 1, 2005).

Center for Food Security and Public Health. 2004. Eastern equine encephalomyelitis, Western equine encephalomyelitis, and Venezuelan equine encephalomyelitis. http://www.cfsph.iastate.edu/Factsheets/pdfs/easter_wester_venezuelan_equine_encephalomyelitis.pdf (accessed March 21, 2006).

Center for Food Security and Public Health, Iowa State University. 2005. Foot and mouth disease. http://www.cfsph.iastate.edu/Factsheets/pdfs/foot_and_mouth_disease.pdf (accessed October 26, 2005).

Center for Food Security and Public Health, Iowa State University. 2006. Highly pathogenic avian influenza. http://www.cfsph.iastate.edu/Factsheets/pdfs/highly_pathogenic_avian_influenza.pdf (accessed April 21, 2006).

Center for Food Security and Public Health, Iowa State University. 2005. Influenza. http://www.cfsph.iastate.edu/Factsheets/pdfs/influenza.pdf (accessed October 21, 2005).

Center for Food Security and Public Health, Iowa State University. 2004. Monkeyoxvirus infection. http://www.cfsph.iastate.edu/Factsheets/pdfs/monkeypox.pdf (accessed January 6, 2004).

Center for Food Security and Public Health, Iowa State University. 2005. Newcastle disease. http://www.cfsph.iastate.edu/Factsheets/pdfs/newcastle_disease.pdf (accessed August 29, 2005).

Center for Food Security and Public Health, Iowa State University. 2004. Nipah. http://www.cfsph.iastate.edu/Factsheets/pdfs/nipah.pdf (accessed August 5, 2005).

Center for Food Security and Public Health, Iowa State University. 2005. Swine vesicular disease. http://www.cfsph.iastate.edu/Factsheets/pdfs/swine_vesicular_disease.pdf (accessed October 29, 2005).

Center for Food Security and Public Health, Iowa State University. 2006. Vesicular Stomatitis. http://www.cfsph.iastate.edu/Factsheets/pdfs/vesicular_stomatitis.pdf (accessed January 28, 2006).

Center for Food Security and Public Health, Iowa State University. 2005. Viral encephalitis. www.cfsph.iastate.edu/TrainTheTrainer/ppts/pptCompanionAnimal.ppt (accessed August 8, 2005).

Center for Food Security and Public Health, Iowa State University. 2004. West Nile encephalitis. http://www.cfsph.iastate.edu/Factsheets/pdfs/west_nile_fever.pdf (accessed August 8, 2005).

Chao, D-Y., K. Chwan-Chuen, W-K Wang, W-J Chen, H-L Wu, and G-J J. Chang. 2005. Strategically examining the full-genome of dengue virus type 3 in clinical isolates reveals its mutation spectra. *Virology Journal* 2:72.

Coffey, L., A.-S. Carrara, S. Paessler, M. Haynie, R. Bradley, R. Tesh, and S. Weaver. 2004. Experimental everglades virus infection of cotton rats (*Sigmodon hispidus*). *Emerging Infectious Disease* 10(12):2182–8.

Couch, R. 2005. Orthomyxoviruses. http://www.gsbs.utmb.edu/microbook/toc.htm (accessed March 3, 2006).

Cowan, M., and K. Talaro. 2006. *Microbiology: A Systems Approach*. New York: McGraw-Hill.

Cunha, B. 2006. Hantavirus pulmonary syndrome. http://www.emedicine.com/med/topic3402.htm (accessed May 2, 2006).

Cuhna, B. 2006. West Nile Encephalitis. http://www.emedicine.com/med/topic3160. htm (accessed December 3, 2006).

Day, J. 2001. Predicting St. Louis encephalitis virus epidemics: lessons from recent, and not so recent, outbreaks. *Annual Review of Entomology* 46(1):545–71.

Derlet, R., and R. Lawrence. 2004. Influenza. http://www.emedicine.com/MED/ topic1170.htm (accessed September 28, 2004).

Donson, D., and M. Lai. 2002. West Nile Virus. http://www.emedicine.com/aaem/ topic542.htm (accessed September 3, 2002).

Edlow, J. 2005. Tick-borne diseases, Lyme. http://www.emedicine.com/EMERG/ topic588.htm (accesed April 25, 2005).

Gillard, J. 2006. West Nile Virus: a mounting threat. *Emergency Medicine* 38(8):40–4.

Hanson, R. 1952. The natural history of vesicular stomatitis. *Bacteriology Review* 3:179–204.

Hawayek, L., and N. Rubeiz. 2006. Contagious ecthyma. http://www.emedicine. com/derm/topic605.htm (accessed February 28, 2006).

Innvista. 2005. Orthomyxoviruses. http://www.innvista.com/health/microbes/ viruses/orthomy.htm (accessed March 3, 2006).

Institutional Animal Care and Use Committee. 1996. Zoonotic diseases. http:// research.ucsb.edu (accessed March 22, 2005).

Jones, T. C., R. D. Hunt, and N. W. King. 1997. *Veterinary Pathology*, 6th ed. Baltimore, MD: Williams & Wilkins.

Klein, J., and M. Tryland. 2005. Characterization of parapoxviruses isolated from Norwegian semi-domesticated reindeer (*Rangifer tarandus tarandus*). *Virology Journal*. http://www.virologhyj.com (accessed June 26, 2005).

Krauss, H., A. Weber, M. Appel, B. Enders, H. D. Isenberg, H. G. Schiefer, W. Slenczka, A. von Graeventiz, and H. Zahner. 2003. *Zoonoses: Infectious Diseases Transmissible from Animals to Humans*, 3rd ed. Washington, DC: American Society of Microbiology Press.

Lazoff, M. 2005. Encephalitis. http://www.emedicine.com/emerg/topic163.htm (accessed September 12, 2005).

Loerke, C. 1998. Cats reovirus. http://virus.stanford.edu/reo/reoviridae.html (accessed September 9, 2005).

Lumumba-Kasongo, M., and M. Gang. 2006. Hantavirus cardiopulmonary syndrome. http://www.emedicine.com/emerg/topic861.htm (accessed April 3, 2006).

Lutwick, L., and R. Deaner. 2006. Herpes B. http://www.emedicine.com/med/ topic3367.htm (accessed June 9, 2006).

Lutwick, L., and Y. Bron. 2006. Picornavirus: overview. http://www.emedicine. com/MED/topic1831.htm (accessed August 31, 2006).

Marfin, A., and G. Campbell. 2005. Colorado tick fever and related coltivirus infections. In *Tick-Borne Diseases of Humans*, edited by J. Goodman, D. Dennis, and D. Sonenshine. Washington, DC: American Society of Microbiology Press.

Marr, J., and C. Calisher. 2003. Alexander the Great and West Nile Virus encephalitis. *Journal Title* 9(12):1599–1603

Medline Plus. 2005. Rabies. http://www.nlm.nih.gov/medlineplus/ency/ article/001334.htm (accessed April 8, 2005).

Merck Veterinary Manual. 2006. Rabies http://www.merckvetmanual.com/mvm/ index.jsp?cfile=htm/bc/102300.htm&word=rabies (accessed May 1, 2006).

Microbiology at Leicester. 2004. Virus replication. http://www.microbiologybytes. com/virology/3035Replication.html (accessed February 3, 2006).

Mills, J., A. Corneli, J. Young, L. Garrison, A. Khan, and T. Ksiazek. 2002. Hantavirus pulmonary syndrome: U.S. updated recommendations for risk reduction. *Morbidity and Mortality Weekly Review* 51(RR09):1–12

Mylonakis, E., and E. Soliman. 2006. California encephalitis. http://www.emedicine.com/med/topic3161.htm (accessed October 18, 2006).

Mylonakis, E., and E. Soliman, 2006. St. Louis encephalitis. http://www.emedicine.com/med/topic3157.htm (accessed October 18, 2006).

National Academic Press. 2004. Learning from SARS: preparing for the next outbreak, workshop summary 2004. http://www.fermat.nap.edu (accessed May 12, 2006).

Obijeski, J., D. Bishop, F. Murphy, and E. Palmer. 1976. Structural proteins of La Crosse virus. *Journal of Virology* 19(3):985–97.

Oehler, R., and Lorenzo, N. 2005. Severe acute respiratory syndrome (SARS). http://www.emedicine.com/med/topic3662.htm (accessed April 27, 2005).

Pfau, C. 2005 Arenaviruses. http://gsbs.utmb.edu/microbook/ch057.htm (accessed April 14, 2006).

Pigott, D., and T. McGovern. 2005. CBRNE: viral hemorrhagic fevers. http://www.emedicine.com/emerg/topic887.htm (accessed December 14, 2005).

Prescott, L. M., D. A. Klein, and J. P. Harley. 2002. *Microbiology*, 5th ed. New York, New York: McGraw-Hill.

Price, D., and S. Wilson. 2005. Dengue fever. http://www.emedicine.com/emerg/topic124.htm (accessed November 9, 2005).

Rasouli, G., and J. King. 2005. Reovirus. http://www.emedicine.com/med/topic2007.htm (accesed July 6, 2005).

Roberts, L., and J. Janovy. 2005. *Foundations of Parasitology*, 7th ed. New York: McGraw-Hill.

Ruan, Y. J., C. L. Wei, A. L. Ee, V. B. Vega, H. Thoreau, S. T. S. Yun, J-M Chia, P. Ng, K. P. Chiu, L. Lim, Z. Tae, C. K. Peng, L. O. L. Ean, N. M. Lee, L. Y. Sin, L. F. P. Ng, R. E. Chee, L. W. Stanton, P. M. Long, and E. T. Liu. 2003. Comparative full-length genome sequence analysis of 14 SARS coronavirus isolates and common mutations associated with putative origins of infection. *Lancet* 361:1779–85.

Sanofi Pasteur. 2005. Rabies. http://www.rabies.com (accessed April 8, 2005).

Schmaljohn, A., and D. McClain. 2005. Alphaviruses (Togaviridae) and flaviviruses (Flaviviridae). http://www.gsbs.utmb.edu/microbook/ch054.htm (accessed May 15, 2006).

Schmidtmann, E., W. Tabachnick, G. Hunt, L. Thompson, and H. Hurd. 1999. Epizootic of vesicular stomatitis (New Jersey serotype) in the Western U.S.: an entomologic perspective. *Journal of Medical Entomology* 36(1):78–82.

Seppa, N. 2005. One in a million. http://www.sciencenews.org (accessed January 29, 2005).

Shanley, J. 2006. Poxviruses. http://www.emedicine.com/med/topic1903.htm (accessed May 26, 2006).

Shepherd, S., and P. Hinfey. 2006. Dengue fever http://www.emedicine.com/MED/topic528.htm (accessed April 6, 2006).

Shimeld, L. A. 1999. *Essentials of Diagnostic Microbiology*. Albany, NY: Delmar Publishers.

Shope, R. 2005. Bunyaviruses. http://www.gsbs.utmb.edu/microbook/ch056.htm (accessed October 8, 2005).

Shroyer, D. 2004. Saint Louis encephalitis: a Florida problem, University of Florida Extension. http://edis.ifas.ufl.edu/BODY_MG337 (accessed March 31, 2006).

Siegel, R. 1999. Rhabdovirus. http://www.stanford.edu/group/virus/1999/sohoni/rhabdovirus.html (accessed April 8, 2005).

Siegel, R. 1999. World of flaviviruses. http://www.stanford.edu/group/virus/1999/asb-flavi/flavivirus.htm (accessed February 3, 2006).

The History Channel. 2006. Viruses. http://www.history.com/encyclopedia.do?vendorId=FWNE.fw..vi032400.a#FWNE.fw..vi032400.a (accessed May 17, 2006).

U.S. Department of Agriculture Animal & Plant Inspection Unit. 2002. Venezuelan Equine Encephalitis http://www.aphis.usda.gov/lpa/pubs/fsheet_faq_notice/fs_ahvee.html (accessed September 8, 2005).

U.S. Department of Agriculture Animal & Plant Inspection Unit. 2003. Exotic Newcastle disease: veterinary services. http://www.aphis.usda.gov/lpa/pubs/fsheet_faq_notice/fs_ahend.html (accessed January 15, 2003).

U.S. Department of Agriculture Animal & Plant Inspection Unit. 2002. Information about vesicular stomatitis for the beef producer. http://www.aphis.usda.gov/lpa/pubs/fsheet_faq_notice/fs_ahvsbeef.pdf (accessed May 4, 2006).

U.S. Department of Agriculture Animal & Plant Inspection Unit. 2003. Summary of selected disease events, October–December 2003, Center for Emerging Issues. http://www.aphis.usda.gov/vs/ceah/cei/taf/iw_2003_files/summary2003/summary_10_12_2003_files/summary_10_to_12_2003.htm (accessed May 1, 2006).

U.S. Department of Agriculture Animal & Plant Inspection Unit. 2004. West Nile Virus. http://www.aphis.usda.gov/lpa/pubs/fsheet_faq_notice/fs_ahwnv.html (accessed October 11, 2004).

Wentworth, D., L. Gillim-Ross, N. Espina, and K. Bernard. 2004. Mice susceptible to SARS coronavirus. *Emerging Infectious Diseases* 10(7):1293–6.

Wikipedia.org. 2006. Severe acute respiratory syndrome. http://en.wikipedia.org/wiki/SARS (accessed May 11, 2006).

World Health Organization. 2004. Avian influenza: asssssing the threat. http://www.who.int/csr/disease/influenza/H5N1-9reduit.pdf (accessed June 9, 2007).

World Health Organization. 2001. Nipah virus. http://www.who.int/mediacentre/factsheets/fs262/en/index.html (accessed May 9, 2006).

World Health Organization. 2006. Rabies. http://www.who.int/mediacentre/factsheets/fs099/en/index.html (accessed September 8, 2006).

Chapter

8

Prion Zoonoses

Objectives

After completing this chapter, the learner should be able to

- Briefly describe the history of prions
- Describe prion structure
- Identify the geographic distribution of prion zoonoses
- Describe the transmission, clinical signs, and diagnostic procedures of prion zoonoses
- Describe methods of controlling prion zoonoses
- Describe protective measures professionals can take to prevent transmission of prion zoonoses

Key Terms

florid plaque
prion
PrP^C

PrP^{CWD}
PrP^{Sc}

transmissible spongiform
encephalopathies

OVERVIEW

Prion diseases, also known as **transmissible spongiform encephalopathies** or TSEs, are a group of animal and human central nervous system (CNS) diseases that have long incubation periods, produce spongiform vacuoles in brain tissue, do not produce an inflammatory response in affected tissues, and are uniformly fatal. **Prions, pro**teinaceous **in**fectious particles, were named from the extraction and transposition of letters from proteinaceous infectious particles in 1982 by Stanley Prusiner to describe self-replicating, infectious proteins. There are a variety of TSEs in animals that have been known for some time. Scrapie, also known as cuddy trot (Scotland), the rubbers or the googles (England), and *la tremblante* or the trembles (France), is a TSE mainly in sheep that produces pruritus and an almost continuous scraping of the skin against stationary objects. Scrapie has been a recognized disease since 1732 when Spanish shepherds reported what they called *la enfermedad trotoria* (the trotting disease) in their flocks. Bovine spongiform encephalopathy (BSE), commonly called mad cow disease, was first observed in Great Britain in April, 1985 and diagnosed in November 1986 on a dairy in Kent, England. The name mad cow disease was coined as a result of the abnormal motor nerve control and aggressiveness seen in cattle with the disease. Another animal TSE is chronic wasting disease (CWD), a condition first detected in mule deer in northeast Colorado in 1967 and determined to be a TSE in 1978 by Dr. Elizabeth Williams of the University of Wyoming. Since that time it has been seen in elk, white-tailed deer, and black-tailed deer in numerous states. Other TSEs in animals include transmissible mink encephalopathy (discovered in farm-raised mink in Wisconsin and Minnesota in 1947) and feline spongiform encephalopathy (discovered in domestic cats in the United Kingdom (U.K.) in 1990).

In the 1920s, Hans Gerhard Creutzfeldt and Alfons Maria Jakob independently observed human cases of a slowly progressive dementing neurologic disease in humans. The disease was named Creutzfeldt-Jakob disease (CJD) and was an extremely rare, sporadic condition of unknown origin in older people. Another form of CJD, iatrogenic CJD, had been known to occur from corneal and dural transplants, contaminated neurosurgical instruments, and growth hormone replacement. In 1996 a new form of CJD was identified and named variant CJD (vCJD) and occurred in younger people (many younger than 30 years old). vCJD has been linked to the consumption of BSE-infected beef and since that time animal management procedures have been altered to prevent this disease. Since the discovery of CJD and vCJD other degenerative neurologic diseases were identified including kuru (a disorder that surfaced among the South Fore people (a tribe of remote highland natives) of New Guinea who practiced mortuary cannibalism reaching epidemic levels in the 1960s), Gerstmann-Sträussler-Scheinker syndrome (GSS) (a familial autosomal dominant disease occurring in the fourth and fifth decade of life), and fatal familial insomnia (FFI) (an autosomal dominant disease discovered by an Italian physician in 1979 and carried in the genome of 28 families worldwide). Table 8-1 summarizes the most common TSEs in people and animals.

Causative Agent

Prions are proteinaceous infections particles that are one-tenth the size of viruses and consist only of protein. Prions are self-replicating yet do not possess a genome. Prions are encoded by a host gene, the prion protein (PrP) gene. Cellular prion protein is a normal component of neuronal cell membranes and is a three-dimensional structure with an α helix configuration. All forms of TSE have a similar alteration of the PrP gene which leads to incorporation of false amino acids into the protein resulting in abnormal folding into a β helix arrangement. The abnormally folded version of the prion protein is extremely stable and seems to have the ability to cause other normally folded prion proteins to adopt the abnormal folding pattern. Through propagation of the abnormal structures, these abnormally folded protein structures accumulate in intracellular vesicles that produce large vacuoles and are deposited as **florid plaques** (focal amyloid protein deposits surrounded by vacuolized cells).

The theory that TSEs are caused by an infectious protein was developed in 1960s by Tikvah Alper and J.S. Griffith. Their theory was based on the fact that the mysterious infectious agent causing diseases such as scrapie and CJD resisted ultraviolet radiation (which breaks down nucleic acids), yet was inactivated by agents that disrupt proteins. In 1981, Patricia Merz at the Institute for Basic Research in Developmental Disabilities at Staten Island, New York, observed strange fibrils in scrapie-infected mouse brains. In 1982 Stanley Prusiner of the University of California, San Francisco purified infectious material and confirmed that the infectious agent was protein (he received the Nobel Prize in Medicine in 1997 for this research on prions).

PrP is found throughout the body on cell membranes in healthy people and animals. The normal form of the protein is called **PrP^C** (C is for cellular), whereas the infectious form is called **PrP^{Sc}** (Sc is for scrapie). It is assumed that PrP^{Sc} directly interacts with PrP^C to cause the normal form of the protein to rearrange its structure. As PrP^{Sc} is introduced into normal healthy cells, it causes the

TSEs are the only diseases that can be sporadic, genetic, or infectious.

The infectious isoform of prion protein is also abbreviated PrP^{res} (protease-resistant) and the normal prion protein is also abbreviated PrP^{sen} (protease-sensitive).

Table 8-1   **Summary of the Most Common Transmissible Spongiform Encephalopathies (TSEs), Their Host, and Mechanism of Transmission**

Disease	Host	Transmission Mechanism
Kuru	Human	Ritualistic cannibalism
Sporadic CJD	Human	Spontaneous PrP^C to PrP^{Sc} conversion or somatic mutation
Iatrogenic CJD	Human	Infection from prion-containing material (dura mater, electrode, surgical equipment)
Familial CJD	Human	Mutations in the PrP gene
vCJD	Human	Ingestion from BSE-contaminated meat
Gerstmann-Sträussler-Scheinker syndrome (GSS)	Human	Mutations in the PrP gene
Fatal familial insomnia (FFI)	Human	Mutation in the PrP gene
Scrapie	Sheep (goats to a lesser degree)	From ewe to offspring and to other lambs through contact with the placenta and placental fluids
BSE	Cattle	Infection from contaminated food that has been improperly processed
TME	Mink	Infection from contaminated food that has been improperly processed
CWD	Mule deer, deer, elk	Unclear
Feline spongiform encephalopathy (FSE)	Felines (domestic cats, wild felines)	Infection from contaminated food that has been improperly processed

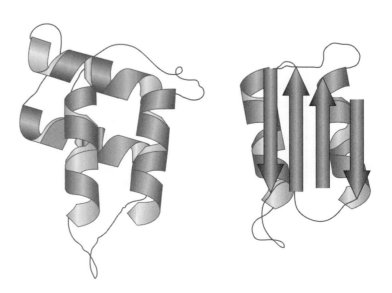

Figure 8-1   Helical pattern of normal (left) and abnormal prion (right).

conversion of PrP^C into PrP^{Sc}, initiating a self-perpetuating vicious cycle of replication of abnormal proteins (Figure 8-1).

EPIZOOTIOLOGY AND PUBLIC HEALTH SIGNIFICANCE

Scrapie

Scrapie was first recognized as a disease of sheep in Great Britain and other countries of Western Europe more than 250 years ago. Since that time scrapie has been reported throughout the world except in Australia and New Zealand. The first case of scrapie in the United States was diagnosed in 1947 in a Michigan flock that had imported sheep of British origin from Canada. Scrapie is uncommon in the United States being diagnosed in only about 1,600 sheep and goats since it was first found here in 1947. In the United States, scrapie has primarily been reported in the Suffolk breed (it also has been diagnosed in Border Leicester, Cheviot, Corriedale, Cotswold, Dorset, Finnsheep, Hampshire, Merino, Montadale, Rambouillet, Shropshire, Southdown, and a number of crossbreeds).

Scrapie costs U.S. producers $20 to $25 million every year mainly from the loss of sales from abroad (U.S. sheep are banned for sale in Europe and China). The Animal and Plant Health Inspection Service (APHIS) estimates that in the United States, scrapie eradication program will cost $100 million over seven years.

BSE

BSE was first described in 1985 in the U.K., and since that time more than 180,000 cattle have died from the disease or have been slaughtered in England. In 2001, a total of 8,516,227 cattle were tested for BSE with 2,153 being positive. Between January 1 and May 31, 2002, a total of 427 new BSE cases were detected in British cattle (most born after the feed ban). In 2002, approximately 40 animals per month were found BSE positive. Approximately 1,500 cattle have died of BSE in other European countries. It is estimated that during the BSE outbreak in England in the 1980s 1 in 200 cattle were infected. Other European countries (Ireland, Netherlands, Germany, France, Denmark, Portugal, Spain, Austria, Belgium, Czech Republic, Finland, Greece, Switzerland, and Italy) also have BSE in cattle. Japan, Israel, Canada, and the United States have also had BSE-positive cattle in their countries.

BSE eradication costs in Great Britain were estimated at $11.6 million in 1986 for countrywide depopulation and were not implemented as a result of cost. In time, funding for the BSE eradication program was shared by the EU (70%) and the U.K. (30%) with the final cost not expected to be known or paid for some time. To date, the total known costs are $181 million for rendering, $1.03 billion for cattle producers, and $166 million for slaughterhouses. In the U.K., expenditures on diagnosis and surveillance totaled approximately $14.4 million from 1988 to 1996. In April 2000, government officials estimated that the total net cost of the BSE epidemic will be $6.9 trillion by the end of the 2001–2002 financial year. In France, the surveillance of cattle at risk represents an average cost of $22,004 per case. The United States is the leading exporter of beef in the world ($3.5 billion for 2003) and the discovery of BSE in the United States has cost producers reven[...] a result of bans on importation of United States beef into countries such [...] Approximately 10% of United States beef is sold outside its borders [...]

Positive BSE cases in the United States have been found on D[...] (downer dairy cow in Washington state that originated f[...]

Alberta, Canada) and in June 2005 (a 12-year-old Brahma cross raised in Texas, sold at a livestock sale in November 2004, and found dead on arrival at a packing plant. This was the first BSE case born in the United States and was probably infected from meat-and-bone meal in feed prior to the 1997 feed ban).

CWD

First seen in 1967, CWD is now considered to be enzootic in the United States. As of February 2006, CWD has been diagnosed in wild deer and/or elk in Colorado, Wyoming, Wisconsin, South Dakota, New Mexico, Utah, Nebraska, Oklahoma, Minnesota, Illinois, Kansas, Montana, New York, West Virginia, Alberta, and Saskatchewan. CWD has been diagnosed in captive herds in Colorado, Nebraska, South Dakota, Montana, Oklahoma, Kansas, Minnesota, New York, Wisconsin, Wyoming, Alberta and Saskatchewan. CWD is estimated to cost $100 billion in lost revenue to areas dependant on hunting and tourism economies. State eradication programs are estimated to cost $4 million dollars annually.

Other Animal TSEs

Transmissible mink encephalopathy (TME) was first detected in the United States in 1947 on one mink ranch in Wisconsin and then on a mink ranch in Minnesota that had received mink from the Wisconsin ranch. In 1961, TME outbreaks occurred on five ranches in Wisconsin and in 1963, outbreaks occurred in Idaho, Minnesota, and Wisconsin. The most recent TME outbreak occurred on one mink ranch in Stetsonville, Wisconsin, in 1985. A herd of 7,300 adult mink suffered a 60% loss of animals during the 5-month-long outbreak. Since then, TME outbreaks have been reported in Canada, Finland, Germany, and the republics of the former Soviet Union.

TME is no longer a major problem in the United States and Canada (last reported outbreak was in 1985). Commercial mink food has replaced raw meat diets and currently less than 1% of mink ranches use dead stock and downer animals for food. Mink are not imported into the United States in large numbers making the risk of importing TME low.

Feline spongiform encephalopathy (FSE) was first reported in the U.K. in 1990. Eighty-seven confirmed FSE cases found only in domestic cats have been reported in the U.K. Other cases have been reported in Norway (1), Northern Ireland (1), and Switzerland (1). Since 1990, other feline species in zoos have been reported to have contracted FSE.

FSE is currently a rare disease. The acquisition of large felines for zoos may become a problem if cases of FSE in these animals increases; however, permits are required by Veterinary Services in many countries that regulate the importation of animals into zoological settings. Many animals are quarantined when they arrive at zoos; however, as a result of the long incubation period of TSEs, the animal may be permitted to intermingle with other animals in a facility prior to demonstration of clinical signs. This is especially a problem with scrapie and CWD in other species.

vCJD

The human disease linked to BSE, vCJD, is a new form of transmissible encephalopathy in Europe. The first cases of vCJD were reported in 1995, when CJD was two British teenagers. As of July 2002, greater than 130 cases of vCJD have been reported since 1995 (128 cases in England, 3 in France, 2 in Ireland, and 1 in Italy). This number was expected to double within 3 years, but case numbers in 2002 (17), 2003 (18), 2004 (9), 2005 (5), and 2006 (2) did not show this expected rapid increase in vCJD cases. As of May 2006 there have been 185 cases of vCJD worldwide with 156 cases in the U.K.

The current risk of acquiring vCJD from eating beef (muscle meat) and beef products produced from cattle in countries with an increased risk of BSE is extremely small (approximately 1 case per 10 billion servings in the U.K.). The risk is lower in other countries since the highest numbers of BSE-infected cattle were found in the U.K. The cost to the Department of Health in the U.K. of staff time spent on BSE/vCJD-related activities from 1988 to 1996 was approximately $1,535,000. The average cost per vCJD patient is about $37,400 in the U.K. (ranging from $12,000 to $75,000 per year). There have been cases of people contracting vCJD from infected blood transfusions. The United States blood donor deferral criteria was put into effect in September 2004 and focuses on the time (cumulatively 3 months or more) that a person lived in the U.K. from 1980 through 1996 or in the rest of Europe (cumulatively 5 years or more) that a person lived in these countries from 1980 through the present.

Kuru

Kuru was only found in isolated cannibalistic tribes in New Guinea beginning in the early 1900s. Between 1957 and 1968 there were 1,100 deaths attributed to the disease occurring mainly in women, children, and the elderly. Kuru is eradicated and does not pose a significant public health risk.

TRANSMISSION

Scrapie

Scrapie is thought to be spread most commonly from the ewe to her offspring and to other lambs through contact with the placenta and placental fluids. Scrapie has been experimentally transmitted to hamsters, mice, rats, voles, gerbils, mink, cattle, and some species of monkeys by inoculation. There is no epidemiologic evidence that scrapie is transmitted to humans. Susceptibility to scrapie in sheep is affected by the amino acid sequence of the sheep's prion protein (codons 136, 154 and 171 of the prion protein play the major role in natural transmission of scrapie in sheep). Genetic selection may be helpful in controlling and possibly eliminating scrapie within flocks.

BSE

BSE is transmitted to cattle via the gastrointestinal tract and the forced feeding of animal offal (cadavers, inferior quality meat, and slaughterhouse scraps) of similar species origin. If feed is heated to 130°C (to inactivate anthrax spores), prions are inactivated. In the 1970s the production of animal feed was changed (as a result of the energy crisis) and feed was processed at a lower temperature. Because cattle feed also contained scrapie-infected sheep offal (at a higher level in the U.K.) cattle ingested prions at substantial level. Prion-infected cattle were then processed into cattle feed which helped the spread of BSE. Milk replacer contained cadaver feed and was the source of BSE in calves.

CWD

The method by which CWD is transmitted is not understood. Evidence suggests that the disease can pass from animal to animal by contact or via contamination of feed or pasture with saliva, urine, and/or feces. Live deer and elk represent the most likely mechanisms for CWD spread. Prions cannot be directly demonstrated in excretions or soil; however, CWD prion protein (**PrP^CWD**) accumulates in gut-associated lymphoid tissues such as the tonsils, Peyer's patches, and mesenteric lymph nodes of infected mule deer which implies there is gastrointestinal shedding of the CWD prion in both feces and saliva. Indirect transmission of CWD in excretions that survive in soil that is spread by animals foraging and soil consumption seems possible. Deer do not actively consume decomposed carcass remains, but they forage in the immediate vicinity of carcass sites where they may become infected. PrP^CWD is believed to be able to persist in contaminated environments for more than 2 years.

Other Animal TSE

Mink acquire TME by eating contaminated feed. There is no evidence of TME transmission by contact between unrelated mink or from mother to nursing young. Cats, both domestic and wild, acquire FSE by eating BSE-contaminated meat.

Humans

Humans contract TSEs by ingestion. Kuru was acquired when deceased family members were ritualistically cooked and eaten following their death. The closest female relatives and children usually consumed the brain, which was the most infectious organ. The women scooped the brain tissue out of the dead person's skull with their bare hands and did not subsequently wash their hands for weeks. During this period, they were handling, caring for, and infecting young children. Kuru has been eradicated through education of tribal people in New Guinea.

vCJD is acquired by ingestion of infected animal products and appears to occur in people that are genetically prone to prion disease (codon 129 of the PrP gene codes for both alleles for methionine in 100% of human patients). Bioassays have identified the BSE agent in the brain, spinal cord, retina, dorsal root ganglia, distal ileum, and bone marrow of cattle experimentally infected by the oral route, suggesting that these tissues represent the highest risk of prion transmission to humans. In 2003, the first case of blood transfusion transmission of vCJD was documented in the U.K.

PATHOGENESIS

A feature of all prion diseases is that they affect the nervous system. Prions tend to affect the gray matter of the CNS, producing loss of neurons and characteristic spongiform change. Spongiform changes are the result of vacuolation of the neurons (Figure 8-2.) Abnormal prions also have the tendency to clump together to form large aggregates called amyloids. Amyloid damages and kills brain cells (usually in the cerebellum or cerebral hemispheres) causing loss of brain function, loss of coordination, and psychological changes. In cattle with BSE, prions are found mainly in the brain but also in extracerebral neural tissues and in lymphatic tissue. After infection, prions are amplified in peripheral organs such as the spleen and lymphatic tissue.

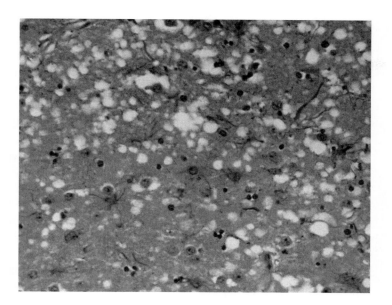

Figure 8-2 The presence of vacuoles (microscopic holes) in the gray matter of affected animals is the classical histologic finding with transmissible spongiform encephalopathies (TSEs). (Courtesy of CDC/Dr. Al Jenny)

Clinical Signs in Animals

The clinical signs of TSE in animals vary somewhat with the species as follows:

- Scrapie in sheep and goats presents at different rates and clinical signs as a result of the slow development of the disease and the age of the animal infected. Damage to nerve cells causes the affected animals to typically show behavioral changes, tremor (especially of head and neck), rubbing as a result of intense pruritus, weakness, weight loss, high-stepping of the front limbs, and ataxia that progresses to recumbency and death. An infected animal may appear normal if left undisturbed at rest; however, when the animal is stimulated by a sudden noise, excessive movement, or the stress of handling, the animal may fall down in a seizure-like state.

- BSE in cattle has an insidious onset and progresses slowly. Clinical signs are neurologic including hyperesthesia, tremors, ataxia, falling, and behavioral changes (apprehension, nervousness, and frenzy) (Figure 8-3). Nonspecific

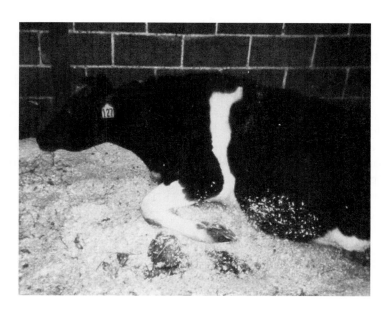

Figure 8-3 Classic signs of bovine spongiform encephalopathy (BSE) in cattle include hyperesthesia, tremors, ataxia, falling, and behavioral changes. (Courtesy of CDC/Dr. Art Davis)

Figure 8-4 Classic signs of chronic wasting disease (CWD) in deer and elk include behavioral changes, staggering, lowered head and ears, decreased food consumption, emaciation, drooling, polyuria, and polydipsia.
(Courtesy of Getty Image)

signs such as loss of condition, weight loss, and decreased milk production may also be seen. Once clinical signs appear BSE is progressive and death occurs weeks to months after an animal becomes recumbent.

- CWD in deer and elk presents with behavioral changes, staggering, poor posture, lowered head and ears, decreased food consumption, emaciation, drooling, polyuria, and polydipsia (increased drinking) (Figure 8-4). Clinical signs of CWD may be more subtle and prolonged in elk than in mule deer. In infected captive animals, they may increase or decrease their interaction with handlers or other members of the herd and show repetitive behaviors (walking set patterns in their pens). The clinical course of CWD varies from a few days to a year with death occurring in all cases.

- TME has an average incubation period of more than 7 months before the onset of clinical signs with signs lasting from 3 days to 6 weeks. Early clinical signs include an increase in nest soiling, difficulty eating, and mink may step into their food. As TME progresses, an infected animal becomes increasingly excited, arches its tail over its back, has ataxia, and displays jerkiness of hind limbs. In advanced cases, signs include rapid circling, compulsive chewing of the tail, and clenching of the jaw. Affected mink become sleepy and unresponsive prior to death.

- FSE presents in felines with early clinical signs of behavioral changes, tremors, ataxia, and increased aggression. As the disease progresses cats develop seizures, excessive salivation, hyperresponsiveness to loud noise, and dilated pupils. Death occurs in 6 to 8 weeks.

Clinical Signs in Humans

Clinical signs of vCJD become apparent after an incubation period of greater than 16 years and typically presents as neurologic and psychomotor deficits. In the early stages of vCJD people have cerebellar signs such as psychiatric abnormalities (depression, annoyance, or fits of rage) and neurologic signs (unsteadiness, difficulty walking, and involuntary movements). As the disease progresses cerebral cortical signs develop such as memory loss, dementia, and myoclonus. As vCJD progresses people become completely immobile and mute prior to death.

Diagnosis in Animals

TSEs in animals are diagnosed as follows:

- Scrapie can be diagnosed in the live animal by clinical signs and by biopsy of the lymphoid tissues on the inside of the third eyelid. This immunohistochemistry test is used by the USDA APHIS to determine whether exposed flocks are infected with PrP^{Sc}. Scrapie is most often diagnosed by microscopic examinations of brain tissue at necropsy or by procedures that detect the presence of the abnormal prion protein in brain tissue.

- BSE is suspected in animals that develop a slowly progressive, fatal neurologic disease. Laboratory tests include detecting PrP^{Sc} in unfixed brain extracts by immunoblotting and in fixed brains by immunohistochemistry. Characteristic fibrils of PrP^{Sc} can be found with electron microscopy. Commercial tests to detect BSE in cattle brain samples include a modified immunoblot, a chemiluminescent ELISA test, an immunoassay, and a two-site noncompetitive immunometric procedure. Since December 2003, at least five rapid diagnostic tests have been licensed for use; however, it is not known how much prion protein needs to be present to give positive results. These tests appear to have good sensitivity and specificity and are still in the testing stages. Samples need to be properly collected and handled as well as state and federal authorities need to be contacted. Serology is not useful for diagnosis of BSE because antibodies are not made against prions.

- CWD cannot be diagnosed based on clinical signs because they are nonspecific. Diagnosis is currently based on brain examination for spongiform lesions and/or accumulation of PrP^{CWD}. The dorsal portion of the medulla oblongata is submitted for immunohistochemical examination for diagnosis of CWD. Supplemental tests include negative-stain electron microscopy or Western blotting for detection of PrP^{CWD} in brain. An ELISA screening test may be used in some states. Demonstration of PrP^{CWD} in lymph nodes, tonsil, and conjunctival lymphoid tissues is currently being tested in deer and elk to determine its sensitivity and specificity.

- TME is diagnosed by identifying spongiform changes in the brains of infected animals.

- FSE is diagnosed by identifying spongiform changes in the brains of infected animals.

Diagnosis in Humans

The clinical presentation, progressive nature of the disease, and rule out of other neurologic diseases help in the diagnosis of vCJD. The diagnosis of vCJD can only be confirmed by brain biopsy (which may not always produce tissue from the affected part of the brain) or autopsy. Brain histopathology reveals multiple microscopic and abnormal aggregates encircled by holes resulting in a daisy-like appearance described as florid plaques are seen with vCJD. Tonsil biopsy samples that are examined by electron microscopy and Western blot methods have been used in a laboratory in London. A test in development involves detecting protein markers in CSF.

> In 2006, prions were isolated from skeletal muscle in deer with CWD.

Treatment in Animals

There is no treatment for TSEs in animals.

Treatment in Humans

There is no treatment for vCJD in people.

MANAGEMENT AND CONTROL IN ANIMALS

Scrapie

The presence of scrapie in the United States prevents the export of breeding stock, semen, and embryos to many other countries. The United States developed the Scrapie Ovine Slaughter Surveillance (SOSS) to estimate the national and regional prevalence of scrapie in mature sheep. The final SOSS study was released in March 2004 and it was determined that 0.2% of sheep in the United States had scrapie (higher rates were found in the Western United States) and the incidence was higher in black faced sheep (0.84%). The USDA also initiated an accelerated scrapie eradication program that includes identification of nonclinical infected sheep through live-animal testing and active slaughter surveillance, effective tracing of infected animals to their flock/herd of origin, and providing effective cleanup strategies for flocks (including indemnity for high risk, suspect, and scrapie positive sheep and goats, which owners agree to destroy, scrapie live-animal testing, genetic testing, and testing of exposed animals that have been sold out of infected and source flocks/herds).

BSE

Controlling BSE revolves around the BSE-status of a particular country. BSE-free countries should have targeted surveillance programs in place to monitor occurrences of clinical neurologic disease, place safeguards on importation of live ruminant species and their products, and determine policy and procedures for importation of embryos. Countries with BSE cases should slaughter affected animals and provide compensation for positive cases, place controls on the use and recycling of mammalian protein, and develop effective identification and tracing of cattle.

Specific ways to prevent the spread of BSE in cattle or to keep BSE prions from the human food source include:

- Enforcing the ruminant feed ban. The inclusion of rendered beef protein, blood and blood byproducts, milk products, pure porcine and pure equine proteins, plate waste, tallow, gelatin, and nonmammalian protein (poultry, marine, vegetable) was prohibited in feed in August 1997.
- Banning the importation of live ruminants and restrictions on most ruminant products from countries where BSE has been diagnosed. This ban began in the United States in 1989 and by December 1997 was expanded to include all European nations. On December 7, 2000, the USDA prohibited importation of all rendered animal protein products from any European country (regardless of species or country of origin).
- Prohibition of downer cattle in human food. Prior to December 30, 2003, downer cattle could be slaughtered for human food once they had passed inspection for signs of disease by a USDA veterinarian, whereas a sample of brain tissue was taken for BSE analysis as part of the surveillance program. Under new rules, cattle that are downed for any reason may not be slaughtered for human food.
- Banning brain and spinal cord tissue of animals 30 months of age or older from the human food supply.
- Banning the use of air-gun stunning to kill cattle, a process which can cause CNS tissue to move into and to contaminate muscle tissue.

- Requiring meat processors to show that tissue is separated from the carcass using advanced meat recovery systems (AMR) and does not contain CNS tissue.
- Removal of all specified risk material (SRM) from entering the human food chain. This material includes skull, brain, trigeminal ganglia, eyes, vertebral column, spinal cord, and dorsal root ganglia of cattle older than 30 months of age.
- Develop a national animal identification plan to trace possibly infected animals.
- Sample all cattle used in the food supply for BSE. After the first U.S. case of BSE was detected in 2003, the government developed an enhanced surveillance program to determine the probability of BSE in U.S. cattle. Under the enhanced surveillance program, 0.0003% of the sampled population was positive for BSE. Currently any suspect cattle (adults with neurologic disease) are held until BSE tests are confirmed negative.

CWD

Control of CWD involves population reduction, testing and removal of affected animals, and intensified surveillance. Control of CWD in free-ranging animals is difficult and varies depending on the location. Control of CWD in farmed cervids involves depopulation with indemnity and herd plan development. Movement of live cervids is banned in the United States. Importation of cervid carcasses into states may be forbidden or limited to certain body parts such as teeth or hides.

CWD can be prevented from possible human contamination by having hunters or people who handle deer and elk wear rubber gloves for field-dressing, wash hands and forearms thoroughly when done handling deer or elk, following state regulations and guidelines for the transportation of harvested game animals (shooting, handling, or consuming an animal that appears sick, wearing rubber or latex gloves when field-dressing, boning out the meat from game, minimizing the handling of brain and spinal tissues, washing hands and instruments after field-dressing, asking deer processors to process meat individually, without meat from other animals, or processing your own meat, and having the animal processed in the area of the state where it was harvested so high-risk body parts can be disposed of properly). If possible, an animal should be tested and not consumed until it tests negative for CWD.

Other Animal TSE

TME and FSE can be prevented in affected species by not feeding infected meat and feeding only properly processed feed.

MANAGEMENT AND CONTROL IN HUMANS

The link between vCJD and BSE has made BSE a notifiable disease in many countries including Great Britain (since June 1988) and the United States. Currently the best way to reduce the risk of vCJD involve preventing BSE in cattle (as described above). Prions are very resistant to disinfectants, heat, ultraviolet radiation, ionizing radiation, and formalin. Infectious tissues or material should be autoclaved at 134°C to 138°C for 18 minutes, incinerated, or treated with 2% sodium hypochlorite or 2 N sodium hypochlorite for more than 1 hour at 20°C and overnight for equipment. In Great Britain infected animal carcasses must be rendered at 133°C at 3 bar pressure for a minimum of 20 minutes. Disposable instruments are recommended for neurosurgical procedures. Since 1991, WHO has held many scientific conferences on issues related to animal and human TSEs. From 1997 to 2000, WHO held a series of training courses worldwide, particularly in developing countries, in order to help individual countries establish national surveillance of CJD and its variants. WHO has developed and revised its training guidelines for surveillance and provides a globally accepted case definition for all forms of human TSEs as follows:

- The clinical presentation, progressive nature of the disease and failure to find any other diagnosis are the hallmarks of vCJD.
- There are no available, completely reliable diagnostic tests for use before the onset of clinical symptoms; however, magnetic resonance scans, tonsillar biopsy, and cerebrospinal fluid tests are useful diagnostic tests.
- The brainwave pattern observed during an electroencephalogram (EEG) was abnormal in most of the vCJD patients (and does not occur in cases of sporadic CJD).
- Currently the diagnosis of vCJD can only be confirmed following pathological examination of the brain.
- No part or product of any animal which has shown signs of a TSE should enter any (human or animal) food chain.
- Countries should not permit tissues that are likely to contain the BSE agent to enter any (human or animal) food chain.
- All countries should ban the use of ruminant tissues in ruminant feed.
- Human and veterinary vaccines prepared from bovine materials may carry the risk of transmission of animal TSE agents; therefore, pharmaceutical companies should avoid the use of bovine materials and materials from other animal species in which TSEs naturally occur.

SUMMARY

Prion diseases, also known as transmissible spongiform encephalopathies or TSEs, are a group of animal and human CNS diseases that have long incubation periods, produce spongiform vacuoles in brain tissue, do not produce an inflammatory response in affected tissues, and are uniformly fatal. Prions are proteinaceous infections particles that are one-tenth the size of viruses and consist only of protein. Cellular prion protein is a normal component of neuronal cell membranes and is a three-dimensional structure with an α helix configuration. All forms of TSE have a similar alteration of the PrP gene resulting in abnormal folding into a β helix arrangement. These abnormally folded protein structures accumulate in intracellular vesicles that produce large vacuoles and are deposited as florid plaques. PrP is found throughout the body on cell membranes in healthy people and animals. The normal form of the protein is called PrP^C, whereas the infectious form is called PrP^{Sc}. It is assumed that PrP^{Sc} directly interacts with PrP^C causing the conversion of PrP^C into PrP^{Sc} and initiating a self-perpetuating vicious cycle of replication of abnormal proteins.

Prion diseases of animals include scrapie, BSE, CWD, TME, and FSE. Human prion diseases that are infectious are vCJD and kuru. Prion diseases are transmitted mainly by ingestion of prions (BSE, TME, FSE, vCJD, and kuru); however, scrapie is spread most commonly from the ewe to her offspring and to other lambs through contact with the placenta and placental fluids, whereas the transmission of CWD is not fully understood. The clinical signs of TSE in animals vary somewhat with the species. Scrapie in sheep and goats presents as behavioral changes, tremors, rubbing as a result of intense pruritus, weakness, weight loss, high-stepping of the front limbs, and ataxia that progresses to recumbency and death. BSE in cattle has an insidious onset and progresses slowly with clinical signs including hyperesthesia, tremors, ataxia, falling, and behavioral changes. CWD in deer and elk presents with behavioral changes, staggering, poor posture, lowered head and ears, decreased food consumption, emaciation, drooling, polyuria, and polydipsia. TME has clinical signs of increased nest soiling, difficulty eating, ataxia, jerkiness of hind limbs, rapid circling, compulsive chewing of the tail, and clenching of the jaw. FSE presents in felines as behavioral changes, tremors, ataxia, increased aggression, seizures, excessive salivation, hyperresponsiveness to loud noise, and dilated pupils. Clinical signs of vCJD in people as psychiatric abnormalities and neurologic signs. As the disease progresses cerebral cortical signs develop such as memory loss, dementia, and myoclonus that leads to immobility and death. TSEs in animals are diagnosed in a variety of ways including biopsy of the lymphoid tissues on the inside of the third eyelid and performing immunohistochemistry (scrapie), detection of PrP^{Sc} in unfixed brain extracts by immunoblotting and in fixed brains by immunohistochemistry (BSE), electron microscopy (all), modified immunoblot (BSE), a chemiluminescent ELISA test (BSE), an immunoassay (BSE), two-site noncompetitive immunometric procedure (BSE), and Western blotting (CWD). TSEs in people are diagnosed by brain biopsy. There is no treatment for TSEs in animals or people. Prevention of prion diseases involves eradication and surveillance programs in animals, feed bans in animals, preventing animals with neurologic disease or downer animals into the food supply, monitoring meat processing techniques, eliminating high-risk products (neurologic tissue) from being processed with meat, proper disinfection of potentially contaminated areas, and coordination of efforts between national and international organizations.

Review Questions

Multiple Choice

1. Creutzfeldt-Jakob disease, kuru, scrapie, BSE, and CWD are caused by
   a. virions.
   b. plasmids.
   c. prions.
   d. florid plaques.

2. In comparison to viruses, prions are
   a. larger.
   b. smaller.
   c. the same size.
   d. mirror images.

3. Transmissible spongiform encephalopathies are described as spongiform because
   a. they absorb extracellular protein.
   b. they are fatal diseases.
   c. they are self-replicating.
   d. they produce vacuoles in affected tissue.

4. What country has the highest percentage of cattle affected with BSE?
   a. United States
   b. France
   c. Japan
   d. U.K.

5. In the United States, what agency tracks cases of BSE?
   a. FDA
   b. USDA
   c. EPA
   d. CDC

True or False

6. Scrapie is found in higher percentages in some breeds of sheep and some genetic lines of sheep.
   a. true
   b. false

7. CWD surveillance is easier than other prion diseases because its transmission if fully understood.
   a. true
   b. false

8. All prions are abnormal.
   a. true
   b. false

9. All TSEs are relatively new diseases, being discovered in the 1960s.
   a. true
   b. false

10. Changes in the rendering process of animal feed has been one of the contributing factors in the development and spread of TSEs.
    a. true
    b. false

Matching

11. _____ CWD

12. _____ Scrapie

13. _____ BSE

14. _____ kuru

15. _____ FSE

16. _____ TME

A. TSE in cattle that was discovered in the 1980s in the U.K.

B. TSE in mink fed raw meat diets

C. TSE in cats fed prion-contaminated feed

D. TSE in sheep known for over 250 years

E. TSE in cannibalistic humans

F. TSE in elk and deer first reported in Colorado

17. _____ vCJD

18. _____ mad cow disease

19. _____ GSS

20. _____ CJD

G. Another name for BSE in cattle

H. autosomal dominant disease occurring in the fourth and fifth decade in humans

I. rare, sporadic, dementing neurologic disease in humans

J. TSE in people linked to BSE-contaminated meat

Short Answer

21. What are the similarities between transmission of kuru in people and BSE in cattle?

22. What is the basic mechanism by which prions cause disease?

23. Why did prion disease become common in some herbivores?

References

Belay, E., and Schonberger, L. 2005. The public health impact of prion disease. *Annual Review of Public Health* 26:319–39.

Black, J. 2005. *Microbiology Principles and Explorations*, 6th ed. New York, NY: John Wiley & Sons, Inc.

Centers for Disease Control and Prevention. 2005–2006. Bovine Spongiform Encephalopathy and Variant Creutzfeldt Jakob Disease: Prevention of Specific Infectious Diseases, Travelers' Health Yellow Book. http://www2.ncid.cdc.gov/travel/yb/utils/ybGet.asp?section=dis&obj=madcow.htm (accessed May 1, 2006).

Center for Food Security & Public Health. 2004. Bovine Spongiform Encephalopathy. http://www.cfsph.iastate.edu/Factsheets/pdfs/transmissible_spongiform_encephalopathy.pdf (accessed January 13, 2006).

Food and Drug Administration. 2000. Bovine Spongiform Encephalopathy (BSE): Estimating Risks for vCJD in Vaccines Using Bovine-Derived Materials. http://www.fda.gov/cber/bse/bse.htm (accessed March 20, 2006).

Haubrich, W. 1997. *Medical Meanings: A Glossary of Word Origins*. Philadelphia, PA: ACP.

Janson, P., and R. Chung. 2005. Kuru. http://www.emedicine.com/med/topic1248.htm (accessed October 15, 2005).

Kahler, S. 2002. *The Rationale for Ridding the United States of Scrapie*. Schaumburg, IL: AVMA.

Kellar, J., and V. Lees. 2003. Risk management of the transmissible spongiform encephalopathies in North America. *Scientific and Technical Review Office International des Epizooties* 22(1):201–25.

Krauss, H., A. Weber, M. Appel, B. Enders, H. D. Isnebergy, H. G. Schiefer, W. Slenczka, A. von Graeventiz, and H. Zahner.. 2003. Zoonoses associated with prions. In: *Zoonoses infectious diseases transmissible from animals to humans*, 3rd ed. Washington, DC: ASM Press, pp. 167–72.

Miller, M., E. Williams, H. Thompson, and L. Wolfe. 2004. Environmental sources of prion transmission in mule deer. *Emerging Infectious Diseases* 10 (6):119–26.

National Institute for Animal Agriculture. 2001. Scrapie Fact Sheet. http://www. animalagriculture.org/scrapie/AboutScrapie/FactSheet.htm (accessed April 13, 2006).

O'Rourke, K., T. Baszler, T. Besser, J. M. Miller, R. C. Cutlip, G. A. H. Wells, S. J. Ryder, S. M. Parish, A. N. Hamir, N. E. Cockett, A. Jenny, and D. P. Knowles2000. Preclinical diagnosis of scrapie by immunohistochemistry of third eyelid lymphoid tissue. *Journal of Clinical Microbiology* 38(9):3254–9.

U.S. Department of Agriculture. 2002. Transmissible mink encephalopathy. http:// www.aphis.usda.gov/lpa/pubs/fsheet_faq_notice/fs_ahtme.html   (accessed April 13, 2006).

William, E., and M. Miller. 2003. Transmissible spongiform encephalopathies in non-domestic animals: origin, transmission and risk factors. *Scientific and Technical Review Office International des Epizooties* 22(1):145–56.

Wisniewski, T., and E. Sigurdsson. 2006. Prion-related diseases. http://www. emedicine.com/neuro/topic662.htm (accessed May 1, 2006).

World Health Organization. 2002. Variant Creutzfeldt-Jakob Disease. http://www. who.int/mediacentre/factsheets/fs180/en/ (accessed May 25, 2006).

Appendices

APPENDIX A: FOODBORNE ZOONOSES

BACTERIA

- *Bacillus anthracis*
- *Brucella* spp.
- *Campylobacter jejuni*
- *Clostridium botulinum*
- *Clostridium perfringens*
- *Escherichia coli*—enteroinvasive (EIEC)
- *Escherichia coli*—enteropathogenic (EPEC)
- *Escherichia coli*—enterotoxigenic (ETEC)
- *Escherichia coli* O157:H7 enterohemorrhagic (EHEC)
- *Listeria monocytogenes*
- *Salmonella enteritidis*
- *Shigella* spp.
- *Staphylococcus aureus*
- *Staphylococcus* spp.
- *Streptococcus*
- *Vibrio cholerae*
- *Vibrio parahaemolyticus*
- *Vibrio vulnificus*
- *Yersinia enterocolitica*
- *Yersinia pseudotuberculosis*

FUNGI

- Most fungal-borne food illness is the result of mycotoxin production, which is not a feature of the zoonotic fungi covered in this textbook.

PARASITES

- *Ascaris lumbricoides*
- *Clonorchis sinensis*
- *Cryptosporidium parvum*
- *Diphyllobothrium* spp.
- *Dracunculus insignis*
- *Echinococcus* spp.
- *Entamoeba histolytica*
- *Fasciola hepatica*
- *Fasciolopsis buski*
- *Giardia duodenalis*
- *Heterophyes heterophyes*
- *Opisthorchis felineus*
- *Opisthorchis viverrini*
- *Paragonimus westermani*
- *Taenia saginata*
- *Taenia solium*
- *Toxoplasma gondii*
- *Trichinella spiralis*
- *Trichuris trichiura*

VIRUSES

- Enteric viruses have host range variants in different species and are not a significant source of food-borne illness in people.

PRIONS

- Bovine spongiform encephalopathy

APPENDIX B: ZOONOSES BY ANIMAL SPECIES

Selected Zoonoses of Dogs and Cats

Zoonosis	Organism	Organism Type
Amebiasis	*Entamoeba histolytic*	Protozoan
Anthrax	*Bacillus anthracis*	Bacterium
Babesiosis	*Babesia canis*	Protozoan
Bite wound infections	*Pasteurella multocida, Capnocytophaga canimorsus, Capnocytophaga cynodegmi, Bergeyella zoohelcum*	Bacteria
Blastomycosis	*Blastomyces dermatitis*	Fungus
Brucellosis	*Brucella canis*	Bacterium
Campylobacteriosis	*Campylobacter jejuni*	Bacterium
Canine granulocytic anaplasmosis	*Anaplasma phagocytophilum* and *Ehrlichia ewingii*	Rickettsia
Canine monocytic ehrlichiosis	*Ehrlichia canis* and *Ehrlichia chaffeensis*	Rickettsia
Cat-scratch disease	*Bartonella henselae* (*Afipia felis* and *Bartonella clarridgeiae*)	Bacteria
Chagas' disease (American trypanosomiasis)	*Trypanosoma cruzi*	Protozoan
Clonorchiasis	*Clonorchis sinensis*	Trematode
Coccidioidomycosis	*Coccidioides immitis*	Fungus
Coenurosis	*Trypanosoma multiceps, Trypanosoma serialis,* and *Trypanosoma brauni*	Cestodes
Cryptococcosis	*Cryptococcus neoformans*	Fungus
Cryptosporidiosis	*Cryptosporidium parvum*	Protozoan
Cutaneous larva migrans	*Ancylostoma caninum* and *Ancylostoma braziliense; Strongyloides stercoralis*	Nematodes
Diphyllobothriasis	*Diphyllobothrium latum*	Cestode
Echinococcosis	*Echinococcus granulosus* and *Echinococcus multilocularis*	Cestode
Filariasis	*Brugia malayi, Dirofilaria immitis,* and *Dirofilaria repens*	Nematodes
Giardiasis	*Giardia* spp.	Protozoan
Gnathostomiasis	*Gnathostoma spinigerum*	Nematode
Histoplasmosis	*Histoplasma capsulatum*	Fungus

Zoonosis	Organism	Organism Type
Leishmaniasis	*Leishmania donovani*	Protozoan
Leptospirosis	*Leptospira interrogans*	Bacterium
Mite dermatitis	*Cheyletiella yasguri*	Mite
Notoedric mange	*Notoedres cati*	Mite
Ornithosis (conjunctivitis)	*Chlamydophila psittaci*	Bacterium
Pasteurellosis	*Pasteurella multocida*	Bacterium
Plague	*Yersinia pestis*	Bacterium
Powassan	Powassan virus	Virus
Poxvirus	Cowpox virus	Virus
Pulmonary capillariasis	*Capillaria aerophila;* renamed *Eucoleus aerophila*	Nematode
Q fever	*Coxiella burnetii*	Bacterium
Rabies	Rabies virus	Virus
Ringworm	*Microsporum* spp. and *Trichophyton* spp.	Fungus
Rocky Mountain spotted fever	*Rickettsia rickettsii*	Rickettsia
Russian spring-summer encephalitis	RSSE virus	Virus
Salmonellosis	*Salmonella enteritidis*	Bacterium
Scabies	*Sarcoptes scabiei*	Mite
Sporotrichosis	*Sporothrix schenckii*	Fungus
Staphylococcus infections	*Staphylococcus intermedius*	Bacterium
Thelaziasis	*Thelazia callipaeda* and *Thelazia californiensis*	Nematodes
Toxoplasmosis	*Toxoplasma gondii*	Protozoan
Trichinosis	*Trichinella spiralis*	Nematode
Tularemia	*Francisella tularensis*	Bacterium
Visceral larva migrans	*Toxocara canis* and *Toxocara cati*	Nematodes
Yersiniosis	*Yersinia* spp.	Bacterium

Selected Zoonoses of Livestock and Horses

Zoonosis	Organism	Organism Type
Amebiasis	*Entamoeba* spp.	Protozoan
Anthrax	*Bacillus anthracis*	Bacterium
Babesiosis	*Babesia bigemina, Babesia bovi* and *Babesia divergens* in cattle *Theileria equi* and *Babesia caballi* in horses	Protozoa
Botulism	*Clostridium botulinum*	Bacterium
Brucellosis	*Brucella abortus, Brucella melitensis, Brucella suis*	Bacteria
Campylobacteriosis	*Campylobacter jejuni*	Bacterium
Chlamydia infection	*Chlamydia trachomatis*	Bacterium
Contagious ecthyma	Orf virus	Virus
Cowpox	Cowpox virus	Virus
Cryptosporidiosis	*Cryptosporidium parvum*	Protozoan
Escherichia coli infection (diarrhea)	*Escherichia coli*	Bacterium
Erysipelas (animals) Erysipeloid (humans)	*Erysipelothrix rhusiopathiae*	Bacterium
European tick-borne encephalitis	TBE virus	Virus
Foot-and-mouth disease	Foot-and-mouth disease virus	Virus
Giardiasis	*Giardia duodenalis*	Protozoan
Glanders	*Burkholderia mallei*	Bacterium
Glanders	*Burkholderia mallei*	Bacterium
Influenza	Swine influenza virus	Virus
Leptospirosis	*Leptospira interrogans*	Bacterium
Louping ill	Louping ill virus	Virus
Pasteurellosis	*Pasteurella multocida*	Bacterium
Pseudocowpox	*Pseudocowpox virus*	Virus
Q fever	*Coxiella burnetii*	Bacterium
Rabies	*Rabies virus*	Virus

Zoonosis	Organism	Organism Type
Ringworm	*Microsporum* spp. and *Trichophyton* spp.	Fungus
Salmonellosis	*Salmonella enteritidis*	Bacterium
Scabies	*Sarcoptes scabiei*	Mite
Streptococcal infection	*Streptococcus* spp.	Bacterium
Swine vesicular disease	Swine vesicular disease virus	Virus
Taeniasis/Cysticercosis	*Taenia solium* and *Taenia saginata*	Cestode
Trichinosis	*Trichinella spiralis*	Nematode
Tuberculosis	*Mycobacterium bovis*	Bacterium
Tularemia	*Francisella tularensis*	Bacterium
Visceral larva migrans	*Ascaris suum*	Nematode
Yersiniosis	*Yersinia enterocolitica* and *Yersinia pseudotuberculosis*	Bacteria

Selected Zoonoses of Birds

Zoonosis	Organism	Organism Type
Campylobacteriosis	*Campylobacter jejuni*	Bacterium
Cryptococcosis	*Cryptococcus neoformans*	Fungus
Cryptosporidiosis	*Cryptosporidium parvum*	Protozoan
Erysipelas (animals) Erysipeloid (humans)	*Erysipelothrix rhusiopathiae*	Bacterium
Histoplasmosis	*Histoplasma capsulatum*	Fungus
Histoplasmosis	*Histoplasma capsulatum*	Fungus
Influenza	*Avian influenza virus*	Virus
Mite dermatitis	*Ornithonyssus bursa, Ornithonyssus sylviarum, Dermanyssus gallinae, Eutrombicula* spp., *Neotrombicula* spp., *Schoengastia* spp., *Euschoengastia* spp., *Acomatacarus* spp., *Siseca* spp., *Blankaartia* spp.	Mites
Newcastle disease	Newcastle disease virus	Virus
Ornithosis (psittacosis)	*Chlamydophila psittaci*	Bacterium
Pasteurellosis	*Pasteurella multocida*	Bacterium

Selected Zoonoses of Birds (Continued)

Zoonosis	Organism	Organism type
Ringworm	*Trichophyton gallinae*	Fungus
Salmonellosis	*Salmonella* spp.	Bacterium
Yersiniosis	*Yersinia pseudotuberculosis*	Bacterium
Yersiniosis	*Yersinia pseudotuberculosis*	Bacterium

Selected Zoonoses of Rodents

Zoonosis	Organism	Organism Type
Campylobacteriosis	*Campylobacter jejuni*	Bacterium
Cryptosporidiosis	*Cryptosporidium parvum*	Bacterium
Escherichia coli infection (enteritis)	*Escherichia coli*	Bacterium
Flea-borne typhus	*Rickettsia typhi, Rickettsia felis*	Rickettsia
Hantavirus pulmonary syndrome	Hantavirus	Virus
Hemorrhagic fever with renal syndrome	Hantavirus	Virus
Leptospirosis	*Leptospira interrogans*	Bacterium
Listeriosis	*Listeria monocytogenes*	Bacterium
Lymphocytic choriomeningitis	Lymphocytic choriomeningitis virus	Virus
Mite dermatitis	*Sarcoptes scabiei, Ornithonyssus bacoti*	Mite
Plague	*Yersinia pestis*	Bacterium
Rat-bite fever	*Streptobacillus moniliformis, Spirillum minus*	Bacterium
Ringworm	*Microsporum* spp. and *Trichophyton* spp.	Fungus
Salmonellosis	*Salmonella* spp.	Bacterium
Tapeworm infection	*Hymenolepis nana, Hymenolepis diminuta*	Cestode
Tick-borne relapsing fever	*Borrelia* spp.	Bacterium
Yersiniosis	*Yersinia enterocolitica*	Bacterium

Selected Zoonoses of Reptiles, Fish, and Wildlife

Zoonosis	Organism	Organism Type
Botulism	*Clostridium botulinum*	Bacterium
Brucellosis	*Brucella suis*	Bacterium
Campylobacteriosis	*Campylobacter* spp.	Bacterium
Clonorchiasis	*Clonorchis sinensis*	Trematode
Cryptosporidiosis	*Cryptosporidium parvum*	Protozoan
Escherichia coli infection (enteritis)	*Escherichia coli*	Bacterium
Erysipelas (animals) Erysipeloid (humans)	*Erysipelothrix rhusiopathiae*	Bacterium
Giardiasis	*Giardia* spp.	Protozoan
Granuloma lesions	*Mycobacterium marinum*	Bacterium
Herpes B	*Herpesvirus simiae*	Virus
Heterophyiasis	*Heterophyes heterophyes*	Trematode
Leprosy	*Mycobacterium leprae*	Bacterium
Mite dermatitis	*Cheyletiella blakei* and *Cheyletiella parasitovorax*	Bacteria
Necrotizing cellulitis	*Vibrio vulnificus*	Bacterium
Paracoccidioidomycosis	*Paracoccidioides brasiliensis*	Fungus
Pasteurellosis	*Pasteurella multocida*	Bacterium
Plague	*Yersinia pestis*	Bacterium
Q fever	*Coxiella burnetii*	Bacterium
Rabies	*Rabies virus*	Virus
Ringworm	*Microsporum* spp., *Trichophyton* spp.	Fungi
Salmonellosis	*Salmonella* spp.	Bacterium
Scabies	*Sarcoptes scabiei*	Mite
Shigellosis	*Shigella* spp.	Bacterium
Trichinosis	*Trichinella spiralis*	Nematode

Selected Zoonoses of Reptiles, Fish, and Wildlife (Continued)

Zoonosis	Organism	Organism Type
Tuberculosis	*Mycobacterium tuberculosis, Mycobacterium bovis*	Bacteria
Tuberculosis	*Mycobacterium bovis*	Bacterium
Tularemia	*Francisella tularensis*	Bacterium
Vibriosis	*Vibrio* spp.	Bacterium
Visceral larva migrans	*Baylisascaris procyonis*	Nematode
Yersiniosis	*Yersinia* spp.	Bacterium

Selected Zoonoses of Nonhuman Primates

Zoonosis	Organism	Organism type
Amebiasis	*Entamoeba histolytica* and *Entamoeba polecki*	Protozoa
Campylobacteriosis	*Campylobacter jejuni*	Bacterium
Giardiasis	*Giardia duodenalis*	Protozoan
Herpes B virus	*Herpesvirus simiae*	Virus
Leprosy	*Mycobacterium leprae*	Bacterium
Marburg hemorrhagic fever	Marburg virus	Virus
Monkeypox	Monkeypox virus	Virus
Salmonellosis	*Salmonella* spp.	Bacterium
Shigellosis	*Shigella* spp.	Bacterium
Tuberculosis	*Mycobacterium bovis* and *mycobacterium tuberculosis*	Bacteria
Tularemia	*Francisella tularensis*	Bacterium

APPENDIX C: BIOLOGIC SAFETY LEVELS

Biosafety levels are described in the document *Biosafety in Microbiological and Biomedical Laboratories* (BMBL). Each biosafety level consists of a combination of laboratory practices and techniques, safety equipment, and laboratory facilities. The higher the biosafety number for a particular organism, the greater the hazard. The Centers for Disease Control and Prevention (CDC) publishes a list of biologic agents classified according to their risk.

Biosafety Level	Types of Organisms	Proper Handling
BSL 1	▪ Organisms are not known to nor have minimal potential hazard. ▪ Appropriate for teaching laboratories and other laboratories that work with defined strains of microorganisms not known to cause disease in healthy adult humans.	▪ Work is conducted on open bench tops with access to an open bench top sink. ▪ Personnel and supervisors should have appropriate training in the procedures performed.
BSL 2	▪ Organisms are of moderate risk. ▪ Appropriate for clinical, diagnostic, teaching, and other facilities.	▪ Work is done in an open bench with precautions for splashes and aerosols (may include gloves, gowns, and biosafety cabinet). ▪ Laboratory personnel have specific training in handling pathogenic agents and are directed by supervisors competent in working with this biosafety level. ▪ Access to the laboratory is limited when work is in progress. ▪ Secondary barriers (hand washing and waste decontaminating facilities) must be available.
BSL 3	▪ Organisms may be a variety of indigenous and exotic agents with a potential for respiratory transmission and that may cause serious and potentially lethal infections. ▪ Appropriate for clinical, diagnostic, teaching, research, or production facilities.	▪ All work in performed in a biosafety cabinet. ▪ Laboratory personnel have specific training in handling pathogenic and potentially lethal agents and are supervised by competent, experienced scientists. ▪ More emphasis is placed on primary and secondary barriers.
BSL 4	▪ Organisms are dangerous and may be exotic with life-threatening potential, which may be transmitted by the aerosol route, and for which there is no available vaccine or therapy. ▪ Appropriate for limited facilities	▪ Work is done in a class III biosafety cabinet or in a biohazard suit that completely surrounds the worker. ▪ Access to the laboratory is strictly controlled. ▪ Laboratory personnel have specific and thorough training in handling extremely hazardous infectious agents, and they understand special containment practices. ▪ This is the maximum level of containment.

The CDC also classifies Zoonotic Agents that are potential Bio-Weapon Agents as Categories A through C.

Category	Definition	Examples
A (highest risk)	▪ easily spread or transmitted from person to person ▪ high death rates and have the potential for major public health impact ▪ might cause public panic and social disruption ▪ require special action for public health preparedness.	▪ Anthrax ▪ Botulism ▪ Plague ▪ Tularemia ▪ Viral hemorrhagic fevers
B (high risk)	▪ moderately easy to spread ▪ moderate illness rates and low death rates ▪ require specific enhancements of laboratory capacity and enhanced disease monitoring.	▪ Brucellosis ▪ Glanders ▪ Ornithosis (Psittacosis) ▪ Q fever ▪ Viral encephalitis ▪ Toxins ▪ Food safety zoonotic agents ▪ Water safety zoonotic agents
C (emerging pathogens with future risk)	▪ easily available ▪ easily produced and spread ▪ potential for high morbidity and mortality rates and major health impact.	▪ Nipah virus ▪ Hantavirus

APPENDIX D: NATIONALLY NOTIFIABLE ZOONOTIC INFECTIOUS DISEASES OF THE UNITED STATES—2007

HUMAN

- Anthrax
- Arboviral neuroinvasive and nonneuroinvasive diseases
  - California serogroup virus disease
  - Eastern equine encephalitis virus disease
  - Powassan virus disease
  - St. Louis encephalitis virus disease
  - West Nile virus disease
  - Western equine encephalitis virus disease
- Botulism
- Brucellosis
- Cholera
- Coccidioidomycosis
- Cryptosporidiosis
- Cyclosporiasis
- Ehrlichiosis/Anaplasmosis
  - Anaplasmosis, human granulocytic
  - Ehrlichiosis, human monocytic
  - Ehrlichiosis, human, other or unspecified agent
- Giardiasis
- Hansen disease (leprosy)
- Hantavirus pulmonary syndrome
- Hemolytic uremic syndrome, postdiarrheal
- Influenza-associated pediatric mortality
- Listeriosis
- Lyme disease
- Novel influenza A virus infections
- Ornithosis (Psittacosis)
- Plague
- Q Fever
- Rabies
- Rocky Mountain spotted fever
- Salmonellosis
- Severe Acute Respiratory Syndrome-associated Coronavirus (SARS-CoV) disease
- Shiga toxin-producing *Escherichia coli* (STEC)
- Shigellosis
- Smallpox
- Streptococcal disease, invasive, Group A
- Streptococcal toxic shock syndrome
- *Streptococcus pneumoniae*, drug resistant, invasive disease
- *Streptococcus pneumoniae*, invasive in children younger than 5 years
- Tetanus
- Trichinellosis (Trichinosis)
- Tuberculosis
- Tularemia
- Typhoid fever
- Vancomycin-intermediate *Staphylococcus aureus* (VISA)
- Vancomycin-resistant *Staphylococcus aureus* (VRSA)
- Vibriosis
- Yellow fever

Notifiable Bacterial Foodborne Zoonotic Diseases

- Anthrax
- Botulism
- Brucellosis
- Cholera
- Enterohemorrhagic *Escherichia coli*
- Hemolytic uremic syndrome, postdiarrheal
- Listeriosis
- Salmonellosis (other than *Salmonella* Typhi)
- Shigellosis
- Typhoid fever (*Salmonella* Typhi and *Salmonella* Paratyphi infections)

Notifiable Parasitic Foodborne Zoonotic Diseases

- Cryptosporidiosis
- Cyclosporiasis
- Giardiasis
- Trichinosis

ANIMAL*

All Species

- Anthrax
- Brucellosis
- Echinococcosis/hydatidosis
- Exotic myiasis (fly larvae)
- Foot-and-mouth disease
- Leptospirosis
- Rabies
- Screwworm
- Transmissible spongiform encephalopathy
- Tuberculosis
- Vesicular stomatitis
- West Nile virus encephalitis

Avian

- Avian influenza (highly pathogenic)
- Avian tuberculosis
- Avian chlamydiosis (*Chlamydophila psittaci*)
- Exotic Newcastle disease (viscerotropic velogenic)
- Fowl cholera (*Pasteurella multocida*)
- Fowl pox (avipoxvirus)
- *Salmonella enteritidis* serovar Enteritidis

Bovine

- Babesiosis
- Bovine spongiform encephalopathy
- Hemorrhagic septicemia (*Pasteurella multocida*)
- Listeriosis
- *Salmonella enteritids* serovar Typhimirium (*Salmonella* DT104)
- Scabies
- Trypanosomiasis

Caprine/Ovine

- Enzootic abortion of ewes (*Chlamydophila abortus*)
- Goat and sheep pox
- Listeriosis (dairy goats)
- Q fever
- Scrapie

Equine

- Equine encephalomyelitis (EEE, WEE, VEE)
- Equine influenza
- Equine piroplasmosis (*Babesia caballi, Babesia equi*)
- Horse mange (*Sarcoptes*)
- Horse pox
- Japanese encephalitis virus
- Surra (*Trypanosoma evansi*)

Porcine

- Atrophic rhinitis (*Pasteurella multocida*)
- Babesiosis
- Cysticercosis (*Taenia solium*)
- Swine vesicular disease
- Trichinellosis

Privately Owned Cervid

- Chronic wasting disease

*There are other nationally notifiable nonzoonotic diseases in animals that can be found at state agriculture websites.

Aberrant host Host in which a parasite cannot complete its development; also known as a dead-end host.

Abnormal host A host that is not the usual one. There are two kinds of abnormal hosts: accidental and aberrant.

Abopercular end The opposite end of an operculum in parasite eggs.

Abortion storm Series of abortions.

Abundance The number of parasites in or on a host including uninfected hosts.

Accidental host An animal/human that harbors an organism that is not usually parasitic in that particular organism. Sometimes the accidental host becomes a dead-end host because the parasite develops but fails to find a portal of exit and cannot continue its life cycle.

Acetabulum A muscular organ of attachment usually associated with the scolex of tapeworms; commonly called a sucker.

Acid fast Property of some microorganisms of resisting destaining of carbolfuchsin stain based on cell wall structure; acid fast organism examples include *Mycobacterium* and *Cryptosporidium*.

Adaptive immunity Portion of the immune system that is activated by exposure to antigen.

Adult stage The stage of a parasite that is sexually mature and in which reproduction occurs.

Agar Solid food source for microorganisms.

Agglutinate To clump together.

Agglutination reaction Antigen-antibody reaction that produces a visible clumping of particles.

Alternate host The host that alternates with another host in the life cycle of a parasite; another term used for intermediate host.

Amastigote Aflagellated, developmental stage found in some hemoflagellated protozoa.

Amoeba Protozoa that move by pseudopodia.

Amoeboid Cells that resemble amoebae by forming pseudopods.

Amplifying vector Designation given to an arthropod that contributes to the buildup and transmission of a disease-causing agent by transmitting the agent among vertebrate hosts; also called amplifying host.

Antibody A large protein molecule produced in response to an antigen; an antibody interacts with a specific antigen.

Antigen A foreign substance; any cell, particle, or chemical that induces a specific immune response.

Antigenic determinant Small region of the antigen that a lymphocyte recognizes; also called the epitope.

Antiseptic A chemical that inhibits the growth of or kills microorganisms on living tissue.

Antitoxin Antibodies that bind to and inactivate toxins.

APHIS Animal and Plant Health Inspection Service.

Artifically acquired active immunity Protection from disease stimulated by intensional means such as a vaccine.

Artifically acquired passive immunity Protection from disease conferred byintensionally administering antibodies fromed by another another or person.

Asexual reproduction Reproduction without the formation and fusion of gametes, for example binary fission, budding, schizogony, etc.

Attenuated vaccine Vaccine composed of weakened, non-pathogenic, live microorganisms.

Autoinfection Reinfection by a parasite from within the host and is not exposed to the outside environment; also known as hyperinfection.

Axostyle A rod-like structure that gives rigidity to the bodies of some flagellates.

B lymphocytes Lymphocytes that differentiate to produce antibodies.

Bacillus Rod-shaped bacterium when used with lower case b; bacilli is plural.

Bacteria Prokaryotic, single-celled microorganism.

Bacteriophage Viruses that infect bacteria.

Baerman technique Fecal test used to recover and identify parasitic larvae from feces using sedimentation principles.

Barophiles Organisms that grow only or more rapidly at pressures greater than 1 atmosphere.

Basal granule The granule-like body from which each cilium arises in cilates.

Binary fission Reproduction by division of an individual that produces two duplicates of the original; occurs in bacteria and some parasites.

Binomial nomenclature A system of naming organisms in which each organism is identified by a genus designation and a species designation.

Biological vector A vector in whose body the infecting organism develops or multiplies before becoming infective; a living obligate host in which morphologic change and/or replication occurs.

Biovar A bacterial strain differentiated by biochemical or non-serological methods.

Blepharoplast A small granule-like body in the cytoplasm of some parasites.

Bothrium An organ of attachment (sucker) in the form of a grove on the scolex of some tapeworms.

Bradykinin A substance released during inflammation that causes vasodilation and increased blood vessel permeability.

Broad-spectrum antibiotics Antibiotics that are effective against both gram-positive and gram-negative bacteria.

Broth Liquid food source for microorganisms.

Budding A form of asexual reproduction common in yeasts in which a bubble forms on the cell surface, grows, and pinches off, forming a new cell.

Caecum A sac-like extension of the intestine that is open only at one end; seen in organisms with a true intestine.

Capnophile An organism requiring CO_2 at a level higher than air for growth.

Carrier Healthy animal or person who are reservoirs of infection.

CD4 cells Lymphocytes that increase immune responsiveness; also known as helper T cells or TH cells.

CD8 cells Cytotoxic lymphocytes that kill infected cells; also known as cytotoxic T cells or TC cells.

CDC Centers for Disease Control and Prevention.

Cell-mediated immunity Immune response carried out by T lymphocytes.

Centers for Disease Control and Prevention The U.S. national agency located in Atlanta, Georgia that does research, collects statistics, and publishes information on infectious disease; also known as the CDC.

Cercaria Free-swimming larva of a trematode that escapes from a sporocyst or redia generation in the intermediate host (mollusk). Cercaria typically have a tail and is the transfer stage to the next host.

Chagoma An erythematous primary lesion of Chagas' disease.

Charcot-Leyden crystals Slender, pointed crystals formed from the breakdown product of eosinophils and are found in stool (typically indicating a parasitic infection); commonly called CL crystals.

Chemotaxis Process by which cells sense certain chemicals and move toward regions that contain optimal concentrations of them.

Chitinous shell The hard shell of nematode eggs that encases the embryo.

Chromatoid body Rod-shaped mass of RNA found in the cysts of amebae.

Cilia Small fibrils that function in movement.

Ciliates A class of protozoa bearing cilia.

Coccus Spherical-shaped bacteria when used with lower case c; cocci is plural.

Coenurus A larval cystic stage of a tapeworm.

Colony A clone of cells all originating from the same parent that is large enough to be visible on solid medium.

Communicable disease Disease that can be transmitted from one host to another.

Communsalism A symbiotic relationship where one organism (commensal) dervies benefit from another (host) without harming or helping the other.

Complement A family of more than 30 different proteins in serum that function together as a nonspecific defense against infection.

Complement fixation assays Tests that detect antigen-antibody reactions by their utilization (fixation) of complement.

Complete metamorphosis Process of growth and development of insects in which an individual insect develops through several larval stages and then a nonfeeding pupal stage before reaching adulthood.

Concentration method A procedure for increasing the strength of or the number of organisms in a medium. Fecal concentration methods are performed on a fecal specimen increasing the number of organisms found per given unit.

Cord factor A mycolic acid found only in virulent strains of *Mycobacterium tuberculosis* that causes these strains to form parallel rows of cells called cords.

Crustacea An arthropod class that is a group of aquatic animals having hard shells, jointed bodies and appendages, and gills for breathing such as crabs, shrimp, and copepods.

Cuticle The epidermis of vertebrates; the outer covering secreted from the hypodermis or subcuticular layer in helminths.

Cyclophyllidiea Tapeworm order in which the worms have four sucker discs encircling the scolex with or without a rostellum (for example *Taenia* spp.).

Cyst The protozoan stage in which the organism is encased in a cyst wall producing a more resistant stage that may be transmitted to a new host; an organism together with the enveloping membrane or wall secreted by that organism.

Cyst wall The outermost protective coating of an encysted protozoan.

Cysticercoid Tapeworm larva in which the scolex is invaginated into a cystic cavity.

Cysticercus Tapeworm larva in which the scolex is invaginated into a bladder filled with fluid.

Cytostome In flagellates and ciliates, the cavity that opens by way of the peristome (lips) to allow solid food particles to enter the mouth.

Dead-end host   Host in which the parasite reaches an end point and is unable to continue its life cycle; also known as an aberrant host.

Decorticated   Loss of characteristic mamillated outer covering of *Ascaris lumbricoides* egg.

Definitive host   Host in which sexual reproduction (fertilization) of a parasite occurs.

Density   The number of organisms per quantity of tissue.

Density gradient   The varying sequence of densities of materials suspended in a liquid in which the most dense material is found at the bottom and the least dense material is found at the top due to gravity sedimentation.

Diecious   Distinctly separate sexes; male reproductive organs are present in one and female organs are present in another.

Direct life cycle   Single-host life cycle; does not have intermediate host.

Disc diffusion method   Determining the sensitivity of a microorganism to antimicrobial drugs by seeding a plate with the microorganism and placing filter paper discs embedded with known quantities of different antimicrobial agents; also known as the Kirby-Bauer method.

Disease   Deviation from normal health or production.

Disinfectant   A chemical that inhibits the growth of or kills microorganisms on inanimate objects.

Doubling time   The period required for a microbial population to produce two new cells for each one that previously existed; also called generation time.

Ectoparasite   Parasite that lives on the host's body surfaces.

Ectoplasm   Granule free cytoplasm of ameba lying immediately under the plasma membrane.

Edema   Swelling of tissue by accumulation of fluid between cells.

Egg   Female reproductive cell after fertilization.

Embryo   The developmental stage following cleavage of the egg up to the first larval or first juvenile stage.

Embryonated   Containing a developing embryo; used to describe eggs that are infective.

Endemic   The ongoing, low-level presence of disease within a group; present in the group at all times.

Endoparasite   Parasite that lives within the body or internal organs of its host.

Endoplasm   Inner, granule-rich cytoplasm of ameba.

Endosome   Nucleolus-type organelle found in some protozoa.

Endospores   Extremely resistant dormant structures that form within the cells of certain genera of bacteria.

Endotoxin   The lipopolysaccharide component of the outer membrane of gram-negative bacteria that is harmful to animals and humans.

Enteric   Intestinal.

Enterotoxins   Compounds that are harmful to the epithelial cells lining the intestinal tract.

Enzootic   The ongoing, low-level presence of disease within an animal group; present in the animal group at all times.

Enzyme   Protein that catalyze specific metabolic reactions.

Enzyme-linked immunosorbent assay   Diagnostic immunological test that contains an enzyme linked to an indicator antibody; abbreviated ELISA.

EPA   Environmental Protection Agency.

Epidemic   Sudden, widespread, and rapidly spreading disease within a group of humans at the same time.

Epidemiology   The study of relationships of the factors determining frequency and distribution of disease.

Epitope   Small region of the antigen that a lymphocyte recognizes; also called the antigenic determinant.

Epizootic   Sudden, widespread, and rapidly spreading disease within a group of animals at the same time.

Epizootiology   The study of distribution and abundance of animal diseases.

Erythema   Abnormal skin redness.

Erythema migrans   A reddish skin rash associated with Lyme disease that is caused by infection and initial spread of *Borrelia burgdorferi* spirochetes in the vicinity of the bite of an infected tick.

Erythrocytes   Red blood cells.

Etiology   The study of causative agents of disease.

Eukaryote   Organisms composed of a membrane-bound nucleus.

Exotoxin   Highly destructive protein produced by some gram-positive and gram-negative bacteria.

Facultative   Capable of adapting to different conditions.

Facultative anaerobe   Organism that uses oxygen to grow when it is available, but can grow without oxygen.

Facultative parasite   A parasite that has free-living stages that do not require hosts.

Fastidious organism   Organism that requires numerous complex nutrients to grow.

FDA   Food and Drug Administration.

Fecal-oral route   A pattern of disease transmission by which pathogens shed in feces enter a new host through the mouth.

Filariasis   Disease caused by a group of parasitic filiarial nematodes.

Filariform   A juvenile postfeeding stage of a nematode; the infective stage of hookworms, filarial worms, and some other nematodes.

Final host   Another term used for definitive host.

Flagellates   A class of protozoa that move by means of flagella.

Flagellum   A filament that usually projects from the body of an organism; functions as an organelle of locomotion or when in the groove in the cytostome causes movement of fluid; flagella is plural.

Florid plaques   Focal amyloid protein deposits surrounded by vacuolized cells.

Flotation methods Fecal test in which feces with worm eggs are suspended in a liquid with a specific gravity greater than that of the eggs making the eggs float to the surface.

Fluke Common name for trematodes.

Fluorescent-labeled antibodies Antibodies chemically bonded to fluorochromes (fluorescent chemicals).

Fomite An object that mechanically transfers disease organisms.

Free-living Living free of a host.

Fungi Group of nonphototrophic eukaryotic organisms that includes yeasts, molds, and mushrooms.

Gas gangrene Infection caused by *Clostridium perfringens*.

Gel electrophoresis Movement of charged molecules through a gel driven by an electric current resulting in molecules of different size and/or charge becoming separated.

Generation time The period required for a microbial population to produce two new cells for each one that previously existed; also called doubling time.

Ghon complexes Calcified caseous tubercles that indicate a past primary tuberculosis infection.

Gingivitis Inflammation of the gums.

Gram stain A differential staining technique that is based on differences in cell wall structure allowing gram-positive bacteria to stain deep blue and gram-negative bacteria to stain light red.

Gram-negative bacteria Bacteria that have a thin cell wall surrounded by an outer membrane.

Gram-positive bacteria Bacteria that have a thick cell wall and no outer membrane.

Gravid Filled with eggs.

Halophile Organism that grows well in environments with high salt concentrations.

Halteres Reduced hind wings of flies.

Helminth Worm; consists of nematodes, trematodes, cestodes, and acanthocephalans (thorny-headed worms).

Hemagglutination Clumping of red blood cells.

Hemagglutinin Protein that clumps red blood cells.

Hematophage Blood feeing arthropod.

Hemocoele A body cavity in mollusks and arthropods through which hemocoeles fluid (blood) circulates carrying nutrients to the organs.

Hemolysin Protein that destroys red blood cells.

Herd immunity Prevention of disease due to the scarcity of new susceptible hosts.

Hermaphroditic Containing both male and female reproductive organs; also known as monecious.

Hooklet The small hook-like organ of attachment present on the rostellum of the tapeworm scolex.

Horizontal transmission Transfer of an infectious agent from one animal to another.

Host An organism that harbors or nourishes another organism.

Humoral immunity Protection conferred by antibodies.

Hydatid cyst A cystic larval stage of *Echinococcus* spp.

Hyperinfection Reinfection by a parasite from within the host and is not exposed to the outside environment; also known as autoinfection.

Hyphae Tubelike filaments in mold that make up a mycelium.

Hypobiotic Dormant.

Immune system Cells (mainly lymphocytes) and organs that extends throughout the body and functions as a defense against infection.

Immunization Artificially stimulating the body's immune defenses.

Immunocompetence The process by which lymphocytes acquire the capability to function fully in the body's defense.

Immunodiffusion test A type of precipitation reaction in which antigens and antibodies are diluted and mixed by diffusion through a gel.

Immunoelectrophoresis assay A type of precipitation reaction in which antigens and antibodies are diluted and mixed by electrophoresis through a gel.

Immunofluorescence assay Tests in which antigen-antibody reactions are detected by fluorescence because one of the reactants is tagged with a fluorescent dye.

Immunoglobulin Antibody.

Immunological memory The ability of memory lymphocytes to recognize an antigen if they encounter it again with better speed and amplification.

Inactivated vaccine Vaccine containing killed microorganisms.

Inapparent infection Subclinical infection; infection that does not produce clinical signs.

Incidence An expression of the rate at which a certain event occurs; the percentage of new infections in a given population during a specific time; number of new infections divided by the number examined.

Indirect life cycle Multiple hosts used in the life cycle.

Infection Colonization of the body with pathogenic organisms.

Infection source The medium, object, or living entity containing an infective stage of an organism.

Infectious disease A disease that is caused by a transmissible agent, such as viral, bacterial, fungal, or parasitic organisms.

Infective stage The stage in the parasitic life cycle during which it is capable of producing infection.

Infestation A host/ectoparasite relationship in which the parasite lives on the surface of the host.

Inflammatory mediators Molecular messengers that initiate inflammation.

Inflammatory response The body's nonspecific reaction to injury or infection; signs include redness, pain, swelling, heat, and loss of function.

Innate immunity Protection that does not depend on exposure to antigens to be active.

Instar The growth period between molts.

Interferons Small glycoproteins produced by host cells in response to viral infection.

Interleukin-1 Cytokine produced by white blood cells that functions to induce fever.

Intermediate host Host necessary for development, asexual reproduction, or transmission of a parasite; also called the alternate host.

Isolate Pure culture derived from a heterogeneous population of microbes.

Juvenile stage Any stage in the development of a helminth parasite between the egg and the mature adult stage that appears similar in shape and structure to the adult.

Karyosome A structure within the nuclue of amoebae having a relatively constant size and location in each species.

Kinetoplast Mass of mitochondrial DNA in flagellate protozoa.

Kinetosome Basal body of cilia (found in cilated protozoa).

Knott's test Concentration procedure used to identify microfilariae.

Larva The postembryonic parasite or arthropod stage in which internal organs are developing and are at least partially functioning; any preadult stage in the life cycle of a parasite that is morphologically distinct from the adult stage (not used for immature nematodes, because they are not morphologically distinct from the adults).

Larviparous Organisms such as flies that deposit first-stage larvae instead of laying eggs.

Leukocidin Enzymes that kill leukocytes.

Leukocytes White blood cells.

Lumen The space within a tubular organ.

Lymphokines Messenger proteins produced by lymphocytes.

Lysis Rupture of the cytoplasmic membrane resulting in cell destruction.

Lysogeny State in which an infecting phage exists as a prophage.

Macrogamete Female gamete of Toxoplasma; the larger of two gametes.

Macronucleus The large kidney-shaped nucleus found in some ciliates.

Mechanical vector A vector which transmits an infective organism from one host to another, but is not essential to the parasite life cycle.

Mesophile Organism that grows best at moderate temperatures (around 37°C).

Metacercaria The encysted stage of a monecious trematode; this stage succeeds the cercaria; metacercariae is plural.

Microgamete Male gamate of *Toxoplasma*; the smaller of two gamete.

Micronucleus A small nucleus closely associated with the macronucleus in ciliates.

Micropyle In coccidia, a pore in the cyst wall closed by a plug of material which is more easily dissolved that the cyst wall; this is where structures formed in the cyst emerge.

Minus strand Single-stranded RNA comprising a viral genome that must be transcribed by RNA-dependent RNA polymerase to act as mRNA; also called negative sense.

Miracidium The larve that emerges from the egg in trematodes.

Molt The process of shedding the outer body covering.

Monecious Containing both male and female reproductive organs; also known as hermaphroditic.

Morbid Afflicted with disease.

Morbidity The sick rate; the ratio of diseased animals/humans to well animals/persons in a population.

Moribund Dying; near death.

Mortal Fatal; causing death.

Mortality The death rate; the ratio of diseased animals/humans to diseased animals/human that die.

Morula The cleaving stage of an egg that forms a solid mass of cells.

Mutualism A symbiotic relationship where both organisms benefit.

Mycelium Mass of hyphae produced by some mold.

Narrow-spectrum antibiotics Antibiotics that are effective against only either gram-positive or gram-negative bacteria.

Natural host The usual host; also known as the typical host and normal host.

Natural killer cells Non-B cells and non-T cells that lyse cells by secreting perforins.

Naturally acquired active immunity Immunological protection that occurs unintentionally following exposure to an organism such as with infectious disease.

Naturally acquired passive immunity Immunological protection that occurs unintentionally following antibodies transferred from mother to fetus across the placenta or to the newborn through colostrum.

Negative sense Single-stranded RNA comprising a viral genome that must be transcribed by RNA-dependent RNA polymerase to act as mRNA; also called minus strand.

Negri bodies Inclusion bodies that develop in the brains of animals or people with rabies.

Neurotoxin Toxic protein that specifically affect nerve function.

Non-infectious disease A disease that is not caused by a transmissible agent, such as nutritional or genetic.

Normal flora Microorganisms that coexist with animals/humans in a stable, nonpathogenic relationship.

Nucleoli Dense masses of RNA and protein in the eukaryotic nucleus that manufactures ribosomes.

Nucleus Cell organelle that contains DNA and controls growth, cell division, and other activities of the cell.

Nuisance biter   A bloodsucking arthropod that causes discomfort when biting, but is not known to be involved in the transmission of a disease-causing agent.

Obligate   Must have; also called strict.

Obligate aerobe   Organism that grows only in the presence of oxygen.

Obligate anaerobe   Organism that grows only in the absence of oxygen.

Obligate barophiles   Organisms that grow only at pressures greater than 1 atmosphere.

Obligate intracellular parasites   Parasites that can only reproduce inside a host cell.

Obligate parasite   A parasite that requires certain hosts to develop and cannot live apart from their host.

Occult blood   Blood present in such small quantities that it is not detectable except by chemical means.

Onchospher   The stage that emerges from the egg shell and later from the embryophore of tapeworms.

Oocyst   Coccidian protozoal stage that is shed with the feces.

Operculum   Cap-like structure at one end of some trematode and cestode eggs through which the embryo or larva emerges.

Opportunistic   An organism that is not typically a parasite or disease-causing organism but may become so under specific conditions.

Opsonin   Protein that facilitates phagocytosis.

Opsonization   Process by which an opsonin faciliatate phagocytosis.

Ovum   An unfertilized egg.

Pandemic   Widely epidemic; occurring over a large region, continent, or geographic area.

Parabasal body   A heavy fiber present in some flagellates.

Parasite   An organism that lives on or within another organism at the expense of that organism.

Parasitism   A symbiotic relationship where one organism (parasite) is metabolically and obligately dependent upon the other (host).

Paratenic host   An animal acting as a substitute intermediate host of a parasite; no additional development of the parasite occurs in the paratenic host; also known as a transport host.

Parthenogenesis   Reproduction without any male element.

Passive immunity   Immunity conferred by administering antibodies.

Passive transmission   A form of transmission in which the host/parastie contact is accidental.

Patent period   Period of time in which the diagnostic stage of a parasite can be demonstrated.

Pathogen   A disease-causing organism.

Pathogenesis   The manner of disease development.

Pathogenic   Disease causing.

Pathology   The study of disease.

Periplast   The limiting, outer memembrane of protozoa.

Peristome   Any parts around the mouth or oral opening of invertebrates (comparable to lips).

Phage   Virus that infects bacteria; shortened name for bacteriophage.

Phage typing   Identifying bacterial strains by their pattern of susceptibility to phages.

Phagocytosis   Engulfment of one cell by another.

Pili   Straight, hair-like appendages that extend from the surface of a bacterial cell.

Piroplasm   Any organism in the class Piroplasmea, while in a circulating red blood cell; for example, Babesia.

Plaque   Circular clear zones on a lawn of cells.

Plaque count   Procedure for determining the number of bacteriophages and phage-infected cells in a sample.

Plaque-forming units   Virions and virus-infected cells; abbreviated PFUs.

Plerocercoid   Tapeworm larva in which the scolex is embedded in an enlarged tail; for example in Diphyllobothrium latum.

Plus strand   Single-stranded RNA constituting a viral genome that can act directly as mRNA; also called positive sense.

Polar plugs   Mucoid plugs located at both ends of eggs of Trichuris worms.

Portal of entry   Anatomic site through which a pathogen enters a host.

Portal of exit   Anatomic site through which a pathogen leaves a host.

Positive sense   Single-stranded RNA constituting a viral genome that can act directly as mRNA; also called plus strand.

Precipitation reaction   An antigen-antibody reaction that forms lattices large enough to precipitate.

Prepatent period   Time from infection until the infection is detectable.

Prevalence   The total number of disease cases in existence at a certain time in a certain location; percentage of infected animals in a given population at a given time.

Primary host   Another term used for definitive host.

Primary immune response   Production of antibody that occurs when a person first encounters a particular antigen.

Prion   An infections agent composed only of protein.

Procercoid   The first larval stage of pseudophyllidian tapeworms which develops from the onchosphere.

Proglottid   One complete unit of a tapeworm below the scolex; also known as a tapeworm segment.

Prokaryotes   Organisms that do not have a membrane bound nucleus; bacteria.

Promastigote   Developmental stage of a protozoan that possesses a prominent flagellum.

Prophage   A phage genome integrated into the chromosome of a host cell.

Protozoa   Nonphotosynthetic, unicellular eukaryotes; protozoan is singular.

Pseudophyllidian   An order of tapeworms in which the scolex has a single terminal bothria (such as in Diphyllobotrium latum).

Pseudopod   Tubelike structures that amoeboid cells use for movement.

Purulent   Pus-producing.

Pus   A mixture of dead leukocytes, microoganisms, and host cells.

Radioimmunoassay   A test to detect antigen-antibody reactions in which one of the reactants is tagged radioactively.

Reservoir host   An animal/human that is infected by a parasite and serves as a source of infection for others.

Reticuloendothelial system   The aggregate of the phagocytic cells (including certain cells of the bone marrow, lymphatic system, liver, and spleen), that function in the immune system's defense against foreign bodies; abbreviated RES.

Reverse zoonosis   A disease that can be naturally transmitted from humans and animals.

Rhabditoid juvenille   The first feeding stage of a juvenile nematode.

Ringworm   Common name for dermatophyte infections.

Rostellum   Apical portion of the scoles of some tapeworms.

Schizogony   Multiplication in protozoa in which repeated division of the nucleus and daughter nuclei occurs within the cell.

Scolex   The attachement end of a tapeworm from which the neck arises and gives rise to proglottids; scolices is plural.

Secondary host   Another term used for intermediate host.

Secondary immune response   Immune response initiated by memory cells.

Selective media   Media that favor that growth of certain microorganisms over others.

Sensilium   On the flea abdomen a structure that detects vibrations and temperature changes for host detection.

Septa   Cross walls in fungi.

Septicemia   Persistent infection of bacteria and their toxins in blood.

Serology   Diagnostic tests using blood serum.

Serotype   Taxonomic classification identified by serology; also called a serovar.

Serovar   A bacterial strain differentiated by serological methods; also called a serotype.

Serum   Liquid portion of blood minus the clotting factors/proteins.

Sign   A characteristic of disease that can be observed by others.

Smear   A thin film spread on a microscope slide.

Sparganum   The second larval stage of pseudophyllidian tapeworms.

Species barrier   Factors that restrict pathogens to certain host species.

Sporadic   Occuring occasionally in a population.

Sporocyst   A cyst that develops within an oocyst of a coccidain protozoa in which sporozoites develop.

Sporogenesis   Development of a spore.

Sporozoa   Nonmotile protozoa.

Sporozoite   The infective unit that develops within an oocyst that penetrates the intestinal wall initiating infection.

Stage   Any particular form in the parasitic life cycle which can be distinguished from its other forms.

Stains   Dyes used to increase contrast.

Strain   A subset of a bacterial species that is different from other bacteria of the same species by a minor, identifiable difference.

Strict   Must have; also called obligate.

Strobila   A complete tapeworm consisting of scolex, neck, and all proglottid stages.

Symbiosis   The close association of two dissimilar organisms.

Syngamy   Multiplication by a sexual process.

Tachyzoite   Actively proliferating trophozoites.

Taxonomy   Science of classifying organisms.

Thermophil   Organism that grows at high temperatures.

Tinea   Common name for dermatophyte infections in humans.

Tissue culture   Cultivation of eukaryotic cells or tissues in vitro.

Titer   Highest dilution of a test solution that is active.

Toxin   Poison.

Toxoid   Treated toxins that have lost their pathogenicity but still stimulate the immune system to produce antitoxin.

Transfer host   An animal/human that serves as a host until the appropriate definitive host is reached (not necessary for the life cycle of the parasite).

Transmission   Transfer from one animal to another.

Transovarial transmission   Passage of infectious microorganisms from one generation of host to the next through their eggs.

Transport host   An animal acting as a substitute intermediate host of a parasite; no additional development of the parasite occurs in the paratenic host; also known as a paratenic host.

Transstadial transmission   Passage of a disease-causing agent from an infected immature stage to the next stage of an arthropod vector.

Trematode   Fluke.

Trophozoite   The active, vegetative stage of a protozoan.

Tubercles   Granulomas produced by tuberculosis infection; dense nodules containing activated macrophages and monocytes.

Turbidity   Cloudiness of a liquid caused by suspended particles.

Typical host Host in which the parasite is usually found and in which it can continue development.

Ubiquitous Existing or being everywhere.

Uncoating The process by which the capsid and envelope of a virion are removed.

Undulating membrane Wavy membranous structure attached to the other portion of some flagellate protozoa.

Unembryonated Not containing a developing embryo; used to describe eggs that do not contain developing embryoes and are not infective.

Unilocular An intermediate larval stage (cyst) of tapeworms having only a single cavity.

Vaccination The use of vaccines to produce artificial active immunity.

Vaccine Agents that confer immunity without causing disease.

Variant Something that differs in its characteristic from the classification to which it belongs.

Vector An animal (usually an arthropod) that transmits disease organisms.

Vegetative cell Cell that grows and reproduces asexually.

Vertical transmission Transfer from one generation to another.

Vesicle Tiny, fluid-filled skin lesion.

Virion Intact, nonreplicating virus particle.

Virulence The degree of pathogenicity; the ability to cause overt disease.

Virus Microscopic packet of nucleic acid typically wrapped in a protein coat.

Vitelline gland The glands in Platyhelminthes that produce yoke material and the egg shell.

Wet mount A drop of liquid containing microorganisms on a microscope slide covered by a cover slip.

Xenodiagnosis A diagnostic technique in which uninfected arthropods are allowed to feed on a suspect individual and then examined in an attempt to recover the parasite.

Yeast A single-celled fungus.

Ziehl-Neelsen stain Special staining techniqe used to indentify *Mycobacterium tuberculosis* and closely related bacteria.

Zoonosis A disease that can be naturally transmitted from animals and humans; plural is zoonoses.